Sulfate-Reducing Bacteria and Archaea

Larry L. Barton • Guy D. Fauque

Sulfate-Reducing Bacteria and Archaea

Larry L. Barton
Department of Biology
167 Castetter Hall
University of New Mexico
Albuquerque, NM, USA

Guy D. Fauque
Mediterranean Inst Oceano
Aix-Marseille Université
Marseille Cedex 09, France

ISBN 978-3-030-96701-7 ISBN 978-3-030-96703-1 (eBook)
https://doi.org/10.1007/978-3-030-96703-1

This Springer imprint is published by the registered company Springer Nature Switzerland AG
The registered company address is: Gewerbestrasse 11, 6330 Cham, Switzerland

Contents

1 Sulfate-Reducing Prokaryotes: Changing Paradigms 1
1.1 Introduction 1
1.2 Nutrients for Growth: Initial Discovery Followed by an Exploration to Assess Diversity 2
1.2.1 Broadening the Scope of Electron Donors 3
1.2.2 Diverse Electron Acceptors 13
1.2.3 Disproportionation of Thiosulfate, Sulfite, and Sulfur 21
1.2.4 Fermentation of Organic Substrates 24
1.2.5 Sulfide/Sulfur Oxidation Coupled to Nitrate, Mn(IV), and O_2 Reduction 25
1.3 Syntrophic Growth 26
1.4 Autotrophic Growth 33
1.5 Geographic Distribution 36
1.6 Perspective 37
References 38

2 Characteristics and Taxonomy 57
2.1 Introduction 57
2.2 Phenotypic Characteristics 57
2.2.1 Cell Anatomy and Morphology 57
2.2.2 Cell Architecture 60
2.2.3 Cytoplasmic Structures 64
2.3 Endospores 69
2.4 Chemotaxonomy 70
2.4.1 Biomarkers 70
2.4.2 FISH Technologies, PhyloChip, and GeoChip 71
2.4.3 Cytochromes 72
2.4.4 Quinones and Lipids 72
2.4.5 DNA G + C Content 74

2.5 Taxonomic Placement . . . 74
2.5.1 Classification . . . 74
2.5.2 Insights from Gene and Genome Analysis . . . 83
2.6 Filamentous or Cable Bacteria . . . 95
2.7 DSR, APS, and Lateral Gene Transfer . . . 100
2.8 Development of Genetic Manipulations . . . 102
2.9 Perspective . . . 103
References . . . 104

3 Reduction of Sulfur and Nitrogen Compounds . . . 121
3.1 Introduction . . . 121
3.2 Sulfate Activation and Bisulfite Production . . . 121
3.2.1 ATP Sulfurylase . . . 123
3.2.2 Inorganic Pyrophosphatase . . . 125
3.2.3 APS Reductase . . . 126
3.3 Assimilatory Sulfate Reduction . . . 131
3.4 Dissimilatory Sulfate Reduction . . . 134
3.4.1 Dissimilatory Bisulfite Reductase . . . 134
3.4.2 Thiosulfate Reductase . . . 137
3.4.3 Trithionate Metabolism . . . 138
3.4.4 Mechanism of Bisulfite Reduction . . . 138
3.5 Sulfate Transport . . . 140
3.6 Elemental Sulfur Reduction . . . 143
3.7 Contributions of SRP to Sulfur Cycling . . . 144
3.8 Enzymology of Nitrogen Respiration . . . 151
3.8.1 Nitrate Reduction . . . 151
3.8.2 Nitrite Reduction . . . 153
3.9 Nitrogen Fixation . . . 155
3.10 Role of SRP in Nitrogen Cycling . . . 156
3.11 Summary and Perspective . . . 157
References . . . 157

4 Electron Transport Proteins and Cytochromes . . . 173
4.1 Introduction . . . 173
4.2 Hydrogenases . . . 174
4.2.1 [FeFe] Hydrogenases . . . 176
4.2.2 [NiFe] Hydrogenases . . . 182
4.2.3 [FeNiSe] Hydrogenases . . . 186
4.3 Formate Dehydrogenase . . . 188
4.3.1 Periplasmic and Membrane . . . 188
4.3.2 Cytoplasmic . . . 190
4.4 Cytoplasmic Proteins with Low Redox Potentials . . . 191
4.4.1 Ferredoxin . . . 191
4.4.2 Flavodoxin . . . 194
4.5 Cytoplasmic Proteins with High Redox Potentials . . . 197
4.5.1 Rubredoxin . . . 197

4.5.2 Rubrerythrin 199
4.5.3 Desulfoferrodoxin 200
4.5.4 Desulforedoxin 201
4.5.5 Neelaredoxin 201
4.5.6 Nigerythrin 202
4.6 Cytochromes 202
4.6.1 C-Type Cytochromes 202
4.6.2 B-Type Cytochromes 211
4.7 Heme-Containing Enzymes 213
4.7.1 Nitrite Reductase 213
4.7.2 Nitrate Reductase 214
4.7.3 Sulfite Reductase: Sirohemes 215
4.7.4 Oxygen Reductases 215
4.7.5 Quinol:Fumarate Oxidoreductase 217
4.7.6 Molybdopterin Oxidoreductase 218
4.7.7 Catalase 218
4.8 Synthesis of Heme 219
4.9 Conclusion and Perspective 221
References 221

5 Systems Contributing to the Energetics of SRBP 245
5.1 Introduction 245
5.2 Growth and Yield Coefficients 246
5.3 Energetic Considerations with Organic Acids and Ethanol 248
5.3.1 Lactate Oxidation 250
5.3.2 Pyruvate Oxidation 254
5.3.3 Formate Dehydrogenases 255
5.3.4 Alcohol Dehydrogenase 257
5.4 Location of Soluble Hydrogenases, Formate Dehydrogenases and Cytochromes 258
5.4.1 Periplasmic Activity 258
5.4.2 Cytoplasmic Activity 258
5.5 Protons and Energetic Considerations 259
5.5.1 Proton Motive Force 259
5.5.2 Proton Translocation Experiments Using Cells 260
5.5.3 Proton-Translocating Pyrophosphatase: HppA Complex 260
5.6 Transmembrane Electron Transport Complexes 261
5.6.1 Quinone Oxidoreductase Complex: Qmo 262
5.6.2 Sulfite Reduction Complex: Dsr 264
5.6.3 Redox Complexes: Hmc, Tmc, Nhc, Ohc 266
5.6.4 Quinone Reductase Complex: Qrc 267
5.6.5 Ion-Translocating NADH Dehydrogenase Complexes: Rnf, Nqr, Nuo 267

5.7 Cytoplasmic Electron Transport Complexes 268
5.7.1 NADH-Ferredoxin Complex: Nfn 268
5.7.2 Heterodisulfide Reductase: Hdr/flox 268
5.8 ATPase (F-Type and V-Type) 269
5.9 Direct Measurement of ATP Production via Anaerobic Oxidative Phosphorylation 270
5.9.1 Phosphorylation Coupled to Dissimilatory Sulfate Reduction: Bisulfite and Associated Reactions 270
5.9.2 Elemental Sulfur Reduction 272
5.9.3 Fumarate as an Electron Acceptor 273
5.9.4 Nitrite as an Electron Acceptor 273
5.9.5 Energy Generated by Heme Biosynthesis 274
5.10 Energy Conservation by Metabolite Cycling 275
5.10.1 Hydrogen Cycling 275
5.10.2 CO Cycling 277
5.10.3 Formate Cycling 278
5.11 Substrate-Level Phosphorylation 278
5.11.1 Fermentation 278
5.11.2 Pyruvate Phosphoroclastic Reaction 279
5.11.3 Succinate-Fumarate Reactions 280
5.12 Summary and Perspective 282
References 282

6 Cell Biology and Metabolism 295
6.1 Introduction 295
6.2 Using Genomic, Proteomic, and Biochemical Analysis 295
6.2.1 Cell Surface 296
6.2.2 Metabolism of Carbon Compounds 297
6.2.3 Transition to Stationary Phase 319
6.2.4 Genomic Islands 325
6.3 Stress Response 326
6.3.1 Oxidative Stress 327
6.3.2 Starvation Response and CO_2 Stress 328
6.3.3 Extreme Temperatures 329
6.3.4 Salt Adaptation 330
6.3.5 Nitrogen Stress 331
6.3.6 pH Extremes 333
6.4 Biofilm 334
6.5 CRISPR-Cas Systems, Proviruses, and Viruses 337
6.6 Perspective 339
References 340

7 Geomicrobiology, Biotechnology, and Industrial Applications 355
7.1 Introduction 355

7.2 Contributions of SRB to Major Nutrient Cycles 355
7.2.1 Sulfur and Nitrogen Cycling 356
7.2.2 Carbon Cycling 356
7.3 H_2S Pollution: Agricultural and Commercial Impact 362
7.4 Oil Technology and SRB 364
7.5 Metabolism of Hydrocarbons 366
7.5.1 Oxidation of Environmentally Relevant Organic Compounds 366
7.5.2 Reductive Dehalogenation 371
7.6 Magnetosomes and Iron Mineralization 371
7.7 Mercury Methylation 373
7.8 Biologically Induced Minerals 377
7.8.1 Iron Sulfide Mineral Precipitation 377
7.8.2 Cu Sulfide Deposits 378
7.8.3 Zn Sulfide Deposits 379
7.8.4 Ni Sulfide Formations 379
7.8.5 Mo Sulfide Minerals 380
7.8.6 Co Sulfides 380
7.8.7 Carbonate Minerals and Dolomite 381
7.9 Reduction of Redox Active Metals and Metalloids Including U and Radionucleotides 382
7.9.1 Reduction of Metal(loid)s: Cr, Mo, Se, and As 382
7.10 Pollutants and Bioremediation Processes 388
7.10.1 Biogenic Hydrogen Sulfide Production 388
7.10.2 Acid Mine and Acid Rock Drainage Bioremediation 389
7.10.3 Uranium Remediation 390
7.10.4 Perchlorate Reduction and Use as an Inhibitor 391
7.10.5 Bioremediation of Petroleum Hydrocarbons 392
7.11 Industrial Applications 392
7.11.1 Production of Metallic Nanoparticles 393
7.11.2 Energy Technology 395
7.11.3 Dye Decolorization 396
7.12 Perspective 396
References 397

8 Biocorrosion 427
8.1 Destructive Effects of Biocorrosion 427
8.1.1 Activities of SRB Leading to Corrosion of Non-metallic Surfaces 427
8.1.2 Association of SRB with Metal Corrosion 429
8.1.3 Mechanisms for Corrosion of Ferrous Metals 431
8.1.4 Biocorrosion of Different Types of Steel Alloys 436
8.1.5 SRB–Metal Interface 437
8.1.6 Control of SRB-Based Corrosion 443

8.2 Perspective 450
References 451

9 Ecology of Dissimilatory Sulfate Reducers: Life in Extreme Conditions and Activities of SRB 463
9.1 Introduction 463
9.2 Microbes in Extreme Environments 464
9.2.1 Hyperthermophiles 464
9.2.2 Thermophiles: Extreme and Moderate 466
9.2.3 Psychrophiles 469
9.2.4 Halophiles 479
9.2.5 Alkaliphiles 486
9.2.6 Acidophiles 490
9.2.7 Piezophiles 492
9.3 Activities and Communities in Extreme, Unique, or Isolated Environments 495
9.3.1 Adaptations to the Environment 495
9.3.2 Soda Lakes and Other Alkaline Environments 499
9.3.3 SRB Growing on Surfaces 501
9.3.4 Hydrothermal Vent Sediments 508
9.3.5 Deep Subsurface and Mines 508
9.3.6 Floodplains and Estuaries 510
9.3.7 Low Nutrient Environment 511
9.4 Summary and Perspective 512
References 512

10 Interactions of SRB with Animals and Plants 529
10.1 Introduction 529
10.2 Symbiosis with Termites and a Gut-Residing Protist 530
10.3 Symbiosis with Root-Feeding Larvae 531
10.4 Symbiosis with Invertebrates 531
10.4.1 Gutless Marine Worm 531
10.4.2 Polychaete Serpulid Worm 531
10.4.3 Sea Cucumber 532
10.5 SRB Presence/Activity in Mice 532
10.6 SRB Present in Rats 533
10.7 SRB in Pigs 534
10.8 SRB Associated with Ruminates 535
10.9 Interactions of SRB with Humans 535
10.9.1 SRB as Flora of the Human Gastrointestinal Tract . . . 536
10.9.2 Oral SRB 536
10.9.3 Are *Desulfovibrio* Human Pathogens? 538
10.9.4 Antibiotic Susceptibility and Resistance 540
10.9.5 Do SRB Have Virulence Factors? 541
10.10 SRB Interactions with Plants 542

10.11 SRB on Surfaces of Living Marine Organisms 544
10.12 SRB with Other Animals 545
10.13 Summary and Perspective 546
References 546

Index 555

Chapter 1
Sulfate-Reducing Prokaryotes: Changing Paradigms

1.1 Introduction

This introductory chapter provides a historical context supporting the study of sulfate-reducing microorganisms (SRMs). Information presented is derived from early publications and reveals the evolution of the field of sulfate reduction by bacteria. While the initial concept was that the growth of sulfate-reducing bacteria (SRB) resulted using a limited number of nutrients, the reports of using a broad range of electron donors and electron acceptors provide the basis for considerable diversity of species. Early reports recognized the broad distribution of sulfate reducers in the environment and growth of these organisms is attributed to the specific electron donors and electron acceptors employed. Growth of these sulfidogenic bacteria and archaea has been summarized in several reviews (Zobell and Rittenberg 1948; Butlin et al. 1949; Postgate 1965; Postgate 1979; Gibson 1990; Odom and Singleton Jr. 1993; Barton 1995; Fauque 1995; Barton and Fauque 2009; Muyzer and Stams 2008; Rabus et al. 2015), and the reader will find these publications highly informative. To evaluate growth by these microorganisms, the energy released from reactions detailing electron donor and electron acceptor is useful, and unless stated otherwise, free energy values in this chapter (and throughout this book) relies on the report by Thauer et al. (1977). By adjusting the culture media with specific electron donors and electron acceptors, the nutrients used to support growth of sulfate-reducing prokaryotes (SRP) became established. Growth studies revealed that considerable difference occurs in the use of electron donors and electron acceptors by specific strains of SRP. The physical parameters for growth and the enzymology of growth-associated reactions are presented in subsequent chapters.

L. L. Barton, G. D. Fauque, *Sulfate-Reducing Bacteria and Archaea*,
https://doi.org/10.1007/978-3-030-96703-1_1

1.2 Nutrients for Growth: Initial Discovery Followed by an Exploration to Assess Diversity

The chemical composition of hydrogen sulfide was discovered by Carl W. Scheele in 1777, and a little over a hundred years later, hydrogen sulfide production was attributed to bacteria. Plauchud (1877) proposed that the conversion of sulfate to hydrogen sulfide was attributed to a "living" process and not to heated organic material. Hoppe-Seyler (1886) demonstrated that the bioproduction of H_2S (smell of "rotten eggs") resulted following the addition of calcium sulfate (gypsum) to anaerobic mud, and he indicated that this event could not be explained as a chemical event. A report in 1893 by N.D. Zelinsky, an organic chemist with the Russian Geographical Society, stated that the origin of hydrogen sulfide in the Black Sea "was associated with biological sulfate reduction in the presence of organic matter." While insightful comments were useful, a major scientific achievement remained, and that was the isolation of a bacterium capable of converting sulfate to sulfide. In 1895, Martinus W. Beijerinck discovered a bacterium that produced H_2S from sulfate, and he called it *Spirillum desulfuricans* (Beijerinck 1895). While Beijerinck was educated as a chemical engineer in Delft and as a botanist in Leiden, he became director of the "Netherlands Yeast and Spirits" laboratory and by occupation assumed activities of a microbiologist. One of the challenges that he addressed was how to improve the quality of canal water so it could be used as boiler feedwater. It was through his investigations addressing removal of sulfate in canal water that he isolated SRB (la Rivière 1997). In Delft, the study of SRB was continued by van Delden who isolated the marine sulfate reducer *Spirillum aestuarii* (van Delden 1903). Later, Elion described the thermophilic *Spirillum thermodesulphuricans* that reduced sulfate to hydrogen sulfide (Elion 1925). Following an examination of the biochemical and physiological activities of sulfate-reducing bacteria, Baars (1930) proposed that all the SRB should be designated as members of a single genus/species which he termed *Vibrio desulfuricans*. Starkey (1938) found that sulfate-reducing bacterial cultures could be induced to produce spores, and this led him to suggest the use of *Sporovibrio* as the genus for all SRB. However, Rittenberg (1941) was unable to induce marine isolates of SRB to sporulate and suggested a separate classification was needed for marine sulfate reducers. As reports revealed the presence of SRB in different geographical regions and environments around the world, sulfate-reducing bacteria became accepted as having global distribution (Tausson and Alioschina 1932; Copenhagen 1934; Bunker 1936, 1939; Starkey 1938; Datta 1943; Morita and Zobell 1955). The activity of SRB in marine microbiology was examined by numerous investigators, and reports by Pankhurst (1968), Güven and Çubukçu (2009), and Fashchuk (2011) provided an insight into the role of SRB in degradation of organic matter in marine sediments. Initially, biochemical, morphological, and physiological characteristics were employed to classify the strains of SRB, and recently, molecular taxonomy has been applied to identify SRB isolates. By 2018, there were 420 species of SRP belonging to 92 gênera (Barton and Fauque 2009 and unpublished information), and new isolates are being added to this number as environmental research proceeds throughout the world.

Over the years, the nomenclature and classification of SRB changed to accommodate new discoveries. Research revealed that SRB were similar to traditional bacteria in several ways, and the SRB were no longer considered as biological curiosities. At the beginning of the twentieth century, SRB were assigned to a single genus, and generic designations for SRB included *Bacillus*, *Bacterium*, *Spirillum*, *Microspira*, *Vibrio*, *Sporovibrio*, or *Desulphovibrio* (Butlin et al. 1949). In the sixth edition of Bergey's Manual of Determinative Bacteriology (Breed et al. 1948), the SRB were assigned to *Desulphovibrio*, and as a result, many of the isolated SRB carried this genus designation. However, Postgate and Campbell (1966)changed the spelling of *Desulphovibrio* to *Desulfovibrio* (substituting "f" for "ph") to conform with Latin (Postgate 1979). In the mid-1960s, sulfate-reducing bacteria were classified into two genera: (i) spore-forming *Desulfotomaculum* (Campbell and Postgate 1965) and (ii) nonspore-forming *Desulfovibrio* (Postgate and Campbell 1966). In contrast to bacteria using sulfate in the environment for the synthesis of cysteine and methionine by an activity referred to as "assimilatory sulfate reduction," the use of sulfate as an electron acceptor by SRB was designated as "dissimilatory sulfate reduction" (Postgate 1959). The enzymatic basis to distinguish dissimilatory sulfate reduction from assimilatory sulfate reduction was provided by Peck Jr. (1961).

With the development of marine microbiology, a new dimension of study resulted when thermophilic archaea were found to grow by dissimilatory sulfate reduction. A euryarchaeon, *Archaeoglobus* (*A.*) *fulgidus* (Fig. 1.1), was isolated from anaerobic submarine hydrothermal systems at Vulcano and Stufe di Nerone, Italy (Stetter 1988). This organism was found to represent a new branch of archaea which contains some characteristics of methanogens and producer of hydrogen sulfide from dissimilatory sulfate reduction (Stetter et al. 1987). Additional sulfate-reducing archaea isolated included *A. profundus in* Fig. 1.1 that was isolated from deep sea sediments of a hydrothermal system at Guaymas, Mexico (Burggraf et al. 1990), and *A. sulfaticallidus* that was isolated from the eastern edge of Juan de Fuca Ridge, eastern Pacific Ocean (Steinsbu et al. 2010). Crenarchaeota organisms isolated from hot marine or spring environments that coupled growth to sulfate reduction included *Thermocladium modestius* (Itoh et al. 1998) and *Caldivirga maquilingensis* (Itoh et al. 1999). *Vulcanisaeta moutnovskia*, a thermoacidophilic anaerobic crenarchaeon, was isolated from a hot spring in Kamchatka, Russia, and a complete set of genes for sulfate reduction pathway have been identified in this organism (Gumerov et al. 2011).

1.2.1 Broadening the Scope of Electron Donors

1.2.1.1 From Acetate to Complex Organic Molecules

In the initial cultures isolated by Beijerinck (1895), malate was the carbon source for sulfate reduction, and in 1903, van Delden introduced the use of lactate for the cultivation of SRB.

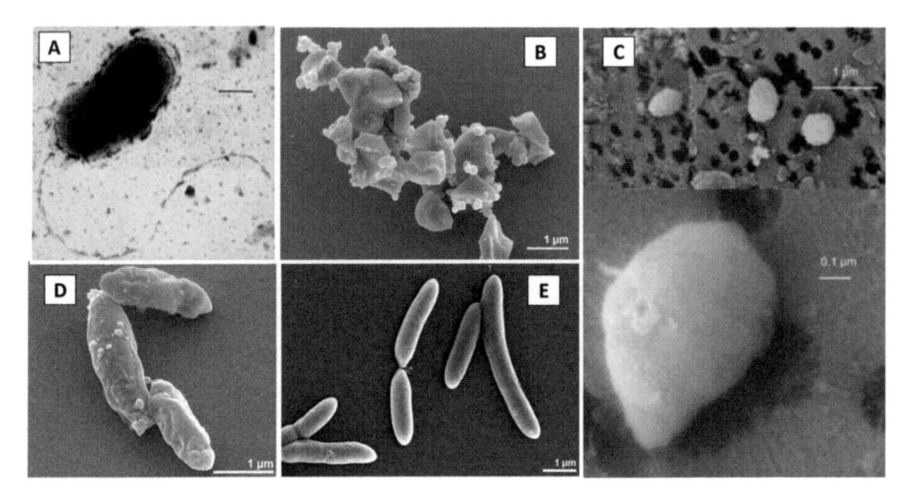

Fig. 1.1 Images of SRP. (**a**) Transmission electron micrograph of bacterium; *D. vulgaris*, bar = 0.5 microns (public domain). (**b**) Scanning electron micrograph of cells of *A. profundus* strain AV18^{T}. (von Jan et al. 2010). (**c**) *A. fulgidus* strain 7324. (**d**) Scanning electron microscopic photograph of *Dst. nigrificans*\DSM 574. (Visser et al. 2014). (**e**) Scanning electron microscopic photograph of *Dst. carboxydivorans* DSM 14880. (Visser et al. 2014)

The growth of SRB coupled to the oxidation of cellulose was described by Rubentschick (1928).

Baars (1930) isolated an acetate-oxidizing SRB which he designated as *D. rubentschikii* but that culture was lost. Following unsuccessful attempts to isolate SRB that grew on acetate and sulfate, Selwyn and Postgate (1959) concluded that the *D. rubentschikii* culture was most likely a mixed population. This conclusion is supported by the observation that polymeric sugar molecules with β-glycosidic linkages are not known to be hydrolyzed by SRB. Lactate became the most common electron donor to grow sulfate reducers, and acetate was commonly reported as the end product of metabolism (Campbell and Postgate 1965; Postgate and Campbell 1966). The reaction describing catabolism of lactate is as follows:

$$2\text{Lactate}^- + SO_4{}^{2-} + H^+ \rightarrow 2\text{acetate}^- + 2CO_2 + 2H_2O + HS^- \quad \Delta G^{0'} = -196.4\ \text{kJ/mol sulfate} \tag{1.1}$$

With acetate as the end product of the reaction, lactate oxidation with sulfate reduction by sulfate reducers represents incomplete oxidation, and many of the sulfate reducers display this reaction. Changes in classification of SRB occurred in the 1960s with *D. desulfuricans* Hildenborough being designated as *D. vulgaris* Hildenborough. Images of *D. vulgaris* Hildenborough and other sulfate reducers are given in Fig. 1.1. This change in designation of species should be recalled when reviewing the older literature. Incomplete oxidation of substrates by SRB with the production of acetate was a concern to Jørgensen and Fenchel (1974) who had proposed that over two-thirds of organic residue in marine sediments would remain if SRB and other anaerobes could not degrade acetate. The importance of SRB in marine environments (Pankhurst 1968; Güven and Çubukçu 2009; Fashchuk 2011) was confirmed with the report that 50% of organic decomposition of marine sediment was attributed to the action of SRB (Jørgensen 1977). The isolation of *Desulfotomaculum* (*Dst.*) *acetoxidans* and *Desulfobacter postgatei* provided evidence for the existence of SRB with the capability of growing on acetate (Widdel and Pfennig 1977; Widdel and Pfennig 1981a). Acetate is oxidized according to the following reaction:

$$CH_3COO^- + SO_4^{2-} \rightarrow HS^- + 2HCO_3{}^- \quad \Delta G^{0'} = -44.5\ \text{kJ/mol acetate} \tag{1.2}$$

Then, with remarkable skill, Friedrich Widdel (1980) isolated several unique SRB capable of growing on higher and lower fatty acids. This development that SRB could be cultured on organic substrates other than lactate or pyruvate (Widdel 1980) provided an impetus for discovering new isolates that grew by oxidizing variety of carbon compounds (Widdel 1988). A selection of sulfate-reducing microorganisms and their electron donors is given in Tables 1.1 and 1.2.

Based on the utilization of organic substrates, SRB were divided into two metabolic groups: (i) incomplete oxidizers of organic substrates producing acetate and CO_2 and (ii) complete oxidizers of organic substrates producing only CO_2 (Widdel 1988). Complete oxidation of lactate by SRB would be according to the following equation:

$$CH_3CHOHCOO^- + 1.5SO_4^{2-} + 4H^+ \rightarrow 3CO_2 + 3H_2O + 1.5HS^- + 1.5H^+ \quad \Delta G^{0'} = -150\ \text{kJ/mol sulfate} \tag{1.3}$$

Table 1.1 Examples of sulfate-reducing prokaryotes indicating the range of electron donors used by these microorganisms

	Electron donors									Substrate oxidation	
Microorganisms	H	Fo	Py	La	Et	Ac	Fu	Ma	FA (C atoms)	(C or I)	Reference
Archaea											
Archaeoglobus fulgidus	+	+	+	+	+	+	+	nr	−	I	Thauer and Kunow (1995)
Bacteria											
Desulfarculus baarsii ATCC 33931	−	+	−	nr	−	+	−	−	3–18	C	Kuever (2014, pp. 42–45)
Desulfatibacillum aliphaticivorans DSM 15576	+	+	+	−	−	+	+	+	3–18	C	Kuever (2014, pp. 46–73)
Desulfobacca acetoxidans DSM 11109	+	−	−	−	−	+	−	nr	−	C	Göker et al. (2011)
Desulfobacter postgatei ATCC 33911	−	−	−	−	−	+	−	−	−	C	Kuever (2014, pp. 46–73)
Desulfobacterium autotrophicum DSMZ 3382	+	+	+	+	+	+	+	+	3–14	I	Brysch et al. (1987)
Desulfobacula toluolica DSM 7467	−	−	+	−	+	−	+	+	3–14	C	Rabus et al. (1993)
Desulfobotulus sapovorans DSM 2055	−	−	−	+	−	−	−	−	4–16	I	Kuever (2014, pp. 46–73)
Desulfobulbus propionicus DSM 2032	+	−	+	+	+	−	−	−	3	I	Widdel and Pfennig (1982)
Desulfocapsa sulfoexigens DSM 10523	+	+	−	−	−	−	−	−	−	nr	Kuever (2014, pp. 75–86)
Desulfocella halophila DSM 11763 T	−	−	+	−	−	−	−	−	4–16	I	Brandt et al. (1999)
Desulfococcus multivorans DSM 2059	−	+	+	+	+	+	−	−	3–16	C	Kuever (2014, pp. 46–73)
Desulfofaba gelida DSM 12344	−	+	+	+	+	−	+	+	3, 4	I	Knoblauch et al. (1999)
Desulfofrigus oceanense DSM 12341	+	+	+	+	+	+	−	+	4, 5	C	Kuever (2014, pp. 46–73)
Desulfofustis glycolicus DSM 9705		−	**nr**	+	−	−	+	+	−	I	Friedrich et al. (1996)
Desulfohalobium retbaense DSM 5692	+	+	+	−	−	−	−	−	−	I	Ollivier et al. (1991)
Desulfomicrobium baculatum DSM 4028	+	+	+	+	−	−	+	+	−	I	Kuever and Galushko (2014)
Desulfomonile limimaris ATCC 700979[T]	+	+	+	+	nr	−	−	nr	3, 4	I	Sun et al. (2001)
Desulfonatronovibrio hydrogenovorans DSM 9292	+	+	−	−	−	−	nr	nr	−	I	Zhilina et al. (1997)
Desulfonatronum thiodismutans DSM 14708[T]	+	+	−	−	+	−	nr	nr	−	I	Pikuta et al. (2003)
Desulfonauticus submarinus DSM 15269[T]	+	+	nr	−	−	−	−	−	nr	C	Audiffrin et al. (2003)

Desulfonema limicola ATCC 33961	+	+	+	+	–	+	+	–	3–14	C	Kuever (2014, pp. 46–73)
Desulforegula conservatrix DSM 13527^{T}	–	–	–	–	–	–	–	–	4–17	I	Rees and Patel (2001)
Desulforhabdus amnigenus DSM 10338	+	+	–	+	+	+	nr	nr	3–4	C	Elferink et al. (1995)
Desulforhopalus vacuolatus DSM 9700	+	–	+	+	+	–	–	nr	3	I	Isaksen and Teske (1996)
Desulfosarcina variabilis DSM 2060	+	+	+	+	+	+	+	–	3–14	C	Kuever (2014, pp. 75–86)
Desulfospira joergensenii DSM	+	+	+	+	nr	–	+	–	4,8,10,12	C	Finster et al. (1997)
Desulfosporosinus orientis DSM 765^{T}	+	+	+	+	+	–	+	–	6, 8	I	Hippe and Stackebrandt (2009)
Desulfotalea psychrophila DSM 12343	+	+	+	+	+	–	+	+	–	I	Kuever (2014, pp. 75–86)
Desulfotignum balticum DSM 7044^{T}	+	+	+	+	–	+	+	+	4,10,12,16,18	C	Kuever et al. (2001)
Desulfotomaculum nigrificans DSM 574	+	+	+	+	–	–	–	–	–	I	Kuever and Rainey (2009)
Desulfovibrio africanus NCIB 8401	+	+	+	+	+	–	–	+	–	I	Campbell et al. (1966)
Desulfovibrio desulfuricans DSM 642	+	+	+	+	+	–	+	+	–	I	Widdel (1988)
Desulfovibrio salexigens DSM 2638	+	+	+	+	+	–	–	+	–	I	Widdel (1988)
Desulfovibrio vulgaris ATCC 29579	+	+	+	+	+	–	+	+	–	I	Widdel (1988)
Thermodesulfovibrio yellowstonii ATCC 51303	+	+	+	+	–	–	nr	–	nr	I	Henry et al. (1994)

Abbreviations: *C* complete, *I* incomplete, *nr* not reported, *H* H_2, *Fo* formate, *Py* pyruvate, *La* lactate, *Et* ethanol, *Ac* acetate, *Fu* fumarate, *Ma* malate, *FA* fatty acid with number of carbon atoms as indicated

Table 1.2 Examples of sulfate-reducing bacteria displaying the use of various electron acceptors in addition to S^0

Bacteria	e^- donors	Habitat	Optimum growth, °C	e^- acceptors[+]	References
Ammonifex thiophilus	H_2, formate	F	75	Thiosulfate	Miroshnichenko et al. (2008)
Ammonifex degensii	H_2, formate	F	*70*	*Nitrate*	Huber et al. (1996)
Desulfofustis glycolicus	Glycolate, OAS, H_2	M	28	Sulfite	Friedrich et al. (1996)
Desulfohalobium retbaense	H_2, lactate, pyruvate, ethanol	M	37–40	Thiosulfate, sulfite	Ollivier et al. (1991)
Desulfomicrobium baculatum Norway 4 DSM 1741	H_2, SCA	F	35	Thiosulfate, sulfite	Biebl and Pfennig (1977)
Desulfomicrobium baculatum DSM 1743	H_2, SCA	F	35	Thiosulfate, sulfite	Biebl and Pfennig (1977)
Desulfonauticus submarinus	H_2, OA	M	45	Thiosulfate, sulfite	Audiffrin et al. (2003)
Desulfosarcina cetonica[a]	Fatty acids	F	30	Thiosulfate	Galushko and Rozanova (1991)
Desulfospira joergensenii	H_2, SCA	M	26–30	Thiosulfate, sulfite	Finster et al. (1997)
Desulfosporosinus meridiei	H_2, SCA	F	10–37	Sulfite, thiosulfate, DMSO, Fe(III)	Robertson et al. (2001)
Desulfovermiculus halophilus	H_2, SCA	M	37	Thiosulfate, sulfite	Beliakova et al. (2006)
Desulfovibrio alcoholovorans	Glycerol, H_2, SCA	F	37	Thiosulfate, sulfite	Qatibi et al. (1991)
Desulfovibrio bastinii	H_2, SCA	F	37	Thiosulfate, sulfite	Magot et al. (2004)
Desulfovibrio biadhensis	H_2, SCA	F	37	Thiosulfate, sulfite	Fadhlaoui et al. (2015)
Desulfovibrio bizertensis	H_2, SCA	M	40	Thiosulfate, sulfite, fumarate	Haouari et al. (2006)
Desulfovibrio burkinensis	Glycerol, H_2, SCA	F	37	Thiosulfate, sulfite, fumarate	Ouattara et al. (1999)
Desulfovibrio fructovorans	Fructose, H_2, SCA	F	35	Thiosulfate, sulfite, fumarate	Ollivier et al. (1988)
Desulfovibrio gabonensis	H_2, SCA	F	30	Thiosulfate, sulfite	Tardy-Jacquenod et al. (1996)

(continued)

Table 1.2 (continued)

Bacteria	e^- donors	Habitat	Optimum growth, °C	e^- acceptors[+]	References
Desulfovibrio gigas	H_2, SCA	F	35	Thiosulfate, sulfite	Biebl and Pfennig (1977)
Desulfovibrio gracilis	H_2, SCA	F	37	Thiosulfate, sulfite, fumarate	Magot et al. (2004)
Desulfovibrio idahonensis	H_2, SCA	F	30	Thiosulfate, sulfite, DMSO, fumarate	Sass et al. (2009)
Desulfovibrio legallis	H_2, SCA	F	35	Thiosulfate, sulfite	Ben Dhia Thabet et al. (2011)
Desulfovibrio longus	H_2, SCA	F	35	Thiosulfate, sulfite, fumarate	Magot et al. (1992)

SCA short-chain organic acids such as lactate, formate, pyruvate, *OA* fatty acids, dicarboxylic acids, oxoacids, hydroxyacids, *OAS* organic acids, sugars, *DMSO* dimethylsulfoxide, *F* fresh water and *M* marine water

[a]Formerly *Desulfobacterium cetonicum* (Stackebrandt et al. 2003)

While there are >70 organic compounds (hydrocarbons, monocarboxylic and dicarboxylic acids, alcohols, amino acids, sugars, and aromatic compounds) that can serve as electron donors for SRB, many of these substrates are utilized by a specific species or strain of SRB. The most commonly used substrates by *Desulfovibrio* species include lactate, pyruvate, malate, fumarate, ethanol, glycerol, and H_2 (Fauque et al. 1991; Hansen 1993). It should be noted that a few species of SRB are capable of using sugars as electron donors. *D. fructosovorans* uses fructose (Ollivier et al. 1988), *D. simplex* grows on glucose (Zellner et al. 1989), while *Dst. carboxydivorans* and *Dst. nigrificans* metabolize glucose and fructose for growth (Parshina et al. 2005b). *A. fulgidus* uses complex peptides (Hocking et al. 2014) and starch (Labes and Schönheit 2001) as electron sources with the reduction of sulfate. Images of some of these SRP are given in Fig. 1.1.

1.2.1.2 Growth Coupled to CO Oxidation

The oxidation of CO to CO_2 by *D. vulgaris* was first observed by Yagi (1958, 1959), and several decades later, the growth of *D. vulgaris* strain Madison on CO in sulfate medium was reported (Lupton et al. 1984). Growth on CO was slow and was attributed to H_2 formation from CO oxidation by a hydrogenase that was insensitive to CO. The source of H_2 in CO oxidation was attributed to "H_2 cycling" (Peck Jr. 1993). Metabolism of CO in *D. vulgaris* Hildenborough is also associated with endogenous "CO cycling" (Voordouw 2002). Utilization of CO as the electron donor with sulfate as electron acceptor was discovered in *Desulfotomaculum* sp. RHT-3 growing in 50% CO and *Dst. nigrificans* growing in 20% CO (Klemps et al. 1985). Reactions associated with CO metabolism are listed in Eqs. 1.4–1.8 (Parshina et al. 2010). When grown in the presence of sulfate, the apparent reaction is in Eq. 1.4. With sulfate, *Dst. kuznetsovii* and *Dst. thermobenzoicum* subsp. *thermosyntrophicum* oxidize CO producing transient levels of acetate (Eq. 1.5), but in co-culture with *Carboxydothermus hydrogenoformans* (Svetlichny et al. 1991), H_2 production is an intermediate (Eq. 1.6) (Parshina et al. 2005a; Parshina et al. 2005b). Acetate can also be produced by the combination of Eqs. 1.6 and 1.7. When *Dst. carboxydivorans* strain CO-1-SRBT grows in 100% CO with sulfate, H_2 is produced and serves as the immediate electron donor (see Eqs. 1.6 and 1.8) (Parshina et al. 2005b):

$$4\mathrm{CO} + \mathrm{SO_4^{2-}} + 4\mathrm{H_2O} \rightarrow 4\mathrm{HCO_3^-} + \mathrm{HS^-} + 3\mathrm{H^+} \quad \Delta G^0 = -37.1\ \mathrm{kJ/mol\ CO} \tag{1.4}$$

$$4\mathrm{CO} + 4\mathrm{H_2O^-} \rightarrow \mathrm{acetate^-} + 2\mathrm{HCO_3^-} + 3\mathrm{H^+} \quad \Delta G^0 = -28.2\ \mathrm{kJ/mol\ CO} \tag{1.5}$$

$$\mathrm{CO} + 2\mathrm{H_2O} \rightarrow \mathrm{HCO_3^-} + \mathrm{H_2} + \mathrm{H^+} \quad \Delta G^0 = -28.2\ \mathrm{kJ/mol\ CO} \tag{1.6}$$

$$4\mathrm{H_2} + 2\mathrm{HCO_3^-} + \mathrm{H^+} \rightarrow \mathrm{acetate^-} + 4\mathrm{H_2O} \quad \Delta G^0 = 36.4\ \mathrm{kJ/mol\ H_2} \tag{1.7}$$

$$4\mathrm{H_2} + \mathrm{SO_4^{2-}} + \mathrm{H^+} \rightarrow \mathrm{HS^-} + 4\mathrm{H_2O} \quad \Delta G^0 = -45.2\ \mathrm{kJ/mol\ H_2} \tag{1.8}$$

When *A. fulgidus* grow in a medium containing sulfate and CO, production of H_2 was not observed, but formate was transiently produced, and formate was used as the immediate electron donor (Henstra et al. 2007).

1.2.1.3 Hydrogen as an Electron Donor

While the reports of Nikitinsky (1907), Kroulik (1913), and Niklewski (1914) provided some of the first suggestions that sulfate-reducing bacteria in mixed cultures oxidized H_2, the utilization of H_2 by a pure culture of sulfate reducer came later. *D. desulfuricans* (now *vulgaris*) Hildenborough was reported by Stephenson and Stickland (1931) to use H_2 to reduce sulfate to hydrogen sulfide, and this observation was confirmed by Postgate (1949). The manometric measurement of H_2 uptake (oxidation) became a useful method to quantify the amount of electron acceptor used to reduce sulfur oxyanions (Postgate 1951a, b). The oxidation of H_2 coupled to reduction of sulfate is expressed in Eq. 1.8.

Hydrogenase activity was considered to be broadly distributed in SRB with only a few SRB lacking the ability to oxidize H_2 (Adams et al. 1951; Sisler and Zobell 1951). Hydrogen oxidation by cell extracts from SRB was coupled to sulfur compounds (Ishimoto et al. 1954) and was also used to examine hydroxylamine as an electron acceptor (Senez and Pichinoty 1958a; Senez and Pichinoty 1958b). The characterization of hydrogenase from *D. vulgaris* was pursued (Sadana and Jagannathan 1954), and the coupling of ATP synthesis to H_2 oxidation established the importance of hydrogenase for energetics of SRB (Peck 1959). H_2 oxidation was recognized as important to energize respiratory reduction of sulfite and sulfate (Postgate 1951b), reduction of elemental sulfur (Stetter and Gaag 1983), methane production (Bryant 1979), nitrate reduction (Wolin et al. 1961), and acetate production (Wieringa 1940). Early reports concerning H_2 oxidation to support the growth of *Desulfovibrio* spp. were discussed in a review by Postgate (1965). Several SRB strains were demonstrated to grow in a mineral salt medium containing acetate, sulfate, and CO_2 with H_2 as the electron source, and these bacteria include *D. vulgaris* strain Madison, *D. vulgaris* strain Marburg, *D. vulgaris* Hildenborough, *D. desulfuricans* strain Essex 6, and *D. gigas* (Badziong et al. 1978, 1979; Badziong and Thauer 1978). It has been proposed that H_2 formation by geological activities is an important energy source in deep environments of the Earth where organic materials are limiting (Hoehler et al. 1998; Chivian et al. 2008; D'Hondt et al. 2009). As reviewed by Barton and McLean (2019; Schrenk et al. 2013; Ménez et al. 2012; Stevens and Mckinley 1995), abiotic production of H_2 can result from several processes: (i) radiolysis of water attributed to U^{238} and Th^{232} in the rocky environment, (ii) addition of water to FeO, or (iii) addition of water to Fe_2SiO_4. Information concerning hydrogenases that have been isolated and purified from different strains of SRB is discussed in Chap. 5.

1.2.1.4 Inorganic Sulfur Compounds as Electron Donors

In an interesting application of sulfur pathway enzymology, several strains of SRB are capable of oxidizing sulfite, thiosulfate, or elemental sulfur with O_2 as the electron acceptor but do not grow with this transfer of electrons (Dannenberg et al. 1992). *D. propionicus* oxidized sulfite and sulfide to sulfate, while elemental sulfur (S^0) was oxidized to thiosulfate, and *D. desulfuricans* oxidizes sulfite, thiosulfate, or S^0 to sulfate. With nitrite as the electron acceptor, *D. desulfuricans* and *D. propionicus* oxidized sulfide to sulfate. Oxidation of S^0 to sulfate was observed with *D. desulfuricans* ATCC 29577, *Desulfomicrobium baculatum* (DSM 1741), *Desulfobacterium autotrophicum* DSN 3382, and *Desulfuromonas acetoxidans* DSM 684 with Mn(IV) as the electron acceptor, but growth did not occur (Lovley and Phillips 1994).

An interesting observation concerning sulfide oxidation coupled to dissimilatory nitrate and nitrite reduction by *D. alkaliphilus* is that the genome of this organism has all the genes used for dissimilatory sulfate reduction but it is unable to reduce sulfate to sulfide (Thorup et al. 2017). The authors propose that the dissimilatory sulfate reductase enzymes are involved in conversion of sulfide to sulfate where nitrate is reduced to ammonium by a periplasmic nitrate reductase and a membrane-bound nitrite reductase.

1.2.1.5 Phosphite as an Electron Donor

Desulfotignum phosphitoxidans oxidizes phosphite to phosphate by a unique process not found in other sulfate-reducing bacteria (Schink et al. 2002). Based on 16S rRNA and *dsrAB* analysis, *Desulfotignum phosphitoxidans* is closely related to *Desulfotignum balticum.* Growth of this *Deltaproteobacteria* is coupled to dissimilatory sulfate reduction according to the following reaction (Schink et al. 2002; Poehlein et al. 2013):

$$4HPO_3^{2-} + SO_4^{2-} + H^+ \rightarrow 4HPO_4^{2-} + HS^- \quad \Delta G^{0'} = -364\ \mathrm{kJ/mol\ sulfate, or} - 91\ \mathrm{kJ/mol\ phosphite} \quad (1.9)$$

Previously, phosphite and hypophosphite have been found to serve as phosphorous sources for aerobic and anaerobic bacterial cultures. Phosphite occurs in marine and patty field water but its origin is unknown.

1.2.1.6 Extracellular Electron Transport Donors

In the anaerobic oxidation of methane, extracellular electron transport between methanogens and uncultured SRB has been proposed to be facilitated by conductive filaments (McGlynn et al. 2015; Wegener et al. 2015; Scheller et al. 2016). Interspecies electron transfer involving redox active proteins (i.e., filaments) present in the extracellular region between anaerobic methane-oxidizing archaea (ANME)-SRB syntrophic consortia was reported (McGlynn et al. 2015). This interspecies transfer of electrons from ANME to SRB was decoupled by soluble artificial electron acceptors (Wegener et al. 2015; Scheller et al. 2016).

D. ferrophilus IS5 acquires electrons directly from elemental iron to support growth and sulfate reduction by a process similar to that in anaerobic corrosion of ferrous iron (Dinh et al. 2004). The mechanism of this extracellular transport of electrons involves outer membrane multiheme cytochromes located on the surface of the cell or on the nanowires (Deng et al. 2018) by a process used by *Shewanella oneidensis* and *Geobacter sulfurreducens* for the extraction of electrons from extracellular elemental iron (Shi et al. 2009) where electrons cross the lipid membranes as characterized by Hartshorne et al. (2009). It is proposed that many uncultivated SRB in the deep marine environment may use extracellular filaments or nanowires to move electrons from MnS or FeS to SRB cells (Deng et al. 2018), and this will be an important research field for the future. While the quantity of electrons moved by extracellular processes is unknown, it is sufficient to energize SRB. When organic electron donors are limiting in anaerobic marine subsurface but sulfur species are available as electron acceptors, inorganic compounds have long been considered to be an important electron source for microbial metabolism (Ravenschlag et al. 2000; Mußmann et al. 2005; Muyzer and Stams 2008).

1.2.2 Diverse Electron Acceptors

SRB responds to nutrient deficiencies in the environment, and alternate electron acceptors become important when sulfate is limiting. The capability of using a specific chemical as an alternate electron acceptor is usually restricted to a few species of SRB. At this time, extracellular electron acceptors have not been reported for SRB, but this may reflect the lack of research initiatives to pursue this topic. With the recently reported presence of outer membrane cytochromes and nanowires in *D. ferrophilus* IS5 (Deng et al. 2018), it will be interesting to determine if some SRB are capable of using humic substances as electron acceptors as has been reported for *Geobacter metallireducens* and *Shewanella alga* (Lovley et al. 1996). A listing of appropriate electron acceptors for several different sulfate-reducing microorganisms is given in Tables 1.1 and 1.2.

1.2.2.1 Sulfur Oxyanions

In the first part of the twentieth century, many considered that members of the genera *Desulfovibrio* and *Desulfotomaculum* represented unique forms of life and these bacteria could only use sulfate as the electron acceptor. By definition, all SRB and archaea use sulfate as the terminal electron acceptor. Early research indicated that *Desulfovibrio desulfuricans* (now *Desulfovibrio vulgaris*) strain Hildenborough coupled H_2 oxidation to the reduction of sulfite, thiosulfate, tetrathionate ($S_4O_6^{2-}$), or dithionate ($S_2O_6^{2-}$) (Postgate 1951b), and trithionate ($S_3O_6^{2-}$) reduction by this same species was characterized by Kim and Akagi (1985). Demonstration of thiosulfate, tetrathionate, or dithionate reductase is not an indication that a bacterium will grow on that substrate because the enzyme may be used for assimilatory sulfate reduction. Trithionate and tetrathionate are polythionates, and the chemical structures of dithionate, thiosulfate, and polythionates are given in Fig. 1.2. Dithionate is a strong reducing agent and is not employed for growing bacteria, while thiosulfate and polythionates are suitable sulfur compounds for cultivation of bacteria. Thiosulfate and polythionates are produced transiently in the environment with thiosulfate most stable at neutral and basic pH, while polythionates are most stable in acidic environments (Le Faou et al. 1990). Most commonly, thiosulfate and sulfite are reported to serve as electron acceptors for dissimilatory sulfate reducers, and examples of microorganisms with this capability are given in Table 1.1.

1.2.2.2 Elemental Sulfur

The solubility of elemental sulfur in water is only 0.16 mmol per liter at 25 °C (Boulègue 1978), and for microorganisms to metabolize elemental sulfur, they must convert sulfur to a "hydrophilic" form which contains oxo compounds such as polythionates. When cultivating bacteria on elemental sulfur, the source of colloidal

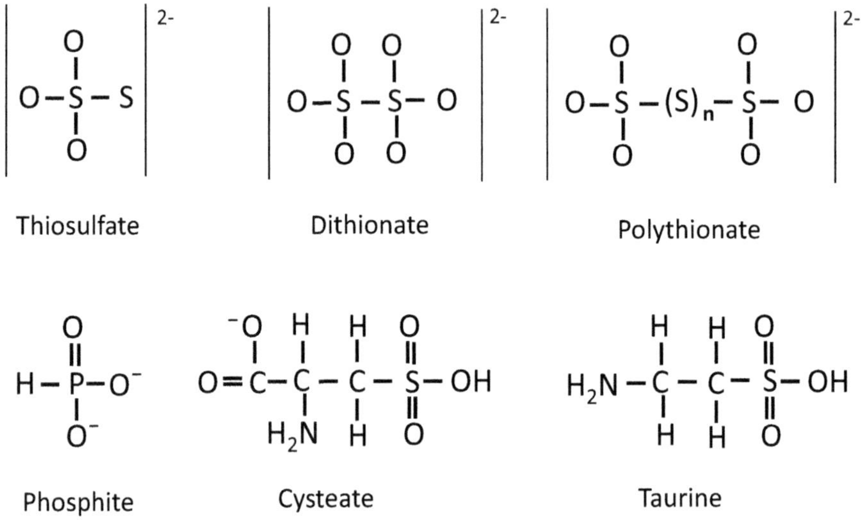

Thiosulfate Dithionate Polythionate

Phosphite Cysteate Taurine

Fig. 1.2 Structures of low molecular weight compounds serving as electron acceptors for sulfate-reducing bacteria and archaea

sulfur is important because the physical form of sulfur may be different from that found in the environment (Le Faou et al. 1990). Sulfur globules from *Thiobacillus ferrooxidans* are proposed to have the chemical formula of $xS_8 \bullet yH_2S_nO_6 \bullet zH_20\text{-}S_8$ (Steudel et al. 1987; Steudel 1989). Sulfur flour contains orthorhombic sulfur associated with chains of sulfur atoms (Domange 1982), while formation of elemental sulfur from the addition of hydrochloric acid to thiosulfate produces chains of sulfur atoms associated with thiosulfuric acid (Pauli 1947). Elemental sulfur (S_8) can be activated in the presence of sulfite (HSO_3^-) to produce $HS_8SO_3^-$ which is soluble in membranes (Le Faou et al. 1990).

Also, elemental sulfur can be dissolved in sulfide at a sulfur/sulfide ratio of 1:10 to produce a polysulfide, and polysulfide was proposed as the substrate for sulfate reduction (Schauder and Müller 1993). The conversion of the S_8-ring of elemental sulfur to a soluble form occurs with the nucleophilic attack of HS^- anion on the sulfur crown to produce polysulfide (Cammack et al. 1984; Steudel et al. 1986; Hedderich et al. 1999). Various microorganisms are capable of using elemental sulfur (S^0) to support growth, and several prokaryotes are capable of reducing S^0 (Le Faou et al. 1990; Schauder and Kröger 1993; Widdel and Hansen 1992). Several SRB are facultative sulfur-reducing bacteria in that they use elemental sulfur as the alternate electron acceptor in respiratory metabolism to support growth (Fauque and Barton 2012) when other appropriate electron acceptors are unavailable. Growth with S^0 reduction has been reported for *D. multispirans* NCIB 12078, *D. carbinolicus* DSM 3852, *D. desulfuricans* strain DK81, *D. sapovorans* strain DKbu 14, *Dst. saponomandens* strain Pato DSM 3223, and *D. vulgaris* strain Woolwich NCIB 8457 (He 1987; Ollivier et al. 1988; Cord-Ruwish and Garcia

1985; Nanninga and Gottschal 1986, 1987; Al-Hitti et al. 1983). Additional examples of sulfate reducers that reduce elemental sulfur are listed in Table 1.2.

Although small levels of sulfide were produced from S^0 by type strains of *D. desulfuricans*, *D. vulgaris*, and *D. pigra*, these cultures were reported to be unable to grow on S^0 (Biebl and Pfennig 1977). When *D. sapovorans*, *D. postageti*, *Dst. acetoxidans*, and several strains of *Desulfonema* are growing with sulfate, S^0 inhibits their growth (Widdel and Pfennig 1981a, b; Widdel 1988). Tetraheme cytochrome c_3 functions as the sulfur reductase for several sulfate-reducing bacteria, and the sensitivity of cytochrome c_3 from *D. vulgaris* Hildenborough (NCIB 8303) to sulfide would account for the inability of this bacterium to grow on S^0 (Fauque et al. 1979).

1.2.2.3 Nitrogen Compounds

Early reports that cells of *D. desulfuricans* reduced nitrite and hydroxylamine (Senez et al. 1956; Senez and Pichinoty 1958a, b) suggested that SRB could have a role in reducing oxidized compounds, but due to the possibility of nonenzymatic interactions attributed to nitrite and hydroxylamine, these observations were not continued. Some microbiologists were reluctant to believe that SRB could also display nitrate respiration because it would reduce the uniqueness of bacteria growing by sulfate respiration. For some time, it was considered that there was a possibility that nitrite reduction by SRB was attributed to the sulfite reductases. In the early 1980s, the use of nitrate as an alternate electron acceptor for the sulfate-reducing bacteria was explored, and dissimilatory nitrate reduction was reported for *Desulfobulbus propionicus* (Widdel and Pfennig 1982), *D. desulfuricans* DT01 (Keith and Herbert 1983), *Desulfovibrio* sp. (McCready et al. 1983), and *D. desulfuricans* strain Essex (Seitz and Cypionka 1986). In subsequent years, growth with nitrate substituted for sulfate was reported for *Desulforhopalus singaporensis* (Lie et al. 1999), *D. desulfuricans* ATCC 27774 (Marietou et al. 2009), *D. profundus* (Bale et al. 1997), *D. termitidis* (Trinkerl et al. 1990), *Thermodesulfovibrio islandicus* (Sonne-Hansen and Ahring 1999), *D. furfuralis* (Folkerts et al. 1989), *D. oxamicus* strain Monticello DSM 1925 (López-Cortés et al. 2006), *Dm. catecholicum* (Szewzyk and Pfennig 1987), *Thermodesulfobium narugense* (Mori et al. 2003), *D. thermobenzoicum* (Tasaki et al. 1991), *Desulfosporosinus acididurans* (Sánchez-Andrea et al. 2015), and *Dst. ferrireducens* (Yang et al. 2016).

While only a few SRB are able to use nitrate as an alternate to sulfate as a final electron acceptor, it should be noted that the nitrate-respiring bacteria are associated with several different phyla (Marietou 2016). The inducibility of nitrate reduction and the preference of sulfate over nitrate as electron acceptor vary with the strain of SRB (Moura et al. 2007). Unlike nitrate reduction in *Escherichia coli* where nitrite accumulates, nitrate is reduced to ammonia by several strains of sulfate-reducing bacteria which indicate that nitrite reduction occurs along with nitrate reduction. *Desulfobulbus propionicus* and *D. desulfuricans* strain Essex 6 and *D. desulfuricans* strain ATCC 27,774 grow with nitrite as the terminal electron acceptor

(Seitz and Cypionka 1986; Marietou et al. 2009). Cyanobacterial mats consist of different zones with an oxygen-generating upper photosynthesizing zone and an anaerobic zone containing SRB. The enzymology of nitrate and nitrite respiration is discussed in Chap. 2.

1.2.2.4 Fumarate

When using a sulfate-free medium, several marine strains of *D. desulfuricans* were reported to use fumarate as an electron acceptor with H_2 oxidation (Sisler and Zobell 1951). Shortly thereafter, Grossman and Postgate (1955) suggested a fumarate-succinate cycle in SRB, and they established that fumarate reduction occurred with H_2 when using cells of *D. desulfuricans* strain El Agheila Z and *D. desulfuricans* strain California 43:63 but not with cells of *D. desulfuricans* strains New Jersey SW8 and Canet 41. This reduction of fumarate by H_2 is by the following reaction:

$$\text{Fumarate}^- + H_2 \rightarrow \text{succinate}^- \quad (1.10)$$

Energy from the reaction of H_2 oxidation with fumarate supports bacterial growth, and experiments with cell-free extracts from *D. gigas* revealed esterification of orthophosphate coupled to the transfer of electrons from H_2 to fumarate (Barton et al. 1970). In the absence of sulfate, hydrogen oxidation was observed with *D. desulfuricans* El Agheila Z when malate was present, and this suggested that malate was converted to fumarate by fumarase (or fumarate hydratase) and fumarate was actually the electron acceptor (several species of *Desulfovibrio* and *Desulfotomaculum* are capable of using fumarate as an alternate electron acceptor).

1.2.2.5 Dismutation of Organic Compounds

In the absence of sulfate, fumarate (Miller and Wakerley 1966) and malate (Miller et al. 1970) served as both electron donor and acceptor by a process termed dismutation or disproportionation. Metabolism of fumarate produces pyruvate which is oxidized to acetate, and electrons from this reaction are directed to additional molecules of fumarate which is reduced to succinate (see the following):

$$\begin{array}{l} 3\ \text{Fumarate} \rightarrow \text{pyruvate} \rightarrow \text{acetate} + CO_2 \\ \quad\quad\quad \downarrow \\ 2\ \text{Succinate} \end{array}$$

D. desulfuricans strain Essex 6 grows on fumarate in the absence of sulfate, and the enzymes required for this metabolism are found in the genomes of *D. desulfuricans* (now *alaskensis*) G20 and *D. vulgaris* Hildenborough (Zaunmüller et al. 2006). Malate is metabolized to pyruvate which is oxidized to acetate, and fumarate is the electron acceptor. See the following reaction:

$$
\begin{array}{l}
3\ \text{Malate} \rightarrow \text{lactate} \rightarrow \text{pyruvate} \rightarrow \text{acetate} + CO_2 \\
\qquad \downarrow \\
2\ \text{Fumarate} \rightarrow 2\ \text{succinate}
\end{array}
$$

1.2.2.6 Oxygen

After many decades of designating SRB as obligate anaerobes, there was the slow acceptance of the concept that oxygen could serve as an electron acceptor for some *Desulfovibrio*. Cell-free extracts of *D. desulfuricans* El Agheila Z and *D. vulgaris* Hildenborough were used to transfer electrons from H_2 oxidation to O_2 (the N_2 gas phase contained ~4%) with cytochrome as an intermediate electron carrier (Postgate 1951c; Grossman and Postgate 1955). Algal mats and stromatolites exposed to the atmosphere were found to harbor SRB that actively reduced sulfate to hydrogen sulfide (Jørgensen and Cohen 1977; Krumbein et al. 1979). These reports were supported by additional metabolic and ecological studies of SRB in cyanobacterial marine and lake mats as well as in oxic marine sediments (Canfield and Des Marais 1991; Fründ and Cohen 1992; Visscher et al. 1992; Teske et al. 1998; Wierenga et al. 2000; Jonkers et al. 2003; Bühring et al. 2005). *D. desulfuricans* remained viable on the surface of an agar plate exposed to atmospheric oxygen for 2 da (Abdollahi and Wimpenny 1990). Five strains of *D. vulgaris* isolated from the North Sea were found to retain viability after a 3-day exposure to atmospheric oxygen (Hardy and Hamilton 1981). After air sparging cultures of SRB for 3 hr, growth of *D. vulgaris*, *D. desulfuricans*, *D. salexigens*, and *Desulfobacter postgatei* was observed, while the viability of *Dst. ruminis*, *Dst. nigrificans*, *Dst. orientis*, and *Desulfococcus multivorans* cultures was greatly diminished (Cypionka et al. 1985).

Another set of experiments demonstrated the capability of SRB to use oxygen as the final electron acceptor. *Desulfobacter postgatei* oxidized acetate to CO_2 with O_2, while *D. desulfuricans* CSN and *Desulfobulbus propionicus* oxidized sulfite or sulfide to sulfate with O_2 as the electron acceptor (Dannenberg et al. 1992). Aerobic metabolism of polyglucose involving reduction of O_2 was demonstrated with *D. gigas* (Santos et al. 1993) and *D. salexigens* (van Niel et al. 1996). Under microaerophilic conditions, *D. desulfuricans* strain CSN, *D. vulgaris*, *D. sulfodismutans*, *Desulfobacterium autotrophicum*, *Desulfobulbus propionicus*, and *Desulfococcus multivorans* displayed aerobic respiration (Dilling and Cypionka 1990). *D. oxyclinae* grew in the absence of sulfate or thiosulfate on lactate with 5% O_2 as the electron acceptor (Sigalevich and Cohen 2000). Growth of *D. vulgaris* occurred slowly at 0.04% (0.48 mM) oxygen, and this culture was completely inhibited at 0.08% (0.95 mM) O_2 (Johnson et al. 1997). Aerotaxis was demonstrated in semisolid media with *D. vulgaris* Hildenborough (Johnson et al. 1997) and with

D. magneticus (Lefèvre et al. 2016). The sensing protein for O_2 in *D. vulgaris* Hildenborough was determined to be a *c*-type heme-containing methyl-accepting protein (Fu et al. 1994). In a carefully crafted experiment, *D. desulfuricans ATCC* 27774 was demonstrated to grow with O_2 as the electron acceptor at near atmospheric levels (Lobo et al. 2007). Several strains of SRB tolerate O2 exposure, and the proteins involved in defense to oxygen toxicity have been discussed in the reviews by Dolla et al. (2006) and Lu and Imlay (2021).

1.2.2.7 Carbon Dioxide

While several strains of sulfate-reducing bacteria grow autotrophically with H_2 and CO_2, *Dst. gibsoniae* DSM 7213 and *Dst. orientis* grow slowly with H_2 and CO_2 in the absence of sulfate (Kuever et al. 2014; Klemps et al. 1985). Reduction of CO_2 to formate is by an apparent CO_2 reductase (Schuchmann and Müller 2013) which is the reverse of formate dehydrogenase, and it can be expressed by the following equation:

$$H_2 + CO_2 \rightarrow HCOO^- + H^+ \quad \Delta G^{0'} = -33.5 \text{ to} - 44.9\ \text{kJ/mol}\ H_2 \qquad (1.11)$$

At 1 Pa H_2 and 10 mM formate, the $\Delta G^{0\prime}$ would be -33.5 kJ mol^{-1}, while at 10 Pa H_2 and 1 mM formate, the $\Delta G^{0\prime}$ would be -44.9 kJ mol^{-1}. With laboratory experiments, the amount of free energy required to support bacterial growth was proposed to be -20 kJ mol^{-1} (Schink 1997), while growth of SRBs in marine sediments was reported to require -19 kJ mol^{-1} (Hoehler et al. 2001). According to the energy yield in Eq. 1.11, electron transfer from H_2 to CO_2 could support bacterial growth. Additionally, the redox potential of $E^{0\prime} = -420$ mV for CO_2/formate and redox potential of -410 mV for H^+/H_2 are relatively close (Reeve et al. 2017; Roger et al. 2018), and Eq. 1.11 would favor formation of formate with slight adjustments in the environment of the enzyme. The reduction of CO_2 by formate dehydrogenases has been proposed for *D. desulfuricans* and *D. vulgaris Hildenborough* (Maia et al. 2016; da Silva et al. 2013).

1.2.2.8 Acrylate

The marine environment contains dimethylsulfoniopropionate (DMSP) which is produced by various algae. DMSP is hydrolyzed to acrylate and dimethylsulfide (DMS) (Eq. 1.12) by an enzyme produced by *D. acrylicus* (van der Maarel et al. 1996, 1998):

$$\underset{\text{DMSP}}{(CH_3)_2S^+CH_2CH_2COO} + H^+ \rightarrow \underset{\text{DMS}}{CH_3SCH_3} + \underset{\text{acrylate}}{H_2C = CHCOO} + H^+ \qquad (1.12)$$

$$\underset{\text{Acrylate}}{H_2C = CHCOO} + H^+ + H_2 \rightarrow \underset{\text{propionate}}{H_3CCH_2COO} + H^+ \tag{1.13}$$

The reaction of H_2 oxidation coupled to acrylate reduction to propionate (Eq. 1.13) is more favorable than the H_2 + sulfate reaction ($\Delta G^{0\prime}$ values of −75 kJ/mol and −38 kJ/mol, respectively) and explains why *D. acrylicus* grows with acrylate as the electron acceptor in the presence of sulfate. Marine microorganisms oxidize DMS to dimethylsulfoxide (DMSO), and DMSO is reduced by a few strains of marine SRB (see Sect. 1.2.2.12). DMS is also metabolized by a marine bacterial consortium which appears to contain SRB.

1.2.2.9 Sulfonates

Several SRB are capable of using sulfonates as alternate electron acceptors. Isethionate, taurine, and cysteate are sulfonates where a sulfur atom is covalently bonded to a carbon atom and sulfur carries an oxidative state of +5. Cultures of sulfate-reducing bacteria that grew with cysteate or isethionate as the electron acceptor included *D. desulfuricans* strain ICI and *Desulfomicrobium baculatum* DSM 1741, while *Desulfobacterium autotrophicans* DSM 3382 was able to grow with cysteate, and *D. desulfuricans* ATCC 29577 grew with isethionate (Lie et al. 1996, 1998). The complete genome of *D. desulfuricans* strain ICI has been sequenced (Day et al. 2019). *D. idahonensis* reduces anthraquinone disulfonate (AQDS) (Sass et al. 2009). *D. oceani* isolated off the coast of Peru Kai uses taurine as a final electron acceptor (Finster and Kjeldsen 2010).

Fermentation of sulfonates occurs when appropriate electron acceptors are unavailable to SRB.

Desulforhopalus singaporensis ferments taurine (Eq. 1.14) but is unable to use taurine as an electron donor in the presence of sulfate (Lie et al. 1999), and the thermodynamics of taurine fermentations would favor growth of bacteria (Thauer et al. 1977). A novel *Desulfovibrio*, strain GRZCYSA, ferments cysteate with the formation of acetate, ammonia, sulfide, and sulfate (Laue et al. 1997):

$$C_2H_7O_3NS \rightarrow 0.5\ C_2H_4O_2 + CO_2 + NH_4{}^+ + HS^- \quad \Delta G^{0'} = -136.99\ \text{kJ/mol}^{-1}\text{taurine} \tag{1.14}$$

1.2.2.10 Dissimilatory Metal Reduction

The reduction of metal ions by SRB is widely reported; however, the growth of SRB using metal respiration is less common even though there is a favorable energy yield with metal ions when coupled to oxidation of butyric acid (Table 1.3). Following the initial report of dissimilatory reduction of Fe(III) with sulfur-respiring

Table 1.3 Thermodynamics of dissimilatory metal reductions by SRB

Reaction	$\Delta G^{0\prime}$ (k·J mol^{-1} electron donor)
Butyrate + $4Fe^{3+}$ + $2H_2O \rightarrow$ 2 acetate +4 Fe^{2+} + 5 H^+	−400[a]
Butyrate + 2/3 $Cr_2O_7^{2-}$ 4/3 H_2O + 1/3 $H^+ \rightarrow$ 2 acetate +4/3 $Cr(OH)_3$	−333[a]
Butyrate + 2 MnO_2 + 3 $H^+ \rightarrow$ 2 acetate +2 Mn^{2+} + 2 H_2O	−291[a]
Butyrate + 2 UO_2^{2+} + 2 $H_2O \rightarrow$ 2 acetate +2 UO_2 5 H^+	−130[a]
$Lactate^-$ + 2 $HAsO_4^{-2}$ + $H^+ \rightarrow$ 3 $acetate^{-1}$ + 2 $H_2AsO_3^-$ + HCO_3^-	−140[b]

[a]Tebo and Obraztsova (1998); [b]Laverman et al. (1995)

bacteria (such as *Desulfuromonas acetoxidans*) (Roden and Lovley 1993), Fe(III) reduction coupled to oxidation of organic acids was found to support the growth of *Desulfofrigus oceanense*, *Desulfofrigus fragile*, *Desulfotalea psychrophila*, *Desulfotalea arctica*, and *Dst. reducens* strain M1–1 (Tebo and Obraztsova 1998; Knoblauch et al. 1999). Metal metabolism by *Dst. reducens* strain M1–1 is unique in that this bacterium is capable of growing with Cr(VI) and Mn(IV) as alternate electron acceptors to sulfate (Tebo and Obraztsova 1998). *Desulfosporosinus auripigmenti* DSM 13351 T (formerly *Dst. auripigmentum*) reduces arsenate (Newman et al. 1997; Stackebrandt et al. 2003). *Desulfohalophilus alkaliarsenatis* is an interesting bacterium in that it grows with dissimilatory reduction of sulfate, arsenate, or nitrate and displays chemoautotrophic growth with nitrate as the final electron acceptor, but with sulfide as the electron donor, it displays lithotrophic growth in a medium containing sulfide and arsenate with lactate as the carbon source (Blum et al. 2012). Growth of SRB coupled to uranium respiration, [U(VI) $\rightarrow$ U(IV)], is thus far limited to only two species. *Dst. reducens* strain M1–1 couples butyrate oxidation to uranyl reduction (Tebo and Obraztsova 1998), while *Desulfovibrio* strain UFZ B 490 grows with lactate as the electron donor and U(VI) as the electron acceptor (Pietzsch et al. 1999).

1.2.2.11 Growth by Dehalorespiration

Anaerobic bacteria capable of growth by coupling electrons from an electron donor to halogenated aromatic compounds have been described as dehalorespirers (Fetzner 1998).

There are two examples of SRB that grow using halogenated aromatic compounds as alternate electron acceptors. With H_2 as the electron donor, *Desulfomonile tiedjei* uses tetraethylene or 3-chlorobenzoate as the terminal electron acceptor (DeWeerd et al. 1990). Additionally, this bacterium is capable of reductive dehalogenation of isomers 3,4–/3,5–/2,5-dichlorobenzoate to benzoate (Holliger et al. 1999). A related bacterium *Desulfomonile limimaris* ATCC 700979 displays growth with lactate oxidation coupled to dehalogenation of 3-chlorobenzoate, 3-bromobenzoate, and 2,3–/2,5–/3,5-dichlorobenzoate (Sun et al. 2001). Both of these bacteria are capable of using benzoate as the electron source with sulfate as the electron acceptor. *D. dechloracetivorans* is another dehalorespirer which dechlorinates 2-chlorophenol or 2,6-dichlorophenol with acetate or lactate as the electron source (Sun et al. 2000).

1.2.2.12 Growth with DMSO

Dimethylsulfoxide (DMSO) is found in fresh and marine waters where it is produced from the oxidation of dimethylsulfide (DMS) by bacterial or chemical oxidation. The reduction of DMSO to DMS has been attributed to hydrogen sulfide and anaerobic bacteria (Jonkers et al. 1996).

DMSO serves as the electron acceptor and supports the growth of *Desulfosporosinus orientis* DSM 765 T, *Desulfosporosinus meridiei* DSM 13257 T, and *D. idahonensis* (Robertson et al. 2001; Sass et al. 2009). SRB isolated from marine environments that grow with DMSO respiration include *D. desulfuricans* strain PA2805, *D. vulgaris*, *D. halophilus*, and *Desulfovibrio* sp. strain HDv (Jonkers et al. 1996). While *Desulfomicrobium salsuginis* grows with DMSO as the terminal electron acceptor, other species of *Desulfomicrobium* have not been tested for this respiratory activity (Dias et al. 2008). Recently, it was reported that *Desulfolutivibrio sulfoxidireducens* grows on DMSO with lactate or ethanol as the electron source (Thiel et al. 2020).

1.2.3 Disproportionation of Thiosulfate, Sulfite, and Sulfur

There is a process employed by some SRB to grow in nutrient-limited environments, and this is substrate disproportionation (dismutation) where the reactant is both oxidized and reduced. Unlike respiration where electrons are transferred from the substrate to an acceptor, disproportionation implies the substrate is the recipient of the electrons, and this could also be referred to as inorganic "fermentation" (Kelly 1987). Le Faou et al. (1990) question if fermentation is an appropriate term for this activity because oxidative phosphorylation coupled to electron transport is responsible for energy production. Examples of bacteria that grow as a result of disproportionation of S^0, sulfite, or thiosulfate are listed in Table 1.4.

Disproportionation of thiosulfate and sulfite was reported for *D. sulfodismutans* (Bak and Pfennig 1987; Bak and Cypionka 1987). The disproportionation of thiosulfate and sulfite is as follows:

Thiosulfate:

$$S_2O_3^{2-} + H_2O \rightarrow SO_4^{2-} + HS^- + H^+ \quad \Delta G^0 = -21.9\,\text{kJ/mol} \tag{1.15}$$

Table 1.4 Sulfate-reducing bacteria that disproportionate S^0, sulfite, or thiosulfate

		Disproportionation with growth (G)		
		Disproportionation without growth (D)		
Bacteria	Thiosulfate	Sulfite	S^0	References
Desulfocapsa Cad626	G	G	G	Peduzzi et al. (2003)
Desulfocapsa sulfoexigens	G	G	G	Finster et al. (1998)
Desulfocapsa thiozymogenes	G	G	G	Janssen et al. (1996)
Desulfobulbus propionicus	G	–	G	Krämer and Cypionka (1989); Lovley and Phillips (1994)
Desulfovibrio oxyclinae	G	G	–	Krekeler et al. (1997)
Desulfovibrio brasiliensis	G	–	–	Warthmann et al. (2005)
Desulfovibrio sulfodismutans[a]	G	D	–	Bak and Pfennig (1987)
Desulfofustis glycolicus	–	–	G	Finster (2008)
Desulfolutivibrio sulfoxidireducens	G	G[b]	–	Thiel et al. (2020)
Desulfotomaculum thermobenzoicum	G	–	–	Jackson and McInerney (2000)
Desulfovibrio aminophilus	D	D	–	Baena et al. (1998)
Desulfovibrio mexicanus	D	D	–	Hernandez-Eugenio et al. (2000)
Desulfobacter curvatus	D	–	–	Krämer and Cypionka (1989)
Desulfobacter hydrogenophilus	D	–	–	Krämer and Cypionka (1989)
Desulfococcus multivorans	D	–	–	Krämer and Cypionka (1989)
Desulfotomaculum nigrificans	D	–	–	Krämer and Cypionka (1989)

[a]Reclassified as *Desulfolutivibrio sulfodismutans* (Thiel et al. 2020)
G[b] metabisulfite used and not sulfite

Sulfite:

$$4SO_3^{2-} + H^+ \rightarrow 3SO_4^{2-} + HS^- \quad \Delta G^0 = -58.9\ \text{kJ/mol} \qquad (1.16)$$

From a survey of 19 cultures of sulfur bacteria, only *D. sulfodismutans*, SRB strains Bra02 and NTA3, and *D. desulfuricans* strain CSN grew with sulfite, while *D. sulfodismutans* and seven additional strains grew with thiosulfate disproportionation . Several strains of bacteria were found to disproportionate thiosulfate and sulfite but unable to couple this activity to growth (Krämer and Cypionka 1989). *Desulfocapsa sulfoexigens* (Finster et al. 1998) and *Dst. thermobenzoicum* grew by thiosulfate disproportionation, but this reaction did not support the growth of *Dst. nigrificans*, *Dst. ruminis*, or *Desulfosporosinus orientis* (Jackson and McInerney 2000). *D. aminophilus* will disproportionate sulfite and thiosulfate to sulfide and sulfate (Baena et al. 1998).

The oxidation of thiosulfate and sulfite to sulfate was proposed to be a reversal of the electron transport system for sulfate reduction to sulfide as indicated in the following reactions:

Sulfite reaction:

$$3HSO_3^- + 3AMP \rightarrow 3APS + 6e^- + 6H^+ \quad (1.17)$$

$$3APS + 3PP_i \rightarrow 3SO_4^{2-} + 3ATP \quad (1.18)$$

$$6e^- + 6H^+ + HSO_3^- \rightarrow HS^- + 3H_2O \quad (1.19)$$

Overall:

$$4HSO_3^- + 3PP_i + 3AMP \rightarrow 3SO_4^{2-} + HS^- + 3ATP + 3H_2O \quad (1.20)$$

Thiosulfate reaction:

$$S_2O_3^{2-} + 2e^- + 2H^+ \rightarrow HSO_3^- + HS^- \quad (1.21)$$

$$HSO_3^- + AMP \rightarrow APS + 2e^- + 2H^+ \quad (1.22)$$

$$APS + PP_i \rightarrow SO_4^{2-} + ATP \quad (1.23)$$

Overall:

$$S_2O_3^{2-} + PP_i + AMP \rightarrow SO_4^{2-} + HS^- + ATP \quad (1.24)$$

It is unresolved if the mechanisms for these disproportionation reactions relies on existing enzymes or if a unique cytoplasmic electron transfer system is involved (Peck Jr. 1993). However, cell extracts of *D. desulfuricans* and *D. propionicus* oxidized sulfite to sulfate with O_2 as the electron acceptor forming 37 nmol ATP per 100 nmol sulfite which was interpreted to indicate that sulfate was formed using APS reductase and ATP sulfurylase with reversal of the sulfate activation pathway (Dannenberg et al. 1992). The disproportionation of thiosulfate is an important contribution to sulfur cycling in freshwater sediment (Jørgensen 1990a), marine sediment (Jørgensen 1990b; Jørgensen and Bak 1991), and cyanobacterial mats (Jørgensen 1994).

The disproportionation of sulfur has been reported to have existed for about 3.5 billion years, and it has been proposed to be one of the oldest biological processes on Earth (Finster 2008; Philippot et al. 2007). The disproportionation of sulfur could be according to the overall reaction (Eq. 1.25) and unable to support growth because it is endergonic (Bak and Cypionka 1987). With the removal of sulfide from solution by precipitation with iron and manganese so that the concentrations of sulfide and sulfate are 10^{-7} and 10^{-2} M, respectively, the $\Delta G^{0\prime}$ is -92 kJ/reaction which is sufficient for production of 1 mol ATP (Thamdrup et al. 1993). Thus, bacterial growth occurs if the sulfide generated by disproportionation of sulfur is removed from solution by chemical reduction of oxidized iron or manganese (Thamdrup et al. 1993).

The first organisms in pure culture reported to display disproportionation of sulfur were *Desulfocapsa thiozymogenes* DSM 7269 (Bak 1993; Janssen et al. 1996) and *Desulfobulbus propionicus* DSM 2032 (Lovley and Phillips 1994). Both of these bacteria were reported to convert S^0 to sulfate and hydrogen sulfide according to Eq. 1.25. The energy of this reaction would not favor bacterial growth; however, when using hydrogen sulfide and sulfate concentrations that reflect environmental concentrations (10^{-7} M and 2.8×10^{-2} M, respectively) the $\Delta G^{0\prime}$ for the reaction would be −120 kJ/reaction (Finster 2008).

Oxidation of S^0 to sulfate was observed by *D. desulfuricans* (ATCC 29577), *Desulfomicrobium baculatum* (DSM 1741), and *Desulfobacterium autotrophicum* (DSM 3382) when MnO_2 was added to the reaction mixture (Eq. 1.26), and even though the reaction is strongly exergonic, growth of these SRB was not observed:

$$4S^0 + 4H_2O \rightarrow SO_4^{2-} + 3HS^- + 5H^+ \quad \Delta G^{0'} = +41\ \mathrm{kJ/reaction} \tag{1.25}$$

$$S^0 + 3MnO_2 + 4H^+ \rightarrow S0_4^{2-} + 3Mn^{2+} + 2H_20 \quad \Delta G^0 = -348\ \mathrm{kJ/reaction} \tag{1.26}$$

Desulfocapsa sulfoexigens (DSM 10523) reduces sulfate with formate as the electron donor, but it is best recognized for its slow growth on S^0 with ferrihydrite as the scavenger of hydrogen sulfide (Finster et al. 1998). Iron was deposited in the growth medium as FeS and pyrite (FeS_2) where pyrite was proposed to be produced according to the following reaction:

$$FeS + S^0 \rightarrow FeS_2 \tag{1.27}$$

When testing bacteria present in marine sediments for S^0 disproportionation, the activity of this reaction was dependent on the presence of red, amorphous ferric hydroxide (FeOOH) or MnO_2, and this activity would account for the oxidation of elemental sulfur to sulfate in the absence of O_2 (Thamdrup et al. 1993). The reaction between S^0 and Mn(IV) would follow Eq. 1.26., and the equation for FeOOH is as follows in Eq. 1.28:

$$3S^0 + 2FeOOH \rightarrow S0_4^{2-} + 2FeS + 2H^+ \tag{1.28}$$

1.2.4 Fermentation of Organic Substrates

In the absence of sulfate, several strains of SRB display fermentative activities, and this metabolism varies with the bacterial species. Additionally, growth supported by fermentation may be slow with generation times of 100–130 hr reported for *D. vulgaris* Hildenborough growing on pyruvate (Voordouw 2002). *D. vulgaris* strain Marburg converts pyruvate to acetate, CO_2, and H_2 according to the following equation (Pankhania et al. 1988; Tasaki et al. 1993):

$$\text{Pyruvate} + 2\text{H}_2\text{O} \rightarrow \text{acetate} + \text{HCO}_3^- + \text{H}_2 + \text{H}^+ \quad \Delta G^{0'} = -47.1\ \text{kJ/mol} \tag{1.29}$$

Nine strains of *Dst. nigrificans*, Teddington Garden NCIB 8351, Delft 74 T NCIB 8395, strain ATCC 3750, strain ATCC 7946, strains 55, 106, 134, "By," and "Dp," were reported to grow by fermentation of pyruvate with production of acetate, H_2 and CO_2 (Postgate 1963). Growth of *Dst. carboxydivorans*, *Dst. nigrificans*, *Dst. putei*, and *Dst. ruminis* was attributed to fermentation of pyruvate (Parshina et al. 2005b), and *Dst. reducens* strain MI-1 ferments pyruvate while Fe(III) is being reduced (Vecchia et al. 2014). Additionally, some strains of SRB are capable of fermenting choline and cysteine. Fermentation of choline by *D. desulfuricans* yields end products of trimethylamine, ethanol, and acetate (Hayward and Stadtman 1959; Senez and Pascal 1961; Baker et al. 1962). SRB have been reported to ferment cysteine (Senez and Leroux-Gilleron 1954).

Fermentation of glucose and fructose is a characteristic of *Dst. carboxydivorans* and *Dst. nigrificans* (Parshina et al. 2005b). In the absence of sulfate, *D. fructosovorans* ferments fructose, malate, and fumarate to succinate and acetate (Ollivier et al. 1988). When sulfate was not provided in the media, *Dst. thermobenzoicum* TSB fermented 4 mol of pyruvate to 5 mol of acetate (Tasaki et al. 1993), while *Desulfobulbus propionicus* MUD fermented 3 mol pyruvate to 2 mol acetate and 1 mol propionate (Tasaki et al. 1993). *Desulfobulbus propionicus* fermented ethanol to propionate and acetate, which indicates the involvement of a randomizing pathway in the formation of propionate (Stams et al. 1984). *D. desulfuricans* strain El Agheila and *D. desulfuricans* strain Canet and *Vibrio cholinicus* (now *D. desulfuricans*) ferment choline in the absence of sulfate (Baker et al. 1962). Fermentation of taurine is discussed in Sect. 1.2.2.9.

1.2.5 Sulfide/Sulfur Oxidation Coupled to Nitrate, Mn(IV), and O_2 Reduction

A coupling of the sulfur and nitrogen cycles occurs with *D. propionicus* as it oxidizes sulfide completely to sulfate with the reduction of nitrate or nitrite to ammonia (Dannenberg et al. 1992). *D. desulfuricans* was found to couple sulfide oxidation to the reduction of nitrite to ammonia. In another unique reaction, *D. desulfuricans* ATCC 29577 coupled S^0 oxidation to sulfate with the reduction of manganese dioxide to Mn(II) according to the following reaction (Lovley and Phillips 1994):

$$\text{S}^0 + 3\text{MnO}_2 + 4\text{H}^+ \rightarrow \text{S0}_4{}^{2-} + 3\text{Mn(II)} + 2\text{H}_20 \quad \Delta G^{0'} = -348\,\text{kJ}(\text{reaction})^{-1} \tag{1.30}$$

The S^0 oxidation–Mn(IV) reduction reaction occurred with *Desulfomicrobium baculatum* DSM 1741, *Desulfobacterium autotrophicum* DSM 3382, *Desulfobulbus propionicus* DSM 2032, and the sulfur-reducing bacterium *Desulfuromonas acetoxidans* DSM 684; however, none of the SRB cultures coupled growth to this electron transfer even though it was an exergonic reaction. When Fe(III) oxide was substituted for MnO, none of the SRB tested would oxidize S^0 to sulfate (Lovley and Phillips 1994). In the presence of O_2, *Desulfobulbus propionicus* oxidizes elemental sulfur to sulfate and sulfide, and in the presence of myxothiazol, an electron transport inhibitor, sulfide oxidation is inhibited with the accumulation of S^0 accumulates (Fuseler and Cypionka 1995).

1.3 Syntrophic Growth

Syntrophy is derived from Greek with "syn" which means "together" and "trophy" "nourishment." Basically, syntrophy describes the relationship between microorganisms of different species where one or both species benefit from the byproducts of the other. Syntrophy is not a synonym for symbiosis because not all forms of symbiosis involve nutrition. This form of cooperative metabolism was first identified when using a mutant of *Escherichia coli* deficient in tryptophan production to grow on tryptophan produced by a strain of *Salmonella typhi* (Fildes 1956). As Selwyn and Postgate (1959) searched for a sulfate-reducing bacterium that used acetate as the electron donor, they pursued the isolation of syntrophs along with their unsuccessful attempts to re-isolate *D. rubentschikii* which was initially reported by Baars (1930). A syntrophic relationship occurred with *Methanobacillus omelianskii* which was considered to be a pure culture (Barker 1939); however, it was in fact a co-culture of two anaerobic microorganisms (Bryant et al. 1967). A bacterium designated as the "S-organism" converted ethanol to acetate and H_2 (Eq. 1.31), and the other organism, *Methanobacterium bryantii* strain M.o.H., used the H_2 to energize the reduction of CO_2 to CH_4 (Eq. 1.32):

Strain S:

$$2CH_3CH_2OH + 2H_2O \rightarrow 2CH_3COO^{-} + 2H^{+} + 4H_2 \quad \Delta G^{0'} = +19\ \text{kJ/mol ethanol} \tag{1.31}$$

Strain M.o.H.:

$$4H_2 + CO_2 \rightarrow CH_4 + 2H_2O \quad \Delta G^{0'} = -131\ \text{kJ/mol methane} \tag{1.32}$$

Co-culture of S and M.o.H. organisms:

$$2CH_3CH_2OH + CO_2 \rightarrow 2CH_3COO^{-} + 2H^{+} + CH_4 \quad \Delta G^{0'} = -112\ \mathrm{kJ/mol\ methane} \tag{1.33}$$

Growth of *Desulfovibrio*-type strains (*D. desulfuricans* strain Essex 6 and *D. vulgaris* Hildenborough) with *Methanobacterium* M.o.H. gave results similar to that obtained with “S-organism” and *Methanobacterium* M.o.H. (Bryant et al. 1977).

The culture known as *Chloropseudomonas ethylica* is a co-culture of an anaerobic green sulfur phototroph and a heterotrophic organism, and there are several cultures of *Chloropseudomonas ethylica*” available with either sulfur or SRB present (Pfennig and Biebl 1976; Biebl and Pfennig 1977). The culture “*Chloropseudomonas ethylica* strain N_2” contains the green sulfur bacterium *Chlorobium limicola* and the sulfate-reducing bacterium *Desulfomicrobium baculatum* DSM 1743 (Biebl and Pfennig 1978). Syntrophic growth with *D. gigas* or *D. desulfuricans* Essex 6 and *Chlorobium limicola* would be characterized by Eqs. 1.34 to 1.36:

Desulfovibrio desulfuricans:

$$24CH_3CH_2OH + 17SO_4^{2-} \rightarrow 34CH_3COO^{-} + 17H_2S + 34H_2O \tag{1.34}$$

Chlorobium limicola:

$$17H_2S + 34CH_3COO^{-} + 28CO_2 + 16H_2O \xrightarrow{\text{light}} 24(C_4H_7O_3) + 17SO_4^{2-} \tag{1.35}$$

Syntrophy:

$$24CH_3CH_2OH + 28CO_2 \xrightarrow{\text{light}} 24(C_4H_7O_3) + 16H_2O \tag{1.36}$$

As reviewed by Morris et al. (2013), there are numerous examples of metabolic cross-feeding between microorganisms by syntrophy. The phylogeny and ecology of syntrophy have been covered in several reviews (Fauque 1995; Schink 1997, 2002; McInerney et al. 2008, 2009; Stams and Plugge 2009; Morris et al. 2013; Timmers et al. 2016). While there may be no obligate syntrophic bacteria (McInerney et al. 2008), a facultative partner relationship is found in many environments. Several reviews involving microorganisms capable of fatty acid oxidation with sulfidogenic sulfate reducers are available (Gieg et al. 2014; McInerney et al. 2009; Plugge et al. 2011; Schink 1997; Sieber et al. 2012). Co-culture of SRB with another microorganism has attracted considerable attention, and the capability of SRB to establish syntrophy has been attributed to genetic polymorphism (Großkopf et al. 2016). While four genome copies are predicted for stationary phase cells of *D. vulgaris* Hildenborough and nine for *D. gigas* (Postgate et al. 1984), the role of polyploidy in SRB remains to be established.

Nutrients shuttled between syntrophic partners commonly include H_2, formate, and acetate. Co-culture of *Desulfovibrio* sp. strain G11 and a hydrogen consumer, *Methanobrevibacter arboriphilus* AZ, grew on formate even though neither culture was capable of growing with formate (Dolfing et al. 2008). The Gibbs free energy change for oxidation of formate by the *Desulfovibrio* is endergonic and is shown in Eq. 1.37, and the production of methane from hydrogen and carbon dioxide by the methanogen is given in Eq. 1.38. By the methanogen maintaining a low partial pressure of H_2, the oxidation of formate by *Desulfovibrio* becomes favorable, and the overall reaction is given in Eq. 1.39:

$$4HCOO^{-} + 4H_2O \rightarrow 4H_2 + 4HCO_3^{-} \quad \Delta G^{0'} = +5.2\ \mathrm{kJ\ mol^{-1}} \tag{1.37}$$

$$4H_2 + HCO_3^{-} + H^{+} \rightarrow CH_4 + 3H_2O \quad \Delta G^{0'} = -135.6\ \mathrm{kJ\ mol^{-1}} \tag{1.38}$$

$$4HCOO^{-} + H2O + H^{+} \rightarrow 3CH_4 + 3HCO_3^{-} \quad \Delta G^{0'} = -130.4\ \mathrm{kJ\ mol^{-1}} \tag{1.39}$$

Hydrocarbon degradation occurs with *Desulfatibacillum alkenivorans* AK-01 mineralizing *n*-alkanes in the presence of sulfate (Westerholm et al. 2011). However, in the absence of sulfate but presence of *Methanospirillum hungatei* IF-1, a H_2 and formate utilizing methanogen enable *Desulfatibacillum alkenivorans* AK-01 to completely oxidize alkanes. The anaerobic degradation of hexadecane and phenanthrene coupled to reduction of sulfate in seafloor sediments has been suggested to involve a syntrophic population (Shin et al. 2019). Interspecies H_2 transfer enables *Methanosarcina barkeri* to couple growth to the oxidation of ethanol to acetate, and H_2 by *Desulfovibrio* sp. oxidizes ethanol to acetate plus H_2 (Laube and Martin 1981). The oxidation of ethanol by *Desulfovibrio* sp. (see Eq. 1.40) is not energetically favorable for growth unless *Methanosarcina barkeri* rapidly consumes H_2 to maintain a low partial pressure of molecular hydrogen. Summation of the *Desulfovibrio* sp. and *Methanosarcina barkeri* metabolism is presented in Eq. 1.41:

$$CH_3CH_2OH + H_2O = CH_3COO^{-} + H^{+} + 2H_2 \quad \Delta G^{0'} = +19.2\ \mathrm{kJ\ mol^{-1}} \tag{1.40}$$

$$2CH_3CH_2OH + HCO_3^{-} = 2CH_3COO^{-} + CH_4 + H_2O \quad \Delta G^{0'} = -116.2\ \mathrm{kJ\ mol^{-1}} \tag{1.41}$$

Multi-culture metabolism was conducted including *Acetivibrio cellulolyticus*, *Desulfovibrio* sp., and *Methanosarcina barkeri* to rapidly oxidize cellulose to carbon dioxide and methane (Laube and Martin 1981).

Interspecies transfer of H_2 accounts for the syntrophic growth of some SRB and methanogens. *D. vulgaris* (DSM 1744) reduces sulfate and grows with H_2 released from *Methanococcus concilii* growing on acetate (Ozuolmez et al. 2015). In another report involving the co-culture of *D. vulgaris* Hildenborough and *Methanococcus maripaludis* strain S2, the methanogen maintains a low H_2 concentration as the SRB ferments lactate with the formation of H_2 (Walker et al. 2009, 2012). The reactions are listed below. If lactate fermentation is adjusted to steady-state concentrations of 4 mM lactate, 26 mM acetate, 2.5×10^{-5} atm H_2, 0.05 atm CO_2, and 0.0006 atm CH_4 in Eqs. 1.42 and 1.44, the free energy becomes -67.3 kJ mol^{-1} and -82.8 kJ mol^{-1}, respectively, which is far more favorable for growth (Walker et al. 2009):

Lactate fermentation:

$$C_3H_5O_3^- + H_2O \rightarrow C_2H_3O_2^- + 2H_2 + CO_2 \quad \Delta G^{0'} = -8.8\ \text{kJ mol}^{-1} \tag{1.42}$$

Methanogenesis:

$$4H_2 + CO_2 \rightarrow CH_4\_2H_2O \quad \Delta G^{0'} = -130.7\ \text{kJ mol}^{-1} \tag{1.43}$$

Syntrophy with lactate:

$$C_3H_5O_3^- \rightarrow C_2H_3O_2^- + 0.5CH_4 + 0.5CO_2 \quad \Delta G^{0'} = -74.2\ \text{kJ mol}^{-1} \tag{1.44}$$

Pyruvate fermentation:

$$C_3H_3O_3^- + H_2O \rightarrow C_2H_3O_2^- + H_2 + CO_2 \quad \Delta G^{0'} = -52.0\ \text{kJ mol}^{-1} \tag{1.45}$$

Syntrophy with pyruvate:

$$C_3H_3O_3^- + 0.5H_2O \rightarrow C_2H_3O_2^- + 0.25CH_4 + 0.75CO_2 \quad \Delta G^{0'} = -84.7\ \text{kJ mol}^{-1} \tag{1.46}$$

The overall reaction with lactate fermentation to methane, acetate, and CO_2 accounts for ~ -82.8 kJ mol^{-1}, and this free energy is partitioned unequally to the SRB (~ -60 kJ mol^{-1}) and the methanogen (~ -20 kJ mol^{-1}) (Walker et al. 2009, 2012). This quantity of free energy available to the methanogen would be in the range appropriate for the thermodynamic requirement of -15 to -20 kJ mol^{-1} required for the synthesis of ATP (Schink and Stams 2006). *D. vulgaris* Hildenborough adjusts to syntrophy with and *Methanococcus maripaludis* strain S2 by enhanced transcription of genes for high molecular weight cytochrome (Hmc), membrane hydrogenase (Coo), and periplasmic hydrogenases (Hyd and Hyn) (Walker et al. 2009). Using physiological and genome analysis of *D. alaskensis* G20 and *D. vulgaris* Hildenborough growing on lactate with *Methanococcus maripaludis*, there was no evidence of a central set of core genes associated with syntrophy in *Desulfovibrio* (Meyer et al. 2013).

The co-culture of a bacterium and archaea results in production of a biofilm that has ridges that are approximately 50 × 300 μm. It was suggested that *Methanococcus maripaludis* in the biofilm with *D. vulgaris* had an altered carbohydrate metabolism as compared to monoculture growth. The close interaction of the co-culture is seen in Fig. 6.3 (Brileya et al. 2014). In a variation of a co-culture involving SRB, a tri-culture consisting of *D. vulgaris* Hildenborough, *Dehalococcoides ethenogenes* strain 195 and *Methanobacterium congolense* were grown with lactate as the energy source. D. vulgaris Hildenborough fermented lactate to H_2 and acetate which was consumed by the syntrophic partners (Men et al. 2012). Robust growth occurred in syntrophic growth with rates of dechlorination being three times faster in tri-culture or di-culture (*D. vulgaris* Hildenborough/ *Dehalococcoides ethenogenes*) as compared to a *Dehalococcoides ethenogenes* monoculture. Transcriptomic analysis and proteomic analysis revealed that 102 genes expressed and 120 proteins produced were differentially impacted with adjustment of *Dehalococcoides ethenogenes* to either di- or tri-culture associations.

Transcriptional analysis has revealed that *D. vulgaris* Hildenborough uses an electron transfer system with lactate syntrophic growth that has some distinct features as compared to an SRB monoculture growing on lactate-sulfate. With syntrophic growth of *D. vulgaris* Hildenborough and *Methanococcus maripaludis*, *D. vulgaris* Hildenborough has upregulated genes for lactate transport and oxidation, membrane-bound Coo hydrogenase, two periplasmic hydrogenases (Hyd, the [Fe] hydrogenase, and Hyn, the [Ni-Fe] hydrogenase) and the transmembrane high molecular weight cytochrome complex Hmc (Walker et al. 2009). Distribution of the bacteria and methanogen in co-culture is given in Fig. 1.3. The shifting of a di-culture of *D. vulgaris* Hildenborough and *Methanosarcina barkeri* to *D. vulgaris* Hildenborough respiring on sulfate results in about 132 genes being differentially expressed (Plugge et al. 2010). As *D. vulgaris* Hildenborough transitioned to sulfidogenic growth, 21 genes upregulated were concerned with energy metabolism and included genes for Ech hydrogenase, ATP synthase, cytochrome c network, thiosulfate reductase, ferrous iron uptake, iron-sulfur cluster binding protein, ferredoxin, and fructose-1,6-bisphosphate aldolase. Downregulated in *D. vulgaris* Hildenborough were several genes for cell surface including several membrane proteins and lipoproteins.

Examination of the syntrophic interactions between *D. alaskensis* G20 grown with *Methanococcus maripaludis* and *Methanospirillum hungatei* revealed 68 genes were differentially expressed for syntrophic growth compared to sulfate respiration (Meyer et al. 2013, 2014). Downregulated were 54 genes, and these included 33 genes of the 11 operons of the Fur regulon, genes for ferrous iron uptake, regulators, and signal transduction. The 14 genes upregulated during syntrophy included those for lactate oxidation and other members of the energy metabolism system.

Anaerobic oxidation of methane (AOM) in marine sediments was proposed by Reeburgh (1976) to proceed by co-metabolism involving sulfate-reducing bacteria

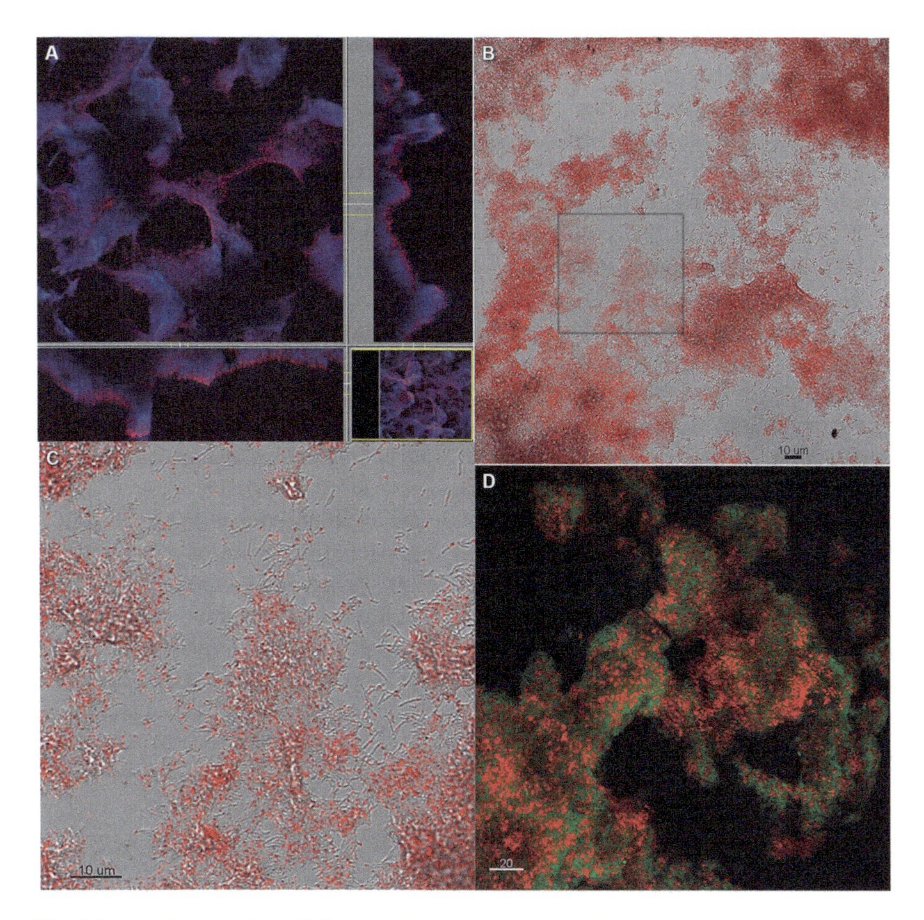

Fig. 1.3 Co-culture biofilm. (Brileya et al. 2014). (**a**) stained with CTC while intact and hydrated and showing all biomass stained with DAPI in blue, and CTC in red, or purple where both DAPI and CTC are present (**b**) scraped from the slide after CTC staining. (**c**) Zoomed in from the inset in (**b**) showing individual grains of red fluorescent CT-formazan in each cell. (**d**) Co-culture biofilm scraped from the substratum, fixed and hybridized with domain-specific probes for *D. vulgaris* (green) and *Methanococcus maripaludis* (red)

and another anaerobic bacterium. Syntrophy between SRB and methanogens was suggested by Hansen et al. (1998; Hallam et al. 2004) to account for AOM by a process which was identified as reverse methanogenesis (see Eq. 1.47):

$$CH_4 + SO_4^{2-} + 2H^+ \rightarrow H_2S + CO_2 + H_2O \tag{1.47}$$

A consortium of anaerobic methanotrophic archaea (ANME) and SRB are considered responsible for AOM. The ANME are uncultured archaea and due to significant differences in the ANME are clustered into three major groups, ANME-1, ANME-2, and ANME-3 (Niemann et al. 2006; Knittel et al. 2005). The bacteria associated with the ANME-2c and ANME-3 ecotypes appear to include members of *Desulfosarcina* and *Desulfobulbus* (Orphan et al. 2001; Niemann et al. 2006; Pernthaler et al. 2008). This activity has been recently reviewed (Conrad 2009; Knittel and Boetius 2009), and a discussion of AOM syntrophy is found in Chap. 7.

The anaerobic oxidation of fatty acids is energetically favorable when metabolism involves two dissimilar species growing in syntrophic association (Worm et al. 2014). While formate and H_2 utilizing strains of methanogens are often encountered as a member of the syntrophic relationship, the oxidation of short-chain fatty acids by syntrophy may include SRB in place of methanogens. The utilization of saturated fatty acids with four to nine carbons with *Syntrophomonas cellicola* oxidizing the hydrocarbons and the H_2 released which is consumed by *Desulfovibrio* strain G11 is an example of a syntrophic relationship involving a sulfate reducer (Wu et al. 2006). *Desulfovibrio* G11 is frequently used as a model hydrogenotrophic partner in syntrophy and has been recently classified as *D. desulfuricans* strain G11 (Sheik et al. 2017). *Dm. thermocisternum* and *Dm. thermobenzoicum* grow on propionate with a hydrogenothrophic methanogen as a syntrophic partner (Nilsen et al. 1996; Plugge et al. 2002). Reactions for this propionate syntrophic activity (Liu et al. 1999) are given in Eqs. 1.48 to 1.50:

Propionate oxidation:

$$CH_3CH_2COO^{\mp}2H_2O \rightarrow CH_3COO^{\mp}CO_2 + 3H_2 \quad \Delta G^{0'} = +71.7\ \text{kJ mol}^{-1} \tag{1.48}$$

Methanogenesis:

$$CO_2 + 4H_2 \rightarrow CH_4 + 2H_2O \quad \Delta G^{0'} = -130.7\ \text{kJ mol}^{-1} \tag{1.49}$$

Syntrophic metabolism:

$$4CH_3CH_2COO^{\mp}2H_2O \rightarrow 4CH_3COO^{\mp}CO_2 + 3CH_4 \quad \Delta G^{0'} = -105.6\ \text{kJ mol}^{-1} \tag{1.50}$$

Members of the genus *Syntrophobacter* are unique SRB that can oxidize propionate and in the absence of sulfate grow with *Methanospirillum hungateii* and other hydrogenotrophic methanogens. Propionate-oxidizing bacteria capable of syntrophic growth include *Syntrophobacter wolinii* (Boone and Bryant 1980; Liu et al. (1999), *Syntrophobacter pfennigii* (Wallrabenstein et al. 1995), *Syntrophobacter fumaroxidans* (Harmsen et al. 1998), and *Syntrophobacter sulfatireducens* (Chen et al. 2005).

Reports concerning anaerobic decomposition of halogenated hydrocarbons highlight the syntrophic association of SRB (Holliger et al. 1999). The dehalogenation of 2- and 4-fluorobenzoate by a soil bacterium is dependent on the metabolic activity of a *Desulfobacterium* sp. (Drzyzga et al. 1994). *Degradation of dichloromethane by Dehalobacterium formicoaceticum requires a Desulfovibrio* sp. *in co-culture* (Mägli et al. 1996). Tetrachloroethene dehalorespiration by *Desulfitobacterium frappieri* TCE1 is dependent on the presence of *D. fructosivorans* (Drzyzga and Gottschal 2002). In sulfate-limiting media with excess fructose, *D. fructosivorans ferments fructose according to the following reaction* (Cord-Ruwisch et al. 1988):

$$C_6H_{12}O_6 + 4H_2O \rightarrow 2CH_3COO^{-}+2HCO_3^- + 4H^+ + 4H_2 \quad (1.51)$$

Research supports the proposal that H_2 produced by *D. fructosivorans is used to energize the dehalogenation system of Dehalobacterium formicoaceticum* (Drzyzga and Gottschal 2002).

In addition to *Dehalobacterium* spp., *species of Dehalococoides* are known to use hexachlorobenzene (HCB) as electron acceptors with growth attributed to coupled electron transfer. Co-culture of *Dehalococcoides mccartyi* strain CBDB1 with either *Desulfovibrio vulgaris* DSM 2119 or *Syntrophobacter fumaroxidans* DSM 10017 enhanced the growth and metabolism of HCB (Chau et al. 2018). With sulfate as the electron acceptor, *Syntrophobacter fumaroxidans* grows with the oxidation of propionate with the production of acetate, CO_2, and H_2. *D. vulgaris* and S. *fumaroxidans* produce H_2 which energizes the respiratory reduction of HCB by *Dehalococcoides mccartyi*. The maximum distance for interspecies cell-to-cell transfer of H_2 between suspended cells of *D. vulgaris* or S. *fumaroxidans* and *Dehalococcoides mccartyi* has been calculated to be 178 and 19 μm, respectively (Chau et al. 2018). The syntrophic partners of *Dehalococcoides mccartyi* consume carbon monoxide which is produced as a byproduct of HCB metabolism and is toxic to *Dehalococcoides mccartyi*.

A nonclassical type of syntrophy (Morris et al. 2013) is the microaerophilic growth of *D. oxyclinae* following detoxification of high levels of O_2 by Marinobacter sp. (Sigalevich et al. 2000). With limiting concentrations of lactate and oxygen and sufficient levels of sulfate, *D. desulfuricans* PA2805 grows when in co-culture with *Thiobacillus thioparus* T5 (van den Ende et al. 1997).

1.4 Autotrophic Growth

The concept of autotrophy in bacteria evolved slowly and included the idea that bacteria had the capability to oxidize and reduce inorganic compounds. While it had been proposed that chemoautotrophic bacteria had a unique energy coupling which distinguished autotrophs from heterotrophs (Koffler and Wilson (1951), Mechalas and Rittenberg (1960) supported the idea that energy obtained from inorganic or organic molecules was equivalent and could be used to support growth-related activities. In the intervening time, the basis of autotrophic growth in SRB was focused on H_2 oxidation, CO_2 providing carbon for biosynthesis and sulfate as the terminal electron acceptor. The oxidation of H_2 has long been considered important for energizing autotrophic growth, and several reports discussed using hydrogen released from rods of mild steel in soil or water environments; however, the cultivation of autotrophic sulfate-reducing bacteria was complicated in that the cultures were not pure (Pont 1939; Starkey and Wright 1945). At that time, hydrogenase activity was being demonstrated in sulfate-reducing bacteria (see Sect. 1.2.1.3). Using mineral salt media with the addition of H_2 and CO_2, autotrophic growth of several strains of sulfate reducers was reported (Butlin and Adams 1947; Butlin et al. 1949; Sisler and Zobell 1951; Senez and Volcani 1951; Sorokin 1954; Postgate 1960). *D. desulfuricans* (now *vulgaris*) Hildenborough growing in media containing acetate plus H_2 and CO_2 was reported by Mechalas and Rittenberg (1960) to derive most of their carbon for growth from acetate and that only 10% to 12% of cellular material was derived from $^{14}CO_2$. In a similar experiment, Postgate reported that 13% of cell carbon was derived from CO_2 and this data was presented in a footnote in the paper by Mechalas and Rittenberg (1960). The potential for sulfate-reducing bacteria to use autotrophic grow or metatropic growth (where small quantities of organics would supplement CO_2 reduction) was extensively reviewed by Postgate (1959). Only a few of the sulfate-reducing bacteria were facultative autotrophs, and this capability would favor these bacteria that are in an environment where biologically available organics would be growth limiting.

Several species of SRB utilize H_2 and CO_2 for growth with acetate to provide carbon for growth (Badziong et al. 1979). With H_2, CO_2, and acetate in growth media, *D. vulgaris* strain Hildenborough, *D. vulgaris* Marburg, *D. desulfuricans* strain Essex 6, and *D. gigas* acquired 68–76% of cell carbon from acetate (Brandis and Thauer 1981). The presence of acetate was also required for H_2/CO_2 growth by several species of *Desulfobulbus* and by *Desulfomonas pigra* (Brysch et al. 1987). Autotrophic growth was reported for *Dst. ruminus* strain DL, *Dst. nigrificans* Delft 74, and *Desulfotomaculum* strains TEW, TWC, and TWP (Klemps et al. 1985; Cypionka and Pfennig 1986). The autotrophic cultivation of *Desulfotomaculum* (reclassified as *Desulfosporosinus*) *orientis* revealed that 96% of cell carbon was from CO_2 (Brysch et al. 1987). Other species of sulfate-reducing bacteria grew autotrophically on H_2 and CO_2 without organic carbon include *Desulfobacterium autotrophicum* (Brysch et al. 1987; Schauder et al. 1987), *Desulfococcus niacini* (Imhoff-Stuckle and Pfennig 1983), *Desulfonema limicola*, *Desulfosarcina variabilis* (Brysch et al. 1987), *Dst. geothermicum* (Daumas et al. 1988), *Desulfobacter hydrogenophilus* (Schauder et al. 1987), *Desulfobacterium vacuolatum* (Brysch et al. 1987), and *Desulfomonile tiedjei* (DeWeerd et al. 1990). While *D. baarsii* and *Desulfococcus multivorans* did not oxidize H_2, these two bacteria grew autotrophically with formate as the electron donor with CO_2 as the carbon source (Jansen et al. 1985). *Desulfomicrobium apsheronum* grows autotrophically with formate as the electron donor with CO_2, but no organic compounds required (Rozanova et al. 1988). *Desulfosporosinus orientis* and *Desulfotomaculum* sp. strain TEP grew chemoautotrophically in sulfate-containing medium with methanol as the electron and carbon source (Klemps et al. 1985).

Sulfate-reducing bacteria use two different metabolic pathways for the fixation of CO_2, and these have been the topic of several reviews (Fauque et al. 1991; Hansen 1993; Fuchs 2011; Berg 2011; Schuchmann and Müller 2014). The reductive citric acid cycle (Fig. 1.4) is employed by *Desulfobacter hydrogenophilus* (Brandis-Heep et al. 1983), while the reductive acetyl-CoA (Wood-Ljungdahl) pathway (Fig. 1.5) (Schauder et al. 1986) is used by *Desulfobacterium autotrophicum* (Schauder et al. 1989) and *D. baarsii* (Jansen et al. 1985). The reductive acetyl-CoA is an ancient metabolic pathway, and it has a unique benefit in that it produces ATP as CO_2 is fixed. The reductive pentose phosphate (Calvin-Benson) cycle employs ribulose 1,5-bisphosphate carboxylase/oxygenase (RubisCO), and the only sulfate reducer reported to have RubisCO (form III) is *A. fulgidus* which does not fix CO_2, but RubisCO III has a high affinity for O_2 (Kreel and Tabita 2007).

Recently, it has been reported that *Desulfovibrio desulfuricans* strain G11 fixes CO_2 by the reductive glycine pathway (Sánchez-Andrea et al. 2020). This strain of *D. desulfuricans* uses H_2 and sulfate as the energy source as it grows autotrophically with CO_2. Using genomic and biochemical analysis, all the enzymes for this unique pathway was demonstrated. In the reductive glycine pathway, CO_2 is reduced to formate, and formate combines with a CO_2 molecule to produce glycine. Using glycine reductase, glycine is converted to acetyl-CoA, and another CO_2 molecule is

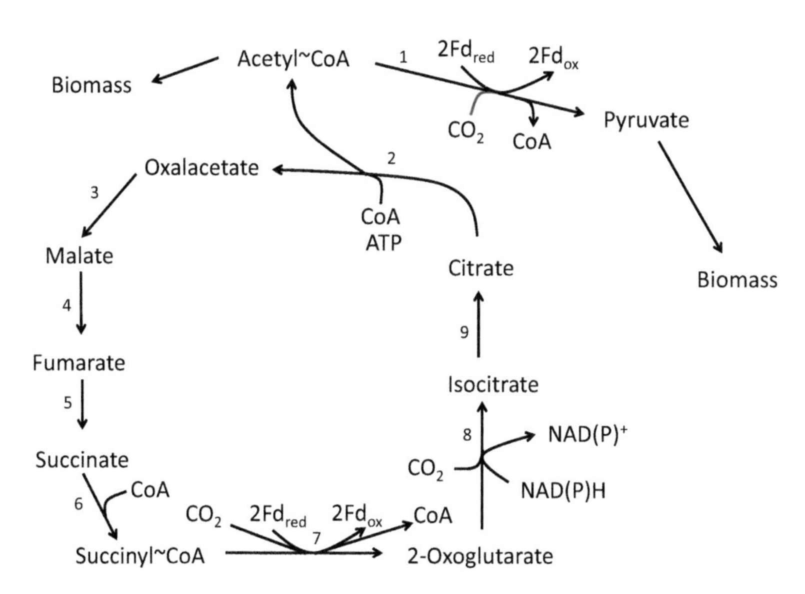

Fig. 1.4 Autotrophic carbon dioxide fixation by the reductive citric acid cycle in SRB. Enzymes are as follows: (1) pyruvate/ferredoxin oxidoreductase, Fd ferredoxin; (2) ATP citrate lyase; (3) malate dehydrogenase; (4) fumarate hydratase (fumarase); (5) fumarate reductase; (6) succinyl-CoA synthetase; (7) 2-oxoglutarate:ferredoxin oxidoreductase; (8) isocitrate dehydrogenase; (9) aconitate hydratase (aconitase)

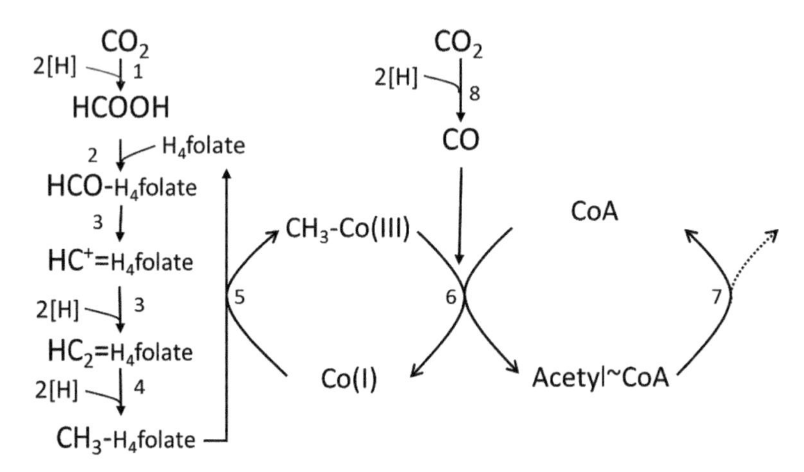

Fig. 1.5 Outline of the Wood-Ljungdahl pathway for autotrophic CO_2 fixation in SRB. Enzymes are (1) formate dehydrogenase, (2) 10-formyl-H_4 folate synthetase, (3) 5,10-methenyl H_4 folate cyclohydrolase and 5,10-methylene-H_4folate dehydrogenase, (4) 5,10-methylene-H_4folate reductase, (5) methyl transferase, (6) acetyl-CoA synthase, (7) corrinoid iron-sulfur protein, (8) CO dehydrogenase. 2[H] = 2 e^- plus 2 H^+. (Enzyme terminology is according to Ragsdale and Pierce (2008))

added to produce pyruvate. Since genes and enzymes for the reductive glycine pathway are present in many bacteria, it remains to be determined if this pathway functions in other SRB.

1.5 Geographic Distribution

With reports of SRB being cultured from different environments, it became evident that SRB were broadly distributed in terrestrial and aquatic environments (Starkey 1948). Cultures of SRB commonly carried the name of where they were isolated. Strains of *Desulfotomaculum* were identified as Teddington Garden, Aberdeen, Holland, Delft, and Singapore. Strains of *Desulfovibrio* and especially *D. desulfuricans* carried the names of Walvis Bay, Benghazi, Essex, Norway, New Jersey, California, Canet, British Guiana, Australia, and Aberdovey (Postgate 1979). Strains of *D. vulgaris* were isolated from many different geographical regions including Teddington, Wandel, Holland, Marseille, Venezuela, Llanelly, Denmark, and Woolwich (Postgate 1979). A considerable amount of research was conducted with *D. vulgaris* strains from Hildenborough, Madison, Marburg, and Miyazaki. With respect to specialized environments, bacteria of the *Desulfotalea*/*Desulforhopalus* cluster were found in the permanently cold Arctic sediments (Knoblauch et al. 1999), Antarctic sediments (Purdy et al. 2003), and sediments in the Japan Trench (Li et al. 1999). The spore-forming SRB, *Desulfotomaculum*, are widely dispersed in nature and have been detected in numerous deep subsurface environments including the Great Artesian Basin, Australia, geothermal groundwater in France, an oilfield in the Paris Basin in France, an oilfield in Russia, an oilfield in the North Sea, a mine in Japan, and a mine South Africa (Aüllo et al. 2013). A highly specialized spore-forming SRB, *Candidatus* Desulforudis audaxviator, was found at a depth of 2.8 km in South Africa where it became adapted to the low nutrient environment, and it has also been detected in deep subsurface samples in Turkey, Germany, and Finland. Information and references for *Candidatus* Desulforudis audaxviator are found in Sect. 2.5.1.2 in Chap. 2. Spores of thermophilic *Desulfotomaculum* spp. have been detected in the cold seabed of the Arctic Ocean (Aüllo et al. 2013). To explain this anomaly, two hypotheses have been raised: (i) ocean floor would be connected to the oceanic crust with the spores being transported to the oceanic ridge and then released into the ocean by way of hydrothermal vents and (ii) the connection of oilfields to the Arctic Sea by oceanic fluids (Aüllo et al. 2013). From a study involving several different marine sites at different geological regions, environmental temperature showed a greater correlation to SRB present than concentration of organic carbon present or the C:N ratio in the environment (Robador et al. 2016). While the biogeography of microorganisms is regulated by selection, drift, dispersal, and mutation, the contribution of each of these processes is difficult to assess for microorganisms (Hanson et al. 2014). A study concerning the influence of geographical isolation on the evolution of laterally transfer of *dsr* genes revealed that the hydrothermal vent sequences were distinguished from hot spring sequences (Fru 2011). There may have been a common ancestor for *Thermodesulfobacterium*, but their gene sequences may have diverged when the bacteria became associated with distinct habitats. However, as observed with some of the D*esulfotomaculum*, the DNA sequences may have had separate origins, but with isolation in the same niche, the genes may have converged (Fru 2011). With creative experiments and new instruments, insights into biogeographical information of the SRP will be obtained.

1.6 Perspective

The isolation of an anaerobic bacterium, *Spirillum desulfuricans*, by Beijerinck in 1895 was the first of hundreds of bacteria and archaea identified as belonging of the physiological group referred to as dissimilatory sulfate reducers. Although these organisms may be expected to have a restricted metabolism, these sulfate reducers are capable of utilizing a broad spectrum of organic and inorganic compounds as alternate electron acceptors. And while each species may use a small number of substrates as electron donors, collectively as a group, the SRP account for the utilization of many different compounds. This breadth of metabolic capability enables the SRP to grow and persist in the environments. Through syntrophic growth studies, cooperative activities of microorganism from different microbial domains have been established. Future studies will expand the metabolic, structural, and cultural diversity of the SRP.

References

Abdollahi H, Wimpenny JWT (1990) Effects of oxygen on the growth of *Desulfovibrio desulfuricans*. J Gen Microbiol 136:1025–1030

Adams ME, Butlin KR, Hollands SJ, Postgate JR (1951) The role of hydrogenase in the autotrophy of *Desulphovibrio*. Research 4:245–246

Al-Hitti IK, Moody GJ, JDR T (1983) Sulphide ion-selective electrode studies concerning *Desulfovibrio* species of sulphate-reducing bacteria. Analyst 108:1209–1220

Audiffrin C, Cayol J-L, Joulian C, Casalot L, Thomas P, Garcia J-L, Ollivier B (2003) *Desulfonauticus submarinus* gen. nov., sp. nov., a novel sulfate-reducing bacterium isolated from a deep-sea hydrothermal vent. Int J Syst Evol Microbiol 53:1585–1590

Aüllo T, Ranchou-Peyruse A, Ollivier B, Magot M (2013) *Desulfotomaculum* spp. and related gram-positive sulfate-reducing bacteria in deep subsurface environments. Front Microbiol. https://doi.org/10.3389/fmicb.2013.00362

Baars JK (1930) Over Sulfaatreductie door Bacteriën. Dissertation. W.D. Meinema, NV, Delft

Badziong W, Ditter B, Thauer RK (1979) Acetate and carbon dioxide assimilation by *Desulfovibrio vulgaris* (Marburg), growing on hydrogen and sulfate as sole energy source. Arch Microbiol 123:301–305

Badziong W, Thauer RK, Zeikus JG (1978) Isolation and characterization of *Desulfovibrio* growing on hydrogen plus sulfate as the sole energy source. Arch Microbiol 116:41–49

Badziong W, Thauer RK (1978) Growth yields and growth rates of *Desulfovibrio vulgaris* (Marburg) growing on hydrogen plus sulfate and hydrogen plus thiosulfate as the sole energy sources. Arch Microbiol 117:209–214

Baena S, Fardeau M-L, Labat M, Ollivier B, Garcia J-L, Patel BKC (1998) *Desulfovibrio aminophilus* sp. nov., a novel amino acid degrading and sulfate-reducing bacterium from an anaerobic dairy wastewater lagoon. Syst Appl Microbiol 21:498–504

Bak F (1993) Fermentation of inorganic sulfur compounds by sulfate-reducing bacteria. In: Guerrero R, Pedrós-Alió C (eds) Trends in microbial ecology. Barcelona, Spanish Society for Microbiology, pp 75–78

Bak F, Cypionka H (1987) A novel type of energy metabolism involving fermentation of inorganic sulphur compounds. Nature 326:891–892

Bak F, Pfennig N (1987) Chemolithotrophic growth of *Desulfovibrio sulfodismutans* sp. nov. by disproportionation of inorganic sulfur compounds. Arch Microbiol 147:184–189

Baker FD, Papiska HR, Campbell LL (1962) Choline fermentation by *Desulfovibrio desulfuricans*. J Bacteriol 84:973–978

Bale SJ, Goodman K, Rochelle PA, Marchesi JR, Fry JC, Weightman AJ, Parkes RJ (1997) *Desulfovibrio profundus* sp. nov., a novel barophilic sulfate-reducing bacterium from deep sediment layers in the Japan Sea. Int J Syst Evol Microbiol 47:515–521

Barker HA (1939) Studies upon methane fermentation: IV The isolation and culture of *Methanobacterium omelianskii*. Antoine Van Leeuwenhoek 6:201–220

Barton LL (1995) Sulfate-reducing bacteria: biotechnology handbooks. Plenum Press, New York

Barton LL, Fauque GD (2009) Biochemistry, physiology and biotechnology of sulfate-reducing bacteria. Adv Appl Microbiol 68:41–98

Barton LL, Le Gall J, Peck HD Jr (1970) Phosphorylation coupled to oxidation of hydrogen with fumarate in extracts of the sulfate-reducing bacterium *Desulfovibrio gigas*. Biochem Biophys Res Commun 41:1036–1042

Barton LL, McLean RJC (2019) Environmental microbiology and microbial ecology. Wiley Publishing Co, New York

Beijerinck WM (1895) Üeber Spirillum desulfuricans als Ursache von Sulfatreduction. Zentbl Bakt ParasitKde (abt. 2) 1:1–9, 49–59, 104–114

Beliakova EV, Rozanova EP, Borzenkov IA, Turova TP, Pusheva MA, Lysenko AM, Kolganov TV (2006) The new facultatively chemolithoautotrophic, moderately halophilic, sulfate-

reducing bacterium *Desulfovermiculus halophilus* gen. nov., sp. nov., isolated from an oil field. Mikrobiologiia 75:201–211
Ben Dhia Thabet O, Wafa T, Eltaief K, Cayol J-L, Hamdi M, Fauque G, Fardeau M-L (2011) *Desulfovibrio legallis* sp. nov.: a moderately halophilic, sulfate-reducing bacterium isolated from a wastewater digestor in Tunisia. Curr Microbiol 62:486–491
Berg IA (2011) Ecological aspects of the distribution of different autotrophic CO_2 fixation pathways. Appl Ennviron Microbiol 77:1925–1936
Biebl H, Pfennig N (1977) Growth of sulfate-reducing bacteria with sulfur as electron acceptor. Arch Microbiol 112:115–117
Biebl H, Pfennig N (1978) Growth yields of green sulfur bacteria in mixed cultures with sulfur and sulfate-reducing bacteria. Arch Microbiol 117:9–16
Birkeland N-K, Schönheit P, Poghosyan L, Anne Fiebig A, Klenk H-P (2017) Complete genome sequence analysis of *Archaeoglobus fulgidus* strain 7324 (DSM 8774), a hyperthermophilic archaeal sulfate reducer from a North Sea oil field. Stand Genomic Sci 12:79. https://doi.org/10.1186/s40793-017-0296-5
Blum JS, Kulp TR, Han S, Lanoil B, Saltikov CW, Stolz JF, Miller LG, Oremland RS (2012) *Desulfohalophilus alkaliarsenatis* gen. nov., sp. nov., an extremely halophilic sulfate- and arsenate-respiring bacterium from Searles Lake, California. Extremophiles 16:727–742
Boone DR, Bryant MP (1980) Propionate-degrading bacterium *Syntrophobacter wolinii* sp. nov., gen. nov., from methanogenic ecosystems. Appl Environ Microbiol 40:626–632
Boulègue J (1978) Solubility of elemental sulfur in water at 298 K. Phosphorus Sulfur Silicon Relat Elem 5:127–128
Brandis A, Thauer R (1981) Growth of *Desulfovibrio* species on hydrogen and sulphate as sole energy source. J Gen Microbiol 126:249–252
Brandis-Heep A, Gebhardt NA, Thauer RK, Widdel F, Pfennig N (1983) Anaerobic acetate oxidation to CO_2 by *Desulfobacter postgatei*. I Demonstration of all enzymes required for the operation of the citric acid cycle. Arch Microbiol 136:222–229
Brandt KK, Patel BK, Ingvorsen K (1999) *Desulfocella halophila* gen. nov., sp. nov., a halophilic, fatty-acid-oxidizing, sulfate-reducing bacterium isolated from sediments of the Great Salt Lake. Int J Syst Bacteriol 49:193–200
Breed RS, Murray EGD, Hitchens AP (1948) Bergey's manual of determinative bacteriology, 6th edn. Williams & Wilkins Co., Baltimore, pp 207–209
Brileya KA, Camilleri LB, Zane GM, Wall JD, Fields MW (2014) Biofilm growth mode promotes maximum carrying capacity and community stability during product inhibition syntrophy. Front Microbiol 5:693, 2. https://doi.org/10.3389/fmicb.2014.00693
Bryant M (1979) Microbial methane production – theoretical aspects. J Animal Sci 48:193–201
Bryant MP, Campbell LL, Reddy CA, Crabill MR (1977) Growth of *Desulfovibrio* in lactate or ethanol media low in sulfate in association with H_2-utilizing methanogenic bacteria. Appl Environ Microbiol 33:1162–1169
Bryant MP, Wolin EA, Wolin MJ, Wolfe RS (1967) *Methanobacillus omelianskii* a symbiotic association of two species of bacteria. Arch Microbiol 59:20–31
Brysch K, Schneider C, Fuchs G, Widdel F (1987) Lithoautrophic growth of sulfate-reducing bacteria and description of *Desulfobacterium autotrophicum* gen. nov., sp. nov. Arch Microbiol 148:264–274
Bühring SI, Elvert M, Witte U (2005) The microbial community structure of different permeable sandy sediments characterized by the investigation of bacterial fatty acids and fluorescence in situ hybridization. Environ Microbiol 7:281–293
Bunker HJ (1936) A review of the physiology and biochemistry of the sulphur bacteria. Dep Sci Industr Res Chem Res Special Rep. no. 3. London, His Majesty's Stationery Office, London. 48 pp
Bunker HJ (1939) Microbial experiments in anaerobic corrosion. J Soc Chem Ind Lond 58:93–100
Burggraf S, Jannasch HW, Nicolaus B, Stetter KO (1990) *Archaeoglobus profundus* sp. nov., represents a new species within the sulfate-reducing archaebacteria. Microbiol. 13:24–28

Butlin KR, Adams ME (1947) Autotrophic growth of sulphate-reducing bacteria. Nature 160: 154–155

Butlin KR, Adams ME, Thomas M (1949) The isolation and cultivation of sulphate-reducing bacteria. J Gen Microbiol 3:46–59

Cammack R, Fauque G, Moura JJ, LeGall J (1984) ESR studies of cytochrome *c* from *Desulfovibrio desulfuricans* strain Norway 4: midpoint potentials of the four haems and interactions with ferredoxin and colloidal sulphur. Biochim Biophys Acta 784:68–74

Campbell LL, Kasprzycki MA, Postgate JR (1966) *Desulfovibrio africanus* sp. n., a new dissimilatory sulfate-reducing bacterium. J Bacteriol 92:1122–1127

Campbell LL, Postgate JR (1965) Classification of the spore-forming sulfate-reducing bacteria. Bacteriol Rev 29:359–363

Canfield DE, Des Marais DJ (1991) Aerobic sulfate reduction in microbial mats. Science 251: 1471–1473

Chau ATT, Lee M, Adrian L, Manefield MJ (2018) Syntrophic partners enhance growth and respiratory dehalogenation of hexachlorobenzene by *Dehalococcoides mccartyi* strain CBDB1. Front Microbiol (22 August 2018). https://doi.org/10.3389/fmicb.2018.01927

Chen S, Liu X, Dong X (2005) *Syntrophobacter sulfatireducens* sp. nov., a novel syntrophic, propionate-oxidizing bacterium isolated from UASB reactors. Int J Sys Evol Microbiol 55: 1319–1324

Chivian D, Brodie EL, Alm EJ, Culley DE, Dehal PS, DeSantis TZ et al (2008) Environmental genomics reveals a single-species ecosystem deep within earth. Science 322:275–278

Conrad R (2009) The global methane cycle: recent advances in understanding the microbial process involved. Env Microbiol Rep 1:285–292

Copenhagen WJ (1934) Occurrence of sulphides in certain areas of the sea bottom on the South African Coast. Fish and Marine Biology Survey Division, Union South African, Rep No 11, pp. 3–18

Cord-Ruwish R, Garcia JL (1985) Isolation and characterization of an anaerobic benzoate-degrading spore-forming sulfate-reducing bacterium: *Desulfotomaculum sapomandens* sp. nov. FEMS Microbiol Lett 29:325–330

Cord-Ruwisch R, Seitz HJ, Conrad R (1988) The capacity of hydrogenotrophic anaerobic bacteria to compete for traces of hydrogen depends on the redox potential of the terminal electron acceptor. Arch Microbiol 149:350–357

Cypionka H, Pfennig N (1986) Growth yields of *Desulfotomaculum orientis* with hydrogen in chemostat culture. Arch Microbiol 143:396–399

Cypionka H, Widdel F, Pfennig N (1985) Survival of sulfate-reducing bacteria after oxygen stress, and growth in sulfate-free oxygen-sulfide gradients. FEMS Microbiol Ecol 31:39–45

da Silva SM, Voordouw J, Leitão C, Martins M, Voordouw G, Pereira IAC (2013) Function of formate dehydrogenases in *Desulfovibrio vulgaris* Hildenborough energy metabolism. Microbiol. 159:1760–1769

Dannenberg S, Kroder M, Dilling W, Cypionka H (1992) Oxidation of H_2, organic compounds and inorganic sulfur compounds coupled to reduction of O_2 or nitrate by sulfate-reducing bacteria. Arch Microbiol 158:93–99

Datta SC (1943) Production of elementary Sulphur by reduction of sulphate through bacterial agency. Curr Sci 12:305

Daumas S, Cord-Ruwisch R, Garcia JL (1988) *Desulfotomaculum geothermicum* sp. nov., a thermophilic, fatty acid-generating, sulfate-reducing bacterium isolated with H_2 from geothermal ground water. Antoine van Leeuwenhoek. 54:165–178

Day LA, De León KB, Kempher ML, Zhou J, Wall JD (2019) Complete genome sequence of *Desulfovibrio desulfuricans* IC1, a sulfonate-respiring anaerobe. Microbiol Resour Announc 8: e00456–e00419. https://doi.org/10.1128/MRA.00456-19

Deng X, Dohmae N, Nealson KH, Hashimoto K, Okamoto A (2018) Multi-heme cytochromes provide a pathway for survival in energy-limited environments. Sci Adv 2018:4. https://doi.org/10.1126/sciadv.aao5682

DeWeerd K, Mandelco L, Tanner RS, Woese CR, Suflita JM (1990) *Desulfomonile tiedjei* gen. nov. and sp. nov., a novel anaerobic dehalogenating, sulfate-reducing bacterium. Arch Microbiol 154:22–30

D'Hondt S, Spivack AJ, Pockalny R, Ferdelman TG, Fischer JP, Kallmeyer J et al (2009) Subseafloor sedimentary life in the South Pacific Gyre. Proc Natl Acad Sci. USA. 106: 11651–11656

Dias M, Salvado JC, Monperrus M, Caumette P, Amouroux D, Duran R, Guyoneaud R (2008) Characterization of *Desulfomicrobium salsuginis* sp. nov. and *Desulfomicrobium aestuarii* sp. nov., two new sulfate-reducing bacteria isolated from the Adour estuary (French Atlantic coast) with specific mercury methylation potentials. Syst Appl Microbiol 31:30–37

Dilling W, Cypionka H (1990) Aerobic respiration in sulfate-reducing bacteria. FEMS Microbiol Lett 71:123–128

Dinh HI, Kuever J, Mußmann M, Hassel AW, Stratmann M, Widdel F (2004) Iron corrosion by novel anaerobic microorganisms. Nature 427:829–832

Dolla A, Fournier M, Dermoun Z (2006) Oxygen defense in sulfate-reducing bacteria. J Biotechnol 126:87–100

Dolfing J, Jiang B, Henstra AM, Stams AJ, Plugge CM (2008) Syntrophic growth on formate: a new microbial niche in anoxic environments. Appl Environ Microbiol 74:6126–6131

Domange L (1982) Préis de Chimie Générale et de Chimie Minérale. Masson, Paris, pp 318–322

Drzyzga O, Gottschal JC (2002) Tetrachloroethene dehalorespiration and growth of *Desulfitobacterium frappieri* TCE1 in strict dependence on the activity of *Desulfovibrio fructosivorans*. Appl Environ Microbiol 68:642–649

Drzyzga O, Jannsen S, Blotevogel K-H (1994) Mineralization of monofluorobenzoate by a diculture under sulfate-reducing conditions. FEMS Microbiol Lett 116:215–219

Elferink SO, Maas RN, Harmsen HJM, Stams AJM (1995) *Desulforhabdus amnigenus* gen. nov. sp. nov., a sulfate reducer isolated from anaerobic granular sludge. Arch Microbiol 164:119–124

Elion L (1925) A thermophilic sulphate-reducing bacterium. Zentbl Bakt ParasitKde (abt.2) 63:58–67

Fadhlaoui K, Hania WB, Postec A, Fauque G, Hamdi M, Ollivier B, Fardeau M-L (2015) Characterization of *Desulfovibrio biadhensis* sp. nov., isolated from a thermal spring. Int J Syst Evol Microbiol 65:1256–1261

Fashchuk DY (2011) Marine ecological geography – theory and experience. Springer, Berlin, p 429

Fauque GD (1995) Ecology of sulfate-reducing bacteria. In: Barton LL (ed) Sulfate-reducing bacteria. Plenum Press, New York, pp 217–242

Fauque GD, Barton LL (2012) Hemoproteins in dissimilatory sulfate-and sulfur-reducing prokaryotes. Adv Microbial Phys 60:2–91

Fauque G, Herve D, Le Gall J (1979) Structure-function relationship in hemoproteins: the role of cytochrome c_3 in the reduction of colloidal sulfur by sulfate-reducing bacteria. Arch Microbiol 121:261–264

Fauque G, LeGall J, Barton LL (1991) Sulfate-reducing and sulfur-reducing bacteria. In: Shively JM, Barton LL (eds) Variations in autotrophic life. Academic Press Limited, San Diego, CA, pp 271–337

Fetzner S (1998) Bacterial dehalogenation. Appl Microbiol Biotechnol 50:633–657

Fildes P (1956) Production of tryptophan by *salmonella typhi* and *Escherichia coli*. J Gen Microbiol 15:636–643

Finster K, Liesack W, Thamdrup B (1998) Elemental sulfur and thiosulfate disproportionation by *Desulfocapsa sulfoexigens* sp. nov., a new anaerobic bacterium isolated from marine surface sediment. Appl Environ Microbiol 64:119–125

Finster K (2008) Microbiological disproportionation of inorganic sulfur compounds. J Sulfur Chem 29:281–292

Finster W, Kjeldsen KU (2010) *Desulfovibrio oceani* subsp. *oceani* sp. nov., subsp. nov. and *Desulfovibrio oceani* subsp. *galateae* subsp. nov., novel sulfate-reducing bacteria isolated from the oxygen minimum zone off the coast of Peru Kai. Antoine van Leeuwenhoek 97:221–229

Finster K, Liesack W, Yindall BJ (1997) *Desulfospira joergensenii*, gen. nov., sp. nov., a new sulfate-reducing bacterium isolated from marine surface sediment. Syst Appl Microbiol 20:201–208

Folkerts M, Ney U, Kneifel H, Stackebrandt E, Witte EG, Förstel H, Schoberth SM, Sahm H (1989) *Desulfovibrio furfuralis* sp. nov., a furfural degrading strictly anaerobic bacterium. Syst Appl Microbiol 11:161–169

Friedrich M, Springer N, Ludwig W, Schink B (1996) Phylogenetic positions of *Desulfofustis glycolicus* gen. nov., sp. nov., and *Syntrophobotulus glycolicus* gen. Nov., sp. nov., two new strict anaerobes growing with glycolic acid. Int J Syst Bacteriol 46:1065–1069

Fru EC (2011) Microbial evolution of sulphate reduction when lateral gene transfer is geographically restricted. Int J Syst Evol Microbiol 61:1725–1735

Fründ C, Cohen Y (1992) Diurnal cycles of sulfate reduction under oxic conditions in cyanobacterial mats. Appl Environ Microbiol 58:70–77

Fu R, Wall JD, Voordouw G (1994) DcrA, a *c*-type heme-containing methyl-accepting protein from *Desulfovibrio vulgaris* Hildenborough, senses the oxygen concentration or redox potential of the environment. J Bacteriol 176:344–350

Fuchs G (2011) Alternative pathway of carbon dioxide fixation insights into the early evolution of life. Annu Rev Microbiol 65:631–658

Fuseler K, Cypionka H (1995) Elemental sulfur as an intermediate of sulfide oxidation with oxygen by *Desulfobulbus propionicus*. Arch Microbiol 164:104–109

Galushko AS, Rozanova EP (1991) *Desulfobacterium cetonicum* sp. nov.: a sulfate-reducing bacterium which oxidizes fatty acids and ketones. Mikrobiologiya 60:102–107

Gibson GR (1990) Physiology and ecology of the sulphate-reducing bacteria. J Appl Bacteriol 69: 769–797

Gieg LM, Fowler SJ, Berdugo-Clavijo C (2014) Syntrophic biodegradation of hydrocarbon contaminants. Curr Opin Biotech 27:21–29

Göker M, Teshima H, Lapidus A, Nolan M, Lucas S, Hammon N et al (2011) Complete genome sequence of the acetate-degrading sulfate reducer *Desulfobacca acetoxidans* type strain (ASPRB2^{T}). Stand Genom Sci 4:393–401

Großkopf T, Zenobi S, Alston M, Folkes L, Swarbreck D, Soyer OS (2016) A stable genetic polymorphism underpinning microbial syntrophy. ISME J 10:2844–2853

Grossman JP, Postgate JR (1955) The metabolism of malate and certain other compounds by *Desulfovibrio desulfuricans*. J Gen Micribiol 12:429–445

Gumerov VN, Mardanov AV, Beletsky AV, Prokofeva MI, Bonch-Osmolovskaya EA, Ravin NV, Skryabin KG (2011) Complete genome sequence of "*Vulcanisaeta moutnovskia*" strain768-28, a novel member of the hyperthermophilic crenarchaeal genus *Vulcanisaeta*. J Bacteriol 193: 2355–2356

Güven KC, Çubukçu N (2009) A method for hydrogen sulfide removal in air of submarine by Lewatit TP 208. J Black Sea/Mediterranean Environ 15:87–98

Hallam SJ, Putnam N, Preston CM, Detter JC, Rokhsar D, Richardson PM, DeLong EF (2004) Reverse methanogenesis: testing the hypothesis with environmental genomics. Science 305: 1457–1462

Hansen TA (1993) Carbon metabolism of sulfate-reducing bacteria. In: Odom JM, Singleton R Jr (eds) The sulfate-reducing bacteria: contemporary perspectives. Springer-Verlag, New York, pp 21–41

Hansen LB, Finster K, Fossing H, Iversen N (1998) Anaerobic methane oxidation in sulfate depleted sediments: effects of sulfate and molybdate additions. Aquat Microbiol Ecol 14:195–204

Hanson CA, Fuhrman JA, Horner-Devine MC, Martiny JBH (2014) Beyond biogeographic patterns: processes shaping the microbial landscape. Nat Rev Microbiol 10:597–506

Haouari O, Fardeau M-L, Casalot L, Tholozan J-L, Hamdi M, Ollivier B (2006) Isolation of sulfate-reducing bacteria from Tunisian marine sediments and description of *Desulfovibrio bizertensis* sp. nov. Int J Syst Evol Microbiol 56:2909–2913

Hardy JA, Hamilton WA (1981) The oxygen tolerance of sulfate-reducing bacteria isolated from North Sea waters. Curr Microbiol 6:259–262

Harmsen HJM, van Kuijk BLM, Plugge CM, Akkermans ADL, de Vos WM, Stams AJM (1998) *Syntrophobacter fumaroxidans* sp. nov., a syntrophic propionate-degrading sulfate-reducing bacterium. Int J Syst Bacteriol 48:1383–1387

Hartshorne HS, Reardon CL, Ross D, Nuester J, Clarke TA, Gates AJ et al (2009) Characterization of an electron conduit between bacteria and the extracellular environment. PNAS, USA 106: 22169–22174

Hayward HR, Stadtman TC (1959) Anaerobic degradation of choline. I. Fermentation of choline by an anaerobic, cytochrome-producing bacterium, *Vibrio cholinicus* N. sp. J Bacteriol 78:557–561

He SH (1987) Isolation of a new species of sulphate-reducing bacterium, *Desulfovibrio multispirans,* and studies on the hydrogenase, cytochromes and fumarate reductase from this bacterium. Ph.D. Thesis, University of Georgia, Athens, 199 pp

Hedderich R, Klimmek O, Kröger A, Dirmeier R, Keller M, Stetter KO (1999) Anaerobic respiration with elemental sulfur and with sulfides. FEMS Microbiol Rev 22:353–381

Henry EA, Devereux R, Maki JS, Gilmour CC, Woese CR, Mandelco L et al (1994) Characterization of a new thermophilic sulfate-reducing bacterium *Thermodesulfovibrio yellowstonii*, gen. nov. and sp. nov.: its phylogenetic relationship to *Thermodesulfobacterium commune* and their origins deep within the bacterial domain. Arch Microbiol 161:62–69

Henstra AM, Dijkema C, Stams AJM (2007) *Archaeoglobus fulgidus* couples CO oxidation to sulfate reduction and acetogenesis with transient formate accumulation. Environ Microbiol 9: 1836–1841

Hernandez-Eugenio G, Fardeau M-L, Patel BKC, Macarie H, Çarcia J-L, Ollivier B (2000) *Desulfovibrio mexicanus* sp. nov., a sulfate-reducing bacterium isolated from an upflow anaerobic sludge blanket (UASB) reactor treating cheese wastewaters. Anaerobe 6:305–312

Hippe H, Stackebrandt E (2009) Genus VI. *Desulfosporosinus*. In: de Vos P, Garrity MG, Jones D, Krieg MR, Ludwig W (eds) Bergey's manual of systematic bacteriology, *The Firmicutes*, vol 3. Springer, New York, pp 983–989

Hocking WP, Stokke R, Roalkvam I, Steen IH (2014) Identification of key components in the energy metabolism of the hyperthermophilic sulfate-reducing archaeon *Archaeoglobus fulgidus* by transcriptome analyses. Front Microbiol 5:95. https://doi.org/10.3389/fmicb.2014.00095. eCollection 2014

Hoehler TM, Alperin MJ, Albert DB, Martens CS (1998) Thermodynamic control on hydrogen concentrations in anoxic sediments. Geochim Cosmochim Acta 62:1745–1756

Hoehler TM, Alperin MJ, Albert DB, Martens CS (2001) Apparent minimum free energy requirements for methanogenic archaea and sulfate-reducing bacteria in an anoxic marine sediment. FEMS Microbiol Ecol 38:33–41

Holliger C, Wohlfarth G, Diekert G (1999) Reductive dechlorination in the energy metabolism of anaerobic bacteria. FEMS Microbiol Rev 22:383–398

Hoppe-Seyler F (1886) Ueber die Gährung der Cellulose mit Bildung von Methan und Kohlensäure: II. Der Zerfall der Cellulose durch Gährung unter Bildung von Methan und Kohlensäure und die Erscheinungen, welche dieser Process veranlasst. Zeitschr. Physiol Chem 10:401–440

Huber H, Rossnagel P, Woese CR, Rachel R, Langworthy TA, Stetter KO (1996) Formation of ammonium from nitrate during chemolithoautotrophic growth of the extremely thermophilic bacterium *Ammonifex degensii* gen. nov. sp. nov. Syst Appl Microbiol 19:40–49

Imhoff-Stuckle D, Pfennig N (1983) Isolation and characterization of a nicotinic acid-degrading sulfate-reducing bacterium, *Desulfococcus niacini* sp nov. Arch Microbiol 136:194–198

Isaksen MF, Teske A (1996) Desulforhopalus vacuolatus gen. nov., sp. nov., a new moderately psychrophilic sulfate-reducing bacterium with gas vacuoles isolated from a temperate estuary. Arch Microbiol 166:160–168

Ishimoto M, Koyama J, Omura T, Nagai Y (1954) Biochemical studies on the sulfate-reducing bacteria: III Sulfate reduction by cell suspensions. J Biochem (Tokyo) 41:537–546

Itoh T, Suzuki K, Nakase T (1998) *Thermocladiurn rnodestius* gen. nov., sp. nov., a new genus of rod-shaped, extremely thermophilic crenarchaeote. Int J Syst Bacteriol 48:879–887
Itoh T, Suzuki K, Sanchez PC, Nakase T (1999) *Caldivirga maquilingensis gen.* nov., sp. nov., a new genus of rod-shaped crenarchaeote isolated from a hot spring in the Philippines. Int J Syst Bacteriol 49:1157–1163
Jackson BE, McInerney MJ (2000) Thiosulfate disproportionation by *Desulfotomaculum thermobenzoicum*. Appl Environ Microbiol 66:3650–3653
Jansen K, Fuchs G, Thauer RK (1985) Autotrophic CO2 fixation by *Desulfovibrio baarsii*: demonstration of enzyme activities characteristics for the acetyl-CoA pathway. FEMS Microbiol Lett 28:311–315
Janssen PH, Schuhmann A, Bak F, Liesack W (1996) Disproportionation of inorganic sulfur compounds by the sulfate-reducing bacterium *Desulfocapsa thiozymogenes gen.* nov., sp. nov. Arch Microbiol 166:184–192
Johnson MS, Zhulin IB, Gapuzan ER, Taylor BL (1997) Oxygen-dependent growth of the obligate anaerobe *Desulfovibrio vulgaris* Hildenborough. J Bacteriol 179:5598–5601
Jonkers HM, Ludwig R, De Wit R, Pringault O, Muyzer G, Niemann H, Finke N, De Beer D (2003) Structural and functional analysis of a microbial mat ecosystem from a unique permanent hypersaline inland lake: "La Salad de Chiprana" (NE Spain). FEMS Microbiol Ecol 44:175–189
Jonkers HM, van der Maarel MJEC, Gemerden H, Hansen TA (1996) Dimethylsulfoxide reduction by marine sulfate-reducing bacteria. FEMS Microbiol Lett 136:283–287
Jørgensen BB (1977) The sulfur cycle of a coastal marine sediment (Limfjorden, Denmark). Limnol Oceanogr 22:814–831
Jørgensen BB (1990a) Sulfur cycle of freshwater sediments: role of thiosulfate. Limnol Oceanogr 35:1329–1342
Jørgensen BB (1990b) Thiosulfate shunt in the sulfur cycle of marine sediments. Nature 249:152–154
Jørgensen BB, Bak F (1991) Pathways and microbiology of thiosulfate transformations and sulfate reduction in a marine sediment (Kattegat, Denmark). Appl Environ Microbiol 57:847–856
Jørgensen BB, Cohen Y (1977) Solar Lake (Sinai): 5. The sulfur cycle of the benthic cyanobacterial mat. Limnol Oceanogr 22:657–666
Jørgensen BB (1994) Sulfate reduction and thiosulfate transformations in a cyanobacterial mat during a diel oxygen cycle. FEMS Microbiol Ecol 13:303–312
Jørgensen BB, Fenchel TM (1974) The sulfur cycle of a marine sediment model system. Mar Biol 24:189–201
Keith SM, Herbert RA (1983) Dissimilatory nitrate reduction by a strain of *Desulfovibrio desulfuricans*. FEMS Microbiol Lett 18:55–59
Kelly DP (1987) Sulphur bacteria first again. Nature 326:830
Kim JH, Akagi JM (1985) Characterization of a trithionate reductase system from *Desulfovibrio vulgaris*. J Bacteriol 163:472–475
Klemps R, Cypionka H, Widdel F, Pfennig N (1985) Growth with hydrogen, and further physiological characteristics of *Desulfotomaculum* species. Arch Microbiol 143:203–208
Knoblauch C, Sahm K, Jørgensen BB (1999) Psychrophilic sulfate-reducing bacteria isolated from permanently cold Arctic marine sediments: description of *Desulfofrigus oceanense* gen. nov., sp. nov., *Desulfofrigus fragile* sp. nov., *Desulfofaba gelida* gen. nov., sp. nov., *Desulfotalea psychrophila* gen. nov., sp. nov. and *Desulfotalea arctica* sp. nov. Int J Sys Bacteriol 49:1631–1643. https://doi.org/10.1099/00207713-49-4-1631
Knittel K, Boetius A (2009) Anaerobic oxidation of methane: progress with an unknown process. Annu Rev Microbiol 63:311–334
Knittel K, Lösekann T, Boetius A, Kort R, Amann R (2005) Diversity and distribution of methanotrophic Archaea at cold seeps. Appl Environ Microbiol 71:467–479
Koffler H, Wilson PW (1951) The comparative biochemistry of molecular hydrogen. In: Werkman CW, Wilson PW (eds) Bacterial physiology. Academic Press, Inc., New York, pp 517–530

Krämer M, Cypionka H (1989) Sulfate formation via ATP sulfurylase in thiosulfate- and sulfite-disproportionating bacteria. Arch Microbiol 151:232–237

Kreel NE, Tabita FR (2007) Substitutions at methionine 295 of *Archaeoglobus fulgidus* ribulose-1,5-bisphosphate carboxylase/oxygenase affect oxygen binding and CO_2/O_2 specificity. J Biol Chem 282:1341–1351

Krekeler D, Sigalevich PA, Teske A, Cypionka H, Cohen Y (1997) A sulfate-reducing bacterium from the oxic layer of a microbial mat from Solar Lake (Sinai), *Desulfovibrio oxyclinae* sp. nov. Arch Microbiol 167:369–375

Kroulik A (1913) Über thermophile Zellulosevergärer. Vorläufige Mitteilung Zbl Bakt II Aby 36: 339–346

Krumbein WE, Buchholz H, Franke P, Giani D, Giele C, Wonneberger K (1979) O_2 and H_2S coexistence in stromatolites. Naturwissenschaften 66:381–389

Kuever J (2014) The family *Desulfarculaceae*. In: Rosenberg E, DeLong EF, Loy S, Stackebrandt E, Thompson F (eds) The prokaryotes deltaproteobacteria and epsilonproteobacteria, 4th edn. Springer, Heidelberg, pp 42–86

Kuever J, Galushko A (2014) The family *Desulfomicrobiaceae*. In: Rosenberg E, DeLong EF, Loy S, Stackebrandt E, Thompson F (eds) The prokaryotes deltaproteobacteria and epsilonproteobacteria, 4th edn. Springer, Heidelberg, pp 75–86

Kuever J, Könneke M, Galushko A, Drzyzga O (2001) Reclassification of *Desulfobacterium phenolicum* as *Desulfobacula phenolica* comb. nov. and description of strain SaxT as *Desulfotignum balticum* gen. nov., sp. nov. Int J Syst Evol Microbiol 51:171–177

Kuever J, Rainey FA (2009) Genus VII. *Desulfotomaculum*. In: de Vos P, Garrity GM, Jones D, Krieg MR, Ludwig W (eds) Bergey's manual of systematic bacteriology, The Firmicutes, vol 3. Springer, New York, pp 469–585

Kuever J, Visser M, Loeffler C, Boll M, Worm P, Sousa DZ et al (2014) Genome analysis of *Desulfotomaculum gibsoniae* strain Groll^T a highly versatile gram-positive sulfate-reducing bacterium. Stand Genom Sci. 9:821–839

Labes A, Schönheit P (2001) Sugar utilization in the hyperthermophilic, sulfate-reducing archaeon *Archaeoglobus fulgidus* strain 7324: starch degradation to acetate and CO_2 via a modified Embden-Meyerhof Pathway and Acetyl-CoA synthetase (Adp-forming). Arch Microbiol 176: 329–338

la Rivière JWM (1997) The Delft School of Microbiology in historical perspective. Antoine van Leeuwenhoek 71:3–13

Laube VM, Martin SM (1981) Conversion of cellulose to methane and carbon dioxide by triculture of *Acetivibrio cellulolyticus*, *Desulfovibrio* sp., and *Methanosarcina barkeri*. Appl Environ Microbiol 42:413–420

Laue H, Denger K, Cook A (1997) Fermentation of cysteate by a sulfate-reducing bacterium. Arch Microbiol 168:210–214

Laverman AM, Blum JS, Schaefer JK, Phillips EJP, Lovley DR, Oremland RS (1995) Growth of strain SES-3 with arsenate and other diverse electron acceptors. Appl Environ Microbiol 61: 3556–3561

Le Faou A, Rajagopal BS, Daniels L, Fauque G (1990) Thiosulfate, polythionates and elemental sulfur assimilation and reduction in the bacterial world. FEMS Microbiol Rev 75:351–382

Lefèvre CT, Howse PA, Schmidt ML, Sabaty M, Menguy N, Luther GW III, Bazylinski DA (2016) Growth of magnetotactic sulfate-reducing bacteria in oxygen concentration gradient medium. Environ Microbiol Rep 8:1003–1015

Li L, Kato C, Horikoshi K (1999) Microbial diversity in sediments collected from the deepest cold-seep area, the Japan Trench. Mar Biotechnol 1:391–400

Lie TJ, Clawson ML, Godchaux W, Leadbetter ER (1999) Sulfidogenesis from 2-aminoethanesulfonate (taurine) fermentation by a morphologically unusual sulfate-reducing bacterium, *Desulforhopalus singaporensis* sp nov. Appl Environ Microbiol 65:3328–3334

Lie TJ, Leadbetter JR, Leadbetter ER (1998) Metabolism of sulfonic acids and other organosulfur compounds by sulfate-reducing bacteria. Geomicrobiol J 15:135–149

Lie TJ, Pitta T, Leadbetter ER, Godchaux W III, Leadbetter JR (1996) Sulfonates: novel electron acceptors in anaerobic respiration. Arch Microbiol 166:204–210

Liu Y, Balkwill DL, Aldrich HC, Drake GR, Boone DR (1999) Characterization of the anaerobic propionate-degrading syntrophs *Smithella propionica* gen. nov, sp. nov. and *Syntrophobacter wolinii*. Int J Syst Bacteriol 49:545–556

Lobo SAL, Melo AMP, Carita JN, Teixeira MT, Saraiva LM (2007) The anaerobe *Desulfovibrio desulfuricans ATCC* 27774 grows at nearly atmospheric oxygen levels. FEBS Lett 581:433–436

López-Cortés A, Fardeau M-L, Fauque G, Joulian C, Ollivier B (2006) Reclassification of the sulfate- and nitrate-reducing bacterium *Desulfovibrio vulgaris* subsp. *oxamicus* as *Desulfovibrio oxamicus* sp. nov., comb. nov. Int J Syst Evol Microbiol 56:1495–1499

Lovley DR, Coates JD, Blunt-Harris EL, Phillips EJP, Woodward JC (1996) Humic substances as electron acceptors for microbial respiration. Nature 382:445–448

Lovley DR, Phillips EJP (1994) Novel processes for anaerobic sulfate production from elemental sulfur by sulfate-reducing bacteria. Appl Environ Microbiol 60:2394–2399

Lu Z, Imlay JA (2021) When anaerobes encounter oxygen: mechanisms of oxygen toxicity, tolerance and defence. Nat Rev Microbiol. https://doi.org/10.1038/s41579-021-00583-y

Lupton FS, Conrad R, Zeikus JG (1984) CO metabolism of *Desulfovibrio vulgaris* strain Madison: physiological function in the absence and presence of exogenous substrates. FEMS Microbiol Lett 23:263–268

Mägli A, Wendt M, Leisinger T (1996) Isolation and characterization of *Dehalobacterium formicoaceticum* gen. nov. sp. nov., a strictly anaerobic bacterium utilizing dichloromethane as source of carbon and energy. Arch Microbiol 166:101–108

Magot M, Basso O, Tardy-Jacquenod C, Caumette P (2004) *Desulfovibrio bastinii* sp. nov. and *Desulfovibrio gracilis* sp. nov., moderately halophilic, sulfate-reducing bacteria isolated from deep subsurface oilfield water. Int J Syst Evol Microbiol 54:1693–1697

Magot M, Caumette P, Desperrier JM, Matheron R, Dauga C, Grimont F, Carreau L (1992) *Desulfovibrio longus* sp. nov., a sulfate-reducing bacterium isolated from an oil-producing well. Int J Syst Bacteriol 42:398–403

Maia LB, Fonseca L, Moura I, Moura JJG (2016) Reduction of carbon dioxide by a molybdenum-containing formate dehydrogenase: a kinetic and mechanistic study. J Am Chem Soc 138:8834–8846

Marietou A, Griffiths L, Cole J (2009) Preferential reduction of the thermodynamically less favorable electron acceptor, sulfate, by a nitrate-reducing strain of the sulfate-reducing bacterium *Desulfovibrio desulfuricans* 27774. J Bacteriol 191:882–889

Marietou A (2016) Nitrate reduction in sulfate-reducing bacteria. FEMS Microbiol Lett 363(15): fnw155. https://doi.org/10.1093/femsle/fnw155

McCready RGL, Gould WD, Cook FD (1983) Respiratory nitrate reduction by Desulfovibrio sp. Arch Microbiol 135:182–185

McGlynn SE, Chadwick GL, Kempes CP, Orphan VJ (2015) Single cell activity reveals direct electron transfer in methanotrophic consortia. Nature 526:531–535

McInerney MJ, Sieber JR, Gunsalus RP (2009) Syntrophy in anaerobic global carbon cycles. Cur Opin Biotechnol 20:623–632

McInerney MJ, Struchtemeyer CG, Sieber J, Mouttaki H, Stams AJM, Schink B, Rholin L, Gunsalus RP (2008) Physiology, ecology, phylogeny, and genomics of microorganisms capable of syntrophic metabolism. Ann N Y Acad Sci 1125:58–72

Mechalas BJ, Rittenberg SC (1960) Energy coupling in *Desulfovibrio desulfuricans*. J Bacteriol 80: 501–507

Men Y, Feil H, VerBerkmoes NC, Shah MB, Johnson DR, Lee PKH et al (2012) Sustainable syntrophic growth of *Dehalococcoides ethenogenes* strain 195 with *Desulfovibrio vulgaris* Hildenborough and *Methanobacterium congolense*: global transcriptomic and proteomic analyses. ISME J 6:410–421

Ménez B, Pasini V, Brunelli D (2012) Life in the hydrated suboceanic mantle. Nat Geosci 5:133–137

Meyer B, Kuehl JV, Deutschbauer AM, Arkin AP, Stahl DA (2014) Flexibility of syntrophic enzyme systems in *Desulfovibrio* species ensures their adaptation capability to environmental changes. J Bacteriol 195:4900–4914
Meyer B, Kuehl J, Deutschbauer AM, Price MN, Arkin AP, Stahl DA (2013) Variation among *Desulfovibrio* species in electron transfer systems used for syntrophic growth. J Bacteriol 195: 990–1004
Miller JDA, Neumann PM, Elford L, Wakerley DS (1970) Malate dismutation by *Desulfovibrio*. Arch Mikrobiol 71:214–219
Miller JDA, Wakerley DS (1966) Growth of sulphate-reducing bacteria by fumarate dismutation. J Gen Microbiol 43:101–107
Miroshnichenko ML, Tourova TP, Kolganova TV, Kostrikina NA, Chernych N, Bonch-Osmolovskaya EA (2008) *Ammonifex thiophilus* sp. nov., a hyperthermophilic anaerobic bacterium from a Kamchatka hot spring. Int J Syst Evol Microbiol 58:2935–2938
Mori K, Kim H, Kakegawa T, Hanada S (2003) A novel lineage of sulfate-reducing microorganisms: *Thermodesulfobiaceae* fam. nov., *Thermodesulfobium narugense*, gen.nov., a new thermophilus isolate from a hot spring. Extremophiles 7:283–290
Morita RY, Zobell CE (1955) Occurrence of bacteria in pelagic sediments collected during the mid-Pacific Expedition. Deep-Sea Res 3:66–73
Morris BEL, Henneberger R, Huber H, Moissl-Eichinger C (2013) Microbial syntrophy: interaction for the common good. FEMS Microbiol Rev 37:384–406
Moura JJG, Gonzalez P, Moura I, Fauque G (2007) Dissimilatory nitrate and nitrite ammonification by sulphate-reducing eubacteria. In: Barton LL, Hamilton WA (eds) Sulphate-reducing bacteria. Cambridge University Press, Cambridge, UK, pp 241–264
Mußmann M, Ishii K, Rabus R, Amann R (2005) Diversity and vertical distribution of cultured and uncultured Deltaproteobacteria in an intertidal mud flat of the Wadden Sea. Environ Microbiol 7:405–418
Muyzer G, Stams AJM (2008) The ecology and biotechnology of sulphate-reducing bacteria. Nat Rev Microbiol 6:441–454
Nanninga HJ, Gottschal JC (1986) Isolation of a sulfate-reducing bacterium growing with methanol. FEMS Microbiol Ecol 38:125–130
Nanninga HJ, Gottschal JC (1987) Properties of *Desulfovibrio carbinolicus* sp. nov. and other sulfate-reducing bacteria isolated from an anaerobic-purification plant. Appl Environ Microbiol 53:802–809
Newman DK, Kennedy EK, Coates JD, Ahmann D, Ellis DJ, Lovley DR, Morel FMM (1997) Dissimilatory arsenate and sulphate reduction in *Desulfotomaculum auripigmentum* sp. nov. Arch Microbiol 168:380–388
Niemann H, Lösekann T, de Beer D, Elvert M, Nadalig T, Knittel K et al (2006) Novel microbial communities of the Haakon Mosby mud volcano and their role as a methane sink. Nature 443: 854–858
Nikitinsky J (1907) Die anaerobe Bindung des Wasserstoffes durch Mikroorganismen. Zbl Bakt II Apt 19:495–499
Niklewski B (1914) Über die Wasserstoffaktivierung durch Bakterien unter besonderer Berücksichtigung der neuen Gattung *Hydrogenomonas agilis*. IV Zbl Bakt II Abt 40:430–433
Nilsen RK, Torsvik T, Lien T (1996) *Desulfotomaculum thermocisternum* sp. nov., a sulfate reducer isolated from a hot North Sea oil reservoir. Int J Syst Bacteriol 46:397–402
Odom JM, Singleton R Jr (1993) The sulfate-reducing bacteria: contemporary perspectives. Springer-Verlag, New York
Ollivier B, Cord-Ruwisch R, Hatchikian EC, Garcia JL (1988) Characterization of *Desulfovibrio fructosovorans* sp. nov. Arch Microbiol 149:447–450
Ollivier B, Hatchikian CE, Prensier G, Guezennec J, Garcia J-L (1991) *Desulfohalobium retbaense* gen. nov., sp. nov., a halophilic sulfate-reducing bacterium from sediments of a hypersaline lake in Senegal. Int J Syst Bacteriol 41:74–81

Orphan VJ, KHinrichs K-U, Ussler W III, Paull DK, Taylor LT, Sylva SP, Hayes JM, Delong EF (2001) Comparative analysis of methane-oxidizing archaea and sulfate-reducing bacteria in anoxic marine sediments. Appl Environ Microbiol 67:1922–1934

Ouattara AS, Patel BKC, Cayol J-L, Cuzin N, Traore AS, Garcia J-L (1999) Isolation and characterization of *Desulfovibrio burkinensis* sp. nov. from an African rice field, and phylogeny of *Desulfovibrio alcoholivorans*. Int J Syst Bacteriol 49:639–643

Ozuolmez D, Na H, Lever MA, Kjeldsen KU, Jørgensen BB, Plugge CM (2015) Methanogenic archaea and SRB co-cultured on acetate: teamwork or coexistence? Front Microbiol. https://doi.org/10.3389/fmicb.2015.00492

Pankhania IP, Spormann AM, Hamilton WA, Thauer RK (1988) Lactate conversion to acetate, CO_2 and H_2 in cell suspensions of *Desulfovibrio vulgaris* (Marburg): indications for the involvement of an energy driven reaction. Arch Microbiol 150:26–31

Pankhurst ES (1968) Significance of sulphate-reducing bacteria in the gas industry: a review. J Appl Bacteriol 31:179–183

Parshina SN, Kijlstra S, Henstra AM, Sipma J, Plugge CM, Stams AJM (2005a) Carbon monoxide conversion by thermophilic sulfate-reducing bacteria in pure culture and in co-culture with *Carboxydothermus hydrogenoformans*. Appl Microbiol Biotechnol 68:390–396

Parshina SN, Sipma J, Nakashimada Y, Henstra AM, Smidt H, Lysenko AM et al (2005b) *Desulfotomaculum carboxydivorans* sp. nov., a novel sulfate-reducing bacterium capable of growth at 100% CO. Int J Syst Evol Microbiol 55:2159–2165

Parshina SN, Sipma J, Henstra AM, Stams AJM (2010) Carbon monoxide as an electron donor for the biological reduction of sulphate. Int J Microbiol 2010:319527, 9 pages. https://doi.org/10.1155/2010/319527

Pauli W (1947) The structure and properties of colloidal sulfur. J Coll Sci 2:333–348

Peck HD (1959) The ATP-dependent reduction of sulfate with hydrogen in extracts of *Desulfovibrio desulfuricans*. Proc Natl Acad Sci U S A 45:701–708

Peck HD Jr (1961) Enzymatic basis for assimilatory and dissimilatory sulfate reduction. J Bacteriol 82:933–939

Peck HD Jr (1993) Bioenergetic strategies of the sulfate-reducing bacteria. In: Odom JM, Singleton R Jr (eds) The sulfate-reducing bacteria: contemporary perspectives. Springer-Verlag, New York, pp 41–87

Peduzzi S, Tonolla M, Hahn D (2003) Isolation and characterization of aggregate-forming sulfate-reducing and purple sulfur bacteria from the chemocline of meromictic Lake Cadagno, Switzerland. FEMS Microbiol Ecol 45:29–37

Pernthaler A, Dekas AE, Brown CT, Goffredi SK, Embaye T, Orphan VJ (2008) Diverse syntrophic partnerships from deep-sea methane vents revealed by direct cell capture and metagenomics. PNAS, USA 105:7052–7057

Pfennig N, Biebl H (1976) *Desulfuromonas acetoxidans* gen. nov. and sp. nov., a new anaerobic, sulfur-reducing, acetate-oxidizing bacterium. Arch Microbiol 110:3–12

Philippot P, Van Zuilen M, Lepot K, Thomazo C, Farquhar J, Van Kranendonk MJ (2007) Early Archaean microorganisms preferred elemental sulfur, not sulfate. Science 317:1534–1537

Pietzsch K, Hard BC, Babel W (1999) A *Desulfovibrio* capable of growing by reducing U(VI). J Basic Microbiol 39:365–372

Pikuta EV, Hoover RB, Bej AK, Marsic D, Whitman WB, Cleland D, Krader P (2003) *Desulfonatronum thiodismutans* sp. nov., a novel alkaliphilic, sulfate-reducing bacterium capable of lithoautotrophic growth. Int J Syst Evol Microbiol 53(Pt 5):1327–1332

Plauchud M (1877) Recherches sur la formation des eaux sulfereuses naturelles. Comptes Rendus LXXXIV:235–238

Plugge CM, Balk M, Stams AJM (2002) *Desulfotomaculum thermobenzoicum* subsp. *thermosyntrophicum* subsp. nov., a thermophilic, syntrophic, propionate-oxidizing, spore-forming bacterium. Int J Syst Evol Microbiol 52:391–399

Plugge CM, Scholten JCM, Culley DE, Nie L, Brockman FJ, Zhang W (2010) Global transcriptomics analysis of *Desulfovibrio vulgaris* lifestyle change from syntrophic growth with *Methanosarcina barkeri* to sulfate reducer. Microbiol. 156:2746–2756

Plugge CM, Zhang W, Scholten JCM, Stams AJM (2011) Metabolic flexibility of sulfate-reducing bacteria. Front Microbiol. https://doi.org/10.3389/fmicb.2011.00081

Poehlein A, Daniel R, Schink B, Simeonova DD (2013) Life based on phosphite: a genome-guided analysis of *Desulfotignum phosphitoxidans*. BMC Genomics 14:753. http://www.biomedcentral.com/1471-2164/14/753BMC

Pont EG (1939) Association of sulphate reduction in the soil with anaerobic iron corrosion. J Aust Inst Agric Sci 5:170

Postgate JR (1949) Inhibition of sulphate-reduction by selenate. Nature Lond 164:670–671

Postgate JR (1951a) On the nutrition of *Desulphovibrio desulphuricans*. J Gen Microbiol 5:714–724

Postgate JR (1951b) The reduction of sulphur compounds by *Desulphovibrio desulphuricans*. J Gen Microbiol 5:725–738

Postgate JR (1951c) Dependence of sulphate reduction and oxygen utilization on a cytochrome in *Desulfovibrio*. Biochemist 58:ix

Postgate J (1959) Sulphate reduction by bacteria. Ann Rev Microbiol. 13:505–520

Postgate JR (1960) On the autotrophy of *Desulphovibrio desulphuricans*. Z Allg Mikrobiol 1:53–56

Postgate JR (1963) Sulfate-free growth of *clostridium nigrificans*. J Bacteriol 85:1450–1451

Postgate JR (1965) Recent advances in the study of the sulfate-reducing bacteria. Bacteriol Rev 29:425–441

Postgate JR (1979) The sulphate-reducing bacteria. Cambridge University Press, Cambridge, UK

Postgate JR, Campbell LL (1966) Classification of *Desulfovibrio* species, the nonsporeforming sulfate-reducing bacteria. Bacteriol Rev 30:732–738

Postgate JR, Kent HM, Robson RL, Chesshyre JA (1984) The genomes of *Desulfovibvio gigas* and *D. vulgaris*. J Gen Microbiol 130:1597–1601

Purdy KJ, Nedwell DB, Embley TM (2003) Analysis of the sulfate-reducing bacterial and methanogenic archaeal populations in contrasting Antarctic sediments. Appl Environ Microbiol 69:3181–3191

Qatibi A-I, Nivière V, Garcia JL (1991) *Desulfovibrio alcoholovorans* sp. nov., a sulfate-reducing bacterium able to grow on glycerol, 1,2- and 1,3-propanediol. Arch Microbiol 155:143–148

Rabus RR, Nordhaus WL, Widdel F (1993) Complete oxidation of toluene under strictly anoxic conditions by a new sulfate-reducing bacterium. Appl Environ Microbiol 59:1444–1451

Rabus R, Venceslau SS, Wöhlbrand L, Voordouw G, Wall JD, Pereira IAC (2015) A post-genomic view of the ecophysiology, catabolism and biotechnological relevance of sulphate-reducing prokaryotes. Adv Microbial Physiol 66:58–321

Ragsdale SW, Pierce E (2008) Acetogenesis and the Wood-Ljungdahl pathway of CO_2 fixation. Biochim Biophys Acta 1784:1873–1898

Ravenschlag K, Sahm K, Knoblauch C, Jørgensen BB, Amann R (2000) Community structure, cellular rRNA content, and activity of sulfate-reducing bacteria in marine Arctic sediments. Appl Environ Microbiol 66:3592–3602

Reeburgh WS (1976) Methane consumption in Cariaco trench waters and sediments. Earth Planetary Sci Lett 28:337–344

Rees GN, Patel BK (2001) *Desulforegula conservatrix* gen. nov., sp. nov., a long-chain fatty acid-oxidizing, sulfate-reducing bacterium isolated from sediments of a freshwater lake. Int J Syst Evol Microbiol 51:1911–1916

Reeve HA, Ash PA, Park H, Huang A, Posidias M, Tomlinson C, Lenz O, Vincent KA (2017) Enzymes as modular catalysts for redox half-reactions in H_2-powered chemical synthesis: from biology to technology. Biochem J 474:215–230

Rittenberg SC (1941) Studies on marine sulphate-reducing bacteria. Dissertation No. 650, University of California, La Jolla

Robador A, Müller AL, Sawicka JE, Berry D, Hubert CRJ, Loy A, Jørgensen BB, Brüchert V (2016) Activity and community structures of sulfate-reducing microorganisms in polar, temperate and tropical marine sediments. ISME J 10:796–809

Robertson WJ, Bowman JP, Franzmann PD, Mee BJ (2001) *Desulfosporosinus meridiei* sp. nov., a spore-forming sulfate-reducing bacterium isolated from gasolene-contaminated groundwater. Int J Syst Evol Microbiol 51:133–140

Roden EE, Lovley DR (1993) Dissimilatory Fe(III) reduction by the marine microorganism *Desulfuromonas acetoxidans*. Appl Environ Microbiol 59:734–742

Roger M, Brown F, Gabrielli W, Sargent F (2018) Efficient hydrogen-dependent carbon dioxide reduction by *Escherichia coli*. Curr Biol 28:140–145

Rozanova EP, Nazina TN, Galushko AS (1988) Isolation of a new genus of sulfate-reducing bacteria and description of a new species of this genus, *Desulfomicrobium apsheronum* gen. nov., sp. nov. Mikrobiologiya 57:634–641

Rubentschick LI (1928) Über Sulfatreduktion durch Bakterien bei Zellulosegärungsprodukten als Energiequelle. Zbl Bakt II Abt 73:483–496

Sadana JC, Jagannathan V (1954) Purification of hydrogenase from *Desulfovibrio desulfuricans*. Biochim Biophys Acta 14:287–288

Sánchez-Andrea I, Guedes IA, Bastian Hornung B, Boeren S, Lawson CE, Sousa DZ et al (2020) The reductive glycine pathway allows autotrophic growth of *Desulfovibrio desulfuricans*. Nat Commun 11:5090. https://doi.org/10.1038/s41467-020-18906-7

Sánchez-Andrea I, Stams AJ, Hedrich S, Ňancucheo I, Johnson DB (2015) *Desulfosporosinus acididurans* sp. nov.: an acidophilic sulfate-reducing bacterium isolated from acidic sediments. Extremophiles 19:39–47

Santos H, Fareleira P, Xavier AV, Chen L, Liu MY, LeGall J (1993) Aerobic metabolism of carbon reserves by the "obligate anaerobe" *Desulfovibrio gigas*. Biochem Biophys Res Commun 195: 551–557

Sass H, Ramamoorthy S, Yarwood C, Langner H, Schumann P, Kroppenstedt RM, Spring S, Rosenzwei RF (2009) *Desulfovibrio idahonensis* sp. nov., sulfate-reducing bacteria isolated from a metal(loid)-contaminated freshwater sediment. Int J Syst Evol Microbiol 59:2208–2214

Schauder R, Eikmanns B, Thauer TK, Widdel F, Fuchs G (1986) Acetate oxidation to CO_2 in anaerobic bacteria via a novel pathway not involving reactions of the citric acid cycle. Arch Microbiol 145:162–172

Schauder R, Kröger A (1993) Bacterial sulphur respiration. Arch Microbiol 159:491–497

Schauder R, Müller E (1993) Polysulfide as a possible substrate for sulfur-reducing bacteria. Arch Microbiol 160:377–382

Schauder R, Preuß A, Jetten M, Fuchs G (1989) Oxidative and reductive acetyl CoA/carbon monoxide dehydrogenase pathway in *Desulfobacterium autotrophicum*. Arch Microbiol 151: 84–89

Schauder R, Widdel F, Fuchs G (1987) Carbon assimilation pathways in sulfate-reducing bacteria. II Enzymes of a reductive citric acid cycle in the autotrophic *Desulfobacter hydrogenophilus*. Arch Microbiol 148:218–225

Scheller S, Yu H, Chadwick GL, McGlynn SE, Orphan VJ (2016) Artificial electron acceptors decouple archaeal methane oxidation from sulfate reduction. Science 351:703–707

Schink B (1997) Energetics of syntrophic cooperation in methanogenic degradation. Microbiol Mol Biol Rev 61:262–280

Schink B (2002) Synergistic interactions in the microbial world. Antoine Van Leeuwenhoek 81: 257–261

Schink B, Stams AJM (2006) Syntrophism among prokaryotes. In: Dworkin M, Falkow S, Rosenberg E, Schliefer K-H, Stackebrandt E (eds) The prokaryotes: an evolving electronic resource for the microbiological community. Springer-Verlag, New York, pp 309–335

Schink R, Thiemann V, Laue H, Friedrich MW (2002) *Desulfotignum phosphitoxidans* sp. nov., a new marine sulfate reducer that oxidizes phosphite to phosphate. Arch Microbiol 177:381–391

Schrenk MO, Brazelton WJ, Lang SQ (2013) Serpentinization, carbon, and deep life. Rev Mineral Geochem 75:575–606

Schuchmann K, Müller V (2013) Direct and reversible hydrogenation of CO_2 to formate by a bacterial carbon dioxide reductase. Science 342:1382–1385

Schuchmann K, Müller V (2014) Autotrophy at the thermodynamic limit of life: a model for energy conservation in acetogenic bacteria. Nat Rev Microbiol 12:809–821

Seitz HJ, Cypionka H (1986) Chemolithotrophic growth of *Desulfovibrio desulfuricans* with hydrogen coupled to ammonification of nitrate and nitrite. Arch Microbiol 146:63–67

Selwyn SC, Postgate JR (1959) A search for the Rubentschikii group of *Desulphovibrio*. Antoine van Leeuwenhoek 25:456–472

Senez JC, Leroux-Gilleron L (1954) Preliminary note on the anaerobic degradation of cysteine and cystine by sulfate-reducing bacteria. Bull Soc Chim Biol Paris 36:553–559

Senez JC, Pascal MC (1961) Dégradation de la choline par les bactéries sulfato-réductrices. Z Allg Mikrobiol 1:142–149

Senez JC, Pichinoty F (1958a) Reduction de l'hydroxylamine liée à l'activité de l'hydrogénase de *Desulfovibrio desulfuricans*. I Activité des cellules et des extraits. Biochim Biophys Acta 27: 569–580

Senez JC, Pichinoty F (1958b) Reduction de l'hydroxylamine liée é l'activité de l'hydrogénase de *Desulfovibrio desulfuricans*. II Nature du système enzymatique et du transporteur d'electrons intervenant dans la réaction. Biochim Biophys Acta 28:355–364

Senez JC, Pichinoty F, Konavaltchikoff-Mazoyer M (1956) Réduction des nitrites et de l'hydroxylamine par les suspensions et les extraits de *Desulfovibrio desulfuricans*. Compt Rend. 242:570–573

Senez J, Volcani B (1951) Utilization de l'hydrogene moleculaire par des souches pures de bacteries sulfato-reductrices d'origine marine. Compt Rend 232:1035–1036

Sheik CS, Sieber JR, Badalamenti JP, Carden K, Olson A (2017) Complete genome sequence of *Desulfovibrio desulfuricans* strain G11, a model sulfate-reducing, hydrogenotrophic, and syntrophic partner organism. Genome Announc 5:e01207–e01217. https://doi.org/10.1128/genomeA.01207-17

Shi L, Richardson DJ, Wang Z, Kerisit SN, Rosso KM, Zachara JM, Fredrickson JK (2009) The roles of outer membrane cytochromes of *Shewanella* and *Geobacter* in extracellular electron transfer. Environ Microbiol Rep 1(4):220–227. https://doi.org/10.1111/j.1758-2229.2009.00035.x

Shin B, Kim M, Zengler K, Chin KJ, Overholt WA, Gieg LM, Konstantinidis KT, Kostka JE (2019) Anaerobic degradation of hexadecane and phenanthrene coupled to sulfate reduction by enriched consortia from northern Gulf of Mexico seafloor sediment. Sci Rep 9(1):1239. https://doi.org/10.1038/s41598-018-36567-x

Sieber JR, McInerney MJ, Gunsalus RP (2012) Genomic insights into syntrophy: the paradigm for anaerobic metabolic cooperation. Ann Rev Microbiol 66:429–452

Sigalevich P, Baev MV, Teske A, Cohen Y (2000) Sulfate reduction and possible aerobic metabolism of the sulfate-reducing bacterium *Desulfovibrio oxyclinae* in a chemostat coculture with *Marinobacter* sp. strain MB under exposure to increasing oxygen concentrations. Appl Environ Microbiol 66:5013–5018

Sigalevich P, Cohen Y (2000) Oxygen-dependent growth of the sulfate-reducing bacterium *Desulfovibrio oxyclinae* in coculture with *Marinobacter* sp. strain MB in an aerated sulfate-depleted chemostat. Appl Environ Microbiol 66:5019–5023

Sisler FD, Zobell CE (1951) Hydrogen utilization by some marine sulfate-reducing bacteria. J Bacteriol 62:117–127

Sonne-Hansen J, Ahring BK (1999) *Thermodesulfobacterium hveragerdense* sp. nov., and *Thermodesulfovibrio islandicus* sp. nov., two thermophilic sulfate reducing bacteria isolated from a Icelandic hot spring. Syst Appl Microbiol 22:559–564

Sorokin YI (1954) Chemistry of the process of hydrogen reduction of sulfates. Truidy Inst Mikrobiol Akad Nauk SSSR 3:21–34

Stackebrandt E, Schumann P, Schüler E, Hippe H (2003) Reclassification of *Desulfotomaculum auripigmentum* as *Desulfosporosinus auripigmenti* corrig., comb. nov. Int J Syst Evol Microbiol 53:1439–1443

Stams AJM, Kremer DR, Nicolay K, Weenk GH, Hansen TA (1984) Pathway of propionate formation in *Desulfobulbus propionicus*. Arch Microbiol 139:167–173

Stams AJM, Plugge CM (2009) Electron transfer in syntrophic communities of anaerobic bacteria and archaea. Nat Rev Microbiol 7:568–577

Starkey RL (1938) A study of spore formation and other morphological characteristics of *Vibrio desulfuricans*. Arch Mikrobiol 9:268–304

Starkey RL (1948) Characteristics and cultivation of sulfate-reducing bacteria. J Am Water Works Assoc 40:1291–1298

Starkey RL, Wright KM (1945) Anaerobic corrosion of iron in soil. Final report of the American Gas Assoc. Iron Corrosion Res. Fellowship, Amer. Gas Assoc. New York

Steinsbu BO, Thorseth IH, Nakagawa S, Inagaki F, Lever MA, Engelen B, Øvreås L, Pedersen RB (2010) *Archaeoglobus sulfaticallidus* sp. nov., a novel thermophilic and facultatively lithoautotrophic sulfate-reducer isolated from black rust exposed to hot ridge flank crustal fluids. Int J Syst Evol Microbiol 60:2745–2752

Stephenson M, Stickland L (1931) The reduction of sulphate to sulphide by molecular hydrogen. Biochem J 25:215–220

Stetter K (1988) *Archaeoglobus fulgidus* gen. nov., sp. nov.: a new taxon of extremely thermophilic archaebacteria. Syst Appl Microbiol 10:172–173

Stetter K, Gaag G (1983) Reduction of molecular sulfur by methanogenic bacteria. Nature 305:309–311

Stetter KO, Laurer G, Thomm M, Neuner A (1987) Isolation of extremely thermophilic sulfate reducers: evidence for a novel branch of archaebacteria. Science 236:822–824

Steudel R (1989) On the nature of the 'elemental sulfur' (S^0) produced by sulfur-oxidizing bacteria – a model for the S^0 globules. In: Schlegel HG, Bowien B (eds) Autotrophic bacteria. Springer-Verlag, Berlin, pp 289–303

Steudel R, Holdt G, Gobel T, Hazeu W (1987) Chromatographic separation of higher polythionatcs $S_nO_6^{2-}$ (n = 3...22) and their detection in cultures of *Thiobacillus ferrooxidans:* molecular composition of bacterial sulfur secretions. Angew Chem Int Ed Engl 26:151–153

Steudel R, Holdt G, Nagorka R (1986) On the autoxidation of aqueous sodium polysulfide. Zeitschrift fur Naturforschung 41b:1519–1522

Stevens TO, Mckinley JP (1995) Lithoautotrophic microbial ecosystems in deep basalt aquifers. Science 270:450–455

Sun B, Cole JR, Sanford RA, Tiedje JM (2000) Isolation and characterization of *Desulfovibrio dechloracetivorans* sp. nov., a marine dechlorinating bacterium growing by coupling the oxidation of acetate to the reductive dechlorination of 2- chlorophenol. Appl Environ Microbiol 66:2408–2413

Sun B, Cole JR, Tiedje JM (2001) *Desulfomonile limimaris* sp. nov., an anaerobic dehalogenating bacterium from marine sediments. J Syst Evol Microbiol 51:365–371

Svetlichny VA, Sokolova TG, Gerhardt M, Ringpfeil M, Kostrikina NA, Zavarzin GA (1991) *Carboxydothermus hydrogenoformans* gen. nov., sp. nov., a CO-utilizing thermophilic anaerobic bacterium from hydrothermal environments of Kunashir Island. Syst Appl Microbiol 14: 254–260

Szewzyk R, Pfennig N (1987) Complete oxidation of catechol by the strictly anaerobic sulfate-reducing *Desulfobacterium catecholicum* sp. nov. Arch Microbiol 147:163–168

Tardy-Jacquenod C, Magot M, Laigret F, Kaghad M, Patel BKC, Guezennec J, Matheron R, Caumette P (1996) *Desulfovibrio gabonensis* sp. nov., a new moderately halophilic sulfate-reducing bacterium isolated from an oil pipeline. Int J Syst Bacteriol 46:710–715

Tasaki M, Kamagata Y, Nakamura K, Mikami E (1991) Isolation and characterization of a thermophilic benzoate degrading, sulfate-reducing bacterium, *Desulfotomaculum thermobenzoicum* sp nov. Arch Microbiol 155:348–352

Tasaki M, Kamagata Y, Nakamura K, Okamura K, Minami K (1993) Acetogenesis from pyruvate by *Desulfotomaculum thermobenzoicum* and differences in pyruvate metabolism among three sulfate-reducing bacteria in the absence of sulfate. FEMS Microbiol Lett 106:259–263
Tausson WO, Alioschina WA (1932) Über die bakterielle sulfatreduktion bei anwesenheit der Kohlenwasserstroffe. Mikrobiol (USSR) 1:229–261
Tebo BM, Obraztsova AY (1998) Sulfate-reducing bacterium with Cr(VI), U(VI), Mn(IV), Fe(III) as electron acceptors. FEMS Microbiol Lett 162:195–198
Teske A, Ramsing NB, Habicht KS, Fukui M, Küver J, Jørgensen BB, Cohen Y (1998) Sulfate-reducing bacteria and their activities in cyanobacterial mats of Solar Lake (Sinai, Egypt). Appl Environ Microbiol 64:2943–2951
Thamdrup B, Finster KM, Hansen JW, Bak F (1993) Bacterial disproportionation of elemental sulfur coupled to chemical reduction of iron or manganese. Appl Environ Microbiol 59:101–108
Thauer RK, Jungermann K, Decker K (1977) Energy conservation in chemotrophic anaerobic bacteria. Bacteriol Rev 41:100–180
Thauer RK, Kunow J (1995) Sulfate-reducing archaea. In: Barton LL (ed) Sulfate-reducing bacteria: biotechnology handbooks. Plenum Press, New York, pp 33–48
Thiel J, Spring S, Tindall BJ, Spröer C, Bunk B, Koeksoy E et al (2020) *Desulfolutivibrio sulfoxidireducens* gen. nov., sp. nov., isolated from a pyrite-forming enrichment culture and reclassification of *Desulfovibrio sulfodismutans* as *Desulfolutivibrio sulfodismutans* comb. nov. Syst Appl Microbiol 43:126105. https://doi.org/10.1016/j.syapm.2020.126105
Thorup C, Schramm A, Findlay AJ, Finster KW, Schreiber L (2017) Disguised as a sulfate reducer: growth of the deltaproteobacterium *Desulfurivibrio alkaliphilus* by sulfide oxidation with nitrate. mBio 8:e00671–e00617. https://doi.org/10.1128/mBio.00671-17
Timmers PHA, Vavourakis CD, Kleerebezem R, Damsté JS, Muyzer G, Stams AJM et al (2016) Metabolism and occurrence of methanogenic and sulfate-reducing syntrophic acetate oxidizing communities in haloalkaline environments. Front Microbiol 9:3039. https://doi.org/10.3389/fmicb.2018.03039
Trinkerl M, Breunig A, Schauder R, König H (1990) *Desulfovibrio termitidis* sp. nov., a carbohydrate-degrading sulfate-reducing bacterium from the hindgut of a termite. Syst Appl Microbiol 13:372–377
van Delden A (1903) Beitrag zur Kenntnis der Sulfat-reduktion durch Bakterien. Zentbl Bakt ParasitKde (abt.2) 11:81–94, 113–119
van den Ende FP, Meier J, van Gemerden H (1997) Syntrophic growth of sulfate-reducing bacteria and colorless sulfur bacteria during oxygen limitation. FEMS Microbiol Ecol 23:65–80
van der Maarel MJEC, van Bergeijk S, van Werkhoven AF, Laverman AM, Meijer WG, Stam WT, Hansen TA (1996) Cleavage of dimethylsulfoniopropionate and reduction of acrylate by *Desulfovibrio acrylicus* sp. nov. Arch Microbiol 166:109–111
van der Maarel MJEC, Jansen M, Jonkers HM, Hansen TA (1998) Demethylation and cleavage of dimethylsulfoniopropionate and reduction of dimethyl sulfoxide by sulfate-reducing bacteria. Geomicrobiol J 15:37–44
van Niel EWJ, Gomes TMP, Willems A, Collins MD, Prins RA, Gottschal JC (1996) The role of polyglucose in oxygen-dependent respiration by a new strain of *Desulfovibrio salexigens*. FEMS Microbiol Ecol 21:243–253
Vecchia D, Suvorova EI, Maillaard J, Bernier-Latmani R (2014) Fe(III) reduction during pyruvate fermentation by *Desulfotomaculum reducens* strain MI-1. Geobiology 12:48–61
Visscher PT, Prins RA, van Gemerden H (1992) Rates of sulfate reduction and thiosulfate consumption in a marine microbial mat. FEMS Microbiol Ecol 86:283–294
Visser M, Parshina SN, Alves JI, Sousa DZ, Pereira IAC, Muyzer G et al (2014) Genome analyses of the carboxydotrophic sulfate-reducers *Desulfotomaculum nigrificans* and *Desulfotomaculum carboxydivorans* and reclassification of *Desulfotomaculum caboxydivorans* as a later synonym of *Desulfotomaculum nigrificans*. Stand Genom Sci. 9:655–675

von Jan M, Lapidus A, Del Rio TG, Copeland A, Tice H, Cheng J-F (2010) Complete genome sequence of *Archaeoglobus profundus* type strain (AV18^T). Stand Genom Sci. 2(3):327–346. https://doi.org/10.4056/sigs.942153

Voordouw G (2002) Carbon monoxide cycling by *Desulfovibrio vulgaris* Hildenborough. J Bacteriol 184:5903–5911

Walker CB, He Z, Yang ZK, Ringbauer JA Jr, He Q, Zhou J et al (2009) The electron transfer system of syntrophically grown *Desulfovibrio vulgaris*. J Bacteriol 191:5793–5801

Walker CB, Redding-Johanson AM, Baidoo EE, Rajeev L, He Z, Hendrickson EL et al (2012) Functional responses of methanogenic archaea to syntrophic growth. ISEM J 6:2045–2055

Wallrabenstein C, Hauschild E, Schink B (1995) *Syntrophobacter pfennigii* sp. nov., a new syntrophically propionate-oxidizing anaerobe growing in pure culture with propionate and sulfate. Arch Microbiol 164:346–352

Warthmann R, Vasconcelos C, Sass H, McKenzie JA (2005) *Desulfovibrio brasiliensis* sp. nov., a moderate halophilic sulfate-reducing bacterium from Lagoa Vermelha (Brazil) mediating dolomite formation. Extremophiles 9:255–261

Wegener G, Krukenberg V, Riedel D, Tegetmeyer HE, Boetius A (2015) Intercellular wiring enables electron transfer between methanotrophic archaea and bacteria. Nature 526:587–590

Westerholm M, Dolfing J, Sherry A, Gray ND, Head IM, Schnurer A (2011) Quantification of syntrophic acetate oxidizing microbial communities in biogas processes. Environ Microbiol Rep 3:500–505

Widdel F (1980) Anaerober Abbau von Fettsäuren und Benzoesäure durch neu isolierte Arten Sulfatreduzierender Bakterien. Thesis, Georg-August-Univ, Göttingen, 443 pp

Widdel F (1988) Microbiology and ecology of sulfate- and sulfur-reducing bacteria. In: Zehnder AJB (ed) Biology of anaerobic organisms. John Wiley, New York, pp 469–585

Widdel F, Hansen TA (1992) The dissimilatory sulfate-and sulfur-reducing bacteria. In: Balows A, Trüper HG, Dworkin M, Harder W, Schleifer K-H (eds) The prokaryotes, vol 1, 2nd edn. Springer-Verlag, New York, pp 583–624

Widdel F, Pfennig N (1977) A new anaerobic sporing acetate-oxidizing sulfate-reducing bacterium, *Desulfotomaculum acetoxidans* (emend.). Arch Microbiol 112:119–122

Widdel F, Pfennig N (1981a) Sporulation and further nutritional characteristics of *Desulfotomaculum acetoxidans*. Arch Microbiol 129:401–402

Widdel F, Pfennig N (1981b) Studies on dissimilatory sulfate-reducing bacteria that decompose fatty acids. I. Isolation of new sulfate-reducing bacteria enriched with acetate from saline environments. Description of *Desulfobacter postgatei* gen. nov., sp. nov. Arch Microbiol 129: 395–400

Widdel F, Pfennig N (1982) Studies on dissimilatory sulfate-reducing bacteria that decompose fatty acids II: incomplete oxidation of propionate by *Desulfovibrio propionicus* gen. nov., sp. nov. Arch Microbiol 131:360–365

Wierenga EBA, Overmann J, Cypionka H (2000) Detection of abundant sulphate-reducing bacteria in marine oxic sediment layers by a combined cultivation and molecular approach. Environ Microbiol 2:417–427

Wieringa KT (1940) The formation of acetic acid from carbon dioxide and hydrogen by anaerobic spore-forming bacteria. Antoine van Leeuwenhoek 6:251–262

Wolin MJ, Wolin EA, Jacobs NI (1961) Cytochrome-producing anaerobic vibrio *Vibrio succinogenes* sp. n. J Bacteriol 81:911–917

Worm P, Koehorst JJ, Visser M, Sedano-Núñez VT, Schaap PJ, Plugge CM, Sousa DZ, Stams AJM (2014) A genomic view on syntrophic versus non-syntrophic lifestyle in anaerobic fatty acid degrading communities. Biochim Biophys Acta 837:2004–2016

Wu C, Liu X, Dong X (2006) *Syntrophomonas cellicola* sp. nov., a spore-forming syntrophic bacterium isolated from a distilled-spirit-fermenting cellar, and assignment of *Syntrophospora bryantii* to *Syntrophomonas bryantii* comb. nov. Int J Syst Evol Microbiol 56:2331–2335

Yagi T (1958) Enzymatic oxidation of carbon monoxide. Biochim Biophys Acta 30:194–195

Yagi T (1959) Enzymatic oxidation of carbon monoxide II. J Biochem (Tokyo) 46:949–955

Yang G, Guo J, Zhuang L, Yuan Y, Zhou S (2016) *Desulfotomaculum ferrireducens* sp. nov., a moderately thermophilic sulfate-reducing and dissimilatory Fe(III)-reducing bacterium isolated from compost. Int J Syst Evol Microbiol 66:3022–3028

Zaunmüller T, Kelly DJ, Glöckner FO, Unden G (2006) Succinate dehydrogenase functioning by a reverse redox loop mechanism and fumarate reductase in sulphate-reducing bacteria. Microbiol 152:2443–2453

Zelinsky ND (1893) On hydrogen sulfide fermentation in the Black Sea and the Odessa estuaries. Proc Russ Phys Chem Soc 25:298–303

Zellner G, Messner P, Kneifel H, Winter J (1989) *Desulfovibrio simplex* spec. nov., a new sulfate-reducing bacterium from a sour whey digester. Arch Microbiol 152:329–334

Zhilina TN, Zavarzin GA, Rainey FA, Pikuta EN, Osipov GA, Kostrikina NA (1997) Desulfonatronovibrio hydrogenovorans gen. nov., sp. nov., an alkaliphilic, sulfate-reducing bacterium. Int J Syst Bacteriol 47:144–149

Zobell CE, Rittenberg SC (1948) Sulfate-reducing bacteria in marine sediments. J Mar Res 7:602–617

Chapter 2
Characteristics and Taxonomy

2.1 Introduction

While the hallmark characteristic which unifies the physiologic group of microorganisms known as dissimilatory-sulfate reducers is respiratory sulfate reduction, these microorganisms can be distinguished by phenotypic and genotypic characteristics. Listed in Table 2.1 are several SRP with their biochemical characteristics. While microorganisms with dissimilatory sulfate reduction are found in fewer taxonomic groups than microorganisms using nitrate for anaerobic respiration, microorganisms with anaerobic sulfate respiration include a sizeable number of species. SRB display cellular characteristics found in heterotrophic bacteria, and while some spore-forming SRB stain as Gram-negative, thin sections of these bacteria examined via electron microscopy display a cell wall characteristic of Gram-positive types of bacteria. Thus, spore-forming SRB do not constitute a unique physiological group where endospore production is by a Gram-negative staining cell. It appears that there is a significant number of SRP in the environment that are uncultivated and these organisms may have some unique biochemical characteristics. This chapter provides a taxonomic overview of SRP as currently reported.

2.2 Phenotypic Characteristics

2.2.1 Cell Anatomy and Morphology

Considerable metabolic diversity occurs throughout the SRP, and a substantial amount of research has focused on *Desulfovibrio* spp. which were relatively easy to cultivate. The cell of *D. desulfuricans* has a vibrio to sigmoid form which is 3 to 5 μm by 0.5 to 1 μm (Postgate and Campbell 1966). Some strains of SRB have larger

L. L. Barton, G. D. Fauque, *Sulfate-Reducing Bacteria and Archaea*,
https://doi.org/10.1007/978-3-030-96703-1_2

Table 2.1 Selected characteristics of representative sulfate-reducing prokaryotes[a]

Physiological group	Genus	GC content of DNA (%)	Cyto-chromes	Oxidation of organics	Characteristic
Mesophilic bacteria	*Desulfobulbus*	50–60	b,c,c_3	I	Cells are oval or lemon-shaped. Single polar flagella in some species. Anaerobic degradation of propionate.
	Desulfomicrobium	52–67	b,c	I	Motile rod shaped cell.
	Desulfovibrio	46–66	b,c,c_3	I	Curved cell with polar flagellum.
	Desulfobacter	44–46	?	C	Rod shaped. Some species motile by single polar flagella.
	Desulfobacterium	37–46	b,c	C	Rod shaped cells. Some species with gas vacuoles.
	Desulfococcus	46–57	b,c	C	Spherical cells.
	Desulfomonile	49	c_3	C	Rod shape. Reductive dichlorination respiration.
	Desulfonema	35–42	b,c	C	Cells form long filaments. Gliding movement.
	Desulfosarcina	51	b,c	C	Spherical cells in packets
Spore-forming bacteria	*Desulfotomaculum*	37–46	b	I/C	Straight or curved rods. Motility. Flagella are peritrichous or polar.
Thermophilic bacteria	*Thermodesulfobacterium*	30–38	b,c	I	Short rods. Thermophile with growth at 70 °C. cell membranes contain ether-linked lipids.
Thermophilic archaea	*Archaeoglobus*	41–46	b,c	I	Irregular coccus cell is about 0.7 μm in diameter flagella. Cell wall is glycoprotein not peptidoglycan. Hyperthermophilic with growth at 83 °C.

[a]Source is Castro et al. 2000; Stackebrandt et al. 1995; Reed and Hartzell 1999

cells, and these include *D. giganteus* with cells 5.0–10.0 μm long and 1.0 μm wide (Esnault et al. 1988), *D. gigas* DSM 1382 and ATCC 19364 with cells 5 to 6 μm by 1 μm (Le Gall 1963), and *Desulfovibrio longus* with cells 0.4 to 0.5 μm wide and 5 to 10 μm long (Magot et al. 1992). While *D. gigas* DSM 1382 was isolated in France, a second strain of *D. gigas* was isolated in Germany, and it is listed as DSM 496 or ATCC 29494. *D. gigas* DSM 1382 has been used extensively in biochemical studies, and when the literature does not identify the strain of *D. gigas* used, it may be assumed that it is the DSM 1382 strain. Since lipid analysis of these two *D. gigas* strains is markedly distinct, reclassification of these two stains would be appropriate (Stackebrandt et al. 1995). Filamentous *Desulfonema magnum*, *Desulfonema limicola*, and *Desulfonema ishimotonii* (members of the *Desulfobacteraceae* and the *Desulfobacterales*) have cells that are 2.3 to 3 by 2–5 μm but form very long filaments that may exceed 1 mm in length (Widdel et al. 1983; Teske et al. 2009). Cable bacteria of the family *Desulfobulbacceae* produce multicellular filaments that are several hundred mm in length (Müller et al. 2020), and cable bacteria are discussed later in this chapter. The straight or slightly curved cells of the spore-forming *Desulfotomaculum* spp. are 0.3–2.5 μm × 2.5–15 μm (Aüllo et al. 2013).

A few of the sulfate-respiring organisms have unusual cell shapes and cellular characteristics. The *Desulfococcus* includes three characterized species (*biacutus*, *multivorans*, and *oleovorans*), and these non-flagellated Gram-negative SRB are members of the *Desulfosarcina-Desulfococcus* clade within the *Desulfobacteraceae* family (Platen et al. 1990; So and Young 1999; Dörries et al. 2016). Cells of *Desulfococcus* spp. are spherical, nonuniform-shaped cells and are commonly reported as short rods that are about 1.4 × 2.3 μm in size. Individual cells of *Desulfosarcina* spp. are attached together to produce short rods (0.5 × 1–1.5 μm), segmented rods (5–10 μm), or segmented chains (>70 μm) (Orphan et al. 2002; Watanabe et al. 2017). Spherical cells have also been reported for SRP. *Archaeoglobus* (*A.*) *fulgidus* has irregular coccoid to dis-shaped cells that are 0.3 to 1 μm wide, and *A. profundus* has irregular cocci or occasionally triangular cells that are 0.7–1.3 μm × 1.4–1.9 μm (Burggraf et al. 1990; Huber and Stetter 2001). Unusual cellular features are reported for a couple of SRB. On the surface of *Desulforhopalus singaporensis*, there are spinae which are 2.5 μm long (Lie et al. 1999). These protein structures of unknown function have been previously reported for bacteria, and detailed analysis of spinin (the protein subunit) has been performed along with molecular arrangement of these subunits (Easterbrook and Coombs 1976; Easterbrook et al. 1976). Cells of *Desulfomonile tiedjei* form a collar by the invagination of the cell wall, and it appears to function for binary fission (Mohn et al. 1990; DeWeerd et al. 1990). The presence of a collar on cells of SRB has also been reported for *Desulfomonile limimaris* (Sun et al. 2001).

2.2.2 *Cell Architecture*

2.2.2.1 Surfaceome and Outer Membrane Protein Complexes

Proteins in the cell envelope of 5 strains of *Desulfotomaculum* and 12 strains of *Desulfovibrio* were examined to assess their immunological relationship (Norqvist and Roffey 1985). Proteins from *Desulfotomaculum* were characteristic of cells with a Gram-positive-type cell wall and no cross-reactions occurred with immune reactions between species of *Desulfotomaculum* and *Desulfovibrio* that were tested. Antisera-based reactions indicated some similarity in proteins from *Desulfotomaculum nigrificans* and *Dst. ruminis*, and antisera to proteins from *Dst. orientis* did not cross react with *Dst. nigrificans* or *Dst. ruminus*. Antisera produced to proteins from *D. salexigens* and from *D. desulfuricans* were species-specific, while there was similarity between *D. vulgaris*, *D. africanus*, and *D. gigas*. It should be noted that after phylogenic analysis, *Dst. orientis* has been reclassified as *Desulfosporosinus orientis* (Stackebrandt et al. 1997).

Dst. reducens is a Gram-positive bacterium that reduces both soluble Fe(III) citrate and insoluble hydrous ferric oxide with lactate as the electron donor; however, significant growth does not result, and this suggests energy is not conserved in the transfer of electrons from lactate to Fe(III) (Vecchia et al. 2014). The surface-exposed proteins on a cell are referred to as the surfaceome (Fig. 2.1), and in *Dst. reducens*, surface proteins have a series of functions including reduction of Fe(III). Using lysozyme treatment or a process used called "cell shaving" to remove exposed proteins by trypsin, released proteins were analyzed by liquid chromatography-

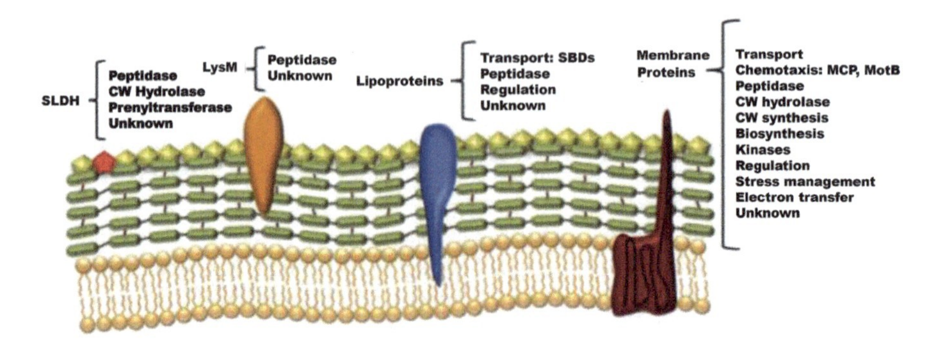

Fig. 2.1 Model of surfaceome of Gram-positive cell wall characteristic of *Desulfotomaculum reducens* (Vecchia et al. 2014). The surfaceome comprises proteins bound to the surface through SLDH (bright red). LysM-like domains (orange) anchor to the membrane lipids (blue) and TMHs (dark red). Different localizations and binding types are associated with different general functions, which are listed in the figure. (*Copyright Permission*: Copyright © 2014 Dalla Vecchia, Shao, Suvorova, Chiappe, Hamelin and Bernier-Latmani. This is an open-access article distributed under the terms of the Creative Commons Attribution License (CC BY). The use, distribution, or reproduction in other forums is permitted, provided the original author(s) or licensor are credited and that the original publication in this journal is cited, in accordance with accepted academic practice)

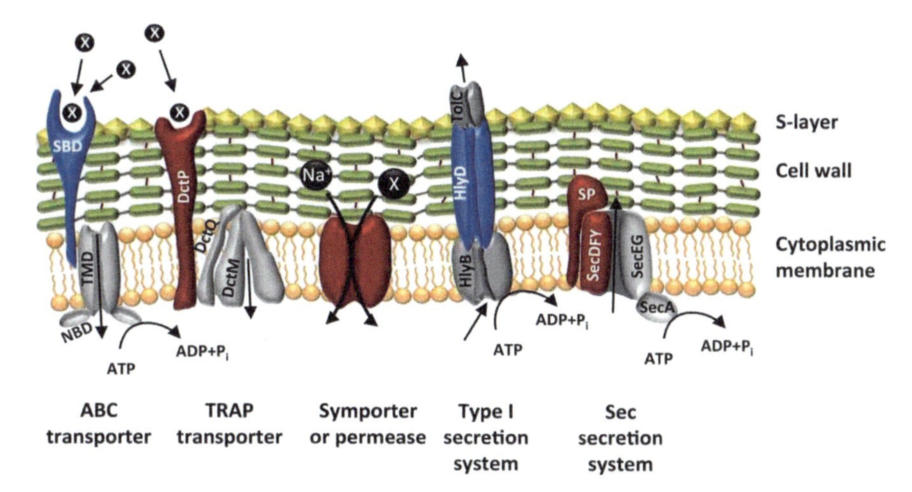

Fig. 2.2 Model of the localization of surface proteins involved in *Desulfotomaculum reducens* (Vecchia et al. 2014). Gray coloring indicates proteins expected to belong to the transport complex, but that were not identified in our extracts, while colored proteins are identified; blue coloring indicates lipoproteins, red indicates membrane-spanning proteins, black circles represent solutes [unspecified, if marked with an ×]. SBP, solute binding protein; NBD, nucleotide binding domain; SP, signal peptidase.

tandem mass spectrometry. A sizeable number of proteins were associated with the cell wall, and about 75% were bound into the plasma membrane (Fig. 2.2). Several proteins associated with the cell surface were identified, and they functioned in peptidoglycan synthesis and modification, flagella structure and related chemotaxis, protein export, and solute transport. The reduction of insoluble Fe(III) (hydrous ferric oxide) required cell contact, and this reduction was attributed to a *c*-type cytochrome bound to the surface of the Gram-positive cell wall of *Thermincola potens* (Carlson et al. 2012). However, genes encoding the two *c*-type cytochromes (NrfA and NrfH) in *Dst. reducens* are not upregulated during Fe(III) reduction (Vecchia et al. 2014), and *c*-type cytochromes were not involved in iron reduction by *Pelobacter carbinolicus* or *Dsm. metallireducens* (Lovley et al. 1995; Finneran et al. 2002). Three surface proteins considered important for Fe(III) reduction by *Dst. reducens* included the following: (i) The gene for protein (Dred_0143) is annotated as a ferredoxin in that it contains a 4Fe-4S cluster and a NADH-binding region and has two heterodisulfide reductase subunit A segments. Potentially, electrons from NADH would be transferred to Fe(III). (ii) The gene for protein (Dred_ 0462) is annotated as a ferredoxin and is a component of a trimeric hydrogenase (Dred_0461–3). Dred_-461 is a putative cytochrome b which may have ten transmembrane helices and would be a cytoplasmic enzyme. (iii) A protein (Dred_1533) annotated as alkyl hydroperoxidase reductase (AhpC) may function as a thio-disulfide oxidoreductase may have a role in Fe(III) reduction as well as for U (VI) reduction. The region of Dred_1527-Dred_1533 contains genes for ferric iron uptake, cadmium resistance, AhpC-like enzyme, and protein associated with production of a *c*-type cytochrome. While there is no mechanism suggested for solid Fe (III) reduction by *Dst. reducens*, it has been proposed that this reduction does not involve nanowires (Vecchia et al. 2014).

Outer membrane preparations from *D. vulgaris* Hildenborough were employed in proteomic studies, and 296 proteins were identified with 70 considered to be associated with the outer membrane (Walian et al. 2012). Since 1–2% of proteins in Gram-negative bacteria is associated with the outer membrane, the predicted number of outer membrane proteins for *D. vulgaris* Hildenborough would be 85% which indicate that about 80% of outer membrane proteins were identified. In comparison, Zhang et al. (2006) identified over 2000 proteins of the *D. vulgaris* proteome, and 68 were predicted to be associated with the outer membrane. Of outer membrane proteins, 35% were lipoproteins, 30% did not have definitive annotations, and 104 of the proteins were involved in heteromeric protein-protein interactions (Walian et al. 2012). The three most abundant outer membrane proteins included a TolC-like protein (DVU1013) and two "conserved hypothetical" proteins (DVU0797 and DVU0799). The TolC-like protein would form a conduit between the inner and outer membrane and would function in the efflux process (Walian et al. 2012). The conserved hypothetical proteins had considerable sequence similarity to bacterial porins with 48% identity to the porin from *D. desulfuricans* (*alaskensis*) G20 (Dde 1011).

2.2.2.2 Nanowires

Extracellular electron transport in anaerobic Gram-negative bacteria involves outer membrane cytochromes and nanostructures referred to as nanowires or nanofilaments extending from the cell. Extracellular uptake of electrons by SRB has been proposed to involve outer membrane heme proteins or nanofilaments (Venzlaff et al. 2013; Enning and Garrelfs 2014). Recently, the presence of electron-conductive nanofilaments on *D. ferrophilus* IS5 has been documented, and this bacterium serves as the model for redox activity of cytochromes in the outer membrane of SRB. Cytochrome content in the outer membrane of *D. ferrophilus* is ten times greater in lactate-starved cultures as compared to lactate-rich media (Deng et al. 2018). Heme proteins in the outer membrane of *D. ferrophilus* were sequenced and determined to be products of genes *DFE_449* which encode a 12-heme cytochrome and *DFE_461* which encode a 14-heme cytochrome. Transcriptome analysis revealed that genes of outer membrane cytochromes of *D. ferrophilus* (*DEF_450* and *DEF_464*) were overexpressed in lactate-limited growth. Lactate-starved cells were shown to obtain electrons from an electrode, and electron microscopy revealed the presence of filaments on the electrode surface (see Fig. 2.3). Nanofilaments were observed on the surface of lactate-starved cells of *D. ferrophilus*, and the nanowires had "bamboo-like" structures with diameters of

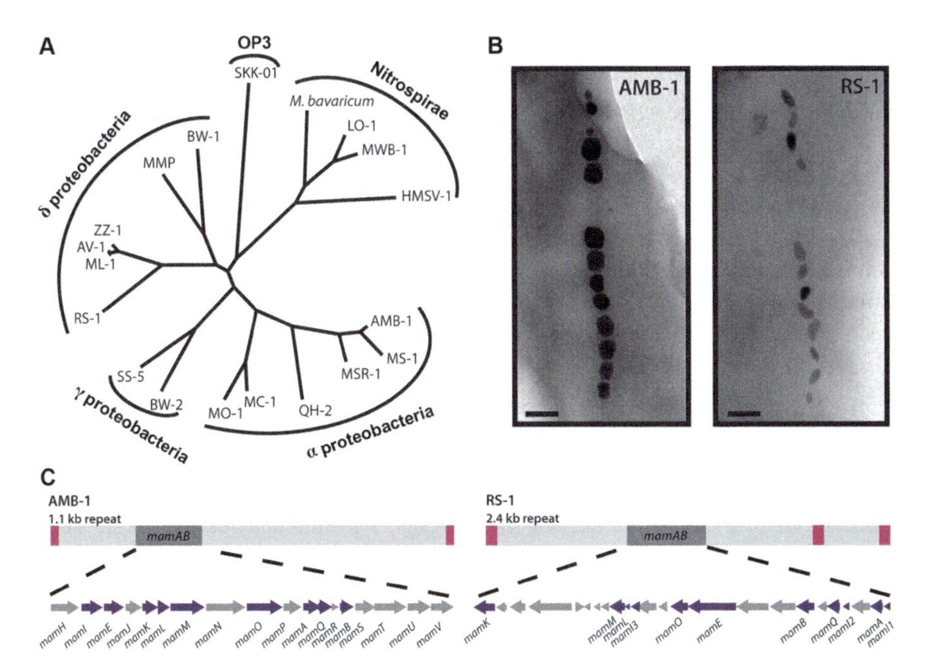

Fig. 2.3 RS-1 is a representative of a group of bacteria that are phylogenetically and phenotypically distinct from the magnetotactic α-proteobacteria (Rahn-Lee et al. 2015). (**a**) 16S phylogenetic tree of magnetotactic bacteria. (**b**) AMB-1 and RS-1 magnetite crystals visualized by TEM. Scale bar 100 nm. (**c**) The *mamAB* gene clusters of AMB-1 and RS-1 shown in the context of the MAI. Pink squares represent the repeats surrounding each island. Purple arrows represent the genes that are conserved between the two. (*Copyright*:)

30 to 50 nm (see Fig. 2.3). Nanowires which appear to be coated with cytochromes are extensions of the cell membrane and are induced in cultures of *D. ferrophilus* when grown in lactate-limited media. These structures appeared to be more similar to those of *Shewanella oneidensis* and not like those of *Geobacter sulfurreducens* (Deng et al. 2018).

Genome analysis of *D. ferrophilus* revealed the presence of 26 genes which encode for multiheme cytochromes and that none of the multiheme cytochromes associated with extracellular electron transfer by *Shewanella oneidensis* and *Geobacter sulfurreducens* were present in *D. ferrophilus* (Deng et al. 2018). However, homologies to genes *DEF_450* and *DFE_450* were found in members of the *Proteobacteria*, *Thermodesulfobacteria*, and *Aquificales* phyla as well as with uncultivable SRB. These nanofilaments and nanowires of SRB may be essential for energy production by sedimentary marine organisms that reduce sulfur compounds and for interspecies electron transfer (syntrophy) in methane oxidation.

2.2.3 Cytoplasmic Structures

2.2.3.1 Polyglucose as a Storage Polymer

Carbohydrate reserves may appear as intracellular granules of polyglucose and polyhydroxybutyrate in SRB cells. In some SRB, accumulation of polyglucose occurs when carbon sources are in excess with favorable growth conditions, and in some SRB, this production of polyglucose is induced when Fe^{2+} and NH_4^+ are limiting in the medium (Stams et al. 1983). The accumulation of polyglucose has been reported in *D. vulgaris* Hildenborough NCIB 8303, *D. gigas* NCIB 9322, *D. desulfuricans* strain HJL, *D. desulfuricans* strain BH, *Desulfovibrio* sp. strain HL21, and *Desulfobulbus propionicus* DSM 2032 (Stams et al. 1983). While these SRB are unable to grow with glucose as the electron donor, polyglucose is used to support the growth of SRB when electron donors are limiting in nutrient media. In the utilization of polyglucose by *D. gigas*, the Embden-Meyerhof pathway and methylglyoxal bypass were demonstrated to operate simultaneously at a rate of 3: 2, while enzymes for the Entner-Doudoroff pathway were not detected (Fareleira et al. 1997). In the presence of sulfate and with limiting electron donor, *Desulfovibrio* strains HL21 and BH couple the conversion of 1 mol glucose to 0.9 mol sulfide with acetate and CO_2 as end products. In the absence of sulfate and electron donor added to the medium, *Desulfovibrio* HL21 ferments glucose from stored polyglucose to acetate and H_2 as major end products with trace levels of ethanol and succinate. *D. propionicus* ferments polyglucose to acetate, propionate, H_2, and CO_2.

Initial studies indicated that endogenous polyglucose in *D. gigas* was fermented using two different metabolic pathways to produce acetate, glycerol, and ethanol as end products (Santos et al. 1993). Based on an analysis of enzymes present in *D. gigas*, fermentation of glucose was not by the Entner-Doudoroff pathway but by the Embden-Meyerhof pathway and the methylglyoxal bypass which contributed 60% and 40%, respectively. In the presence of sulfate, acetate and CO_2 were produced from polyglucose, and NADH generated from glucose catabolism participated in sulfate reduction. This metabolism of glucose from endogenous polyglucose by *D. gigas* and *Desulfobulbus propionicus* is in contrast to the fact that these SRB are unable to grow with sugars as an energy source. With the exception of *D. fructosovorans* (Ollivier et al. 1988) and *D. simplex* (Zellner et al. 1989), *Desulfovibrio* spp. lacks the enzymology to use sugars as electron donors.

2.2.3.2 Polyhydroxyalkanoate

Intracellular accumulation of polyhydroxyalkanoate (PHA) have been reported for several SRB with cellular PHA consisting primarily of poly-3-hydroxybutyric acid (3HB) and 3-hydroxyvaleric acid with minor quantities of 3-hydroxyhexanoic acid, 3-hydroxyoctanoic acid, and 3-hydroxytetradecanoic acid (Hai et al. 2004). The amount of PHA as 3HB purified from *Desulfococcus multivorans* and *Desulfonema magnum* when grown on benzoate was 8% and 88% cell dry wt., while for *Desulfobotulus sapovorans* (formerly *D. sapovorans*), 13.5% (wt/wt) was purified from cells with caproate as the carbon source. With *Desulfobacterium autotrophicum* grown on valerate, the quantity of PHA isolated was 5.4%, and of this, 48% was 3HB, and 52% was 3-hydroxyvaleric acid (Hai et al. 2004). In related studies, an estimation of PHA in *Desulfosarcina variabilis* when grown on benzoate was 27% and 22% of cell dry wt., respectively, while *Desulfobotulus sapovorans* was estimated to produce 43% PHA with caproate as the carbon source (Hai et al. 2004). The quantity of PHA estimated to be produced by *Desulfobacterium autotrophicum* grown on valerate or caproate was 8% and 11%, respectively, while no PHA was produced when grown on lactate (Hai et al. 2004). With taurine and malate as the carbon source, *Desulforhopalus singaporensis* was reported to produce 25% and 7% of cell mass, respectively, as PHA (Lie et al. 1999).

The synthesis of poly-PHA involves three steps, and the best characterized pathway for polymer synthesis is for production of poly-PH,B and this pathway is considered to function in SRB. Initially, two acetyl-CoA molecules are condensed to produce acetoacetyl-CoA by acetyl-CoA acetyltransferase (β-ketothiolase, PhaA). Acetoacetyl-CoA is converted to (R)-3-HB-CoA by acetoacetyl-CoA reductase (PhaB), and (R)-3-HB-CoA is polymerized into the PHB chain by PHA synthase (PhaC) (Sagong et al. 2018). The gene (*phaC*) for a subunit of PHA synthesis was obtained from *Desulfococcus multivorans*, and predicted amino acids synthesized by this DNA compared favorably to PhaC subunits of class III PHA synthesis found in other bacteria. The PhaF subunit for PHA synthase from *Desulfococcus multivorans* contains the amino acid motif characteristic of class IIIPHA synthesis (Hai et al. 2004). The detection of genes with similarity to *phaC* and *phaF* was identified in the genome of *Desulfobacterium autotrophicum* but was absent in *D. vulgaris* Hildenborough (Hai et al. 2004).

2.2.3.3 Iron Inclusions and Magnetosomes

Magnetotactic sulfate-reducing bacteria are found in two orders (*Desulfovibrionales* and *Desulfobacterales*) of the *Deltaproteobacteria* class. Bacteria identified as strain RS-1 (Sakaguchi et al. 2002), strain FH-1 (Lefèvre et al. 2013), and strains ML-1, AV-1, and ZZ-1 (Lefèvre et al. 2011a) are associated with *Desulfovibrionales*, while *Candidatus* Magnetoglobus multicellularis (Abreu et al. 2014), *Candidatus* Desulfamplus magnetomortis strain BW-1 (Lefèvre et al. 2011a), *Candidatus* Magnetananas tsingtaoensis and *Candidatus* Magnetananas drummondensis (Zhou et al. 2012; Chen et al. 2016), *Candidatus* Magnetomorum litorale, *Candidatus* Magnetoglobus multicellularis, *Candidatus* Magnetananas rongchenensis, and *Desulfamplus magnetovallimortis* strain BW-1 (Abreu et al. 2011; Chen et al. 2015; Descamps et al. 2017) are with *Desulfobacterales*. In a recent survey of 168 MTB draft genomes including 26 genomes from the *Desulfobacterales*, evidence for magnetosomes has been ascribed to numerous bacteria with putative *man* genes (Lin et al. 2020).

D. magneticus RS-1 is a freshwater SRB isolate that produces intracellular magnetite (Fe_3O_4) particles (Sakaguchi et al. 1993: Sakaguchi et al. 2002). While cells of *D. magneticus* RS-1 usually have cytoplasmic magnetosomes (~ 40 nm) of magnetite, some cells contain hematite (Fe_2O_3) on the cell surface (Pósfai et al. 2006). *D. magneticus* RS-1 cells contain on average six irregular bullet-shaped magnetosomes and are weakly magnetotactic (Sakaguchi et al. 1993; Sakaguchi et al. 2002; Pósfai et al. 2006). In comparison, *Magnetospirillum* strains have intracellular compartments (magnetosomes) of ferrimagnetic iron oxide (magnetite) enclosed in an organic membrane and are strongly magnetotactic (Gorby et al. 1988).

Based on the reports in the literature, *D. magneticus* RS-1 is the model organism for magnetotactic activity in SRB. Genome analysis of *D. magneticus* RS-1 indicated the conservation of genes associated with magnetotactic bacteria, and these include separate regions on the chromosome for *nuo* and *mamAB*-like gene clusters with another gene region on a cryptic plasmid (pDMC1) (Nakazawa et al. 2009). While *D. magneticus* RS-1 genome contains the core genes for synthesis of magnetosomes, multiple genome transfer events would have provided this bacterium with the magnetotactic character. The shape of the "tooth-shaped" magnetite crystals found in *D. magneticus* RS-1 is under genetic control (Fig. 2.3), and these magnetic crystals may be enclosed in a membrane. Rahn-Lee et al. (2015) suggests that there is a single site on the inner side of the plasma membrane for the production of these magnetic crystals and these crystals are self-arranged due to the magnetic field of the crystals (see Sect. 2.5.2.2 for more genomic information).

A study of the formation of magnetite in *D. magneticus* RS-1 revealed that a solid ferrous iron phase is a precursor in the mineralization of magnetite. When ferrous iron was provided to iron-starved cells, small amorphous Fe-P granules were apparent in the cytoplasm as was described earlier by Byrne et al. (2010), and spectral near-edge structure (XANES) analysis indicated the presence of Fe (II) triphosphate (68%), Fe(III) triphosphate (21%), and Fe-S (11%) (Baumgartner et al. 2016). XANES analysis suggested the presence of green rust which is a mixture of ferrous and ferric hydroxides which reflects the unique Fe(II)/Fe(III) redox chemistry of this bacterium, and the bullet-shaped magnetite granule results from solid-state growth of the magnetite crystal.

D. magneticus sp. RS1 contains two intracellular granules which are either magnetite or polyphosphate (Byrne et al. 2010). These two granules are considered to develop independently and have distinct functions. The Fe-P organelle is amorphous and unlike the magnetite granule is surrounded by a membrane. The Fe-P granule is rapidly formed in cells following the addition of Fe(II) to iron-limited cells and is proposed to function to detoxify the cell from iron overload. While there are four distinct pools of iron (i.e., Fe-S granules, magnetite, ferritin, and FeS) in *D. magneticus* RS-1, future studies are needed to clarify the mineralization of magnetosomes and other iron pools.

Recently, there have been reports of magnetotactic bacteria in several anaerobic environments. The SRB *Desulfamplus magnetovallimortis* strain BW-1 was isolated from a brackish spring, and it is unique in that it produces a chain of greigite (Fe_3S_4) and/or magnetite nanocrystals in the cytoplasm (Lefèvre et al. 2011b; Descamps et al. 2017). An uncultured magnetosome-containing bacterium, *Candidatus* Magnetoglobus multicellularis, was reported to contain several genes characteristic of sulfate-reducing bacteria, and these genes encode ATP-sulfurylase, sulfate permease, adenylyl-sulfate reductase (*apr*, α and β subunits), and sulfite reductase (*dsr*, α and β subunits) (Abreu et al. 2014). A magnetotactic *Deltaproteobacteria* strain WYHR-1 was isolated from Weiyang Lake in China, and the bean-shaped cells have an average length of 4.3 μm and a width of 1.4 μm (Li et al. 2019). A single cell contains an average of 62 bullet-shaped magnetosomes containing magnetite which range in length from a few nanometers to ~180 nm which are arranged along the long axis of the cell.

In addition to *D. magneticus* strain RS-1 T and *Desulfamplus magnetovallimortis* strain BW-1 T, *Desulfovibrio* sp. strain FSS-1 becomes the third member of *Deltaproteobacteria* which has been cultivated (Shimoshige et al. 2021). Figure 2.4 provides images of the tooth-shaped magnetic crystals. With the use of EDS spectra and STEM-EDS elemental maps, Fe and O were determined to be present in the nanoparticles, and the particles were proposed to be either magnetite or maghemite. Although the genome of *Desulfovibrio* sp. strain FSS-1 is not completely finished, this magnetotactic SRB has a magnetosome gene cluster of 13 *mam* genes and 16 *mad* genes which differ from the gene cluster in *D. magneticus* strain RS-1^T. The number of magnetotactic granules inside the FSS-1 cells was 9.4, and the length of the bullet-shaped nanoparticles was 53.9 nm. It is unknown how strain FSS-1 regulates the production of hydrogen sulfide from sulfate reduction because sulfide would interfere with iron metabolism in the formation of the magnetite nanoparticles.

2.2.3.4 Phosphorus Inclusions

Phosphorus inclusion granules have been identified in a few SRP. Spherical granules 0.2 to 0.8 μm in diameter were observed in cells of *D. gigas* by light microscopy with basic dyes. Using transmission electron microscopy, one to three granules per cell were identified, and the structure was close to amorphous with a fine-grain size (Jones and Chambers 1975). The phosphorus inclusion granules were purified from disrupted cells of *D. gigas*, and emission spectroscopy indicated the presence of phosphorus and magnesium as the major elements with trace levels of Na, Ni, Ca, Co, Si, Fe, and Cu. When isolated, these granules had a white color and contained 6.3% (wt/wt) organic carbon. The publication indicated the presence of short chains of magnesium polyphosphate with a molar ratio of P to Mg of 1.17 (Jones and Chambers 1975). Using NMR spectroscopy, a subsequent report of electron-dense bodies in *D. gigas* was not polyphosphate but α-glucose 1,2,3,4,6-pentakis (diphosphate) which has five discrete diphosphate groups attached to glucose (Hensgens

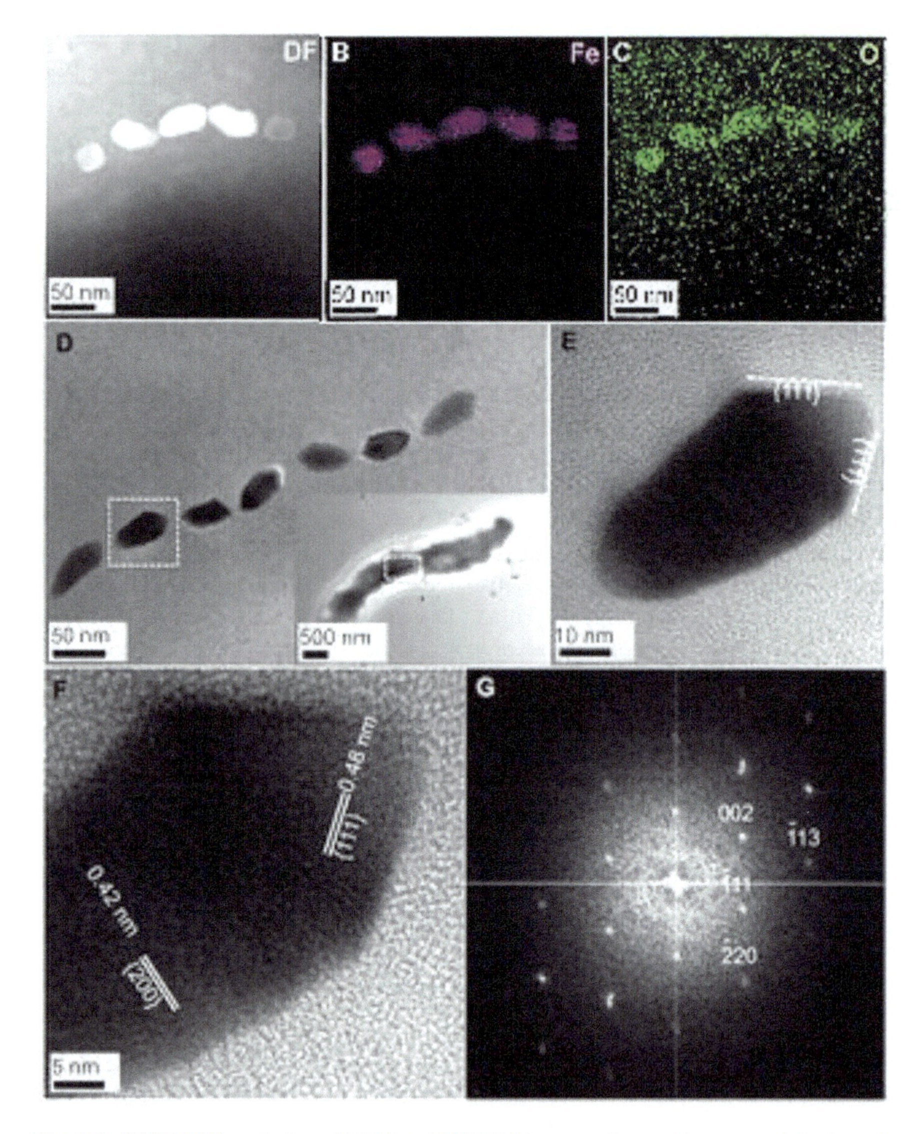

Fig. 2.4 STEM-EDS analysis and TEM and HRTEM images of magnetic nanoparticles in each FSS-1 cell (Shimoshige et al. 2021). (**a**) STEM image of a magnetosome. (**b**, **c**) STEM-EDS elemental maps corresponding to iron (Fe) and oxygen (O). (**d**) TEM image of a magnetosome in a cell corresponding to the dashed line box indicated in the inset. (**e**) HRTEM image of a magnetic nanoparticle indicated by the dashed line box in panel (**d**). (**f**) HRTEM image of the same nanoparticle as shown in panel (**e**). (**g**) Fast-Fourier transform (FFT) pattern of the same nanoparticle as shown in the panel (**f**). (*Copyright*: © 2021 Shimoshige et al. This is an open-access article distributed under the terms of the Creative Commons Attribution License, which permits unrestricted use, distribution, and reproduction in any medium, provided the original author and source are credited)

et al. 1996). The function of this novel phosphate compound and electron-dense bodies is unresolved but perhaps is a response to nutrient stress (Jones and Chambers 1975; Hensgens et al. 1996). *D. desulfuricans* ATCC 27774 produces a phosphorous-containing metabolite identified as methyl-1,2,3,4-tetrahydroxybutane-1,3-cyclic bisphosphate, but this compound is not associated with the formation of electron-dense granules (Santos et al. 1991).

Polyphosphate granules have been reported in the cells of *A. fulgidus* VC16 where they coexist with iron sulfide granules (Toso et al. 2016). The polyphosphate granules are about 220–240 nm in diameter, occupy about 1.4% of the cytoplasm, and consist of P and O at a ratio of 1:2 with minor quantities of Mg, Ca, Cu, and Al. These granules are suggested to be used for energy storage and may function in sequestering cationic metals including those that are toxic to cells.

2.2.3.5 Gas Vacuoles

Gas vacuoles have been demonstrated in *Desulforhopalus vacuolatus* (Isaksen and Teske 1996; Isaksen and Jørgensen 1996), *Dst. geothermicum* (Dauma et al. 1988), *and Dst. sapomandens* (Cord-Ruwisch and Garcia 1985). In addition to these anaerobic SRB, gas vacuoles have also been reported in *Clostridium loretii* (Oren 1983), in *Methanosarcina vacuolata* (Zhilina and Zavarzin 1987), in bacteria found in sea ice and waters of the Antarctic (Staley et al. 1989), and in the SRB in the surface layer of the Kattegat sediment (Jørgensen and Bak 1991). It has been suggested that a function of gas vacuoles in anaerobic bacteria, including the SRB, growing in cold environments may be to increase the surface area of vegetative cells without increasing the volume of the cytoplasm (Isaksen and Jørgensen 1996). In spore-forming bacteria, gas vacuoles attached to endospores would contribute to the dispersal of these endospores.

2.3 Endospores

Endospores produced by sulfate-reducing genera are either round or oblong, located central to terminal, and in some species the spore expands the cell (Table 2.2). While the presence of bacteria endospores in amber (Cano and Borucki 1995), salt crystals (Vreeland et al. 2000), manganese nodules (Nealson and Ford 1980), and deep subseafloor sediments (Lomstein et al. 2012) suggests the viability of bacterial endospores for thousands to millions of years, a concern has been raised that activation and germination of spores for extended periods of time may require DNA maintenance (Lindahl 1993; Johnson et al. 2007). Based on a study of the Baltic Sea sediments, the half-life of spores from *Desulfotomaculum* spp. is about 400 years (De Rezende et al. 2013).

An interesting feature of sporulating SRB is that many species give a Gram-negative stain (Table 2.2), but thin sections of these cells indicate a typical Gram-

Table 2.2 Characteristics of spore-forming bacteria that use sulfate as electron acceptor

Family Genus species	Cell motility	Gram stain	Spore & location	Reference
Thermoanaerobacteraceae				
Ammonifex thiophilus	Peritrichous	Positive	R,T	Miroshnichenko et al. (2008)
Peptococcaceae				
Desulfotomaculum nigrificans	Peritrichous	Negative	O,T,St	Campbell and Postgate (1965)
Desulfosporsinus orientis	Peritrichous	Negative	O,C,pa	Campbell and Postgate (1965)
Desulfosporosinus meridiei	Lateral flagella	Negative	E,St	Robertson et al. (2001)
Desulfurispora thermophila	Motile	Positive	O,C,St	Kaksonen et al. (2007)
Veillonellaceae				
Desulfosporomusa polytropa	Lateral flagella	Negative	O,T,St	Sass et al. (2004)

R round, *O* oval, *T* terminal, *E* elliptical, *C* central, *St* subterminal, *T* terminal, *Pa* paracentral

positive-type cell wall (Haouari et al. 2008; Spring et al. 2009). When examining cells of *Desulfosporosinus orientis*, it was established that the cell wall appeared typical for Gram-positive bacteria in that it was about 15 nm thick and structurally unlike the outer membrane of *D. desulfuricans* (Stanley and Southam 2018). The cell wall of *Desulfosporosinus orientis* serves as nucleation site for the deposition of iron sulfide (mackinawite) as subhedral plate-like structures that are about 300 nm in size which totally enclose the cell, but the cell avoids fossilization by lysis associated with release of the endospore. In contrast, Gram-negative SRB such as *D. desulfuricans* promotes the formation of iron sulfide precipitate in the medium and not associated with the cell periphery (Stanley and Southam 2018).

2.4 Chemotaxonomy

2.4.1 Biomarkers

Taxonomic markers for SRB are used to assist in the identification of cultures isolated from various environments. Species-specific and genus-specific markers used in the study of SRB have been reviewed by Odom (1993). The "desulfoviridin test" was developed by Postgate (1959) to identify bacteria that produce bisulfite reductase, the key enzyme in dissimilatory sulfate reductase. Fluorescence is attributed to the alkaline treatment of the prosthetic group of bisulfite reductase, and with careful standards employed, this test can be used to quantitate SRB that produce desulfoviridin (Barton and Carpenter 2013). Due to the abundance of dissimilatory bisulfite reductase in sulfate-reducing microorganisms, cell extracts can be surveyed to determine the enzyme type with desulfoviridin, P-582, desulfofuscidin, desulforubidin, and archaeal having maximum absorption at 630 nm (Oliveira et al. 2008), 582 nm (Akagi and Adams 1973), 578 nm (Fauque et al. 1990), 545 nm (Fauque et al. 1991), and 591 nm (Schiffer et al. 2008), respectively. *Desulfovibrio* spp. have highly active hydrogenases, and a species-specific probe was developed to identify *Desulfovibrio* subgroups (Voordouw et al. 1990a), but the application of this test to environmental samples was unable to reliably detect SRB present. The use of antibodies to cytochrome c_3 and to the surface of SRB cells has been pursued (Singleton Jr. et al. 1984, 1985), but future studies are needed to evaluate the application of these immune probes to detect SRB in the environment. Studies to evaluate antibodies to adenosine 5′-phosphate (APS) reductase in species of *Desulfovibrio*, *Desulfotomaculum*, *Desulfobulbus*, *Desulfosarcina*, and several sulfide oxidizers indicated a potential for detection of SRB in field environments. Using immunological studies of APS reductase, SRB could be distinguished from anaerobic sulfide-oxidizing phototrophic bacteria (Odom et al. 1991).

2.4.2 FISH Technologies, PhyloChip, and GeoChip

DNA probes are of great specificity for SRB species, and a detailed listing of 16S rRNA gene-targeted primers for different taxa of SRB has been compiled (Castro et al. 2000; Stahl et al. 2007; Lücker et al. 2007). Fluorescence in situ hybridization (FISH) provides for identification of SRB species in environmental settings, and several FISH-related techniques for this detection are reviewed by Stahl et al. (2007). Included in the publications illustrating the application of FISH technology to SRB are several examples of its use in the location of SRB in biofilms and mats (Ito et al. 2002; Labrenz and Banfield 2004: Baumgartner et al. 2006; Lücker et al. 2007; Probst et al. 2013).

To examine the presence of SRP in an environment, the 16S RNA gene-targeted oligonucleotide microarray called the SRP-PhyloChip was developed (Loy et al. 2002). This chip carries over 200,118-mer probes which include SRB from a broad range of environmental and clinical settings. The sensitivity and accuracy in the detection of SRB are enhanced when the SRP-PhyloChip is combined with other tests such as denaturing gradient gel electrophoresis and polar lipid-derived fatty acid analysis (Miletto et al. 2008). A GeoChip was developed that contained 4243 oligonucleotide probes which included 410,000 genes to detect groups of bacteria, including SRB, associated with cycling of nutrients, metal resistance, and metal reduction (He et al. 2007). The microbial community including SRB present on the inner and outer side of a chimney at the Juan de Fuca Ridge was evaluated using the GeoChip (Wang et al. 2009). Additionally, the microbial composition in water (Van Nostrand et al. 2009) and in marine sediment (Wu et al. 2008) was evaluated using GeoChip technology. A review critiquing the advantages of using GeoChip to evaluate bacterial community structure has been provided by Van Nostrand et al. (2012).

2.4.3 Cytochromes

The presence of cytochrome in *Desulfovibrio* was discovered by independent studies of Ishimoto et al. (1954) and Postgate (1954). This hallmark discovery became an important phenotypic characteristic for identification of Gram-negative SRB. Due to the quantity of cytochrome present in SRB cells, bacteria in the *Deltaproteobacteria* and *Nitrospira* groups are referred to as being cytochrome-rich, while SRB in the taxonomic groups of *Clostridia* and *Archaea* are cytochrome-poor (Pereira et al. 2011). For example, *D. gigas* has 1.2 nmole cytochrome *c*/mg cell protein with 36% of the total cytochrome *c* associated with the membrane (Odom et al. 1981). Cytochrome-rich SRP use periplasmic cytochrome c_3 (Tplc_3) to shuttle electrons from periplasmic hydrogenases and formate dehydrogenase to membrane complexes containing multiheme cytochrome *c*. While cytochrome-poor SRP lack a periplasm, hydrogenases and formate dehydrogenases are localized in the plasma membrane (Pereira et al. 1998; Matias et al. 2005; Louro 2007; Pereira et al. 2008; Pereira et al. 2011; da Silva et al. 2012; Romão et al. 2012;). Additionally, membrane-bound cytochrome of *b*-type is found in several SRP, but its concentration is markedly less than the concentration of *c*-type cytochromes. As found in *D. gigas*, cytochrome *b* is 0.1–0.15 nmol/mg cell protein which is about one-tenth the quantity of cytochrome *c*. Cytochrome presence and type have been used to differentiate SRB (*see* Table 2.1).

2.4.4 Quinones and Lipids

Menaquinones (MK) with a narrow variation of isoprenoid side chains are present in the membranes of SRP (Table 2.3), and ubiquinones were not detected in any of these sulfate-reducing microorganisms. MK-7 is the dominant quinone found in species of *Desulfotomaculum*, *Desulfococcus*, *Desulfobacter*, and *Thermodesulfobacterium* (Collins and Widdel 1986). The major quinone is *Desulfobulbus propionicus* has MK-5, *Desulfotalea psychrophila* and *Desulfovibrio* species which oxidize lactate and H_2 commonly have MK-6 as the most abundant quinone, while *D. thermophilus*, *Desulfonema limicola*, *Desulfobacterium autotrophicum*, and *A. fulgidus* have MK-7 (Hemmi et al. 2005; Kuever et al. 2005a, b, c, d). The distribution of these menaquinones in SRP is not sufficiently specific for species identification without additional character assessments.

From studies of fatty acids present in SRP, it appears that the type of fatty acids present is of taxonomic value. Isomonoenoic and anteisomonoenoic fatty acids and branched β-hydroxy acids are present in *D. desulfuricans* (Boon et al. 1977). Ueki

Table 2.3 Distribution of quinones in sulfate-reducing bacteria[a]

Menaquinone	Genus	Species
MK-5 (H_2)	*Desulfobulbus*	*propionicus, marinus, elongatus*
MK-6	*Desulfovibrio*	*desulfuricans, vulgaris, gigas, piger*
MK-6 (H_2)	*Desulfovibrio*	*africanus, salexigens*
MK-7	*Desulfobotulus*	*sapovorans*
	Desulfobacter	*postgatei, hydrogenophilus, curvatus*
	Desulfococcus	*multivorans*
	Desulfosarcina	*variabilis*
	Desulfobacterium	*autotrophicum, niacini*
	Desulfonema	*limicola*
	Desulfotomaculum	*nigrificans, orientis, ruminis, acetoxidans*
	Thermodesulfobacterium	*commune, mobile*
MK-7 (H_2)	*Desulfobacterium*	*vacuolatum, phenolicum, indolicum*
	Desulfoarculus	*baarsii*
MK-9	*Desulfonema*	*magnum*

[a]Source is Devereux and Stahl 1993

and Suto (1979) found that the major fatty acid in *Dst. nigrificans* was saturated branched-chain fatty acids (*iso*-$C_{15:0}$ and *anteiso*-$C_{15:0}$), while *Dst. orientis* and *Dst. ruminis* contained straight-chain fatty acids ($C_{16:0}$ and $C_{16:1}$). Using cellular fatty acids in SRB as biomarkers, the following observations were reported (Taylor and Parkes 1983). When grown on acetate, 79% of fatty acids of *Desulfobacter* sp. were even numbered straight chain (C_{14}–C_{16}), while propionate growth of *Desulfobulbus* sp. produced 83% of fatty acids as odd number carbon atoms (C_{15}–C_{17}). With H_2/CO_2 or lactate, branched iso- and anteiso-fatty acids of C_{15} and C_{17} were dominant in *D. desulfuricans*. These fatty acid biomarkers for SRB were effective in evaluating the changing populations of SRB in marine sediments (Taylor and Parkes 1983, 1985). A review of 40 strains of SRB indicated that iso-branched 17:1 FA was a biomarker for most species of *Desulfovibrio*, while *Desulfobacterium autotrophicum* contained 10-methyl 16:0 FA, but no iso- or anteiso-FA (Vainshtein et al. 1992). Variability between species of *D. desulfuricans* was reported with intestinal isolates having a higher ratio of saturated-to-unsaturated fatty acids than isolates from soil (Dzierżewicz et al. 1996). The location of uncultured *Desulfosarcina/Desulfococcus* species in a *Beggiatoa* mat near an active methane vent at the crest of the Hydrate Ridge was established using a genus-specific fatty acid probe (Elvert et al. 2003).

Polar lipid-derived fatty acids (PLFA) of SRB have been examined, and specific PLFA have been associated with specific strains of SRB (Dowling et al. 1986). PLFA biomarkers for SRB are as follows: *Desulfobacter* and *Desulfobacula* cy17:0 and 10Me16:0, *Desulfobulbus* and *Desulforhabdus* 15:1g6c and 17:1g6c, and *Desulfovibrio* and *Syntrophobacter* i17:1g7c (Pelz et al. 2001). Studies using PLFA have provided information on the SRB community in oil reservoirs (Fan et al. 2017) and uranium deposits (Jiang et al. 2012).

Membrane lipids in SRB include ester-linked fatty acids, and in members of the archaeon *A. fulgidus*, isoprenoid lipids ($D_{40}H_{72-80}$) are linked by ether bonds. Lipids in *A. fulgidus* have sn-2,3-diphytanylglycerol diether or sn-2,3-dibiphytanyldiglycerol tetraether, and the ratio of tetraether to diether lipid differs with growth conditions and with isolates (Lai et al. 2008). *Thermodesulfobacterium commune* is a unique bacterium in that it contains phospholipids that are ether-linked in the membrane (Langworth et al. 1983).

2.4.5 *DNA G + C Content*

A property of taxonomic use is the calculation of guanine plus cytosine (G + C) in DNA calculated as guanine + cytosine/adenine + thiamine + guanine + cytosine and expressed as a percent. The G + C content in DNA from SRP ranges from ~30% to ~65% (see Table 2.1). Members of the *Clostridia* group characteristically have low G + C% values and are in contrast to *Deltaproteobacteria* members that are commonly of high G + C% content. Organisms with G + C ratios that differ by more than 5% are considered to be unrelated. This is in comparison with the 95% species-delimiting value for nucleotide identity (Goris et al. 2007; Richter and Rosselló-Móra 2009).

2.5 Taxonomic Placement

Several different tests and measurements have been employed to assess the characteristics of microorganisms that are used for taxonomy. Background on these methods used for taxonomic purposes have been reviewed by Busse et al. (1996) and Tindall et al. (2010). Classifications based on these and related tests have resulted in classification of SRP. It should be noted that classification and taxonomy of SRP are subject to adjustments as new information on recent isolates is produced.

2.5.1 *Classification*

The initial approaches to classify SRB were the publications addressing *Desulfotomaculum* (Campbell and Postgate 1965) and *Desulfovibrio* (Postgate and Campbell 1966). With an increasing number of isolates of SRP, classifications of SRB (Stackebrandt et al. 1995) and Archaea (Thauer and Kunow 1995) were discussed with respect to relationships of SRP. Additional reviews of SRP classification were published by Castro et al. (2000), Huber and Stetter (2001), Kuever et al. (2005a, b, c, d), Rabus et al. (2007), Thauer et al. (2007), Rabus and Strittmatter (2007), and Rabus et al. (2013). The most recent classification was by Rabus et al. (2015), and it along with the publication of Castro et al. (2000) serves as the basis for the arrangements in Table 2.4. As of 2018, there are over 420 species and over

Table 2.4 Taxonomy of sulfate-reducing prokaryotes and relevant characteristics

Designations	mol% G + C of DNA	Shape and form	Oxidation of e^- donor	Optimum growth		Reference
				pH	°C	
Domain: Archaea **Phylum: Euryarchaeota** **Class: *Archaeoglobi*** **Order: *Archaeoglobales*** **Family: Archaeoglobaceae**						
Archaeoglobus fulgidus	48	Irregular coccus	C	6.5	83	1
Phylum: *Crenarchaeota* **Class: *Thermoprotei*** **Order: *Thermoproteales*** **Family: *Thermoproteaceae***						
Caldivirga maquilingensis	43	Straight, curved, branched rods	I	4.0	85	2
Thermocladium tenax	55	Straight, curved cell	C	4.0	75	3
Domain: Bacteria **Phylum Proteobacteria** **Class: Deltaproteobacteria** **Order: Desulfovibrionales** **Family: Desulfovibrionaceae**						
Desulfovibrio vulgaris	61.3	Vibrio	I	7.5	35	4
Desulfovibrio gigas	60.2	Spirilloid	I	7.5	35	5
Desulfovibrio desulfuricans	55.1	Vibrio	I	7.5	35	6
Family: *Desulfomicrobiaceae*						
Desulfomicrobium baculatum	56.8	Short rod	I	7.2	33	7
Family: *Desulfohalobiacea*						
Desulfohalobium retbaense	57.1	Rod	I	6.7	38	8
Desulfonatronovibrio hydrogenovorans	48.6	Vibrio	I	9.6	37	9

(continued)

Table 2.4 (continued)

Designations	mol% G + C of DNA	Shape and form	Oxidation of e$^-$ donor	Optimum growth		Reference
				pH	°C	
Desulfonauticus submarinus	34.3	Curved rod	Nr	7.0	45	10
Desulfothermus okinawensis	34.9	Rods	Nr	6.2	50	11
Family: Desulfonatronumaceae						
Desulfonatronum thiodismutans	63.0	Vibrio	Nr	9.5	37	12
Order: *Desulfobacterales* **Family: Desulfobacteraceae**						
Desulfobacterium autotrophicum	48.9	Rod	I	6.7	26	13
Desulfobotulus sapovorans	53	Vibrio	I	7.7	34	13
Desulfococcus multivorans	57	Spherical	C	7.3	35	14
Desulfonema limicola	35	Multicellular filaments	C	7.6	30	15
Desulfosarcina variabilis	52.1	Irregular cells in packets	C	7.3	30	16
Family: Desulfobulbaceae						
Desulfobulbus propionicus	60	Lemon-shaped cell	I	7.2	39	17
Desulfofustis glycolicus	56	Straight to curved rods	I	7.3	28	18
Desulforhopalus vacuolatus	48	Oval	I	7.2	18	19
Desulfotalea psychrophila	46.8	Rod	I	7.2	10	20
Order: Desulfoarcales **Family: Desulfoarculaceae**						
Desulfarculus baarsii	65.7	Vibrio	C	7.3	37	21
Order: Syntrophobacteriales **Family: *Syntrophobacteraceae***						

Thermodesulforhabdus norvegica	51	Rod	C	6.9	60	22
Desulforhabdus amnigenus	52.5	Rod	C	7.4	37	**23**
Syntrophobacter fumaroxidans	60.6	Rod, lemon shape	I	7.3	37	**24**
Family: Syntrophaceae						
Desulfobacca acetoxidans	51.1	Oval	C	7.6	37	**25**
Desulfomonile limimaris	49.0	Rod	I	7.5	37	**26**
Phylum: Nitrospirae **Class: Nitrospira** **Order: Nitrospirales** **Family*: Nitrospiraceae***						
Thermodesulfovibrio yellowstonii	29.5	Curved rod	I	6.9	65	**27**
Phylum: Firmicutes **Class: Clostridia** **Order: Thermoanaerobacterales** **Family: Thermoanaerobiaceae**						
Ammonifex thiophilus	56	Rod		6.8	75	**28**
Thermacetogenium phaeum	53.5	Rod		6.8	58	**29**
Family: *Thermodesulfobacteriaceae*						
Thermodesulfobacterium commune	34.4	Rod	I	7.0	70	**30**
Thermodesulfatator autotrophicus	43.1	Rod		6.8	68	**31**
Family: Thermodesulfobiaceae						
Thermodesulfobium narugense	33.7	Rod		4.9	55	**32**
Thermodesulfobium acidiphilum						**33**

(continued)

Table 2.4 (continued)

Designations	mol% G + C of DNA	Shape and form	Oxidation of e^- donor	Optimum growth		Reference
				pH	°C	
Order: Clostridiales **Family: Peptococcaceae**						
Desulfotomaculum nigrificans	44.7	Rod	I	7.0	55	**34**
Desulfosporsinus orientis	41.7	Curved rod	I	6.6	35	**35**
Desulfurispora thermophila	53.5	Rod	I	7.2	60	**36**
Candidatus Desulforudis audaxviator	60.9	Rod		9.3	60	**37**
Class: Negativicutes **Order: *Selenomonadeles*** **Family:*Veillonellaceae***						
Desulfosporomusa polytropa	45–46	Rod, tapered ends	I	6.1–8.2	28	**38**

1. Klenk et al. 1997; Birkeland et al. 2017. 2. Itoh et al. 1999. 3. Itoh et al. 1998; Siebers et al. 2011. 4. Heidelberg et al. 2004. 5. Morais-Silva et al. 2014. 6. Postgate and Campbell 1966. 7. Copeland et al. 2009. 8. Olliver et al. 1991. 8. Spring et al. 2010. 9. Zhilina et al. 1997. 10. Audiffrin et al. 2003. 11. Nunoura et al. 2007. 12. Pikuta et al. 2003. 13. Brysch et al. 1987. 14. Dörries et al. 2016. 15. Fukui et al. 1999. 16. Widdel 1980. 17. Pagani et al. 2011. 18. Friedrich et al. 1996. 19. Isaksen and Teske 1996. 20. Rabus et al. 2004. 21. Sun et al. 2010. 22. Beeder et al. 1995. 23. Oude Elferink et al. 1995. 34. Harmsen et al. 1998. 35. Oude Elferink et al. 1999. 26. Sun et al. 2001. 27. Bhatnagar et al. 2015. 28. Miroshnichenko et al. 2008. *29.* Hattori et al. 2002. 30. Bhatnagar et al. 2015. 31. Lai et al. 2016. 32. Mori et al. 2003. 33. Frolov et al. 2017. 34. Visser et al. 2014. 35. Pester et al. 2012. 36. Kaksonen et al. 2007. 37. Chivian et al. 2008. 38. Sass et al. 2004

92 genera of SRP (Barton and Fauque 2009 and unpublished information) and new isolates are being evaluated in laboratories through the world. Following the 16S rRNA phylogeny of Meyer and Kuever (2007), the SRB can be assigned to three phyla (*Deltaproteobacteria* subdivision within the *Proteobacteria*, *Firmicutes*, and *Nitrospira*), while the sulfate-reducing archaea are in two phyla (*Crenarchaeota* and *Euryarchaeota*) (see Table 2.4).

2.5.1.1 Archaea

The euryarchaeotal microorganism, *A. fulgidus*, has robust growth with sulfate as the electron acceptor and has been extensively characterized (Stetter et al. 1987; Stetter 1988). *A. profundus* grows lithotrophically on H_2, CO_2, and sulfate (Burggraf et al. 1990). Another sulfate-reducing strain of *Archaeoglobus* is reported as *Archaeoglobus lithotrophicus* (Burggraf et al. 1990); however, this name was not validly published under the rules of the International Code of Nomenclature of Bacteria. Two members of the *Crenarchaeota* that have been proposed as dissimilatory sulfate reducers are *Thermocladium tenax* (Zillig et al. 1981) and *Caldivirga maquilingensis* (Itoh et al. 1999). There is a difference of opinion if *Thermocladium tenax* and *Caldivirga maquilingensis* are indeed dissimilatory sulfate-reducing microorganisms because they were reported to lack *qmo* genes considered essential for sulfate reduction (Pires et al. 2003; Zane et al. 2010), and these archaea also lack the *aprM* gene which functions to anchor APS reductase to the plasma membrane. However, Siebers et al. (2011) report the presence of *sat*, *apsAB*, and *dsrABCGK* which respectively encode ATP sulfurylase, APS reductase, dissimilatory (bi)sulfite reductase, and a protein which anchors APS reductase to the membrane. Additionally, Siebers et al. (2011) report the growth of *Thermocladium tenax* by dissimilatory sulfate reduction. The inclusion of *Thermocladium tenax* and *Caldivirga maquilingensis* in classification of SRP (Table 2.1) should be considered provisional until additional information establishing growth on sulfate is reported.

2.5.1.2 Endospore-Producing SRB

Sporulation in sulfate reducers was initially associated with *Vibrio desulfuricans* (Starkey 1938), *Sporovibrio desulfuricans* (Senez 1951), *Clostridium nigrificans* (Campbell Jr. et al. 1957), and *Desulfovibrio orientis* (Adams and Postgate 1959). Spore-forming SRB were assigned to the *Desulfotomaculum* genus (Campbell and Postgate 1965; Postgate and Campbell 1966). With the isolation of new cultures of SRB, the classification system with only two genera (*Desulfovibrio* and *Desulfotomaculum*) was soon expanded. *Dst. antarcticum* displayed interesting characteristics (Iizuka et al. 1969); however, it appears that this SRB culture has been lost (Stackebrandt et al. 1997). Spore-forming SRB have been isolated from a variety of ecosystems, and 12 *Desulfotomaculum* sp. obtained from deep subsurfaces are described (Aüllo et al. 2013).

Thermophilic SRB endospores are found in the cold marine sediments of Aarhus Bay, and the distribution of these endospores over several thousand years are proposed to serve as an indication of passive dispersion in marine environments (De Rezende et al. 2013).

The classification of spore-forming SRB is evolving with *Dst. orientis* (previously *D. orientis*) reclassified as *Desulfosporosinus orientis* (Stackebrandt et al. 1997), and based on reexamination of genomic analysis, *Dst. carboxydivorans* is to be considered a synonym for *Dst. nigrificans* (Visser et al. 2014). The classification of *Dst. guttoidenum* has been questioned because of its similarity to several *Clostridia* species (Stackebrandt et al. 1997; Spring et al. 2009). As presented in Table 2.4, spore-forming SRB are assigned to two *Thermoanaerobacterales* and *Clostridiales* in the *Clostridia* class, while *Desulfosporomusa polytropa* (Sass et al. 2004) is proposed to be assigned to the *Selenomonadeles* order in the *Negativicutes* class. *Candidatus* Desulforudis audaxviator is an interesting bacterium found at a depth of 2.8 km in the Mponeng gold mine in South Africa (Chivian et al. 2008) and remains uncultivated. The quantitation of spore-producing SRB in environmental samples often results in an underreporting, and a method has been reported that induces germination and exponential growth (de Rezende et al. 2017). From molecular analysis of deep subsurface samples, bacteria with a high genomic similarity to *Candidatus* Desulforudis audaxviator have been detected in sediments of Lake Maslak in Turkey (Balci et al. 2012), in the Outokumpu borehole in Finland (Itävaara et al. 2011), and in a sample taken near a geothermal plant in Germany (Alawi et al. 2011). Based on microbial characteristics (Sass et al. 2004), it would follow that *Desulfosporomusa polytropa* would be of the *Negativicutes* class, *Selenomonadeles* order, and *Veillonellaceae* family. At this time, there is only one SRB genus in this family. A unique SRB recently isolated is *Desulfosporomusa polytropa* a spore-forming SRB which is tentatively placed in the family *Veillonellaceae*, and this placement awaits approval (Sass et al. 2004).

There are 30 recognized species of *Desulfotomaculum* with 17 thermophilic, 3 halophilic, and 1 alkaliphilic species (Stackebrandt et al. 1997; Aüllo et al. 2013). Analysis of phylogenic information indicated that these organisms could be assigned to three clusters. Cluster I contained most of the *Desulfotomaculum* species, cluster II contained *Dst. orientis*, and cluster III contained *Dst. guttoidenum.* Based on 16S rRNA gene sequence, the *Desulfotomaculum* species from Cluster I were resolved into 6 subclusters (Ia, Ib, Ic, Id, Ie, If), and members of these subclasses were assigned to five different genera (see Table 2.5). Subcluster 1 g contain *Desulfocucumis palustris*, and subcluster Ih contain several *Pelotomaculum* species which were identified by Watanabe et al. (2018). It is significant to note that *Sporotomaculum* sp. of *Desulfotomaculum* subspecies Ib and *Pelotomaculum thermopropionicum* of subspecies 1 h lack the ability to grow by dissimilatory sulfate reduction due to a loss of *dsrAB* as these organisms adopted a syntrophic lifestyle (Imachi et al. 2000, 2002, 2006). Methanogenic communities contain propionate-oxidizing *Desulfotomaculum* living syntrophically with methanogens, and over time, these syntrophic *Desulfotomaculum* have lost the genes for dissimilatory sulfate reduction (Imachi et al. 2006).

Table 2.5 Members of *Desulfotomaculum* Cluster I with proposed new genera[a]

Cluster I	Proposed genus	Proposed species	Description is the same as
Subcluster 1a	*Desulfotomaculum*	*aeronauticum*	*Dst. aeronauticum*
		aquiferis	*Dst. aquiferis*
		defluvii	*Dst. defluvii*
		ferrireducens	*Dst. ferrireducens*
		hydrothermale	*Dst. hydrothermale*
		nigrificans	*Dst. nigrificans*
		profundi	*Dst. nigrificans*
		pitei	*Dst. pitei*
		ruminis	*Dst. ruminis*
Subcluster 1b	*Desulfallas*	*sapomandens*	*Dst. pitei*
		geothermicus	*Dst. pitei*
		thermosapovorans	*Dst. pitei*
		gibsoniae	*Dst. pitei*
		arcticus	*Dst. pitei*
		alcoholivorax	*Dst. pitei*
Subcluster Ic/Id	*Desulfofundulus*	*kuznetsovii*	*Dst. pitei*
		thermobenzoicus	*Dst. pitei*
		australicus	*Dst. pitei*
		thermoacetoxidans	*Dst. pitei*
		thermocisternus	*Dst. pitei*
		luciae	*Dst. pitei*
		solfataricus	*Dst. pitei*
		thermosubterraneus	*Dst. pitei*
Subcluster Ie	*Desulfofarcimen*	*acetoxidans*	*Dst. pitei*
		intricatum	*Dst. pitei*
Subcluster If	*Desulfohalotomaculum*	*halophium*	*Dst. pitei*
		alkaliphilum	*Dst. pitei*
		peckii	*Dst. pitei*
		tongense	*Dst. pitei*

Dst. = *Desulfotomaculum*
[a] Watanabe et al. 2018

2.5.1.3 *Deltaproteobacteria* and Non-sporing Bacteria

The proteobacterial class *Deltaproteobacteria* contains a highly diverse group of bacteria including numerous species of mesophilic and thermophilic SRB. The *Desulfovibrionaceae* family in *Deltaproteobacteria* embodies over 35 species in the *Desulfovibrio* genus (Kuever et al. 2005a), and 60 species of *Desulfovibrio* were commonly reported (Schoch et al. 2020). The 16S rRNA gene identity value is used with a cut-off level of 94.5% for the delineation of genera (Yarza et al. 2014). A comparative analysis of 16S rRNA gene sequences of *Desulfovibrionaceae* family members was resolved into six groups, and these are presented in Table 2.6. Using

Table 2.6 Groups of Gram-negative SRB based on 16S rRNA gene sequences[a]

Group A	*D. alaskensis*	*D. oxamicus*	*Halodesulfovibrio aestuarii*
	D. cuneatus	*D. piger*	*Halodesulfovibrio marinisediminis*
	D. desulfuricans	*D. psychrotolerans*	*Halodesulfovibrio oceani*
	'D. fairfieldensis'	*D. simplex*	*Halodesulfovibrio spirochaetisodalis*
	D. intestinalis	*D. termitidis*	
	D. legallii	*D. vietnamensis*	
	D. litoralis	*D. vulgaris*	
	D. longreachensis		
Group B	*D. biadhensis*		
	D. gabonensis		
	D. giganteus		
	D. gigas		
	D. indonesiensis		
	D. marinus		
	D. paquesii		
Group C	*Desulfobaculum xiamenense*		
	Desulfocurvus thunnarius		
	Desulfocurvus vexinensis		
	D. bizertensis		
	D. senezii		
Group D	*D. aminophilus*	*D. hydrothermalis*	*Pseudodesulfovibrio aespoeensis*
	D. arcticus	*D. idahonensis*	*Pseudodesulfovibrio hydrargyri*
	D. bastinii	*D. longus*	*Pseudodesulfovibrio indicus*
	'D. brasiliensis'	*D. mexicanus*	*Pseudodesulfovibrio piezophilus*
	D. capillatus	*D. oxyclinae*	*Pseudodesulfovibrio portus*
	D. ferrireducens	*D. salexigens*	*Pseudodesulfovibrio profundus*
	D. frigidus	*D. tunisiensis*	
	D. gracilis	*D. zosterae*	
	D. halophilus		
Group E	*D. aerotolerans*	*D. carbinoliphilus*	
	D. alcoholivorans	*D. inopinatus*	
	D. Burkinensis	*D. magneticus*	
	D. butyratiphilus	*D. sulfodismutans*	
	D. carbinolicus		
Group F	*D. africanus*		
	D. alkalitolerans		
	D. cavernae		
	D. reitneri		
	D. sp. X2		

[a]Source of information is Spring et al. 2019

the universal RpoB protein (β-subunit of the DNA-directed RNAS polymerase) and DsrAB (subunits of dissimilatory sulfite reductase) as alternative phylogenic markers, six groups were again identified (Spring et al. 2019).

There are 21 genera in the *Desulfobacteraceae* family which include *Desulfobacter*, *Desulfatibacillum*, *Desulfatiferula*, *Desulfatirhabdium*, *Desulfatitalea*, *Desulfobacterium*, *Desulfobacula*, *Desulfobotulus*, *Desulfocella*, *Desulfococcus*, *Desulfoconvexum*, *Desulfofaba*, *Desulfofrigus*, *Desulfoluna*, *Desulfonatronobacter*, *Desulfonema*, *Desulforegula*, *Desulfosalsimonas*, *Desulfosarcina*, *Desulfospira*, and *Desulfotignum* (Kuever 2014a). The *Desulfobulbaceae* family (Kuever 2014b; Trojan et al. 2016) contains *Desulfobulbus*, *Desulfocapsa*, *Desulfofustis*, *Desulforhopalus*, *Candidatus* Electrothrix, and *Candidatus* Electronema as genera.

Some interesting organizational initiatives involving classification are being pursued. There is a proposal to reclassify the proteobacterial classes *Deltaproteobacteria* and *Oligoflexia* and the phylum *Thermodesulfobacteria* into four phyla (Waite et al. 2020). An isolate of SRB which grows under high pressure has been reported as *Pseudodesulfovibrio indicus* and is proposed to be classified along with four species of *D.* (*piezophilus*, *profundus*, *portus*, and a*espoeensis*) as *Pseudodesulfovibrio* (Cao et al. 2016a). There has been a proposal to modify the family *Thermodesulfobiaceae* to delete the genus *Coprothermobacter* but to retain the genus *Thermodesulfobium* (Pavan et al. 2018).The taxonomic framework of cable bacteria is being constructed (Trojan et al. 2016).

2.5.2 *Insights from Gene and Genome Analysis*

Following the isolation, purification, and sequencing of enzymes and proteins from different strains of *Desulfovibrio*, several enzymes were cloned (Table 2.7). Most of the enzymes cloned were from *D. vulgaris* Hildenborough, and this was several years before the genome of this SRB became available (Heidelberg et al. 2004).

2.5.2.1 Archaea Domain

A. fulgidus VC-16 DSM 4304 is the type strain of the genus, and its genome contains several repetitive elements of unknown function. One region of the genome contains 48 copies of a 30 bp segment, another region contains 60 copies of the 30 bp, and a third region contains 42 copies of a 37 bp segment. Additionally, there were 9 instances where long coding repeats of 628–1886 bp were present with copy numbers varying from 2 to 7. The versatility of *A. fulgidus* VC-16 is seen in that the genome contains 57 β-oxidation enzymes, 5 types of ferredoxin-dependent oxido-reductases, 14 acyl-CoA ligases, and 4 tungsten-containing oxidoreductases. To provide iron for the numerous iron-containing redox proteins, the systems for uptake of Fe^{3+} and Fe^{2+} have been identified. The transport of Fe^{+} would enable hemin and

Table 2.7 *Desulfovibrio* spp. genes cloned and expressed from 1985 to 1992

Enzymes/proteins	Name	Bacteria	Reference
[Fe] hydrogenase	*hydA,B*	*D. vulgaris* H.[a]	Voordouw et al. (1985)
		D. *vulgaris* H.	Voordouw and Brenner (1985)
		D. vulgaris H.	Prickril et al. (1986)
		D. vulgaris H.	Voordouw et al. (1987a)
		D. vulgaris H.	Voordouw et al. (1987b)
		D.vulgaris H.	van Dongen et al. (1988)
		D. vulgaris subsp. *oxamicus*	Voordouw et al. (1989a)
[NiFe] hydrogenase	*hynA, hynB*	*D. gigas*	Li et al. (1987)
		D. gigas and *D. baculatus*	Voordouw et al. (1989b)
		D. vulgaris Miyazaki	Deckers et al. (1990)
		D. fructosovorans	Rousset et al. (1990)
[NiFeSe] hydrogenase	*hysA, hysB*	*D. baculatus*	Menon et al. (1987)
		D. gigas and *D. baculatus*	Voordouw et al. (1989b)
Cytochrome c_3	*cyc*	*D. vulgaris* H.	Voordouw and Brenner (1986)
		D. vulgaris H.	Voordouw et al. (1987b)
		D. vulgaris H.	Pollock et al. (1991)
		D. vulgaris H.	Voordouw et al. (1990a, b, c)
Cytochrome c_{553}	*cyf*	*D. vulgaris* H.	van Rooijen et al. (1989)
Assimilatory sulfite reductase	*asr*	*D. vulgaris* H.	Tan et al. (1991)
High molecular weight cyto-chrome (Hmc)	*hmc*	*D. vulgaris* H.	Pollock et al. (1991)
Flavodoxin	*fla*	*D. vulgaris* H.	Curley and Voordouw (1988)
		D. vulgaris H.	Krey et al. (1988)
		D. vulgaris H.	Carr et al. (1990)
Rubredoxin	*rub*	*D. vulgaris* H.	Voordouw (1988)
Desulforedoxin	*dsr*	*D. gigas*	Brumlik et al. (1990)
Nitrogenase, Fe protein	*nifH*	*D. gigas*	Kent et al. (1989)
Rubrerythrin	*rbr*	*D. vulgaris* H.	Prickril et al. (1991)
Methyl-accepting chemotaxis protein	*dcrA*	*D. vulgaris* H.	Dolla et al. (1992)

[a] *H* Hildenborough

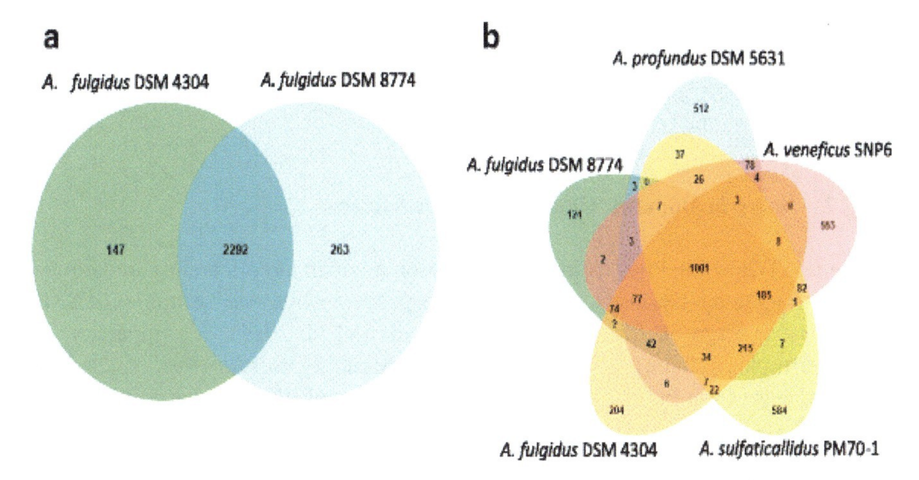

Fig. 2.5 Venn diagrams showing the distribution of orthologous and unique genes for (**a**) *A. fulgidus* strains DSM 4304 and DSM 8774 and (**b**) for all the completely genome sequenced *Archaeoglobus* representatives: *A. fulgidus* DSM 4305, *A. fulgidus* DSM 8774, *Archaeoglobus profundus* DSM 5631, *Archaeoglobus veneficus* DSM 11195, and *Archaeoglobus sulfaticallidus* DSM 19444 (Birkeland et al. 2017).

hemin-like compounds to provide iron, while the FeoB of the Feo system would transport Fe^{2+} into the cell (Klenk et al. 1997). Genome sequence analysis of *A. fulgidus* strain 7324 DSM 8774 (Birkeland et al. 2017) revealed several differences when compared to *A. fulgidus* VC-16. There is a 93.5% similarity between these two strains with both sharing 2292 genes (see Fig. 2.5). When comparing 5 strains of *Archaeoglobus*, all strains shared 1001 genes (see Fig. 2.5), and most of these genes encoded for energy production, biosynthetic activities, and regulatory functions. An unusual characteristic of *Caldivirga maquilingensis* is the presence of introns in the 16S rDNA (Itoh et al. 1999). These rRNA introns are only found in the Crenarchaeotic families of *Desulfurococcaceae* and Thermoproteaceae, and these introns are considered to be more mobile than 16S rRNA exons.

It was reported that the genome of *Thermoproteus tenax* (Siebers et al. 2011) has a well-developed electron transport system involving membrane proteins with two copies of genes for the NADH:quinone oxidoreductase (complex I), genes for succinate dehydrogenase (complex II), and genes encoding the bc_1 complex (complex III). Repeat DNA segments with unknown function were observed in the genome and included five type I clusters and two type II clusters. Systems for energy production would be attributed to genes encoding for A_0A_1-ATP synthase and a membrane-bound pyrophosphatase. While A_0A_1 have a function similar to F_0F_1 in bacteria, their protein subunits are unique to archaea. With respect to dissimilatory sulfate reduction, genes for the formation of APS from sulfate + ATP (*sat*), sulfite from APS (*apsAB*), and sulfide production from sulfite (*dsrABCGK*) were identified.

2.5.2.2 *Bacteria* Domain: Class *Deltaproteobacteria*

D. gigas ATCC 19364 has a circular chromosome which has apparently undergone little rearrangement because it has only17 transposases while other genomes of SRB contain on average 34 transposases (Morais-Silva et al. 2014). A comparison of protein orthology among genomes from four species of *Desulfovibrio* (including *D. gigas)* is given in Fig. 2.6. There is only one operon of rRNA which is in contrast to three to six operons in other *Desulfovibrio* spp., and this the single rRNA operon may contribute to the slow growth rate with a division time of 8 hr. Also, the chromosome contains genes that encode for nine selenocysteine containing proteins. In *D. gigas*, genes for three c_3-type cytochromes and two hydrogenase genes are present. Genes encoding for the [NiFe]-type hydrogenases were identified as (i) HynAB which is periplasmic and (ii) Ech which is an energy-conserving

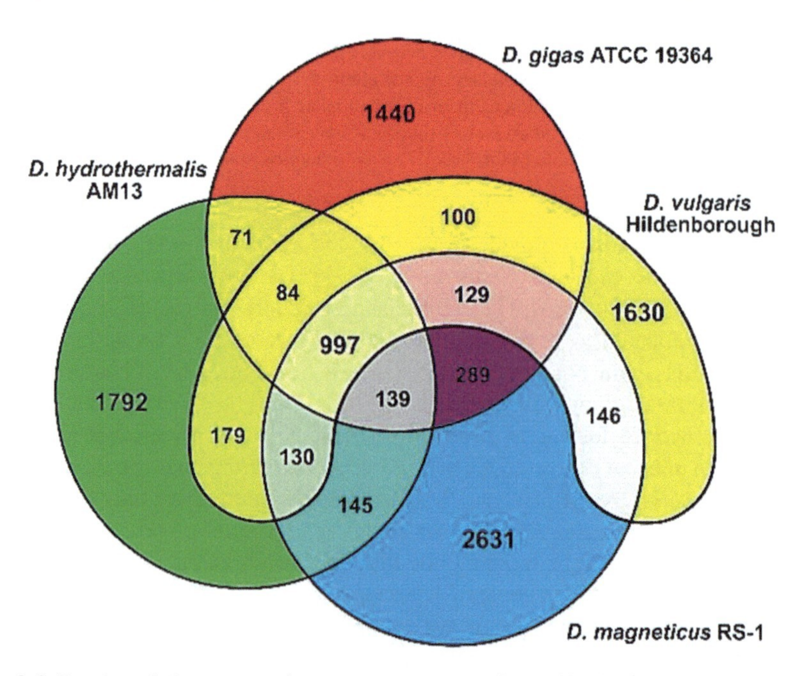

Fig. 2.6 Protein orthology comparison among genomes of *Desulfovibrio gigas*, *D. magneticus* RS-1, *D. hydrothermalis* AM 13, and *D. vulgaris* Hildenborough (Morais-Silva et al. 2014). The Venn diagram shows shared ortholog groups for any given gene of each species under analysis. (*Copyright*: 2014 The Authors. MicrobiologyOpen published by John Wiley & Sons Ltd. This is an open-access article under the terms of the Creative Commons Attribution License, which permits the use, distribution, and reproduction in any medium, provided the original work is properly cited)

hydrogenase. With respect to energy production, *D. gigas* has genes encoding for the F_1F_0-ATP synthase module, and unlike other *Desulfovibrio* spp., genes encoding the vacuolar-type ATPase (V_0V_1) were also present giving rise to the speculation that it may function as a sodium pump. Genes involved in cell size, response to O_2 exposure, carbon metabolism, and energy transfer complexes are discussed in other parts of this book. *D. gigas* has a single plasmid (accession number CP006586) with 101,949 bp and 72 protein coding genes. Of the proteins encoded by plasmid DNA, 12 proteins encoding acetyl, methyl, and glycosyl transferases were identified along with an operon for a type II secretory system, an operon for capsule polysaccharide synthesis, and the apsK gene which encodes for a bifunctional enzyme with activities characteristic of sulfate adenylyltransferase and adenylylsulfate kinase activities.

D. magneticus sp. RS1 which is classified as a *δ-proteobacteria* is the only magnetotactic sulfate-reducing bacterium (Sakaguchi et al. 1993, 2002). The genome of *D. magneticus* sp. RS1 is significantly larger than for most *Desulfovibrio* spp. (Table 2.8) and genome analysis reveals that this bacterium has all the genes characteristic of *Desulfovibrio* spp. (Nakazawa et al. 2009). Multiple genes were identified for several enzymes and activities, and these include three sets of genes for ferrous ion uptake (*feoAB*), two gene sets for sulfate adenylyltransferases, two sets of fumarate reductase, four Fe-only hydrogenases, two FeNi-hydrogenases, ten *c*-type cytochromes, two gene sets for NADH:quinone oxidoreductases (complex I), and numerous genes for quinone biosynthesis. While *D. vulgaris* Hildenborough and *D. vulgaris* DP4 and *D. alaskensis* G20 have 204, 218, and 210 genes, respectively, for signal transduction, *D. magneticus* RS-1 has 383 genes. This large number of signal transduction genes is suggested to enable *D. magneticus* to change from magnetic to nonmagnetic forms as the cells grow (Nakazawa et al. 2009). Three gene regions found in *D. magneticus* that are characteristic of magnetotactic bacteria include the *nuo* gene subset which encodes for NADH:quinone oxidoreductase (complex I), the *mamAB*-like gene cluster, and the gene region in the plasmid DMC1. These gene regions have G + C values that ware lower than adjacent genes, and this has been taken to suggest that these genes were acquired by horizontal gene transfer. The *mamABEKQMOPT* genes encode magnetosome membrane proteins in *D. magneticus*, and these genes appear to be contained in a single operon. The cryptic plasmid (pDMC1) contains three genes of unknown function, but these three are found in other BLAST searches.

Syntrophobacter fumaroxidans is a Gram-negative, propionate-oxidizing member of the class *Deltaproteobacteria.* This bacterium produces H_2 for a syntrophic partner, D*ehalococcoides mccartyi* strain CBDB1, which uses hexachlorobenzene as the electron acceptor. Genome analysis of *Syntrophobacter fumaroxidans* strain (MPOBT) revealed the presence of 4.990 Mbp, 4179 genes with 98.06% encoding for protein, no plasmids, and the G + C % is 59.95 (Plugge et al. 2012). Based on genes present, the conversion of propionate to acetate and CO_2 is by the methylmalonyl-CoA pathway. All enzymes for dissimilatory sulfate reduction with production of sulfide are present in *Syntrophobacter fumaroxidans* as well as membrane complexes for generating proton motive force. Regulation and signal

Table 2.8 Genomic analysis of a selection of SRP

Domain/Family/Genus/species		Size (Mb)	Total gene	RNA gene	Protein coding	COG	Signal peptides	Trans-membrane helices	CRISPR repeats	Accession number[a]	Reference (available)
Domain *Archaea* Family: *Archaeoglobaceae*											
Archaeoglobus fulgidus	DSM 8774	2.316	2615	56	2558	1759	80	490	5	CP006577.1	Birkeland et al. (2017)
Archaeoglobus fulgidus	DSM 4304	2.178	2436	49	2407	1817	382		3	NC_000917	Klenk et al. (1997)
Archaeoglobus profundus	DSM 5631	1.563	1901	52	1858	1267	141	128	0	CP001857 NC_013741	von Jan et al. (2010)
Family: *Thermoproteaceae*											
Thermocladium tenax	Kra1 DSM 2078	1.841		50	2051	1572	56	412	7	Gc01285	Siebers et al. (2011)
Domain: Bacteria Family: Desulfovibrionaceae											
Desulfovibrio gigas	ATCC 19364	3.693	3370	51	3273	2273	Nr	Nr	6	CP006585	Morais-Silva et al. (2014)
Desulfovibrio fructosovorans	JJ	4.675	4210	51	4159	2440	359	1103	2	Gs0012061	
Desulfovibrio *alaskensis*	G20	3.730	3258	78					1	CP000112	Hauser et al. (2011)
D. desulfuricans	ATCC 27774	2.873	2356	61					1	NC_011883	
D. piger	FI11049	2.807	2535	97					Nd	LT630450	Wegmann et al. (2017)

D. vulgaris Miyazaki		4. 040	3178	80					2	NC_011769	
D. vulgaris DP4		3.661	3295	88	3149				1	NC_008741	
D. vulgaris Hildenborough		3.773	3535	82	2824					NC_002937	Heidelberg et al. (2004)
Desulfovibrio aminophilus	Strain DSM 12254	3.419	3299	60	3202				Nr	GCF_000422565.1	
Pseudodesulfovibrio indicus		3.966	3611	60	3461				1	CP014206.1	Cao et al. (2016b)
Pseudodesulfovibrio aespoeensis	Aspo-2, DSM 10631	3.629	3405	55	3269				Nd	CP002431.1	Pedersen et al. (2014)
Ca. Desulfovibrio trichonymphae		1.410	1082	46	1982						Kuwahara et al. (2017)
Family: *Desulfomicrobiaceae*											
Desulfomicrobium baculatum	Strain XT (DSM 4028)	3.942	3565	71	3494	2689	723	892	0	Gc01026	Copeland et al. (2009)
Family: *Desulfohalobiacea*											
Desulfohalobium retbaense	DSM 5692	2.909	2609	57	2552	1976	456	634	1	CP001734	Spring et al. (2010)
Desulfonatronospira thiodismutans	AS03-1	3.971	3791	50	3741	1943	132	711	5	Go0003396	Sorokin et al. (2008)
Family: Desulfobacteraceae											
Desulfobacterium autotrophicum	HRM2, DSM 3382	5.598	4917	70	4776				1	NC_012108	Strittmatter et al. (2009)

(continued)

Table 2.8 (continued)

Domain/Family/Genus/ species		Size (Mb)	Total gene	RNA gene	Protein coding	COG	Signal peptides	Trans-membrane helices	CRISPR repeats	Accession number[a]	Reference (available)
Desulfococcus multivorans	1be1, DSM 2059	4.455	3942	57					6	CP015381	Dörries et al. (2016)
Desulfococcus oleovorans Hxd3	DSM 6200	3.944	3265	53	3255	2036	191	872	2	CP000859	
Desulfococcus biactus	KMRActS	4.646	4773	65	4708			1128	8		Gutiérrez Acosta et al. (2014)
Desulfotignum phosphitoxidans	FiPS-3, DSM 13687	4.998	4699	53	4646	3648	354		2	APJX01-000000	Poehlein et al. (2013)
Desulfonema ishimotonii	Tokyo 01, DSM 9680	6.638	5197	88	5109		361		6	BEXT-01000000	Watanabe et al. (2019)
Family: Desulfobulbaceae											
Desulfobulbus propionicus	DSM 2032	3.851	3408	57	3351	2502	1073	812	1	CP002364	Pagani et al. (2011)
Desulfotalea psychrophile LSv54	DSM 12343	3.523	3232	91	3076		2023			NC_006138	Rabus et al. (2004)
Family: Desulfoarculaceae											
Desulfarculus baarsii	DSM 2075	3.655	3355	52	3303	2466	686	768	3	CP002085	Sun et al. (2010)
Family: ***Syntrophobacteraceae***											
Syntrophobacter fumaroxidan	DSM 10017	4.990	4176	81	4098	2959	741	1035	4	CP000478	Plugge et al. (2012)
Family: Syntrophaceae											
Desulfobacca acetoxidans	ASRB2^T	3.282	3023	51	2969	2106	488	726	4	CP002629	Göker et al. (2011)

Family: *Nitrospiraceae*											
Thermodesulfovibrio yellowstonii	DSM	11,347	2.003	2058	49	1996			4	NC_011296	Bhatnagar et al. (2015)
Family: Thermoanaerobiaceae											
Thermacetogenium phaeum	DSM 12270	2.939	3215	51					11	CP003732	Oehler et al. (2012)
Family: *Thermodesulfobacteriaceae*											
Thermodesulfobacterium commune	DSM 2178	2.003	1532	46	1453	Nr	Nr	Nr	2	CP008796	Bhatnagar et al. (2015)
Family: Peptococcaceae											
Syntrophobotulus glycolicus	strain FlGlyRT, DSM 8271	3.406	3439	69	3370	2399	463	848	2	CP002547	Han et al. (2011)
Desulfotomaculum kuznetsovii	strain 17 T, DSM 6115	3.601	3625	58	3567	2560	258	748	Nr	NC_015573	Visser et al. (2013)
Desulfotomaculum nigrificans	DSM 574	3.052	3112	98	3014	2340	582	721		Gi03933	Visser et al. (2014)
Desulfotomaculum reducens	strain MI-1, *DSM* 100696	3.608	3424	71	3324			510		NC_009253	Junier et al. (2010)
Desulfotomaculum carboxydivorans	strain CO-1-SRB, DSM 14880	2.982	2844	97	2747	2174	504	647		Gc01783	Visser et al. (2014)
Desulfotomaculum gibsoniae	strain Groll^T, DSM 7213	4.855	4762	96	4666	3345	737	966	Nr	CP003273	Kuever et al. (2014)
Desulfotomaculum acetoxidans	DSM 5575	4.545	4470	100	4370	2702	701	791	11	CP001720 NC_013216	Spring et al. (2009)

(continued)

Table 2.8 (continued)

Domain/Family/Genus/species		Size (Mb)	Total gene	RNA gene	Protein coding	COG	Signal peptides	Trans-membrane helices	CRISPR repeats	Accession number[a]	Reference (available)
Desulfotomaculum ruminis	DSM 2154	3.969	3959	87	3783				10	NC_015589	Spring et al. (2012)
Candidatus Desulforudis audaxviator	MP104C	2.349	2293	45	2239				4	NC_010424	Jungbluth et al. (2017)
Candidatus Desulfopertinax cowenii	modA32[b]	1.778	1778	44	1842				1		Jungbluth et al. (2017)

[a] GenBank; Genomes Online Database (GOLD, www.genomesonline.org)
[b] Draft 91.67% complete

transduction are achieved by 12 sigma factors including 1 housekeeping sigma 70 factor, 2 additional genes similar to sigma 70 factors, 1 sigma 28 factors, 1 sigma 54 factors for global regulation, and 7 sigma 24 type stress-related factors.

2.5.2.3 Plasmids

Plasmids are found in some of the SRP. The only archaeal dissimilatory sulfate reducer reported to contain a plasmid is *A. profundus*, and it has a 2.8 kb plasmid (pGS5) that is negatively supercoiled (López-García et al. 2000). As indicated in Table 2.9, some SRB contain one or two plasmids, and the size of plasmids varies with the species of SRB. The functions of many of the genes encoded in SRB plasmids are unclear; however, some general statements about potential activities of genes in plasmids have been provided. Synthesis of magnetosomes is proposed for some gene products of the plasmids in *D. magneticus* (Nakazawa et al. 2009); modification of cell surface components, nitrogen fixation, a type III protein secretion system, and changes of cell surface have been indicated for *D. vulgaris* Hildenborough (Heidelberg et al. 2004); and the *nif* genes for nitrogen fixation are

Table 2.9 Characteristics of plasmids in sulfate-reducing bacteria

Bacteria	Chromosome	Plasmid 1	Plasmid 2	Total	References
	Mbp	Kbp	Kbp	Mbp	
	G + C %	G + C %	G + C %	G + C %	
	[ORFs]	[ORFs]	[ORFs]	[ORFs]	
Desulfovibrio magneticus	4.630	58.70	8867	5.315	Nakazawa et al. (2009)
Strain RS-1	62.77%	58.03%	37.17%	62.67%	
	[4630]	[65]	[10]	[4.705]	
Desulfovibrio gigas	3.69	101,9		3.791	Morais-Silva et al. (2014)
	63.4%	64.1%			
	[3273]	[75]		[3348]	
Desulfovibrio vulgaris	3.570	202.3		3.772	Heidelberg et al. (2004)
Hildenborough	63.2%	65.7%		65%	
	[3370]	[152]		[3522]	
Desulfotalea psychrophila	3.523	121.6	14.66	3.659	Rabus et al. (2004)
LSv54	46%	43%	28%		
	[3116]	[101]	[17]	[3224]	
Desulfobacterium	5.589	62.96		5.651	Strittmatter et al. (2009)
Autotrophicum	48,9%	42.0%		48%	
HRM2	[4871]	[76]		[4.947]	

encoded on plasmids of some strains of *D. vulgaris* (Postgate et al. 1986). Annotated ORFs of the *D. gigas* plasmid suggest the presence of genes encoding acetyl, glycosyl, and methyl transferases, a type II secretory system for folded exoproteins, capsule polysaccharide biosynthesis, and transporter proteins (Morais-Silva et al. 2014). SRB such as *D. desulfuricans* G100A are capable of receiving IncQ plasmids from *Escherichia coli* by conjugation (Argyle et al. 1992). *D. desulfuricans* strain G100A contains the plasmid (pBGI) with 2303 bp, and there are 20 plasmid copies per genome (Wall et al. 1993). When the pBG1 plasmid was cured from *D. desulfuricans* strain G100A, the resulting strain was identified as *D. desulfuricans* G20, and subsequently, this strain was renamed *D. alaskensis* G20.

2.5.2.4 Uncultured Bacteria

C. Desulfovibrio trichonymphae is an uncultured parasite of *Trichonympha.* An unusual tripartite relationship occurs with *Trichonympha agilis*, a protist in the gut of the cellulolytic termite *Reticulitermes speratus.* This relationship involves two bacteria (*C.* Desulfovibrio trichonymphae and *C.* Endomicrobium trichonymphae) and a hydrogenosome. The genome of the uncultivated bacterium *C.* Desulfovibrio trichonymphae was sequence and analyzed and revealed interesting information about the lifestyle of this bacterium (Kuwahara et al. 2017). *C.* Desulfovibrio trichonymphae has a circular chromosome containing 1.410 Mbp and no plasmids and a G + C of 54.8%. In comparison, the chromosome of *Lawsonia intracellularis* has 1.457 Mbp, while the chromosome of *C.* Endomicrobium trichonymphae has 1.125 Mbp. Based on comparison sequences of 30 ribosomal proteins to existing bacterial genomes, *C.* Desulfovibrio trichonymphae was determined to be most closely related to *D. desulfuricans* 27,774. As is the case for *D. desulfuricans* ATCC 27,774, *C.* Desulfovibrio trichonymphae lacks genes for lipid A biosynthetic pathway. The genome of *C.* Desulfovibrio trichonymphae contains genes for gluconeogenesis and pentose phosphate biosynthesis and some of the enzymes in the tricarboxylic acid cycle. Present are genes for [NiFe] hydrogenase, *c*-type cytochromes, F0F1-ATPase, all enzymes associated with dissimilatory sulfate reduction, fumarate dehydrogenase, and numerous biosynthetic pathways. Absent are genes for [FeFe] and [NiFeSe] hydrogenases, lactate dehydrogenase, formate dehydrogenase, and alcohol dehydrogenase. In *Trichonympha* collaris, three endosymbiont bacteria are present: *C.* Desulfovibrio trichonymphae, an *Endomicrobium*, and "*C.* Adiutrix intracellularis" (Kuwahara et al. 2017). Further genome studies will provide useful information to understand the metabolism and ecology of uncultured SRB and these endosymbionts.

Deep subsurface uncultured members of the *Firmicutes* include the terrestrial *C.* Desulforudis audaxviator and the marine *C.* Desulfopertinax cowenii. There is considerable similarity between these bacteria with the marine bacterium having a GC content of 60.2%, while with the terrestrial bacterium, the GC content is 60.9%. With 98.09% of the genome analyzed for *C.* Desulforudis audaxviator and 97.61% of the genome analyzed for *C.* Desulfopertinax cowenii, the genomes contained 2.35

Mbp and 1.78 Mbp, respectively (Chivian et al. 2008; Jungbluth et al. 2017). However, the predicted function of encoded proteins is similar with 1518 for *C.* Desulfopertinax cowenii and 1587 for *C.* Desulforudis audaxviator. When comparing these two bacteria, there is 76.9% nucleotide similarity, 74.2% average amino acid identity, and 73% of genes shared between. The small genome in *C.* Desulfopertinax cowenii, a single CRISPR element, absence of pseudogenes, and few paralogs suggest that this bacterium has undergone some type of genome streamlining. Both genomes of these subsurface terrestrial and marine bacteria contain genes for dissimilatory sulfate reduction, carbon dioxide fixation by the Wood-Ljungdahl (reductive acetyl-CoA) pathway, and pathways for synthesis of essential amino acids. Unlike *C.* Desulforudis audaxviator, *C.* Desulfopertinax cowenii has a large number of sugar transporters which may reflect a difference of organic matter in marine and terrestrial deep subsurfaces. Comparison of these two genomes to metagenome from around the world indicates ta wide distribution of bacteria similar to *C.* Desulforudis audaxviator and *C.* Desulfopertinax cowenii.

Isolates of *Desulfatiglans* from marine sediments couple the oxidization of aromatic hydrocarbons to dissimilatory sulfate reduction. Based on the analysis of seven *Desulfatiglans*-related single-cell genomes (SAGs) from Aarhus Bay sediment, the genome size was estimated to be about 1.3 Mbp, and these SAGs represented Group 1 which had the capability of sulfate respiration, while Group 2 lacked genes for dissimilatory sulfate reduction but displayed the capability of conserving energy by fermentation and acetogenesis (Jochum et al. 2018). SAGs from both groups had genes for aromatic hydrocarbon degradation and sulfatase activity. A model reflecting these two activities is shown in Fig. 2.7. The presence of sulfatase is unique for cultured SRB and would enable these SAGs to utilize organic compounds with sulfate esters as a carbon source and release of inorganic sulfate to be used for respiration.

2.6 Filamentous or Cable Bacteria

A unique type of multicellular bacteria forming long filaments with the capability of transporting electrons over a distance of several centimeters (Pfeffer et al. 2012) has been designated as "cable bacteria." These bacteria are found in fresh water and marine environments (Malkin et al. 2014; Risgaard-Petersen et al. 2014; Seitaj et al. 2015). Cable bacteria form linear filaments consisting of hundreds of cells which can vertically align in the water column across a redox gradient (Teske 2019). Filaments of cable bacteria are shown in Fig. 2.7. Electrons are transported along the filament starting with the oxidation of sulfide in anoxic sediments to oxygen reduction in the aerobic layer (Nielsen et al. 2010; Bjerg et al. 2018; Meysman et al. 2019) or to nitrate-rich region (Marzocchi et al. 2014). Cable bacteria have also been reported in the rhizosphere of aquatic plants that release O_2 from their roots (Scholz et al. 2019). Because cable bacteria remain uncultivated, the initial reports of the presence of cable bacteria relied on short segments of the 16S rRNA gene and fluorescence

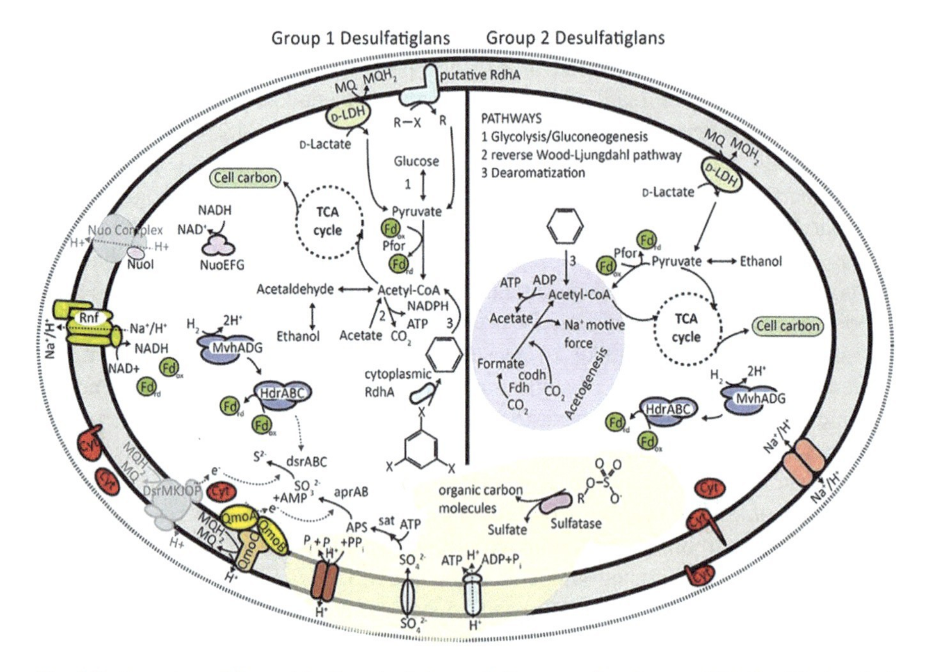

Fig. 2.7 Overview of the metabolic potential encoded in group 1 and group 2 SAGs. All chemical compounds shown are representative, the aromatic ring structure may represent any monoaromatic compound, and halogenated organic compounds may represent any organohalide (Jochum et al. 2018). Sulfate produced from sulfatases may be used in respiration in group 1 cells. Pfor, pyruvate: ferredoxin oxidoreductase; Fdh, NAD-dependent formate hydrogenase; Codh, carbon monoxide dehydrogenase; Cyt, multiheme cytochrome; D-LDH, D-lactate dehydrogenase; MQ/MQH2, menaquinone pool; Fd, ferredoxin; Nuo, NADH:quinone oxidoreductase; Rnf, ferredoxin-dependent transmembrane proton pump; Hdr, heterodisulfide reductase; Mvh, methyl viologen reducing hydrogenase; RdhA, reductive dehalogenase subunit A; PPi, pyrophosphate. (*Copyright* © 2018 Jochum, Schreiber, Marshall, Jørgensen, Schramm, and Kjeldsen. This is an open-access article distributed under the terms of the Creative Commons Attribution License (CC BY). The use, distribution, or reproduction in other forums is permitted, provided the original author(s) and the copyright owner(s) are credited and that the original publication in this journal is cited, in accordance with accepted academic practice. No use, distribution, or reproduction is permitted which does not comply with these terms)

(FISH) analysis targeting 16S rRNA using a *Desulfobulbaceae*-specific probe (Malkin et al. 2014; Marzocchi et al. 2014; Vasquez-Cardenas et al. 2015; Larsen et al. 2015). Using almost full-length 16S rRNA gene sequences combined with an evaluation of full or partial *dsrAB* genes, 16 samples containing cable bacteria were examined and assigned to candidate genera in the family *Desulfobulbaceae*: "*Candidatus* Electrothrix" which contains mostly marine isolates and "*Candidatus* Electronema" which contains freshwater isolates (Trojan et al. 2016). Six species have been proposed ("*C.* Electronema *nielsenii*," "*C.* Electronema *palustris*," "*C.* Electrothrix *marina*," "*C.* Electrothrix *communis*," "*C.* Electrothrix *japonica*," "C. Electrothrix *aarhusiensis*"), and at least five undesignated species were suggested (Trojan et al. 2016). Images of cable bacteria are presented in Figs. 2.8 and 2.9 where

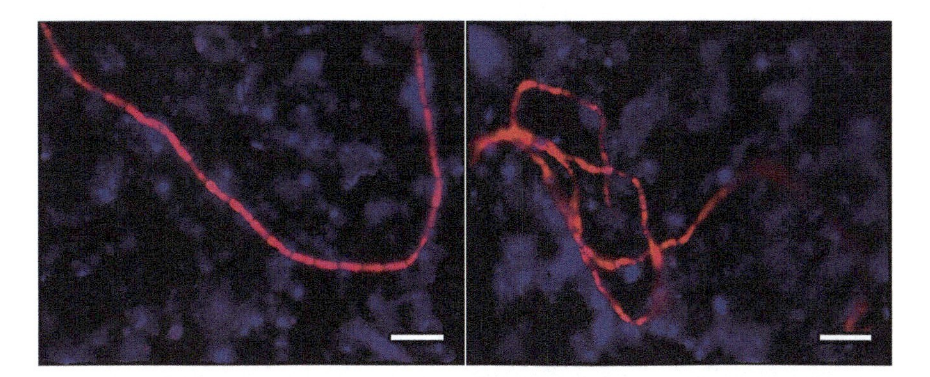

Fig. 2.8 FISH images of filamentous bacteria hybridized with DSB706 from sediment of core N1 of the unamended incubation (Geelhoed et al. 2020). The images show the overlay of false colored images for the DSB706 (red) and DAPI (blue) signal. The scale bars denote 10 μm. (*Copyright* © 2020 Geelhoed, van de Velde and Meysman. This is an open-access article distributed under the terms of the Creative Commons Attribution License (CC BY). The use, distribution, or reproduction in other forums is permitted, provided the original author(s) and the copyright owner(s) are credited and that the original publication in this journal is cited, in accordance with accepted academic practice)

longitudinal ridges along the cell are visible (Fig. 2.9). The physiology and ecology of these cable bacteria are extremely interesting since the filaments are about a thousand times the length of an individual cell. Evidence for the transport of electrons over long distances by filaments produced by these cable bacteria was based on Raman microscopy which measured the cytochrome redox state along the filament with one end of the filament in sulfide and the other in O_2 (Bjerg et al. 2018). A cytochrome redox potential gradient along the filament was observed when sulfide was oxidized and O_2 was reduced, but with the removal of O_2, the redox potential was lost (Bjerg et al. 2018). Periplasmic fibers provided for high conductivity of intact filaments 10.1 mm long which consisted of over 2000 adjacent cells, and this magnitude of conductivity is greater than that observed with nanowires of *Shewanella* or *Geobacter* pili (Meysman et al. 2019). It remains to be tested if electrons from cable bacteria can be transferred to other bacteria.

Even though septa are produced between the bacterial cells in the linear filament, structures are present which secure the cell-to-cell arrangement. Cross sections of cells reveal a ridge formation which extends from cell to cell with filaments imbedded on the ridges, and this ridge formation is not reported for other bacteria. Using atomic force microscopy (AFM), "strings" along the length of the cable including through the cell-cell junctions were observed, and these strings are imbedded in ridges in the outer membrane (Jiang et al. 2018). Using a combination of several different types of electron microscopy, the architecture of the outer membrane was probed, and fibers of ~50 nm in diameter were observed in the periplasmic region extending across the cell-to-cell junctions to stabilize the cable bacteria filament against fragmentation (Cornelissen et al. 2018). Examination of cell envelopes from different samples revealed 52–62 ridges per cell, while others had 15 or

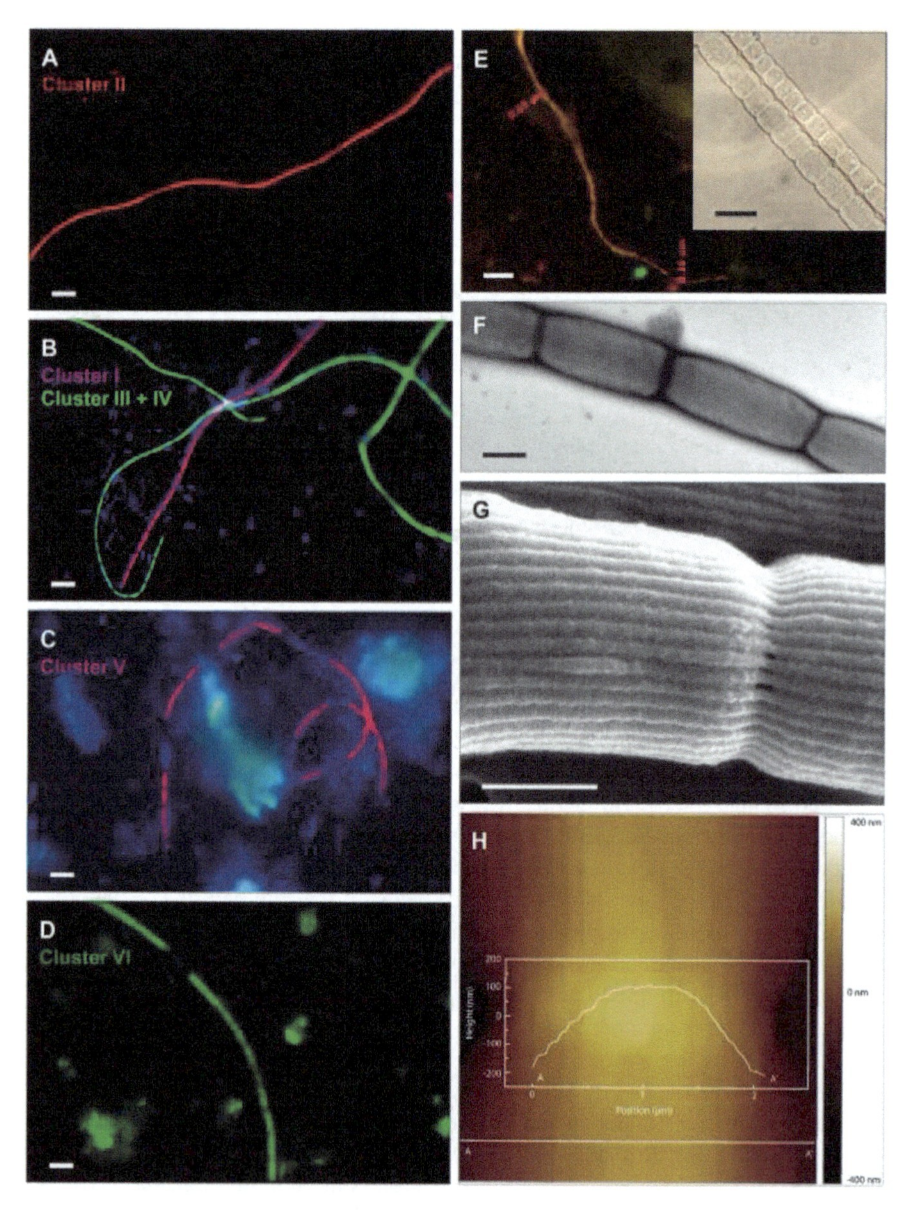

Fig. 2.9 Micrographs of cable bacteria (Trojan et al. 2016). (**a–d**) FISH identification of cable bacteria: (**a**) "*C.* Electrothrix *aarhusiensis*," hybridized with probe EXaa430-CY3 (red); (**b**) "*C.* Electrothrix *marina*," hybridized with probe EXma1271-CY3 (magenta from overlay with DAPI); and "*C.* Electrothrix *communis/japonica*," hybridized with probe EXco1016-FAM (blue-green from overlay with DAPI); (**c**) "*C.* Electronema *nielsenii*," hybridized with probe ENni1437-CY3 (magenta from overlay with DAPI); (**d**) "*C.* Electronema *palustris*," hybridized with probe ENpa1421-FAM (green). Scale bars for panels **a–d**, 5 μm. (**e**) FISH-identified cable bacteria filament (orange from overlay of DSB706-CY3 and EUB338-FAM) showing differences in filament diameter over a short distance. Scale bar, 5 μm. Insert: phase contrast image of two distant, differently sized sections of the same cable bacteria filament. Scale bar, 10 μm (for complete filament image, see Fig. S1). (**f**) TEM micrograph of "*C.* Electronema *nielsenii*" after

16 ridges. A series of highly organized parallel ridge compartments extend along the length of the cell, and the ridge was 205–231 nm thick as measured by focused ion beam scanning electron microscopy (FIB-SEM) or about 100 nm when samples were dried prior to examination by AFM (Cornelissen et al. 2018).

Genome analysis of single filaments and metagenomics of cable bacteria has enabled scientists to understand the physiology and metabolism of these microorganisms. Using metagenomics with an enrichment containing cable bacteria, the genome of MAG Dsb_1MN was predicted to contain 2740 protein coding genes (compared to 2649 genes predicted for *Candidatus* Electronema sp. GS (Kjeldsena et al. 2019), and 38% of these genes in MAG Dsb_1MN coded for uncharacterized proteins (Müller et al. 2020). This cable bacterium contains genes for dissimilatory sulfate reduction, thiosulfate reductase and sulfide-quinone reductase, 4 different types of hydrogenases, CO_2 fixation genes by the Wood-Ljungdahl pathway, genes for nitrate and nitrite reductase, and 15 different genes for *c*-type cytochromes including a gene for cytochrome *bd* oxidase. MAG Dsb_1MN lacked a complete TCA cycle with the absence of genes for succinate synthase and fumarate reductase (Müller et al. 2020).

Draft genomes of Ca. Electrothrix and Ca. Electronema were assembled from individual filaments. While about half of these genes are found only in cable bacteria, 781 of the 806 core gene families of the *Desulfobulbaceae* were detected, and 197 gene families were unique and found only in cable bacteria (Kjeldsena et al. 2019). Missing in these draft genome sequences were genes encoding for flagella, enolase, subunits DsrOP of the DsrKMJOP complex, and NADH-quinone oxidoreductase. Since the cable bacteria contained genes for the dissimilatory sulfate reduction pathway but not genes for the SOX pathway (reverse-type dissimilatory sulfite reductase or flavocytochrome *c* sulfide dehydrogenase), it is proposed that dissimilatory sulfate-reducing bacteria were the ancestors for cable bacteria. It is proposed that type III SQR (sulfide:quinone oxidoreductase) which is encoded in the cable bacteria genes is responsible for the oxidation of sulfide to sulfur, and a reaction between sulfide and sulfur results in the formation of polysulfide. The polysulfide reductase (PSR) in the plasma membrane of cable bacteria would acquire electrons from the quinone pool to produce sulfide with concomitant pumping of protons across the membrane to conserve energy. PSR combined with sulfur oxidation could result in sulfur disproportionation and enable cells in the middle of the cable to obtain energy when there is no electron acceptor for the cable (Kjeldsena et al. 2019).

Fig. 2.9 (continued) phosphotungstic acid staining to visualize the longitudinal ridges. Scale bar, 1 μm. (**g**) SEM micrograph of gold-sputtered "*C.* Electronema *nielsenii*," showing that the longitudinal ridges are continuous over a cell-cell junction. Scale bar, 0.5 μm. (**h**) AFM height image of an unfixed, hydrated "*C.* Electronema *nielsenii*." The height profile along line A–A′ is overlaid onto the image, illustrating the 3D structure of the ridges. (*Copyright*: © 2016 The Author(s). Published by Elsevier GmbH. This is an open-access article under the CCBY-NC-ND license (http://creativecommons.org/licenses/by-nc-nd/4.0/))

Energetics of cable bacteria is an important topic which is being addressed in several research programs. Growth of cells in cable bacteria by the use of energy conservation of sulfur disproportionation has been proposed for groundwater cable bacteria (Müller et al. 2020). Thus, a "cryptic sulfur cycle" may provide energy for sulfate reduction by internal cells in the cable. The proposal that cells in the middle of the cable have the potential for growth (Kjeldsena et al. 2019; Müller et al. 2020) is contrasted by the suggestion that there is a significant division of labor in the cable with cell growth in the sulfidic zone and is attributed to energy obtained only from sulfide oxidation (Geerlings et al. 2020). The transport of electrons from the anoxic zone to the aerobic region is proposed to be mediated by periplasmic cytochromes, and the primary function of this process may be for detoxification of accumulated reducing equivalents (Geerlings et al. 2020). While discharging electrons to O_2, cable bacteria could acquire energy from utilization of polyphosphate and polyglucose granules because both of these storage materials have been demonstrated in cable bacteria and genes encoding for enzymes for their utilization have been identified in their genomes (Kjeldsena et al. 2019).

The growth of cable bacteria requires both sulfidogenic and aerobic (or nitrate) zones, and metabolic activities impact the geochemistry of sediments. Acidification of the nano-environment adjacent to cable bacteria has been proposed to account for increased concentrations of Ca, Fe, and Mn in pore water (Geelhoed et al. 2020). Acidification of the environment converts iron sulfide deposits to H_2S which is a positive feedback in providing additional sulfide for the cable bacteria (Seitaj et al. 2015). The conversion of iron monosulfides to iron oxides in the Gulf of Finland has contributed to the activity of cable bacteria (Hermans et al. 2019). Cable bacteria migrate in the aquatic environment by gliding mobility without the presence of flagella, and at this time, only three chemotaxis operons have been identified in cable bacteria genomes (Kjeldsena et al. 2019). Genes for polysaccharide export have been identified in cable bacteria and are proposed to contribute to excretion-based gliding movement (Kjeldsena et al. 2019). Movement of filaments involves the formation of loops which are often twisted, and this suggests a helical rotation of filaments (Bjerg et al. 2016).

2.7 DSR, APS, and Lateral Gene Transfer

The transfer of genes between microorganisms accounts for increased metabolic versatility for microorganisms, and there are numerous examples where lateral gene transfer (LGR) involving SRB has resulted in new physiological characteristics for the recipient bacteria. Dissimilatory (bi)sulfite reductase (DSR) is found in all bacteria and archaea that grow with sulfate as the electron acceptor. DSR consists of two dissimilar polypeptides with a $\alpha_2\beta_2$ structure and may have resulted from gene duplication of an early gene (Dahl et al. 1993). Adjacent genes encoding DSR (*dsrAB*) are highly conserved throughout the microbial world. The presence of a ferredoxin-like domain is present in all sequenced DSR, and the location of this domain is identical in all DSR thus far examined. A single form of DSR would suggest its presence in cells prior to divergence of *Bacteria* and *Archaea*. It remains to be explained why the DSR is currently found in a small number of taxonomic groups of *Bacteria* and *Archaea*. Genes for DSR could have been lost through microbial evolution, or the natural diversity of DSR genes has yet to be isolated (Wagner et al. 1998). Based on a comparison of sequences of a large number of proteins, it is considered that the domains diverged about 3.1 to 3.6 billion years ago (Feng et al. 1997), and this compares to isotope analysis that supports 2.8 to 3.1 billion years ago as a starting date for dissimilatory sulfate reduction (Schidlowski et al. 1983; Shen et al. 2001). In general, there was high similarity between 16S rRNA and DsrAB (amino acids encoded by *dsrAB*) across distinct microbial lineages (Wagner et al. 1998); however, a difference was found when comparing DsrAB and 16S rRNA analysis of several SRB species (Klein et al. 2001). While *Thermodesulfobacterium commune* and *Thermodesulfobacterium mobile* are members of a separate division by 16S rRNA sequence analysis, these two bacteria are grouped within the *Deltaproteobacteria* when using DsrAB analysis. With 16S rRNA sequence analysis, all species of *Desulfotomaculum* are aligned with the low G + C Gram-positive bacteria, but with DsrAB analysis, three species (*aeronauticum*, *ruminis*, and *putei*) group with *Desulfosporosinus orientis*. Based on DsrAB analysis, several species of *Desulfotomaculum* (*acetoxidans*, *thermoacetoxidans*, *thermobenzoicum*, *thermocisternum*, *kuznetsovii*, *thermosapovorans*) and *geothermicum* group with *Desulfobacula toluolica* within the *Deltaproteobacteria* group (Klein et al. 2001). This difference in classification would reflect LGT of *dsrAB*. The distribution of dissimilatory sulfate reduction genes throughout prokaryotes is now attributed to both vertical inheritance and lateral (horizontal) gene transfer (Klein et al. 2001; Zverlov et al. 2005). A review of primers for detection of *dsrAB* genes in the environment reveals that many *dsrAB* may be considered hidden because the sequences do not align with known organisms (Müller et al. 2015). In expanding on this limitation of detecting *dsrAB* in the environment, new probes were used, and the number of sulfate–/sulfite-respiring bacteria and archaea detected was found to represent members of 13 phyla which include 8 candidate phyla (Anantharaman et al. 2018).

An examination of insertions and deletions in the amino acid sequences of DSR was examined to assess lateral gene transfer. Specific to organisms of the *Deltaproteobacteria* is the presence of one insertion sequence in the α-subunit and two insertions in the β-subunit. As discussed earlier, these insertions were found in the seven *Desulfotomaculum* species and the two *Thermodesulfobacterium* species, and this identification of inserts is used to support the proposal that these bacteria acquired *dsrAB* from an unidentified ancestor of the *Deltaproteobacteria* (Klein et al. 2001). There is a considerable evidence to support the proposal that DSR genes in *Archaeoglobus* sp. have also been acquired by lateral transfer from Gram-positive bacteria (Hipp et al. 1997; Klenk et al. 1997; Wagner et al. 1998).

Adenosine 5’-phosphosulfate (APS) reductase converts APS to sulfite plus adenosine monophosphate (AMP) and is encoded on the genes *aprBA* (also known as *apsBA*). The APS reductase has a heterodimer $\alpha\beta$ structure which is highly conserved throughout the SRP, and a single copy of the *apsA* gene occurs in each cell (Fritz et al. 2000). The proposition that *apsBA* (also known as *aprBA*) genes have been transferred from bacteria to *Archaeoglobus* sp. remains under discussion (Hipp et al. 1997). Comparison of ApsA with 16S rRNA of SRB indicated multiple exchanges of *apsA* between SRP (Friedrich 2002). The LGT of *apsAA* and *dsrAB* genes involves thermophilic bacteria and transfer of apsA genes from Gram-positive SRB to bacteria belonging to *Nitrospinaceae* and *Syntrophobacteraceae* has been reported (Friedrich 2002). The involvement of Gram-positive bacteria with LTR may reflect the increased survival of these bacteria as they adjust to oxygen and changing water levels. Since many of the SRB can grow by fermentation of carbon sources, recipient bacteria have a period of adjustment of gene organization and regulation before a fully functional sulfate-respiring system supports growth. There is a marked similarity between the ApsA (also known as AprA)-based tree and the rRNA-based tree for members of the *Desulfovibrionales*, *Desulfobacteraceae*, and *Desulfobulbaceae* (Friedrich 2002). The tree topologies for AprB and AprA were identical which indicates these genes have a shared evolutionary development (Hattori et al. 2002; Meyer and Kuever 2007). It has been suggested that the genes for dissimilatory sulfate reduction could move by a genomic island (Friedrich 2002; Mussmann et al. 2005); however, evidence for such a mobilizing of genes for LGT was not apparent in a report by Meyer and Kuever (2007).

There are several reports detailing the LGT of genes to dissimilatory sulfate-reducing prokaryotes. The transfer of *aprBA* from a *Thermodesulfovibrio* species to *Thermacetogenium phaeum* resulted from co-cultivation of these two microorganisms (Meyer and Kuever 2007). Genome analysis of *Desulfonema ishimotonii* Tokyo 01 T suggests that there were multiple gene transfers to this bacterium from gliding bacteria (Watanabe et al. 2019). Genes unique to cable bacteria may have been acquired by LGT (Kjeldsena et al. 2019). Genomic analysis of anaerobic methane-oxidizing archaea indicates that a number of respiratory strategies unique to these members were acquired via LGT involving bacteria and archaea (Leu et al. 2020).

2.8 Development of Genetic Manipulations

While the dissimilatory sulfate reducers are anaerobic and this presents some challenge in their growth, this problem was quickly overcome with improved cultivation techniques, and members of the *Desulfovibrio* species became models for genetic evaluations. A historical review of the use of plasmids for genetic experiments and early reports of conjugation in *Desulfovibrio* species focused on *Desulfovibrio* spp. are provided by Wall et al. (1993).

The sensitivity of *D. vulgaris* Hildenborough, *D. desulfuricans* ATCC 27774, and *D. fructosovorans* to ampicillin, streptomycin, kanamycin, tetracycline, rifampicin, spectinomycin, gentamycin, and nalidixic acid varied with the species, but all those strains were sensitive to chloramphenicol (Wall 1993). The high level of antibiotic resistance by *Desulfovibrio* greatly complicated the use of selective markers for genetic studies. Initially, the target for transfer of broad-host range of plasmids in *Desulfovibrio* focused on redox-active metalloproteins (van Dongen 1995), and the first vector transfers to *D. vulgaris* Hildenborough, *Desulfovibrio* sp. Holland SH-1, and *D. desulfuricans* Norway 4 were reported (Powell et al. 1989; van den Berg et al. 1989). Shortly thereafter, plasmid transfer in *D fructosovorans* and *D. desulfuricans* G200 was reported (Voordouw et al. 1990a, b, c; Rousset et al. 1991). With the cloning and sequencing of genes from several species of *Desulfovibrio* (Voordouw 1993), the transfer of these genes by conjugation and transformation was developed. Some of the first genes cloned from *Desulfovibrio* species are given in Table 2.7, and this enabled taxonomic evaluation of the various bacterial cultures. In a review by Wall et al. (2003), a series of cloning vectors commonly used in *Desulfovibrio* studies are discussed as well as the construction of mutants and potential use of reporter genes. Over the years, there have been a considerable number of advances in genetics as applied to SRB, and these have been reviewed by Wall et al. (2008). The most extensive genetic studies have been with *D. vulgaris* Hildenborough, and a review of genetics applied to understand energy conservation and pathways of electron flow was published by Keller and Wall (2011). Details of constructing marker-exchange deletions, site-directed mutations, in-frame unmarked deletions, and transposon mutant libraries are reviewed by Rabus et al. (2015).

2.9 Perspective

There continues to be a large number of bacteria isolated from the environment with the metabolic characteristic of dissimilatory sulfate reduction. When considering the taxonomic array of bacteria and archaea, the use of sulfate as a final electron acceptor is characteristic of only a few phyla. It is unclear why a limited number of taxonomic groups use sulfate as a final electron acceptor while bacteria and archaea of many phyla couple growth to respiratory nitrate reduction. Perhaps environmental factors limit successful horizontal gene transfer for dissimilatory sulfate reduction. The SRB most commonly cultivated from the environment include members of the genus *Desulfovibrio*, and this may reflect their tolerance to atmospheric O_2. With improved techniques for cultivation of anaerobic sulfate reducers and the sampling from remote or unique environments, a number of new isolates have provided an interesting array of bacteria which reveal the abundance of sulfate reducers. An interesting observation is the large number of uncultivated SRB found in environmental studies because these uncultivated microorganisms have unique gene sequences that distinguish them from the cultivated microorganisms. Clearly, the taxonomy of SRP is a fluid process.

References

Abreu F, Cantao ME, Nicolas MF, Barcellos FG, Morillo V, Almeida LGP et al (2011) Common ancestry of iron oxide- and iron-sulfide-based biomineralization in magnetotactic bacteria. ISME J 5:1634–1640

Abreu F, Morillo V, Nascimento FF, Werneck C, Cantao ME, Ciapina LP et al (2014) Deciphering unusual uncultured magnetotactic multicellular prokaryotes through genomics. ISME J 8:1055–1068

Adams ME, Postgate JR (1959) A new sulphate-reducing vibrio. J Gen Microbiol 20:252–257

Akagi JM, Adams V (1973) Isolation of a bisulfite reductase activity from *Desulfotomaculum nigrificans* and its identification as the carbon monoxide-binding pigment P582. J Bacteriol 116: 372–396

Alawi M, Lerm S, Vetter A, Wolfgramm M, Seibt A, Würdemann H (2011) Diversity of sulfate-reducing bacteria in a plant using deep geothermal energy. Grundwasser. https://doi.org/10.1007/s00767-011-0164-y

Anantharaman K, Hausmann B, Jungbluth SP, Kantor RS, Lavy A, Warren LA et al (2018) Expanded diversity of microbial groups that shape the dissimilatory sulfur cycle. ISME J 12: 1715–1728

Argyle JL, Rapp-Giles BJ, Wall JD (1992) Plasmid transfer by conjugation in *Desulfovibrio desulfuricans*. FEMS Microbiol Lett 94:255–262

Audiffrin C, Cayol JL, Joulian C, Casalot L, Thomas P, Garcia JL, Ollivier B (2003) *Desulfonauticus submarinus* gen. nov., sp. nov., a novel sulfate-reducing bacterium isolated from a deep-sea hydrothermal vent. Int J Syst Evol Microbiol 53:1585–1590

Aüllo T, Ranchou-Peyruse A, Ollivier B, Magot M (2013) *Desulfotomaculum* spp. and related gram-positive sulfate-reducing bacteria in deep subsurface environments. Front Microbiol. https://doi.org/10.3389/fmicb.2013.00362

Balci N, Vardar N, Yelboga E, Karaguler N (2012) Bacterial community composition of sediments from artificial Lake Maslak. Environ Monitor Assess, Istanbul. https://doi.org/10.1007/s10661-011-2368-0

Barton LL, Carpenter CM (2013) Suitability of fluorescence measurements to quantify sulfate-reducing bacteria. J Microbiol Meth. 93:192–197

Barton LL, Fauque GD (2009) Biochemistry, physiology and biotechnology of sulfate-reducing bacteria. Adv Appl Microbiol 68:41–98

Baumgartner LK, Reid RP, Dupraz C, Decho AW, Buckley DH, Spear JR et al (2006) Sulfate-reducing bacteria in microbial mats: changing paradigms, new discoveries. Sed Geol 185:131–145

Baumgartner J, Menguy N, Gonzalez TP, Morin G, Widdrat M, Faivre D (2016) Elongated magnetite nanoparticle formation from a solid ferrous precursor in a magnetotactic bacterium. J R Soc Interface 13(124):20160665. https://doi.org/10.1098/rsif.2016.0665. PMID: 27881802 Free PMC article

Beeder J, Torsvik T, Lien T (1995) *Thermodesulforhabdus norvegicus* gen. nov., sp. nov., a novel thermophilic sulfate-reducing bacterium from oil field water. Arch Microbiol 164:331–336

Bhatnagar S, Badger JH, Madupu R, Khouri HM, O'Connor EM, Robb FT, Ward NL, Eisen JA (2015) Genome sequence of the sulfate-reducing thermophilic bacterium *Thermodesulfovibrio yellowstonii* strain DSM 11347T (phylum *Nitrospirae*). Genome Announc 3(1):e01489–e01414. https://doi.org/10.1128/genomeA.01489-14

Birkeland N-L, Schönheit P, Poghosyan L, Fiebig A, Klenk H-P (2017) Complete genome sequence analysis of *Archaeoglobus fulgidus* strain 7324 (DSM 8774), a hyperthermophilic archaeal sulfate reducer from a North Sea oil field. Stand Gen Sci. 12(1):79. https://doi.org/10.1186/s40793-017-0296-5

Bjerg JT, Damgaard LR, Holm SA, Schramm A, Nielsen LP (2016) Motility of electric cable bacteria. Appl Environ Microbiol 82:3816–3821

Bjerg JT, Boschker HTC, Larsen S, Berry D, Schmid M, Millo D et al (2018) Long-distance electron transport in individual, living cable bacteria. PNAS, USA. 115:5786–5791

Boon JJ, de Leeuw JW, Hoek GJD, Vosjan JH (1977) Significance and taxonomic value of iso and anteiso monoenoic fatty acids and branched β-hydroxy acids in *Desulfovibrio desulfuricans*. J Bacteriol 129:1183–1191

Brumlik MJ, Leroy G, Bruschi M, Voordouw G (1990) The nucleotide sequence of the *Desulfovibrio gigas* desulforedoxin gene indicates that the *Desulfovibrio vulgaris rbo* gene originated from a gene fusion event. J Bacteriol 172:7289–7292

Brysch K, Schneider C, Fuchs G, Widdel F (1987) Lithoautotrophic growth of sulfate-reducing bacteria, and description of *Desulfobacterium autotrophicum* gen. nov., sp. nov. Arch Microbiol 148:264–274

Burggraf S, Jannasch HW, Nicolaus B, Stetter KO (1990) Archaeoglobus profundus sp. nov., represents a new species within the sulfate-reducing archaebacteria. Sys Appl Microbiol 13:24–28

Busse H-J, Denner EBM, Lubitz W (1996) Classification and identification of bacteria: current approaches to an old problem. Overview of methods used in bacterial systematics. J Biotechnol 47:3–38

Byrne ME, Ball DA, Guerquin-Kern J-L, Rouiller I, Wu T-D, Downing KH, Vali H, Komeili A (2010) *Desulfovibrio magneticus* RS-1 contains an iron- and phosphorus-rich organelle distinct from its bullet-shaped magnetosomes. Proc Natl Acad Sci. USA. 107:12263–12268

Campbell LL Jr, Frank HA, Hall ER (1957) Studies on the thermophilic sulfate reducing bacteria. I identification of *Sporovibrio desulfuricans* as *Clostridium nigrificans*. J Bacteriol 73:516–521

Campbell LL, Postgate JR (1965) Classification of the spore-forming sulfate-reducing bacteria. Bacteriol Rev 29:359–363

Cano RJ, Borucki MK (1995) Revival and identification of bacterial spores in 25-to 40-million-year-old Dominican amber. Science 268:1060–1064

Cao J, Gayet N, Zeng X, Shao Z, Jebbar M, Alain K (2016a) *Pseudodesulfovibrio indicus* gen. nov., sp. nov., a piezophilic sulfate-reducing bacterium from the Indian Ocean and reclassification of four species of the genus *Desulfovibrio*. Int J Syst Evol Microbiol 66:3904–3911

Cao J, Maignien L, Shao Z, Alain K, Jebbar M (2016b) Genome sequence of the piezophilic, mesophilic sulfate-reducing bacterium *Desulfovibrio indicus* J2^{T}. Genome Announc 4(2): e00214–e00216. https://doi.org/10.1128/genomeA.00214-16

Carlson HK, Iavarone AT, Gorur A, Yeo BS, Tran R, Melnyk RA et al (2012) Surface multiheme *c*-type cytochromes from *Thermincola potens* and implications for respiratory metal reduction by Gram-positive bacteria. Proc Natl Acad Sci U S A 109:1702–1707

Carr MC, Curley GP, Mayhew SG, Voordouw G (1990) Effects of substituting asparagine for glycine-61 in flavodoxin from *Desulfovibrio vulgaris* (Hildenborough). Biochem Int 20:1025–1032

Castro HF, Williams NH, Ogram A (2000) Phylogeny of sulfate-reducing bacteria. FEMS Microbiol Ecol 31:1–9

Chen YR, Zhang R, Du HJ, Pan HM, Zhang WY, Zhou K, Li JH, Xiao T, Wu LF (2015) A novel species of ellipsoidal multicellular magnetotactic prokaryotes from Lake Yuehu in China. Environ Microbiol 17:637–647

Chen YR, Zhang WY, Zhou K, Pan HM, Du HJ, Xu C et al (2016) Novel species and expanded distribution of ellipsoidal multicellular magnetotactic prokaryotes. Environ Microbiol Rep 8: 218–226

Chivian D, Brodie E, Alm E, Culley D, Dehal P, DeSantis T et al (2008) Environmental genomics reveals a single-species ecosystem deep within Earth. Science 322:275–278

Collins MD, Widdel F (1986) Respiratory quinones of sulphate-reducing and sulphur-reducing bacteria: a systematic investigation. Syst Appl Microbiol 8:8–18

Copeland A, Spring S, Göker M, Schneider S, Lapidus A, Del Rio TG et al (2009) Complete genome sequence of *Desulfomicrobium baculatum* type strain (XT). Stand Genomic Sci 1:29–37

Cord-Ruwisch R, Garcia JL (1985) Isolation and characterization of an anaerobic benzoate-degrading spore-forming sulfate-reducing bacterium, *Desulfotomaculum sapomandens* sp. nov. FEMS Microbiol Lett 29:325–330

Cornelissen R, Bøggild A, Thiruvallur Eachambadi R, Koning RI, Kremer A, Hidalgo-Martinez S et al (2018) The cell envelope structure of cable bacteria. Front Microbiol 9:3044. https://doi.org/10.3389/fmicb.2018.03044

Curley GP, Voordouw G (1988) Cloning and sequencing of the gene encoding flavodoxin from *Desulfovibrio vulgaris* Hildenborough. FEMS Microbiol Lett 49:295–299

da Silva SM, Pacheco I, Pereira IAC (2012) Electron transfer between periplasmic formate dehydrogenase and cytochromes *c* in *Desulfovibrio desulfuricans* ATCC 27774. J Biol Inorg Chem 17:831–838

Dahl C, Kredich NM, Deutzmann R, Trüper HG (1993) Dissimilatory sulfite reductase from *Archaeoglobus fulgidus*: physico-chemical properties of the enzyme and cloning, sequencing and analysis of the reductase gene. J Gen Microbiol 139:1817–1828

Daumas S, Cord-Ruwisch R, Garcia JL (1988) *Desulfotomaculum geothermicum* sp. nov., a thermophilic, fatty acid-degrading, sulfate-reducing bacterium isolated with H_2 from geothermal ground water. Antonie Van Leeuwenhoek 54:165–178

De Rezende JR, Kjeldsen KU, Hubert CR, Finster K, Loy A, Jørgensen BB (2013) Dispersal of thermophilic *Desulfotomaculum* endospores into Baltic Sea sediments over thousands of years. ISME J 7:72–84. https://doi.org/10.1038/ismej.2012.83

de Rezende JR, Hubert CRJ, Røy H, Kjeldsen KU, Jørgensen BB (2017) Estimating the abundance of endospores of sulfate-reducing bacteria in environmental samples by inducing germination and exponential growth. Geomicrobiol J 34:338–345

Deckers HM, Wilson FR, Voordouw G (1990) Cloning and sequencing of a [NiFe] hydrogenase operon from *Desulfovibrio vulgaris* Miyazaki F. J Gen Microbiol 136:2021–2028

Deng X, Dohmae N, Nealson KH, Hashimoto K, Okamoto A (2018) Multi-heme cytochromes provide a pathway for survival in energy-limited environments. Sci Adv 2018(4):eaao5682

Descamps ECT, Monteil CL, Menguy N, Ginet N, Pignol D, Bazylinski DA, Lefèvre CT (2017) *Desulfamplus magnetovallimortis* gen. nov., sp. nov., a magnetotactic bacterium from a brackish desert spring able to biomineralize greigite and magnetite, that represents a novel lineage in the *Desulfobacteraceae*. Syst Appl Microbiol 40:280–289

Devereux R, Stahl DA (1993) Phylogeny of sulfate-reducing bacteria and a perspective for analyzing their natural communities. In: Odom JM, Singleton R Jr (eds) The Sulfate-reducing bacteria: contemporary perspectives. Brock/Springer Series in Contemporary Bioscience. Springer, New York, pp 131–160

DeWeerd K, Mandelco L, Tanner RS, Woese CR, Suflita JM (1990) *Desulfomonile tiedjei* gen. nov. and sp. nov., a novel anaerobic dehalogenating, sulfate-reducing bacterium. Arch Microbiol 154:22–30

Dolla A, Fu R, Brumlik MJ, Voordouw G (1992) Nucleotide sequence of dcrA, a *Desulfovibrio vulgaris* Hildenborough chemoreceptor gene, and its expression in *Escherichia coli*. J Bacteriol 174:1726–1733

Dörries M, Wöhlbrand L, Kube M, Reinhardt R, Rabus R (2016) Genome and catabolic subproteomes of the marine, nutritionally versatile, sulfate-reducing bacterium *Desulfococcus multivorans* DSM 2059. BMC Genomics 17:918. https://doi.org/10.1186/s12864-016-3236-7

Dowling NJE, Widdel F, White DC (1986) Phospholipid ester-linked fatty acid biomarkers of acetate-oxidizing sulphate-reducers and other sulphide-forming bacteria. J Gen Microbiol 132: 1815–1825

Dzierżewicz Z, Cwalina B, Kurkiewicz S, Chodurek E, Wilczok T (1996) Intraspecies variability of cellular fatty acids among soil and intestinal strains of *Desulfovibrio desulfuricans*. Appl Environ Microbiol 62:3360–3365

Easterbrook KB, Coombs RW (1976) Spinin: the subunit protein of bacterial spinae. Can J Microbiol 22:438–440

Easterbrook KB, Willison JHM, Coombs RW (1976) Arrangement of morphological subunits in bacterial spinae. Can J Microbiol 22:619–629

Elvert M, Boetius A, Knittel K, Jørgensen BB (2003) Characterization of specific membrane fatty acids as chemotaxonomic markers for sulfate-reducing bacteria involved in anaerobic oxidation of methane. Geomicrobiol J 20:403–419

Enning D, Garrelfs J (2014) Corrosion of iron by sulfate-reducing bacteria: new views of an old problem. Appl Environ Microbiol 80:1226–1236

Esnault G, Caumeette P, Garcia J-L (1988) Characterization of *Desulfovibrio giganteus* sp. nov., a sulfate-reducing bacterium isolated from a Brackish Coastal Lagoon System. Appl Microbiol 10:147–151

Fan F, Zhang B, Morrill PL, Husain T (2017) Profiling of sulfate-reducing bacteria in an offshore oil reservoir using phospholipid fatty acid (PLFA) biomarkers. Water Air Soil Poll 228:410. https://doi.org/10.1007/s11270-017-3595-y

Fareleira P, LeGall J, Xavier AV, Santos H (1997) Pathways for utilization of carbon reserves in *Desulfovibrio gigas* under fermentative and respiratory conditions. J Bacteriol 179:3972–3980

Fauque G, Lino AR, Czechowski M, Kang L, DerVartanian DV, Moura JJG, LeGall J, Moura I (1990) Purification and characterization of bisulfite reductase (desulfofuscidin) from *Desulfovibrio thermophilus* and its complexes with exogenous ligands. Biochim Biophys Acta 1040:112–118

Fauque G, LeGall J, Barton LL (1991) Sulfate-reducing and sulfur-reducing bacteria. In: Shively JM, Barton LL (eds) Variations in autotrophic life. Academic Press Limited, London, pp 271–337

Feng D-F, Cho G, Doolittle RF (1997) Determining divergence times with a protein clock: update and reevaluation. PNAS, USA. 94:13028–13033

Finneran KT, Forbush HM, VanPraagh CVG, Lovley DR (2002) *Desulfitobacterium metallireducens* sp nov., an anaerobic bacterium that couples growth to the reduction of metals and humic acids as well as chlorinated compounds. Int J Syst Evol Microbiol 52:1929–1935

Friedrich MW (2002) Phylogenetic analysis reveals multiple lateral transfers of adenosine-5-′-phosphosulfate reductase genes among sulfate-reducing microorganisms. J Bacteriol 184: 278–289

Friedrich M, Springer NM, Ludwig W, Schnink B (1996) Phylogenetic positions of *Desulfofustis glycolicus* gen. nov., sp. nov. and *Syntrophobotulus glycolicus* gen. nov., sp. nov., two new strict anaerobes growing with glycolic acid. J Syst Bacteriol 46:1065–1069

Fritz G, Buchert T, Huber H, Stetter KO, Kroneck PMH (2000) Adenylylsulfate reductases from archaea and bacteria are 1:1 alpha beta-heterodimeric iron-sulfur flavoenzymes—high similarity of molecular properties emphasizes their central role in sulfur metabolism. FEBS Lett 473:63–66

Frolov EN, Kublanov IV, Toshchakov SV, Samarov NI, Novikov AA, Lebedinsky AV et al (2017) *Thermodesulfobium acidiphilum* sp. nov., a thermoacidophilic, sulfate-reducing, chemoautotrophic bacterium from a thermal site. Int J Syst Evol Microbiol 67:1482–1485

Fukui M, Teske A, Assmus B, Muyzer G, Widdel F (1999) Physiology, phylogenetic relationships, and ecology of filamentous sulfate-reducing bacteria (genus desulfonema). Arch Microbiol 172: 193–203

Geelhoed JS, van de Velde SJ, Meysman FJR (2020) Quantification of cable bacteria in marine sediments via qPCR. Front Microbiol 11:1506. https://doi.org/10.3389/fmicb.2020.01506

Geerlings NMJ, Karman C, Trashin S, As KS, Kienhuis MVM, Hidalgo-Martinez S et al (2020) Division of labour and growth during electrical cooperation in multicellular cable bacteria. PNAS USA. https://doi.org/10.1073/pnas.1916244117

Göker M, Teshima H, Lapidus A, Nolan M, Lucas S, Hammon N et al (2011) Complete genome sequence of the acetate-degrading sulfate reducer *Desulfobacca acetoxidans* type strain ($ASRB2^T$). Stand Genomic Sci 4:393–401. https://doi.org/10.4056/sigs.2064705

Gorby YA, Beveridge TJ, Blakemore RP (1988) Characterization of the bacterial magnetosome membrane. J Bacteriol 170:834–841

Goris J, Konstantinidis KT, Klappenbach JA, Coenye T, Vandamme P, Tiedje JM (2007) DNA-DNA hybridization values and their relationship to whole-genome sequence similarities. Int J Syst Evol Microbiol 57:81–91

Gutiérrez Acosta OB, Schleheck D, Schink B (2014) Acetone utilization by sulfate-reducing bacteria: draft genome sequence of *Desulfococcus biacutus* and a proteomic survey of acetone-inducible proteins. BMC Genomics 15:584. http://www.biomedcentral.com/1471-2164/15/584

Hai T, Lange D, Rabus R, Steinbuchel A (2004) Polyhydroxyalkanoate (PHA) accumulation in sulfate-reducing bacteria and identification of a class III PHA synthase (PhaEC) in *Desulfococcus multivorans*. Appl Environ Microbiol 70:4440–4448

Han C, Mwirichia R, Chertkov O, Held B, Lapidus A, Nolan M et al (2011) Complete genome sequence of *Syntrophobotulus glycolicus* type strain (FlGlyRT). Stand Gen Sci 4:371–380

Haouari O, Fardeau M-L, Cayol J-L, Casiot C, Elbaz-Poulichet F, Hamdi M, Joseph M, Ollivier B (2008) *Desulfotomaculum hydrothermale* sp. nov., a thermophilic sulfate-reducing bacterium isolated from a terrestrial Tunisian hot spring. Int J Syst Evol Microbiol 58:2529–2535

Harmsen HJM, Van Kuijk BLM, Plugge DM, Akkermans ADL, De Vos WM, Stams AJM (1998) *Syntrophobacter furnaroxidans* sp. nov., a syntrophic propionate-degrading sulfate-reducing bacterium. Int J Syst Bacteriol 48:1383–1387

Hattori S, Kamagata Y, Hanada S, Shoun H (2002) *Thermacetogenium phaeum* gen. nov., sp. nov., a strictly anaerobic, thermophilic, syntrophic acetate-oxidizing bacterium. Int J Syst Evol Microbiol 50:1601–1609

Hauser LJ, Land ML, Brown SD, Larimer F, Keller KL, Rapp-Giles BJ et al (2011) Complete genome sequence and updated annotation of *Desulfovibrio alaskensis* G20. J Bacteriol 193: 4268–4269

He Z, Gentry TJ, Schadt CW, Wu L, Liebich J, Chong SC et al (2007) GeoChip: a comprehensive microarray for investigating biogeochemical, ecological and environmental processes. ISME J 1:67–77

Heidelberg JF, Seshadri R, Haveman SA, Hemme CH, Paulsen IT, Kolonay JF, Eisen JA et al (2004) The genome sequence of the anaerobic, sulfate-reducing bacterium *Desulfovibrio vulgaris* Hildenborough: consequences for its energy metabolism and reductive metal bioremediation. Nature Biotech 22:554–559

Hemmi H, Takahashi Y, Shibuya K, Nakayama T, Nishino T (2005) Menaquinone-specific prenyl reductase from the hyperthermophilic Archaeon *Archaeoglobus fulgidus*. J Bacteriol 187:1937–1944

Hensgens CHM, Santos H, Zhang C, Kruizinga WH, Hansen TA (1996) Electron-dense granules in *Desulfovibrio gigas* do not consist of inorganic triphosphate but of a glucose pentakis (diphosphate). Eur J Biochem 242:327–331

Hermans M, Lenstra WK, Hidalgo-Martinez S, van Helmond NAGM, Witbaard R, Meysman FJR, Gonzalez S, Slomp CP (2019) Abundance and biogeochemical impact of cable bacteria in Baltic Sea sediments. Environ Sci Technol 53:7494–7503

Hipp WM, Pott AS, Thum-Schmitz N, Faath I, Dahl C, Trüper HG (1997) Towards the phylogeny of APS reductases and sirohaem sulfite reductases in sulfate-reducing and sulfur-oxidizing prokaryotes. Microbiology (Reading) 143:2891–2902

Huber H, Stetter KO (2001) Order I Archaeoglobales. In: Brenner DJ, Krieg NR, Staley ST (eds) Bergey's mannual of systematic bacteriology, 2nd ed, vol 1, The archaea and deeply branching and phototrophic bacteria. Springer, New York, pp 349–352

Iizuka H, Okazaki H, Seto N (1969) A new sulfate-reducing bacterium. J Gen Appl Microbiol 15: 11–18

Imachi H, Sekiguchi Y, Kamagata Y, Ohashi A, Harada H (2000) Cultivation and in situ detection of a thermophilic bacterium capable of oxidizing propionate in syntrophic association with hydrogenotrophic methanogens in a thermophilic methanogenic granular sludge. Appl Environ Microbiol 66:3608–3615

Imachi H, Sekiguchi Y, Kamagata Y, Hanada S, Ohashi A, Harada H (2002) *Pelotomaculum thermopropionicum* gen. nov. sp. nov., an anaerobic, thermophilic, syntrophic propionate-oxidizing bacterium. Int J Syst Evol Microbiol 52:1729–1735

Imachi H, Sekiguchi Y, Kamagata Y, Loy A, Qiu Y-L, Hugenholtz P et al (2006) Non-sulfate-reducing, syntrophic bacteria affiliated with *Desulfotomaculum* Cluster I are widely distributed in methanogenic environments. Appl Environ Microbiol 72:2080–2091

Isaksen MF, Jørgensen BB (1996) Adaptation of psychrophilic and psychrotrophic sulfate-reducing bacteria to permanently cold marine environments. Appl Environ Microbiol 62:408–414

Isaksen MF, Teske A (1996) *Desulforhopalus vacuolatus* gen. nov., sp. nov., a new moderately psychrophilic sulfate-reducing bacterium with gas vacuoles isolated from a temperate estuary. Arch Microbiol 166:60–168

Ishimoto M, Koyama J, Nagai Y (1954) A cytochrome and a green pigment of sulfate-reducing bacteria. Bull Chem Soc Jpn 27:564–565

Itävaara M, Nyyssönen M, Kapanen A, Nousiainen A, Ahonen L, Kukkonen I (2011) Characterization of bacterial diversity to a depth of 1500m in the Outokumpu deep borehole, Fennoscandian Shield. FEMS Microbiol Ecol. https://doi.org/10.1111/j.1574-6941.2011.01111.x

Ito T, Nielsen JL, Okabe S, Watanabe Y, Nielsen PH (2002) Phylogenetic identification and substrate uptake patterns of sulfate-reducing bacteria inhabiting an oxic-anoxic sewer biofilm determined by combining microautoradiography and fluorescent in situ hybridization. Appl Environ Microbiol 68:356–364

Itoh T, Suzuki K-I, Nakase T (1998) *Therrnocladiurn rnodestius* gen. nov., sp. nov., a new genus of rod-shaped, extremely thermophilic crenarchaeote. Int J Syst Bacteriol 48:879–887

Itoh T, Suzuki K-i, Sanchez PC, Nakase T (1999) *Caldivirga maquilingensis* gen. nov., sp. nov., a new genus of rod-shaped crenarchaeote isolated from a hot spring in the Philippines. Int J Syst Bacteriol 49:1157–1163

Jiang L, Cai C, Zhang Y, Mao SY, Sun YG, Li K, Lei Xiang L, Zhang CM (2012) Lipids of sulfate-reducing bacteria and sulfur-oxidizing bacteria found in the Dongsheng uranium deposit. Chin Sci Bull 57:1311–1319. https://doi.org/10.1007/s11434-011-4955-4

Jiang Z, Zhang S, Klausen LH, Song J, Li Q, Wang Z et al (2018) *In vitro* single-cell dissection revealing the interior structure of cable bacteria. PNAS, USA 115:8517–8522

Jochum LM, Schreiber L, Marshall IPG, Jørgensen BB, Schramm A, Kjeldsen KU (2018) Single-cell genomics reveals a diverse metabolic potential of uncultivated *Desulfatiglans*-related deltaproteobacteria widely distributed in marine sediment. Front Microbiol. https://doi.org/10.3389/fmicb.2018.02038

Johnson SS, Hebsgaard MB, Christensen TR, Mastepanov M, Nielsen R, Munch K et al (2007) Ancient bacteria show evidence of DNA repair. Proc Natl Acad Sci U S A 104:14401–14405

Jones HE, Chambers LA (1975) Localized intracellular polyphosphate formation by *Desulfovibrio gigas*. J Gen Microbiol 89:67–72

Jørgensen BB, Bak F (1991) Pathways and microbiology of thiosulfate transformations and sulfate reduction in marine sediment (Kattegat, Denmark). Appl Environ Microbiol 57:847–856

Jungbluth SP, del Rio TG, Tringe SG, Stepanauskas R, Rappé MS (2017) Genomic comparisons of a bacterial lineage that inhabits both marine and terrestrial deep subsurface systems. PeerJ 5: e3134. https://doi.org/10.7717/peerj.3134

Junier P, Junier Y, Podell S, Sims DR, Detter JC, Lykidis A et al (2010) The genome of the Gram-positive metal- and sulfate-reducing bacterium *Desulfotomaculum reducens* strain MI-1. Environ Microbiol 12:2738–2754

Kaksonen AH, Spring S, Schumann P, Kroppenstedt RM, Puhakka JA (2007) *Desulfurispora thermophila* gen. Nov., sp. nov., a thermophilic, spore-forming sulfate-reducer isolated from a sulfidogenic fluidized-bed reactor. Int J Syst Evol Microbiol 57:1089–1094

Keller KL, Wall JD (2011) Genetics and molecular biology of the electron flow for sulfate respiration in *Desulfovibrio*. Front Microbiol 2011(2):135. https://doi.org/10.3389/fmicb.2011.00135

Kent HM, Buck M, Evans DJ (1989) Cloning and sequencing of the *nifH* gene of *Desulfovibrio gigas*. FEMS Microbiol Lett 61:73–78
Kjeldsena KU, Schreibera L, Thorupa CA, Boesen T, Bjerg JT, Yanga T et al (2019) On the evolution and physiology of cable bacteria. PNAS 116:19116–19125
Klein M, Friedrich M, Roger AJ, Hugenholtz P, Fishbain S, Abicht H et al (2001) Multiple lateral transfers of dissimilatory sulfite reductase genes between major lineages of sulfate-reducing prokaryotes. J Bacteriol 183:6028–6035
Klenk HP, Clayton RA, Tomb JF, White O, Nelson KE, Ketchum KA et al (1997) The complete genome sequence of the hyperthermophilic, sulphate-reducing archaeon *Archaeoglobus fulgidus*. Nature 390:364–370
Krey GD, Vanin EF, Swenson RP (1988) Cloning, nucleotide sequence, and expression of the flavodoxin gene from *Desulfovibrio vulgaris* (Hildenborough). J Biol Chem 263:15436–15443
Kuever J (2014a) The family *Desulfobacteraceae*. In: Rosenberg E, DeLong EF, Lory S, Stackebrandt E, Thompson F (eds) The prokaryotes. Springer, Berlin, pp 45–73. https://doi.org/10.1007/978-3-642-39044-9_266
Kuever J (2014b) The family *Desulfobulbaceae*. In: Rosenberg E, DeLong EF, Lory S, Stackebrandt E, Thompson F (eds) The prokaryotes. Springer, Berlin, pp 75–86. https://doi.org/10.1007/978-3-642-39044-9_267
Kuever J, Rainey FA, Widdel F (2005a) Order II. *Desulfovibionales*. In: Brenner DJ, Krieg NR, Staley ST (eds) Bergey's mannual of systematic bacteriology, 2nd ed, vol 2, the Proteobactgeria Part C. The Alpha-, Beta-. Delta- and Epsilonproteobacteria. Springer, New York, pp 925–956
Kuever J, Rainey FA, Widdel F (2005b) Order III. *Desulfobacteriales*. In: Brenner DJ, Krieg NR, Staley ST (eds) Bergey's mannual of systematic bacteriology, 2nd ed, vol 2, the Proteobactgeria Part C. The Alpha-, Beta-. Delta- and Epsilonproteobacteria. Springer, New York, pp 959–999
Kuever J, Rainey FA, Widdel F (2005c) Order V. *Desulfuromonales*. In: Brenner DJ, Krieg NR, Staley ST (eds) Bergey's mannual of systematic bacteriology, 2nd ed, vol 2, the Proteobactgeria Part C. The Alpha-, Beta-. Delta- and Epsilonproteobacteria. Springer, New York, pp 1005–1020
Kuever J, Rainey FA, Widdel F (2005d) Order VI. *Syntrophobacterales*. In: Brenner DJ, Krieg NR, Staley ST (eds) Bergey's mannual of systematic bacteriology, 2nd ed, vol 2, The Proteobactgeria Part C. The Alpha-, Beta-. Delta- and Epsilonproteobacteria. Springer, New York, pp 11021–11039
Kuever J, Visser M, Loeffler C, Boll M, Worm P, Sousa DZ et al (2014) Genome analysis of *Desulfotomaculum gibsoniae* strain Groll^T a highly versatile Gram-positive sulfate-reducing bacterium. Stand Genomic Sci 9:821–839
Kuwahara H, Yuki M, Izawa K, Ohkuma M, Hongoh Y (2017) Genome of '*Ca. Desulfovibrio trichonymphae*', an H_2-oxidizing bacterium in a tripartite symbiotic system within a protist cell in the termite gut. The ISME J. 11:766–776
Labrenz M, Banfield JF (2004) Sulfate-reducing bacteria-dominated biofilms that precipitate ZnS in a subsurface circumneutral-pH mine drainage system. Microbial Ecol 47:205–217
Lai D, Springstead JR, Monbouquette HG (2008) Effect of growth temperature on ether lipid biochemistry in *Archaeoglobus fulgidus*. Extremophiles 12:271–278
Lai O, Cao J, Dupont S, Shao Z, Jebbar M, Alain K (2016) *Thermodesulfatator autotrophicus* sp. nov., a thermophilic sulfate-reducing bacterium from the Indian Ocean. Int J Syst Evol Microbiol 66:3978–3982
Langworth TA, Holzer G, Zeikus JG, Torabene T (1983) Iso- and anteiso-branched glycerol diethers of the thermophilic anaerobe *Thermodesulfotobacterium commune*. Syst Appl Microbiol 4:1–17
Larsen S, Nielsen LP, Schramm A (2015) Cable bacteria associated with long-distance electron transport in New England salt marsh sediment. Environ Microbiol Rep 7:175–179
Le Gall J (1963) A new species of *Desulfovibrio*. J Bacteriol 86:1120

Lefèvre CT, Frankel RB, Pósfai M, Prozorov T, Bazylinski DA (2011a) Isolation of obligately alkaliphilic magnetotactic bacteria from extremely alkaline environments. Environ Microbiol 13:2342–2350

Lefèvre CT, Menguy N, Abreu F, Lins U, Pósfai M, Prozorov T et al (2011b) A cultured greigite-producing magnetotactic bacterium in a novel group of sulfate-reducing bacteria. Science 334: 1720–1723

Lefèvre CT, Trubitsyn D, Abreu F, Kolinko S, Jogler C, de Almeida LG et al (2013) Comparative genomic analysis of magnetotactic bacteria from the *Deltaproteobacteria* provides new insights into magnetite and greigite magnetosome genes required for magnetotaxis. Environ Microbiol 15:2712–2735

Leu AO, McIlroy SJ, Ye J, Parks DH, Orphan VJ, Tyson GW (2020). Lateral gene transfer drives metabolic flexibility in the anaerobic methane-oxidizing archaeal family *Methanoperedenaceae*. mBio 11:e01325-20. https://doi.org/10.1128/mBio.01325-20

Li C, Peck HD Jr, LeGall J, Przybyla AE (1987) Cloning sequencing characterization of the genes encoding hydrogenase of *Desulfovibrio gigas*. DNA 6:539–551

Li J, Zhang H, Liu P, Menguy N,Roberts AP, Chen H, Wang Y, Pan Y (2019) Phylogenetic and structural identification of a novel magnetotactic Deltaproteobacteria strain,WYHR-1, from a freshwater lake. Appl Environ Microbiol 85:e00731-19. https://doi.org/10.1128/AEM.00731-19

Lie TJ, Clawson ML, Godchaux W, Leadbetter ER (1999) Sulfidogenesis from 2-aminoethanesulfonate (taurine) fermentation by a morphologically unusual sulfate-reducing bacterium, *Desulforhopalus singaporensis* sp. nov. Appl Environ Microbiol 65:3328–3334

Lin W, Zhang W, Paterson GA, Zhu Q, Zhao X, Knight R et al (2020) Expanding magnetic organelle biogenesis in the domain *Bacteria*. Microbiome 8:152. https://doi.org/10.1186/s40168-020-00931-9

Lindahl T (1993) Instability and decay of the primary structure of DNA. Nature 362:709–715

Lomstein BA, Langerhuus AT, D'hondt S, Jørgensen BB, Spivack AJ (2012) Endospore abundance, microbial growth and necromass turnover in deep sub-seafloor sediment. Nature 484: 101–104

López-García P, Forterre P, van der Oost J, Erauso G (2000) Plasmid pGS5 from the hyperthermophilic archaeon *Archaeoglobus profundus* is negatively supercoiled. J Bacteriol 182:4998–5000

Louro RO (2007) Proton thrusters: overview of the structural and functional features of soluble tetrahaem cytochromes c_3. J Biol Inorg Chem 12:1–10

Lovley DR, Phillips EJP, Lonergan DJ, Widman PK (1995) Fe(III) and S^0 reduction by *Pelobacter carbinolicus*. Appl Environ Microbiol 61:2132–2138

Loy A, Lehner A, Lee N, Adamczyk J, Meier H, Ernst J, Schleifer K-H, Wagner M (2002) Oligonucleotide microarray for 16S rRNA gene-based detection of all recognized lineages of sulfate-reducing prokaryotes in the environment. Appl Environ Microbiol 68:5064–5081

Lücker S, Steger D, Kjeldsen KU, MacGregor BJ, Wagner M, Loy A (2007) Improved 16S rRNA-targeted probe set for analysis of sulfate-reducing bacteria by fluorescence in situ hybridization. J Microbiol Meth 69:523–528

Magot M, Caumette P, Desperrier JM, Matheron R, Dauga C, Grimont F, Carreau L (1992) *Desulfovibrio longus* sp. nov., a sulfate-reducing bacterium isolated from an oil-producing well. Int J Syst Bacteriol 42:398–403

Malkin SY, Rao AMF, Seitaj D, Vasquez-Cardenas D, Zetsche E-M, Hidalgo-Martinez S, Boschker HTS, Meysman FJR (2014) Natural occurrence of microbial sulphur oxidation by long-range electron transport in the seafloor. ISME J 8:1843–1854

Marzocchi U, Trojan D, Larsen S, Meyer RL, Revsbech NP, Schramm A, Nielsen LP, Risgaard-Petersen N (2014) Electric coupling between distant nitrate reduction and sulfide oxidation in marine sediment. ISME J 8:1682–1690

Matias PM, Pereira IAC, Soares CM, Carrondo MA (2005) Sulphate respiration from hydrogen in *Desulfovibrio* bacteria: a structural biology overview. Prog Biophys Mol Biol 89:292–329

Menon NK, Peck HD Jr, Le Gall J, Przybyla AE (1987) Cloning and sequencing of the genes encoding the large and small subunits of the periplasmic (NiFeSe) hydrogenase of *Desulfovibrio baculatus*. J Bacteriol 169:5401–5407

Meyer B, Kuever J (2007) Phylogeny of the alpha and beta subunits of the dissimilatory adenosine-5'-phosphosulfate (APS) reductase from sulfate-reducing prokaryotes – origin and evolution of the dissimilatory sulfate-reduction pathway. Microbiology 153:2026–2044

Meysman FJR, Cornelissen R, Trashin S, Bonné R, Martinez SH, van der Veen J et al (2019) A highly conductive fibre network enables centimetre-scale electron transport in multicellular cable bacteria. Nat Commun 10:4120. https://doi.org/10.1038/s41467-019-12115-7

Miletto M, Loy A, Antheunisse AM, Loeb R, Bodelier PLE, Laanbroek HJ (2008) Biogeography of sulfate-reducing prokaryotes in river foodplains. FEMS Microbiol Ecol 64:395–406

Miroshnichenko ML, Tourova TP, Kolganova TV, Kostrikina NA, Chernych N, Bonch-Osmolovskaya EA (2008) Ammonifex thiophilus sp. nov., a hyperthermophilic anaerobic bacterium from a Kamchatka hot spring. Int J Syst Evol Microbiol 58:2935–2938

Mohn WW, Linkfield TG, Pamlratz HS, Tiedje JM (1990) Involvement of a collar structure in polar growth and cell division of strain DCB-1. Appl Environ Microbiol 56:1206–1211

Morais-Silva FO, Rezende AM, Pimentel C, Santos CI, Clemente C, Varela-Raposo A et al (2014) Genome sequence of the model sulfate reducer *Desulfovibrio gigas*: a comparative analysis within the *Desulfovibrio* genus. Microbiology Open 3:513–530

Mori K, Kim H, Kakegawa T, Hanada S (2003) A novel lineage of sulfate-reducing microorganisms: *Thermodesulfobiaceae* fam. nov., *Thermodesulfobium narugense*, gen. nov., sp. nov., a new thermophilic isolate from a hot spring. Extremophiles 7:283–290

Müller AL, Kjeldsen KU, Rattei T, Pester M, Loy A (2015) Phylogenetic and environmental diversity of DsrAB-type dissimilatory (bi) sulfite reductases. ISME J 2015(9):1152–1165

Müller H, Marozava S, Probst AJ, Meckenstock RU (2020) Groundwater cable bacteria conserve energy by sulfur disproportionation. ISME J 14:623–634

Mussmann M, Richter M, Lombardot T, Meyerdierks A, Kuever J, Kube M, Glöckner FO, Amann R (2005) Clustered genes related to sulfate respiration in uncultured prokaryotes support the theory of their concomitant horizontal transfer. J Bacteriol 187:7126–7137

Nakazawa H, Arakaki A, Narita-Yamada S, Yashiro I, Jinno K, Aoki N et al (2009) Whole genome sequence of *Desulfovibrio magneticus* strain RS-1 revealed common gene clusters in magnetotactic bacteria. Genome Res 19:1801–1808

Nealson KH, Ford J (1980) Surface enhancement of bacterial manganese oxidation: implications for aquatic environments. Geomicrobiol J 2:21–37

Nielsen LP, Risgaard-Petersen N, Fossing H, Christensen PB, Sayama M (2010) Electric currents couple spatially separated biogeochemical processes in marine sediment. Nature 463:1071–1074

Norqvist A, Roffey R (1985) Biochemical and immunological study of cell envelope proteins in sulfate-reducing bacteria. Appl Environ Microbiol 50:31–37

Nunoura T, Oida H, Miyazaki M, Suzuki Y, Takai K, Horikoshi K (2007) *Desulfothermus okinawensis* sp. nov., a thermophilic and heterotrophic sulfate-reducing bacterium isolated from a deep-sea hydrothermal field. Int J Syst Evol Microbiol 57:2360–2364

Odom JM (1993) Industrial and environmental activities of sulfate-reducing bacteria. In: Odom JM, Singleton R Jr (eds) The Sulfate-reducing bacteria: contemporary perspectives. Springer, New York, pp 189–210

Odom JM, Peck HD, JR. (1981) Localization of dehydrogenases, reductases, and electron transfer components in the sulfate-reducing bacterium *Desulfovibrio gigas*. J Bacteriol 147:1161–1169

Odom JM, Jesse K, Knodel E, Emptage M (1991) Immunological cross-reactivities of adenosine 5'-phosphate reductases from sulfate-reducing and sulfide-oxidizing bacteria. Appl Environ Microbiol 57:7727–7733

Oehler D, Poehlein A, Leimbach A, Müller N, Daniel R, Gottschalk G, Schink B (2012) Genome-guided analysis of physiological and morphological traits of the fermentative acetate oxidizer

Thermacetogenium phaeum. BMC Genomics 13:723. https://doi.org/10.1186/1471-2164-13-723

Oliveira TF, Vornhein C, Matias PM, Venceslau SS, Pereira IAC, Archer M (2008) Purification, crystallization and preliminary crystallographic analysis of a dissimilatory DsrAB sulfite reductase in complex with DsrC. J Struct Biol 164:236–239

Olliver B, Hatchikian CE, Prensier G, Gunezekkec J, Garcia J-L (1991) *Desulfohalobium retbaense* gen. nov. sp. nov. a halophilic sulfate-reducing bacterium from sediments of a hypersaline lake in Senegal. Int J Syst Bacteriol 41:74–81

Ollivier B, Cord-Ruwisch R, Hatchikian EC, Garcia JL (1988) Characterization of *Desulfovibrio fructosovorans* sp. nov. Arch Microbiol 149:447–450

Oren A (1983) *Clostridium lortetii* sp. nov., a halophilic obligatory anaerobic bacterium producing endospores with attached gas vacuoles. Arch Microbiol 136:42–48

Orphan VJ, House CH, Hinrichs K-U, McKeegan KD, DeLong EF (2002) Multiple archaeal groups mediate methane oxidation in anoxic cold seep sediments. Proc Natl Acad Sci U S A 99:7663–7668

Oude Elferink SJ, Maas RN, Harmsen HJ, Stams AJ (1995) *Desulforhabdus amnigenus* gen. nov. sp. nov., a sulfate reducer isolated from anaerobic granular sludge. Arch Microbiol 164:119–124

Oude Elferink SJ, Akkermans-van Vliet WM, Bogte JJ, Stams AJ (1999) *Desulfobacca acetoxidans* gen. nov., sp. nov., a novel acetate-degrading sulfate reducer isolated from sulfidogenic granular sludge. Int J Syst Bacteriol 49:345–350

Pagani I, Lapidus A, Nolan M, Lucas S, Hammon N, Deshpande S et al (2011) Complete genome sequence of *Desulfobulbus propionicus* type strain (1pr3T). Genomic Sci 4:100–110

Pavan ME, Pavan EE, Glaeser SP, Etchebehere C, Kampfer P, Pettinari MJ, López NI (2018) Proposal for a new classification of a deep branching bacterial phylogenetic lineage: transfer of *Coprothermobacter proteolyticus* and *Coprothermobacter platensis* to *Coprothermobacteraceae* fam. nov., within *Coprothermobacterales* ord. nov., *Coprothermobacteria classis* nov. and *Coprothermobacterota* phyl. nov. and emended description of the family *Thermodesulfobiaceae*. Int J Syst Evol Microbiol 68:1627–1632

Pedersen K, Bengtsson A, Edlund J, Rabe L, Hazen T, Chakraborty R, Goodwin L, Shapiro N (2014) Complete genome sequence of the subsurface, mesophilic sulfate-reducing bacterium *Desulfovibrio aespoeensis* Aspo-2. Genome Announc. 2(3):e00509–e00514. https://doi.org/10.1128/genomeA.00509-14

Pelz O, Chatzinotas A, Zarda-Hess A, Abraham W-R, Zeyer J (2001) Tracing toluene-assimilating sulfate-reducing bacteria using 13C-incorporation in fatty acids and whole-cell hybridization. FEMS Microbiol Ecol 38:123–131

Pereira IAC, Romão CV, Xavier AV, LeGall J, Teixeira M (1998) Electron transfer between hydrogenases and mono and multiheme cytochromes in *Desulfovibrio* spp. J Biol Inorg Chem 3:494–498

Pereira PM, He Q, Xavier AV, Zhou JZ, Pereira IAC, Louro RO (2008) Transcriptional response of *Desulfovibrio vulgaris* Hildenborough to oxidative stress mimicking environmental conditions. Arch Microbiol 189:451–461

Pereira IAC, Ramos AR, Grein F, Marques MC, da Silva SM, Venceslau SS (2011) A comparative genomic analysis of energy metabolism in sulfate-reducing bacteria and archaea. Front Microbiol 2:69. https://doi.org/10.3389/fmicb.2011.00069

Pester M, Brambilla E, Alazard D, Rattei T, Weinmaier T, Han J et al (2012) Complete genome sequences of *Desulfosporosinus orientis* DSM765^T, *Desulfosporosinus youngiae* DSM17734^T, *Desulfosporosinus meridiei* DSM13257^T, and *Desulfosporosinus acidiphilus* DSM22704^T. J Bacteriol 194:6300–6301

Pfeffer C, Larsen L, Song J, Dong M, Besenbacher F, Meyer RL et al (2012) Filamentous bacteria transport electrons over centimetre distances. Nature 491:218–221

Pikuta EV, Hoover RB, Bej AK, Marsic D, Whitman WB, Cleland D, Krader P (2003) *Desulfonatronum thiodismutans* sp. nov., a novel alkaliphilic, sulfate-reducing bacterium capable of lithoautotrophic growth. Int J Syst Evol Microbiol 53:1327–1332

Pires RH, Lourenco AIC, Morais F, Teixeira M, Xavier AV, Saraiva LM, Pereira IA (2003) A novel membrane-bound respiratory complex from *Desulfovibrio desulfuricans* ATCC 27774. Biochim Biophys Acta 1605:67–82

Platen H, Temmes A, Schink B (1990) Anaerobic degradation of acetone by *Desulfococcus biacutus* spec. nov. Arch Microbiol 154:355–361

Plugge CM, Henstra AM, Worm P, Swarts DC, Paulitsch-Fuchs AH, Scholten JCM (2012) Complete genome sequence of *Syntrophobacter fumaroxidans* strain (MPOBT). Stand Genomic Sci 7:91–106

Poehlein A, Daniel R, Schink B, Simeonova DD (2013) Life based on phosphite: a genome-guided analysis of *Desulfotignum phosphitoxidans*. BMC Genomics 14:753. https://doi.org/10.1186/1471-2164-14-753

Pollock WBR, Loutfi M, Bruschi M, Rapp-Giles BJ, Wall JD, Voordouw G (1991) Cloning, sequencing, and expression of the gene encoding the high-molecular-weight cytochrome *c* from *Desulfovibrio vulgaris* Hildenborough. J Bacteriol 173:220–228

Pósfai M, Moskowitz BM, Arató B, Schüler D, Flies C, Bazylinski DA, Frankel RB (2006) Properties of intracellular magnetite crystals produced by *Desulfovibrio magneticus* strain RS-1. Earth Planet Sci Lett 249:444–455

Postgate JR (1954) Presence of cytochrome in an obligate anaerobe. Biochem J 56:xi–xii

Postgate J (1959) A diagnostic reaction of *Desulphovibrio desulphuricans*. Nature 183:481–482

Postgate JR, Campbell LL (1966) Classification of *DesulJovibrio* species, the nonsporulating sulfate-reducing bacteria. Bacteriol Rev 30:732–738

Postgate JR, Kent HM, Robson RL (1986) DNA from diazotrophic *Desulfovibrio* strains is homologous to *Klebsiella pneumoniae* structural *nif* DNA and can be chromosomal or plasmid-borne. FEMS Microbiol Lett 33:159–165

Powell B, Mergeay M, Christofi N (1989) Transfer of broad-range plasmids to sulphate-reducing bactgeria. FEMS Microbiol Lett 59:269–274

Prickril BC, Czechowski MH, Przybyla AE, Peck HD Jr, LeGall J (1986) Putative signal peptide on the small subunit of the periplasmic hydrogenase from *Desulfovibrio vulgaris*. J Bacteriol 167: 722–725

Prickril BC, Kurtz DM Jr, LeGall J, Voordouw G (1991) Cloning and sequencing of the gene for rubrerythrin from *Desulfovibrio vulgaris* (Hildenborough). Biochemist 30:11118–11123

Probst AJ, Holman H-YN, DeSantis TZ, Andersen GL, Birarda G, Bechtel HA et al (2013) Tackling the minority: sulfate-reducing bacteria in an archaea-dominated subsurface biofilm. The ISME J 7:635–651

Rabus R, Strittmatter A (2007) Functional genomics of sulphate-reducing prokaryotes. In: Barton LL, Hamilton WA (eds) Sulphate-reducing bacteria environmental and engineered systems. Cambridge University Press, Cambridge, pp 117–140

Rabus R, Ruepp A, Frickey T, Rattei T, Fartmann B, Stark M et al (2004) *Desulfotalea psychrophila*, a sulfate-reducing bacterium from permanently cold Arctic sediments. Environ Microbiol 6:887–902

Rabus R, Hansen T, Widdel F (2007) Dissimilatory sulfate- and sulfur-reducing prokaryotes. In: Dworkin M, Falkow S, Rosenberg E, Schleifer KH, Stackebrandt E (eds) The prokaryotes, vol 2. Springer, New York, pp 659–768

Rabus R, Hansen TA, Widdel F (2013) Dissimilatory sulfate- and sulfur-reducing prokaryotes. In: Rosenberg E, DeLong EF, Lory S, Stackebrandt E, Thompson F (eds) The prokaryotes. Springer, Berlin, pp 309–404

Rabus R, Venceslau SS, Wöhlbrand L, Voordouw G, Wall JD, Pereira IAC (2015) A post-genomic view of the ecophysiology, catabolism and biotechnological relevance of sulphate-reducing prokaryotes. Adv Microbial Physiol 66:58–321

Rahn-Lee L, Byrne ME, Zhang M, Le Sage D, Glenn DR, Milbourne T et al (2015) A genetic strategy for probing the functional diversity of magnetosome formation. PLoS Genet 11(1): e1004811. https://doi.org/10.1371/journal.pgen.1004811

Reed DW, Hartzell PL (1999) The *Archaeoglobus fulgidus* D-lactate dehydrogenase is a Zn^{2+} flavoprotein. J Bacteriol 181:7580–7587

Richter M, Rosselló-Móra R (2009) Shifting the genomic gold standard for the prokaryotic species definition. Proc Natl Acad Sci U S A 106:19126–19131

Risgaard-Petersen N, Damgaard LR, Revil A, Nielsen LP (2014) Mapping electron sources and sinks in a marine biogeobattery. JGR Biogeosci 119:1475–1486

Robertson WJ, Bowman JP, Franzmann PD, Mee BJ (2001) *Desulfosporosinus meridiei* sp. nov., a spore-forming sulfate-reducing bacterium isolated from gasolene-contaminated groundwater. Int J Syst Evol Microbiol 51:133–140

Romão CV, Archer M, Lobo SA, Louro RO, Pereira IAC, Saraiva LM, Teixeira M, Matias PM (2012) Diversity of heme proteins in sulfate-reducing bacteria. In: Kadish KM, Smith KM, Guilard R (eds) Handbook of porphyrin science. World Scientific Publishing Company, Singapore, vol 19, pp 139–230

Rousset M, Dermoun Z, Hatchikian CE, Bélaich J-P (1990) Cloning and sequencing of the locus encoding the large and small subunit genes of the periplasmic [NiFe]hydrogenase from *Desulfovibrio fructosovorans*. Gene 94:95–101

Rousset M, Dermoun Z, Chippaux M, Bélaich JP (1991) Marker exchange mutagenesis of the *hydN* genes in *Desulfovibrio fructosovorans*. Mol Microbiol 5:1735–1740

Sagong H-Y, Son HF, Choi SY, Lee SY, Kim K-J (2018) Structural insights into polyhydroxyalkanoates biosynthesis. Trends Biochem Sci 43:790–805

Sakaguchi T, Arakaki A, Matsunaga T (2002) *Desulfovibrio magneticus* sp. nov., a novel sulfate-reducing bacterium that produces intracellular single-domain-sized magnetite particles. Int J Syst Evol Microbiol 52:215–221

Sakaguchi T, Burgess JG, Matsunaga T (1993) Magnetite formation by a sulphate-reducing bacterium. Nature 365:47–49

Santos H, Fareleira P, Pedegral C, LeGall J, Xavier AV (1991) *In vivo* 3'P-NMR studies of *Desuljovibrio* species. Detection of a novel phosphorus-containing compound. Eur JBiochem 201:283–287

Santos H, Fareleira P, Xavier AV, Chen L, Liu M-Y, LeGall J (1993) Aerobic metabolism of carbon reserves by the "obligate anaerobe" *Desulfovibrio gigas*. Biochem Biophys Res Commun 195: 551–557

Sass H, Overmann J, Rütters H, Babenzien H-D, Cypionka H (2004) *Desulfosporomusa polytropa* gen. nov., sp. nov., a novel sulfate-reducing bacterium from sediments of an oligotrophic lake. Arch Microbiol 182:204–211

Schidlowski M, Hayes JM, Kaplan IR (1983) Isotopic inferences of ancient biochemistries: carbon, sulfur, hydrogen, and nitrogen. In: Schopf JW (ed) Earth's earliest biosphere, its origin and evolution. Princeton University Press, Princeton, pp 149–186

Schiffer A, Parey K, Warkentin E, Diederichs K, Huber H, Stetter KO, Kroneck PM, Ermler U (2008) Structure of the dissimilatory sulfite reductase from the hyperthermophilic archaeon *Archaeoglobus fulgidus*. J Mol Biol 379:1063–1074

Schoch CL, Ciufo S, Domrachev M, Hotton CL, Kannan S, Khovanskaya R et al (2020) NCBI Taxonomy: a comprehensive update on curation, resources and tools. Database (Oxford). 2020: baaa062. https://doi.org/10.1093/database/baaa062

Scholz VV, Müller H, Koren K, Nielsen LP, Meckenstock RU (2019) The rhizosphere of aquatic plants is a habitat for cable bacteria. FEMS Microbiol Ecol 95:fiz062. https://doi.org/10.1093/femsec/fiz062. PMID: 31054245; PMCID: PMC6510695

Seitaj D, Schauer R, Sulu-Gambari F, Hidalgo-Martinez S, Malkin SY, Burdorf LDW, Slomp CP, Meysman FJR (2015) Cable bacteria generate a firewall against euxinia in seasonally hypoxic basins. PNAS, USA. 112:13278–13283

Senez JC (1951) Étude comparative de la croissance de *Sporovibrio desulfuricans* sur pyruvate et sur lactate de soude. Ann Inst Pasteur 80:395–409

Shen YA, Buick R, Canfield DE (2001) Isotopic evidence for microbial sulphate reduction in the early Archaean era. Nature 410:77–81

Shimoshige H, Kobayashi H, Shimamura S, Mizuki T, Inoue A, Maekawa T (2021) Isolation and cultivation of a novel sulfate-reducing magnetotactic bacterium belonging to the genus *Desulfovibrio*. PLoS One 16(3):e0248313. https://doi.org/10.1371/journal.pone.0248313

Siebers B, Zaparty M, Raddatz G, Tjaden B, Albers S-V, Bell SD et al (2011) The complete genome sequence of *Thermoproteus tenax*: a physiologically versatile member of the *Crenarchaeota*. PLoS One. 6(10):e24222. https://doi.org/10.1371/journal.pone.0024222

Singleton R Jr, Denis J, Campbell LL (1984) Antigenic diversity of cytochrome c_3 from the anaerobic, sulfate reducing bacteria. Arch Microbiol 139:91–95

Singleton R Jr, Denis J, Campbell LL (1985) Whole-cell antigens of the members of the sulfate-reducing genus *Desulfovibrio*. Arch Microbiol 141:195–197

So CM, Young LY (1999) Isolation and characterization of a sulfate-reducing bacterium that anaerobically degrades alkanes. Appl Environ Microbiol 65:2969–2976

Sorokin DY, Tourova TP, Henstra AM, Stams AJM, Galinski EA, Muyzer G (2008) Sulfidogenesis under extremely haloalkaline conditions by *Desulfonatronospira thiodismutans* gen. nov., sp. nov., and *Desulfonatronospira delicata* sp. nov. – a novel lineage of Deltaproteobacteria from hypersaline soda lakes. Microbiology 154:1444–1453

Spring S, Lapidus A, Schröder M, Gleim D, David Sims D, Meincke L et al (2009) Complete genome sequence of *Desulfotomaculum acetoxidans* type strain (5575T). Stand Genomic Sci 1: 242–253

Spring S, Nolan M, Lapidus A, Del Rio TG, Copeland A, Tice H et al (2010) Complete genome sequence of *Desulfohalobium retbaense* type strain (HR100^T). Stand Genomic Sci 2:38–48

Spring S, Sorokin DY, Verbarg S, Rohde M, Woyke T, Kyrpides NC (2019) Sulfate-reducing bacteria that produce exopolymers thrive in the calcifying zone of a hypersaline cyanobacterial mat. Front Microbiol. https://doi.org/10.3389/fmicb.2019.00862

Spring S, Visser M, Lu M, Copeland A, Lapidus A, Lucas S et al (2012) Complete genome sequence of the sulfate-reducing firmicute *Desulfotomaculum ruminis* type strain (DLT). Stand Genomic Sci 7:304–319

Stackebrandt E, Stahl DA, Devereux R (1995) Taxonomic relationships. In: Barton LL (ed) Sulfate-reducing bacteria. Plenum Press Inc, New York, pp 49–88

Stackebrandt E, Sproer C, Rainey FA, Burghardt J, Päuker O, Hippe H (1997) Phylogenetic analysis of the genus *Desulfotomaculum*: evidence for the misclassification of *Desulfotomaculum guttoideum* and description of *Desulfotomaculum orientis* as *Desulfosporosinus orientis* gen. nov., comb. nov. Int J Syst Bacteriol. 47:1134–1139

Staley JT, Irgens RL, Herwig PP (1989) Gas vacuolate bacteria from the sea ice of Antarctic. Appl Environ Microbiol 55:1033–1036

Stahl DA, Loy A, Wagner N (2007) Molecular strategies for studies of natural populations of sulphate-reducing microorganisms. In: Barton LL, Hamilton WA (eds) Sulphate-reducing bacteria-environmental and engineered systems. Cambridge University Press, Cambridge, UK, pp 55–116

Stams FJM, Veenhuis M, Weenk GH, Hansen TA (1983) Occurrence of polyglucose as a storage polymer in *Desulfovibrio* species and *Desulfobulbus propionicus*. Arch Microbiol 136:54–59

Stanley W, Southam G (2018) The effect of Gram-positive (*Desulfosporsinus orientis*) and Gram-negative (*Desulfovibrio desulfuricans*) sulfate-reducing bacteria on iron sulfide mineral precipitation. Can J Microbiol 64:629–637

Starkey RL (1938) A study of spore formation and other morphological characteristics of *Vibrio desulfuricans*. Arch Microbiol 8:268–304

Stetter KO (1988) *Archaeoglobus fulgidus* gen. nov., sp. nov. a new taxon of extremely thermophilic archaebacteria. Syst Appl Microbiol 10:172–173

Stetter KO, Laurer G, Thomm M, Neuner A (1987) Isolation of extremely thermophilic sulfate reducers: evidence for a novel branch of archaebacteria. Science 236:822–824

Strittmatter AW, Liesegang H, Rabus R, Decker I, Amann J, Andres S et al (2009) Genome sequence of *Desulfobacterium autotrophicum* HRM2, a marine sulfate reducer oxidizing organic carbon completely to carbon dioxide. Environ Microbiol 11:1038–1055

Sun B, Cole JR, Tiedje JM (2001) *Desulfomonile limimaris* sp. nov., an anaerobic dehalogenating bacterium from marine sediments. Int J Syst Evol Microbiol 51:365–371

Sun H, Spring S, Lapidus A, Davenport K, Del Rio TG, Tice H et al (2010) Complete genome sequence of *Desulfarculus baarsii* type strain (2st14T). Stand Genomic Sci 3:276–284

Tan J, Helms LR, Swenson RP, Cowan JA (1991) Primary structure of the assimilatory-type sulfite reductase from *Desulfovibrio vulgaris* (Hildenborough): cloning and nucleotide sequence of the reductase gene. Biochemist 30:9900–9907

Taylor J, Parkes RJ (1983) The cellular fatty acids of the sulphate-reducing bacteria, *Desulfobacter* sp., *Desulfobulbus* sp. and *Desulfovibvio desulfuvicans*. J Gen Microbiol 129:3303–3309

Taylor J, Parkes RJ (1985) Identifying different populations of sulphate-reducing bacteria within marine sediment systems, using fatty acid biomarkers. J Gen Microbiol 131:631–642

Teske A (2019) Cable bacteria, living electrical conduits in the microbial world. PNAS, USA 116: 18759–18761

Teske A, Jørgensen BB, Gallardo VA (2009) Filamentous bacteria inhabiting the sheaths of marine *Thioploca* spp. on the Chilean continental shelf. FEMS Microbiol Ecol 68:164–172

Thauer RK, Kunow J (1995) Sulfate reducing archaea. In: Barton LL (ed) Sulfate-reducing bacteria. Plenum Press Inc, New York, pp 33–48

Thauer RF, Stackebrandt E, Hamilton WA (2007) Energy metabolism and phylogenic diversity of sulphate-reducing bacteria. In: Barton LL, Hamilton WA (eds) Sulphate-reducing bacteria environmental and engineered systems. Cambridge University Press, Cambridge, pp 1–38

Tindall BJ, Rosselló-Móra R, Busse H-J, Ludwig W, Kämpfer P (2010) Notes on the characterization of prokaryote strains for taxonomic purposes. Int J Syst Evol Microbiol 60:249–266

Toso DB, Javed MM, Czornyj E, Gunsalus RP, Zhou ZH (2016) Discovery and characterization of iron sulfide and polyphosphate bodies coexisting in *Archaeoglobus fulgidus* cells. Archaea. 2016: Article ID 4706532, 11 pages https://doi.org/10.1155/2016/4706532

Trojan D, Schreiber L, Bjerg JT, Bøggild A, Yang T, Kjeldsen KU, Schramm A (2016) A taxonomic framework for cable bacteria and proposal of the candidate genera *Electrothrix* and *Electronema*. Syst Appl Microbiol 39:297–306

Ueki A, Suto T (1979) Cellular fatty acid composition of sulfate-reducing bacteria. J Gen Appl Microbiol 25:185–196

Vainshtein M, Hippe H, Kroppenstedt RM (1992) Cellular fatty acid composition of *Desulfovibrio* species and its use in classification of sulfate-reducing bacteria. Syst Appl Microbiol 15:554–566

van den Berg WAM, Stokkermans JPWG, van Dongen WMAM (1989) Development of a plasmid transfer system for the anaerobic sulphate reducer, *Desulfovibrio vulgaris*. J Biotechnol 12:173–184

van Dongen WMAM (1995) Molecular biology of redox-active metal proteins from *Desulfovibrio*. In: Barton LL (ed) Sulfate-reducing bacteria. Plenum Press, New York, pp 185–216

van Dongen W, Hagen W, van den Berg W, Veeger C (1988) Evidence for an unusual mechanism of membrane translocation of the periplasmic hydrogenase of *Desulfovibrio vulgaris* as derived from expression in *E. coli*. FEMS Microbial Ecol 50:5–9

Van Nostrand JD, He Z, Zhou J (2012) Use of functional gene arrays for elucidating in situ biodegradation. Front Microbiol 3:Article 339. https://doi.org/10.3389/fmicb.2012.00339

Van Nostrand JD, Wu W-M, Wu L, Deng Y, Carley J, Carroll S et al (2009) GeoChip-based analysis of functional microbial communities during the reoxidation of a bioreduced uranium contaminated aquifer. Environ Microbiol 11:2611–2626

van Rooijen GJH, Bruschi M, Voordouw F (1989) Cloning and sequencing of the gene encoding cytochrome c_{553} from *Desulfovibrio vulgaris* Hildenborough. J Bacteriol 171:3575–3578

Vasquez-Cardenas D, van de Vossenberg J, Polerecky L, Malkin SY, Schauer R, Hidalgo-Martinez S et al (2015) Microbial carbon metabolism associated with electrogenic sulphur oxidation in coastal sediments. ISME J 9:1966–1978

Vecchia ED, Shao PP, Suvorova E, Chiappe D, Hamelin R, Bernier-Latmani R (2014) Characterization of the surfaceome of the metal-reducing bacterium *Desulfotomaculum reducens*. Front Microbiol 5:432. https://doi.org/10.3389/fmicb.2014.00432

Venzlaff H, Enning D, Srinivasan J, Mayrhofer KJJ, Hassel AW, Widdel F, Stratmann M (2013) Accelerated cathodic reaction in microbial corrosion of iron due to direct electron uptake by sulfate-reducing bacteria. Corros Sci 66:88–96

Visser M, Parshina SN, Alves JI, Sousa DZ, Pereira IAC, Muyzer G et al (2014) Genome analyses of the carboxydotrophic sulfate-reducers *Desulfotomaculum nigrificans* and *Desulfotomaculum carboxydivorans* and reclassification of *Desulfotomaculum caboxydivorans* as a later synonym of *Desulfotomaculum nigrificans*. Stand Genomic Sci 9:655–675

Visser M, Worm P, Muyzer G, Pereira IAC, Schaap PJ, Plugge CM et al (2013) Genome analysis of *Desulfotomaculum kuznetsovii* strain 17T reveals a physiological similarity with *Pelotomaculum thermopropionicum* strain SIT. Stand Genomic Sci 8:69–87

von Jan M, Lapidus A, Del Rio TG, Copeland A, Tice H, Cheng J-F et al (2010) Complete genome sequence of *Archaeoglobus profundus* type strain (AV18^T). Stand Genomic Sci 2:327–346

Voordouw G (1993) Molecular biology of sulfate-reducing bacteria. In: Odom JM, Singleton R (eds) The sulfate-reducing bacteria: contemporary perspectives. Brock/Springer Series in Contemporary Bioscience, Springer, New York, pp 88–130

Voordouw G, Brenner S (1985) Nucleotide sequence of the gene encoding the hydrogenase from *Desulfovibrio vulgaris* (Hildenborough). Eur J Biochem 148:515–520

Voordouw G, Brenner S (1986) Cloning and sequencing of the gene encoding cytochrome c_3 from *Desulfovibrio vulgaris* (Hildenborough). Eur J Biochem 159:347–351

Voordouw G, Hagen WR, Krüse-Wolters KM, van Berkel-Arts A, Veegler C (1987a) Purification and characterization of *Desulfovibrio vulgaris* (Hildenborough) hydrogenase expressed in *Escherichia coli*. Eur J Biochem 162:31–36

Voordouw G, Kent HM, Postgate JR (1987b) Identification of the genes for hydrogenase and cytochrome c_3 in *Desulfovibrio*. Can J Microbiol 33:1006–1010

Voordouw G, Menon NK, LeGall J, Choi ES, Peck HD Jr, Przybyla AE (1989b) Analysis and comparison of nucleotide sequences encoding the genes for [NiFe] and [NiFeSe] hydrogenases from *Desulfovibrio gigas* and *Desulfovibrio baculatus*. J Bacteriol 171:2894–2899

Voordouw G, Niviere V, Ferris FG, Fedorak PM, Westlake DWS (1990a) Distribution of hydrogenase genes in *Desulfovibio* spp. and their use in identification of species from the oil field environment. Appl Environ Microbiol 56:3748–3754

Voordouw G, Pollock WB, Bruschi M, Guerlesquin F, Rapp-Giles B-J, Wall JD (1990b) Functional expression of *Desulfovibrio vulgaris* Hildenborough cytochrome c_3 in *Desulfovibrio desulfuricans* G200 after conjugational gene transfer from *Escherichia coli*. J Bacteriol 172: 6122–6126

Voordouw G, Strang JD, Wilson FR (1989a) Organization of the genes encoding [Fe] hydrogenase in *Desulfovibrio vulgaris* subsp. *oxamicus* Monticello. J Bacteriol 171:3881–3889

Voordouw G, Walker JE, Brenner S (1985) Cloning of the gene encoding the hydrogenase from *Desulfovibrio vulgaris* (Hildenborough) and determination of the NH_2-terminal sequence. Eur J Biochem 148:509–514

Voordouw G (1988) Cloning of genes encoding redox proteins of known amino acid sequence from a library of the *Desulfovibrio vulgaris* (Hildenborough) genome. Gene 67:75–83

Voordouw G, Niviere V, Ferris FG, Fedorak PM, Westlake DW (1990c) Distribution of hydrogenase genes in *Desulfovibrio* spp. and their use in identification of species from the oil field environment. Appl Environ Microbiol 56:3748–3754

Vreeland RH, Rosenzweig WD, Powers DW (2000) Isolation of a 250 million-year-old halotolerant bacterium from a primary salt crystal. Nature 407:897–900

Wagner M, Roger AJ, Flax JL, Brusseau GA, Stahl DA (1998) Phylogeny of dissimilatory sulfite reductases supports an early origin of sulfate respiration. J Bacteriol 180:2975–2982

Walian PJ, Allen S, Shatsky M, Zeng L, Szakal ED, Haichuan Liu H et al (2012) High-throughput isolation and characterization of untagged membrane protein complexes: outer membrane complexes of *Desulfovibrio vulgaris*. J Proteome Res 11:5720–5735

Wall JD (1993) Genetics of the sulfate-reducing bacteria. In: Odom JM, Singleton R (eds) The sulfate-reducing bacteria: contemporary perspectives. Brock/Springer Series in Contemporary Bioscience, Springer, New York, pp 77–87

Wall JD, Arkin AP, Balci NC, Rapp-Giles B (2008) Genetics and genomics of sulfate respiration in *Desulfovibrio*. In: Dahl C, Friedrich CG (eds) Microbial sulfur metabolism. Springer, Berlin, pp 1–12

Wall JD, Hemme CL, Rapp-Giles B, Ringbauer JA, Casalot L, Giblin T (2003) Genes and genetic manipulations of *Desulfovibrio*. In: Ljungdahl LG, Adams MW, Barton LL, Ferry JG, Johnson MJ (eds) Biochemistry and physiology of anaerobic bacteria. Springer, New York, pp 85–98

Wall JD, Rapp-Giles BJ, Rousset M (1993) Characterization of a small plasmid from *Desulfovibrio desulfuricans* and its use for shuttle vector construction. J Bacteriol 175:4121–4128

Wang F, Zhou H, Meng J, Peng X, Jiang L, Sun P et al (2009) GeoChip-based analysis of metabolic diversity of microbial communities at the Juan de Fuca Ridge hydrothermal vent. PNAS, USA. 106:4840–4845

Waite DW, Chuvochina M, Pelikan C, Parks DH, Yilmaz P, Wagner M et al (2020) Proposal to reclassify the proteobacterial classes *Deltaproteobacteria* and *Oligoflexia*, and the phylum *Thermodesulfobacteria* into four phyla reflecting major functional capabilities. Int J Syst Evol Microbiol 70:5972–6016

Watanabe M, Higashioka Y, Kojima H, Fukui M (2017) *Desulfosarcina widdelii* sp. nov. and *Desulfosarcina alkanivorans* sp. nov., hydrocarbon-degrading sulfate-reducing bacteria isolated from marine sediment and emended description of the genus *Desulfosarcina*. Int J Syst Evol Microbiol 67:2994–2997

Watanabe M, Kojima H, Fukui M (2018) Review of *Desulfotomaculum* species and proposal of the genera *Desulfallas* gen. nov., *Desulfofundulus* gen. nov., *Desulfofarcimen* gen. nov. and *Desulfohalotomaculum* gen. nov. Int J Syst Evol Microbiol 68:2891–2899

Watanabe M, Kojima H, Umezawa K, Fukui M (2019) Genomic characteristics of *Desulfonema ishimotonii* Tokyo 01T implying horizontal gene transfer among phylogenetically dispersed filamentous gliding bacteria. Front Microbiol 10:227. https://doi.org/10.3389/fmicb.2019.00227

Wegmann U, Nueno Palop C, Mayer MJ, Crost E, Narbad A (2017) Complete genome sequence of *Desulfovibrio piger* FI11049. Genome Announc 5:e01528–e01516. https://doi.org/10.1128/genomeA.01528-16

Widdel F (1980) Anaerober Abbau von Fettsäuren und Benzoesäure durch neu isolierte Arten Sulfat-reduzierender Bakterien. Dissertation. Georg-August-Universität zu Göttingen, Lindhorst/Schaumburg-Lippe, Göttingen

Widdel F, Kohring GW, Mayer F (1983) Studies on dissimilatory sulfate-reducing bacteria that decompose fatty acids. III. Characterization of the filamentous gliding *Desulfonema limicola* gen. nov. sp. nov., and *Desulfonema magnum* sp. nov. Arch Microbiol 134:286–294

Wu L, Kellogg L, Devol AH, Tiedje JM, Zhou J (2008) Microarray based characterization of microbial community functional structure and heterogeneity in marine sediments from the Gulf of Mexico. Appl Environ Microbiol 74:4516–4529

Yarza P, Yilmaz P, Pruesse E, Glöckner FO, Ludwig W, Schleifer K-H et al (2014) Uniting the classification of cultured and uncultured bacteria and archaea using 16S rRNA gene sequences. Nat Rev Microbiol 12:635–645

Zane GM, Yen HC, Wall JD (2010) Effect of the deletion of qmoABC and the promoter-distal gene encoding a hypothetical protein on sulfate reduction in *Desulfovibrio vulgaris* Hildenborough. Appl Environ Microbiol 76:5500–5509

Zellner G, Messner P, Kneifel H, Winter J (1989) *Desulfovibrio simplex* spec. nov., a new sulfate-reducing bacterium from a sour whey digester. Arch Microbiol 152:329–334

Zhang W, Gritsenko MA, Moore RJ, Culley DE, Nie L, Petritis K et al (2006) A proteomic view of *Desulfovibrio vulgaris* metabolism as determined by liquid chromatography coupled with tandem mass spectrometry. Proteomics 6:4286–4299

Zhilina TN, Zavarzin GA (1987) *Methanosarcina vacuolate* sp. nov., a vacuolated *Methanosarcina*. Inter J Syst Bacteriol 37:281–283

Zhilina TN, Zavarzin GA, Rainey FA, Pikuta EM, Osipov GA, Kostrikina NA (1997) *Desulfonatronovibrio hydrogenovorans* gen. nov., sp. nov., an alkaliphilic, sulfate-reducing bacterium. Int J Syst Bacteriol 47:144–149

Zhou K, Zhang WY, Yu-Zhang K, Pan HM, Zhang SD, Zhang WJ et al (2012) A novel genus of multicellular magnetotactic prokaryotes from the Yellow Sea. Environ Microbiol 14:405–413

Zillig W, Stetter KO, Schaefer W, Janekovic D, Wunderl S, Holz I, Palm P (1981) *Thermoproteales*: a novel type of extremely thermoacidophilic anaerobic archaebacteria isolated from Icelandic solfataras. Zentralbl Mikrobiol Parasitenkd Infektionskr Hyg Abt 1 Orig 2: 205–227

Zverlov V, Klein M, Lücker S, Friedrich MW, Kellermann J, Stahl DA, Loy A, Wagner M (2005) Lateral gene transfer of dissimilatory (bi)sulfite reductase revisited. J Bacteriol 187:2203–2208

Chapter 3
Reduction of Sulfur and Nitrogen Compounds

3.1 Introduction

The use of sulfate as an electron acceptor by bacteria and archaea not only requires a unique set of enzymes but also requires an input of energy to activate sulfate. In fact, respiratory reduction of sulfate is the only process in microbial physiology where the electron acceptor requires an input of ATP before electrons are consumed. Since sulfur chemistry is relatively complex, the enzymology concerning the reduction of sulfur-oxy anions and elemental sulfur involves novel biochemistry. The enzymes for the reduction of sulf-oxy molecules have been characterized and are discussed in this chapter. As discussed in Chap. 1, several sulfate-reducing bacteria (SRB) grow by the disproportionation of thiosulfate, sulfite, and sulfur, and these dismutation reactions are assumed to be attributed to dissimilatory sulfate reduction. Nitrate is one of the more common alternate electron acceptors used by SRB, and this anaerobic reduction of nitrate to ammonium is also characteristic of other anaerobes. From a standpoint of energetics, dissimilatory nitrate reduction provides more energy than dissimilatory sulfate reduction. In freshwater sediments where sulfate concentrations are low, nitrate is an important electron acceptor for SRB. Thus, from an overall perspective, sulfate-reducing prokaryotes (SRP) provide important contributions to both the sulfur and nitrogen cycles.

3.2 Sulfate Activation and Bisulfite Production

The production of sulfide by the physiological group of microorganisms uses sulfate as the final electron acceptor for growth. While sulfate respiration is a hallmark characteristic of SRP, these microorganisms also assimilate sulfate for the synthesis of sulfur amino acids. A generalized view of dissimilatory and assimilatory sulfate reduction is given in Fig. 3.1. It is useful here to retrace the publications that

L. L. Barton, G. D. Fauque, *Sulfate-Reducing Bacteria and Archaea*,
https://doi.org/10.1007/978-3-030-96703-1_3

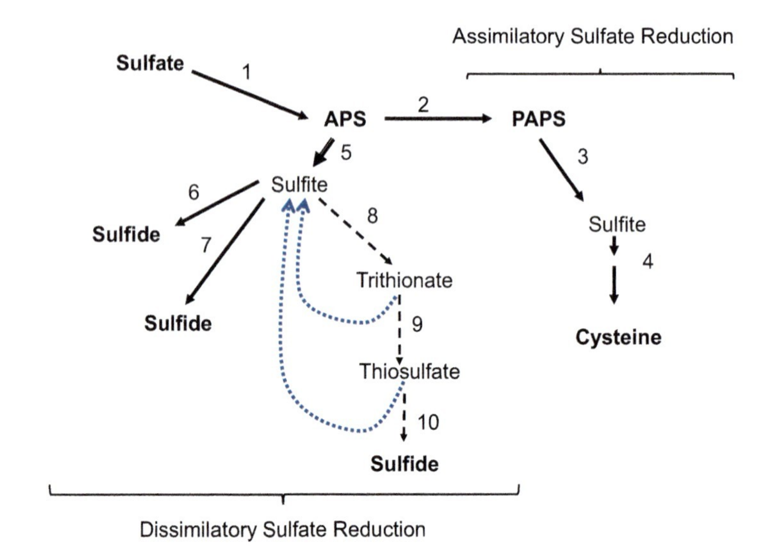

Fig. 3.1 Sulfur pathway indicating sulfate activation and reduction to sulfide. The dissimilatory (respiratory) sulfate reduction pathway is unique to the SRP and is distinct from the assimilatory sulfate reduction pathway. Enzymes are as follows: (1) ATP sulfurylase, (2) APS kinase, (3) PAPS reductase, (4) multienzyme step resulting in the synthesis of cysteine, (5) APS reductase, (6) dissimilatory sulfite reductase (Dsr type), (7) assimilatory sulfite reductase (dSiR type), (8) trithionate-forming enzyme, (9) trithionate reductase, and (10) thiosulfate reductase

established the processes leading to sulfate reduction in plants, animals, and microorganisms. Sulfate is a stable anion with a redox potential of −516 mV for the SO_4^{2-}/HSO_3^- couple (Thauer et al. 2007), which is too electronegative for biological systems. Sulfate must be activated by reacting with ATP to produce adenosine phosphosulfate (APS). With the redox potential of −60 mV for the APS/HSO_3^- couple, metabolism of APS can be achieved by appropriate enzymes. As shown in Eq. 3.1, this activation of sulfate to APS is energetically expensive in that it has a net requirement of two ATP molecules for each sulfate anion activated.

Using extracts from lamb liver or from *Neurospora*, Hilz and Lipmann (1955) reported the novel reaction with the production of AMP-sulfate (adenosine 5′-phosphosulfate, APS); see Eq. 3.6, and the structure of APS is given in Fig. 3.2. This enzymatic activity catalyzes pyrophosphate (PP_i) elimination, and this was the first report demonstrating the activation of an inorganic ion other than phosphate. Furthermore, APS was identified as the sulfate donor with phosphorylation in the 3′-position of the ribose moiety forming 3′-phosphoadenosine-5′-phosphosulfate (PAPS) according to Eq. 3.4 (Bandurski et al. 1956). The structure of PAPS was established by Robbins and Lipmann (1957) and is shown in Fig. 3.2. Peck Jr. (1959), Ishimoto (1959), and Ishimoto and Fujimoto (1959) demonstrated independently that cell-free extracts of *Desulfovibrio* production of sulfide from sulfate were dependent on stoichiometric consumption of ATP with APS but not PAPS as an intermediate. Sulfate metabolism in *Desulfovibrio* was resolved into assimilatory

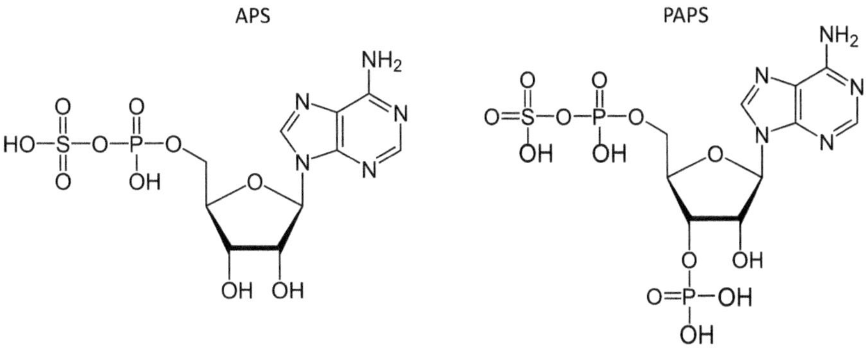

Fig. 3.2 Structures of APS and PAPS. APS adenosine 5′-phosphosulfate, PAPS 3′-phosphoadenosine-5′-phosphosulfate

reduction for biosynthesis of sulfur-containing compounds and dissimilatory reduction for respiration (Peck Jr. 1961a). For the dissimilatory reduction of sulfate to sulfide, bacteria and archaea reduce APS to bisulfite (Eq. 3.3) with bisulfite being reduced to hydrogen sulfide (Eq. 3.4). Components of these pathways were initially discussed (Peck Jr. 1960; Ishimoto et al. 1961; Ishimoto and Yagi 1961; Peck Jr. 1962b) and more recently reviewed by Akagi (1981, 1995) and by Peck Jr. and Lissolo (1988).

$$\mathrm{ATP} + \mathrm{SO_4^{2-}} = \mathrm{APS} + \mathrm{PP_i} \quad \text{ATP sulfurylase} \quad \Delta G^{0'} = +46.0\ \mathrm{kJ/mol} \tag{3.1}$$

$$\mathrm{PP_i} + \mathrm{H_2O} \rightarrow 2\mathrm{P_i} \quad \text{Inorganic pyrophosphatase} \quad \Delta G^{0'} = -21.9\ \mathrm{kJ/mol} \tag{3.2}$$

$$\mathrm{APS} + \mathrm{H_2} = \mathrm{AMP} + \mathrm{HSO_3^-} + \mathrm{H^+} \quad \text{APS reductase} \quad \Delta G^{0'} = -68.6\ \mathrm{kJ/mol} \tag{3.3}$$

$$\mathrm{HSO_3^-} + 3\mathrm{H_2} + \mathrm{H^+} = \mathrm{H_2S} + 3\mathrm{H_2O} \quad \text{Bisulfite reductase} \quad \Delta G^{0'} = -171.7\ \mathrm{kJ/mol} \tag{3.4}$$

3.2.1 ATP Sulfurylase

The production of AMP-sulfate or adenosine 5′-phosphosulfate (APS) in this reaction is catalyzed by sulfate adenylyltransferase (Sat) (EC 2.7.7.4), and this enzyme is alternately known ATP sulfurylase (ATPS), adenosine-5′-triphosphate sulfurylase, or adenylylsulfate pyrophosphorylase. Using partially purified Sat from *Desulfotomaculum* (formerly *Clostridium*) *nigrificans* strain 8351 and from *Desulfovibrio vulgaris* (formerly *desulfuricans*) strain 8303, the equilibrium constant (Eq. 3.5) for the sulfurylase reaction (Eq. 3.1) was calculated to be K_{eq} of 1.8 ×

10^{-8} and 6.2×10^{-9}, respectively (Akagi and Campbell 1962). This equilibrium constant was similar to the 4–5 $\times$ 10^{-8} value calculated by Robbins and Lipmann (1958) using Sat from yeast:

$$K_{eq} = [APS] \times [PP_i]/[ATP] \times \left[SO_4^{2-}\right] \tag{3.5}$$

The reaction (Eq. 3.1) catalyzed by Sat would not favor the production of APS and PP_i unless one or both of these products are consumed. The presence of an inorganic pyrophosphatase would shift the equilibrium of the reaction and enable the activation of sulfate with APS formation to proceed.

Several characteristics of Sat obtained from *Desulfotomaculum* (*Dst.*) *nigrificans* and *Desulfovibrio* (*D.*) *vulgaris* have broad application to the enzymatic activity of other sulfate reducers. Inhibitors and stimulators of ATP sulfurylase are given in Fig. 3.3. Sat is a cytoplasmic enzyme that uses ATP but not GTP, CTP, ICP, or UTP, and the inhibition of sulfate reduction by molybdate, tungstate, or chromate is attributed to the formation of an unstable intermediate resulting in the release of the anion and AMP. Selenate is a strong inhibitor of APS formation, and the intermediate selenium compound is more stable than when Sat is exposed to group VI anions (Baliga et al. 1961; Akagi and Campbell 1962). With Sat isolated from *D. desulfuricans* ATCC 27774 and *D. gigas*, evaluation using extended X-ray

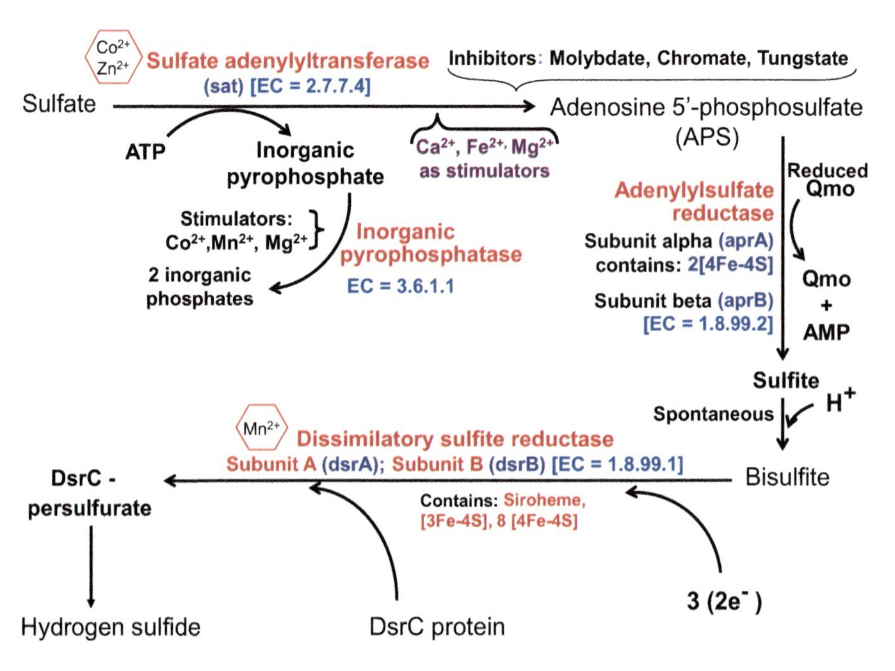

Fig. 3.3 Pathway for sulfide production from dissimilatory sulfate reduction by *Desulfovibrio* species. (Reproduced by permission from Barton et al. (2014), Copyright 2014, with permission from Springer Science & Business Media BV)

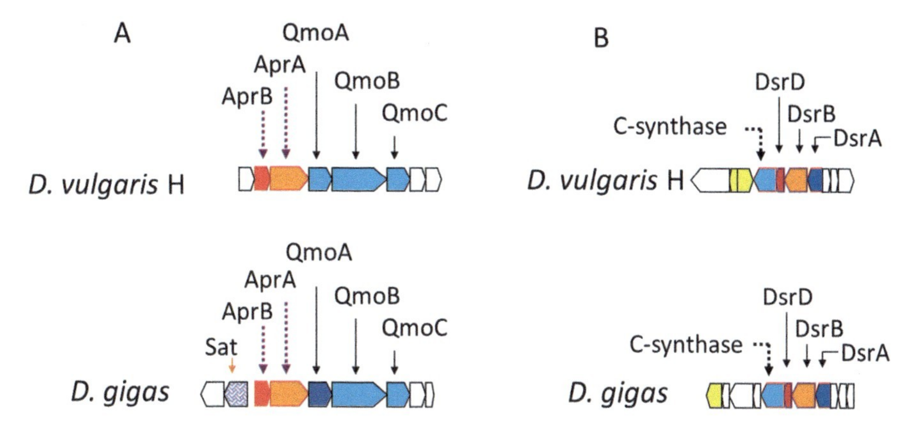

Fig. 3.4 Role of Qmo and Apr proteins associated with sulfate reduction. Qmo quinone-interacting membrane-bound oxidoreductase, Apr adenosine 5′-phosphosulfate reductase. (Reproduced by permission from Barton et al. (2014), Copyright 2014, with permission from Springer Science & Business Media BV)

absorption fine structure (EXAFS) and electron paramagnetic resonance (EPR) revealed that Co^{2+} and Zn^{2+} ions are coordinated to three cysteine residues and one histidine residue in a tetrahedral coordination (Gavel et al. 1998). The Co^{2+} and Zn^{2+} ions are not involved in catalysis, but they are proposed to contribute to structural stability (Gavel et al. 2008). Crystal structure analysis of ATP sulfurylase from sulfur-oxidizing bacteria (*Thermus thermophilus* and *Allochromatium vinosum*) revealed the presence of tetrahedral coordination sites for Zn^{2+} ions (Taguchi et al. 2004;Parey et al. 2013). Sat purified from *D. desulfuricans* ATCC 27774 and *D. gigas* is a homotrimer of 141 kDa (47 kDa monomers) and 147 kDa (49 kDa monomers), respectively (Gavel et al. 1998). Characteristic of the dissimilatory sulfate reduction pathway is a gene cluster (see Fig. 3.4), which includes *sat, aprBA,* and *qmoABC* (Meyer and Kuever 2007a, b).

3.2.2 *Inorganic Pyrophosphatase*

In yeast, the coupling of inorganic pyrophosphatase (PPase, EC 3.6.1.1) in Eq. 3.2 to the ATP sulfurylase reaction as given in Eq. 3.1 accounts for the production of APS from sulfate and ATP (Wilson and Bandurski 1958). A strong pyrophosphatase was reported in extracts of *D. vulgaris* by Peck Jr. (1959), and Baliga et al. (1961) proposed the presence of two PPase in this bacterium with a soluble enzyme activated by Mg^{2+} and an insoluble enzyme activated by Co^{2+}. However, Akagi and Campbell (1963) reported the presence of the only soluble PPase in *D. vulgaris*. The PPase purified from *D. vulgaris* strain 8303 had a molecular mass of 41.6 kDa with a metallic cofactor of Mn^{2+} (Ware and Postgate 1971). Activation of PPase from *D. vulgaris* was achieved with the addition of cysteine, sodium sulfide, sodium sulfite, or a reductant with a half-cell activity of less than $E^{0\prime} = -150$ mV. Extracts of cultures from *D. desulfuricans* Essex 6, *D. desulfuricans* strain Norway 4, *D. gigas*, and *Clostridium pasteurianum* strain W produced PPase, which also required activation (Ware and Postgate 1971). PPase are also present in *Desulfotomaculum* spp. and account for APS formation. High levels of PPase activity were reported for *Dst. orientis*, while lower levels were found with extracts of *Dst. ruminis, Dst. acetoxidans*, and *Dst. nigrificans* (Thebrath et al. 1989). PPase is encoded on the *ppa*C gene.

3.2.3 APS Reductase

Following the demonstration of APS as an intermediate in the sulfate reduction pathway, APS reductase was reported for SRP using *D. vulgaris* as the model bacterium (Peck Jr. 1961b; Ishimoto and Fujimoto 1961). In the presence of APS and a suitable electron source, APS reductase (EC 1.8.99.2) produces AMP and bisulfite (Eq. 3.3). Neither NADH nor NADPH directly serve as electron donors, and reduced methyl viologen (MV) generated by a H_2-hydrogenase coupled reaction is commonly used in the APS reductase assay (Peck Jr and LeGall 1982). Using APS reductase reaction prepared from *D. vulgaris,* the mechanism of APS reduction was examined and was found to involve a FAD-adduct (Michaels et al. 1970; Peck Jr and Bramlett 1982; Peck Jr and LeGall 1982). The APS reductase purified from *D. vulgaris* had a molecular mass of 220 kDa with subunits of 20 kDa and 72 kDa, one FAD, and 12 atoms of nonheme iron per molecule (Bramlett and Peck Jr. 1975).

While most of the early research on APS reductase had focused on species of *Desulfovibrio,* this enzyme has also been demonstrated in *Desulfobacter postgatei, Desulfococcus multivorans, Desulfobulbus propionicus, Desulfosarcina variabilis, Desulfotomaculum* sp., and *Archaeoglobus fulgidus* (Skyring and Trudinger 1973; Stille and Trüper 1984; Speich and Trüper 1988). The physicochemical properties of APS reductase from several sulfate-reducing organisms are similar, and some of the properties are summarized in Table 3.1. When using the bacteria and archaea listed in Table 3.2 as a reference, the APS reductases have the following characteristics: pH optimum 7.4 to 8.0 and K_m for sulfite of 0.34 to 1.3 mM (Lampreia et al. 1994). Additionally, the APS reductase in *Desulfovibrio* has been demonstrated to be readily reversable (Peck Jr. 1961b), and this supports the close relationship between the sulfate-reducing and the sulfur-oxidizing bacteria.

It has been estimated that 2–3% of cytoplasmic protein in *D. vulgaris* is APS reductase (Peck Jr and LeGall 1982). Using immunocytochemical localization, APS reductase in *D. thermophilus* is membrane associated (Kremer et al. 1988). The APS reductase in *Chromatiaceae* functions as in the synthesis of APS from bisulfite plus AMP, and this enzyme is associated with the membranes of the chromatophores (Trüper and Fischer 1982). A membrane complex has been isolated from *D. vulgaris* and *D. alaskensis* G20, and this protein complex contains APS reductase and Qmo

Table 3.1 Characteristics of selected bacterial and archaeal high-spin and low-spin sulfite reductases. Iron compounds are listed for intact enzyme structure

Protein	λmax (nm)	Organism	Mr (kDa)	Non-heme iron	Siroheme (sirohydro-chlorin)	[4Fe-4S] cluster	References
Low-spin "assimilatory-type" sulfite reductase							
	590	*D. vulgaris* Hildenborough	27.2	5	1	1	Lee et al. (1973b); Huynh et al. (1984)
	587	*Drm. acetoxidans*[e]	23.6	5	1	1	Moura et al. (1982)
	590	*Ms. barkeri* [e]	23	5	1	1	Moura and Lino (1994); Moura et al. (1986)
High-spin "dissimilatory-type" bisulfite reductase							
Desulfoviridin	628	*D. gigas*	200	34	2(2)	8	Lee and Peck Jr. (1971); Lee et al. (1973a); Hsieh et al. (2010); Oliveira et al. (2011)
			~400				Lampreia et al. (1990)
	630	*D. vulgaris* Hildenborough	200	34	2(2)	8	Oliveira et al. (2008a, b)
P-582	582	*Dst. nigrificans*	194	16	1.3	4	Akagi and Adams (1973); Trudinger 1970
	582	*Dst. thermocisternum*[a]	196	16	2	4	Larsen et al. (1999)
Desulfofuscidin	576	*T. commune*[b]	167	20–21	4	4	Hatchikian and Zeikus (1983); Hatchikian (1994)
	578	*T. mobile*	190	32	4	8	Fauque et al. (1990)
Desulforubidin	545	*Dsm. novegicum*[c]	225	36	4	8	Fauque et al. (1991); Oliveira et al. (2011); Moura et al. (1988); DerVartanian (1994); Lee et al. (1973a)
	545	*Ds. variabilis*	208	15	2	8	Arendsen et al. (1993)
Archaeal	593	*A. fulgidus*	218	36	4	8	Schiffer et al. (2008); Dahl et al. (1993)
	NR[d]	*A. profundus*	198	24	2	6	Larsen et al. (1999)

[a]Calculated, [b]from gene analysis. [c]Formerly known as *D. thermophilus*. Formerly known as *D. desulfuricans* Norway 4 and *D. baculatus* Norway 4. [d]*NR* not reported. [e]Sulfur reducers and not sulfate reducers

Table 3.2 Homologous genes for dissimilatory sulfate reduction[a]

Bacteria/Archaea___	*sat*	*aprB*	*aprA*	*qmoA*	*qmoB*	*qmoC*
Desulfovibrio vulgaris 8303	DVU1295	DVU0846	DVU0847	DVU0848	DVU0849	DVU0850
Desulfovibrio alaskensis G20	Dde_2265	Dde_1109	Dde_1110	Dde_1111	Dde_1112	Dde_1113
Desulfotalea psychrophilia LSv54	DP1 4 72	DP1104	DP1105	DP1106	DP1107	DP1108
Desulfotomaculum reducens MI-1	Dred-DRAFT_3079	Dred-DRAFT_3080	Dred-DRAFT_3081	Dred-DRAFT_3082	Dred-DRAFT_3083	NA
Archaeoglobus fulgidus	AF1667	AF1669	AF1670	AF0663	AF0662	AF0611

[a]Meyer and Kuever (2007b, 2008). *NA* not apparent

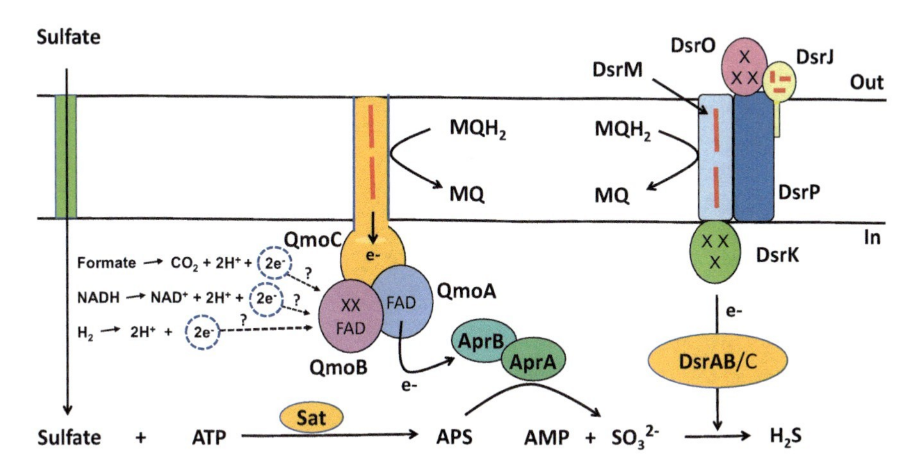

Fig. 3.5 Schematic representation of the QmoABC-AprAB interaction and the proposed involvement of third partners. (**a**) In the hypothesis of an electron bifurcation process, the putative electron acceptor of QmoB with a high redox potential is represented by a question mark. (**b**) In the hypothesis of an electron configuration mechanism, several possible co-electron donors for the Qmo complex are considered: Fd ferredoxin, Hase hydrogenase, Fdh formate dehydrogenase, or Nox NADH dehydrogenase. The soluble HdrABC-MvhGAD complex (**c**) and the membrane-bound HdrED (D) of methanogens are shown for comparison. The gray-dashed arrows represent electron bifurcation in (**a**, **c**) or electron configuration in (**b**). The gray boxes represent the cytoplasmic membrane with "+" indicating the periplasm and "−" the cytoplasm. (Source of information used in this figure is Canfield et al. 2010; Keller and Wall 2011 and Ramos et al. 2012)

(quinone-interacting membrane-bound oxidoreductase) (Krumholz et al. 2013). This membrane complex of APS reductase and Qmo is consistent with the proposition that Qmo provides electrons for APS reduction (Ramos et al. 2012). Using cyclic voltammetry, experiments demonstrated the transfer of electrons from Qmo to APS reductase, and this strengthened the proposal that Qmo is the physiological electron donor for APS reductase (Duarte et al. 2016). A model showing the interaction of Qmo with APS reductase is presented in Fig. 3.5.

The dissimilatory APS reductase (Apr) is a heterodimer consisting of an alpha subunit, which is 75–80 kDa, and a beta subunit, which is 18–23 kDa encoded on the *aprBA* gene (Lampreia et al. 1994; Speich et al. 1994; Hipp et al. 1997; Molitor et al. 1998; Fritz et al. 2000). The alpha and beta subunits (AprA and AprB) of APS reductase are present in a 1:1 configuration (Fritz et al. 2000), and a FAD molecule is bound onto the alpha subunit, while two [4Fe-4S] centers are bound onto the beta subunit. Analysis of crystalline APS reductase from *A. fulgidus* revealed that the FAD-adduct was formed following the nucleophilic attack of the N5 atom of reduced FAD on the sulfite moiety of APS (Fritz et al. 2002). A three-dimensional structure of APS reductase from *A. fulgidus* is presented in Fig. 3.6.

Electrons from the surface of the APS reductase molecule would be transferred initially to the [4Fe-4S] clusters and finally to the FAD moiety associated with the enzyme. Analysis of the APS reductase purified from *D. gigas* revealed that the enzyme has two subunits (70 kDa and 23 kDa) and a native form of about 400 kDa

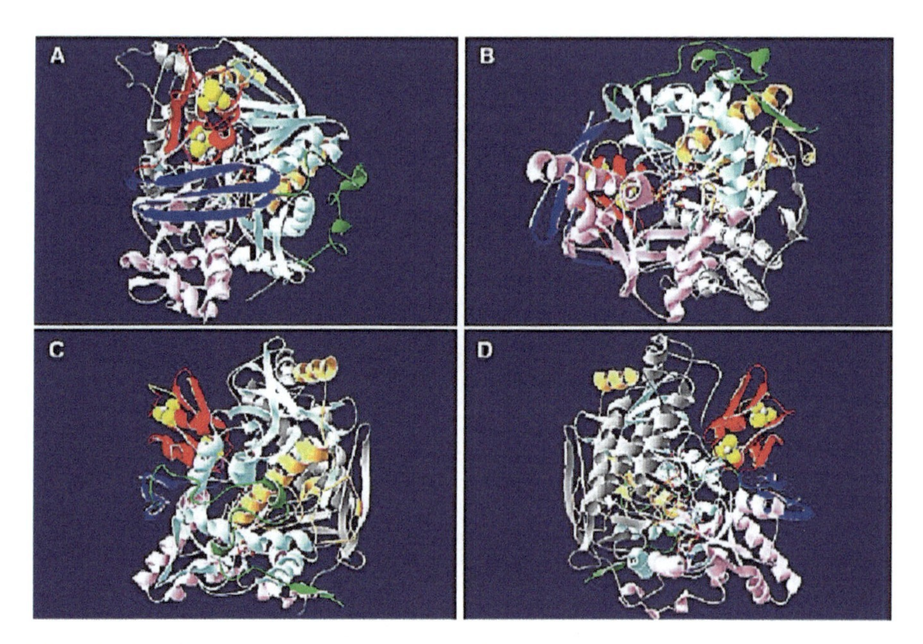

Fig. 3.6 Three-dimensional ribbon structure of APS reductase from *A. fulgidus* (Meyer and Kuever 2008). The beta-subunit segments are colored red (ferredoxin segment), blue (3 antiparallel beta-sheets segment), and green (tail segment); the alpha-subunit domains are colored light blue and orange (FAD-binding domain I and II), pink (capping domain), and gray (helical domain). The [4Fe-4S] clusters, FAD, and substrate APS are shown as ball-and-stick representations; tryptophan Trp-B48 of AprB is highlighted by the color violet. Ribbon structure is shown from (**a**) top view, (**b**) bottom view (substrate channel), (**c**) front view, and (**d**) back view. (https://doi.org/10.1371/journal.pone.0001514.g001. Meyer and Kuever (2008).

with center I (FAD plus one [4Fe-4S] center) having a redox potential of 0 mV and center II (the remaining Fe-S center) having a redox potential of <−400 mV (Lampreia et al. 1990, 1994). A reaction cycle of APS reductase from *A. fulgidus* is presented in Fig. 3.7.

The crystalline structure of APS reductase purified from *D. gigas* revealed an asymmetric unit consisting of a hexamer structure made up of six αβ-heterodimers, but the unit contains the $\alpha_2\beta_2$-heterotetramer found with *A. fulgidus* (Chiang et al. 2009). The choice of buffers (phosphate vs. Tris) is suggested to be important in the molecular mass of APR reductase in *A. fulgidus* (Chiang et al. 2009). Crystalline analysis indicated that the C-terminal segment of the β-subunit attaches to the α-subunit in a fashion where the C-terminus resides in the active site channel of an adjoining α-subunit of another αβ-heterodimer and thereby provides a type of self-regulation (Chiang et al. 2009).

When comparing the AprA subunits from *D. vulgaris, A. fulgidus,* and *Chromatium vinosum*, there are several similarities, but the AprB subunits for

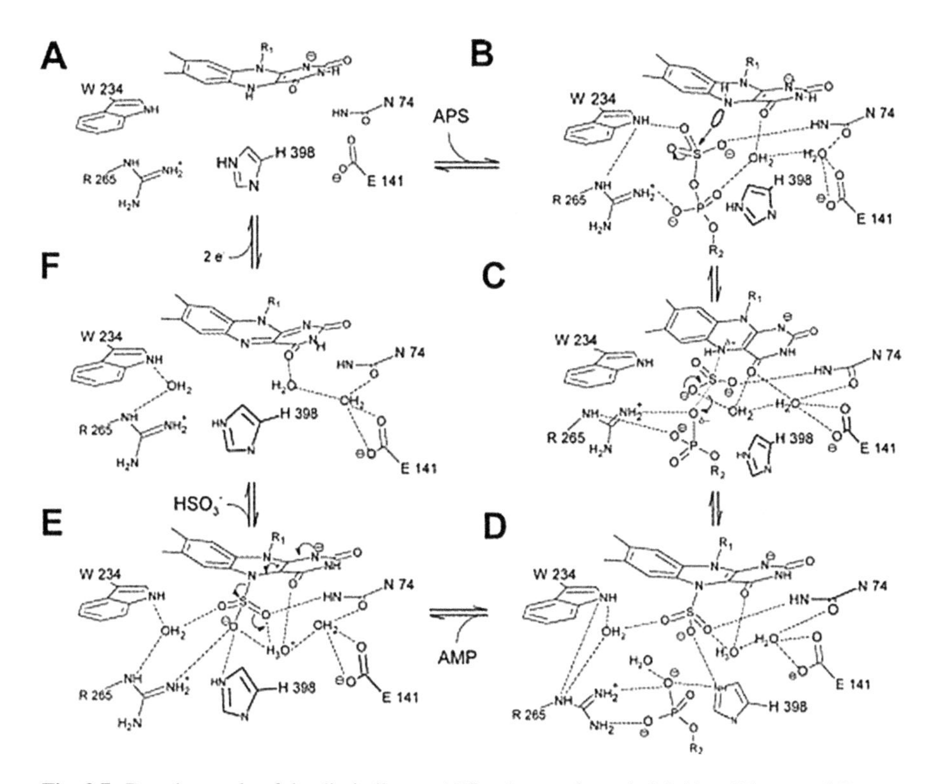

Fig. 3.7 Reaction cycle of the dissimilatory APS reductase from *A. fulgidus*. (Meyer and Kuever 2008. Copyright: 2008 Meyer, Kuever. This is an open-access article distributed under the terms of the Creative Commons Attribution License, which permits unrestricted use, distribution, and reproduction in any medium, provided the original author and source are credited)

these same prokaryotes contain apparent differences and suggest that a probe for AprA to identify dissimilatory sulfate reducers is better than using a probe for AprB (Hipp et al. 1997). In an evaluation of the full-length amino acid sequence of AprA from 14 sulfur-oxidizing and sulfate-reducing prokaryotes, six different clusters were obtained (Meyer and Kuever 2007b). The *aprAB* and *qmoABC* gene arrangements for prokaryotes in several clusters are provided in Fig. 3.8.

3.3 Assimilatory Sulfate Reduction

In most bacteria and yeast, the assimilation of sulfate into cysteine and other sulfur-containing compounds utilizes PAPS as a sulfur intermediate (Eqs. 3.6–3.8). APS kinase (EC 2.7.1.25) catalyzes the formation of PAPS from APS, and this enzyme is alternately called ATP:adenylyl-sulfate 3′-phosphotransferase, adenylylsulfate kinase, adenosine 5′-phosphosulfate kinase, adenosine phosphosulfate kinase, or adenosine-5′-phosphosulfate-3′-phosphokinase. The pathway for assimilatory

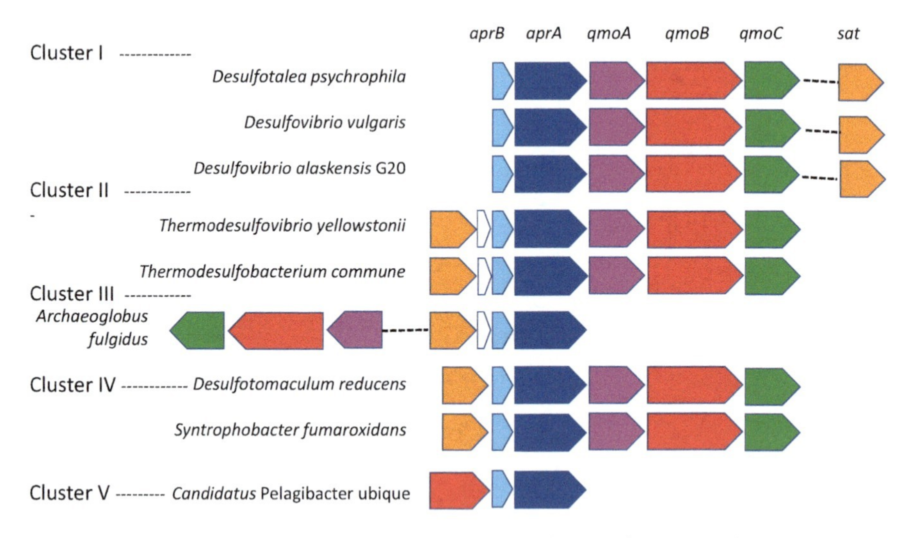

Fig. 3.8 Arrangement of *aprAB* and *qmoABC* genes in a selection of microorganisms. (Meyer and Kuever 2007b)

sulfate reduction by non-oxygenic photosynthetic microorganisms and some cyanobacteria involves PAPS as the sulfur intermediate (Eq. 3.5). PAPS is formed from APS by APS kinase, and sulfite is produced from PAPS by PAPS reductase, which requires NADPH (Eq. 3.7). A sulfite reductase reduces sulfite to hydrogen sulfide according to Eq. 3.8. Sulfide is used for the synthesis of cysteine and in bacteria practicing assimilatory sulfate reduction, there is a close regulation of sulfate reduced with the synthesis of organic sulfur compounds so there is no production of unused sulfide. Two unique bacteria that release hydrogen sulfide into the medium following assimilatory sulfate reduction include a strain of *Bacillus megaterium* (Bromfield 1953) and a strain of *Pseudomonas zelinskii* (Shturm 1948).

$$\mathrm{APS} + \mathrm{ATP} = \mathrm{ADP} + \mathrm{PAPS} \quad \text{APS kinase} \tag{3.6}$$

$$\mathrm{NADPH} + \mathrm{PAPS} = \mathrm{NADP}^+ + \mathrm{PAP} + \mathrm{HSO}_3^- \quad \text{PAPS reductase} \tag{3.7}$$

$$\mathrm{HSO}_3^- + 3\mathrm{NADPH} + 4\mathrm{H}^+ \rightarrow \mathrm{H_2S} + 3\mathrm{NADP}^+ + 3\mathrm{H_2O} \quad \text{Assimilatory sulfite reductase} \tag{3.8}$$

While the PAPS pathway is commonly associated with heterotrophic bacteria, yeast, and plants for the assimilatory synthesis of cysteine from sulfate, SRB use a different pathway. *Desulfovibrio*, *Desulfotomaculum,* and presumedly other sulfate-reducing microorganisms employ an assimilatory sulfite-reducing enzyme according to Eqs. 3.9 and 3.10 (Peck Jr. 1961a, b).

$$APS + 2e^- = AMP + S0_3^{2-} \quad \text{APS reductase} \tag{3.9}$$

$$SO_3^{2-} + 6e^- + 6H^+ = S^{2-} + 3H_2O \quad \text{Assimilatory sulfite reductase} \tag{3.10}$$

One of the two classes of sulfite reductase in sulfate-respiring microorganisms is the assimilatory sulfite reductase (aSiR) (EC 1.8.99.1), which has a low molecular mass and a siroheme prosthetic group in a low-spin state (Moura and Lino 1994). The other class of sulfite reductase (dSiR) (EC 1.8.99.3) has two subunits (DsrA and DsrB) and a siroheme cofactor in a high-spin state (Barton et al. 2014).

The assimilatory sulfite reductase had been demonstrated in *D. vulgaris, D. gigas,* and *D. desulfuricans* (Peck Jr. 1962b). The six-electron reduction step in Eq. 3.10 does not produce any intermediates with reduced methyl viologen as an artificial electron donor with purified sulfite reductase. Ferredoxin and flavodoxin were also effective electron shuttle molecules with crude enzyme fractions (LeGall and Dragoni 1966; LeGall and Hatchikian 1967). While NADPH was an effective electron donor with the *Escherichia coli* sulfite reductase (Siegel and Kamin 1968), NADH and NADPH were not appropriate electron acceptors for the sulfite reductase obtained from *D. vulgaris* Hildenborough (Lee et al. 1973b). The sulfite reductase from *D. vulgaris* Hildenborough contains a single polypeptide chain with one siroheme, one [4Fe-4S], and a molecular mass of 27.2 kDa (Huynh et al. 1984). A siroheme prosthetic group is a tetrahydroporphyrin with Fe-chelated, and the trivial name is sirotetrahydroporphyrin (Akagi 1995). As measured by EPR, a distinctive characteristic of this assimilatory sulfite reductase is the siroheme, which is in a low-spin ferric state with g values of 2.40 and 2.33 (Moura et al. 1986). An assimilatory sulfite reductase was also isolated from *Desulfuromonas acetoxidans* strain 5071 and *Methanosarcina barkeri* DSM 800 (Moura et al. 1986), and the molecular characteristics were similar to those of *D. vulgaris* Hildenborough; see Table 3.1. It has been suggested that this low-spin sulfite reductase is important in methanogens (Johnson and Mukhopadhyay 2008; Susanti and Mukhopadhyay 2012) and numerous other anaerobes (Canfield et al. 2005). In *D. vulgaris*, cysteine synthesis from sulfide is caused by the following reactions (Gevertz et al. 1980):

$$\text{Acetyl-CoA} + \text{L-serine} \rightarrow \text{O-acetyl-L-serine} + \text{CoA} \tag{3.11}$$

$$\text{O-acetyl-L-serine} + \text{sulfide} \rightarrow \text{L-cysteine} + \text{acetate} \tag{3.12}$$

Serine transacetylase (EC 2.3.1.30) catalyzes Eq. 3.11, and cysteine synthase (EC 4.2.99.8) is responsible for Eq. 3.12. It may seem unnecessary to have a special assimilatory sulfate reduction enzyme in *Desulfovibrio* since there is a large amount of sulfide released from the dissimilatory sulfate reduction pathway. However, many *Desulfovibrio* grow in environments where the sulfate concentration is low and electron acceptors other than sulfate are used to support growth.

3.4 Dissimilatory Sulfate Reduction

Dissimilatory sulfate reduction is a unique physiological process, and the enzymology of this sulfur pathway is the hallmark characteristic of several species of anaerobic bacteria and archaea. In addition to sulfate respiration, there is the coupling of sulfide production from sulfate with the generation of ATP. As in all instances where sulfate reduction occurs, the initial enzyme reaction for sulfate reduction is the ATP-dependent activation of sulfate to APS by ATP sulfurylase as given in Eq. 3.1. Dissimilatory sulfite reduction starts with sulfite produced by APS reductase. As indicated in Eqs. 3.13 and 3.14, the reduction of bisulfite with electrons from the oxidation of H_2 is highly exergonic and serves to provide energy for cells growing on sulfate as the electron acceptor.

$$HSO_3^- + 3H_2 + H^+ \rightarrow HS^- + 3H_2O + H^+ \quad \text{Bisulfite reductase} \quad \Delta G^{0'} = -171.7\ \text{kJ/mol} \tag{3.13}$$

$$HSO_3^- + 6e^- + 6H^+ \rightarrow HS^- + 3H_2O \quad E^{0'} = -116\ \text{mV} \tag{3.14}$$

As reported by Peck Jr. (1961b, a) and Peck Jr. and Lissolo (1988), the enzyme associated with dissimilatory sulfate reduction is not the same as that required for assimilatory sulfate reduction. The production of bisulfite ($HS0_3^-$) results from the action of APS reduction, which requires APS and two reducing equivalents. As shown in Fig. 3.1, the pathway for the production of APS from sulfate is the same in both the assimilatory and dissimilatory sulfate reduction. The essential enzyme for dissimilatory sulfate enzyme involves the reduction of $HS0_3^-$ to HS^-, and it is designated as bisulfite reductase, while the enzyme for assimilatory sulfate reduction is referred to as sulfite ($S0_3^{2-}$) reductase. The use of bisulfite reductase and sulfite reductase was proposed by Lee and Peck Jr. (1971) to identify the dissimilatory and assimilatory sulfate pathways; however, reports often use sulfite reductase for both pathways. The enzymes for dissimilatory sulfate reduction (ATP sulfurylase, APS reductase, and bisulfite reductase) are broadly distributed in SRB but can be distinguished by differences of electrophoretic mobilities (Skyring and Trudinger 1973).

3.4.1 Dissimilatory Bisulfite Reductase

The protein responsible for bisulfite reduction in *D. vulgaris* strain Hildenborough was established by Lee and Peck Jr. (1971) to be desulfoviridin. Years earlier, Postgate (1956) reported the isolation of an emerald green protein, which he designated as desulfoviridin, but at that time he was unable to identify a physiological role for this porphyroprotein. A tetrahydroporphyrin prosthetic group is also found in the assimilatory sulfate reductase of several SRB. The dissimilatory sulfite reductase is distinguished from assimilatory sulfite reductase in that the assimilatory reductase is a member of the low-spin class while the dissimilatory sulfite reductases belong to the high-spin group (Fauque et al. 1991; LeGall and Fauque 1988; Barton and Fauque 2009; Fauque and Barton 2012). Based on their siroheme moieties, EPR spectra, and major wave length for optical absorption, the dissimilatory sulfite reductases are assigned to five different groups (Table 3.2). Following an extensive in silico review of dissimilatory sulfite-reduction genes, several clusters of microorganisms based on the arrangement of dissimilatory sulfite reductase Dsr (dSiR) genes were reported (Fig. 3.4) (Meyer and Kuever 2007a, b).

The bisulfite reductase known as desulfoviridin is found in species of *Desulfococcus, Desulfomonile, Desulforegula, Desulfonema,* and *Desulfovibrio,* and the enzyme isolated from *Desulfovibrio* has been extensively studied (Moura et al. 1988; Fauque et al. 1991; Steuber et al. 1994; Wolfe et al. 1994; Steuber et al. 1995; Steuber and Kroneck 1998). The reaction of dissimilatory sulfite reduction requires the action of bisulfite reductase (DSR, also known as dSiR), and the central proteins involved are DsrA and DsrB, which are encoded on the *dsrAB* genes. Generally, the dissimilatory bisulfite reductase has the $\alpha_2\beta_2$ structure containing siroheme-[Fe-S] clusters and [4Fe-4S] clusters located some distance from the siroheme-[Fe-S] cluster. In the dissimilatory sulfite reductase (DsrAB), siroheme-[Fe-S] clusters are found in each of the α-subunit and β-subunit structures, but only one of these siroheme-[Fe-S] clusters is active (Simon and Kroneck 2013). It has been suggested that the DsrAB structure attributed to *dsrA* and *dsrB* resulted from an early gene duplication (Grein et al. 2013).

The binding of a siroheme-[Fe-S] cluster is attributed to the (Cys-X_5 – Cys) -X_n-Cys-X_3-Cys) motif, and the [4Fe-4S] clusters on the α-subunit and β-subunit structure are attributed to the (Cys-X_2-Cys-X_2-Cys) sequence (Adman et al. 1976; George et al. 1985; Bruschi and Guerlesquin 1988; Ostrowski et al. 1989). Sirohemes are also associated with bacterial nitrite reductases, and the structural features of sirohemes were reviewed by Simon and Kroneck (2013). While the siroheme contains eight carboxylate side chains, it has been reported that in some *Desulfovibrio*, one of the carboxylate side chains is replaced at the 2'acetate with an amide (Matthews et al. 1995). Sirohydrochlorin (iron-depleted siroheme) is reported for *D. vulgaris* and *D. gigas*; see Table 3.2 for references.

The $\alpha_2\beta_2\gamma_2$ polypeptide structure is reported for the Dsr from *D. vulgaris* Hildenborough, *D. vulgaris subspecies oxamicus* (Monticello), *D. gigas,* and *D. desulfuricans* ATCC 27774 (Lee et al. 1973b; Pierik et al. 1992). The γ polypeptide subunit is 11 kDa and is encoded on *dsrC,* which is located some distance from the *dsrAB* operon. Expression of the *dsrC* gene is reported to be expressed at a high level (Canfield et al. 2010; Keller and Wall 2011), and *dsrC* is found in all organisms that have the *dsrAB* operon (Venceslau et al. 2014). It has been proposed that electron donors for the Qmo complex could be hydrogenase, formate dehydrogenase, or NADH dehydrogenase (Ramos et al. 2012), and this is presented in a model in Fig. 3.5.

Alternately, the reduction of sulfite to sulfide has been proposed to be a sequential two-electron reduction (Fig. 3.1), and in *A. fulgidus,* the intermediate sulfur compounds would be formed at the DsrABC active site (Parey et al. 2010). This would account for the formation of trithionate and thiosulfate produced with the reduction of sulfite by what has often been termed the "trithionate pathway" and is presented as follows (Akagi 1995) with free energy of the Eqs. 3.15–3.17 as reported by Broco et al. (2005).

$$3HSO_3^- + 2e^- + 3H^+ \rightarrow S_3O_6^{2-} + 3H_2O \quad \text{trithionate-forming system} \quad \Delta G^{0'} = -46.3\ \text{kJ/mol} \tag{3.15}$$

$$S_3O_6^{2-} + 2e^- + H^+ \rightarrow S_2O_3^{2-} + HSO_3^- \quad \text{trithionate reductase} \quad \Delta G^{0'} = -123\ \text{kJ/mol} \tag{3.16}$$

$$S_2O_3^{2-} + 2e^- + 2H^+ \rightarrow HS^- + HSO_3^- \quad \text{thiosulfate reductase} \quad \Delta G^{0'} = -2/1\ \text{kJ/mol} \tag{3.17}$$

Using *D. vulgaris* Miyazaki, the above reactions were proposed by Kobayashi et al. (1969), and the thiosulfate-forming system was reported with *D. vulgaris* Hildenborough (Suh and Akagi 1969). Reports by Findley and Akagi (1970) and Drake and Akagi (1978) supported the trithionate pathway. Under specified conditions, the dissimilatory sulfite reductase from *D. vulgaris* Hildenborough produces only trithionate as the product (Lee and Peck Jr. 1971). It was reported from several laboratories that bisulfite reductase catalyzed the formation of trithionate, thiosulfate, and sulfide from bisulfite, and the quantities of sulf-oxy anions produced varied with assay conditions (Kobayashi et al. 1972; Kobayashi et al. 1974; Jones and Skyring 1975). The three end products of bisulfite reduction (trithionate, thiosulfate, and sulfide) have been reported for *Desulfovibrio* sp. (Jones and Skyring 1975; Drake and Akagi 1976; Fitz and Cypionka 1990; Broco et al. 2005). The potential of thiosulfate and trithionate as intermediates in the reduction of sulfate to sulfide has been reviewed by Chambers and Trudinger (1975). *Thermodesulfobacterium commune* (Hatchikian and Zeikus 1983) and *Desulfomicrobium baculatum* (Liu et al. 1979). The in vivo products of the dissimilatory sulfite reductase remain unknown, but it is evident that the dissimilatory sulfite reductase enzyme is extremely complex.

Using polyacrylamide gel electrophoresis, multiple forms of dissimilatory sulfite reductase have been reported for SRB, which were reported as major and minor bands using reduced methyl viologen as the electron donor (Suh and Akagi 1969; Jones and Skyring 1975; Seki et al. 1979). When the two forms of dissimilatory sulfite reductase were resolved from extracts of *D. vulgaris* Miyazaki, both forms produced trithionate, thiosulfate, and sulfide from sulfite, and the two forms were considered to differ in charge and conformational properties (Seki et al. 1979). More recently, three dissimilatory sulfite reductases with $\alpha_2\beta_2\gamma_2$ polypeptide structure were isolated from *D. gigas* with Dsr-I and Dsr-II having enzymatic activity, while Dsr-III is inactive (Hsieh et al. 2010). Dsr-I has eight [4Fe-4S] clusters, two saddle-shaped sirohemes, and two sirohydrochlorins, but Dsr-II has six [4Fe-4S] clusters, two sirohemes, and two hydrochlorins. Dsr-III has cofactors similar to Dsr-II; however, Dsr-III has a demetallated siroheme, and a [3Fe-4S] cluster replaces the [4Fe-4S] cluster found in Dsr-II. These three Dsr enzyme forms would appear to have resulted from variations in the assembly of the cofactors on the polypeptide chains.

3.4.2 Thiosulfate Reductase

The first report of thiosulfate and tetrathionate as electron acceptors for *D. desulfuricans* (now *vulgaris*) Hildenborough with H_2 as the electron donor was by Postgate (1951). While the enzyme associated with tetrathionate reduction in SRB has not been pursued (Barrett and Clark 1987), considerable attention has been given to the enzymes for thiosulfate reduction. Using the *D. vulgaris* strain Miyazaki, washed cells were shown to couple lactate oxidation to thiosulfate reduction (Ishimoto et al. 1954). The reduction of thiosulfate to sulfide by *D. vulgaris* has long been considered to be a two-step process with the outer sulfane sulfur reduced to sulfide in the initial step, and this is followed by the reduction of sulfite to sulfide (Ishimoto et al. 1955; Ishimoto and Koyama 1957). This reaction sequence with thiosulfate reductase is described by Akagi (1995) and is provided below (Eqs. 3.18 and 3.19):

$$^{*}SSO_3^{2-} + 2e^- + 2H^+ \rightarrow H^{*}S^- + HSO_3^- \quad \text{thiosulfate reduction} \tag{3.18}$$

$$HSO_3^- + 6e^- + 6H^+ \rightarrow HS^- + 3H_2O \quad \text{process} \tag{3.19}$$

Thiosulfate reductase (EC 1.8.2.5) has been purified from *D. vulgaris* Hildenborough (Haschke and Campbell 1971), *D. vulgaris* Miyazaki (Aketagawa et al. 1985), *D. gigas* (Hatchikian 1975), and *Dst. nigrificans* (Nakatsukasa and Akagi 1969). The molecular weight of thiosulfate reductase from *D. vulgaris* Hildenborough was determined to be 16.3 kDa (Haschke and Campbell 1971), while the same enzyme from *D. vulgaris* Miyazaki was reported to be 87 kDa (Aketagawa et al. 1985) and from *D. gigas* was 220 kDa (Hatchikian 1975). *A. fulgidus* is unique in that the purified sulfite reductase does not have a separate thiosulfate reductase because the dissimilatory sulfite reductase displays broad substrate specificity in that it reduces thiosulfate as well as sulfite (Parey et al. 2010).

The characteristic of using thiosulfate as an electron acceptor in lactate-containing medium was widely distributed in the 92 SRB isolates (Skyring et al. 1977), and this suggests that thiosulfate reductase is found in many SRB. Thiosulfate serves as an electron acceptor with lactate as the electron donor for the growth of *D. baculatus* (Rozanova and Nazina 1976), *D. thermophilus* (Rozanova and Khudyakova 1974), *D. africanus* (Jones 1971), *Dst. ruminis* (Coleman 1960), and *Thermodesulfobacterium commune* (Zeikus et al. 1983). *Desulfobulbus propionicus* grows with thiosulfate with propionate as the electron source (Widdel and Pfennig 1982), while *Desulfobacter postgatei* grows with thiosulfate when acetate is the energy source (Thauer 1982).

While NADH or NADPH are not effective electron donors for thiosulfate reductase (Nakatsukasa and Akagi 1969; Hatchikian 1975), reduced methyl viologen interacts with thiosulfate reductase to produce sulfide from thiosulfate. While in vitro reactions indicate that cytochrome c_3 and cytochrome cc_3 are highly efficient in electron transfer between hydrogenase and thiosulfate reductase (Postgate 1956; Hatchikian 1975; Bruschi et al. 1977; Aketagawa et al. 1985), physiological reactions would not involve these cytochromes because thiosulfate reductase resides in the cytoplasm, while the cytochromes are in the periplasm. Flavodoxin and ferredoxin enhance the reduction of thiosulfate by hydrogen in cell-free reactions in *D. gigas* (Hatchikian et al. 1972) but not with cell-free extracts of *D. vulgaris* (Aketagawa et al. 1985). It has been proposed that flavoredoxin (a 40 kDa FMN-containing flavoprotein) mediates the transfer of electrons to dissimilatory sulfite reductase from hydrogenase in *D. gigas* (Chen et al. 1993). However, using a mutant with the gene for flavoredoxin deleted, thiosulfate reduction was also eliminated, and this suggests that flavoredoxin is the physiological electron donor for thiosulfate reductase in *D. gigas* (Broco et al. 2005).

3.4.3 Trithionate Metabolism

A prerequisite for the trithionate pathway in dissimilatory sulfate reduction is the production of trithionate by SRB. Kobayashi et al. (1969, 1974) proposed that trithionate formation was an intermediate in dissimilatory sulfite reduction; see Eq. 3.14. When using purified bisulfite reductase (desulfoviridin) isolated from *D. gigas,* the formation of trithionate was reported by Lee and Peck Jr. (1971). While a trithionate reducing system resulting in the formation of thiosulfate has been reported for *D. vulgaris* Hildenborough, it is not present in *D. gigas* (Kim and Akagi 1985). With washed cells of *D. desulfuricans* strain Essex 6, Fitz and Cypionka (1990) demonstrated that trithionate and thiosulfate were produced from sulfite with formate or H_2 as the electron donor. Conditions for this formation of trithionate included the presence of the energy uncoupler carbonyl cyanide m-chlorophenylhydrozone (CCCP), low concentrations of electron donor, and an excess of sulfite. With sulfite as the electron acceptor, *D. gigas* has been reported to produce trithionate, thiosulfate, and sulfide; however, with sulfate, no intermediates are detected in the reduction of sulfite to sulfide (Broco et al. 2005).

3.4.4 Mechanism of Bisulfite Reduction

Several investigators have examined sulfate reduction to determine if intermediates occur in the pathway as sulfite is reduced to sulfide. As reviewed in previous sections, several reports indicate the presence of trithionate and thiosulfate as intermediates in sulfate respiration (Kobayashi et al. 1969; Suh and Akagi 1969;

Findley and Akagi 1970; Drake and Akagi 1978). The production of multiple products with sulfite reduction and the mechanism for these reactions have been discussed by Akagi (1995), Oliveira et al. (2008b), Hsieh et al. (2010), and Venceslau et al. (2014). The reduction of sulfite using three sequential steps requiring a total of six electrons appears to be highly regulated. With *D. gigas* growing on sulfite, the first intermediate formed from sulfite reduction is trithionate (see Eq. 3.14), which accumulates until sulfite is consumed, and in the absence of sulfite, trithionate is reduced with the production of thiosulfate and sulfite (see Eq. 3.15) (Broco et al. 2005). With *D. gigas* growing on sulfate, no intermediates are formed, and this may be attributed to a low intracellular concentration of sulfite attributed to the slow rate of producing sulfite from sulfate. A mechanism accounting for the presence of intermediates in sulfite reduction in *D. gigas* considers the handling of multiple sulfite residues by the dissimilatory sulfite reductase (Hsieh et al. 2010). There is only one binding site for sulfite on bisulfite reductase for the formation of trithionate and that is to the Fe^{3+} of the siroheme. The first sulfite is bound to Fe, where it is reduced to a sulfide and remains bound to the Fe. With the addition of a second sulfite, the bound sulfide is oxidized partially to produce thiosulfate, and with the addition of a third sulfite, trithionate is produced. An important protein that interacts with sulfites added to DsrAB is the γ-subunit, which can assume several different conformations (Hsieh et al. 2010).

Analysis of crystalline DsrAB crystal from *D. vulgaris* (Oliveira et al. 2008b) and *A. fulgidus* (Schiffer et al. 2008) revealed that there are two siroheme-[Fe-S] clusters per α-subunit and β-subunit: the cofactor bound onto the DsrB subunit is catalytic for sulfite reduction, and the second siroheme-[Fe-S] cofactor bound onto DsrA is inactive. Basic amino acids surround the catalytically active siroheme-[Fe-S] cluster, and sulfite travels through a positively charged channel to the active site. This positively charged channel is absent in the DsrA and is proposed to account for the inactive siroheme-[Fe-S] cluster on DsrA. In *D. vulgaris* Hildenborough, the DsrC (γ-subunit) subunit is positioned along the interface of the DsrA and DsrB subunits, and the DsrC participates in numerous functions (Venceslau et al. 2014) including the reduction of sulfite. The expression of the γ-subunit gene is not under the same regulation as those of the α-subunit and β-subunit genes (Karkhoff-Schweizer et al. 1993). Crystalline analysis of DsrC from *Desulfovibrio vulgaris* provides an understanding of the interaction of DsrC with the DsrAB subunits (Oliveira et al. 2008a, b). The DsrC subunit is usually isolated along with the DsrAB protein; however, the number of DsrC molecules associated with the DsrAB protein when isolated from disrupted cells may vary from 0 to 2 (Pierik et al. 1992). It has been proposed that the initial reduction of sulfite is a four-electron reduction step resulting in the formation of an unstable sulfur intermediate. The sulfur atom from sulfite would associate with two terminal cysteines on the DsrC subunit to produce a trisulfide, and with the addition of two electrons, the sulfur atom is released as sulfide. Electrons for the reduction of the DsrC-trisulfide are proposed to be from the quinone pool in the membrane and are delivered to the DsrC by the membrane-bound Dsr-MKJOP complex (Oliveira et al. 2008b; Santos et al. 2015). Sulfide and the three H_2O molecules (Eqs. 3.12 and 3.13) would exit through a water channel, which has a diameter that is too restrictive for sulfite to traverse (Hsieh et al. 2010). Also associated with the *dsrAB* operon is the *dsrD* gene, and DsrD is suggested to interact with DNA and may have a regulatory activity (Mizuno et al. 2003).

Initially the activity of bisulfite reductase was suggested by Drake and Akagi (1978) to involve membranes and the association of bisulfite reductase was discussed by Badziong and Thauer (1980), Kremer et al. (1988), Thauer (1989), and Fitz and Cypionka (1990). A membrane-associated bisulfite reductase was purified from *D. desulfuricans* (Steuber et al. 1994), and the membrane activity was reported by Pires et al. (2006). The Dsr-MKJOP proteins were purified from *D. desulfuricans* ATCC 27774, and the corresponding genes were identified in the *D. vulgaris* genome (Pires et al. 2006). This transmembrane complex has *b* and *c*-type hemes and iron sulfur centers for electron transfer, and the composition of the membrane-associated proteins is as follows: DsrM is an integral membrane protein, which contains cytochrome *b*, DsrK is on the cytoplasm side of the membrane and is similar to the subunit HdrD of heterodisulfide reductase, DsrJ subunit is a cytoplasmic triheme cytochrome *c*, DsrO is a periplasmic-located Fe-S protein, and DsrP is an integral protein. The association of each heme into subunit DsrJ is distinct and consists of His/Cys, His/His, and His/Met coordination (Pires et al. 2006). A discussion of the model of Dsr association with the membrane as proposed for SRB is given in Chap. 5, Sect. 5.6.2.

3.5 Sulfate Transport

Characteristically SRB do not have sulfatases to release sulfate from organic-sulfate compounds, but these bacteria rely on inorganic sulfate from the environment. With sulfate employed as the final electron acceptor for growth, SRB have a considerable demand for sulfate, and the medium C, which is widely employed for the cultivation of these bacteria, has 4.5 g/L of sodium sulfate (Postgate 1979). In nature, SRB grow in sulfate concentrations ranging from 28 mM in marine waters to a micromolar level in freshwater environments. The saturation constant (K_s) of cells for sulfate and apparent half-saturation constant (K_m) of cell-free extracts for sulfate for several species of SRB were determined to be in the micromolar range. The chemostat growth of *D. vulgaris* Marburg with H_2 as the energy source displayed a K_s of 10 μM for sulfate (Nethe-Jaenchen and Thauer 1984). An apparent K_m of 200 μM for sulfate was observed with *Desulfobacter postgatei* in the range of 5 to 20 μM sulfate (Ingvorsen et al. 1984). With the use of ^{35}S-sulfate as the substrate, *D. salexigens,* a marine isolate, displayed an apparent K_m of 77 10 μM sulfate and *D. vulgaris* Marburg, a species originally from fresh water, had a K_m of 5 10 μM sulfate (Ingvorsen and Jørgensen 1984). With *D. desulfuricans* (Cypionka 1987) and *D. propionicus* (Kreke and Cypionka 1992) are exposed to 1.25 μM sulfate, an internal concentration of about 500 μM sulfate was observed, which is a thousand-fold concentration. *D. salexigens* concentrated sulfate 20,000 fold when sulfate in the environment was 1.25 μM (Kreke and Cypionka 1995). Exposure of *Desulfococcus multivorans* to 2.5 μM sulfate resulted in significantly high levels of accumulated sulfate, but in the presence of 25 mM sulfate, the internal sulfate concentration accumulation was only 1.4 times outside the cells (Warthmann and Cypionka 1990). Thus, if sulfate concentrations in the environment is growth limiting, SRB concentrate sulfate inside the cell several fold; however, in environments where sulfate concentrations are adequate to support optimal growth, little to not intracellular concentration of sulfate occurs. As reviewed by Cypionka (1995), the uptake of sulfate by SRB is dependent on proton motive force involving both membrane charge ($\Delta\psi$) and difference in pH across the membrane (ΔpH). With *D. desulfuricans* suspended in solutions of different acidity, the following transmembrane pH gradients were obtained: ΔpH 0.5 units (intracellular alkaline) was measured when the solution was at pH 7, ΔpH 1.2 units was observed when the solution pH was 5.9, and no pH change occurred with the solution at pH 7.7 (Cypionka 1989; Warthmann and Cypionka 1990; Kroder et al. 1991; Kreke and Cypionka 1992). Using 10 strains of freshwater SRB, the transmembrane pH measurements of these freshly isolated SRB were observed to be 0.25 to 0.8 ΔpH (Kreke and Cypionka 1992). While there is a transmembrane pH gradient in SRB, this ΔpH would not be sufficient to concentrate sulfate a thousandfold (Cypionka 1995). The $\Delta\psi$ values of these freshwater isolates of SRB were −80 to −140 mV, and when including ΔpH values, the proton motive force of these SRB was determined to be −110 to −155 mV (Kreke and Cypionka 1992).

Sulfate transport in several SRB was attributed, in part, to membrane charge and not to a transport system requiring ATP. Thiosulfate is structurally similar to sulfate, and thiosulfate appears to be transported by sulfate transporters in marine SRB (Stahlmann et al. 1991). With the freshwater isolate *Desulfobulbus propionicus*, sulfate uptake was by proton symport with at least two or three protons coupled with each sulfate ion imported (Warthmann and Cypionka 1990; Kreke and Cypionka 1994). Multiple sulfate transport systems occur as SRB. For example, the genome of *Desulfococcus multivorans* DSM 2059 has six genes predicted for Na^+/sulfate symporters and two genes proposed for H^+/sulfate symporters (Dörries et al. 2016). With high levels of sulfate in the environment, the electrogenic symport system uses two protons to drive each sulfate ion by a proton gradient, and at extremely low sulfate concentrations, the electrogenic symport systems use three cations for each sulfate ion (Cypionka 1989). Electrogenic sulfate transport in *Desulfobulbus propionicus* and *Desulfococcus multivorans* by proton and Na^+ ion symport, respectively, was reported to be sensitive to thiocyanate and carbonyl cyanide m-chlorophenylhydrazone (CCCP) with *Desulfococcus multivorans* sensitive to the Na^+/H^+ antiporter monensin and amiloride (Warthmann and Cypionka 1990). With *D. desulfuricans* CSN, sulfate transport was sensitive to CCCP and thiocyanate but not the ATP synthase inhibitor N,N′-dicyclohexylcarbodiimide (DCCD) or the phosphate analog arsenate (Cypionka 1989). Thus, sulfate transport in *D. desulfuricans* CSN is not by an ATP-dependent transport process but by an electrogenic membrane. Further characterization of the sulfate uptake in *D. desulfuricans* CSN revealed that nigericin and monensin, inhibitors of the K^+/H^+

and Na^+/H^+ antiporters, accounted for the partial inhibition of sulfate uptake. While thiocyanate accounted for the inhibition of sulfate transport, valinomycin and amiloride, which function as K^+ transporters and inhibitor of the Na^+/H^+ exchange reaction, respectively, were without effect on sulfate transport in *Desulfovibrio desulfuricans* CSN. In a unique freshwater sulfate reducer, *Desulfomicrobium baculatum*, sulfate transport was attributed to Na^+ ion symport and not to proton symport (Kreke and Cypionka 1995). With *Desulfomicrobium baculatum* displaying a membrane potential of −145 mV, Na^+ ion motive force of −199 mV, and an eightfold gradient of Na^+ (out)/Na^+ (in), two Na^+ ions were calculated to be required for each sulfate ion imported.

An interesting in silico evaluation of sulfate transport by SRB and archaea was presented by Marietou et al. (2018). The review included 44 genomes of sulfate reducers, which included six strains of archaea from *Crenarchaeota* and *Euryarchaeota* phyla and 38 strains of bacteria from *Firmicutes, Nitrospira, Proteobacteria,* and *Thermodesulfobacteria* phyla. None of the 44 genomes of the sulfate reducers examined used the ATP-binding cassette (ABC) system, also known as the SulT sulfate transporters, which were employed by many assimilatory SRB and eukaryotes for sulfate transport sulfate. The molybdate transport system (Mod ABC) in *Escherichia coli* is an ABC-transporter demonstrated to transport sulfate in addition to molybdate; however, molybdate was reported to be ineffective against the sulfate transport uptake of *D. desulfuricans* (Cypionka 1989). Thus, it appears unlikely that the molybdate transporter system would be an important sulfate transporter in SRB. Marietou et al. (2018) suggest that the molybdate permeases are misannotated in the SRB genomes. This lack of genes for SulT would support the discussion by Cypionka (1989) and Marietou et al. (2018) that an ATP requirement for a SulT transport system could not be supported in SRB on the basis of energetics because SRB have an ATP requirement for the activation of sulfate to bisulfite.

As a result of the in silico examination of genomes from SRB, an active transporter system for sulfate was proposed to include the following (Marietou et al. 2018): (i) the SulP sulfate transporters which uses solute:cation and solute: solute antiporters; (ii) the CysT solute transporters of the inorganic phosphate transporter family, which uses H^+ or Na^+ symporters as phosphate and sulfate transporters; (iii) the DASS (divalent anion:sodium symporter) sulfate transporter system; and (iv) the CysZ transporter, which includes the toluene sulfonate uptake permease family and diverse groups of transmembrane proteins. Members of archaea (*Caldivirga maquilingensis* IC-167, *Vulcanisaeta moutnovskia* 768-28, *A. fulgidus* 7324, *Archaeoglobus fulgidus* VC-16, and *Archaeoglobus profundus* Avl8) had proteins associated with the CysP and CysZ transporter family but not the SulP or DASS family, while *A. sulfaticallidus* PM70-1 had putative proteins associated with DASS, CysP, and CysZ transporter family, but not with the SulP family. Multiple sulfate transporters were found in many of the SRB, and each SRB strain examined contained on average 2, 3, 1, and 8 putative proteins associated with the SulP, DASS, CysP, and CysZ families. The SRB frequently used as model organisms (*Dst. reducens* MI-l, *D. gigas*, *D. alaskensis* G20, *Desulfotalea psychrophila, D. desulfuricans*) all contained proteins associated with the SulP, DASS, CysP, and CysZ families. It is interesting to note that in *D. vulgaris*, strains RCH1, DP4, and Hildenborough contained proteins of the SulP, CysP, and CysZ families but not the DASS family, while strain Miyazaki contained proteins of all four sulfate transport families. Several thermophilic SRB (*Dst. kuznetsovii* 17, *Thermodesulfovibrio yellowstonii, Thermodesulfatator indicus* CIR29812, *Thermodesulfobacterium geofontis* OPF15, *Thermodesulfobacterium commune*) lack proteins associated with the SulP family but have proteins of the DASS, CysP, and CysZ families. Three SRB (*Ammonifex degensii* KC4, *Thermodesulfobium narugense* Na82, *Desulfohalobium retbaense* HR100) are unique in that they contain proteins only associated with the CysZ family. Specific sulfate transport families were not correlated with freshwater or marine environments of isolates. With multiple sulfate transport systems potentially present in sulfate reducers, it will be important to determine which is the preferred sulfate uptake system and which are alternates (Keller and Wall 2011).

3.6 Elemental Sulfur Reduction

The report of (Pfennig and Biebl 1976) provided the first information that elemental sulfur (S^0) could be used as the terminal electron acceptor by the anaerobic acetate-oxidizing bacterium, *Desulfuromonas acetoxidans.* Subsequently, *D. desulfuricans* Norway strain (now *Desulfomicrobium baculatum*) and *D. gigas* were reported to use S^0 as an alternate electron acceptor (Biebl and Pfennig 1977). Like *Wolinella succinogenes* DSM 1740, *Desulfuromonas acetoxidans* DSM 1675, and *Sulfurospirillum deleyianum* DSM 6946, the sulfur reductase (EC 1.97.1.3) is constitutive in *Desulfovibrio* and *Desulfomicrobium* species (Fauque 1994). While the reduction of S^0 by cytochrome c_3 from *D. vulgaris* Miyazaki was considered to be nonphysiological (Ishimoto et al. 1958), the sulfur reductase in *Desulfomicrobium norvegicum* strain Norway 4 (NCIB 8310) and *Desulfomicrobium norvegicum* DSM 1743 was demonstrated to be cytochrome c_3 (Fauque et al. 1979; Fauque et al. 1980; Fauque 1994). The redox potential of the four hemes of cytochrome c_3 from *Desulfomicrobium norvegicum* strain Norway 4 was reported to be −150 mV, −300 mV, −330 mV, and − 355 mV with low potential heme being implicated in the reduction of S^0 (Cammack et al. 1984). Due to the action of an exposed heme from cytochrome c_3, sulfide is released from colloidal sulfur. As sulfide opens the sulfur ring containing eight sulfur atoms, the production of new polysulfide molecules occurs, and reduction produces additional hydrogen sulfide (Fauque 1994). With *D. gigas*, membrane preparations catalyzed the reduction of S^0 with H_2 to hydrogen sulfide with concomitant synthesis of ATP (Fauque et al. 1980). Energy from oxidation of H_2 coupled to the reduction of S^0 has been reported to support growth of *D. desulfuricans* (Escobar et al. 2007), and environmental SRB have been observed to reduce polysulfide by H_2 oxidation (Takahashi et al. 2008). *Desulfurivibrio alkaliphilus* reduces polysulfide but not sulfate, and it contains the gene (*psrA*) for molybdenum-polysulfide reductase

(Melton et al. 2016), which is similar to the gene in polysulfide-reducing bacterium *Wolinella succinogenes* (Krafft et al. 1992; Baar et al. 2003). While a few SRB are capable of dissimilatory reduction of S^0 to sulfide, this type of respiration is a feature of numerous bacteria and archaea (Pfennig and Biebl 1976; Zöphel et al. 1988, 1991; Le Faou et al. 1990; Fauque et al. 1991, 1994; Widdel and Pfennig 1992; Widdel and Hansen 1992; Schauder and Kröger 1993; Schauder and Müller 1993; Hedderich et al. 1999; Rabus et al. 2006; Barton and Fauque 2009). Relatively few phyla of the *Bacteria* (i.e., *Clostridia, Nitrospira, Deltaproteobacteria,* and *Thermodesulfobacterium*) and the *Archaea* (i.e., *Archaeoglobus*) have members capable of dissimilatory sulfate reduction. In comparison, *Thermotoga, Gammaproteobacteria, Deltaproteobacteria,* and *Epsilonproteobacteria* of the *Bacteria* and seven phyla of the *Archaea* have members capable of the reduction of elemental sulfur (Fauque and Barton 2012). It is interesting that elemental sulfur in the presence of sulfate has been reported to inhibit growth of some *Desulfonema* species, *Desulfobacter postgatei, D. sapovorans,* and *Desulfomicrobium acetoxidans* (Le Faou et al. 1990). Examples of SRB displaying respiration of sulfur are provided in Tables 3.1 and 3.3.

3.7 Contributions of SRP to Sulfur Cycling

While all life forms assimilate sulfur compounds, SRP use sulfate and a few other sulfur compounds as the electron acceptor for their anaerobic metabolism. Sulfur metabolism associated with sulfur cycling has been the subject of specific research activities focusing on marine and lake sediments (Sass et al. 1997; Wasmund et al. 2017; Jørgensen et al. 2019.), aquifers (Knöller et al. 2008), methane seeps (Sivan et al. 2014; Latour et al. 2018), wetlands (Fortin et al. 2000; Pester et al. 2012; Blodau et al. 2007), costal marshes (Steudler and Peterson 1984), estuaries (Purdy et al. 2002), biofilms (Okabe et al. 1999), lakes with low-sulfate waters (Berg et al. 2019), terrestrial subsurface (Ulrich et al. 1998), acid mine drainage (Tuttle et al. 1969), and human microbiome (Barton et al. 2017). The global sulfur cycle (Fig. 3.9) has been the subject of several reviews (Bharathi 2008; Fike et al. 2016; Anantharaman et al. 2018), and the evolutionary importance of this process has been discussed (Canfield and Raiswell 1999). It is difficult to understand the dynamics of sulfur components of the sulfur cycle because of the intertwining activities of the carbon, nitrogen iron, and manganese cycles with the sulfur cycle (Van Cappellen and Wang 1996). The geochemistry of the anaerobic environments where decomposition of organic matter occurs is highly complex, and a "cryptic sulfur cycle" reflects the interaction of sulfide with Fe (II) (Thamdrup et al. 1993; Sivan et al. 2014; Latour et al. 2018; Berg et al. 2019; Jørgensen et al. 2019). The term cryptic sulfur cycle indicates that the reoxidation of sulfur is hidden and not observed from pore water chemistry.

The sulfate concentration in seawater is ~28 mM, and it is generally considered that the rate of sulfate reduction is diminished when the sulfate concentration is

Table 3.3 Examples of SRB using elemental sulfur as electron acceptor

Bacteria	e⁻ donors	Habitat	Optimum growth, °C	e⁻ acceptors[b]	Reference
Ammonifex thiophilus	H_2, formate	F	75	Thiosulfate	Miroshnichenko et al. (2008)
Ammonifex degensii	H_2, formate	F	70	Nitrate	Huber et al. (1996)
Desulfofustis glycolicus	Glycolate, OAS, H_2,	M	28	Sulfite	Friedrich et al. (1996)
Desulfohalobium retbaense	H_2, lactate, pyruvate, ethanol	M	37--40	Thiosulfate, sulfite	Ollivier et al. (1991)
Desulfomicrobium baculatum Norway 4 DSM 1741	H_2, SCA	F	35	Thiosulfate, sulfite	Biebl and Pfen-nig (1977)
Desulfomicrobium baculatum DSM 1743	H_2, SCA	F	35	Thiosulfate, sulfite	Biebl and Pfennig (1977)
Desulfonauticus submarinus	H_2, OA	M	45	Thiosulfate, sulfite	Audiffrin et al. (2003)
Desulfosarcina cetonica[a]	Fatty acids	F	30	Thiosulfate	Galushko and Rozanova (1991)
Desulfospira joergensenii	H_2, SCA	M	26–30	Thiosulfate, sulfite	Finster et al. (1997)
Desulfosporosinus meridiei	H_2, SCA	F	10–37	Sulfite, thio-sulfate, DMSO, Fe (III)	Robertson et al. (2001)
Desulfovermiculus halophilus	H_2, SCA	M	37	Thiosulfate, sulfite	Beliakova et al. (2006)
Desulfovibrio alcoholovorans	Glycerol, H_2, SCA	F	37	Thiosulfate, sulfite	Qatibi et al. (1991)
Desulfovibrio bastinii	H_2, SCA	F	37	Thiosulfate, sulfite	Magot et al. (2004)
Desulfovibrio biadhensis	H_2, SCA	F	37	Thiosulfate, sulfite	Fadhlaoui et al. (2015)
Desulfovibrio bizertensis	H_2, SCA	M	40	Thiosulfate, sulfite, fumarate	Haouari et al. (2006)
Desulfovibrio burkinensis	Glycerol, H_2, SCA	F	37	Thiosulfate, sulfite, fumarate	Ouattara et al. (1999)
Desulfovibrio fructovorans	Fructose, H_2, SCA	F	35	Thiosulfate, sulfite fumarate	Ollivier et al. (1988)
Desulfovibrio gabonensis	H_2, SCA	F	30	Thiosulfate, sulfite	Tardy-Jacquenod et al. (1996)

(continued)

Table 3.3 (continued)

Bacteria	e⁻ donors	Habitat	Optimum growth, °C	e⁻ acceptors[b]	Reference
Desulfovibrio gigas	H_2, SCA	F	35	Thiosulfate, sulfite	Biebl and Pfennig (1977)
Desulfovibrio gracilis	H_2, SCA	F	37	Thiosulfate, sulfite, fumarate	Magot et al. (2004)
Desulfovibrio idahonensis	H_2, SCA	F	30	Thiosulfate, sulfite, DMSO, fumarate	Sass et al. (2009)
Desulfovibrio legallis	H_2, SCA	F	35	Tthiosulfate, sulfite	Ben Dhia Thabet et al. (2011)
Desulfovibrio longus	H_2, SCA	F	35	Thiosulfate, sulfite fumarate	Magot et al. (1992)

SCA short- chain organic acids such as lactate, formate, pyruvate, *OA* fatty acids, dicarboxylic acids, oxoacids, hydroxyacids, *OAS* organic acids, sugars, *DMSO* dimethylsulfoxide, *F* fresh water, *M* marine water

[a]Formerly identified as *Desulfobacterium cetonicum* (Stackebrandt et al. 2003)

[b]Electron acceptors in addition to sulfate and elemental sulfur

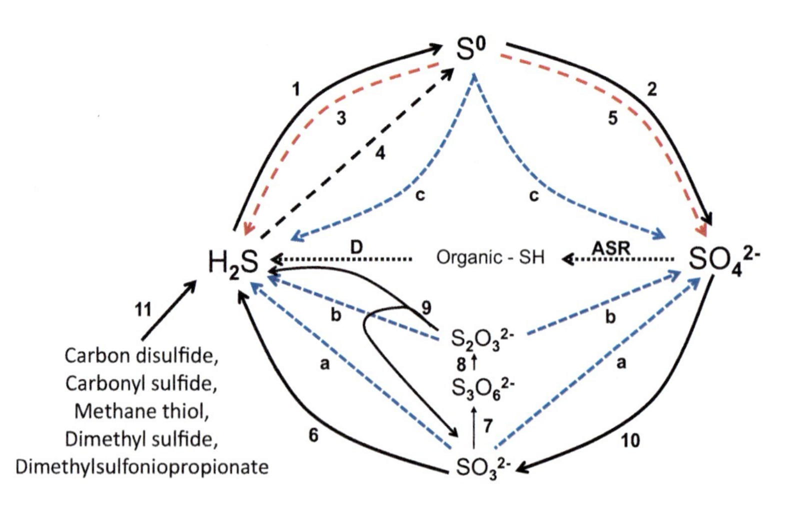

Fig. 3.9 The biological sulfur cycle with roles of bacteria identified. Solid lines indicate aerobic reactions, dashed lines indicate anaerobic reactions, and dotted lines indicate both aerobic and anaerobic activity. (D): Desulfurylation by many aerobic and anaerobic prokaryotes, (ASR): assimilatory sulfate reduction by many aerobic and anaerobic microorganisms. Blue broken lines (a, b, and c): disproportionation by *Desulfovibrio* and *Desulfocapsa* and other SRB. (1 and 2) Sulfide and sulfur oxidation by aerobic colorless sulfur bacteria. (3) Sulfur reduction by anaerobic microorganisms including SRP. (4 and 5) Anaerobic sulfide and sulfur oxidation by purple sulfur bacteria and green sulfur bacteria. (6) Sulfite reduction by microorganisms including SRB. (6 and 10) Dissimilatory sulfate reduction by SRP. (7) Trithionate-forming enzyme. (8) Trithionate reductase. (9) Thiosulfate reductase. (11) Various bacteria including some SRB

<0.2 mM. However, appreciable levels of sulfate are reduced in freshwater sediments and at the sulfate-methane interface region in marine sediments where the sulfate concentrations are in the range of 1–100 μM. A review of reported values for sulfate metabolism by SRB indicates that there are two physiological types of SRB. One population of SRB functions at a low affinity for sulfate ($K_m = 430$ μM), and another population has a high affinity for sulfate ($K_m = 2.6$ μM) (Tarpgaard et al. 2011). It is not known if different species of SRB are required for these two different sulfate concentrations or if individual species of SRB adjust and grow in either concentration of sulfate. Pure cultures of *Desulfobacterium autotrophicum* growing in marine concentrations of sulfate with an apparent K_m of 0.5 mM can grow with sulfate concentrations below 0.5 mM by the induction of a high affinity sulfate uptake system with an apparent $K_m = 0.008$ mM (Tarpgaard et al. 2017). In the methane zone of marine sediment, a steady state of 0.01 mM sulfate has been reported, and this produces a "cryptic sulfur cycle" (Pellerin et al. 2018). The presence of cryptic sulfur cycles beneath the sulfate-methane transition zone proposed for anaerobic sediments worldwide have been reviewed (Brunner et al. 2016; Wasmund et al. 2017). Under marine environmental conditions where the concentrations of sulfate in the sediment is at the μM level, the free energy released would be about -17.9 to -11.9 kJ mol^{-1} sulfate, which would be near or below the energy level needed to support growth of SRB (Pellerin et al. 2018).

Sulfate reduction occurs in anaerobic environments worldwide, and the rate of sulfate reduction reflects the amount of sulfate present and the environmental conditions where the SRB are growing. The daily rate of sulfate reduced in wetlands of the Prairie Pothole Region of North America has been reported to reach a high of 22 μmol cm^{-3} day^{-1} (Martins et al. 2018). In peatlands, everglades, river floodplains, rice paddies, and lake sediments where the sulfate concentration is in the μM range, the amount of daily sulfate reduction was a fraction of the amount present and was measured in nmol cm^{-3} day^{-1} (Pester et al. 2012). While the daily amount of sulfate reduced in sediments of the Gulf of Gdańsk was 1.9 to 31 nM sulfate/g sediment (Mudryk et al. 2000), the daily rate of sulfate reduction with the production of hydrogen sulfide in the Black Sea was reported to be about 21 μM (Albert et al. 1995). Robador et al. (2016) surveyed the sediments of the nine seas and found that the quantity of sulfate reduced daily ranged broadly from 0.5 nmol cm^{-3} observed at the Southern Ocean to 2233 nmol cm^{-3} in the Arctic Ocean. Salt marshes release considerable amounts of hydrogen sulfite into the atmosphere with a total annual release of 1.7 Tg S (1 Tg = 10^{12} g) (Steudler and Peterson 1984). Following an extensive evaluation of reported values for sulfate reduction worldwide, it was estimated that the global quantity of sulfate reduced annually would be 11.3 teramoles (Bowles et al. 2014).

In addition to sulfide being produced by the SRB, various sulfur cycle intermediates including sulfite, elemental sulfur, polysulfides, thiosulfate, and tetrathionate were also formed due to biotic and abiotic action (Wasmund et al. 2017; Jørgensen et al. 2019; Vigneron et al. 2021). As indicated in Fig. 3.10, manganese oxides and especially iron oxides interact with sulfide to produce pyrite (FeS_2) by the polysulfide or the hydrogen sulfide pathways. Elemental sulfur (also known as zero-valent sulfur) is present in the marine environment as polysulfide (${S_n}^{2-}$), cyclic ring of eight atoms (S_8), and polymeric sulfur (S_n) (Lichtschlag et al. 2013; Berg et al. 2014), and

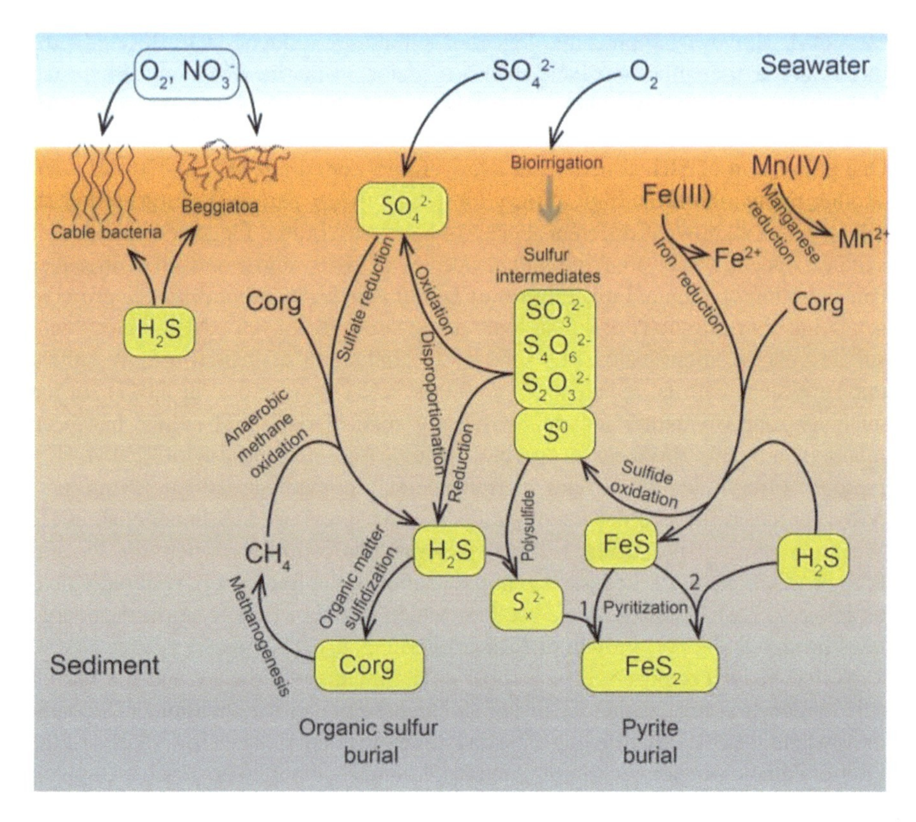

Fig. 3.10 The biogeochemical sulfur cycle of marine sediments. The schematic presentation includes many of the processes discussed in this review. With an increase in depth, the water becomes anaerobic due to the production of H_2S by the bacteria. Arrows indicate fluxes and pathways of biological or chemical processes. (Jørgensen et al. 2019.)

elemental sulfur nanoparticles with a diameter of <0.2 μm are readily isolated from the marine environment (Jørgensen et al. 2019). Sulfur-oxidizing bacteria produce elemental sulfur by oxidation of thiosulfate (Meyer et al. 2007; Ghosh and Dam 2009; Duzs et al. 2018) using the SOX pathway (Sakurai et al. 2010), the tetrathionate intermediate pathway (Dam et al. 2007), the Sox-S_4I interaction system (Hensen et al. 2006), and a pathway involving both thiosulfate dehydrogenase and thiosulfohydrolase (Zhang et al. 2020; Liu et al. 2021). There is a strong possibility that cable bacteria may participate in the oxidation of sulfur intermediates including sulfur disproportionation in the sulfur cycle (Kjeldsen et al. 2019; Müller et al. 2019).

Elemental sulfur reacts with sulfide to produce polysulfides, which are important for pyrite formation, and pyrite serves as the marine sulfur reservoir. A summary of

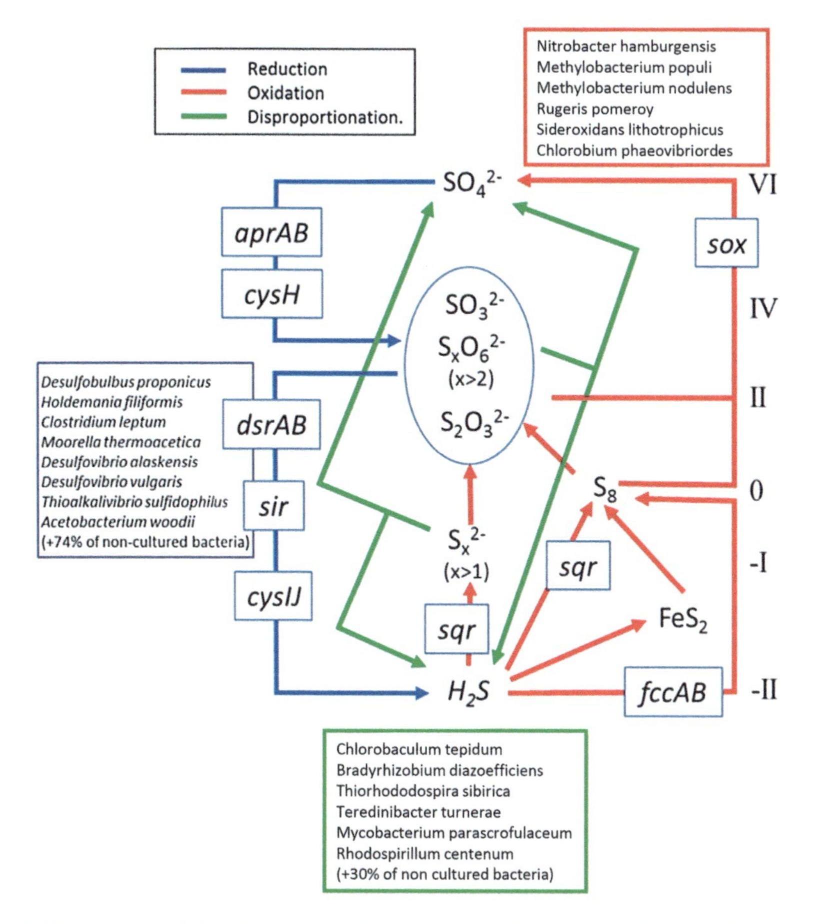

Fig. 3.11 Summary of the sulfur redox cycle in marine environments with the microbial species utilizing each metabolic pathway (sulfate reduction, sulfide oxidation, sulfite, and thiosulfate disproportionation) identified from ALWC samples. Roman numerals indicate the S oxidation state at each stage. (Smith et al. 2019. This figure has been modified to provide the current spelling for the bacteria listed.

the marine sulfur cycle indicating the metabolic pathways and microorganisms involved are presented in Fig. 3.11. As reviewed by Jørgensen et al. (2019), biologically produced sulfur is distinguished from inorganic α-S_8 in terms of geochemical reactions and bacterial metabolism. Sulfite, thiosulfate, and tetrathionate are transiently produced in the anaerobic sediment, and often the concentrations of these sulf-oxy compounds are below the chemical detection level (Jørgensen et al. 2019). Additionally, SRB display the capability of disproportionation reactions involving S^0, thiosulfate, and sulfite (Finster 2008). The cultivation of *Desulfocapsa thiozymogenes* under conditions to reduce polysulfide formation displayed a sulfur disproportionation rate of $10^{-15.4}$ mol S° cell^{-1} h^{-1} (Böttcher et al. 2001). The importance of thiosulfate in the anaerobic sulfur cycle has been called the "thiosulfate shunt," and it functions in marine sediments as well as in the human gut (Jørgensen 1990; Barton et al. 2017; Zhang et al. 2020).

SRB are considered to have an important role in the release of sulfur molecules into the atmosphere. Salt marshes may be especially important because the marshes are shallow and anaerobic sediments are exposed to the atmosphere (Steudler and Peterson 1984). Sulfur- containing gases emitted from salt marshes include hydrogen sulfide, methanethiol, dimethyl sulfide (DMS), carbonyl sulfide (COS), carbon disulfide (CS_2), and dimethyl disulfide (DMDS). Of the gasses from a *Spartina* salt marsh, 49% was DMS, and 35% was hydrogen sulfide (Steudler and Peterson 1984; Bharathi 2008). DMS production is important because it accounts for 90% of the biogenic volatile sulfur compounds emitted from marine environments (Bharathi 2008). The volatile sulfur compounds are photochemically oxidized, and its products contribute to acid rain and production of sulfate particles, which impact the climate by decreasing solar radiation (Bharathi 2008). Many of the phytoplankton and macroalgae produce dimethylsulfoniopropionate (DMSP) as an osmolyte, and one of the precursor molecules for DMSP is methionine (Kirst 1996). SRB not only demethylate DMRSP to produce methylmercaptopropionate (MMPA), carbonate, and sulfide, but SRB also oxidize DMS with the production of sulfide and bicarbonate (Bharathi 2008).

New isolates are found in anaerobic sediments with the capability of sulfate reduction, and these bacteria have unique physiological capabilities (Cao et al. 2016; Zheng and Sun 2020). Evaluations of SRB associated with sulfur cycling in marine sediments and associated environments have provided information suggesting the presence of many uncultivated bacteria. Using metagenomics to evaluate wetland sediments, the SRB populations were determined to be metabolically diverse and capable of using a range of compounds as electron donors (Martins et al. 2018). Two marker genes were assessed in the metagenome data, and they included the *dsrA* gene and *dsrD* gene, which is generally absent in sulfur-oxidizing bacteria. The presence of *dsrD* genes in the metagenome data sets were associated with *Deltaproteobacteria, Nitrospirae, Acidobacteria, Planctomycetes, Firmicutes,* and the candidate phyla *Armatimonadetes, Gemmatimonadetes, Aminicenantes,* and *Schekmanbacteria.* Examination of bacteria present in Lake A of the Canadian High Arctic indicated that the abundance of uncultivatable SRB was only ~4% at the surface of the lake but ~40% in the sediment (Vigneron et al. 2021). From samples taken from a diverse set of sites, metagenome analysis revealed numerous bacteria present with genes for sulfate/sulfite reduction including 20 different phyla including 13 genomes previously not known to have *dsr* genes (Anantharaman et al. 2018). Clearly, the contributions of uncultivated bacteria and archaea to the sulfur cycle are underestimated, and with continued research, a better insight into the dynamics of the global sulfur cycle will be attained.

3.8 Enzymology of Nitrogen Respiration

For some time, it was considered that there was one physiological group of bacteria that reduced sulfate and another that reduced nitrate. Relatively few isolates of bacteria grow by sulfate respiration, while a sizeable number of bacteria grow by nitrate respiration. The reason why there is a limited number of sulfate-reducing microorganism in nature remains unexplained. While there was a slow acceptance to the idea that SRB could also be nitrate-respiring bacteria, it is well documented now that numerous SRB can grow using either sulfate or nitrate as the terminal electron acceptor. The reduction of nitrate by SRB does not result in the production of N_2, but nitrate is reduced to nitrite, which is reduced to ammonium. Thus, nitrogen is not removed from the anaerobic environment as volatile N_2 but is transformed to another nitrogen compound. The half-cell reactions of nitrate and nitrite are given in Eqs. 3.20 and 3.21. Energetically, both nitrate and nitrite reduction with H2 would provide energy for cell growth, and in fact the energy released is greater than that for sulfate reduction (Thauer et al. 1977). Hydroxylamine is reduced by nitrite reductase, and it too could support growth. The enzymology of nitrate and nitrite reduction by sulfate reducers is covered in the following segments (Eqs. 3.20–3.24):

$$NO_3^- + 2e^- + 2H^+ = NO_2^- + H_2O \quad E^{0'} = +420\ \text{mV} \tag{3.20}$$

$$NO_2^- + 6e^- + 8H^+ = NH_4^+ + 2H_2O \quad E^{0'} = +340\ \text{mV} \tag{3.21}$$

$$NO_3^- + 4H_2 + 2H^+ = NH_4^+ + 3H_2O \quad \Delta G^{0'} = -591\ \text{kJ/mol} \tag{3.22}$$

$$NO_2^- + 3H_2 + 2H^+ = NH_4^+ + 2H_2O \quad \Delta G^{0'} = -436.4\ \text{kJ/mol} \tag{3.23}$$

$$H_2 + NH_2OH + H^+ = NH_4^+ + H_2O \quad \Delta G^{0'} = -248.5\ \text{kJ/mol} \tag{3.24}$$

3.8.1 Nitrate Reduction

Based on the molecular structure, operon organization, and cellular location, bacterial nitrate reductases are classified as respiratory membrane-bound nitrate reductase (Nar), assimilatory nitrate reductase (Nas), and periplasmic nitrate reductase (Nap) (Sparacino-Watkins et al. 2014). The periplasmic nitrate reductase would be associated with gram-negative SRB, and the periplasmic nitrate reductase of *Desulfovibrio* has been extensively studied. There are few gram-positive SRB that have been reported to use nitrate as a terminal electron acceptor. Since these bacteria would not have a periplasm, it would be of interest to determine which type of nitrate reductase they employ.

The periplasmic nitrate reductase (Nap, EC. 1.7.99.4) of *D. desulfuricans* ATCC 27774 is a monomeric enzyme with four domains and a molecular mass of 74 kDa. Nap has a molybdopterin guanine dinucleotide (MGD) cofactor, which is ~15 Å from the protein surface, and it is bound to each of the protein domains. A cysteine motif of (Cys-X_2-Cys-X_3-Cys-$X_{24–26}$-Cys) (Sparacino-Watkins et al. 2014) is at the N-terminus, and it coordinates the [4Fe-4S] cluster at the periphery of the molecule, which is about 20 Å from the molybdenum atom of the MGD (Bursakov et al. 1995, 1997; Dias et al. 1999). A single channel within the Nap molecule is proposed to enable nitrate to enter and for nitrite to exit the active site. Coordination of the MGD cofactor was attributed to four dithiolene sulfurs, the fifth ligand is sulfur from a cysteine residue, and the sixth ligand is a sulfur atom (Najmudin et al. 2008). It has been suggested that there is considerable flexibility in binding the Mo complex into the Nap molecule and that there may be two MGD cofactors in the Nap of *D. desulfuricans* ATCC 27774, but only one is active (González et al. 2006; Moura et al. 2007; Sparacino-Watkins et al. 2014). A model indicating the location of enzymes for nitrate and nitrite reduction is given in Fig. 3.12. The mechanism of nitrate reductase in *D. desulfuricans* ATCC 27774 is exceedingly complex and has been reviewed by Moura et al. (2007) and Sparacino-Watkins et al. (2014). A summary of the nitrate reduction process is as follows: Mo^{6+} in the MGD cofactor

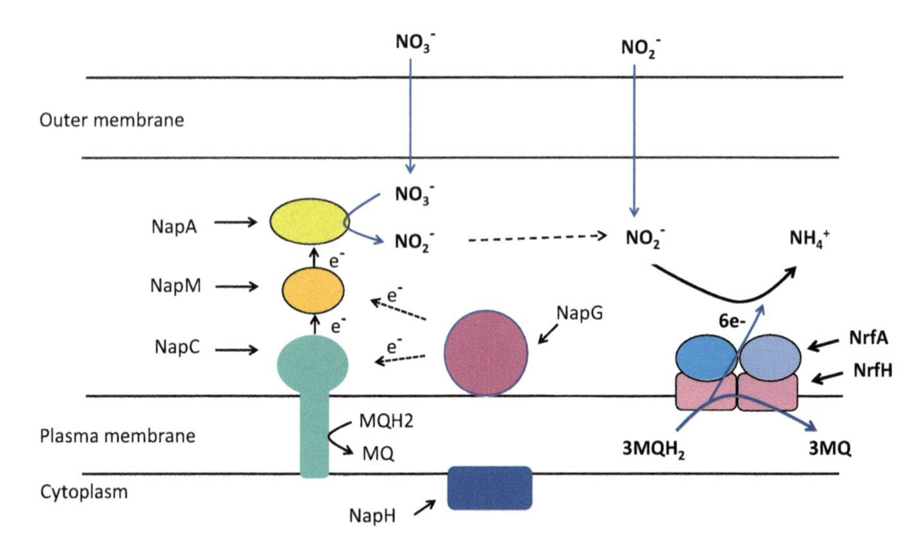

Fig. 3.12 Model of dissimilatory nitrate reduction with the production of ammonium. Nap enzymes are associated with nitrate reduction, and Nrf indicates nitrite reduction. NapA contains Mo-MGD and a [4Fe-4S] cluster; NapM (referred to as Nap B in some bacteria) has two hemes present, and NapC has four hemes present. NrfA is a dimer where each subunit contains five hemes, while NrfH is also a dimer where each subunit contains four hemes

is reduced to Mo^{4+}, and nitrate is added and reduced to produce nitrite with the oxidation of Mo^{4+} to Mo^{6+}.

The genes associated with dissimilatory nitrate reduction in *D. desulfuricans* are localized in a single operon, and this cluster of six genes consists of *napCMADGH* (Marietou et al. 2005). In terms of function, these genes are arranged into two groups (Sparacino-Watkins et al. 2014): redox active proteins (NapA, NapC, NapM, NapG, and NapH) and non-redox active protein (NapD). NapA contains the MGD cofactor and is the catalytic subunit of nitrate reductase. NapA has the twin-arginine motif used to recognize proteins for transport to the periplasm by the TAT (twin arginine transport) exporter system. NapC contains four Cys-X-X-Cys-His, which is the binding motif for the four covalently attached hemes and a transmembrane helix characteristic of quinol dehydrogenase. The function of the NapC polypeptide of 22 kDa is to transfer electrons from the menaquinone pool to the periplasmic dehydrogenase. The NapM is a tetraheme cytochrome *c* with a mass of ~12.5 kDa that resides on the periplasmic side of the membrane, and it carries electrons from NapC or another electron source to NapA. NapD has a molecular mass of 11.9 kDa, and it functions as a chaperon to recognize the TAT leader sequence of NapA and regulates the export of mature NapA once it has been properly folded with assembled cofactors. NapH is a membrane protein of 31.8 kDa with two domains in the protein that extends into the cytoplasm that binds two [4Fe-4S] clusters. NapG has a mass of 18 kDa; it is located on the outer side of the plasma membrane and has four [4Fe-4S] clusters. The function of NapG and NapH appears to be an alternate system for directing electrons from the menaquinone pool to the NapA.

3.8.2 Nitrite Reduction

A model used to study nitrite reduction in sulfate reducers has been the enzyme isolated from *D. desulfuricans* ATCC 27774. The cytochrome *c* nitrite reductase (ccNir; EC. 1.7.2.2.) was initially isolated by Liu and Peck Jr. (1981), and as reviewed by Moura et al. (2007), the unique molecular configuration accounted for difficulty in determining the number of hemes per protein subunit. Not only was the ccNir of *D. desulfuricans* reported to have six hemes (Liu and Peck Jr. 1981), but also it was reported that *Escherichia coli*, (Kajie and Anraku 1985), *Vibrio* sp. (Rehr and Klemme 1986; Liu et al. 1988), *Wolinella succinogenes* (Liu et al. 1983), and "*Spirillum*" strain 5175 (Schumacher and Kroneck 1992) had a hexaheme nitrite reductase. After a reevaluation of the ccNir from *Sulfospirillum deleyianum* and *Wolinella succinogenes,* Schumacher et al. (1994) concluded that it was a tetraheme protein. Finally, the X-ray structure of ccNir from *Sulfospirillum deleyianum* was solved and revealed that the ccNir contained five hemes per subunit (Einsle et al. 1999). A reevaluation of spectral properties and redox values revealed that the ccNir of *D. desulfuricans* ATCC 27774 consisted of two subunits: NrfA, formerly called NirA (68 kDa), and NrfH formerly referred to as NirH (19 kDa) with a quaternary structure of $\alpha_4\beta_2$ (Almeida et al. 2003). NrfA has five hemes with four hemes

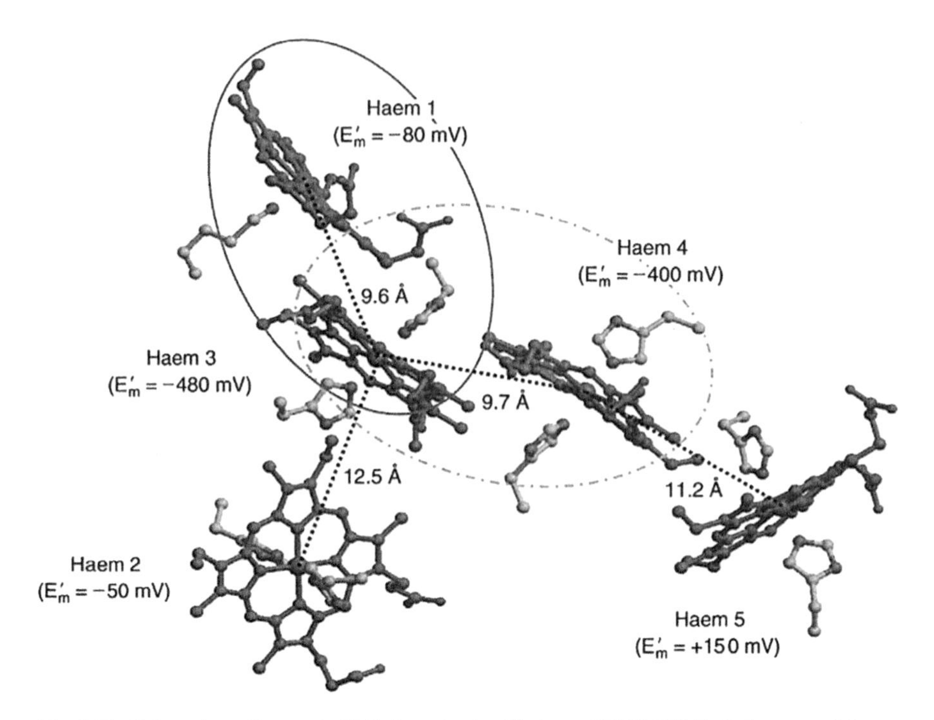

Fig. 3.13 Orientation of hemes in NrfA from *D. desulfuricans* ATCC 27774. Redox potentials are provided for each heme, and distance between hemes is given in angstroms (Å). (Reproduced by permission from Fauque and Barton (2012), Copyright 2012, with permission from Elsevier)

coordinated by *bis*-histidine, and the fifth heme is coordinated by lysine and not histidine. The orientation of the five hemes in NrfA is given in Fig. 3.13. NfrH has four hemes, which were coordinated by *bis*-histidine, and the NrfH has a hydrophobic helix, which is integrated into the membrane facing the periplasm. The active site for NrfA and molecular structure is discussed by Cunha et al. (2003), and this discussion includes the role of Ca^{2+} binding to stabilize the structure of this subunit. The nitrite reductase produced by *D. vulgaris* Hildenborough was isolated as a complex of 760 kDa, and it was resolved into two subunits of 56 kDa and 18 kDa (Pereira et al. 2000; Cunha et al. 2003). The NrfA subunit contains four heme-binding motifs of CXXCH, and the fifth binding motif of CXXCK is unique for heme binding. There is an export signal peptide with a LXXC motif at the leading end of NrfA that is recognized by signal peptidase II. The NrfH subunit with four hemes bound by the CXXCH motif and contains a transmembrane helix. Structural analysis of NrfH revealed a menaquinone binding site, and this would enable electrons to be transferred from the membrane to the periplasm (Rodrigues et al. 2008).

Unlike *D. desulfuricans* ATCC 27774, *D. vulgaris* Hildenborough is unable to grow as it reduces nitrite to ammonia (Rodrigues et al. 2006a). The role of the nitrite reductase in *D. vulgaris* may be to remove the nitrite produced by bacteria in the environment and thereby avoid inhibition of the dissimilatory sulfite reductase (Greene et al. 2003). The reduction of nitrite by DsrAB in *D. vulgaris* has been proposed for the bacterium to use nitrite as a nitrogen source for growth (Korte et al. 2015). Nitrite has been shown to induce the *nrfHA* genes in *D. vulgaris* and downregulate the production of enzymes for dissimilatory sulfate reduction (Haveman et al. 2004). The production of trace levels of nitric oxide (NO) associated with nitrite reduction are characteristically present in SRB cultures, which reduce nitrate. In the natural environment, NO may arise from several biological and chemical based reactions. In *D. desulfuricans* ATCC 27774, the detoxification of NO is, in part, attributed to the reduction of NO by NrfA (Costa et al. 1990). The presence of NO induces a stress response in *D. desulfuricans* ATCC 27774 with an induction of 31 genes and a repression of 26 genes (Cadby et al. 2017). The cluster of genes induced by NO to protect proteins against nitrosative stress includes a flavodiiron protein (Rodrigues et al. 2006b), and another set of genes encodes the hybrid cluster proteins (Hcp) (da Silva et al. 2015; Wang et al. 2016).

3.9 Nitrogen Fixation

Nitrogen fixation by marine isolates of SRB using standard growth techniques was initially reported by Sisler and Zobell (1951), and a few years later, LeGall et al. (1959) demonstrated nitrogen fixation in *D. desulfuricans* Berre Sol (NCIB 8388) and *D. desulfuricans* Berre Eau (NCIB 8387). Using $^{15}N_2$ uptake, *D. desulfuricans* Norway 4 (NCIB 8310), *D. vulgaris* Hildenborough (NCIB 8303), and *D. gigas* (NCIB 9332) were positive for nitrogen fixation (Riederer-Henderson and Wilson 1970). Using the acetylene reduction test to indicate the presence of nitrogenase, strains of *Dst. ruminis* (NCIB 8542 and NCIB 10,149), *D. desulfuricans* strain Essex 6 (NCIB 8307), and *Dst. orientis* (NCIB 8382) showed nitrogen-fixing activity, while *Dst. nigrificans* (NCIB 8395), *D. africanus* strain Benghazi (NCIB 8401), and *D. desulfuricans* strain El Agheila A (NCIB 8309) had no acetylene reduction activity (Postgate 1970). A considerable amount of energy is required for nitrogen fixation (Eq. 3.25; Postgate 1982), and this was apparent from the decreased growth rate of *D. desulfuricans* growing on N_2 (LeGall and Senez 1960). The nitrogen-fixing activity of *Desulfovibrio* has been reviewed by Lespinat et al. (1987) and Postgate et al. (1988).

$$16\text{ATP} + 8\text{e}^- + 8\text{H}^+ + \text{N}_2 \rightarrow 16\text{ADP} + 16\text{P}_\text{i} + \text{H}^2 + 2\text{NH}_3 \tag{3.25}$$

The structural operon (*nifHDKY*) of the nitrogen fixation genes for *Klebsiella pneumoniae* are used to assess intergeneric homologies diazotrophic bacteria. The *nifD* gene of the *nif* cluster has been shown to have homology to 13 strains of *Desulfovibrio* of five species. Six strains of *D. vulgaris*, four strains of *D. desulfuricans*, two strains of *D. salexigens*, two strains of *D. africanus*, and one strain of *D. gigas* and *D. thermophilus* had homology to nifD, a polypeptide of the MoFe protein of nitrogenase (Postgate et al. 1986). The *nif* gene was found on the chromosome of 10 SRB strains and on 130 MDa plasmids of three strains of *D. vulgaris*. New strains of SRB are being isolated with the capability of nitrogen fixation, and *nif* genes are detected in numerous recent genome analysis of SRB.

3.10 Role of SRP in Nitrogen Cycling

Nitrogen is a macronutrient required by all life forms including SRP for the synthesis of essential nitrogen-containing mono- and polymeric molecules. Additional processes associated with SRP include the use of nitrate or nitrite as alternate electron acceptors when sulfate concentration is minimal and the production of ammonia by nitrogen fixation. Several reviews highlight the nitrogen cycle in marine environments (Blackburn and Henriksen 1983; Jensen et al. 1990; Herbert 1999; Pajares and Ramos 2019), and a couple of papers have focused on nitrogen fixation (diazotrophy) associated with sulfate reduction (Sørensen et al. 1979; Lespinat et al. 1987; Bertics et al. 2013). When SRP reduces nitrate as the terminal electron acceptor, the ammonification pathway is used and not the denitrification pathway, which results in N_2 production. The nitric oxide reductase demonstrated in *D. vulgaris* is associated with oxygen stress response and not with denitrification (Silaghi-Dumitrescu et al. 2005). As summarized in Fig. 3.14, the association of SRP with the global nitrogen cycle is restricted to anaerobic zones. Although not included in this global nitrogen cycle, a few SRB have the capability of reducing the nitrate moiety on organic-nitrate molecules to an

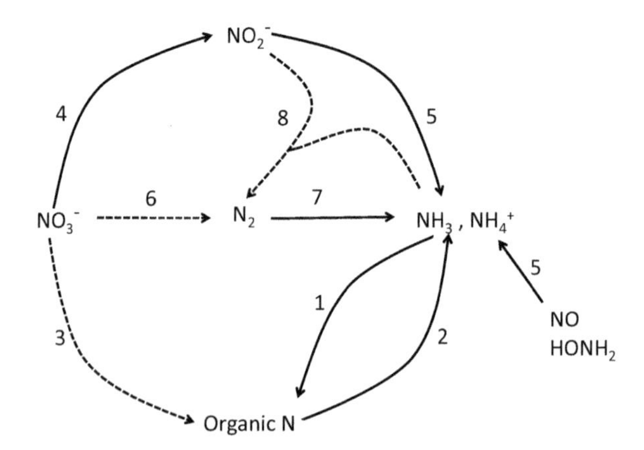

Fig. 3.14 Generalized global nitrogen cycle. Numbers indicate enzymes as follows: (1) assimilation of ammonia by glutamine synthetase and glutamate synthase into amino acids, (2) mineralization, (3) plant uptake for assimilatory process, (4) nitrate reductase, (5) nitrite reductase, (6) denitrification, (7) nitrogen fixation, and (8) anammox process. Dashed line indicates reactions not associated with sulfate-reducing microorganisms. Sulfate-reducing and other microorganisms participating in reactions with a solid line

organic amine, and some reduce hydroxylamine to ammonium. Nitrate and nitrite reductases were discussed earlier in this chapter (Sects. 3.8.1 and 3.8.2).

3.11 Summary and Perspective

The metabolic pathways associated with the reduction of inorganic sulfur compounds have been established, and the enzymes for these reductions have been characterized. A unique group of prokaryotes have the ability to use inorganic sulfur compounds as electron acceptors with concomitant production of ATP. The enzyme common to all prokaryotes that display sulfate respiration is the siroheme-containing (bi)sulfite reductase, which is distinguished from the assimilatory sulfite reductase. The diversity of siroheme type, EPR spectra, and optical properties is used to arrange the dissimilatory bisulfite reductases into five different groups. It will be interesting to see if additional diversities are present in uncultivated SRP. As discussed in Chap. 1, some SRB are assumed to use enzymes of dissimilatory sulfate reduction for the disproportionation of thiosulfate and sulfite with the production of sulfate and sulfide, and information is needed to support this suggestion. Dissimilatory nitrite (and in some species nitrate) reduction with the production of ammonia by SRB is similar to the bisulfite to sulfide reduction in that six electrons are required and heme moieties are involved. However, unlike sulfate activation, which requires ATP for the production of sulfite, nitrate reduction to nitrite is not dependent on ATP input. It appears that enzymes for dissimilatory nitrate/nitrite reduction would have been acquired by SRB to support growth when sulfate concentrations are limiting. Due to the enzymatic capabilities of SRB, their growth reflects an intertwining of the sulfur and nitrogen cycles.

References

Adman ET, Sieker L, Jensen L (1976) Structure of *Peptococcus aerogenes* Ferredoxin – Refinement at 2 Å resolution. J Biol Chem 248:3987–3996

Akagi JM (1981) Dissimilatory sulfate reduction, mechanistic aspects. In: Bothe H, Trebst A (eds) Biology of inorganic nitrogen and sulfur. Springer-Verlag, New York, pp 178–187

Akagi JM (1995) Respiratory sulfate reduction. In: Barton LL (ed) Sulfate-reducing bacteria. Plenum Press, New York, pp 89–111

Akagi JM, Adams V (1973) Isolation of a bisulfite reductase activity from *Desulfotomaculum nigrificans* and its identification as the carbon monoxide-binding pigment P582. J Bacteriol 116: 392–396

Akagi JM, Campbell LL (1962) Studies on thermophilic sulfate-reducing bacteria. III. Adenosine triphosphate-sulfurylase of *Clostridium nigrificans* and *Desulfovibrio desulfuricans*. J Bacteriol 84:1194–1201

Akagi JM, Campbell LL (1963) Inorganic pyrophosphatase of *Desulfovibrio desulfuricans*. J Bacteriol 86:563–568

Aketagawa J, Kobayashi K, Ishimoto M (1985) Purification and properties of thiosulfate reductase from *Desulfovibrio vulgaris*, Miyazaki F. J Biochem 97:1025–1032

Albert DB, Taylor C, Martens CS (1995) Sulfate reduction rates and low molecular weight fatty acid concentrations in the water column and surficial sediments of the Black Sea. Deep-Sea Res I Oceanogr Res Pap 42:1239–1260

Almeida MG, Macieira S, Luisa L, Gonçalves LL, Huber R, Cunha CA, Romão MJ et al (2003) The isolation and characterization of cytochrome *c* nitrite reductase subunits (NrfA and NrfH) from *Desulfovibrio desulfuricans* ATCC 27774 re-evaluation of the spectroscopic data and redox properties. Eur J Biochem 270:3904–3915

Anantharaman K, Hausmann B, Jungbluth SP, Kantor RS, Lavy A, Warren LA et al (2018) Expanded diversity of microbial groups that shape the dissimilatory sulfur cycle. ISME J 12: 1715–1728

Arendsen AF, Verhagen MF, Wolbert RB, Pierik AJ, Stams AJ, Jetten MS, Hagen WR (1993) The dissimilatory sulfite reductase from *Desulfosarcina variabilis* is a desulforubidin containing uncoupled metalated sirohemes and S = 9/2 iron-sulfur clusters. Biochemist 32:10323–10330

Audiffrin C, Cayol J-L, Joulian C, Casalot L, Thomas P, Garcia J-L, Ollivier B (2003) *Desulfonauticus submarinus* gen. nov., sp. nov., a novel sulfate-reducing bacterium isolated from a deep-sea hydrothermal vent. Int J Syst Evol Microbiol 53:1585–1590

Baar C, Eppinger M, Raddatz G, Simon J, Lanz C, Klimmek O et al (2003) Complete genome sequence and analysis of *Wolinella succinogenes*. Proc Natl Acad Sci, USA 100:11690–11695

Badziong W, Thauer RK (1980) Vectoral electron transport in *Desulfovibrio vulgaris* (Marburg), growing on hydrogen plus sulfate as sole energy source. Arch Microbiol 125:167–174

Baliga BS, Vartak HC, Jagannathan V (1961) Purification and properties of *Desulfovibrio desulfuricans*. J Sci Ind Res (India) 20C:33–40

Bandurski RS, Wilson LG, Squires CL (1956) The mechanism of "active sulfate" formation. J Am Chem Soc 78:6408–6409

Barrett EL, Clark MA (1987) Tetrathionate reduction and production of hydrogen sulfide from thiosulfate. Microbiol Rev 51:192–205

Barton LL, Fauque GD (2009) Biochemistry, physiology and biotechnology of sulfate-reducing bacteria. Adv Appl Microbiol 68:41–98

Barton L, Fardeau M-L, Fauque GD (2014) Hydrogen sulfide: a toxic gas produced by dissimilatory sulfate and sulfur reduction and consumed by microbial oxidation. In: Sigel A, Sigel H, Sigel RKO (eds) Metal ions in life sciences. Dordrecht, The Netherlands, Springer Science & Business Media B.V., pp 237–278

Barton LL, Ritz NL, Fauque GD, Lin HC (2017) Sulfur cycling and the intestinal microbiome. Dig Dis Sci 62:2241–2257

Beliakova EV, Rozanova EP, Borzenkov IA, Turova TP, Pusheva MA, Lysenko AM, Kolganov TV (2006, 75) The new facultatively chemolithoautotrophic, moderately halophilic, sulfate-reducing bacterium *Desulfovermiculus halophilus* gen. nov., sp. nov., isolated from an oil field. Mikrobiologiia 201:–211

Ben Dhia Thabet O, Wafa T, Eltaief K, Cayol J-L, Hamdi M, Fauque G, Fardeau M-L (2011) *Desulfovibrio legallis* sp. nov.: a moderately halophilic, sulfate-reducing bacterium isolated from a wastewater digestor in Tunisia. Curr Microbiol 62:486–491

Berg JS, Jézéquel D, Duverger A, Lamy D, Laberty-Robert C, Miot J (2019) Microbial diversity involved in iron and cryptic sulfur cycling in the ferruginous, low-sulfate waters of Lake Pavin. PLoS One 14(2):e0212787. https://doi.org/10.1371/journal.pone.0212787

Berg JS, Schwedt A, Kreutzmann AC, Kuypers MMM, Milucka J (2014) Polysulfides as intermediates in the oxidation of sulfide to sulfate by *Beggiatoa* spp. Appl Environ Microbiol 80:629–636

Bertics VJ, Löscher CR, Salonen I, Dale AW, Gier J, Schmitz RA, Treude T (2013) Occurrence of benthic microbial nitrogen fixation coupled to sulfate reduction in the seasonally hypoxic Eckernforde Bay, Baltic Sea. Biogeosciences 10:1243–1258

Bharathi PAL (2008) Sulfur cycle. In: Jørgensen SE, Fath BD (eds) Encyclopedia of ecology, vol 4. Elsevier, Amsterdam, The Netherlands, pp 3424–3431
Biebl H, Pfennig N (1977) Growth of sulfate-reducing bacteria with sulfur as electron acceptor. Arch Microbiol 112:115–117
Blackburn TH, Henriksen K (1983) Nitrogen cycling in different types of sediment from Danish waters. Limnol Oceanogr 28:477–493
Blodau C, Mayer B, Peiffer S, Moore T (2007) Support for an anaerobic sulfur cycle in two Canadian peatland soils. J Geophys Res 112:G0200
Böttcher ME, Thamdrup B, Vennemann TW (2001) Oxygen and sulfur isotope fractionation during anaerobic bacterial disproportionation of elemental sulfur. Geochim Cosmochim Acta 65:1601–1609
Bowles MW, Mogollon JM, Kasten S, Zabel M, Hinrichs KU (2014) Global rates of marine sulfate reduction and implications for sub-sea-floor metabolic activities. Science 344:889–891
Bramlett RN, Peck HD Jr (1975) Some physical and kinetic properties of adenylyl sulfate reductase from *Desulfovibrio vulgaris*. J Biol Chem 250:2979–2986
Broco M, Rousset M, Oliveira S, Rodrigues-Pousada C (2005) Deletion of flavoredoxin gene in *Desulfovibrio gigas* reveals its participation in thiosulfate reduction. FEBS Lett 579:4803–4807
Bromfield SM (1953) Sulfate reduction in partially sterilized soil exposed to air. J Gen Microbiol 8: 378–390
Brunner B, Arnold GL, Røy H, Müller IA, Jørgensen BB (2016) Off limits: sulfate below the sulfate-methane transition. Front Earth Sci 4:75. https://doi.org/10.3389/feart.2016.00075
Bruschi M, Hatchikian CE, Golovleva LA, Le Gall J (1977) Purification and characterization of cytochrome c3, ferredoxin, and rubredoxin isolated from *Desulfovibrio desulfuricans* Norway. J Bacteriol 129:30–38
Bruschi M, Guerlesquin F (1988) Structure, function and evolution of bacterial ferredoxins. FEMS Microbiol Rev 54:155–176
Bursakov S, Liu MY, Payne WJ, LeGall J, Moura I, Moura JJ (1995) Isolation and preliminary characterization of a soluble nitrate reductase from the sulfate-reducing organism *Desulfovibrio desulfuricans* ATCC 27774. Anaerobe 1:55–60
Bursakov SA, Carneiro C, Almendra MJ, Duarte RO, Caldeira J, Moura I, Moura JJ (1997) Enzymatic properties and effect of ionic strength on periplasmic nitrate reductase (NAP) from *Desulfovibrio desulfuricans* ATCC 27774. Biochem Biophys Res Commun 239:816–822
Cadby IT, Faulkner M, Cheneby J, Long J, van Helden J, Dolla A, Cole JA (2017) Coordinated response of the *Desulfovibrio desulfuricans* 27774 transcriptome to nitrate, nitrite and nitric oxide. Sci Rep 7:16228. https://doi.org/10.1038/s41598-017-16403-4
da Silva SM, Amaral C, Neves SS, Santos C, Pimentel C, Rodrigues-Pousada C (2015) An HcpR paralog of *Desulfovibrio gigas* provides protection against nitrosative stress. FEBS Open Bio 5: 594–604
Cammack R, Fauque G, Moura JJG, LeGall J (1984) ESR studies of cytochrome c from *Desulfovibrio desulfuricans* strain Norway 4: midpoint potentials of the four haems and interactions with ferredoxin and colloidal sulphur. Biochim Biophys Acta 784:68–74
Canfield DE, Kristensen E, Thamdrup B (2005) The sulfur cycle. Adv Mar Biol 48:313–383
Canfield DE, Stewart FJ, Thamdrup B, De Brabandere L, Dalsgaard T, Delong EF (2010) A cryptic sulfur cycle in oxygen-minimum-zone waters off the Chilean coast. Science 330:1375–1378
Canfield DE, Raiswell R (1999) The evolution of the sulfur cycle. Am J Sci 299:697–723
Cao J, Gayet N, Zeng X, Shao Z, Jebbar M, Alain K (2016) *Pseudodesulfovibrio indicus* gen. nov., sp. nov., a piezophilic sulfate-reducing bacterium from the Indian Ocean and reclassification of four species of the genus *Desulfovibrio*. Int J Syst Evol Microbiol 66:3904–3911
Chambers LA, Trudinger PA (1975) Are thiosulfate and trithionate intermediates in dissimilatory sulfate reduction? J Bacteriol 123:36–40
Chen L, Liu MY, LeGall J (1993) Isolation and characterization of flavoredoxin, a new flavoprotein that permits in vitro reconstruction of an electron transfer chain from molecular hydrogen to sulfite reduction in the bacterium *Desulfovibrio gigas*. Arch Biochem Biophys 303:44–50

Chiang Y-L, Hsieh Y-C, Fang J-Y, Liu E-H, Huang Y-C, Chuankhayan P et al (2009) Crystal structure of adenylylsulfate reductase from *Desulfovibrio gigas* suggests a potential self-regulation mechanism involving the C terminus of the β-subunit. J Bacteriol 191:7597–7608
Coleman GS (1960) A sulphate-reducing bacterium from the sheep rumen. J Gen Microbiol 22: 423–436
Costa C, Macedo A, Moura I, Moura JJG, Le Gall J, Berlier Y, Liu M-Y, Payne WJ (1990) Regulation of the hexaheme nitrite/nitric oxide reductase of *Desulfovibrio desulfuricans, Wolinella succinogenes* and *Escherichia coli*: a mass spectrometric study. FEBS 276:67–70
Cunha CA, Macieira S, Dias JM, Almeida G, Gonçalvesa LL, Costa C et al (2003) Cytochrome c nitrite reductase from *Desulfovibrio desulfuricans* ATCC 27774: the relevance of the two calcium sites in the structure of the catalytic subunit (NrfA). J Biol Chem 278:17455–17465
Cypionka H (1987) Uptake of sulfate, sulfite and thiosulfate by proton-anion symport in *Desulfovibrio desulfuricans*. Arch Microbiol 148:144–149
Cypionka H (1989) Characterization of sulfate transport in *Desulfovibrio desulfuricans*. Arch Microbiol 152:237–243
Cypionka H (1995) Sulfate transport and cell energetics. In: Barton LL (ed) Sulfate-reducing bacteria. Plenum Press, New York, pp 151–184
Dahl C, Kredich NM, Deutzmann R, Trüper HG (1993) Dissimilatory sulphite reductase from *Archaeoglobus fulgidus*: physico-chemical properties of the enzyme and cloning, sequencing and analysis of the reductase genes. J Gen Microbiol 139:1817–1828
Dam B, Mandal S, Ghosh W, Das Gupta SK, Roy P (2007) The S4-intermediate pathway for the oxidation of thiosulfate by the chemolithoautotroph *Tetrathiobacter kashmirensis* and inhibition of tetrathionate oxidation by sulfite. Res Microbiol 158:330–338
DerVartanian DV (1994) Desulforubidin: dissimilatory, high-spin sulfite reductase of *Desulfotomaculum* species. Meth Enzymol 243:270–276
Dias JM, Than ME, Humm A, Huber R, Bourenkov GB, Bartunik HD et al (1999) Crystal structure of the first dissimilatory nitrate reductase at 1.9 Å solved by MAD methods. Structure 7:65–79
Dörries M, Wöhlbrand L, Kube M, Reinhardt R, Rabus R (2016) Genome and catabolic subproteomes of the marine, nutritionally versatile, sulfate-reducing bacterium *Desulfococcus multivorans* DSM 2059. BMC Genomics 17:918. https://doi.org/10.1186/s12864-016-3236-7
Drake HL, Akagi JM (1978) Dissimilatory reduction of bisulfite by *Desulfovibrio vulgaris*. J Bacteriol 136:916–923
Drake HL, Akagi JM (1976) Product analysis of bisulfite reductase activity isolated from *Desulfovibrio vulgaris*. J Bacteriol 126:733–738
Duarte AG, Santos AA, Pereira IAC (2016) Electron transfer between the QmoABC membrane complex and adenosine 5′-phosphosulfate reductase. Biochim Biophys Acta (BBA) – Bioenerget 1857:380–386
Duzs Á, Tóth A, Németh B, Balogh T, Kós PB, Rákhely G (2018) A novel enzyme of type VI sulfide: quinone oxidoreductases in purple sulfur photosynthetic bacteria. Appl Microbiol Biotechnol 102:5133–5147
Escobar C, Bravo L, Hernandez J, Herrera L (2007) Hydrogen sulfide production from elemental sulfur by *Desulfovibrio desulfuricans* in an anaerobic bioreactor. Biotechnol Bioeng 98:569–577
Einsle O, Messerschmidt A, Stach P, Bourenkov GP, Bartunik HD, Huber R, Kroneck PM (1999) Structure of cytochrome c nitrite reductase. Nature 400:476–480
Fadhlaoui K, Ben Hania W, Postec A, Fauque G, Hamdi M, Ollivier B, Fardeau ML (2015) Characterization of *Desulfovibrio biadhensis* sp. nov., isolated from a thermal spring. Int J Syst Evol Microbiol 65:1256–1261
Fauque GD (1994) Sulfur reductase from thiophilic sulfate-reducing bacteria. Meth Enzymol 243: 353–367
Fauque GD, Barton LL (2012) Hemoproteins in dissimilatory sulfate- and sulfur-reducing prokaryotes. Adv Microbial Phys 60:2–91

Fauque GD, Barton LL, LeGall J (1980) Oxidative phosphorylation linked to the dissimilatory reduction of elemental sulfur by *Desulfovibrio*. Ciba Found Symp 72:71–86

Fauque G, Hervé D, LeGall J (1979) Structure-function relationship in hemoproteins: the role of cytochrome c3 in the reduction of colloidal sulphur by sulfate-reducing bacteria. Arch Microbiol 121:261–264

Fauque GD, Klimmek O, Kröger A (1994) Sulfur reductases from spirilloid mesophilic sulfur-reducing bacteria. Meth Enzymol 243:367–383

Fauque G, LeGall J, Barton LL (1991) Sulfate-reducing and sulfur-reducing bacteria. In: Shively JM, Barton LL (eds) Variations in autotrophic life. Academic Press, London, pp 271–337

Fauque G, Lino AR, Czechowski M, Kang L, DerVartanian DV, Moura JJ, LeGall J, Moura I (1990) Purification and characterization of bisulfite reductase (desulfofuscidin) from *Desulfovibrio thermophilus* and its complexes with exogenous ligands. Biochim Biophys Acta 1040:112–118

Fike DA, Bradley AS, Leavitt WD (2016) Geomicrobiology of sulfur. In: Ehrlich HL, Newman DK, Kappler A (eds) Ehrlich's geomicribiology, 6th edn. Taylor & Francis Group, LLC, CRC Press, Boca Raton, FL, pp 480–515

Finster K (2008) Microbial disproportionation of inorganic sulfur compounds. J Sulfur Chem 29: 281–292

Finster K, Liesack W, Yindall BJ (1997) *Desulfospira joergensenii*, gen. nov., sp. nov., a new sulfate-reducing bacterium isolated from marine surface sediment. Syst Appl Microbiol 20:201–208

Findley JE, Akagi JM (1970) Role of thiosulfate in bisulfite reduction as catalyzed by *Desulfovibrio vulgaris*. J Bacteriol 103:741–744

Fitz RM, Cypionka H (1990) Formation of thiosulfate and trithionate during sulfite reduction by washed cells of *Desulfovibrio desulfuricans*. Arch Microbiol 154:400–406

Fortin D, Goulet R, Roy M (2000) Seasonal cycling of Fe and S in a constructed wetland: the role of sulfate-reducing bacteria. Geomicrobiol J 17:221–235

Friedrich M, Springer N, Ludwig W, Schink B (1996) Phylogenetic positions of *Desulfofustis glycolicus* gen. nov., sp. nov., and *Syntrophobotulus glycolicus* gen. nov., sp. nov., two new strict anaerobes growing with glycolic acid. Int J Syst Bacteriol 46:1065–1069

Fritz G, Buchert T, Huber H, Stetter KO, Kroneck PMH (2000) Adenylylsulfate reductases from archaea and bacteria are 1: 1 alpha beta-heterodimeric iron-sulfur flavoenzymes – high similarity of molecular properties emphasizes their central role in sulfur metabolism. FEBS Lett 473:63–66

Fritz G, Roth A, Schiffer A, Büchert T, Bourenkov G, Bartunik HD (2002) Structure of adenylylsulfate reductase from the hyperthermophilic *Archaeoglobus fulgidus* at 1.6-Å resolution. PNAS 99:1836–1841

Galushko AS, Rozanova EP (1991) *Desulfobacterium cetonicum* sp. nov. – a sulfate-reducing bacterium which oxidizes fatty acids and ketones. Microbiology (Engl Transl.). Mikrobiologiya (Russia) 60:742–746

Gavel OY, Bursakov SA, Rocco GD, Trincão J, Pickering IJ, George GN et al (2008) A new type of metal-binding site in cobalt- and zinc-containing adenylate kinases isolated from sulfate-reducers *Desulfovibrio gigas* and *Desulfovibrio desulfuricans* ATCC 27774. J Inorg Biochem 102:1380–1395

Gavel OY, Bursakov SA, Calvete JJ, George GN, Moura JJ, Moura I (1998) ATP sulfurylases from sulfate-reducing bacteria of the genus *Desulfovibrio*. A novel metalloprotein containing cobalt and zinc. Biochem 37:16225–16232

Greene EA, Hubert C, Nemati M, Jenneman G, Voordouw G (2003) Nitrite reductase activity of sulphate-reducing bacteria prevents their inhibition by nitrate-reducing, sulphide-oxidizing bacteria. Environ Microbiol 5:607–617

George DH, Hunt LT, L. Yeh LSL, Barker WC. (1985) New perspectives on bacterial ferredoxin evolution. J Mol Evol 22:117–143

Gevertz D, Amelunxen R, Akagi JM (1980) Cysteine synthesis by *Desulfovibrio vulgaris* extracts. J Bacteriol 141:1460–1462

Ghosh W, Dam B (2009) Biochemistry and molecular biology of lithotrophic sulfur oxidation by taxonomically and ecologically diverse bacteria and archaea. FEMS Microbiol Rev 33:999–1043

González PJ, Rivas MG, Brondino CD, Bursakov SA, Moura I, Moura JJ (2006) EPR and redox properties of periplasmic nitrate reductase from *Desulfovibrio desulfuricans* ATCC 27774. J Biol Inorg Chem 11:609–616

Grein F, Ramos AR, Venceslau SS, Pereira IAC (2013) Unifying concepts in anaerobic respiration: insights from dissimilatory sulfur metabolism. Biochim Biophys Acta (BBA) – Bioenerg 1827: 145–160

Haouari O, Fardeau ML, Casalot L, Tholozan JL, Hamdi M, Ollivier B (2006) Isolation of sulfate-reducing bacteria from Tunisian marine sediments and description of *Desulfovibrio bizertensis* sp. nov. Int J Syst Evol Microbiol 56:2909–2913

Haschke RH, Campbell LL (1971) Thiosulfate reductase of *Desulfovibrio vulgaris*. J Bacteriol 106: 603–607

Hatchikian EC (1975) Purification and properties of thiosulfate reductase from *Desulfovibrio gigas*. Arch Microbiol 105:249–256

Hatchikian EC (1994) Desulfofuscidin: dissimilatory, high-spin sulfite reductase of thermophilic, sulfate-reducing bacteria. Meth Enzymol. 243:276–295

Hatchikian EC, Le Gall J, Bruschi M, Dubourdieu M (1972) Regulation of the reduction of sulfite and thiosulfate by ferredoxin, flavodoxin and cytochrome cc'_3 in extracts of the sulfate reducer *Desulfovibrio gigas*. Biochim Biophys Acta 258:701–708

Hatchikian EC, Zeikus JG (1983) Characterization of a new type of dissimilatory sulfite reductase present in *Thermodesulfobacterium commune*. J Bacteriol 153:2111–1220

Haveman SA, Greene EA, Stilwell CP, Voordouw JK, Voordouw G (2004) Physiological and gene expression analysis of inhibition of *Desulfovibrio vulgaris* Hildenborough by nitrite. J Bacteriol 186:7944–7950

Hedderich R, Klimmek O, Kröger A, Dirmeier R, Keller M, Stetter KO (1999) Anaerobic respiration with elemental sulfur and with sulfides. FEMS Microbiol Rev 22:353–381

Hensen D, Sperling D, Truper HG, Brune DC, Dahl C (2006) Thiosulphate oxidation in the phototrophic sulphur bacterium *Allochromatium vinosum*. Mol Microbiol 62:794–810

Herbert RA (1999) Nitrogen cycling in coastal marine ecosystems. FEMS Microbiol Rev 23:563–590

Hilz H, Lipmann F (1955) The enzymatic activation of sulfate. Proc Natl Acad Sci U S A 41:880–890

Hipp WM, Pott AS, Thum-Schmitz N, Faath I, Dahl C, Trüper HG (1997) Towards the phylogeny of APS reductases and sirohaem sulfite reductases in sulfate-reducing and sulfur-oxidizing prokaryotes. Microbiol 143:2891–2902

Hsieh Y-C, Liu M-L, Wang VC-C, Chiang LE-H, Wu WG, Shan SI, Chen C-J (2010) Structural insights into the enzyme catalysis from comparison of three forms of dissimilatory sulphite reductase from *Desulfovibrio gigas*. Mol Microbiol 78:1101–1116

Huber R, Rossnagel P, Woese CR, Rachel R, Langworthy TA, Stetter KO (1996) Formation of ammonium from nitrate during chemolithoautotrophic growth of the extremely thermophilic bacterium *Ammonifex degensii* gen. nov. sp. nov. Syst Appl Microbiol 19:40–49

Huynh BH, Kang L, DerVartanian DV, Peck HD Jr, LeGall J (1984) Characterization of a sulfite reductase from *Desulfovibrio vulgaris*. Evidence for the presence of a low-spin siroheme and an exchange-coupled siroheme-[4Fe-4S] unit. J Biol Chem 259:15373–15376

Ingvorsen K, Jørgensen BB (1984) Kinetics of sulfate uptake by freshwater and marine species of *Desulfovibrio*. Arch Microbiol 139:61–66

Ingvorsen K, Zehnder AJB, Jørgensen BB (1984) Kinetics of sulfate and acetate uptake by *Desulfobacter postgatei*. Appl Environ Microbiol 47:403–408

Ishimoto M (1959) Sulfate reduction in cell-free extracts of *Desulfovibrio*. J Biochem (Tokyo) 46: 105–106
Ishimoto M, Fujimoto D (1959) Adenosine-5′-phosphosulfate as an intermediate in the reduction of sulfate by a sulfate-reducing bacterium. Proc Japan Acad 35:243–245
Ishimoto M, Fujimoto D (1961) Sulfate reducing bacteria. X. Adenosine-5′- phosphosulfate reductase. J Biochem (Tokyo) 50:299–304
Ishimoto M, Fujimoto D, Kishimoto Y (1961) Studies in sulfate reduction in micro-organisms. Proc Intern Congr Biochem, 5th, Moscow 14:281–284
Ishimoto M, Koyama J (1957) Biochemical studies on sulfate-reducing bacteria. VI Separation of hydrogenase and thiosulfate reductase and partial separation of cytochrome and green pigment. J Biochem (Tokyo) 44:233–242
Ishimoto M, Kondo Y, Kameyama T, Yagi T, Shiraz M (1958) The role of cytochrome in the enzyme system of sulphate-reducing bacteria. In: Science Council of Japan (ed) Proceedings of the International Symposium on Enzyme Chemistry. Marüzen, Tokyo and Kyoto, pp 229–234
Ishimoto M, Koyama J, Omura T, Nagi Y (1954) Biochemical studies on sulfate-reducing bacteria. Sulfate reduction by cell suspensions. J Biochem (Tokyo) 41:537–546
Ishimoto M, Koyama J, Nagai Y (1955) Biochemical studies on sulfate-reducing bacteria: IV. Reduction of thiosulfate by cell-free extracts. J Biochem 42:41–53
Ishimoto M, Yagi T (1961) Sulfate-reducing bacteria. IX Sulfite reductase. J Biochem (Tokyo) 49: 103–109
Jensen HM, Lomstein E, Sörensen J (1990) Benthic NH_4^+ and NO_3^- flux following sedimentation of a spring phytoplankton bloom in Aarhus Bight, Denmark. Mar Ecol Prog Ser 61:87–96
Johnson EF, Mukhopadhyay B (2008) A novel coenzyme F420 dependent sulfite reductase and a small sulfite reductase in methanogenic archaea. In: Dahl C, Friedrich CG (eds) Microbial sulfur metabolism. Springer, Berlin, pp 202–206
Jones HE (1971) A re-examination of *Desulfovibrio africanus*. Arch Mikrobiol 80:78–86
Jones HE, Skyring GW (1975) Effect of enzyme assay on sulphite reduction catalyzed by *Desulfovibrio gigas*. Biochem Biophys Acta 377:52–60
Jørgensen BB (1990) A thiosulfate shunt in the sulfur cycle of marine sediments. Science 249:152–154
Jørgensen BB, Findlay AJ, Pellerin A (2019) The biogeochemical sulfur cycle of marine sediments. Front Microbiol 10:849. https://doi.org/10.3389/fmicb.2019.00849. PMID: 31105660; PMCID: PMC6492693
Kajie S, Anraku Y (1985) Purification of a hexaheme cytochrome c_{552} from *Escherichia coli* K 12 and its properties as a nitrite reductase. Eur J Biochem 154:457–463
Karkhoff-Schweizer RR, Bruschi M, Voordouw G (1993) Expression of the γ-subunit gene of desulfoviridin-type dissimilatory sulfite reductase and of the α- and β-subunit genes is not coordinately regulated. Eur J Biochem 211:501–507
Keller KL, Wall JD (2011) Genetics and molecular biology of the electron flow for sulfate respiration in *Desulfovibrio*. Front Microbiol 2:135. https://doi.org/10.3389/fmicb.2011.00135
Kim JH, Akagi JM (1985) Characterization of trithionate reductase system from *Desulfovibrio vulgaris*. J Bacteriol 163:472–475
Kirst GO (1996) Osmotic adjustment in phytoplankton and MacroAlgae. In: Kiene RP, Visscher PT, Keller MD, Kirst GO (eds) Biological and environmental chemistry of DMSP and related sulfonium compounds. Springer, Boston, MA, pp 121–129
Kjeldsen KU, Schreiber L, Thorup CA, Boesen T, Bjerg JT, Yang T et al (2019) On the evolution and physiology of cable bacteria. PNAS USA 116:19116–19125
Knöller K, Vogt C, Feisthauer S, Weise SM, Weiss H, Richnow H-H (2008) Sulfur cycling and biodegradation in contaminated aquifers: insights from stable isotope investigations. Environ Sci Technol 42:7807–7812
Kobayashi K, Seki Y, Ishimoto M (1974) Biochemical studies on sulfate-reducing bacteria. XII. Sulfite reductase from *Desulfoviridin vulgaris* – mechanism of trithionate , thiosulfate and sulfide formation by a sulfate-reducing bacterium. J Biochem 75:519–529

Kobayashi K, Takahashi E, Ishimoto M (1972) Biochemical studies on sulfate reducing bacteria. XI. Purification and properties of sulfite reductase, desulfoviridin. J Biochem 72:879–887

Kobayashi K, Tachibana S, Ishimoto M (1969) Intermediary formation of trithionate in sulfite reduction by a sulfate-reducing bacterium. J Biochem 65:155–157

Korte HL, Saini A, Trotter VV, Butland GP, Arkin AP, Wall JD (2015) Independence of nitrate and nitrite inhibition of *Desulfovibrio vulgaris* Hildenborough and use of nitrite as a substrate for growth. Environ Sci Technol 49:924–931

Krafft T, Bokranz M, Klimmek O, Schroeder I, Fahrenholz F, Kojro E, Kroeger A (1992) Cloning and nucleotide sequence of the *psrA* gene of *Wolinella succinogenes* polysulphide reductase. Eur J Biochem 206:503–510

Kreke B, Cypionka H (1992) Proton motive force in freshwater sulfate-reducing bacteria, and its role in sulfate accumulation in *Desulfobulbus propionicus*. Arch Microbiol 158:183–187

Kreke B, Cypionka H (1994) Role of sodium ions for sulfate transport and energy metabolism in *Desulfovibrio salexigens*. Arch Microbiol 161:55–61

Kreke B, Cypionka H (1995) Energetics of sulfate transport in *Desulfomicrobium baculatum*. Arch Microbiol 163:307–309

Kremer DR, Veenhuis M, Fauque G, Peck Jr HD Jr, LeGall J, Lampreia J, Moura JJG, Hansen TA (1988) Immunocytochemical localization of APS reductase and bisulfite reductase in three *Desulfovibrio* species. Arch Microbiol 150:296–301

Kroder M, Kroneck PMH, Cypionka H (1991) Determination of the transmembrane proton gradient in the anaerobic bacterium *Desulfovibrio desulfuricans* by ^{31}P nuclear magnetic resonance. Arch Microbiol 156:145–147

Krumholz LR, Wang L, Beck DAC, Wang T, Hackett M, Mooney B et al (2013) Membrane protein complex of APS reductase and Qmo is present in *Desulfovibrio vulgaris* and *Desulfovibrio alaskensis*. Microbiol 159:2162–2168

Lampreia J, Moura I, Teixeira M, Peck HD Jr, LeGall J, Huynh BH, Moura JJG (1990) The active centers of adenylylsulfate reductase from *Desulfovibrio gigas* – characterization and spectroscopic studies. Eur J Biochem 188:653–664

Lampreia JL, Pereira AS, Moura JJG (1994) Adenylylsulfate reductases from sulfate-reducing bacteria. Meth Enzymol 243:241–260

Larsen Ø, Lien T, Birkeland NK (1999) Dissimilatory sulfite reductase from *Archaeoglobus profundus* and *Desulfotomaculum thermocisternum*: phylogenetic and structural implications from gene sequences. Extremophiles 3:63–70

Latour P, Hong W-L, Sauer S, Sen A, Gilhooly WP III, Lepland A, Fouskas F (2018) Dynamic interactions between iron and sulfur cycles from Arctic methane seeps. Biogeosci Discuss. https://doi.org/10.5194/bg-2018-223

Lee JP, Peck HD Jr (1971) Purification of the enzyme reducing bisulfite to trithionate from *Desulfovibrio gigas* and its identification as desulfoviridin. Biochem Biophys Res Commun 45:583–589

Lee JP, Yi C, LeGall J, Peck HD Jr (1973a) Isolation of a new pigment, desulforubidin, from *Desulfovibrio desulfuricans* (Norway strain) and its role in sulfite reduction. J Bacteriol 115: 453–455

Lee JP, LeGall J, Peck HD Jr (1973b) Isolation of assimilatory- and dissimilatory-type sulfite reductases from *Desulfovibrio vulgaris*. J Bacteriol 115:529–542

Le Faou A, Rajagopal BS, Daniels L, Fauque G (1990) Thiosulfate, polythionates and elemental sulfur assimilation and reduction in the bacterial world. FEMS Microbiol Rev 75:351–382

LeGall J, Dragoni N (1966) Dependence of sulfite reduction on a crystallized ferredoxin from *Desulfovibrio gigas*. Biochem Biophys Res Commun 23:145–149

LeGall J, Fauque G (1988) Dissimilatory reduction of sulfur compounds. In: Zehnder AJB (ed) Biology of anaerobic microorganisms. Wiley, New York, pp 587–639

LeGall J, Hatchikian EC (1967) Purification et proprietes d'une flavoproteine intervenant dans la reduction du sulfite par *Desulfovibrio gigas*. C R Acad Sci 264:385–387

LeGall J, Senez JC (1960) Influence de la fixation de l'azote sur la croissance de *Desulfovibrio desulfuricans*. Comp Rend Seances Acad Sci 250:404–406

LeGall J, Senez JC, Pichinoty F (1959) Fixation de l'azote sur la croissance de *Desulfovibrio desulfuricans*. Ann Inst Pasteur 96:223–230

Lespinat PA, Berlier YM, Fauque GD, Toci R, Denariaz G, Jean LeGall J (1987) The relationship between hydrogen metabolism, sulfate reduction and nitrogen fixation in sulfate reducers. J Indust Microbiol 1:383–388

Lichtschlag A, Kamyshny A, Ferdelman TG, deBeer D (2013) Intermediate sulfur oxidation state compounds in the euxinic surface sediments of the Dvurechenskii mud volcano (Black Sea). Geochim Cosmochim Acta 105:130–145

Liu R, Shan Y, Xi S, Zhang X, Sun C (2021) A deep-sea sulfate reducing bacterium directs the formation of zero-valent sulfur via sulfide oxidation. bioRxiv. https://doi.org/10.1101/2021.03.23.436689

Liu M-C, Bakel BW, Liu M-Y, Dao TN (1988) Purification of *Vibrio fischeri* nitrite reductase and its characterization as a hexaheme *c*-type cytochrome. Arch Biochem Biophys 262:259–265

Liu CL, DerVartanian DV, Peck HD Jr (1979) On the redox properties of three bisulfite reductases from the sulfate-reducing bacteria. Biochem Biophys Res Commun 91:962–970

Liu M-C, Liu M-Y, Payne WJ, Peck HD Jr, LeGall J (1983) *Wolinella succinogenes* nitrite reductase: purification and properties. FEMS Microbiol Lett 19:201–206

Liu MC, Peck HD Jr (1981) The isolation of a hexaheme cytochrome from *Desulfovibrio desulfuricans* and its identification as a new type of nitrite reductase. J Biol Chem 256:13159–13164

Magot M, Basso O, Tardy-Jacquenod C, Caumette P (2004) *Desulfovibrio bastinii* sp. nov. and *Desulfovibrio gracilis* sp. nov., moderately halophilic, sulfate-reducing bacteria isolated from deep subsurface oilfield water. Int J Syst Evol Microbiol 54:1693–1697

Magot M, Caumette P, Desperrier JM, Mat R (1992) *Desulfovibrio longus* sp. nov., a sulfate-reducing bacterium isolated from an oil-producing well. Int J Microbiol Syst Bacteriol 42:398–403

Marietou A, Richardson D, Cole J, Mohan S (2005) Nitrate reduction by *Desulfovibrio desulfuricans*: a periplasmic nitrate reductase system that lacks NapB, but includes a unique tetraheme *c*-type cytochrome. NapM FEMS Microbiol Lett 248:217–225

Marietou A, Røy H, Jørgensen BB, Kjeldsen KU (2018) Sulfate transporters in dissimilatory sulfate-reducing microorganisms: a comparative genomics analysis. Front Microbiol 9:309. https://doi.org/10.3389/fmicb.2018.0030

Martins PD, Danczak RE, Roux S, Frank J, Borton MA, Wolfe RA, Burris MN, Wilkins MJ (2018) Viral and metabolic controls on high rates of microbial sulfur and carbon cycling in wetland ecosystems. Microbiome 6:138. https://doi.org/10.1186/s40168-018-0522-4

Matthews JC, Timkovich R, Liu MY, Le Gall J (1995) Siroamide: a prosthetic group isolated from sulfite reductases in the genus *Desulfovibrio*. Biochemist 34:5248–5251

Melton ED, Sorokin DY, Overmars L, Chertkov O, Clum A, Pillay M et al (2016) Complete genome sequence of *Desulfurivibrio alkaliphilus* strain AHT2^T, a haloalkaliphilic sulfidogen from Egyptian hypersaline alkaline lakes. Stand Genomic Sci 11(1):67. https://doi.org/10.1186/s40793-016-0184-4

Meyer B, Imhoff JF, Kuever J (2007) Molecular analysis of the distribution and phylogeny of the soxB gene among sulfur-oxidizing bacteria – evolution of the Sox sulfur oxidation enzyme system. Environ Microbiol 9:2957–2977

Meyer B, Kuever J (2007a) Phylogeny of the alpha and beta subunits of the dissimilatory adenosine-5′-phosphate (APS) reductase from sulfate-reducing prokaryotes – origin and evolution of the dissimilatory sulfate-reduction pathway. Microbiol 153:2026–2044

Meyer B, Kuever J (2007b) Molecular analysis of the distribution and phylogeny of dissimilatory adenosine-5′-phosphosulfate reductase-encoding genes (*aprBA*) among sulfur-oxidizing prokaryotes. Microbiol 153:3478–3498

Meyer B, Kuever J (2008) Homology modeling of dissimilatory APS rReductases (AprBA) of sulfur-oxidizing and sulfate-reducing prokaryotes. PLoS One 3(1):e1514. https://doi.org/10.1371/journal.pone.0001514

Michaels GB, Davidson JT, Peck HD Jr (1970) A flavin-sulfite adduct as an intermediate in the reaction catalyzed by adenylyl sulfate reductase from *Desulfovibrio vulgaris*. Biochem Biophys Res Commun 39:321–328

Miroshnichenko ML, Tourova TP, Kolganova TV, Kostrikina NA, Chernych N, Bonch-Osmolovskaya EA (2008) *Ammonifex thiophilus* sp. nov., a hyperthermophilic anaerobic bacterium from a Kamchatka hot spring. Int J Syst Evol Microbiol 58:2935–2938

Mizuno N, Voordouw G, Miki K, Sarai A, Higuchi Y (2003) Crystal structure of dissimilatory sulfite reductase D (DsrD) protein-possible interaction with B- and Z-DNA by its winged-helix motif. Structure 11:1133–1140

Molitor M, Dahl C, Molitor I, Schafer U, Speich N, Huber R, Deutzmann R, Trüper HG (1998) A dissimilatory sirohaem-sulfite-reductase-type protein from the hyperthermophilic archaeon *Pyrobaculum islandicum*. Microbiol 144:529–541

Moura I, LeGall J, Lino AR, Peck HD Jr, Fauque G, Xavier AV et al (1988) Characterization of two dissimilatory sulfite reductases (desulforubidin and desulfoviridin) from the sulfate-reducing bacteria. Mössbauer and EPR studies. J Am Chem Soc 110:1075–1082

Moura I, Lino AR (1994) Low-spin sulfite reductase. Meth Enzymol 243:296–303

Moura I, Lino AR, Moura JJG, Xavier AV, Fauque G, Peck HD Jr, LeGall J (1986) Low-spin sulfite reductases: a new homologous group of non-heme iron-siroheme proteins in anaerobic bacteria. Biochem Biophys Res Commun 141:1032–1041

Moura JJG, González PJ, Moura I, Fauque G (2007) Dissimilatory nitrate and nitrite ammonification by sulphate-reducing bacteria. In: Barton LL, Hamilton WA (eds) Sulphate-reducing bacteria – environmental and engineered systems. Cambridge University Press, Cambridge, UK, pp 241–264

Moura JJG, Moura I, Santos H, Xavier AV, Scandellari M, LeGall J (1982) Isolation of P590 from *Methanosarcina barkeri*: evidence for the presence of sulfite reductase activity. Biochem Biophy Res Commun 108:1002–1009

Mudryk ZJ, Podgórska B, Ameryk A, Bolalek J (2000) The occurrence and activity of sulphate-reducing bacteria in the bottom sediments of the Gulf of Gdańsk. Oceanol 42:105–117

Müller H, Marozava S, Probst AJ, Meckenstock RU (2019) Groundwater cable bacteria conserve energy by sulfur disproportionation. ISME J 14:1–12

Najmudin S, González PJ, Trincão J, Coelho C, Mukhopadhyay A, Cerqueira NM et al (2008) Periplasmic nitrate reductase revisited: a sulfur atom completes the sixth coordination of the catalytic molybdenum. J Biol Inorg Chem 13:737–753

Nakatsukasa W, Akagi JM (1969) Thiosulfate reductase isolated from *Desulfotomaculum nigrificans*. J Bacteriol 98:429–433

Nethe-Jaenchen R, Thauer RK (1984) Growth yields and saturation constant of *Desulfovibrio vulgaris* in chemostat culture. Arch Microbiol 137:236–240

Okabe S, Itoh T, Satoh H, Watanabe Y (1999) Analyses of spatial distributions of sulfate-reducing bacteria and their activity in aerobic wastewater biofilms. Appl Environ Microbiol 65:5107–5116

Ollivier B, Cord-Ruwisch R, Hatchikian EC, Garcia JL (1988) Characterization of *Desulfovibrio fructosovorans* sp. nov. Arch Microbiol 149:447–450

Ollivier B, Hatchikian CE, Prensier G, Guezennec J, Garcia JL (1991) *Desulfohalobium retbaense* gen. nov., sp. nov., a halophilic sulfate-reducing bacterium from sediments of a hypersaline lake in Senegal. Int J Syst Bacteriol 41:74–81

Oliveira TF, Franklin E, Afonso JP, Khan AR, Oldham NJ, Pereira IAC, Archer M (2011) Structural insights into dissimilatory sulfite reductases: structure of desulforubidin from *Desulfomicrobium norvegicum*. Front Microbiol 2011(2):71. https://doi.org/10.3389/fmicb.2011.00071

Oliveira TF, Vornhein C, Matias PM, Venceslau SS, Pereira IAC, Archer M (2008a) Purification, crystallization and preliminary crystallographic analysis of a dissimilatory DsrAB sulfite reductase in complex with DsrC. J Struct Biol 164:236–239

Oliveira TF, Vonrhein C, Matias PM, Venceslau SS, Pereira IAC, Archer M (2008b, 283) The crystal structure of *Desulfovibrio vulgaris* dissimilatory sulfite reductase bound to DsrC provides novel insights into the mechanism of sulfate respiration. J Biol Chem:34141–34149

Ostrowski J, Wu J-Y, Rueger DC, Miller BE, Siegel LM, Kredich NM (1989) Characterization of the cysJIH regions of *Salmonella typhimurium* and *Escherichia coli* B: DNA sequences of cysI and cysH and a model for the siroheme-Fe_4-S_4 active center of sulfite reductase hemoprotein based on amino acid and homology with spinach nitrite reductase. J Biol Chem 264:15726–15737

Ouattara AS, Patel BKC, Cayol J-L, Cuzin N, Traore AS, Garcia J-L (1999) Isolation and characterization of *Desulfovibrio burkinensis* sp. nov. from an African rice field, and phylogeny of *Desulfovibrio alcoholivorans*. Int J Syst Bacteriol 49:639–643

Pajares S, Ramos R (2019) Processes and microorganisms involved in the marine nitrogen cycle: knowledge and gaps. Front Mar Sci 6:739. https://doi.org/10.3389/fmars.2019.00739

Parey K, Demmer U, Warkentin E, Wynen A, Ermler U, Dahl C (2013) Structural, biochemical and genetic characterization of dissimilatory ATP sulfurylase from *Allochromatium vinosum*. PLoS One 8(9):e74707. https://doi.org/10.1371/journal.pone.0074707

Parey K, Warkentin E, Kroneck PMH, Ermler U (2010) Reaction cycle of the dissimilatory sulfite reductase from *Archaeoglobus fulgidus*. Biochemist 49:8912–8921

Peck HD Jr (1959) The ATP-dependent reduction of sulfate with hydrogen in extracts of *Desulfovibrio desulfuricans*. Proc Natl Acad Sci U S A 45:701–708

Peck HD Jr (1960) Evidence for oxidative phosphorylation during the reduction of sulfate with hydrogen by *Desulfovibrio desulfuricans*. J Biol Chem 235:2734–2738

Peck HD Jr (1961a) Enzymatic basis for assimilatory and dissimilatory sulfate reduction. J Bacteriol 82:933–939

Peck HD Jr (1961b) Evidence for the reversibility of the reaction catalysed by adenosine5-′-phosphosulfate reductase. Biochim Biophys Acta 49:621–624

Peck HD Jr (1962a) Symposium on metabolism of inorganic compounds. V Comparative metabolism of inorganic sulfur compounds in microorganisms. Bacteriol Rev 26:67–94

Peck HD Jr (1962b) The role of adenosine-5′- phosphosulfate in the reduction of sulfate to sulfite by *Desulfovibrio desulfuricans*. J Biol Chem 237:198–203

Peck HD Jr, Bramlett RN (1982) Flavoproteins in sulfur metabolism. In: Massey V, Williams V (eds) 7th International Symposium on Flavin and Flavoproteins. University Park Press, Tokyo, pp 851–858

Peck HD Jr, Lissolo T (1988) Assimilatory and dissimilatory sulfate reduction. In: Cole J, Ferguson S (eds) The nitrogen and sulfur cycles. Cambridge University Press, Cambridge, pp 99–132

Peck HD Jr, LeGall J (1982) Biochemistry of dissimilatory sulphate reduction. Philos Trans R Soc Lond Ser B Biol Sci 298:443–466

Pellerin A, Antler G, Røy H, Findlay A, Beulig F, Scholze C, Turchyn AV, Jørgensen BB (2018) The sulfur cycle below the sulfate-methane transition of marine sediments. Geochim Cosmochim Acta 239:74–89

Pereira IA, LeGall J, Xavier AV, Teixeira M (2000, 1481) Characterization of a heme *c* nitrite reductase from a non-ammonifying microorganism, *Desulfovibrio vulgaris* Hildenborough. Biochim Biophys Acta:119–130

Pester M, Knorr K-H, Friedrich MW, Wagner M, Loy A (2012) Sulfate-reducing microorganisms in wetlands – fameless actors in carbon cycling and climate change. Front Microbiol 3:72. https://doi.org/10.3389/fmicb.2012.00072

Pfennig N, Biebl H (1976) *Desulfuromonas acetoxidans* gen. nov. and sp. nov., a new anaerobic, sulfur-reducing, acetate-oxidizing bacterium. Arch Microbiol 110:3–12

Pierik AJ, Duyvis MG, van Helvoort JM, Wolbert RB, Hagen WR (1992) The third subunit of desulfoviridin-type dissimilatory sulfite reductases. Eur J Biochem 205:111–115

Pires RH, Venceslao SS, Morais F, Teixeira M, Xavier AV, Pereira IA (2006) Characterization of the *Desulfovibrio desulfuricans* ATCC 27774 DsrMKJOP complex–a membrane-bound redox complex involved in the sulfate respiratory pathway. Biochemist 45:249–262

Postgate JR (1951) The reduction of sulphur compounds by *Desulfovibrio desulfuricans*. J Gen Microbiol 5:725–738

Postgate JR (1956) Cytochrome c_3 and desulfoviridin, pigments of the anaerobe *Desulfovibrio desulfuricans*. J Gen Microbiol 37L:545–572

Postgate JR (1970) Nitrogen fixation by sporulating sulphate-reducing bacteria including rumen strains. J Gen Microbiol 63:137–139

Postgate JR (1979) The sulphate-reducing bacteria. Cambridge University Press, Cambridge

Postgate JR (1982) The fundamentals of nitrogen fixation. Cambridge University Press, Cambridge

Postgate JR, Kent HM, Robson RL (1986) DNA from diazotrophic *Desulfovibrio* strains is homologous to *Klebsiella pneumoniae* structural *nif* DNA and can be chromosomal or plasmid-borne. FEMS Microbiol Lett 33:159–165

Postgate JR, Kent HM, Robson RL (1988) Nitrogen fixation by *Desulfovibrio*. In: Cole JA, Ferguson SJ (eds) The nitrogen and sulphur cycles. Cambridge University Press, Cambridge, UK, pp 457–471

Purdy KJ, Embley TM, Nedwell DB (2002) The distribution and activity of sulphate reducing bacteria in estuarine and coastal marine sediments. Antonie Van Leeuwenhoek 81:181–187

Qatibi AI, Nivière V, Garcia JL (1991) *Desulfovibrio alcoholovorans* sp. nov., a sulfate-reducing bacterium able to grow on glycerol, 1,2- and 1,3-propanediol. Arch Microbiol 155:143–148

Rabus R, Hansen TA, Widdel F (2006) Dissimilatory sulfate- and sulfur-reducing prokaryotes. In: Dworkin M, Falkow S, Rosenberg E, Schleifer K-H, Stackebrandt E (eds) The prokaryotes, vol 2. Springer, Berlin, pp 659–768

Ramos AR, Keller KL, Wall JD, Pereira IAC (2012) The membrane QmoABC complex interacts directly with the dissimilatory adenosine 5 – phosphosulfate reductase in sulfate reducing bacteria. Front Microbio 3:137. https://doi.org/10.3389/fmicb.2012.00137

Rehr B, Klemme J-H (1986) Metabolic role and properties of nitrite reductase of nitrate-ammonifying marine *Vibrio* species. FEMS Microbiol Lett 35:325–328

Riederer-Henderson MA, Wilson PW (1970) Nitrogen fixation by sulphate-reducing bacteria. J Gen Microbiol 61:27–31

Robador A, Müller AI, Sawicka JE, Berry D, Hubert CRJ, Loy A, Jørgensen BB, Brüchert V (2016) Activity and community structures of sulfate-reducing microorganisms in polar, temperate and tropical marine sediments. ISME J 10:796–809

Robbins PW, Lipmann F (1957) Isolation and identification of active sulfate. J Biol Chem 229:837–851

Robbins PW, Lipmann F (1958) Enzymatic synthesis of adenosine-5'-phosphosulfate. J Biol Chem 233:686–690

Robertson WJ, Bowman JP, Franzmann PD, Mee BJ (2001) *Desulfosporosinus meridiei* sp. nov., a spore-forming sulfate-reducing bacterium isolated from gasoline-contaminated groundwater. Int J Syst Evol Microbiol 51:133–140

Rodrigues ML, Oliveira TF, Pereira IA, Archer M (2006a) X-ray structure of the membrane-bound cytochrome *c* quinol dehydrogenase NrfH reveals novel haem coordination. EMBO J 25:5951–5960

Rodrigues R, Vicente JB, Félix R, Oliveira S, Teixeira M, Rodrigues-Pousada C (2006b) *Desulfovibrio gigas* flavodiiron protein affords protection against nitrosative stress *in vivo*. J Bacteriol 188:2751

Rodrigues ML, Scott KA, Sansom MS, Pereira IA, Archer M (2008) Quinol oxidation by *c*-type cytochromes: structural characterization of the menaquinol binding site of NrfHA. J Mol Biol 381:341–350

Rozanova EP, Khudyakova AI (1974) A new nonsporeforming thermophilic sulfate-reducing organism, *Desulfovibrio thermophilus* nov. sp. Microbiol (USSR) 43:908–912

Rozanova EP, Nazina TN (1976) A mesophilic, sulfate-reducing, rod shaped, nonsporeforming bacterium. Microbiol (USSR) 45:711–716
Sakurai H, Ogawa T, Shiga M, Inoue K (2010) Inorganic sulfur oxidizing system in green sulfur bacteria. Photosyn Res 104:163–176
Santos AA, Venceslau SS, Grein F, Leavitt WD, Dahl C, Johnston DT, Pereira IAC (2015) A protein trisulfide couples dissimilatory sulfate reduction to energy conservation. Science 350: 1541–1545
Sass H, Cypionka H, Babenzien H-D (1997) Vertical distribution of sulfate-reducing bacteria at the oxic-anoxic interface in sediments of the oligotrophic Lake Stechlin. FEMS Microbiol Ecol 22: 245–255
Sass H, Ramamoorthy S, Yarwood C, Langner H, Schumann P, Kroppenstedt RM, Spring S, Rosenzweig RF (2009) *Desulfovibrio idahonensis* sp. nov., sulfate-reducing bacteria isolated from a metal(loid)- contaminated freshwater sediment. Int J Syst Evol Microbiol 59:2208–2214
Schauder R, Kröger A (1993) Bacterial sulphur respiration. Arch Microbiol 159:491–497
Schauder R, Müller E (1993) Polysulfide as a possible substrate for sulfur-reducing bacteria. Arch Microbiol 160:377–382
Schiffer A, Parey K, Warkentin E, Diederichs K, Huber H, Stetter KO, Kroneck PM, Ermler U (2008) Structure of the dissimilatory sulfite reductase from the hyperthermophilic archaeon *Archaeoglobus fulgidus*. J Mol Biol 379:1063–1074
Schumacher W, Hole U, Kroneck PMH (1994) Ammonia-forming cytochrome *c* nitrite reductase from *Sulfurospirillum deleyianum* is a tetraheme protein: new aspects of the molecular composition and spectroscopic properties. Biochem Biophys Res Commun 205:911–916
Schumacher W, Kroneck PMH (1992) Anaerobic energy metabolism of the sulfur-reducing bacterium "*Spirillum*" 5175 during dissimilatory nitrate reduction to ammonia. Arch Microbiol 157:464–470
Seki Y, Kobayashi K, Ishimoto M (1979) Biochemical studies on sulfate-reducing bacteria: XV. Separation and comparison of two forms of desulfoviridin. J Biochem 85:705–711
Shturm LD (1948) Sulfate reduction by facultative aerobic bacteria. Mikrobiologiya 17:415–418
Siegel LM, Kamin H (1968) *E. coli* TPNH-sulfite reductase. In: Yaki K (ed) Flavins and flavoproteins. University of Tokyo Press, Tokyo, pp 15–40
Silaghi-Dumitrescu R, Ng KY, Viswanathan R, Kurtz DM Jr (2005) A flavo-diiron protein from *Desulfovibrio vulgaris* with oxidase and nitric oxide reductase activities. Evidence for an *in vivo* nitric oxide scavenging function. Biochemist 44:3572–3579
Simon J, Kroneck PMH (2013) Microbial sulfite respiration. Adv Microbial Physiol 62:45–117
Susanti D, Mukhopadhyay B (2012) An intertwined evolutionary history of methanogenic archaea and sulfate reduction. PLoS One 7(9):e45313. https://doi.org/10.1371/journal.pone.0045313
Sisler FD, Zobell CE (1951) Nitrogen fixation by sulfate-reducing bacteria indicated by nitrogen argon ratios. Science 113:511–514
Sivan O, Antler G, Turchyn AV, Marlow JJ, Orphan VJ (2014) Iron oxides stimulate sulfate-driven anaerobic methane oxidation in seeps. Proc Natl Acad Sci U S A 111(40):E4139–E4147
Skyring GW, Jones HE, Goodchild D (1977) The taxonomy of some new isolates of dissimilatory sulfate-reducing bacteria. Can J Microbiol 23:1415–1425
Skyring GW, Trudinger PA (1973) A comparison of the electrophoretic properties of the ATP-sulfurylases, APS reductases and sulfite reductases from cultures of dissimilatory sulfate-reducing bacteria. Can J Microbiol 19:375–380
Smith M, Bardiau M, Brennan R, Burgess H, Caplin J, Santanu Ray S, Urios T (2019) Accelerated low water corrosion: the microbial sulfur cycle in microcosm. NPJ Mater Degrad 3:37. https://doi.org/10.1038/s41529-019-0099-9
Sørensen J, Jørgensen BB, Revsbech NP (1979) A comparison of oxygen, nitrate, and sulfate respiration in coastal marine sediments. Microbiol Ecol 5:105–115
Sparacino-Watkins C, Stolz JF, Basu P (2014) Nitrate and periplasmic nitrate reductases. Chem Soc Rev 43:676–706

Speich N, Dahl C, Heisig P, Klein A, Lottspeich F, Stetter KO, Trüper HG (1994) Adenylylsulphate reductase from the sulfate-reducing archaeon *Archaeoglobus fulgidus* – cloning and characterization of the genes and comparison of the enzyme with other iron-sulfur flavoproteins. Microbiol 140:1273–1284

Speich N, Trüper HD (1988) Adenylylsulphate reductase in a dissimilatory sulphate-reducing archaebacterium. J Gen Microbiol 134:1419–1425

Stackebrandt E, Schumann P, Schüler E, Hippe H (2003) Reclassification of *Desulfotomaculum auripigmentum* as *Desulfosporosinus auripigmenti* corrig., comb. nov. Int J Syst Evol Microbiol 53:1439–1443

Stahlmann J, Warthmann R, Cypionka H (1991) Na^+-dependent accumulation of sulfate and thiosulfate in marine sulfate-reducing bacteria. Arch Microbiol 155:554–558

Steudler P, Peterson B (1984) Contribution of gaseous sulphur from salt marshes to the global sulphur cycle. Nature 311:455–457

Steuber J, Arendsen AF, Hagen WR, Kroneck PMH (1995) Molecular properties of the dissimilatory sulfite reductase from *Desulfovibrio desulfuricans* (Essex) and comparison with the enzyme from *Desulfovibrio vulgaris* (Hildenborough). Eur J Biochem 233:873–879

Steuber J, Cypionka H, Kroneck PMM (1994) Mechanism of dissimilatory sulfite reductase by *Desulfovibrio desulfuricans*: purification of a membrane-bound sulfite reductase and coupling to cytochrome c_3 and hydrogenase. Arch Microbiol 162:255–260

Steuber J, Kroneck PMM (1998) Desulfoviridin, the dissimilatory sulfite reductase from *Desulfovibrio desulfuricans* (Essex): new structural and functional aspects of the membranous enzyme. Inorg Chim Acta 276:52–57

Stille W, Trüper HD (1984) Adenylylsulfate reductase in some new sulfate-reducing bacteria. Arch Microbiol 137:145–150

Suh B, Akagi JM (1969) Formation of thiosulfate from sulfite by *Desulfovibrio desulfuricans*. J Bacteriol 103:741–744

Taguchi Y, Sugishima M, Fukuyama K (2004) Crystal structure of a novel zinc-binding ATP sulfurylase from *Thermus thermophilus* HB8. Biochemist 43:4111–4118

Takahashi Y, Suto K, Inoue C, Chida T (2008) Polysulfide reduction using sulfate-reducing bacteria in a photocatalytic hydrogen generation system. J Biosci Bioeng 106:219–225

Tardy-Jacquenod C, Magot M, Laigret F, Kaghad M, Patel BKC, Guezennec J, Matheron R, Caumette P (1996) *Desulfovibrio gabonensis* sp. nov., a new moderately halophilic sulfate-reducing bacterium isolated from an oil pipeline. Int J Syst Bacteriol 46:710–715

Tarpgaard IH, Jørgensen BB, Kjeldsen KU, Røy H (2017, 2017) The marine sulfate reducer *Desulfobacterium autotrophicum* HRM2 can switch between low and high apparent half-saturation constants for dissimilatory sulfate reduction. FEMS Microbiol Ecol 93(4). https://doi.org/10.1093/femsec/fix012

Tarpgaard IH, Røy H, Jørgensen BB (2011) Concurrent low- and high-affinity sulfate reduction kinetics in marine sediment. Geochim Cosmochim Acta 75:2997–3010

Thamdrup B, Finster K, Hansen JW, Bak F (1993) Bacterial disproportionation of elemental sulfur coupled to chemical reduction of iron or manganese. Appl Environ Microbiol 59:101–108

Thauer RK (1982) Dissimilatory sulfate reduction with acetate as the electron donor. Philos Trans R Soc London Ser B 298:467–471

Thauer RK (1989) Energy metabolism of sulfate-reducing bacteria. In: Schlegel HG, Bowien B (eds) Autotrophic bacteria. Science Tech Publishers, Madison, WI, pp 397–413

Thauer RK, Jungermann K, Decker K (1977) Energy conservation in chemotrophic anaerobic bacteria. Bacteriol Rev 41:100–180

Thauer RK, Stackebrandt E, Hamilton WA (2007) Energy metabolism and phylogenetic diversity of sulphate-reducing bacteria. In: Barton LL, Hamilton WA (eds) Sulphate-reducing bacteria. Cambridge University Press, Cambridge, UK, pp 1–38

Thebrath B, Dilling W, Cypionka H (1989) Sulfate activation in *Desulfotomaculum*. Arch Microbiol 152:296–301

Trudinger PA (1970) Carbon monoxide-reacting pigment from *Desulfotomaculum nigrificans* and its possible relevance to sulfite reduction. J Bacteriol 104:158–170

Trüper HG, Fischer U (1982) Anaerobic oxidation of sulfur compounds as electron donors for bacterial photosynthesis. Phil Trans R Soc Lond B 298:529–542
Tuttle JH, Dugan PR, Macmillan CB, Randles CI (1969) Microbial dissimilatory sulfur cycle in acid mine water. J Bacteriol 97:594–602
Ulrich GA, Martion D, Burger K, Grossman EL, Ammerman JW, Suflita JM (1998) Sulfur cycling in the terrestrial subsurface: commensal interactions, spatial scales and microbial heterogeneity. Micro Ecol 36:141–151
Van Cappellen P, Wang Y (1996) Cycling of iron and manganese in surface sediments; a general theory for the coupled transport and reaction of carbon, oxygen, nitrogen, sulfur, iron, and manganese. Am J Sci 296:197
Venceslau SS, Stockdreher Y, Dahl C, Pereira IAC (2014) The "bacterial heterodisulfide" DsrC is a key protein in dissimilatory sulfur metabolism. Biochim Biophys Acta 1837:1148–1164
Vigneron A, Cruaud P, Culley AI, Couture RM, Lovejoy C, Vincent WF (2021) Genomic evidence for sulfur intermediates as new biogeochemical hubs in a model aquatic microbial ecosystem. Microbiome 9(1):46. https://doi.org/10.1186/s40168-021-00999-x. PMID: 33593438; PMCID: PMC7887784
Wang J, Vine CE, Balasiny BK, Rizk J, Bradley CL, Tinajero-Trejo M et al (2016) The roles of the hybrid cluster protein, Hcp and its reductase, Hcr, in high affinity nitric oxide reduction that protects anaerobic cultures of *Escherichia coli* against nitrosative stress. Mol Microbiol 100: 877–892
Ware DA, Postgate JR (1971) Physiological and chemical properties of a reductant-activated inorganic pyrophosphatase from *Desulfovibrio desulfuricans*. J Gen Microbiol 67:145–160
Warthmann R, Cypionka H (1990) Sulfate transport in *Desulfobulbus propionicus* and *Desulfococcus multivorans*. Arch Microbiol 154:144–149
Wasmund K, Mußmann M, Loy A (2017) The life sulfuric: microbial ecology of sulfur cycling in marine sediments. Environ Microbiol Repts 9:323–344
Widdel F, Hansen TA (1992) The dissimilatory sulfate-and sulfur-reducing bacteria. In: Balows A, Trüper HG, Dworkin M, Harder W, Schleifer K-H (eds) The prokaryotes, vol 1, 2nd edn. Springer, New York, pp 583–624
Widdel F, Pfennig N (1982) Studies on dissimilatory sulfate-reducing bacteria that decompose fatty acids. II. Incomplete oxidation of propionate by *Desulfobulbus propionicus* gen. nov., sp. nov. Arch Mikrobiol 131:360–365
Widdel F, Pfennig N (1992) The genus *Desulfuromonas* and other gram-negative sulfur-reducing eubacteria. In: Balows A, Trüper HG, Dworkin M, Harder W, Schleifer KH (eds) The prokaryotes, vol 4, 2nd edn. Springer, New York, pp 3379–3389
Wilson LG, Bandurski RS (1958) Enzymatic reactions involving sulfate, sulfite, selenate and molybdate. J Biol Chem. 233:975–981
Wolfe BM, Lui SM, Cowan JA (1994) Desulfoviridin, a multimeric-dissimilatory sulfite reductase from *Desulfovibrio vulgaris* (Hildenborough). Purification, characterization, kinetics and EPR studies. Eur J Biochem 223:79–89
Zeikus JG, Dawson MA, Thompson TE, Ingvorsen K, Hatchikan EC (1983) Microbial ecology of volcanic sulphidogenesis isolation and characterization of *Thermodesulfobacterium commune* gen.nov. and sp. nov. J Gen Microbiol. 129:1159–1169
Zhang J, Liu R, Xi S, Cai R, Zhang X, Sun C (2020) A novel bacterial thiosulfate oxidation pathway provides a new clue about the formation of zero-valent sulfur in deep sea. ISME J 14:2261–2274
Zheng R, Sun C (2020) *Pseudodesulfovibrio cashew* sp. nov., a novel deep-sea sulfate-reducing bacterium, 2 linking heavy metal resistance and sulfur cycle. bioRxiv. https://doi.org/10.1101/2020.12.08.417204
Zöphel A, Kennedy MC, Beinert H, Kroneck PMH (1988) Investigations on microbial sulfur respiration. 1. Activation and reduction of elemental sulfur in several strains of Eubacteria. Arch Microbiol. 150:72–77
Zöphel A, Kennedy MC, Beinert H, Kroneck PMH (1991) Investigations on microbial sulfur respiration: isolation, purification, and characterization of cellular components from *Spirillum* 5175. Eur J Biochem 195:849–856

Chapter 4
Electron Transport Proteins and Cytochromes

4.1 Introduction

The components of electron transport in SRB have been a subject pursued by numerous chemists, biochemists, and microbiologists, and this chapter examines key structural properties of these proteins. While SRB are anaerobes, their hydrogenases are not destroyed by exposure to air, which enabled these proteins to be initially purified without the use of anaerobic chambers. This ease of protein purification, the localization of hydrogenases within the cell (periplasm, plasma membrane, or cytoplasmic), and the range of metal cofactors (Fe, Ni-Fe, or Ni-Se-Fe) provided a great challenge in understanding the roles for hydrogenases in SRB. With the development of molecular biology techniques, gene isolation and genome analysis provided an additional dimension for understanding hydrogenases in SRB. Formate is used by many of the SRB as an electron donor, and the formate dehydrogenases from SRB have been characterized. The multiple forms of formate dehydrogenase found in a species, along with the distribution in the cell (periplasm, membrane, or cytoplasm), indicate that formate dehydrogenases have special functions in various SRB. Molecular characteristics of ferredoxin and flavodoxin as electron carriers with low redox potentials are discussed along with their role in metabolism. Molecular characteristics of electron carriers with high potential are discussed, and some of the physiological activities associated with these unique proteins are analyzed. At one time, cytochromes were considered to be absent in anaerobic bacteria; however, it is not apparent that SRB have an abundance of cytochromes *c* and most characteristically multiheme cytochromes, including those in transmembrane complexes, to facilitate the movement of electrons across the plasma membrane. Additionally, SRB have cytochromes b, which interface with membrane-associated electron transport processes. The protein chemistry of nitrite, nitrate, and bisulfite reductases has been examined to assess the mechanisms of

L. L. Barton, G. D. Fauque, *Sulfate-Reducing Bacteria and Archaea*,
https://doi.org/10.1007/978-3-030-96703-1_4

interaction for terminal electron acceptors in SRB. Most recently, certain SRB use O_2 respiration to support growth, and characteristics of this terminal respiratory process are discussed.

4.2 Hydrogenases

Hydrogenases are commonly distributed throughout biology, and the classes of hydrogenases, including those from SRP, have been the subject of several reviews (Vignais et al. 2001; Vignais and Billoud 2007; Greening et al. 2016). The SRP have hydrogenases that belong to the [FeFe] and [NiFe] classes, which have the reaction as indicated in Eq. 4.1.

$$H_2 = 2H^+ + 2e^- \tag{4.1}$$

Characteristically, hydrogenases of SRP are members of the [FeFe] class or [NiFe] class with [NiFeSe] as a subclass of [NiFe]. A new "metal-free" hydrogenase had been proposed for a methanogen (Lyon et al. 2004); however, additional examination revealed that it contained a single iron atom in the active site (Shima and Thauer 2006). The distribution of this [Fe] hydrogenase (Hmd), often referred to as "iron-only" hydrogenase, appears to be limited to methanogenic archaea (Shima and Ermler 2011).

While hydrogenases were discovered in *Escherichia coli* and related intestinal bacteria in 1931 by Stephenson and Strickland, it was not until several decades later that hydrogenase research was pursued in sulfate-reducing microorganisms. Dyes have been employed as electron acceptors to demonstrated hydrogenases in cells and cell-free extracts with dichlorophenol indophenol, Janus green, and indigo carmine used with *Desulfovibrio* spp.; however, methyl viologen and benzyl viologen are commonly employed as the electron acceptors for hydrogenase (Sadana and Jagannathan 1956; Postgate 1984). Even though methylene blue is used where hydrogenase reactions occur, methylene blue may also inhibit some hydrogenases, and this has limited its use in hydrogenase assays (King and Winfield 1955; Littlewood and Postgate 1956). Studies of H_2 oxidation with sulfate reducers provided early information on the kinetics of sulfate reduction by strains of *Desulfovibrio* (Postgate 1951a, b). Additionally, even though hydrogenases are typically associated with SRP, it is not an absolute requirement for dissimilatory sulfate reduction, because at least one SRB (*D. sapovorans*) lacks hydrogenase (Postgate 1984) and Δ*hyn*/Δ*ech* mutant strains of *D. gigas* readily grow in lactate-sulfate or pyruvate-sulfate media even though they lack hydrogenase activity (Morais-Silva et al. 2013).

Over the years, molecular characteristics of hydrogenases isolated from SRB had been reviewed (LeGall et al. 1979; Peck Jr. and LeGall 1982; Odom and Peck Jr. 1984; LeGall and Fauque 1988; Fauque et al. 1988). Features of the hydrogenases

Table 4.1 Distribution of hydrogenases in cells of *Desulfovibrio*

	Location in cell[b]		
Classes of hydrogenases	Periplasm	Membrane	Cytoplasm[a]
[FeFe]	Hyd	–	Hnd
[NiFe]	Hyn	Ech and Coo	–
[NiFeSe]	Hys	Hys	–

[a]Includes monomeric, trimeric, tetrameric, and sensor hydrogenases
[b]Reference is Baffert et al. (2019)

with different metal centers are presented in Table 4.2 for those bacteria often used as model organism for hydrogenases. More recently, reports containing genomic information for hydrogenases in *Desulfovibrio* (Baffert et al. 2019), *Desulfotomaculum* (Junier et al. 2010; Kuever et al. 2014; Visser et al. 2014; Sousa et al. 2018), and *Archaeoglobus* (Hocking et al. 2014) have been published. The presence of multiple forms for hydrogenase (i.e., [FeFe], [NiFe], and [NiFeSe]) within a cell and the location of these hydrogenases in the periplasm, membrane, and cytoplasm (see Table 4.1) indicate a specific function for each enzyme form. Considering the activity of hydrogenases, those in the cytoplasm are generally associated with H_2 evolution, while those in the membrane or periplasm are associated with H_2 uptake (Vignais et al. 2001). The report by Sadana and Rittenberg (1963) revealed that hydrogenase activity was found in the soluble protein fraction of the cell as well as in the membrane fraction of *D. desulfuricans* (now *vulgaris*) Hildenborough NCIB 8303. Using standard cell-fractionation techniques, it was determined that a hydrogenase in *D. gigas* was localized in the periplasm (Odom and Peck Jr. 1981). Using biochemical techniques, Ackrell et al. (1966) resolved the location of hydrogenases in *D. desulfuricans* ATCC 7757 into two distinct areas: one soluble and the other with the membrane.

Subsequently, considerable attention has been given to the forms of hydrogenase and their location in the SRB cell using genomic analysis. To predict if a hydrogenase is transported into the plasma membrane or periplasm, the presence of a signal peptide located at the N-terminus of the small subunit is assessed. A specific motif in the signal peptide ((S/T)RRxFxK) is recognized by the membrane targeting and translocation (Mtt) pathway or twin-arginine translocation (Tat) pathway to direct the assembled and folded heterodimer hydrogenase to the membrane or the periplasm (Vignais and Colbeau 2004). While multiple forms of hydrogenase in the cell is advantageous to the organism, it complicates the resolution of the role for each hydrogenase. Based on composition of metal centers, hydrogenases from SRB had been organized [FeFe], [NiFe], or [NiFeSe] groups (Fauque et al. 1988). Currently, hydrogenases such as those found in SRP are assigned to either of two classes: [FeFe] or [NiFe]. The [NiFeSe] hydrogenases of SRB are a subclass of [NiFe] hydrogenase (Vignais and Billoud 2007). The active centers of the three groups of hydrogenases found in SRB are in Fig. 4.1.

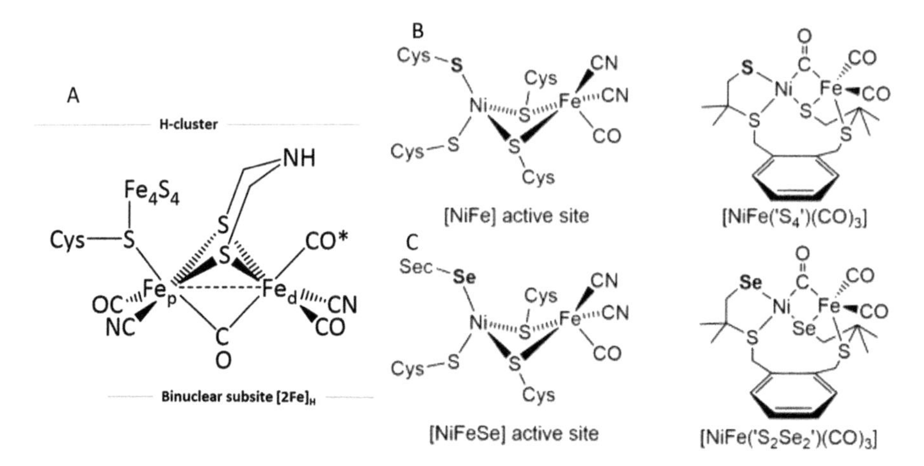

Fig. 4.1 H-cluster models for the different classes of hydrogenases. The [Fe] hydrogenase is not found in sulfate-reducing prokaryotes (SRP), but the [FeFe] and [NiFe] hydrogenases are associated with different members of the SRP. In the [FeFe] and the [NiFe] hydrogenases, metal atoms are coordinated with CO (carbon monoxide) and CN^- (cyanide) ligands. The[NiFeSe] hydrogenase found in some SRB has Se (selenocysteine) replacing S (from cysteine), which is coordinated to the Ni atom (Baffert et al. 2019). These active sites from the three hydrogenases is from Wikimedia Commons, and this file is licensed under the Creative Commons Attribution-Share Alike 4.0 International license

4.2.1 [FeFe] Hydrogenases

The [FeFe] hydrogenase were initially purified and characterized from *D. vulgaris* Hildenborough (van der Westen et al. 1978; Huynh et al. 1984; Hagen et al. 1986; Patil et al. 1988; Huynh 1995) and from *D. vulgaris* Miyazaki F (Yagi et al. 1976, 1985; van der Westen et al. 1978) and served as an impetus for future investigations. The hydrogenase gene from *D. vulgaris* Hildenborough was the first to be cloned, and sequencing of the gene encoding this hydrogenase was achieved (Voordouw et al. 1985; Voordouw and Brenner 1985). Over the years, many papers have reported [FeFe] hydrogenase in SRB, including from *D. desulfuricans* ATCC 7757 (Hatchikian et al. 1992) and *D. fructovorans* (Casalot et al. 1998). Characteristics of [FeFe] hydrogenase from *D. gigas* are provided in Table 4.2 and are compared to hydrogenases produced by SRB with different metal centers. Using *D. vulgaris* Hildenborough as a model, the [FeFe] enzyme has a mass of 56.6 kDa with two dissimilar subunits of 46 kDa and 10 kDa with the catalytic center in the large subunit. Present in the large subunit are two ferredoxin-like [4Fe-4S] clusters, which transfer electrons to an atypical [4Fe-4S] cluster, with which is ligated to a diiron subcluster to produce a 6 Fe system that constitutes the catalytic site referred to as the H-cluster (Peters et al. 1998; Nicolet et al. 1999). The [FeFe] hydrogenases have a greater catalytic activity than the [NiFe] hydrogenases (Frey 2002), and the [FeFe] hydrogenases are reversibly inhibited by 5 nM CO but irreversibly inhibited by O_2 and NO (Goldet et al. 2009).

Table 4.2 General physiochemical and catalytic properties of hydrogenases from Gram-negative bacteria with different metal centers[a]

Characteristic	[FeFe] *Desulfovibrio vulgaris*	[NiFe] *Desulfovibrio gigas*	[NiFeSe] *Desulfomicrobium* (formerly *Desulfovibrio*)
Molecular mass (kDa)	56.6	89.5	81
Subunits (kDa)	2(46 and 10)	2(62 and 26)	2(54 and 27)
Nonheme iron	16[b]	12	8
[4Fe-4S] cluster	3	2	2
[3Fe-4S] cluster	0	1	0
Nickel atoms/molecule	0	1	1
Selenium atoms/ molecule	0	0	1
H_2 evolution[c]	4800	440	467
H_2 specific activity (uptake)[c]	50,000	1500	120
H_2 uptake binding affinity K_m[d]	100 μM	1 μM	1 μM
D_2-H_2 exchange	2700	267	350

[a]Unless specified, data is adapted from Fauque et al. (1988)
[b]Hagen et al. (1986). Twelve of the Fe atoms would be associated with the 3[4Fe-4S] clusters, and the remaining would form the iron cluster active site
[c]Micromol H_2 produced or consumed/min/mg protein
[d]Caffrey et al. (2007)

The [FeFe] hydrogenases are widely distributed in SRB and may be located in the periplasm and the cytoplasm. The *hydAB* genes, which encode for the dimeric [FeFe] periplasmic hydrogenase, were found in 25 of the 32 genomes of *Desulfovibrio* identified to the species level (Table 4.3), and *D. alaskensis* G20 was the only genome of SRB with two sets of *hydAB* genes (Baffert et al. 2019). Some [FeFe] hydrogenases have multiple subunits and are found as trimers and tetramers. As listed in Table 4.3, four species of *Desulfovibrio* carry genes of [FeFe] hydrogenase trimers and tetramers, and one species has only a [FeFe] hydrogenase tetramer without a trimer (Baffert et al. 2019). The genome of *D. alaskensis* G20 and *D. piger* Fl-11,049 contains information for monomeric [FeFe] hydrogenase (Baffert et al. 2019).

The [FeFe] hydrogenases are also found in members of the *Desulfotomaculum* genera, and since these Gram-positive bacteria have no periplasm, all hydrogenases are either associated with the plasma membrane or are in the cytoplasm (Table 4.4). Six [FeFe] hydrogenases are found in *Dst. gibsoniae*: one hydrogenase located on the outer side of the plasma membrane, one monomeric hydrogenase, two trimeric NAD(P)-dependent bifurcating hydrogenases, one hydrogenase with expression associated with flavin-dependent oxidoreductases, and one HsfB-type hydrogenase involved in sensing (Kuever et al. 2014). Genome analysis of *Dst. kuznetsovii* reveals the presence of at least four [FeFe] hydrogenases: two trimeric confurcating hydrogenases and two hydrogenases lacking NAD-binding sites (Visser et al. 2013).

Table 4.3 Distribution of [FeFe] hydrogenases in *Desulfovibrio*[a]

Present in the periplasm	Present in the cytoplasm		
	Trimeric (possible bifurcating)	Tetrameric (possible bifurcating)	Sensory
D. alaskensis G20	*D. alcoholivorans* DSM 5433	*D. alcoholivorans* DSM 5433	*D. alcoholivorans* DSM 5433
D. alcoholivorans DSM 5433	*D. carbinolicus* DSM 3852	*D. carbinolicus* DSM 3852	*D. carbinolicus* DSM 3852
D. aminophilus DSM 12254	*D. fructosovorans* JJ	*D. fructosovorans* JJ	*D. fructosovorans* JJ
D. bastinii DSM 16055	*D. magneticus* RS-1	*D. magneticus* RS-1	*D. magneticus* RS-1
D. bizertensis DSM 18034		*D. magneticus* MBC34	*D. magneticus* MBC34
D. carbinolicus DSM 3852			
D. desulfuricans ATCC 27774			
D. fairfieldensis CCUG 45958			
D. ferrireducens DSM 17176			
D. fructosovorans JJ			
D. gracilis DSM 16080			
D. hydrothermalis DSM 14728			
D. inopinatus DSM 10711			
D. longus DSM 6739			
D. magneticus MBC34			
D. magneticus RS-1			
D. mexicanus DSM 13116			
D. piger fl 11,049			
D. putealis DSM 16056			
D. salexgens DSM 2638			
D. termitidus l#1			
D. vulgaris Hildenborough			
D. zosterae DSM 11974			

[a]Based on Pereira et al. (2011)

Table 4.4 Types of [FeFe] hydrogenases found in *Desulfotomaculum* spp.

	Trimeric (bifurcating?)	Monomeric	Membrane associated	Sensing	Reference
Dst. nigrificans DSM574	3	2	3	1	Visser et al. (2014)
Dst. ruminis DSM 2154	3	1	2	1	Spring et al. (2012)
Dst. kuznetsovii 17^T DSM 6115	2	2	0		Visser et al. (2013)
Dst., reducens MI-1	3	2	1		Junier et al. (2010)

Genomic evaluation of *Dst. nigrificans* and *Dst. carboxydivorans*, which is considered to be a heterotypic synonym of *Dst. nigrificans*, indicated that these two bacteria have nine [FeFe] hydrogenases: three are trimeric bifurcating hydrogenases, two are monomeric hydrogenases, one is a HsfB-type hydrogenase involved in sensing, two are membrane associated which appear to be located on the exterior of the plasma membrane, and one hydrogenase gene is located in a multigene membrane protein with oxidoreductase activity (Visser et al. 2014).

4.2.1.1 Electron-Bifurcating Hydrogenases

The reduction of ferredoxin by electron donors, which have a higher potential, is catalyzed in bacteria and archaea, provided there is a concomitant oxidation of the same donor by a higher potential electron acceptor. Several of the [FeFe] hydrogenases are electron-bifurcating hydrogenases, where the transport of electrons from hydrogen is used to reduce ferredoxin, an endergonic reaction, by coupling it to an exergonic reaction such as the reduction of NAD^+. This activity is summarized in Eq. 4.2.

$$2H_2 + 2Fd_{Ox} + NAD^+ = 2Fd_{Red} + NADH + 3H^+ \tag{4.2}$$

The coupling reaction of the electron-bifurcating hydrogenase was initially reported for *Thermotoga maritima* (Verhagen et al. 1999; Schut and Adams 2009) and subsequently demonstrated in *Acetobacterium woodii* (Schuchmann and Müller 2012), *Moorella thermoacetica* (Wang et al. 2013a), and *D. fructosovorans* (Kpebe et al. 2018). Electron-bifurcating ferredoxin reduction reactions from anaerobic bacteria and archaea are proposed to be important in energy conservation (Buckel and Thauer 2013). It is suggested that bifurcating hydrogenases are important for SRB, since they are found in several *Desulfovibrio* and *Desulfotomaculum.* A model indicating the ferredoxin and pyridine nucleotide activity of a bifurcating hydrogenase is given in Kpebe et al. (2018). While bifurcating hydrogenases are commonly ferredoxin and NAD^+ coupled, a $NADP^+$-specific bifurcating hydrogenase has been isolated from *Clostridium autoethanogenum* (Wang et al. 2013b). A model of

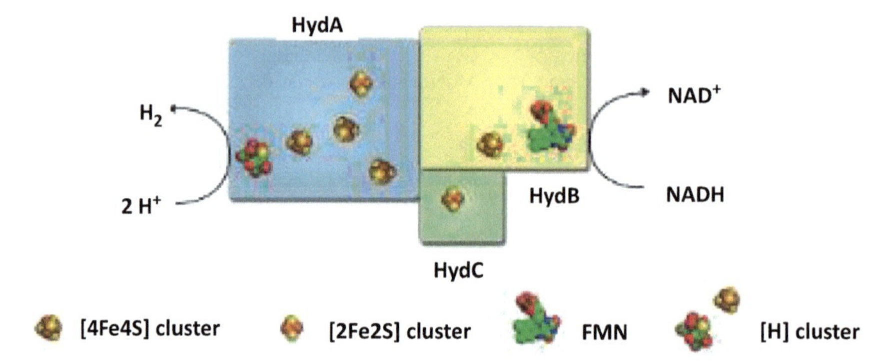

Fig. 4.2 Subunit architecture and cofactor content of the NADH-dependent hydrogenase of *S. wolfei.* This enzyme resembles in the overall subunit architecture electron bifurcating/ confurcating hydrogenases that utilize NADH and reduced ferredoxin in an energetically coupled reaction to produce H_2 (or vice versa). The hydrogenase of *S. wolfei* putatively lacks three iron-sulfur clusters in the subunit HydB and utilizes NADH alone for H_2 production, thus not using FBEB. (Schuchmann et al. 2018. *Copyright* © 2018 Schuchmann, Chowdhury and Müller. This is an open-access article distributed under the terms of the Creative Commons Attribution License (CC BY). The use, distribution, or reproduction in other forums is permitted, provided the original author(s) and the copyright owner(s) are credited and the original publication in this journal is cited, in accordance with accepted academic practice

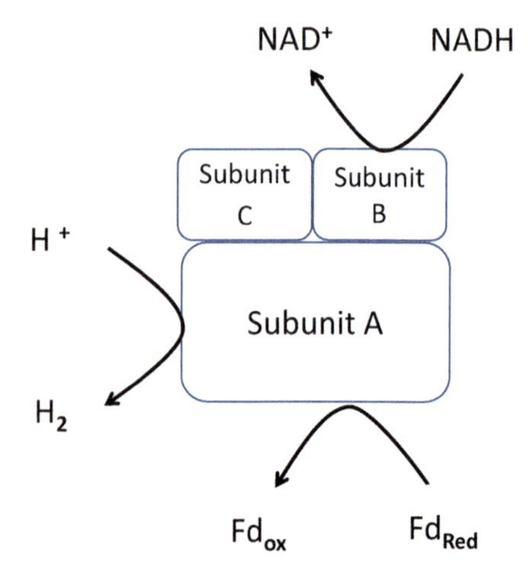

Fig. 4.3 Model of confurcating hydrogenase in bacteria. Fd_{Ox} oxidized ferredoxin; Fd_{Red} reduced ferredoxin, hydrogenase subunits as indicated

subunit architecture of bifurcating hydrogenase using NADH and reduced ferredoxin from *Syntrophomonas wolfei* is given in Fig. 4.2, and a model of configurating hydrogenases in bacteria is given in Fig. 4.3. Future studies should examine SRB for the presence of ferredoxin and $NADP^+$-coupled bifurcating hydrogenases. Bifurcating hydrogenases are either trimeric or tetrameric cytoplasmic [FeFe] hydrogenases, which contain the catalytic subunit, a [2Fe2S] cluster subunit, and a flavin and FeS-containing subunit with the capability of reducing $NAD(P)^+$. The mechanism of the bifurcating hydrogenase is not resolved but may involve flavin with crossed-redox potential (Baymann et al. 2018) or the H-cluster (Peters et al. 2018a, b).

4.2.1.2 H_2-Sensing Hydrogenase

The expression of hydrogenase genes in response to the presence of H_2 results in production of a [NiFe] hydrogenase in *Ralstonia eutropha*, which is regulated by a two-component signal transduction system (Friedrich et al. 2005). At this time, the genomes of *Desulfovibrio* do not contain genes for a sensory [NiFe] hydrogenase (Baffert et al. 2019), but several species of *Desulfovibrio* do contain genes with a similarity to that of the putative H_2-sensory [FeFe] hydrogenase of *Thermotoga maritima* (see Table 4.3) (Chongdar et al. 2018; Baffert et al. 2019). The H_2-sensory [FeFe] hydrogenase has been referred to as the HydS or Hfs, where "S" implies signaling. The H_2-sensing hydrogenase could be important for *Desulfovibrio* in that the synthesis of hydrogenase could be regulated by the H_2 partial pressure inside the cell.

4.2.1.3 Hydrogenase Maturation and Architecture

Several architectural features are important to consider in the assembling of a functional [FeFe] hydrogenase. The catalytic center of the hydrogenase is buried deep within the large subunit, and the three iron-sulfur centers are spatially arranged in the subunit to provide for electrons transferred from the catalytic center to the surface of the molecule. The large subunit can be divided into three domains with secondary structures containing various α-helices and β-sheets (Nicolet et al. 1999). As the proteins in the large subunit are configured, a hydrophobic gas channel is established, through which H_2 diffuses to the active site, and these channels provide an avenue for protons to migrate from the active site to the surface of the molecule. This gas channel also enables O_2 or CO to gain access to the active center, where inhibition of H_2 oxidation would occur. A model indicating the presence of gas channels in the large subunit assembly of hydrogenase was published by Mulder et al. (2011). Currently, there is an interest in using molecular biology to change amino acid residues in the hydrogenase large subunit to change enzyme specificity and kinetic activity (Winkler et al. 2013).

Maturation of the [FeFe] hydrogenase involves the formation of the H-cluster by insertion of the diiron subcluster, which requires three assessor proteins (i.e., HydE, HydF, and HydG) that have been referred to as maturases (Kuchenreuther et al. 2012; Broderick et al. 2014; Shepard et al. 2014; Peters et al. 2015; Bortolus et al. 2018). Reports indicate that a functional hydrogenase can be produced when the genes for the apoprotein of [FeFe] hydrogenase and the maturases are introduced into a recombinant host (Avilan et al. 2018) or the $(Et_4N)_2[Fe_2(adt)(CO)_4(CN)_2]$ complex is added to the apo-form (Birrell et al. 2016). A model (Fig. 4.1) of the iron center of the [FeFe] enzyme contains a ferredoxin-like [4Fe4S] cluster bridged by cysteine to the Fe-Fe center with ligands of one carbonyl (CO) and two cyano (CN) groups, and the diiron are bridged by a dithiolate linkage (Nicolet et al. 1999). The sequence of events leading to the natural formation of the active center of the [FeFe] hydrogenase using the maturases was proposed by Mulder et al. (2011). The small subunit contains α-helical protein structures, which form a belt-like structure on the surface of the large subunit (Nicolet et al. 1999).

4.2.2 [NiFe] Hydrogenases

The dimeric soluble [NiFe] hydrogenase (hyn) from *D. gigas* was the first hyn characterized (Bell et al. 1978; Hatchikian et al. 1978), and the biochemical properties of this enzyme were reviewed by Cammack et al. (1994). The crystalline structure of the *D. gigas* hyn was established (Volbeda et al. 1995), and the active site was identified to contain one carbonyl (CO) and two cyano (CN) ligands (van der Spek et al. 1996; Volbeda et al. 1996; Higuchi et al. 1997). The characteristics of this hyn from *D. gigas* is given in Table 4.5, and a model of the active site is in Fig. 4.1. For the isolation and evaluation of physical properties of the [NiFe] hydrogenase from *D. gigas*, the reader would find the report by Cammack et al. (1994) as highly useful. A series of redox active states of nickel and iron-sulfur clusters have been recorded for the [NiFe] hydrogenase, and midpoint potentials of these states are provided in Table 4.4. The X-ray crystal analysis and related studies of [NiFe] hydrogenase structures have been determined for *D. gigas* (Volbeda et al. 1995), *D. desulfuricans* ATCC 27774 (Matias et al. 2001), *D. fructosovorans* (Montet et al. 1997; Volbeda et al. 2002, 2005), and *D. vulgaris* Miyazaki F (Higuchi et al. 1997, 1999).

The maturation of [NiFe] hydrogenases follows a specific sequence of events with the insertion of Fe, CN^-, and CO, followed by the insertion of Ni in the catalytic center of the enzyme. In *Archaeoglobus fulgidus*, the delivery of nickel is by a maturation factor (HypB), which is a metal-binding GTPase (Chan et al. 2012a). The CN and CO ligands of the [NiFe] hydrogenase of *D. vulgaris* Miyazaki F were identified (Ogata et al. 2002), and all [NiFe] hydrogenases contain [NiFe]–(CN)2CO at the active site (Shafaat et al. 2013). The dimeric hyn hydrogenase of

Table 4.5 Midpoint potential for activities associated with [NiFe] hydrogenase from *Desulfovibrio gigas*[a]

Event or activity	Midpoint potential (mV)
[3Fe-4S] cluster	−35
Nickel (Ni-A signal)	−150
Nickel (NiC/EPR-silent)	−270
Nickel (EPR-silent/Ni-C)	−390
[4Fe-4S] clusters	−290, −340
Reductive activity	−310
Catalytic activity (H_2 evolution)	−360

[a]Data from Cammack et al. (1994)

D. fructosovorans consists of 60 kDa and 25.8 kDa subunits (Hatchikian et al. 1990), and the enzyme was determined to contain 1 nickel atom, 11 iron atoms, and 12 sulfur atoms. This enzyme contains a [NiFe] active site with two [4Fe4S] clusters and one [3Fe4S] cluster forming an electron transfer pathway (Rousset et al. 1998), and a model of the [NiFe] active center is given in Fig. 4.1. Unlike the [FeFe] hydrogenase, which has the iron-sulfur clusters in the large subunit, the [NiFe] hydrogenases have the iron-sulfur clusters spatially positioned across the small subunit. The distal [4Fe4S] cluster remained unchanged when the histidine ligand was replaced with a glycine or cysteine, but the medial [3Fe4S] cluster was changed to [4Fe4S] with cysteine replacing a proline (Rousset et al. 1998; Dementin et al. 2011). The [NiFe] hydrogenases of *Desulfovibrio* are located in the periplasm, and in the genome review, these hydrogenases were found in all genomes examined (Table 4.6) and are more commonly found in the periplasm than [FeFe] hydrogenase (Baffert et al. 2019). Also, the [NiFe] hydrogenases are found in the plasma membrane as Ech or Coo with Ech found in almost all *Desulfovibrio* with only a few species producing Coo hydrogenase (Table 4.7) (Baffert et al. 2019). *D. piger* and *D. litoralis* do not have genes for the production of Ech hydrogenase, but these organisms have genes for Coo hydrogenases. *D. alaskensis* and *D. aminophilus* lack genes encoding for both Ech and Coo hydrogenases.

Table 4.6 Number of hydrogenases reported for a selection of sulfate-reducing prokaryotes[a]

	[NiFe] hydrogenases								
Microorganisms	Hyn	Hys	Hdr	Ech	Coo	Sens	Hox	Mnh	[FeFe]
D. aespoeensis	1			1					1
D. desulfuricans G20	1	1							4
D. desulfuricans ATCC 27774	2			1	1				1
D. gigas	1			1					
D. magneticus RS-1	2			1	1				1
D. piger	1	1			1				1
D. salexigens	1	1		1					2
D. vulgaris Hildenborough	2	1		1	1				2
Desulfomicrobium baculatum	1	1							
Desulfatibacillum alkenivorans	1		1			1			3
Desulfobacterium autotrophicum HRM2	1	1	1						3
Desulfohalobium retbaense DSM 5692	1		1						
Desulfonatronospira thiodismutans	2			1					
Desulfotalea psychrophile	2		1				1		2
Syntrophobacter fumaroxidans MPOB	1		2				1	1	4

[a]References: Pereira et al. (2011); Morais-Silva et al. (2013, 2014)

Table 4.7 Distribution and location of [NiFe] hydrogenases in *Desulfovibrio*[a]

		Number present in the membrane	
	Number present in the periplasm	Ech	Coo
D. africanus Walvis Bay	1	1	0
D. alaskensis G20	1	0	0
D. alcoholivorans DSM 5433	1	1	0
D. alkalitolerans DSM 16529	1	1	0
D. aminophilus DSM 12254	1	0	0
D. bastinii DSM 16055	1	2	1
D. bizertensis DSM 18034	1	1	2
D. carbinolicus DSM 3852	1	1	0
D. cuneatus DSM 11391	1	1	1
D. desulfuricans ATCC 27774	1	1	1
D. fairfieldensis CCUG 45958	1	1	1
D. ferrireducens DSM 16995	1	1	0
D. fructosovorans JJ	1	1	0
D. frigidus DSM 17176	1	1	0
D. gigas DGl	1	1	1
D. gracilis DSM 16080	1	1	0
D. hydrothermalis DSM 14728	1	21	
D. inopinatus DSM 10711	1	2	0
D. legallii KHC7	1	1	1
D. litoralis DSM 111393	1	0	1
D. longus DSM 6739	1	1	0
D. magneticus MBC34	1	1	1
D. magneticus RS-1	1	1	1
D. mexicanus DSM 13116	1	1	1
D. oxyclinae DSM 11498	1	0	1
D. piger Fl 11049	1	1	0
D. putealis DSM 16056	1	1	1
D. salexigens DSM 2638	1	1	1
D. termitidus Hl1	1	1	1
D. vulgaris Hildenborough	1	1	1
D. vulgaris Miyasaki	1	0	1
D. zosterae DSM 11974	1	1	1

[a]Based on Pereira et al. (2011)

The Ech (energy-conserving hydrogenase) in *D. fructovorans* has six subunits, which are closely related to the subunits of the NADH:quinone oxidoreductase (Baffert et al. 2019). The NADH:quinone oxidoreductase has been demonstrated to pump protons, and in this regard, is similar to the mitochondrial complex I (Hedderich 2004; Hedderich and Forzi 2005; Efremov and Sazanov 2012). Ech is bound to the membrane with an active site on the inner face of the membrane (Hedderich and Forzi 2005); it is energized by a low potential ferredoxin (Meuer et al. 1999), and this hydrogenase will translocate protons to the outer side of the membrane (Welte et al. 2010). Under autotrophic growth, Ech catalyzes the reduction of ferredoxin with H_2 with reduced ferredoxin used to drive CO_2 fixation and biosynthetic activities (Grahame and DeMoll 1995; Meuer et al. 2002; Pereira et al. 2011). When cultivated as a biofilm, there is an increased expression of *ech* genes in *D. vulgaris* Hildenborough as compared to planktonic growth (Clark et al. 2012).

The Coo hydrogenase is related to Ech hydrogenase, and the Coo enzyme is also located in the plasma membrane, where it is involved in translocating ions across the membrane using reduced ferredoxin as the electron donor (Hedderich and Forzi 2005). In *D. vulgaris* Hildenborough under syntrophic growth, the Coo hydrogenase is responsible for H_2 production with lactate containing medium, and Coo hydrogenase production is diminished when H_2 is used to support growth (Pereira et al. 2008; Walker et al. 2009). With *D. vulgaris* Hildenborough growing on lactate-sulfate medium, Coo hydrogenase is required for growth, where H_2 production is coupled to reduced ferredoxin coupled to oxidation of menaquinol (Keller and Wall 2011). Homologs of the gene for NADH:quinone oxidoreductase have been reported for *Desulfovibrio* spp. (Pereira et al. 2011; Baffert et al. 2019), and the hypothesis has been presented that Coo hydrogenase displaying electron bifurcation could catalyze the following reversible reaction (Baffert et al. 2019):

$$2H_2 + 2Fd_{ox} + Quinone = 2Fd_{red} + Quinone_{red} + 2H^+ \tag{4.3}$$

4.2.2.1 Maturation

Since the structure of the H-cluster of [NiFe] hydrogenase is highly conserved through the bacteria, it is considered that maturation of the H-cluster would use the same procedure for all bacteria. Maturation of the H-cluster of the [NiFe] hydrogenase is relatively complex as evidenced by the requirement of seven core proteins, ATP, GTP, and carbamoyl phosphate for the synthesis of the *Escherichia coli* metallocenter (Böck et al. 2006), and a peptidase removes a peptide from the large subunit after Ni has been inserted (Rossmann et al. 1995). In *Escherichia coli*, the genes for the five maturation proteins (i.e., HypA, HypB, HypC, HypD, and HypE) are contained in a transcriptional unit, while the gene for the HypF protein is at a different site. As designated in Table 4.8, distinct proteins are used for the insertion of Ni and Fe (Blokesch et al. 2002, 2004). As reported, the crystal structure of *A. fulgidus* HypB (AfHypB) is a metal-binding GTPase (Chan et al. 2012a).

Table 4.8 Proteins associated with maturation of [NiFe] hydrogenase using *Escherichia coli* as a model

Protein	Function	Reference
HypA/HypF	Ni addition	Chan et al. (2012a, b); Olson et al. (2001); Hube et al. (2002)
HypB	Ni insertion, GTPase	Chan et al. (2012b); Maier et al. (1995); Blokesch et al. (2002)
HypC/HypG	Chaperone, Fe insertion	Blokesch et al. (2004); Magalon and Böck (2000)
HypD	Fe insertion, Fe-S protein	Blokesch et al. (2004)
HypE	ATPase, synthesis of CO/CN ligands	Lenz et al. (2007)
HypF	Carbamoyl phosphatase, ATP pyrophosphatase, Synthesis of CO/CN ligands	Paschos et al. (2002)

4.2.3 [FeNiSe] Hydrogenases

While [FeFe] and [NiFe] hydrogenases have been found in numerous microorganisms including SRP, the distribution of [NiFeSe] hydrogenase appears to be limited to a few genera of *Desulfovibrio* (Table 4.7). [NiFeSe] hydrogenases have been purified from *Desulfomicrobium baculatum* (Teixeira et al. 1987), *D. salexigens* (Teixeira et al. 1986), *D. vulgaris* Hildenborough (Valente et al. 2005), and *D. vulgaris* Miyazaki F (Nonaka et al. 2013). The catalytic center of [NiFeSe] hydrogenases has been examined from crystal structures of this enzyme from *Desulfomicrobium baculatum* (Garcin et al. 1999) and *D. vulgaris* Hildenborough (Marques et al. 2010), and a selenocysteine moiety replaces one of the four cysteine residues that bind nickel and iron into the active site. Characteristics of [NiFeSe] hydrogenases from SRB have been summarized previously (Fauque et al. 1988; Zacarias et al. 2018) (Table 4.2), and the chemistry of [NiFeSe] hydrogenases has been reviewed (Wombwell et al. 2015).

The presence of selenium in this hydrogenase prompted numerous studies. Selenium substitution for sulfur in the [NiFeSe] hydrogenase of *Desulfomicrobium* (formerly *Desulfovibrio*) *baculatum* was only in cysteine, since there is no Se-methionine. This selenium substitution for sulfur in coordination of nickel accounted for only a small difference in redox potential, and it was suggested that this selenium substitution for sulfur provided a fine-tuning as the enzyme interacted with various physiological requirements (He et al. 1989). Selenocysteine was demonstrated to be coordinated to nickel in the active site of the [NiFeSe] hydrogenases from *Desulfomicrobium baculatum* (Eidsness et al. 1989), and a model of the active site of a [NiFeSe] hydrogenases is in Fig. 4.1.

A unique feature of the [NiFeSe] hydrogenase is the presence of selenium, which is co-translationally inserted into the protein at an in-frame opal (stop) codon, TGA (Voordouw et al. 1989). In order to read through the opal codon (Böck et al. 1991), an activity encoded in four genes (selA, selB, selC, selD) is required and constitutes the complex machinery for selenocysteine, which has been called the selenosome (Hatfield and Gladyshev 2002). With the construction of *L*-selenocysteine-tRNA^{UGA}, selenocysteine is added at the in-frame UGA. The analysis of [NiFeSe] hydrogenase gene from *D. gigas* and *Desulfomicrobium baculatum* revealed that a TGA codon located near the 3′-end of the protein is near a regular cysteine codon (TGC), and this TGA is the site of selenocysteine insertion (Voordouw et al. 1989). This incorporation of selenocysteine in [NiFeSe] hydrogenase occurs in the large subunit and the ten cysteine residues, which coordinates three iron-sulfur clusters that are present in the small subunit, where they form an electronic relay. In contrast, all iron-sulfur clusters in SRB [NiFe] hydrogenase are in the large subunit. The [NiFeSe] hydrogenase contains a [4Fe-4S] medial cluster, while the [NiFe] hydrogenase of SRB has a [3Fe-4S] cluster (Garcin et al. 1999). In *D. vulgaris* Hildenborough, nutrition is important for synthesis, because when selenium is present in culture media, the production of [FeFe] and [NiFe] hydrogenases is downregulated while [NiFeSe] hydrogenase production is upregulated and becomes the dominant hydrogenase (Valente et al. 2006). With 25 μM iron in the growth medium, the [FeFe] hydrogenase is produced at the greatest levels. With 25 μM iron and 1 μM nickel in the culture medium, the [NiFe] hydrogenase is dominant, and with 25 μM iron, 1 μM nickel, and 1 μM selenium, the [NiFeSe] hydrogenase is the most abundant (Caffrey et al. 2007).

In comparison to [FeFe] hydrogenase, the [NiFeSe] hydrogenase from SRB characteristically displays lower activity and higher affinity for H_2; see Table 4.2. Thus, hydrogenase activity at high concentrations of H_2 is greatest with [FeFe] hydrogenase, while at low levels of H_2, the [NiFe] and [NiFeSe] hydrogenases have greater activity. While the [FeFe] hydrogenase is primarily an uptake hydrogenase, [NiFeSe] has a bias for H_2 production (Wombwell et al. 2015). A recombinant *D. vulgaris* Hildenborough [NiFeSe] hydrogenase with cysteine replacing the selenocysteine residue displayed diminished protection against oxidative damage, while nickel insertion in the H-site was dramatically diminished and altered catalytic activity (Marques et al. 2017).

While [NiFeSe] hydrogenases have been commonly found in the periplasm (Table 4.9), a membrane-bound [NiFeSe] hydrogenase was isolated from *Desulfovibrio vulgaris Hildenborough* (Valente et al. 2005, 2007). This enzyme is translocated to the membrane by the Tat pathway and not by the Sec pathway used by other lipoproteins. It is unusual that both subunits are lipophilic. The large subunit is bound to the membrane by lipophilic groups and lacks the signal peptide, which is characteristic of bacterial lipoproteins. It will be important to assess the distribution of membrane-bound [NiFeSe] hydrogenase in other *Desulfovibrio*.

The movement of H_2 and protons to or from the surface of the [NiFeSe] hydrogenase large subunit to the active center of the enzyme may be by hydrophobic channels. By reconstructing 3D gas cavities in the [NiFeSe] hydrogenase with the established genome libraries, Tamura et al. (2016) proposed that gas channels developed in ancient SRB strains are grown in absolute anaerobic conditions to enable a rapid exchange of H_2 in the hydrogenase large subunit. Therefore, they suggested that SRB present in anaerobic deep-sea environments would have prominent gas channels to connect the active site to the surface of the molecule. With the

Table 4.9 Properties of ferredoxins from sulfate-reducing bacteria[a]

Bacteria	Ferredoxin	Amino acid residues	Structure	Cysteine per monomer	Cluster type	Redox potential (mV)
Desulfovibrio gigas	I	58	Trimer	6	[4Fe]	−450
	II	58	Tetramer	6	[3Fe]	−130
Desulfovibrio vulgaris M	I	61	Dimer	7	[3Fe], [4Fe]	−430
	II	63	Dimer	7	[4Fe]	−405
Desulfovibrio africanus	I	60	Dimer	4	[4Fe]	nd[b]
	III	61	Dimer	7	[3Fe], [4Fe]	−140, −410
Desulfomicrobium baculatum	I	59	Dimer	6	[4Fe]	−374
	II	59	Dimer	8	[4Fe] plus [4Fe]	−500

[a]Moura et al. (1994a, b), and References within Sect. 4.2.2
[b]*nd* not determined

periplasmic location of [NiFeSe] hydrogenase in sulfate reducers, transient exposure to O_2 would have selected those SRB with that harbored silent mutations of proteins in the large subunit to restrict or block the gas channels and thereby prevent O_2 inhibition of the hydrogenase activity.

The SRB, which inhabit anaerobic aquatic environments (i.e., *D. africanus*, *D. salexigens*, *D. hydrothermalis*, and *Desulfomicrobium baculatum*), have well-defined tunnellike structures connecting the H-site to the surface of the [NiFeSe] hydrogenase subunit, while *D. piger* and *Bilophila wadsworthia* have gas cavities that are poorly developed, which is attributed to O_2 diffusion into the gut from the intestinal lining (Tamura et al. 2016).

4.3 Formate Dehydrogenase

4.3.1 Periplasmic and Membrane

Soluble Fdh enzymes in the periplasm may consist of a small subunit and a catalytic subunit (FhdAB) (Almendra et al. 1999) or an enzyme that consists of the small subunit, catalytic subunit, and a cytochrome c_3 subunit ($FdhABC_3$) (Sebban et al. 1995). The membrane-associated Fdh are periplasmic-facing enzymes containing a subunit for quinone reduction and occur as FdhABC or FdhABD, which has a large protein subunit (Pereira et al. 2011). In a survey of genomes from several sulfate-reducing bacteria (Pereira et al. 2011), one or more genes encoding Fdh were detected with genes encoding FdhAB occurring more frequently than.

FdhABC$_3$. Genes for membrane-associated Fdh (FdhABC and FdhABD) were less frequently observed than genes for soluble Fdh.

Fdh has been characterized from *D. gigas* NCIB 9332 (Riederer-Henderson and Peck Jr. 1986a, b; Almendra et al. 1999), *D. vulgaris* Hildenborough (Sebban et al. 1995), *D. vulgaris* Miyazaki (Yagi 1979), *D. desulfuricans* ATCC 27774 (Costa et al. 1997), and *D. alaskensis* NCIMB 13491 (Brondino et al. 2004). *D. vulgaris* Hildenborough has three formate dehydrogenase enzymes (Pereira et al. 2007), with FdhAB and FdhABC$_3$ as soluble proteins in the periplasm and FdhM (also referred to as FdhABD) (*fdnG*, DVU2482) in the periplasm but associated with the plasma membrane. The FdhABC$_3$ (DVU2809–11) enzyme contains molybdenum, while the FdhAB (DVU0587–88) binds either molybdate or tungstate. *D. alaskensis* has three periplasmic formate dehydrogenases with one enzyme binding molybdate and the other binding tungstate (Brondino et al. 2004; Martins et al. 2015, 2016). Sulfide produced from SRB respiration will precipitate molybdenum (Tucker et al. 1997); however, molybdenum sulfide is markedly more stable than tungsten sulfide, and this may reflect the necessity of having multiple forms of formate dehydrogenases (da Silva et al. 2011). In the presence of molybdate, *D. vulgaris* Hildenborough produces primarily FdhABC$_3$, and in the presence of tungstate, the primary formate dehydrogenase produced is FdhAB (da Silva et al. 2013).

Ongoing research efforts have provided useful information concerning sulfate-reducing microorganisms, and some of these reports are summarized as follows:

- The *Dst. kuznetsovii* genome contains two Fdh complexes: Desku_0187–0192 and Desku_2987–2991. The Desku_0187–0192 complex is considered to be associated with a new type of energy conservation and consists of Desku_0187, a FAD nucleotide disulfide oxidoreductase; Desku_0189, a methyl viologen-dependent hydrogenase; Desku_0190, a formate dehydrogenase; and Desku_0192, a protein predicted to be translocated across the plasma membrane (Sousa et al. 2018).
- The genomes of *Dst. nigrificans* (DesniDRAFT) and *D. carboxydivorans* (Desca) have three sets of genes for Fdhs, and they share the following characteristics. The first Fdh gene cluster is as follows: DesniDRAFT_0989 and Desca_1018, an alpha subunit of Fdh; DesniDRAFT_0990 and Desca_1017, a hydrogenase; and DesniDRAFT_0988 and Desca_1019, a flavoprotein with a predicted transmembrane helix. The second Fdh gene cluster: DesniDRAFT_1389–1392 and Desca_2053–2055 are for a putative cytoplasmic enzyme. The third Fdh gene cluster: DesniDRAFT_1396–1397 and Desca_2059–2060 are proposed to be a periplasmic-facing Fdh associated with a protein containing ten transmembrane helices (DesniDRAFT_1395, Desca_2058), which may be a cytochrome *b* (Visser et al. 2014).
- *The genome of Pelotomaculum* candidate BPL, a benzene-degrading bacterium that has all genes for dissimilatory sulfate reduction, contains three genes putatively encoding formate dehydrogenases (Ga0073689_2617920664, Ga0073689_2617920666, and Ga0073689_2617922225), and these genes may be important for syntrophic fermentation (Dong et al. 2017).

- *Dst. gibsoniae* has two putative Fdhs: The first Fdh is (Desgi_1522–1523), which is bound to a polysulfide reductase with 10 transmembrane helices (Desgi_1524). The second Fdh (Desgi_2136–2139) is considered to be a confurcating enzyme with Desgi_2138 similar to NADH binding of NADH:ubiquinone oxidoreductase and iron-sulfur cluster binding motifs (Kuever et al. 2014).
- *Thermodesulfovibrio* sp. N1 contains genes for a periplasmic formate dehydrogenase FdhAB, and this formate is used as an electron source. The Fdh contains a catalytic subunit A (THER_1543) and a B (THER_1541) but no membrane component. Electrons from Fdh are suggested to be transferred to the membrane by periplasmic cytochromes and the Hmc complex (Frank et al. 2016).
- *Syntrophobacter fumaroxidans* produces two Fdhs, and both contain selenium and tungsten but no molybdenum. Fdh-1 has three subunits (89 kDa, 56 kDa, and 19 kDa) and at least 4[2Fe-2S] clusters per molecule. Fdh-2 is a heterodimer, consisting of a 92 kDa and a 33 kDa subunit, and contains at least 2[4Fe-4S] per molecule (de Bok et al. 2003). Examination of Fdh genes under different growth conditions revealed that the transcription of *fdh*-1 and *fdh*-2 (encoding Fdh-1 and Fdh-2, respectively) was greater than the transcription of *fdh*-3 and *fdh*-4 (encoding for Fdh-3 and Fdh-4, respectively) under most growth conditions. However, the greatest transcription of *fdh*-3 and *fdh*-4 was under coculture with *Methanospirillum hungatei.* Fdh-1 is found in the cytoplasm, while Fdh-2, Fdh-3, and Fdh-4 are localized in the periplasm (Worm et al. 2011).

4.3.2 Cytoplasmic

A putative cytoplasmic enzyme, using NAD^+ as the electron acceptor with formate oxidation (NfdABC), has been proposed to be present in SRB (Pereira et al. 2007). The reaction catalyzed would be as follows:

$$NAD^+ + \text{formate} \rightarrow CO_2 + NADH \tag{4.4}$$

Genes encoding the NAD^+-reducing formate dehydrogenase have been detected in *Desulfomicrobium baculatum, Desulfonatronospira thiodismutans* AS03-1, *Desulfotalea psychrophila, Dst. acetoxidans* DSM771, *Dst. reducens*, and *Ammonifex degensii* KC4 (Pereira et al. 2011). A membrane-bound NADH dehydrogenase could contribute to cell energetics and reoxidize the reduced pyridine nucleotide.

A cytoplasmic enzyme that may be involved in formate production is the formate dehydrogenase/hydrogenase (FhcABCD), which would catalyze the following reaction:

$$H_2 + CO_2 \rightarrow \text{formate} + H^+ \quad (4.5)$$

While genes encoding FhcABCD have been detected in *D. desulfuricans* (now *alaskensis*) G20 (Dde0473–0476) and in *Desulfotalea psychrophila* (DP0481–0478), this enzyme remains to be isolated and characterized.

A formate hydrogen lyase (FHL) complex has long been examined in bacteria, where oxidation of formate is coupled to proton reduction. In *Escherichia coli*, the FHL complex consists of a Ni-Fe hydrogenase subunit (Hyd-3), molybdenum-dependent formate dehydrogenase (FdhF), and three iron-sulfur proteins (HycB, HycF, and HycG) (McDowall et al. 2014). Although FHL (FhcABCD) has not been isolated from SRB, there is some suggestion that the genes for this enzyme complex are present in *D. alaskensis* G20, *D. magneticus* RS-1, *D. salexigens, Desulfobacterium autotrophicum* HRM2, and *Desulfotalea psychrophila* (Pereira et al. 2011). Further research is needed to understand the contribution of cytoplasmic formate dehydrogenases to the energetics of sulfate reducers.

4.4 Cytoplasmic Proteins with Low Redox Potentials

4.4.1 Ferredoxin

Ferredoxin was initially isolated and characterized from *Clostridium pasteurianum*, reported by Mortenson and colleagues in 1962, and subsequently, several reviews have been published on bacterial ferredoxins (Valentine 1964; Yoch and Valentine 1972; Yoch and Carithers 1979; LeGall et al. 1979; LeGall et al. 1982; Peck Jr. and LeGall 1982; Odom and Peck Jr. 1984; Fauque and LeGall 1988; Bruschi and Guerlesquin 1988; Moura et al. 1994a, b; Chen et al. 1995; Meyer 2001). The general properties of bacterial ferredoxins include (i) a low molecular weight of about 6 kDa, (ii) one or more iron-sulfur clusters, (iii) a participation in a variety of reactions to transfer electrons, (iv) appreciable numbers of acidic amino acid residue to make it an acidic protein, and (v) a low oxidation-reduction potential. The bacterial ferredoxins are nonheme proteins, and the SRB ferredoxins can be classified according to the type of iron-sulfur clusters present: [3Fe-4Fe], [4Fe-4Fe], [3Fe-4Fe] plus [4Fe-4Fe], and 2x[4Fe-4Fe]. The properties of ferredoxins from several SRB are given in Table 4.9, and the models representing [3Fe-4Fe] and [4Fe-4Fe] clusters are presented in Fig. 4.4.

The first report of ferredoxin isolated from SRB was by LeGall and Dragoni (1966), and the cytoplasmic ferredoxin from this bacterium, *D. gigas,* has been the subject of numerous investigations. The soluble iron-sulfur proteins of *D. gigas* were resolved into two molecular forms: ferredoxin (DgFd) I, a trimer, and (DgFd) II, a tetramer, are found in *D. gigas* (Bruschi et al. 1976a, b; Cammack et al. 1977) with monomeric units containing an identical amino acid composition (Bruschi 1979). While DgFdI is designated as a trimer, Moura et al. (1994a, b) suggest that DgFdI

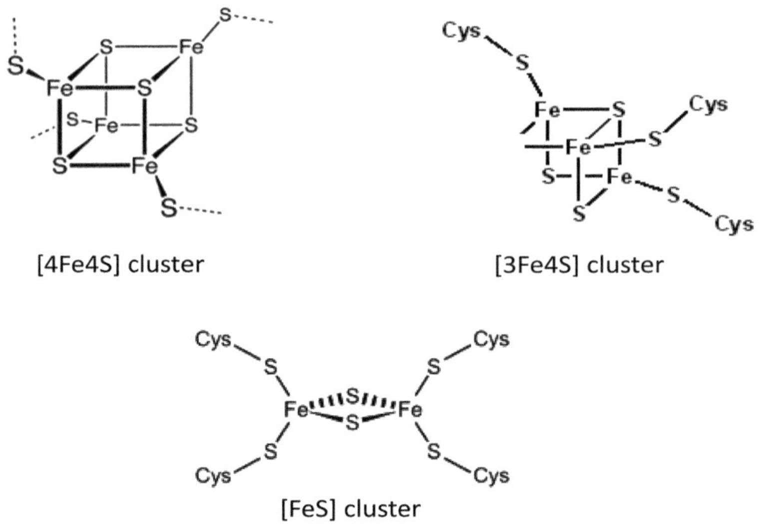

[4Fe4S] cluster

[3Fe4S] cluster

[FeS] cluster

Fig. 4.4 Models indicating Fe and S arrangements in ferredoxins. The source for [4Fe4S] and [3Fe4S] clusters is Wikimedia Commons, and the copyright holder of this work has released this work into the public domain, which applies worldwide. The source of the models for the [2Fe2S] cluster is Wikimedia Commons, and this image is a simple structural formula, which is ineligible for copyright and therefore in the public domain

contains a dimer. In this chapter, DgFdI is referred to as a trimer. A third ferredoxin (DgFdI') has been reported for *D. gigas* (LeGall et al. 1979); however, it remains to be characterized. Based on iron-sulfur stoichiometry and spectroscopic analysis, DgFdII from *D. gigas* was determined to contain a [3Fe-4S] cluster and not the traditional [4Fe-4S] cluster (Huynh et al. 1980; Beinert and Thomson 1983; Beinert et al. 1983). Also, it was determined that the [3Fe-4S] and the [4Fe-4S] clusters were bound to the same polypeptide chain and that the [3Fe-4S] cluster could be converted to an [4Fe-4S] cluster under appropriate conditions (LeGall et al. 1982; Kent et al. 1982; Moura et al. 1984). As indicated in Table 4.10, the [4Fe-4S] ferredoxin from *D. gigas* has an $E^{o'}$ of -455 mV and participates transferring electrons in the phosphoroclastic reaction, while the [3Fe-4S] ferredoxin has an E_o' of -130 mV and transfers electrons from cytochrome c_3 to the sulfite reductase (Moura et al. 1978a, b). Amino acid analysis of ferredoxin from *D. gigas* revealed an appreciable number of aspartic acid and glutamic acid residues with few aromatic amino acids (Laishley et al. 1969). Based on crystallographic data, three-dimensional structures have been constructed for DgFdII (Kissinger et al. 1989, 1991). A distinctive motif of CysXXCysXXCys implies cysteine residues binding iron as the cubane-type clusters in DgFdII (Moura et al. 1994a, b). While each iron atom in the iron-sulfur cluster is tetrahedrally coordinated and generally binds to cysteine, O or N ligands of aspartic acid (or other amino acids in the cysteine position) may also bind iron atoms. DgFdI can accommodate both [4Fe-4S] and [3Fe-4S] clusters with most of the isolated DgFdI molecules containing only.

Table 4.10 Periplasmic [NiFeSe] hydrogenases of *Desulfovibrio*

Bacteria	Strain	Locus	Number
D. africanus	Walvis Bay	Desfo_	3130–3131
D. alaskensis	G20	Dde_	2134–2135
D. ferrireducens	DSM 16995	SAMN05660337	3449–3450
D. frigidus	DSM 17176	WP_	031485205, 084154232
D. inopinatus	DSM 10711	WP_	027182710–027182711
D. piger	FI-11049	DESPIGER_	00296–00297
D. salexigens	DSM 2638	Desal_	2049–2050
D. termitidis	HI1	WP_	35064337, 084559314
D. vulgaris	Miyasaki	DvMF_	0273–0274
D. vulgaris	Hildenborough	DVU_	1917–1918
D. zosterae	DSM 11974	WP_	027722758, 084147016

[4Fe-4S] clusters and less than 25% of isolated DgFdI containing both [4Fe-4S] and [3Fe-4S] clusters (Xavier et al. 1981).

Ferredoxins have been isolated from other sulfate-reducing bacteria (Table 4.10), and these are briefly characterized:

- *Desulfomicrobium baculatum* Norway 4 has ferredoxin I (DmbFdI), which contains one [4Fe-4S] cluster with a midpoint potential of −374 mV (Bruschi et al. 1977; Guerlesquin et al. 1980), and a DmbFdII, which is oxygen-sensitive and a midpoint potential of −500 mV (Bruschi et al. 1985; Guerlesquin et al. 1985). Amino acid sequence has been reported for DmbFdI, which has 59 amino acids with 6 cysteine residues (Bruschi et al. 1985).
- *D. vulgaris* Miyazaki has two ferredoxins: *Dv*MFdI and *Dv*MFdII (Ogata et al. 1988) (Table 4.10). The amino acid sequence for *Dv*MFdI has been reported (Okawara et al. 1988) with midpoint redox potentials of −10 to −140 mV and − 340 mV, which are assumed to reflect the presence of [3Fe-4S] and [4Fe-4S] clusters, respectively (Moura et al. 1994a, b).
- *D. africanus* Benghazi has three ferredoxins: DaFdI, DaFdII, and DaFdIII. Amino acid analysis has been reported for DaFdI, and it only contains four cysteine residues, which are the minimum for binding a [4Fe-4S] cluster (Bruschi and Hatchikian 1982). DaFdII is in small concentration, and little is known about its characteristics. Amino acid analysis revealed that DaFdIII contains seven cysteine residues, which binds both a [3Fe-4S] cluster and a [4Fe-4S] to the protein with midpoint redox potentials of −140 mV and −410 mV, respectively (Bouvier-Lapierre et al. 1987; Armstrong et al. 1989).
- Ferredoxin from *D. vulgaris* Hildenborough has been detected in relatively low amounts as either soluble or bound to nucleic acid (Arendsen et al. 1995). The *D. vulgaris* Hildenborough ferredoxin nonspecifically binds RNA but not DNA and is proposed to regulate protein synthesis at the RNA level. In cell-free preparations, ferredoxin from *Dv*H participates in the phosphoroclastic reaction by shuttling electrons from cytochrome c_3 to pyruvate dehydrogenase (Akagi 1967). This interaction between cytochrome c_3 and ferredoxin within the cell has

been questioned, because cytochrome c_3 is in the periplasm while ferredoxin is in the cytoplasm (Arendsen et al. 1995).

- Ferredoxins have been demonstrated in *D. salexigens*, *D. desulfuricans* ATCC 277, *Desulfomicrobium baculatum* 9974, and *Desulfotomaculum* sp.; however, these iron-sulfur proteins have not been characterized (Moura et al. 1994b).
- A [4Fe-4S] ferredoxin containing 57 amino acid residues was isolated from *D. desulfuricans* strain Berre S (NCIB 8388), and it displayed a redox potential of −330 mV at pH 7.0 (Zubieta et al. 1973).

The cluster interconversion with [3Fe-4S] converted to [4Fe-4S] in *D. gigas* has received considerable attention (LeGall et al. 1982; Kent et al. 1982; Moura et al. 1984; Moura et al. 1994a, b), and it raises the question if this interconversion occurs in vivo and is in response to redox status of the cytoplasm. Interconversion of [3Fe-4S] and [4Fe-4S] has also been reported for FdIII from *D. africanus* (Busch et al. 1997). It has been proposed that the interconversion of [3Fe-4S] clusters to [4Fe-4S] clusters in biological systems may reflect control of protein synthesis at the mRNA level (Klausner and Harford 1989; Rouault et al. 1991).

An interesting feature of the SRB ferredoxin is the formation of heterometal clusters (Moura et al. 1994a, b). The [3Fe-4S] cluster in ferredoxin of *D. gigas* was the first to be converted to the [Co,3Fe-4S] cluster (Thomson et al. 1981). FdIII from *D. africanus* was reported to be converted from [3Fe-4S] to [M,3Fe-4S], where M = Fe, Zn, Co, or Cd (Butt et al. 1997). With the addition of metal to the [3Fe-4S] cluster, the [M,3Fe-4S] assumes the cubane-type structure, and an appropriate amino acid is coordinated to the metal atom. Using FdII from *D. gigas*, the [M,3Fe-4S] clusters were measured by voltammetry, and the following redox values were obtained: [Cd,3Fe-4S] = −495 mV, [Fe,3Fe-4S] = −420 mV, [Ni,3Fe-4S] = −360 mV, and [Co,3Fe-4S] = −245 mV (Moura et al. 1994a, b). With *D. africanus*, the heterometal clusters were reported to have the following redox measurements: [Cd,3Fe-4S] = 500 mV, [Zn,3Fe-4S] = −490 mV, and [Fe3Fe-4S] = −400 mV (Butt et al. 1991, 1997). While a physiological role for heterometal clusters is not readily apparent, it may indicate the proclivity of a [3Fe-4S] cluster to acquire an appropriate metal ion to form the cubane structure.

4.4.2 Flavodoxin

Flavodoxins are low-molecular-weight electron transfer protein found in many microorganisms. Flavin mononucleotide (FMN) is bound into the proteins, and this prosthetic group carries both electrons and protons. Characterization of flavodoxins from SRB concerning structure and physiological activities has been the subject of numerous reviews (LeGall et al. 1979; Peck Jr. and LeGall 1982; LeGall and Fauque 1988; Fauque et al. 1991; VerVoot et al. 1994; Chen et al. 1995). For the purification of flavodoxin from SRB, the procedures established by LeGall and Hatchikian (1967), Irie et al. (1973) and Moura et al. (1980a) may be employed.

Flavodoxin has been isolated from several SRB, including *D. gigas* (LeGall and Hatchikian 1967), *D. vulgaris* Hildenborough (Dubourdieu et al. 1973), *D. desulfuricans* (Palma et al. 1994), and *D. salexigens* (Moura et al. 1980a). Amino acid composition of flavodoxin from *D. vulgaris* Hildenborough indicates an abundance of acid residues, which accounts for the acidic character of flavodoxins (Dubourdieu et al. 1973). Genes encoding flavodoxin production from *D. vulgaris* Hildenborough (Krey et al. 1988; Curley and Voordouw 1988; Curley et al. 1991), *D. gigas* (Helms and Swenson 1992), *D. desulfuricans* Essex 6 (Helms and Swenson 1991), and *D. salexigens* (Helms et al. 1990) have been cloned and characterized. While Knight Jr and Hardy (1966) report the production of flavodoxin by *Clostridium pasteurianum* under iron limited growth, where flavodoxin is used in place of the 2x[4Fe-4S] ferredoxin, the synthesis of flavodoxin by SRB is not always induced by iron limitation. Flavodoxin is produced in large amounts by DvH, which is an iron-sufficient medium (Dubourdieu and Le Gall 1970; Mayhew et al. 1978; LeGall and Peck Jr. 1987). On the other hand, *Desulfomicrobium baculatum* Norway 4 and *Desulfomicrobium baculatum* DSM 1743 do not produce flavodoxin (Bruschi et al. 1977; Fauque et al. 1991). From a comparison of flavodoxin structures from *D. vulgaris* Hildenborough, *D. desulfuricans* ATCC 29577, *D. salexigens*, *D. gigas* ATCC 193, and *D. gigas* ATCC 29494, a flavodoxin signature sequence of ILFGSSTGNTESIAQKL was determined at positions 6–22 on the protein (Caldeira et al. 1994). Analysis of flavodoxin structures from *D. vulgaris* Hildenborough, *D. desulfuricans* ATCC 29577, *D. desulfuricans* ATCC 27774, *D. salexigens*, *D. gigas* ATCC 193, and *D. gigas* ATCC 29494 revealed that similarities between these species ranged from 43.5% to 73.3% (Caldeira et al. 1994). It is interesting that flavodoxin proteins from *D. desulfuricans* ATCC 29577 and *D. desulfuricans* ATCC 27774 are 75% similar, while the flavodoxin proteins of *D. gigas* ATCC 19364/NCIB 9332 and *D. gigas* ATCC 29494/DSM 496 have a similarity of 66% (Helms and Swenson 1992; Caldeira et al. 1994).

Genomic studies of a large diverse set of genomes across the biological spectrum revealed that the production of both ferredoxin and flavodoxin by a species is relatively common, and at least in the case of cyanobacteria, flavodoxin evolution occurred after ferredoxin (Campbell et al. 2019). Flavodoxins vary in mass and range from 15 to 23 kDa, and two subclasses of flavodoxins are designated based on molecular mass. The difference in molecular mass reflects the presence or absence of 20–30 amino acids in the middle of a β-strand (Romero et al. 1996; Hsieh et al. 2013). Members of the "short-chain" flavodoxin do not contain the 20–30 amino acid segment and include flavodoxins from *Desulfovibrio*, while flavodoxin from cyanobacteria have the "long chain" and a greater mass. While the 20–30 amino acid segment has no effect on binding FMN, the additional amino acids do have an influence on the E_2 (semiquinone-quinone) midpoint potentials of flavodoxin with cyanobacterial potentials being more negative than those of the *Desulfovibrio* (Romero et al. 1996). Reaction pH has a greater impact midpoint potential of E_2 than on E_1 (Caldeira et al. 1994).

Midpoint potentials of several flavodoxins from *Desulfovibrio* are given in Table 4.11. The difference in E_2 midpoint potentials of *D. desulfuricans*

Table 4.11 Characteristics of flavodoxins from sulfate reducers

Desulfovibrio	Molecular weight	Midpoint potential (E_1) semiquinone-hydroquinone	(E_2) semiquinone-quinone	References
D. desulfuricans ATCC 27774	15.3 kDa[a]	−387 mV	−40 mV	Romero et al. (1996) Caldeira et al. (1994)
D. vulgaris Hildenborough	15.9 kDa[b]	−438 mV −435 mV −440 mV	−103 mV −143 mV −140 mV	VerVoort et al. (1994) Dubourdieu et al. (1975) Chang and Swenson (1997)
D. vulgaris (Miyazaki F)	15.9 kDa	−434 mV	−151 mV	Kitamura et al. (1998)

[a]Apo-flavodoxin as determined by mass spectrometry was 15.4 kDa (Romero et al. 1996) but 15.3 kDa as calculated from amino acid sequence (Caldeira et al. 1994)
[b]The value of 15.9 was reported following sizing using Sephadex chromatography (Dubourdien and Fox 1977), which is in agreement with gene sequence analysis (Curley and Voordouw 1988; Krey et al. 1988)

(−40 mV) and *Dv*H (−140 mV) reflects a difference of amino acids in the FMN environment. Midpoint potentials for flavodoxin from *D. vulgaris* Hildenborough revealed −440 mV for the semiquinone-hydroquinone form and −140 mV for the semiquinone-quinone form (Dubourdieu et al. 1975). These redox potentials are comparable to ferredoxin from *D. gigas*, where DgFdI has an $E^{o\prime} = -440$ mV and DgFdII has an $E^{o\prime} = -130$ mV (Moura et al. 1978a, b). Based on the similarities of the similar midpoint potentials, it is often assumed that flavodoxin and ferredoxin are interchangeable metabolic reactions.

While there are no established reactions with an obligate requirement for flavodoxin, there are two proposed reactions where flavodoxin replaces ferredoxin and functions in SRB metabolism. The first reaction involves flavodoxin to transfer electrons from pyruvate oxidation by the phosphoroclastic reaction to hydrogenase (Kim and Akagi 1985), and the second reaction of notable importance is the transfer of electrons from the plasma membrane to the sulfate-reduction system (Chen et al. 1993). Another reaction proposed for electron transfer by flavodoxin involves H_2 production from aldehyde oxidoreductase (Barta et al. 1993). In the *Desulfovibrio* that simultaneously produces flavodoxin and ferredoxin, it is unclear what influences which electron carrier is used. In the reaction between hydrogenase and bisulfite reductase, flavodoxin can substitute for ferredoxin in electron transfer; however, phosphorylation coupled to electron transfer in cell-free preparations of *D. gigas* occurs only with ferredoxin and not with flavodoxin (Barton and Peck Jr. 1970; Barton et al. 1972; Peck Jr. and LeGall 1982). It would appear that ferredoxin, which carries electrons and not protons, interfaces with the plasma membrane to pump the proton across the plasma membrane, thereby creating a proton charge across the membrane. Since flavodoxin carries both electrons and protons, no export of protons across the plasma membrane occurs, resulting in no ATP synthesis coupled to electron transport.

An interest in the structure of flavodoxin of SRB was initiated by the initial paper by Watenpaugh et al. (1972) and more recently by Hsieh et al. (2013), describing the molecule consisting of five parallel pleated sheet peptide segments flanked by helical segments with FMN bound to the protein by ionic interactions. This flavoprotein, as is the case with other flavodoxins, is a member of α/β proteins found in at least nine superfamilies with diverse physiological functions associated with members of these families (Brenner 1997). Structural analysis of flavodoxin from *D. desulfuricans* ATCC 27774 reveals a three-dimensional structure similar to other flavodoxins with deviations occurring around the isoalloxazine ring of FMN (Romero et al. 1996). In *D. gigas*, flavodoxin occurs in solution and in crystal as a head-to-head dimer at a distance of 17 Å, which creates an extended negatively charged region (Hsieh et al. 2013). A flavoprotein isolated from *D. gigas* designated as "flavoredoxin" as a homodimer with two FMN residues and a redox potential of −348 mV should be reexamined, since the flavoredoxin has spectral properties and enzymatic activities, characteristic of flavodoxin (Chen et al. 1993). FMN is retained within the flavodoxin by noncovalent bonds, and in *D. gigas*, the exposed FMN residue has an accessible electrostatic area of ~32.5 Å (Hsieh et al. 2013). As observed in different species of bacteria producing flavodoxin, the folding of apo-flavodoxin proceeds prior to the addition of FMN, and a highly stable flavodoxin molecule is produced. The mechanisms for folding of the flavodoxin apoprotein and binding of FMN to produce a molecule with a functional conformation continue to be an area of interest (Sancho 2006).

4.5 Cytoplasmic Proteins with High Redox Potentials

4.5.1 Rubredoxin

Rubredoxins are found in sulfur-metabolizing microorganisms, and each rubredoxin molecule has a single iron atom held by sulfur atoms of four cysteinyl residues. Unlike ferredoxin or flavodoxin, rubredoxin does not have acid labile sulfur atoms. With a mass of 5–6 kDa, rubredoxins are the smallest electron transfer protein in SRB. Rubredoxin has been isolated and amino acid sequences provided for *D. gigas* (LeGall and Dragoni 1966; Le Gall 1968; Laishley et al. 1969; Bruschi 1976a; Vogel et al. 1977), *D. vulgaris* Hildenborough (Bruschi and LeGall 1972; Bruschi 1976b; Vogel et al. 1977), *D. desulfuricans* Norway NCIB 8310 (Bruschi et al. 1977), *D. desulfuricans* ATCC 27774 (Sieker et al. 1983), *D. desulfuricans* var. *azotovorans* (Newman and Postgate 1968), *Desulfovibrio desulfuricans* Berre-Eau NCIB 8387 (Fauque et al. 1987), *D. salexigens* British Guiana NICB 8403 (Moura et al. 1980a), and *Thermodesulfobacterium commune* (Papavassiliou and Hatchikian

1985). Rubredoxin gene (*rub*) has been cloned and expressed from several *Desulfovibrio* species, including from *D. vulgaris* Hildenborough, using amino acid sequence (Voordouw 1988), and the *rub* gene from *Desulfovibrio vulgaris* Miyazaki F was expressed in *Escherichia coli* with the production of a tetrahedral geometry of the iron atom with the cysteine residues and an active electron carrier with a redox potential of −5 mV (Kitamura et al. 1997).

While there is considerable variation in the charged amino acid residues in rubredoxin from different *Desulfovibrio* (Vogel et al. 1977), all rubredoxins have two sets of the sequence -C-x-y-C-G-z- near the iron center (Sieker et al. 1994). The first set has a tyrosine residue as "z" with variations in "x" and "y," while the second set has a proline residue as "x." Additionally, there is always an aromatic amino acid residue at position 4 of either tyrosine, phenylalanine, or tyrosine before the first cysteine. Near the second set, there is a tryptophan residue at position 37 (Sieker et al. 1994). An unusual rubredoxin has been isolated from *D. desulfuricans* 27,774, which is a monomer of 8.5 kDa, and has four amino acid residues between the first two cysteine residues (-C-xxxx-C-), coordinating the iron atom and a redox potential of +25 mV at pH 7.6 (LeGall et al. 1998). A rubredoxin with four amino acids between the cysteine residues has also been demonstrated in *D. vulgaris* Hildenborough (Lumppio et al. 1997). The terminology used to describe rubredoxin with a -C-xx-C- motif is rubredoxin type 1 (Rd-1), and a -C-xxxx-C- motif is characteristic of rubredoxin type 2 (Rd-2) (Lumppio et al. 1997; LeGall et al. 1998; da Costa et al. 2001).

The redox potential of most rubredoxins from *Desulfovibrio* spp. is from −5 to 0 mV (Moura et al. 1979), which is too electropositive for dissimilatory sulfate reduction. One possible function for rubredoxin is shuttling electrons for a cytoplasmic lactate dehydrogenase in *D. desulfuricans* Miyazaki (Shimizu et al. 1989). Rubredoxin-2 has been reported to donate electrons to bacterioferritin in *D. desulfuricans* ATCC 27774 and therefore has a role in iron metabolism (da Costa et al. 2001). In the archaeon *A. fulgidus*, rubredoxin donates electrons to neelaredoxin (Rodrigues et al. 2005). Another activity of potential physiological importance is the transfer of electrons from NADH to oxygen by rubredoxin-1 (Chen et al. 1993a; Gomes et al. 2000) according to the following proposed electron transfer scheme, where NRO is a NADH-rubredoxin oxidoreductase and ROO is a rubredoxin-oxygen oxidoreductase:

$$\text{NADH} \rightarrow \text{NRO} \rightarrow \text{Rubredoxin} \rightarrow \text{ROO} \rightarrow O_2 \qquad (4.6)$$

NRO was isolated from *D. gigas* (Chen et al. 1993a) and demonstrated to donate electrons to ROO, which was also isolated from the same SRB (Chen et al. 1993b). The role of rubredoxin in response to oxygen stress on sulfate-reducing microorganisms is discussed in Chap. 2.

Several reports reveal the analysis of *Desulfovibrio* rubredoxins by X-ray examination of protein crystals. A comparison of rubredoxin from *D. vulgaris* with rubredoxin from *Clostridium pasteurianum* revealed the differences of amino acids at the surface of the molecule but not with bonding at the central core

(Adman et al. 1977). The analysis of crystalline ferredoxin from *D. gigas* indicated a similarity to the ferredoxins from *D. vulgaris* and *Clostridium pasteurianum* with the coordination of an iron atom with four cysteine sulfur atoms (Fe-Cys_4 cluster) forms a regular tetrahedron with a ring of acidic amino acid residues suggested to be important in the docking of electron transfer proteins to ferredoxin (Frey et al. 1987). From crystalline analysis of rubredoxin from *D. vulgaris* Miyazaki F., a structure was revealed with a core region consisting of aromatic amino acid residues and the iron center (Misaki et al. 1999). More recently, Chen et al. (2006) found that the rubredoxin from *D. gigas* displayed hydrophobic and pi-pi interactions for the internal folding of the ferredoxin molecule with numerous conformations of amino acid residues and the Fe-Cys_4 cluster.

With the iron center exposed in rubredoxin, several manipulations have been conducted to modify the protein, and even though the products are not important, physiologically these activities are of interest. The coordinated iron atom in rubredoxin has been replaced with ionic cobalt and nickel (Moura et al. 1991a, b), and it has been reported that the nickel-substituted rubredoxin has hydrogenase-like activity (Saint-Martin et al. 1988). Through the use of modeling programs, a complex was constructed between rubredoxin and a heme moiety of cytochrome c_3 (Stewart et al. 1988; Stewart et al. 1989; Stewart and Wampler 1991).

4.5.2 Rubrerythrin

A nonheme iron protein was isolated from *D. vulgaris* Hildenborough and was designated rubrerythrin (Rr), because it contained a rubredoxin-like iron center and a hemerythrin-type iron cluster (LeGall et al. 1988). A review of rubredoxin from *D. vulgaris* Hildenborough has been provided by Moura et al. (1994a, b). From amino acid sequencing (Van Beeumen et al. 1991) and gene analysis (Prickril et al. 1991), rubrerythrin from *D. vulgaris* Hildenborough was determined to contain 191 amino acid residues. The molecular mass of 22 kDa for rubrerythrin, as initially described (LeGall et al. 1988), is in good agreement with the 21.5 kDa reported using amino acid content analysis (Van Beeumen et al. 1991). At the C-terminal part of the rubrerythrin molecule, there are four cysteine residues in the amino acid sequence of -Cys-X-Y-Cys-Gly-Try- X_{12} – Cys-Pro-X-Cys-, which is characteristic of iron-binding sites in rubredoxin (Van Beeumen et al. 1991). Using recombinant rubrerythrin, Gupta et al. (1995) determined that the 44 kDa homodimer contained six iron atoms with two FeS_4 (Fe-Cys_4) and two diiron sites. As determined by X-ray crystallography, rubrerythrin from *D. vulgaris* is a tetramer of two-domain subunits, where each subunit has a C-terminal rubredoxin-like FeS_4 domain and a diiron site surrounded by a four-helix bundle (Sieker et al. 1988; deMaré et al. 1996). This four-helix bundle is similar to that found in bacterioferritin. Physicochemical and biochemical characteristics of rubrerythrin have been summarized, and each homodimer contains two mononuclear iron centers and two dinuclear iron clusters for a total of six iron atoms for each dimer (Pieiuk et al. 1993).

Using a truncated rubrerythrin gene, which encoded for only 152 amino acids, a 35 kDa homodimer was created that contained two diiron-binding sites. The in vitro addition of ferrous ion accounted for the reconstruction of the two diiron sites and when tested by EPR-based redox titrations produced redox potentials of +215 mV and 154 mV (Gupta et al. 1995), which is lower than the +280 mV reported for recombinant rubrerythrin or +281 mV at pH 7.0 in HEPES buffer (Pieiuk et al. 1993) but close to the +230 mV initially reported at pH 7.6 in phosphate buffer (LeGall et al. 1988). With this high redox potential for rubrerythrin, it is difficult to construct a redox scheme for sulfate reduction involving rubrerythrin in the cytoplasm of an anaerobic bacterium. One function proposed for rubrerythrin in *D. vulgaris* Hildenborough concerns hydrolysis of inorganic pyrophosphate (Liu and LeGall 1990; Van Beeumen et al. 1991); however, this hydrolytic activity remains to be reproduced (Pieiuk et al. 1993). Rubrerythrin from *D. vulgaris* Hildenborough has ferroxidase activity (Bonomi et al. 1996); however, this is not a physiological activity. Several publications propose the involvement of rubrerythrin in the protection of *D. vulgaris* against oxidative stress (Lumppio et al. 2000; Coulter and Kurtz Jr. 2001). The origin of rubrerythrin may be a fusion of genes for rubredoxin and hemerythrin (Van Beeumen et al. 1991).

4.5.3 Desulfoferrodoxin

Desulfoferrodoxin (Dfx) is a nonheme iron protein that was initially isolated from *D. vulgaris* Hildenborough and *D. desulfuricans* ATCC 27774 (Moura et al. 1990). This protein occurs as a monomer with a mass of 14 kDa, which is unusual in that it has no labile sulfide but has a desulforedoxin-like FeS_4 (center I) and a mononuclear iron site (center II) (Moura et al. 1990; Devreese et al. 1996). When isolated, the desulfoferrodoxin is pink with center I in the ferric state with a distorted tetrahedral sulfur coordination similar to that of desulforedoxin and a ferrous state at center II (Verhagen et al. 1993; Tavares et al. 1994). A molecular form of desulfoferrodoxin was also isolated that was not pink but gray, where iron in both centers was in the ferric state and the redox potential of this gray form for center I and center II was +4 mV and +240 mV, respectively (Tavares et al. 1994). Examination of DNA fragments containing genetic material corresponding to the N-terminal and C-terminal domains expressed in *Escherichia coli* revealed a $FeCys_4$ binding motif at the N-terminal domain and a $Fe(N_{\varepsilon}\text{-His})_3(N_{\delta}\text{-His})(S\text{-Cys})$ center in the C-terminal fragment (Devreese et al. 1996; Ascenso et al. 2000). MAD phasing and refinement of desulfoferrodoxin molecule to 1.9-Å resolution confirmed the distorted tetrahedral form of a rubredoxin-type center for iron domain I, and iron domain II has a square pyramidal coordination to nitrogen atoms of four histidine residues and one sulfur from cysteine as the axial ligand (Coelho et al. 1997).

4.5.4 Desulforedoxin

A small nonheme iron protein initially isolated by Moura et al. (1977) from *D. gigas* has been named desulforedoxin (Dx). This dimeric molecule had a mass of 7.9 kDa, two iron atoms, eight cysteine residues, no acid labile sulfur atoms, and a midpoint redox potential of −35 mV (Moura et al. 1977; Moura et al. 1978a, b; Bruschi et al. 1979; Moura et al. 1980b). The monomer has 36 amino acid residues which forms an incomplete β-barrel consisting of antiparallel strands with two iron atoms placed about 16 Å apart at opposite ends of the dimeric molecule (Archer et al. 1995). The iron atom in each monomer is coordinated to four cysteine residues as found in the rubredoxin-like sequence of -Cys-XX-Cys-X_n-Cys-Cys-. The sequence of -Cys-XX-Cys- occurs at the amino terminus of desulforedoxin, while the sequence of -Cys-Cys- is near the carboxyl terminus. The iron atom is bound by four cysteine residues in desulforedoxin, forming a distorted tetrahedral arrangement unlike the tetrahedral structure found in rubredoxin. This difference in iron-binding structure is attributed to the differences in binding motifs in rubredoxin and desulforedoxin. The desulforedoxin gene (*dsr*) from *D. gigas* was cloned and expressed in *Escherichia coli*, where two isoforms of desulforedoxin were produced: one was identical to desulforedoxin from *D. gigas* with two iron atoms and another which had one atom of iron and one atom of zinc (Czaja et al. 1995). In another study, iron replacement in desulforedoxin was readily achieved using In^{3+}, Ga^{3+}, Cd^{2+}, Hg^{2+}, and Ni^{2+} salts (Archer et al. 1999). Metal ion substitution for iron was discussed with rubredoxin in Sect. 4.4.1. While the metal ion that substituted desulforedoxin has similar molecular folding and structural arrangements around the metal center, it supports the suggestion that metal ion availability, and not molecular structure, is key for assembly of certain metalloproteins (Silva and Williams 1991). This underscores the importance of metal ion homeostasis in SRB and microorganisms in general. The origin of desulforedoxin is unknown, but it has been suggested that it is a product of gene fusion of desulforedoxin and neelaredoxin (Devreese et al. 1996).

4.5.5 Neelaredoxin

A blue protein isolated from *D. gigas* was named neelaredoxin (blue in Sanskrit is neela) (Chen et al. 1994a, b). Neelaredoxin (Nlr) is a monomeric protein with a mass of 15 kDa that contains two iron atoms, no labile sulfur atoms, and a redox potential of +190 mV at pH 7.5. The oxidized form of neelaredoxin is blue, while the reduced form is bleached. There is a single iron site with coordination attributed to four histidine and one cysteine residue (Silva et al. 2001). Neelaredoxin is constitutively produced and is proposed to be associated with defense to oxygen stress in *D. gigas* (Chen et al. 1994a, b; Silva et al. 1999). Neelaredoxin has also been demonstrated in *A. fulgidus* (Abreu et al. 2001) and is proposed to participate in superoxide detoxification according to the following reaction scheme:

$$NAD(P)H \rightarrow NRO \rightarrow Rd \rightarrow Nlr \rightarrow O_2^{\cdot} \tag{4.7}$$

where NAD(P)H provided electrons for NADPH:rubredoxin oxidoreductase (NRO) for the reduction of rubredoxin. Reduced rubredoxin transfers electrons to neelaredoxin, which reacts with superoxide ($O_2^{\cdot}$) to produce hydrogen peroxide (Rodrigues et al. 2005).

4.5.6 *Nigerythrin*

Nigerythrin (Ngr) was isolated from *D. vulgaris* Hildenborough and is a homodimer with a molecular mass of 54 kDa (Pieiuk et al. 1993). Nigerythrin is a homodimer that contains two dinuclear clusters and two mononuclear centers with midpoint redox potential of > + 200 mV. Nigerythrin is closely related to rubredoxin, with an amino acid sequence in nigerythrin that displays a 33% identity to rubredoxin of DvH and amino acids functioning as iron ligands in FeS_4, and diiron sites are the same for both nigerythrin and rubredoxin (Lumppio et al. 1997). The high resolution of crystalline nigerythrin revealed a Glu ↔ His ligand toggling of one iron atom of the diiron site (Iyer et al. 2005). With the conversion of Fe1(III) to a diferrous state, there is a 2-Å movement of Fe1 from a carboxylate to a histidine ligand, which is proposed to account for nigerythrin having an affinity for hydrogen peroxide over dioxygen. This iron toggling between protein ligands is also observed with rubredoxin (Iyer et al. 2005).

4.6 Cytochromes

4.6.1 *C-Type Cytochromes*

A characteristic of cytochromes is the transferring of electrons in the electron transport chains, and cytochromes *c* are found in most but not all sulfate-reducing microorganisms. The reports of Postgate (1954) and Ishimoto et al. (1954a, b) provided the initial information of cytochrome *c* in sulfate-reducing bacteria and provided the impetus for many subsequent investigations. The sulfate-reducing microorganisms are unique in that they may have cytochromes *c* with 1, 2, 4, 8, or 16 heme groups per molecule, and with this molecular diversity, LeGall and Fauque et al. (1988) suggested that these cytochromes should be grouped according to number of heme residues per molecule and not by molecular mass. A review by Moura et al. (1991b) summarizes detailed information on cytochromes isolated from sulfate reducers, and the classification of bacterial cytochromes *c* from bacteria is based on the review by Ambler (1991). The electrochemical and biochemical properties of various *c*-type cytochromes from sulfate-reducing bacteria have been reviewed (Yagi 1994; Romão et al. 2012; Fauque and Barton 2012). Physiological activities have been correlated with the abundance of genes for cytochrome *c* in microbial genomes, and the following groups are suggested: (i) members of deltaproteobacteria sulfate reducers with a high number of multiheme cytochrome *c* genes and (ii) members of *Archaea* and *Clostridia* that are sulfate reducers and carry few or no cytochrome *c* genes (Pereira et al. 2011).

4.6.1.1 Monoheme Cytochrome c_{553}

Cytochrome c_{553} has a pronounced absorption at 553 nm in the reduced form and is found in a few SRB, including *D. vulgaris* Hildenborough (Bruschi et al. 1970), *D. vulgaris* Miyazaki (Yagi 1979), *D. desulfuricans* Berre-Eau NCIMB 8387 (Moura et al. 1987), *D. desulfuricans* NCIMB 8372 (Eng and Neujahr 1989), *D. salexigens* British Guiana (Moura et al. 1987), and *D. desulfuricans* G201 (Aubert et al. 1998a, b). Cytochrome c_{553} (cyf) has the motif of -C-X-X-C-H-, where the two cysteine residues of the polypeptide bind to the vinyl side chains of heme c by thioether bonds and histidine serves as an axial fifth ligand in coordination of iron in heme with methionine as the sixth ligand for the iron atom. The heme group is located near the N-terminus of the protein chain, and according to the classification of Ambler (1991), cytochrome c_{553} belongs to the class I of cytochromes c. With respect to amino acid sequences, the cytochromes c_{553} from *D. vulgaris* Hildenborough and *D. vulgaris* Miyazaki are homologous (Van Rooijen et al. 1989) but markedly distinct from cytochrome c_3 obtained from SRB. Some of the relevant electrochemical properties of cytochrome c_{553} from sulfate reducers are presented in Table 4.12.

Table 4.12 Comparison of molecular properties of cytochrome c_{553} and cytochrome $c_{553(550)}$ from sulfate-reducing bacteria

Bacteria	Molecular mass	pI	Redox potential (mV)	References
D. vulgaris Hildenborough cyt-c_{553}	9 kDa	8.0	+18, +20	Bertrand et al. (1982) Bianco et al. (1983) Koller et al. (1987)
D. vulgaris Miyazaki cyt-c_{553}	8 kDa	10.2	+26	Yagi (1979)
D. desulfuricans NCIMB 8372 cyt-c_{553}	7.2 kDa	>9	0	Moura et al. (1987) Eng et al. (1989)
D. desulfuricans NCIMB 8387 cyt-c_{553}		9.2	> −50	Moura et al. (1987)
Desulfomicrobium baculatum 9974 DSM 1743 cyt- $c_{553(550)}$			> −50	Moura et al. (1987)
Desulfomicrobium norvegicum DSM 1741[a] cyt- $c_{553(550)}$	9.2 kDa	6.6	+40, +50	Fauque et al. (1979a, b) Bianco et al. (1983)

[a]Formerly classified as *Desulfovibrio desulfuricans* Norway 4

The structure of cytochrome c_{553} from *D. vulgaris* Miyazaki F as determined by X-ray crystallography revealed alpha helices at each end of the molecule with molecular folding characteristic of cytochrome *c* molecules (Nakagawa et al. 1990), while the 3D structure of cytochrome c_{553} from *D. vulgaris* Hildenborough examined using NMR spectroscopy indicated a conformationally flexible loop, including residues 50–53 (Blackledge et al. 1995). A comparison of the cytochrome c_{553} structures from the two strains of *D. vulgaris* (Hildenborough and Miyazaki F) revealed considerable similarity (Matias et al. 2005).

The physiological role of cytochrome c_{553} has been a subject of considerable debate, and several enzymatic activities have been suggested to involve cytochrome c_{553}. In *D. vulgaris* Miyazaki and *D. vulgaris* Hildenborough, cytochrome c_{553} has been proposed to serve as an electron acceptor for formate dehydrogenase (Yagi 1979; Sebban-Kreuzer et al. 1998a, b) and electron acceptor from lactate dehydrogenase in *D. vulgaris* Miyazaki (Ogata et al. 1981). With reactions using proteins from *D. vulgaris* Hildenborough, cytochrome c_{553} has been shown to accept electrons from [FeFe] hydrogenase (Pereira et al. 1998).

Docking studies between cytochrome c_{553} from *D. vulgaris* Hildenborough and ferredoxin I from *D. norvegicum* revealed a potential for electron transfer across a distance of 4.1 Å between a cysteine residue attached to the distal [4Fe4S] cluster of ferredoxin I and a cysteine bonded to the heme of cytochrome c_{553} (Morelli and Guerlesquin 1999; Morelli et al. 2000). *D. norvegicum* ferredoxin I was used as a model for formate dehydrogenase and [FeFe] hydrogenase, because both molecules had a ferredoxin-like domain with 30% identity to *D. norvegicum* ferredoxin I. Also, crystallographic data for the experiments was available for *D. norvegicum* ferredoxin I but not for formate dehydrogenase and [FeFe] hydrogenase. A mapping of the interaction between cytochrome c_{553} from *D. vulgaris* Hildenborough and ferredoxin I from *D. norvegicum* reveals the heme pocket is surrounded by positively charged amino acid residues while the surface of the distal [4Fe4S] cluster of ferredoxin I is negatively charged. Using NMR data, the interaction between *D. vulgaris* Hildenborough cytochrome c_{553} and the distal [4Fe4S] cluster of *D. vulgaris* Hildenborough [FeFe] hydrogenase revealed a distance of 3.8 Â between the cysteine residues covalently linking heme and the iron-sulfur center.

Using molecular biology techniques, the interaction between cytochrome c_{553} and formate dehydrogenase from *D. vulgaris* Hildenborough was examined. When the Lys62-Lys63- Tyr64 residues located near the region of the heme group of cytochromes c_{553} were modified, electron transfer from cytochrome c_{553} and formate dehydrogenase electron transfer was lost (Sebban-Kreuzer et al. 1998a, b). Examination of genomes of *D. vulgaris* Hildenborough, *D. alaskensis* G20, and other SRB indicates that cytochrome c_{553} is associated with aerobic respiration, since the *cyf* gene is clustered with genes coding for a heme-copper oxygen reductase (Kitamura et al. 1995; Pereira et al. 2011). Additionally, electrons are transferred from cytochrome c_{553} through the membrane of *D. vulgaris* Hildenborough to molecular oxygen (Lobo et al. 2008a, b).

Cytochromes isolated from *Desulfomicrobium baculatum* DSM 1743 and *Desulfomicrobium norvegicum* DSM 1741 (formerly known as *D. desulfuricans*

Norway 4) (Fauque et al. 1979a, b) are unique, in that in the reduced state, they have the characteristic absorption at 553 nm of a cytochrome *c*, but it also has an absorption at 550 nm (Bianco et al. 1983). This asymmetric alpha-band absorption accounts for the designation as cytochrome $c_{553(550)}$. While these bacteria are the only SRB reported to have a positive redox potential cytochrome c_{553} with a split alpha band, this property has also been reported for *Chromatium vinosum* (Cusanovich and Bartsch 1969), *Thiocapsa pfennigii* (Meyer et al. 1973), *Rhodopseudomonas gelatinosa* (Bartsch 1978), and *Ectothiorhodospira shaposhnikovii* (Kusche and Trüper 1984). The cytochrome $c_{553(549)}$ from *Ectothiorhodospira shaposhnikovii* has a mass of 10.4 kDa, an isoelectric point at pH 5.1, and a redox potential of +248 mV (Kusche and Trüper 1984), while the cytochrome $c_{553(550)}$ from *Chromatium vinosum* has a mass of 12.9 kDa, an isoelectric point at pH 4.38, and a redox potential of +300 mV (Cusanovich and Bartsch 1969). Additional research on cytochrome $c_{553(550)}$ from SRB may provide an interesting relationship to anoxygenic photosynthetic bacteria.

4.6.1.2 Homodimeric Diheme Split-Soret Cytochrome *c*

A multiheme cytochrome *c* isolated from *D. desulfuricans* 27,774 displayed two regions of absorption in the Soret region when reduced. Unlike reduced cytochrome *c,* which displayed a single absorption band in the Soret region, this reduced cytochrome *c* had a split-Soret band with an absorption at 424 nm and 415 nm (Liu et al. 1988), and this split-Soret band was proposed to reflect on the interaction between the transition dipole movements of the two heme groups (Matias et al. 1997). The reduced split-Soret cytochrome (SSC) from *D. desulfuricans* 27,774 has two absorption regions: an α-band at 550 nm and a β-band at 520 nm.

This SSC is a homodimer containing identical subunits of 26.3 kDa, with each subunit containing two heme moieties with redox potentials of −168 mV (heme I) and − 330 mV (heme II) (Moura et al. 1991a, b). A three-dimensional structure of SSC displays a stacking of the two heme groups separated by a distance of 3.8 Â and the two hemes in the dimer stacked at a 45° angle with 3.5 Â between these hemes (Matias et al. 1997). The monomers of the SSC dimer display considerable intimacy with a histidine residue from one monomer serving as an axial coordinator for a heme in the other monomer. While the two heme groups are positioned toward the C-terminus of the monomer, a [2Fe2S] cluster at the other end of the molecule is held by cysteine residues (Devreese et al. 1997; Abreu et al. 2003; Rodriguez and Abreu 2005). While the function of the SSC remains a mystery, Rodriguez and Abreu (2005) suggest that a suitable substrate may be a small molecule such as CO or NO.

A survey of genomes of sulfur- and sulfate-reducing bacteria and archaea available in 2011 revealed that genes for SSC are broadly distributed (Pereira et al. 2011). Sulfate reducers with genomes with genes appropriate for SSC include the following: *A. veneficus* DSM 11195, *A. fulgidus* (ATCC 49558), *Thermodesulfovibrio yellowstonii* ATCC 51303, *Desulfococcus oleovorans* DSM 6200, *Desulfurivibrio alkaliphilus* DSM 19089, *Desulfobulbus propionicus* DSM 2032 and

D. desulfuricans ATCC 2774, *D. piger* ATCC 29098, *D. alaskensis* G20, *D. africanus* Walvis Bay, *D. vulgaris* Miyazaki F, *Desulfohalobium retbaense* DSM 5692, and *Desulfonatronospira thiodismutans* AS03–1. The following sulfate reducers had genomes which contained genes suggestive of SSC production: *Pelobacter propionicus* DSM 2379, *Desulfurobacterium thermolithotrophum* DSM 11699, *Desulfurobacterium thermolithotrophum* DSM 11699, *Dethiobacter alkaliphilus* AHT 1, *Desulfitobacterium hafniense* Y51, and *Bilophila wadsworthia* 3_1_6. Pereira et al. (2011) also report that SSC genes are found in other environmentally important bacteria, including *Ferrimonas*, *Shewanella*, and *Geobacter.* The origin of SSC is unknown, but some propose it may be the result of gene fusion involving two unrelated monoheme cytochromes or an unknown Fe-S protein and the ancestor of cytochrome c_3 (Devreese et al. 1997).

4.6.1.3 Tetraheme Cytochrome c_3 (TpI-c_3/TpII-c_3)

Cytochrome c_3 was the first cytochrome discovered in sulfate-reducing bacteria (Ishimoto et al. 1954a, b; Postgate 1954). All *Desulfovibrio* spp. thus far examined contain cytochrome c_3, and this cytochrome has been considered a hallmark characteristic of *Desulfovibrio* (Postgate 1984; LeGall and Fauque 1988). Other sulfate reducers that have been reported to produce tetraheme cytochrome c_3 in members of the *Delta-proteobacteria* and *Nitrospira* taxonomic groups include *Thermodesulfobacterium commune* (Zeikus et al. 1983; Hatchikian et al. 1984), *Desulfobulbus elongatus* (Samain et al. 1986), *Desulfomicrobium baculatum* (Fauque et al. 1979a, b; Fauque et al. 1994), and *Thermodesulfobacterium mobile* (formerly identified as *D. thermophilus*) (Fauque et al. 1988; Rozanova and Pivovarova 1988). A review of genomes of sulfate-reducing bacteria and archaea available in 2011 revealed the presence of tetraheme cytochrome c_3 genes in *A. fulgidus, A. profundus*, *Thermodesulfovibrio yellowstonii*, *Syntrophobacter fumaroxidans* MPOB, *Desulfonatronospira thiodismutans* AS03–1, *Desulfohalobium retbaense* DSM 5692, *Desulfococcus oleovorans* Hxd3, *Desulfobacterium autotrophicum* HRM2, and *Desulfatibacillum alkenivorans* (Pereira et al. 2011). Sulfate-reducing bacteria that do not contain genes for tetraheme cytochrome c_3 include *Caldivirga maquilingensis* and *C. Desulforudis audaxviator* MP104C (Pereira et al. 2011). With internal similarity between the first portion and second portion of the cytochrome c_3, it has been suggested this reflects gene duplication of small cytochromes *c* with production of multiheme cytochromes (Bruschi 1981).

The oxidation-reduction potentials for the four hemes in cytochrome c_3 from *D. vulgaris* Hildenborough are all different and are reported as −284 mV (heme I), −310 mV (heme II), −324 mV (heme III), and − 319 mV (heme IV) (Dervartanian et al. 1978). In general, the redox potential of each of the four hemes in tetraheme cytochrome c_3 is in the range of −200 to −400 mV range (Coutinho and Xavier 1994). Using X-ray diffraction analysis and electron paramagnetic resonance studies, the structure of tetraheme cytochrome c_3 has been established for the following bacteria: *D. vulgaris* Miyazaki (Higuchi et al. 1984), *D. gigas* (Sieker et al. 1986; Matias et al. 1996), *Desulfomicrobium baculatum* (Haser et al. 1979; Pierrot et al. 1982), *D. desulfuricans* ATCC 27774 (Frazão et al. 1994; Morais et al. 1995), *D. africanus* (Nørager et al. 1999; Pereira et al. 2002), and *D. alaskensis* (formerly *desulfuricans*) G20 (Pattarkine et al. 2006). While there is only about 20% identity of amino acid sequences in the various tetraheme cytochromes c_3, certain basic properties are displayed by all these cytochromes, and they include the following: a fold of the polypeptide chain, perpendicular neighboring hemes, heme to heme distances, heme angles, dihedral angles of His to His axial heme ligands, and positions of the aromatic residues (Morais et al. 1995; Brennan et al. 2000; Matias et al. 2005).

The tetraheme cytochrome c_3 of *Desulfovibrio* occurs in two distinct types: type I (TpI-c_3) and type II (TpII-c_3). These two types of cytochromes c_3 are distinguished by location in the cell, physiological activity, genes and genetic features, and molecular structure. TpI-c_3 is soluble, is located in the periplasm, and is reduced by periplasmic hydrogenases (Coutinho and Xavier 1994; Louro et al. 2001; Matias et al. 2005). Although this classification of TpI-c_3 classification is relatively recent, it should be noted that reports on tetraheme cytochromes c_3 from 1956 up through 2000, where no type was designated, would be TpI-c_3. The TpII-c_3 type is associated with the plasma membrane, is reduced slowly by periplasmic hydrogenases but rapidly reduced by electrons from TpI-c_3, and shuttles electrons across the plasma membrane (Valente et al. 2001; Pereira et al. 2002; Pieulle et al. 2005; Paquete et al. 2007). Cloning and sequencing of the gene encoding for TpI-c_3 (Voordouw and Brenner 1986; da Costa et al. 2000) provided information to survey genomes of sulfate-reducing bacteria for tetraheme cytochromes. From genome surveys, *D. vulgaris* Hildenborough was found to have at least 17 tetraheme cytochromes c_3, while only one TpI-c_3 was found in the genome of *Desulfotalea psychrophila* (Pereira et al. 2007). TpII-c_3 has been isolated from *D. desulfuricans* ATCC 27774 (Pires et al. 2003), *D. africanus, D. vulgaris*, and *D. gigas* (Di Paolo et al. 2006) with cloning and sequencing of gene for TpII-c_3 conducted by Pires et al. (2003) and Pereira et al. (2007). The properties of TpII-c_3 are found in domains of the Nhc and HmcA membrane complexes.

4.6.1.4 Octaheme Cytochrome c_3 (M_r 26,000)

A multiheme cytochrome c_3 was isolated from *D. gigas* (Bruschi et al. 1969) containing eight hemes, and it has been called various names, including cytochrome cc_3, di-tetraheme cytochrome c_3, octaheme cytochrome c_3, and cytochrome c_3 (M_r 26,000). Octaheme has also been isolated from *Desulfomicrobium baculatum* Norway 4 (formerly *D. desulfuricans*) (Guerlesquin et al. 1982) and demonstrated in *D. desulfuricans* El Agheila Z and *D. salexigens* Benghazi. The octaheme of *Desulfomicrobium baculatum* is a dimer, where each subunit contains four hemes with redox potential for each of the four hemes reported from cyclic voltammetry as −210 mV, −270 mV, −325 mV, and − 365 mV. Based on the amino acid composition, N-terminal analysis, and electron paramagnetic resonance spectra, the octaheme cytochrome subunits are distinct from tetraheme cytochrome c_3 (Bruschi 1994). Not only is there a difference in amino acids residues between octaheme cytochrome *c* and tetraheme cytochrome c_3, but the octaheme cytochrome has two intermolecular disulfide bridges to secure the identical subunits (Bruschi et al. 1996). Crystalline analysis of octaheme cytochrome c_3 supports suggestions that the octaheme cytochrome is different from the tetraheme cytochrome in *Desulfovibrio* (Sieker et al. 1986; Czjzek et al. 1996; Aubert et al. 1998a, b). The octaheme cytochrome gene (*cycD*) from *D. desulfuricans* Norway was expressed in *D. desulfuricans* G201 and determined to be periplasmic due to the signal sequence of 24 amino acids at the N-terminal segment of the polypeptide (Aubert et al. 1997). Octaheme cytochrome *c3* has been reported to interact with periplasmic [NiFeSe] hydrogenase from *D. desulfuricans* Norway (LeGall et al. 1994). In cell-free extracts of *D. gigas* and purified enzyme fractions, thiosulfate reductase reduces octaheme; however, this is a nonphysiological reaction, because the octaheme cytochrome is periplasmic, while thiosulfate reductase is cytoplasmic (Bruschi et al. 1969; Hatchikian 1975).

4.6.1.5 Nonaheme Cytochrome c_3: NhcA

A periplasmic multiheme cytochrome isolated from *D. desulfuricans* ATCC 27774 had a mass of 40.8 kDa with nine hemes in each molecule (Liu et al. 1988). This cytochrome has been described as nonaheme cytochrome c_3 or NhcA. The amino acid composition of NhcA was distinct from the tetraheme cytochrome c_3, and the split-Soret cytochrome *c* was isolated from the same organism. NhcA was isolated from *D. gigas*, and it has a molecular weight of 67 kDa with redox potentials that ranged from −50 mV to −315 mV (Chen et al. 1994a, b). The gene encoding for the nine-heme cytochrome from *D. desulfuricans* Essex was isolated, and the primary structure of this cytochrome was homologous to the NhcA from *D. desulfuricans* ATCC 27774 and *D. vulgaris* Hildenborough (Fritz et al. 2001). An operon structure was observed in *D. desulfuricans* Essex, and downstream from the gene encoding NhcA was a gene with HmcB homologue. NhcA was reported to be an electron acceptor for [NiFe] hydrogenase (Fritz et al. 2001). Sequencing of the *nhcA* gene from *D. desulfuricans* ATCC 27774 revealed a four-gene operon for the Mhc (or Dd27k 9Hc) complex: *nhcA*; *nhcB*, which is homologous to *mhcB* with binding sites for four [Fe4S] clusters; *nhcC*, which is homologous to *mhcC* and is proposed to interact with quinones and cytochrome *b*; and *nhcD*, which encodes a small protein and is homologous to *mhcD* (Saraiva et al. 1999, 2001).

An evaluation of the polypeptide chain from *D. desulfuricans* ATCC 27774, which consisted of 296 amino acid residues, revealed that the N- and C-terminal domains are physiologically distinct (Bento et al. 2003). The hemes in the N-terminal domain have a lower redox potential than the hemes in the C-terminal domain and it was proposed that electrons would flow from the N-terminal to the C-terminal domain. Structural analysis of NhcA from *D. desulfuricans* ATCC 27774 reveals two cytochrome c_3 motifs. The N-terminal domain contains hemes I, II, III, and V, while the C-terminal domain contains hemes VI, VII, VIII, and IX, with heme IV between the C- and N-terminal domains (Matias et al. 1999a, b; Bento et al. 2003). Structural analysis of NhcA from *D. desulfuricans* Essex revealed cytochrome c_3-like domains at the N-terminal domain and C-terminal domain with the additional heme buried between the two domains. While the two tetraheme domains exhibit heme arrangements similar to cytochrome c_3, there is sufficient difference in these domains to account for the N-terminal domain associated with a transmembrane unit, while the C-terminal domain transfers electrons to hydrogenase (Umhau et al. 2001). In *D. desulfuricans* ATCC 27774, the Nhc complex occurs in both nitrate- and sulfate-grown cells with greater quantity of Nhc complex produced with cultivation on sulfate.

4.6.1.6 Hexadecaheme Cytochrome c_3: HmcA

The cytochrome *c* with 16 hemes was first isolated from *D. vulgaris* Miyazaki (Yagi 1979) and later was found in *D. vulgaris* Hildenborough (Higuchi et al. 1987; Pollock et al. 1991) and *D. gigas* (Chen et al. 1994a, b). The high-molecular-weight cytochrome system consists of several different polypeptides, and HmcA designates a specific polypeptide consisting of about 500 amino acid residues, which is associated with the periplasmic facing side of the plasma membrane. The HmcA isolated from *D. vulgaris* Hildenborough was reported to have a mass of 75 kDa (Higuchi et al. 1987), while the molecular weight calculated from gene sequencing was 65.5 kDa (Pollock et al. 1991). The redox potential of the 16 hemes in the HmcA from *D. vulgaris* Miyazaki was estimated to be 60, 15, −135 (seven hemes), −190 to −205 (five hemes), and −260 (two hemes) mV (Ogata et al. 1993). The hemes are numbered according to the sequence of covalent bonded cysteines to a heme moiety, starting at the N-terminal segment. There is a similarity of amino acid content of HmcA between *D. vulgaris* Hildenborough and *D. vulgaris* Miyazaki, but the amino acid composition from the HmcA is distinct from cytochrome c_3 (Higuchi et al. 1994).

A modular domain structure was proposed for HmcA by Pollock et al. (1991), and three-dimensional structures reported by Czjzek et al. (2002), Matias et al. (2002), Sato et al. (2004) and Santos-Silva et al. (2007) confirmed this modular composition. Near the N-terminal are three hemes, which resemble the cytochrome c_7 from *Desulfuromonas acetoxidans* (Czjzek et al. 2001). Domain II contains three hemes and is similar to cytochrome c_3, while the third domain near the C-terminal resembles the nonaheme cytochrome (Hcc) (Matias et al. 1999a). Additionally, Hcc is constituted from two sub-domains (Matias et al. 1999a, b; Umhau et al. 2001). HmcA has a pronounced fold, which is assumed to enable cytochrome-cytochrome interactions, and a detailed discussion concerning the transmembrane orientation of HmcA has been presented by Romão et al. (2012).

The *hmc* operon from *D. vulgaris* Hildenborough was found to contain eight open reading frames (Orfs) with Orf1 to Orf6 proposed to encode for the Hmc complex, while Rrf1 and Rrf2 were associated with the family of response regulator proteins (Rossi et al. 1993). The six genes of the *hmc* operon encode for the following proteins (Matias et al. 2002): (i) HmcA is a hydrophobic 16-heme cytochrome. (ii) HmcB transcends the membrane with four [4Fe4S] cluster binding motifs near the periplasmic N-terminal domain and has a mass of 40.1 kDa (Keon and Voordouw 1996). (iii) Predicted integral membrane proteins included HmcC (943.2 kDa), HmcD (5.8 kDa), and HmcE (25.3 kDa), where HmcC and HmcF may bind cytochrome *b*. (iv) HmeF (52.7 kDa) is a cytoplasmic protein with two [4Fe4S] cluster binding motifs. The Hmc complex has been considered to be associated with H_2 metabolism, and when the *hmc* operon of *Dv*H was deleted, hydrogen metabolism was impaired (Dolla et al. 2000).

Using the HMC from *Dv*H for a BLAST search of protein sequence in databases, several bacteria were identified as having some similarity to that of *D. vulgaris* Hildenborough (Romão et al. 2012), and these bacteria included the following: *Desulfohalobium retbaense* DSM 5692, *Desulfobacterium* sp. (uncultured), *Desulfococcus oleovorans* DSM 6200, *D. aespoeensis* ATCC 700646, *D. alaskensis* G20, *D. baculatum* DSM 4028, *D. desulfuricans* ND132, *D. fructosovorans* JJ, *D. magneticus* ATCC 700980, *D. salexigens* ATCC 14822, *Thermodesulfobacterium* sp. OPB45, and *Thermodesulfovibrio yellowstonii* ATCC 51303. Some of the bacteria listed above lacked high similarity values, and fewer than 16 hemes could be associated with the HmcA orthologs. Only 15 binding sites for hemes *c* were identified in *D. baculatum* DSM 4028 and *D. aespoeensis* ATCC 700646 with heme V lacking in *D. baculatum* DSM 4028 and heme IX lacking in *D. aespoeensis* ATCC 700646. Only 14 binding sites were identified in *D. desulfuricans* ND 132, where both heme V and heme IX binding sites were absent (Romão et al. 2012).

4.6.1.7 Molecular Docking, Bohr Effect, Proton Thrustor

Metalloproteins may be redox active, and electron transfer by way of electron tunneling in biological systems is usually over a distance range of 4–14 Å. However, if protons are associated with the electron movement, then proton tunneling would be limited to a range of 1–2 Å, because of the greater size of proton (hydride) as compared to an electron (Moser et al. 2010). Exposed redox centers in a protein could result in electron transfer to unwanted redox partners, which would result in a loss in energy and generated superfluous products. Cytochromes *c* with multiple hemes could catalyze redox reactions with several different substrates within a range of ~10 Å. The distribution of cytochromes c_3 in the periplasm and not in the cytoplasm may indicate that nonspecific electron transfer is a greater problem in the cytoplasm than in the periplasm (Moser et al. 2010). The molecular architecture of redox active proteins provides specificity for oxidation-reduction partners with docking sites. For example, with multiheme cytochromes, the most electronegative heme is at the N-terminal region of the protein, and in general, electrons would move toward hemes at the C-terminal segment, where the electronegativity of the heme is not as great. The interaction of TpI-c_3 with hydrogenases is attributed to the positive charge in the environment of heme IV of the cytochrome with the negative surface charge near the [4Fe4S] center of the hydrogenase. The docking of TpI-c_3 with TpII-c_3 is considered to involve heme IV of TpI-c_3 and heme I of TpII-c_3, because heme IV has a positive surface while heme I has a negative surface charge (Valente et al. 2001; Teixeira et al. 2004). The transfer of electrons from TpI-c_3 to HmcA and Nhc is considered to be similar to the electron transfer from TpI-c_3 to TpII-c_3. Surface charge in the environment of hemes is critical for substrate interaction as exemplified with purified reduced cytochrome c_3 from *D. alaskensis* (formerly *desulfuricans*) G20 being oxidized by molybdate and uranyl ions. The negatively charged ions of molybdate and uranyl interact with heme IV to oxidize the reduced tetraheme cytochrome c_3 (Pattarkine et al. 2006).

Interheme interactions in tetraheme cytochrome c_3 have been examined, and a level of cooperativity has been reported that accounts for this cytochrome to accept electrons released from periplasmic hydrogenase and to function as a proton activator (Brennan et al. 2000). The four heme groups in tetraheme cytochrome c_3 display redox cooperativity, and pH influences their redox potentials (Brennan et al. 2000). With changes in redox state, two lysine residues (Lys80 and Lys90) are rearranged, and this contributes to the stabilization of heme II and heme III. This cooperativity accounts for a two-electron step as the cytochrome c_3 interfaces with hydrogenase as a coupling partner. With reduction, propionate 13 of heme I rotates toward the protein interior producing a positive redox-Bohr effect and accounts for proton activation. With the completion of a two-electron redox cycle, the tetraheme cytochrome c_3 performs as a proton thruster (Luro et al. 1998; Brennan et al. 2000; Louro 2006).

4.6.2 B-Type Cytochromes

4.6.2.1 Heme *b* Distribution in Sulfate Reducers

In addition to the cytochromes *c* found in the periplasm and membranes of Gram-negative SRB, the cytochromes *b* are present in the membrane in lesser amounts as compared to cytochromes *c*. To visualize light absorption of reduced cytochromes, ascorbate, and not dithionate, has been used, which reduces cytochromes other than cytochrome *c* and produces absorption at 630 nm for cytochrome *c* and 560 nm for *b*/*o* cytochrome. Additionally, specific hemes can be identified by pyridine-hemochromogen spectra. Cytochromes with heme *b* are not covalently bonded into the protein making isolation of cytochromes *b* extremely difficult. Therefore, the presence of *b*-type cytochromes is based on spectral evidence and comparison of SRB genes to genes in bacteria, where the presence of cytochrome *b* is well-established. A membrane-bound subunit contains two *b* types, which are orientated for optimal electron exchange with menaquinones and transmembrane electron transfer between periplasmic and cytoplasmic subunits. An important function of membrane complexes with protein subunits containing hemes *b* is to proton pump to energize the membrane. Two *b*-type hemes were reported for the TpII-c_3 complex (Matias et al. 2002). The redox potentials of the *b*-type hemes are of a suitable range to receive electrons from menaquinol, which has a midpoint potential menaquinone/menaquinol of −70 mV. The approximate values of −20 mV and +75 mV were reported for the two hemes *b* in the Qmo complex (Pereira 2008). The quinol: fumarate reductase contains two hemes *b* with reduction potentials of −45 and −175 mV (Guan et al. 2018).

In addition to heme *b*, heme *o* and heme *d* are associated with Gram-negative sulfate reducers. Cytochrome oxidase (*cc*/(*o*/*b*)o_3) from *D. vulgaris* Hildenborough contains both heme *b* and heme *o* (Lamrabet et al. 2011), and cytochrome *d* is found in the *bd*-type quinol oxidase (Lemos et al. 2001). Both of these cytochromes are discussed in Sect. 4.7.4. *D. africanus* contains *b*-, *c*-, and *d*-type cytochromes (Jones 1971). It has long been recognized that cytochrome *b* is the dominant cytochrome in members of the genus *Desulfotomaculum* (Postgate 1979).

4.6.2.2 Presence in Electron Transport Complexes

Electron transport complexes facilitate the transfer of electrons across the plasma membrane, and several protein complexes have been demonstrated in the SRB. While the Qmo and Dsr complexes are considered to be present in all dissimilatory sulfate reducers, many of the transmembrane complexes may have a limited distribution throughout the *Desulfovibrio*. A pair of hemes *b* present in a transmembrane subunit is used to shuttle electrons from the membrane complex to the menaquinone pool. A list of some of the transmembrane electron transport complexes found in SRB is provided in Table 4.13, and information about these complexes is in Chap. 5.

Table 4.13 Transmembrane electron transport complexes containing heme *b*[a]

Complex name	Complex subunits	Subunit containing *b*-type heme	Proposed electron donors/acceptors
Qmo	QmoABC	C	MKH_2, APS reductase
Dsr	DsrMKJOP	M	MKH_2, DsrC
Qrc	QrcABCD	D (?)	TpI-c_3, MK
Tmc	TmcABCD	C	TpI-c_3, DsrC
Hmc	HmcABCDE	E	TpI-c_3, DsrC
Nhc (9hc)	NhcABCD	C	TpI-c_3, MK
Ohc	OhcABC	C	TpI-c_3, MK

[a]References: Romão et al. (2012); Rabus et al. (2015)

4.7 Heme-Containing Enzymes

4.7.1 Nitrite Reductase

A multiheme protein with nitrite reducing activity was isolated from *D. desulfuricans* ATCC 27774 (Liu and Peck Jr 1981) and from *D. vulgaris* Hildenborough (Pereira et al. 2000). Nitrite ammonification by SRB is reviewed by Moura et al. (2007). The multiheme nitrite reductase (ccNiR or NiR) from *D. desulfuricans* 27,774, which reduces nitrite to ammonia, was initially considered to have six hemes per molecule (Liu and Peck Jr 1981; Costa et al. 1996). The reaction of nitrite reductase is as follows:

$$NO_2^- + 8H^+ + 6e^- \rightarrow NH_4^+ + 2H_2O \qquad E^0 = +330\ \text{mV} \tag{4.8}$$

The nitrite reductase (cytochrome c_{552}) isolated from *Escherichia coli* was also proposed to be a hexaheme cytochrome, but when reexamined, it was reported to contain four hemes per molecule based on the presence of four Cys-X-X-His cytochrome *c* binding motifs (Darwin et al. 1993), and this was in agreement with the report of the four hemes in the nitrite reductase of *Sulfurospirillum deleyianum* (Schumacher et al. 1994). Subsequently, it was proposed that nitrite reductase contained five cytochromes when the heme-binding motif (CXXCK) using lysine to replace histidine was identified in the genome of *Escherichia coli*. The three-dimensional analysis of nitrite reductase from *Sulfurospirillum deleyianum* (Einsle et al. 1999), *Wolinella succinogenes* (Einsle et al. 2000), *Escherichia coli* (Bamford et al. 2002), and *D. desulfuricans* ATCC 27774 (Cunha et al. 2003) provided the final proof of five hemes per NrfA molecule.

The multiheme nitrite reductase in *D. desulfuricans* ATCC 27774 was determined to consist of NrfA and NrfH subunits, which are encoded by *nrfA* and *nrfH* genes (Almeida et al. 2003). The pentahemic NrfA has a mass of 61 kDa, while NrfH has four hemes and a mass of 19 kDa. These subunits form a complex with a stoichiometry of 2NfrA:1NrfH, although a physiological heterooligomeric unit may be a $\alpha_2\beta_2$ complex. NrfA is the catalytic unit, which is positioned on the periplasmic side of the membrane by a thioether bond between a N-terminal cysteine and a plasma membrane lipid. Additionally, there is a tight association between NrfA and the membrane-bound NrfH subunit. The redox potentials of hemes in NrfA are reported as follows: heme I, −80 mV; heme II, −50 mV; heme III, −480 mV; heme IV, −400 mV; and heme V, +150 mV. The NrfH consists of a soluble domain with hemes 2–4 containing *c*-type hemes and a hydrophobic membrane associated domain containing heme I. The redox potentials of heme I (H_1) of NrfH is >0 mV, and the other hemes (H_2, H_3, H_4) are −300 mV (Almeida et al. 2003). The NrfH has the structure of a cytochrome *c* quinol dehydrogenase with a binding motif of CXXCHXM, where a methionine replaces histidine as the sixth ligand to heme I (Rodrigues et al. 2006). Since the nitrite reductase (NrfA/H) lacks heme *b*, this cytochrome *c* quinol dehydrogenase is suggested to transfer electrons to menaquinone.

It has been proposed that *D. piezophilus* has an octaheme oxidoreductase gene that may be a nitrite reductase (Tikhonova et al. 2012). Two copies of the *nrfHA* gene are present in *D. vulgaris* Miyazaki, *D. termitidis*, and *Desulfovibrio* sp. A2, and single copies of *nrfHA* gene were detected in genomes of bacteria belonging to the *Desulfovibrionales* family (Rajeev et al. 2015). Due to the chemical similarity between sulfite and nitrite, the nitrite reductase from *D. desulfuricans* ATCC 27774 also reduced sulfite (Pereira et al. 1996a, b). It has been suggested that the presence of nitrite reductase in *D. vulgaris* Hildenborough, a non-ammonifying bacterium, may function to detoxify nitrite without coupling nitrite reduction of growth (Pereira et al. 2000). In *D. vulgaris* Hildenborough, NrfR is a sigma-54-dependent two-component response regulator, which under nitrite stress activates the *nrfHA* operon, and NrfR may also influence the nitrate reductase genes (Rajeev et al. 2015). Hydroxylamine is reduced by *Desulfovibrio* strains (Senez and Pichinoty 1958), membranes of *D. gigas* (Barton et al. 1983), and purified nitrite reductase from *D. desulfuricans* ATCC 27774 (Costa et al. 1990). It is not known if SRB contain a hydroxylamine oxidoreductase similar to the hemoprotein from *Nitrosomonas europaea* (Igarashi et al. 1997).

4.7.2 Nitrate Reductase

Nitrite reduction is associated with many different bacterial strains; however, a relatively few strains of SRB are capable of nitrate ammonification with nitrite as an intermediate (Mitchell et al. 1986; Moura et al. 1997). This nitrate reduction in SRB follows the following reaction:

$$NO_3^- + 2H^+ + 2e^- \rightarrow NO_2^- + H_2O \qquad E^0 = +420\ \text{mV} \tag{4.9}$$

Nitrate reduction has been reported for numerous SRB including the following: *D. desulfuricans* (Keith and Herbert 1983; Seitz and Cypionka 1986; Dalsgaard and Bak 1994), *Desulfovibrio* sp. (McCready et al. 1983), *D. furfuralis* (Mitchell et al. 1986), *D. profundus* (Mitchell et al. 1986), *D. fairfieldensis* (Warren et al. 2005), *D. oxamicus* (Lopez-Cortès et al. 2006), *D. simplex* (Zellner et al. 1989), *D. termitidis* (Trinkerl et al. 1990), *Dst. thermobenzoicum* (Plugge et al. 2002), *Thermodesulfobium narugense* (Mori et al. 2003), *Desulforhopalus singaporensis* (Lie et al. 1999), *Thermodesulfovibrio islandicus* (Sonne-Hansen and Ahring 1999), and *Desulfomicrobium* sp. CrR3 (Dorosh et al. 2016).

The dissimilatory reduction of nitrate and nitrate ammonification by SRB has been reviewed by Moura et al. (2007). The genes for nitrate reduction are not present in the genome of *D. vulgaris* Hildenborough, and the current knowledge of nitrate reductase in SRB is based on the system of *D. desulfuricans* ATCC 27774. While it had long been proposed that the nitrate reductase operon consisted of four genes (*napD, napA, napB,* and *napC*) (Potter et al. 2001), subsequent experimentation has revealed six genes in a cluster (*napCMADGH*) associated with nitrate reduction in *D. desulfuricans* 27,774 (Marietou et al. 2005).

NapA was the first protein of the nitrate reductase complex isolated from *D. desulfuricans* 27,774 (Dias et al. 1999). NapA is a monomeric protein that has a mass of 80 kDa, which is divided into four domains, and the binding of two molybdopterin guanine dinucleotide (MGD) cofactors transcend all four domains. While the catalytic activity is buried in NapA, a [4Fe4S] cluster is located at the periphery of NadA. NapA is the catalytic subunit receiving electrons from the quinol pool by way of the tetraheme *c*-type cytochrome NapC and potentially NapG. NapD is suggested to be involved in the maturation of NapA, and NapM contains a tetraheme *c*-type cytochrome. Noteworthy is the absence of the diheme subunit NapB in SRB. The NapA from *D. desulfuricans* 27,774 is inhibited by cyanide, azide, and perchlorate, and crystals of the NapA molecule-containing inhibitors were examined by X-ray analysis. X-ray crystallography confirmed that sulfur of cysteine and not OH/OH_2 was the sixth ligand to coordinate Mo(VI) and the mechanisms describing the reduction of nitrate bonded to Mo(VI) must include the activity of cysteine (Najmudin et al. 2008). The cytoplasmic reduction of nitrate (as well as nitrite reduction) in SRB is highly regulated in response to NO stress by a two-component response regulation, which is suggested to be attributed to involve Rex and sigma-54 (Cadby et al. 2017).

4.7.3 Sulfite Reductase: Sirohemes

The hallmark characteristic of dissimilatory sulfate-reducing bacteria and archaea is the respiratory reduction sulfate with the release of hydrogen sulfide. Sulfite reduction is the enzyme instrumental in this reduction process, and these enzymes are unique in that they contain siroheme as a cofactor. In fact, there are different types of sulfite reductases in SRP with one used for respiration and the second for biosynthesis. Both of these sulfite reductases are discussed in Chap. 3.

4.7.4 Oxygen Reductases

Several systems are used by sulfate-reducing microorganisms for the production of bioenergy or to detoxify the cellular environment from O_2 and include the following reductive processes: (i) a membrane-bound *bd*-type quinol oxidase, (ii) a membrane-bound cc(*b*/*o*)o_3 cytochrome oxidase, and (iii) cytoplasmic systems. This chapter will characterize the *bd*-type quinol oxidase and cc(*b*/*o*)o_3 cytochrome oxidase, which have been examined in only a few SRB.

4.7.4.1 Quinol *bd*-Type Oxidase

With membranes from *D. gigas* grown in lactate/sulfate medium, NADH and succinate provided electrons for O_2 uptake. A terminal oxygenase with subunits of 40 and 29 kDa was isolated and characterized as a member of the cytochrome *bd* family (Lemos et al. 2001). This quinol oxidase *bd* is encoded on *cydAB* genes which has been observed in the genomes of *D. vulgaris* Hildenborough, *D. vulgaris* Miyazaki, *D. alaskensis* (*desulfuricans*) G20, *D. magneticus*, *D. fructosovorans*, *D. piger*, and *Desulfovibrio* sp. FW1012B (Machado et al. 2006; Lobo et al. 2008a, b; Lamrabet et al. 2011). In *D. vulgaris* Hildenborough, the genes *DVU3270-DVU3271* encode for cydAB. The cydA subunit (49 kDa) encoded on *DVU3271* is predicted to contain nine transmembrane helices and to bind two *b*-type hemes. The cydB subunit encoded on *DVU32710* is predicted to contain eight transmembrane helices and to bind one *d*-type heme. Reduction of quinol oxidase *bd* is achieved by menaquinol, and physiologically menaquinone is reduced by type II NADH dehydrogenase, succinate oxidation (Lemos et al. 2001), the Qrc complex (Venceslau et al. 2010), and the Tmc complex (Pereira et al. 2006). Inhibitors of the quinol oxidase *bd* include cyanide, antimycin A, and 2-n-heptyl-4-hydroxyquinole-N-oxide. Results obtained by Ramel et al. (2013) suggest that one or several systems exist in SRB for quinol-dependent oxygen reduction in addition to the quinol oxidase *bd* process. While the quinol oxidase *bd* system is considered to provide defense against exposure to O_2 in SRB and other bacteria, this *bd* oxidase has been suggested to function as a defense to NO stress (Giuffrè et al. 2014). With the identification of the operon for *cydAB*, including the promoter region and transcriptional regulator-binding sites, the mechanisms whereby environmental stress influences cydAB production can be evaluated (Machado et al. 2006).

4.7.4.2 Cytochrome *cc*(*b*/*o*)o_3 Oxidase

A heme-copper oxidase isolated from *D. vulgaris* (Lobo et al. 2008a, b) is a *cc(b/o)*o_3-type oxidase (Lamrabet et al. 2011) and not a *ccaa*$_3$-type, as initially suggested. The *cc(b/o)*o_3 cytochrome oxidase (cox) is unrelated to quinol oxidase *bd*, and both oxidases contribute to a proton charge on the membrane and detoxify cells from O_2. In *D. vulgaris* Hildenborough, the seven genes (DVU1816 - DVU1810) associated with cox production are in a cluster (see Fig. 4.5) adjacent to DVU1817, which

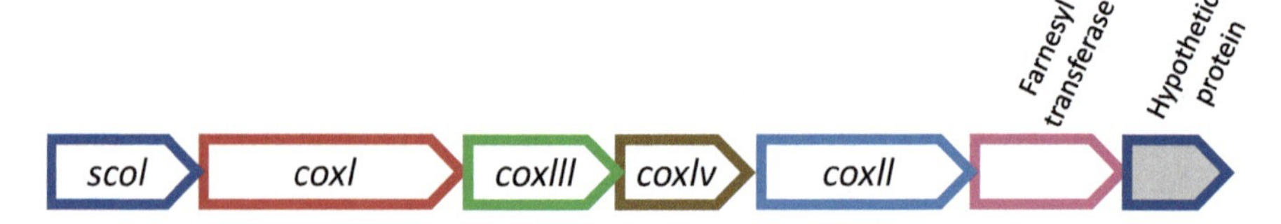

Fig. 4.5 Gene cluster encoding for cytochrome *c* oxidase in *D. vulgaris* Hildenborough. Genes are drawn for position and do not reflect the size

encodes for cytochrome c_{553} (Lamrabet et al. 2011). The gene *scol* encodes a protein, which assists in maturation of copper Cu_A center in coxII. Subunit I (coxI) is the catalytic protein subunit, which contains a low-spin heme and a binuclear center of Cu_B and a high-spin heme, which is the site for O_2 reduction. Subunit II (coxII) contains the Cu_A center with two copper atoms and two *c*-type hemes. Subunit III (coxIII) is the transmembrane component of 23 kDa, and subunit IV (coxIV) is a 23 kDa with three transmembrane helices. Adjacent to the genes for coxI–IV is the gene which encodes for protoheme IX farnesyltransferase, which adds the farnesyl group onto heme *b* to make heme *o*. Finally, the hypothetical protein with four transmembrane helices may function for maturation of subunit II (Lamrabet et al. 2011).

4.7.5 Quinol:Fumarate Oxidoreductase

Many SRB can grow with either fumarate or malate under sulfate-reducing conditions. Fumarate is important in growth on malate, with malate being converted to fumarate by fumarase. Anaerobic growth on fumarate may be attributed to fumarate disproportionation with the production of succinate, acetate, and CO_2 or as an electron acceptor by fumarate respiration with succinate production. With the growth of *D. alaskensis* (formerly *desulfuricans*) G20 on fumarate and sulfate, a "succinate burst" was observed in mid-log phase (Price et al. 2014), and the depletion of this succinate by succinate:menaquinone reductase (Sdh) was proposed to be, by a reversal of the quinol:fumarate reductase (Frd), according to the following reaction:

$$\text{Succinate}^- + \text{menaquinone} \rightarrow \text{fumarate}^- + \text{menaquinol} \qquad (4.10)$$

Since the redox of succinate/fumarate is +30 mV and the redox reaction of menaquinol/menaquinone is −80 mV, anaerobic oxidation of succinate by menaquinol requires a proton potential, and this is referred to as a "reverse redox loop mechanism" (Zaunmüller et al. 2006). Sulfate-reducing bacteria have menaquinone as the respiratory quinone, while aerobic bacteria usually have ubiquinone. With the redox of the ubiquinol/quinone reaction being +100 mV, a proton potential is not required to energize aerobic oxidation of succinate. Unlike most SRB, *D. alaskensis* G20 does not created a proton gradient when fumarate is reduced (Price et al. 2014). The Sdh and Frd enzymes appear similar based on amino acid sequences and prosthetic groups (Romã0 et al. 2012). The genes for quinol:fumarate reductase (*frd*CAB) are found in *D. vulgaris*, *D. alaskensis* G20, *Desulfobacterium autotrophicum*, and *Desulfotalea psychrophila*, but the genes for *frd*D or *sdh*D are absent (Zaunmüller et al. 2006). In *D. desulfuricans* strain Essex 6, the Frd/Suh enzyme functions to reduce fumarate, while in *D. vulgaris*, it functions as a succinate dehydrogenase (Zaunmüller et al. 2006).

A fumarate reductase isolated from *D. multispirans* interacts with menaquinol for the fumarate reduction and reported to contain FAD with four protein subunits

(27, 30, 32, and 45 kDa) (He et al. 1986). Quinol:fumarate reductase (QFR) has been purified from *D. gigas*, and it contains three subunits of 22, 31, and 71 kDa (Lemos et al. 2002). The QFR contains two *b*-type hemes with reduction potentials of −45 and −175 mV and three iron-sulfur clusters ([2Fe-2S], [4Fe-4S], and [3Fe-4S]). This enzyme catalyzes both fumarate reduction and succinate oxidation with reduction occurring 30 times faster than oxidation. The QFR structure consists of a dimer of three subunits ($A_2B_2C_2$), which is secured by three hydrogen bonds between the C subunits. Subunit A contains the FAD-binding site. At the C-terminal domain of the subunit B are the [3Fe4S] and [4Fe4S] clusters with the [2Fe2S] cluster in the N-terminal domain. The subunit C is imbedded in the membrane with five trans-membrane helices and contains two *b*-type hemes. Structural comparisons between the QFR from *D. gigas, Wolinella succinogenes*, and *Escherichia coli* reveal that QFR from *D. gigas* is unique in that it has a menaquinone molecule bound near one of the heme *b* (Guan et al. 2018).

4.7.6 Molybdopterin Oxidoreductase

The molybdopterin oxidoreductase (Mop) of *D. alaskensis* (formerly *desulfuricans*) G20 is a membrane complex that is involved in the transfer of electrons from periplasmic cytochrome c_3 to the menaquinone (Li et al. 2009). The *mop* operon encodes for MopABCD (genes Dde_2932 to Dd_2935) with MopABC subunits located in the periplasm, while MopD is a transmembrane protein subunit. The MopA subunit (23.8 kDa and 209 amino acids) has six (CXXCH) motifs for the binding of six *c*-type hemes. The MopB subunit (73.2 kDa and 689 amino acids) contains a putative binding site for molybdopterin. The MopC subunit (28.8 kDa and 255 amino acids) includes three motifs characteristically used to bind iron-sulfur clusters and one (CXXCH) motif for cytochrome c_3. The transmembrane MopD subunit (47 kDa and 418 amino acids) has 10 transmembrane helices, and this subunit may bind cytochrome *b* for the interaction with menaquinone. The presence of heme *b* in MopD awaits confirmation. It is proposed that Mop transfers electrons from periplasmic TpI-c_3 to menaquinone while the Qmo and Dsr complexes would complete the transfer electrons from reduced menaquinone to adenosine phosphosulfate and sulfite.

4.7.7 Catalase

Catalase is a cytoplasmic enzyme that degrades hydrogen peroxide into O_2 and H_2O. An activity has been demonstrated in a few SRB, including *D. fairfieldensis* (Warren et al. 2005), *D. vulgaris* Hildenborough (Hatchikian et al. 1977), *Desulfomicrobium norvegicum* (Hatchikian et al. 1977), *D. oxyclinae* (Krekeler et al. 1998), *D. gigas* (Dos Santos et al. 2000), *D. vulgaris* Miyazaki F (Kitamura et al. 2001), *D. piger* Vib-7 (Kushkevych et al. 2014), and *Desulfomicrobium* sp. Rod-9 (Kushkevych et al. 2014). The catalase gene from *D. vulgaris* Miyazaki F has been cloned and expressed in *Escherichia coli* (Kitamura et al. 2001). The catalase purified from *D. gigas* is a homotrimer of three subunits of 61 kDa with one high-spin ferric heme per molecule (Dos Santos et al. 2000).

4.8 Synthesis of Heme

The synthesis of heme has been examined in several eukaryotic systems, aerobic microorganisms, photosynthetic bacteria, and SRB. The general metabolic pathway for heme synthesis of bacteria, where 5-aminolevulinic acid (ALA) is the precursor for tetrapyrrole-derived cofactors, has been reviewed by Fauque and Barton (2012) and by Lobo et al. (2012). In many bacteria and eukaryotes, ALA is produced from succinyl-CoA combined with glycine, but with *D. vulgaris* Hildenborough (Ishida et al. 1998) and numerous other anaerobes, the precursor for ALA is glutamyl-tRNA with glutamate-1-semialdehyde (GSA) as an intermediate. As indicated in Fig. 4.6,

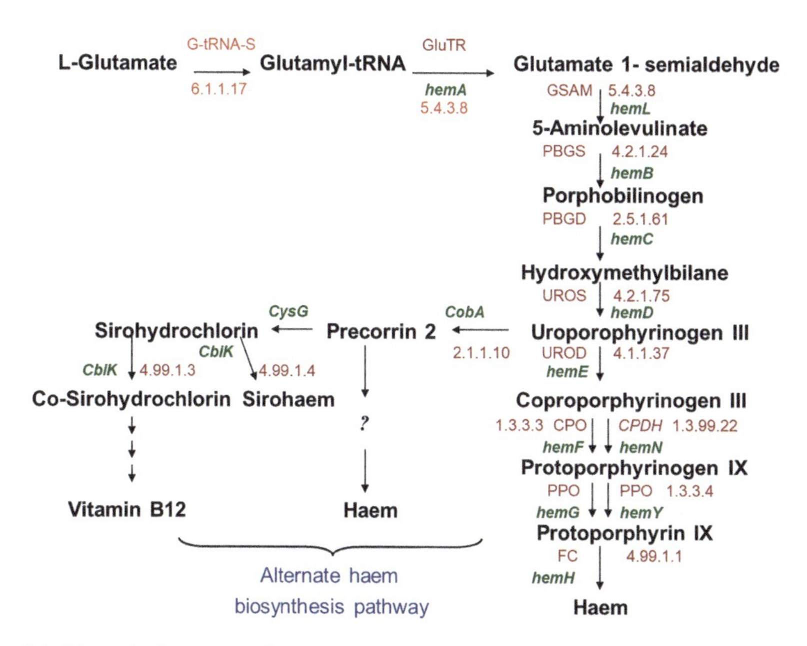

Fig. 4.6 Biosynthetic pathway for heme biosynthesis in prokaryotes. Genes and corresponding enzymes are as follows: hemA glutamyl-t-RNA reductase, hemL glutamate-1-semialdehyde-2,1-aminomutase, hemB porphobilinogen synthase, hemC porphobilinogen deaminase, hemD uroporphyrinogen III synthase, hemE uroporphyrinogen III decarboxylase, hemN and hemF coproporphyrinogen III dehydrogenase, hemG (oxygen-dependent) and hemY (oxygen-independent) protoporphyrinogen IX oxidase, hemH ferrochelatase, cobA uroporphyrinogen III C-methyltransferase, cysG precorrin-2 dehydrogenase, and cbiK cobaltochelatase. Enzyme Commission numbers are listed for each catalytic step. (Reproduced by permission from Fauque and Barton (2012), *Copyright 2012, with permission from Elsevier*)

the product of *hemA* (GSA-1-semialdehyde reductase) and *hemL* (GSA aminotransferase) participates in the sequential conversion of glutamyl-tRNA to ALA. In *D. vulgaris* Hildenborough, the product of *hemB* (ALA dehydratase, also called porphobilinogen synthase) produces porphobilinogen from the condensation of two ALA molecules (Lobo et al. 2009). With the production of uroporphyrinogen III (Fig. 4.6), heme synthesis may result from the classic pathway with protoporphyrinogen IX as an intermediate or by an alternate pathway involving precorrin-2. A report detailing the synthesis of heme *b*, heme *c*, and heme *d* from precorrin-2 in *D. vulgaris* Hildenborough has been published by Lobo et al. (2012). The production of protoporphyrin IX from protoporphyrinogen IX in *D. gigas* has been reported by Klemm and Barton (1985, 1987). As discussed by Fauque and Barton (2012), several of the enzymes for heme synthesis are bifunctional, and isolated enzymes must be sequenced to identify the genes responsible for their production.

Other complex cofactors present in SRP include siroheme and cobalamin (vitamin B_{12}). It is considered that in SRB, the production of siroheme is derived from precorrin-2 with sirohydrochlorin as an intermediate. Several heme-containing proteins have been isolated from SRB, and they include the following: Sulfite reductases in SRB contain siroheme, and these enzymes are referred to as desulforubidin in *D. desulfuricans* (Lee et al. 1973a); desulfoviridin in *D. gigas, D. vulgaris*, and *D. salexigens* (Lee and Peck Jr 1971; Lee et al. 1973b; Czechowski et al. 1986); P-582 in *Dst. nigrificans* (Akagi and Adams 1973; Trudinger 1970); and desulfofuscidin in *D. thermophilus, Thermodesulfobacterium commune*, and *Thermodesulfobacterium mobile* (Hatchikian and Zeikus 1983; Fauque et al. 1990). Another sulfite reductase in *Desulfovibrio* contains siroamide (Matthews et al. 1995). Other heme-containing enzymes include oxygen reductase in *D. vulgaris* (Lobo et al. 2008a), methyl-accepting protein involved with O_2 sensing in *D. vulgaris* (Fu et al. 1994) cobaltochelatase in *D. vulgaris* (Lobo et al. 2008b), and catalase in *D. vulgaris (*Dos Santos et al. 2000). Additional heme-containing proteins include rubredoxin in *D. gigas* (Timkovich et al. 1994) and bacterioferritin in *D. desulfuricans* (Romão et al. 2000).

The synthesis of cobalamin from sirohydrochlorin requires the insertion of cobalt to produce Co-sirohydrochlorin. This insertion of Co employs cobaltochelatase (CbiK), and *D. vulgaris* Hildenborough has two CbiKs. One cobaltochelatase ($CbiK^C$) functions in the cytoplasmic biosynthesis of cobalamin, and the other cobaltochelatase ($CbiK^P$) resides in the periplasm and is proposed to have a role in iron transport (Lobo et al. 2008b). The role of chelatases in the synthesis of heme proteins has been reviewed by Romão et al. (2011). As discussed in the review by Fauque and Barton (2012), there are several unresolved issues pertaining to heme synthesis in SRP, and continued research is needed to complete this story.

4.9 Conclusion and Perspective

The SRB have a wealth of iron-sulfur and heme proteins, which facilitate the movement of electrons, and the number of these proteins varies with the SRB species. While each species has multiple forms of hydrogenases and different classes ([FeFe], [NiFe], and [NiFeSe]), each hydrogenase would appear to have a specific role in metabolism, including activities other than hydrolysis. While the [NiFeSe] hydrogenase may be considered to be a type of [NiFe] hydrogenase, this use of selenium in hydrogenase appears to be unique to the SRB. Although hydrogenase activity is a hallmark characteristic of SRB, hydrogenase is not an absolute requirement for sulfate respiration, since it is absent in one species (*D. sapovorans*). SRB frequently possess multiple formate dehydrogenases, and the specific role in metabolism involves the location of this enzyme in the cell. While ferredoxin and flavodoxin are low-potential redox proteins, they are not redundant electron carriers, but each has a specific role in metabolism. Rubredoxin and other high-potential redox proteins have been isolated from SRB, and a physiological activity of these proteins may include oxidative stress response. Each of the An abundance of cytochromes *c* is characteristic of SRB, and the presence of multiheme cytochromes enables these bacteria to interact with numerous substrates. While some cytochromes *c* may have resulted from partial gene duplication, the mechanisms accounting for gene development is unresolved. The presence of cytochrome *b* associated with the reduction of fumarate reflects a similarity to the electron transport complex II in mitochondrial-like systems. In aerobic respiration, SRB employ cytochromes *b* complexed with other cytochromes in the membrane to transfer electrons from diverse electron donors to O_2. Just as the isoeflation of nitrite reductase provided definitive proof for the enzyme and that nitrite wasn't reduced by a sulfite reductase, the isolation of nitrate reductase from SRB clarified the range of the final electron acceptors in addition to sulfur compounds. Thus, some SRB strains were demonstrated to have enzymes necessary for the reduction of nitrate and nitrite when sulfate in the environment was limiting.

References

Abreu IA, Lourenço A, Xavier AY, LeGall J, Coelho AY, Matias PM et al (2003) A novel iron centre in the split-Soret cytochrome *c* from *Desulfovibrio desulfuricans* ATCC 27774. Bio Inorg Chem 8:360–370

Abreu IA, Saraiva LM, Soares CM, Teixeira M, Cabelli DE (2001) The mechanism of superoxide scavenging by *Archaeoglobus fulgidus* neelaredoxin. J Biol Chem 276:38995–39001

Ackrell BAC, Asato RN, Mower HF (1966) Multiple forms of bacterial hydrogenases. J Bacteriol 92:828–838

Adman ET, Sieker LC, Jensen LH, Bruschi M, Le Gall J (1977) A structural model of rubredoxin from *Desulfovibrio vulgaris* at 2 A resolution. J Mol Biol 112:113–120

Akagi JM (1967) Electron carriers for the phosphoroclastic reaction of *Desulfovibrio desulfuricans*. J Biol Chem 242:2478–2483

Akagi JM, Adams V (1973) Isolation of a bisulfite reductase activity from *Desulfotomaculum nigrificans* and its identification as the carbon monoxide-binding pigment P582. J Bacteriol 116: 372–396

Almendra MJ, Brondino CD, Gavel O, Pereira AS, Tavares P, Bursakov S et al (1999) Purification and characterization of a tungsten-containing formate dehydrogenase from *Desulfovibrio gigas*. Biochemist 38:16366–16372

Almeida MG, Macieira S, Gonçalves LL, Huber R, Cunha CA, Romão MJ et al (2003) The isolation and characterization of cytochrome *c* nitrite reductase subunits (NrfA and NrfH) from *Desulfovibrio desulfuricans* ATCC 27774 – Re-evaluation of the spectroscopic data and redox properties. Eur J Biochem 270:3904–3915

Ambler RP (1991) Sequence variability in bacterial cytochromes *c*. Biochim Biophys Acta 1058: 42–47

Archer M, Carvalho AL, Teixeira S, Moura I, JJG M, Rusnak F, Romão MJ (1999) Structural studies by X-ray diffraction on metal substituted desulforedoxin, a rubredoxin-type protein. Protein Sci 8:1536–1545

Archer M, Huber R, Tavares P, Moura I, Moura JJG, Carrondo MA et al (1995) Crystal structure of desulforedoxin from *Desulfovibrio gigas* determined at 1.8 Å resolution: a novel non-heme iron protein structure. J Mol Biol 251:690–702

Arendsen SJ, Van Dongen WMAM, Hagen WR (1995) Characterization of a ferredoxin from *Desulfovibrio vulgaris* (Hildenborough) that interacts with RNA. Eur J Biochem 231:352–357

Armstrong FA, George SJ, Cammack R, Hatchikian EC, Thomson AJ (1989) Electrochemical and spectroscopic characterization of the 7Fe form of ferredoxin III from *Desulfovibrio africanus*. Biochem J 264:265–273

Ascenso C, Rusnak F, Cabrito I, Lima MJ, Naylor S, Moura I, Moura JJ (2000) Desulfoferrodoxin: a modular protein. J Biol Inorg Chem 5:720–729

Aubert C, Giudici-Orticoni M-T, Czjzek M, Haser R, Bruschi M, Dolla A (1998a) Structural and kinetic studies of the Y73E mutant of octaheme cytochrome c_3 (M_r = 26 000) from *Desulfovibrio desulfuricans* Norway. Biochemist 37:2120–2130

Aubert C, Leroy G, Bianco P, Forest E, Bruschi M, Dolla A (1998b) Characterization of the cytochromes *c* from *Desulfovibrio desulfuricans* G201. Biochem Biophys Res Commun 242: 213–218

Aubert C, Leroy G, Bruschi M, Wall JD, Dolla A (1997) A single mutation in the Heme 4 environment of *Desulfovibrio desulfuricans* Norway cytochrome c_3 (Mr 26,000) greatly affects the molecule reactivity. J Biol Chem 272:15128–15134

Avilan L, Roumezi B, Risoul V, Bernard CS, Kpebe A, Belhadjhassine M et al (2018) Phototrophic hydrogen production from a clostridial [FeFe] hydrogenase expressed in the heterocysts of the cyanobacterium *Nostoc* PCC 7120. Appl Microbiol Biotechnol 102:5775–5783

Baffert C, Kpebe A, Avilan L, Brugna M (2019) Hydrogenases and H_2 metabolism in sulfate-reducing bacteria of the *Desulfovibrio* genus. Adv Microbial Physiol. 74:143–189

Bamford VA, Angove HC, Seward HE, Thomson AJ, Cole JA, Butt JN, Hemmings AM, Richardson DJ (2002) Structure and spectroscopy of the periplasmic cytochrome *c* nitrite reductase from *Escherichia coli*. Biochemist 41:2921–2931

Barta BA, LeGall J, Moura JJ (1993) Aldehyde oxidoreductase activity in *Desulfovibrio gigas*: in vitro reconstitution of an electron-transfer chain from aldehydes to the production of molecular hydrogen. Biochemist 32:11559–11568

Barton LL, LeGall J, Odom JM, Peck HD Jr (1983) Energy coupling to nitrite respiration in the sulfate-reducing bacterium *Desulfovibrio gigas*. J Bacteriol 153:867–871

Barton LL, LeGall J, Peck HD Jr (1972) Oxidative phosphorylation in the obligate anaerobe, *Desulfovibrio gigas*. In: San Pietro A, Gest H (eds) Horizons of bioenergetics. Academic Press, New York, pp 35–51

Barton LL, Peck HD Jr (1970) Role of ferredoxin and flavodoxin in the oxidative phosphorylation catalyzed cell-free preparations of the anaerobe *Desulfovibrio gigas*. Bacteriol Proc. American Society for Microbiology, Washington, DC, p 134

Baymann F, Schoepp-Cothenet B, Duval S, Guiral M, Brugna M, Baffert C, Russell MJ, Nitschke W (2018) On the natural history of flavin-based electron bifurcation. Front Microbiol. https://doi.org/10.3389/fmicb.2018.01357

Bartsch RG (1978) Cytochromes. In: Clayton RK, Sistrom WR (eds) The photosynthetic bacteria. Plenum Press, New York, pp 249–279

Beinert H, Emptage MH, Dreyer JL, Scott RA, Hahn JE, Hodgson KO, Thomson A (1983) Iron-sulfur stoichiometry and structure of iron-sulfur clusters in three iron proteins. Evidence for (3Fe-4S) clusters. Proc Natl Acad Sci U S A 80:393–396

Beinert H, Thomson AJ (1983) Three iron clusters in iron-sulfur proteins. Arch Biochem Biophys 222:333–361

Bell RG, Lee JP, Peck HD Jr, LeGall. (1978) Reactivity of *Desulfovibrio gigas* toward artificial and natural electron donors or acceptors. Biochimie 60:315–320

Bento I, Teixeira VH, Baptista AM, Soares CM, Matias PM, Carrondo MA (2003) Redox-Bohr and other cooperativity effects in the nine-heme cytochrome c from *Desulfovibrio desulfuricans* ATCC 27774 – crystallograhic and modeling studies. J Biol Chem 278:36455–36469

Bertrand P, BruschiM, Denis M, Gayda J P, Manca F (1982) Cytochrome c-553 from *Desulfovibrio vulgaris*: Potentiometric characterization by optical and EPR studies. Biochem Biophys Res Commun 106:756–760

Bianco P, Haladjian J, Loutfi M, Bruschi M (1983) Comparative studies of monohemic bacterial C-type cytochromes. Redox and optical properties of *Desulfovibrio desulfuricans* Norway cytochrome $c_{553(550)}$ and *Pseudomonas aeruginosa* cytochrome c_{551}. Biochem Biophys Res Commun 113:526–530

Birrell JA, Wrede K, Pawlak K, Rodriguez-Maciá P, Rüdiger O, Reijerse EJ, Lubitz W (2016) Artificial maturation of the highly active heterodimeric [FeFe] hydrogenase from *Desulfovibrio desulfuricans* ATCC 7757. Israel J Chem 56:852

Blackledge M, Medvedeva S, Poncin M, Guerlesquin F, Bruschi M, Marion D (1995) Structure and dynamics of ferrocytochrome c_{553} from *Desulfovibrio vulgaris* studied by NMR spectroscopy and restrained molecular dynamics. J Mol Biol 245:661–681

Blokesch M, Paschos A, Theodoratou E, Bauer A, Hube M, Huth S, Böck A (2002) Metal insertion into NiFe-hydrogenases. Biochem Soc Trans 30:674–680

Blokesch M, Albracht SP, Matzanke BF, Drapal NM, Jacobi A, Böck A (2004) The complex between hydrogenase-maturation proteins HypC and HypD is an intermediate in the supply of cyanide to the active site iron of [NiFe]-hydrogenases. J Mol Biol 344:155–167

Böck A, Forchhammer K, Heider J, Leinfelder W, Sawers G, Veprek B, Zinoni F (1991) Selenocysteine: the 21st amino acid. Mol Microbiol 5:515–520

Böck A, King PW, Blokesch M, Posewitz MC (2006) Maturation of hydrogenases. Adv Microbiol Physiol 51:1–71

Bonomi F, Kurtz DM Jr, Cui X (1996) Ferroxidase activity of recombinant *Desulfovibrio vulgaris* rubrerythrin. J Biol Inorg Chem 1:67–72

Bortolus M, Costantini P, Doni D, Carbonera D (2018) Overview of the maturation machinery of the H-cluster of [FeFe]-hydrogenases with a locus on HydF. Int J Mol Sci 19:3118–3137

Bouvier-Lapierre G, Bruschi M, Bonicel L, Hatchikian EC (1987) Amino acid sequence of *Desulfovibrio africanus* ferredoxin III: a unique structural feature for accommodating iron-sulfur cluster. Biochem Biophys Acta. 913:20–26

Brennan L, Turner DL, Messias AC, Teodoro ML, LeGall J, Santos H, Xavier AV (2000) Structural basis for the network of functional cooperativities in cytochrome c_3 from *Desulfovibrio gigas*: solution structures of the oxidized and reduced states. J Mol Biol 298:61–82

Brenner SE (1997) Population statistics of protein structures: lessons from structural classifications. Curr Opin Struct Biol 7:369–376

Broderick JB, Byer AS, Duschene KS, Duffus BR, Betz JN, Shepard EM, Peters JW (2014) H-cluster assembly during maturation of the [FeFe]-hydrogenase. J Biol Inorg Chem 19:747–757

Brondino CD, Passeggi MC, Caldeira J, Almendra MJ, Feio MJ, Moura JJG, Moura I (2004) Incorporation of either molybdenum or tungsten into formate dehydrogenase from *Desulfovibrio alaskensis* NCIMB 13491: EPR assignment of the proximal iron-sulfur cluster to the pterin cofactor in formate dehydrogenases from sulfate-reducing bacteria. J Biol Inorg Chem 9:145–151

Bruschi M (1976a) The amino acid sequence of rubredoxin from the sulfate-reducing bacterium, *Desulfovibrio gigas*. Biochem Biophys Res Commun 70:615–621

Bruschi M (1976b) Non-heme iron proteins. The amino acid sequence of rubredoxin from *Desulfovibrio vulgaris*. Biochim Biophys Acta 434:4–17

Bruschi M (1979) Amino acid sequence of *Desulfovibrio gigas* ferredoxin: revisions. Biochem Biophys Res Commun 91:623–628

Bruschi M (1981) The primary structure of the tetraheme cytochrome c_3 from *Desulfovibrio desulfuricans* (strain Norway 4). Biochim Biophys Acta 671:219–224

Bruschi M (1994) Cytochrome c_3 (M_r 26,000) isolated from sulfate-reducing bacteria and its relationships to other polyhemic cytochromes from *Desulfovibrio*. Meth Enzymol. 243:140–165

Bruschi M, Guerlesquin F (1988) Structure, function and evolution of bacterial. FEMS Microbiol Rev 54:155–176

Bruschi MH, Guerlesquin FA, Bovier-Lapierre GE, Bonicel JJ, Couchoud PM (1985) Amino acid sequence of the [4Fe-4S] ferredoxin isolated from *Desulfovibrio desulfuricans* Norway. J Biol Chem 260:8292–8296

Bruschi M, Hatchikian EC (1982) Non heme iron proteins of *Desulfovibrio*. The primary structure of ferredoxin I from *D. africanus*. Biochimie 64:503–507

Bruschi M, Hatchikian EC, Golovela L, LeGall J (1977) Purification and characterization of cytochrome c_3, ferredoxin, and rubredoxin isolated from *D. desulfuricans* Norway. J Bacteriol 129:30–38

Bruschi M, Hatchikian EC, Le Gall J, Moura JJG, Xavier AV (1976a) Purification, characterization and biological activity of three forms of ferredoxin from the sulfate-reducing bacterium *D. gigas*. Biochim Biophys Acta 449:275–284

Bruschi M, Hatchikian EC, LeGall J, Moura JJG, Xavier AV (1976b) Purification and characterization of cytochrome c_3, ferredoxin, and rubredoxin isolated from *D. desulfuricans* Norway. J Bacteriol 129:30–38

Bruschi M, LeGall J (1972) Purification et propriétés d'une rubrédoxine isolée à partir *D. vulgaris* (souche NCIB 8380). Biochim Biophys Acta 263:279–284

Bruschi M, LeGall J, Dus K (1970) C-Type cytochromes of *Desulfovibrio vulgaris* amino acid composition and end groups of cytochrome c_{553}. Biochem Biophys Res Commun 38:607–616

Bruschi M, Le Gall J, Hatchikian CE, Dubourdieu M (1969) Cristallisation et propriétées d'un cytochrome intervenant dans la réduction du thiosulfate par *Desulfovibrio gigas*. Bull Soc Fr Physiol Veg 15:381–390

Bruschi M, Leroy G, Bonicel J, Campese D, Dolla A (1996) The cytochrome c_3 superfamily: amino acid sequence of a dimeric octahaem cytochrome c_3 (M_r 26,000) isolated from *Desulfovibrio gigas*. Biochem J 320:933–938

Bruschi M, Moura I, Le Gall J, Xavier AV, Sieker LC (1979) The amino acid sequence of desulforedoxin, a new type of non-heme iron protein from D*esulfovibrio gigas*. Biochem Biophys Res Commun 90:596–605

Buckel W, Thauer RK (2013) Energy conservation via electron bifurcating ferredoxin reduction and proton/Na+ translocating ferredoxin oxidation. Biochim Biophys Acta Bioenerg 1827:94–113

Busch JLH, Breton JL, Bartlett BM, Armstrong FA, James R, Thomson AJ (1997) [3Fe-4S] $\leftrightarrow$ [4Fe-4S] cluster interconversion in *Desulfovibrio africanus* ferredoxin III: properties of an Asp14-->Cys mutant. Biochem J 323:95–102

Butt JN, Armstrong F, Breton J, George SJ, Thomson AJ, Hatchikian EC (1991) Investigation of metal ion uptake reactivities of [3Fe-4S] clusters in proteins: voltammetry of co-adsorbed ferredoxin-aminocyclitol films at graphite electrodes and spectroscopic identification of transformed clusters. J Amer Chem Soc 113:6663–6670

Butt JN, Fawcett SEJ, Breton J, Thomson AJ, Armstrong FA (1997) Electrochemical potential and pH dependences of [3Fe-4S] ↔ [M3Fe-4S] cluster transformations (M = Fe, Zn, Co, and Cd) in ferredoxin III from *Desulfovibrio africanus* and detection of a cluster with M = Pb. J Am Chem Soc 119:9729–9737

Cadby IT, Faulkner M, Cheneby J, Long J, van Helden J, Dolla A, Cole JA (2017) Coordinated response of the *Desulfovibrio desulfuricans* 27774 transcriptome to nitrate, nitrite and nitric oxide. Sci Rep 2017(7):16228. https://doi.org/10.1038/s41598-017-16403-4

Caffrey SM, Park H-S, Voordouw JK, He Z, Zhou J, Voordouw G (2007) Function of periplasmic hydrogenases in the sulfate-reducing bacterium *Desulfovibrio vulgaris* Hildenborough. J Bacteriol 189:6159–6167

Caldeira J, Palma PN, Regalla M, Lampreia J, Calvete J, Schäfer W et al (1994) Primary sequence, oxidation-reduction potentials and tertiary-structure prediction of *Desulfovibrio desulfuricans* ATCC 27774 flavodoxin. Eur J Biochem 220:987–995

Cammack R, Fernandez VM, Hatchikian EC (1994) Nickel-iron hydrogenase. Meth Enzymol 243: 43–68

Cammack R, Rao KK, Hall DO, Moura JJG, Xavier A, Bruschi M et al (1977) Spectroscopic studies of the oxidation-reduction properties of three forms of ferredoxin from *Desulfovibrio gigas*. Biochim Biophys Acta 490:311–321

Campbell IJ, Bennett GN, Silberg JJ (2019) Evolutionary relationships between low potential ferredoxin and flavodoxin electron carriers. Front Energy Res 7:79. https://doi.org/10.3389/fenrg.2019.00079

Casalot L, Hatchikian CE, Forget N, de Philip P, Dermoun Z, Bélaïch JP, Rousset M (1998) Molecular study and partial characterization of iron-only hydrogenase in *Desulfovibrio fructosovorans*. Anaerobe 4:45–55

Chan K-H, Li T, Wong C-O, Wong K-B (2012a) Structural basis for GTP-dependent dimerization of hydrogenase maturation factor HypB. PLoS One 7(1):e30547. https://doi.org/10.1371/journal.pone.0030547

Chan K-H, Lee KM, Wong K-B (2012b) Interaction between hydrogenase maturation factors HypA and HypB is required for [NiFe]-hydrogenase maturation. PLoS One 7(2):e32592. https://doi.org/10.1371/journal.pone.0032592. Epub 2012 Feb 27

Chang FC, Swenson RP (1997) Regulation of oxidation–reduction potentials through redox-linked ionization in the Y98H mutant of the *Desulfovibrio vulgaris* [Hildenborough] flavodoxin: direct proton nuclear magnetic resonance spectroscopic evidence for the redox-dependent shift in the pK_a of histidine-98. Biochemist 6:9013–9021

Chen L, Chen MY, LeGall L (1993) Isolation and characterization of flavoredoxin, a new flavoprotein that permits in vitro reconstitution of an electron transfer chain from molecular hydrogen to sulfite reduction in the bacterium *Desulfovibrio gigas*. Arch Biochem Biophys 303:44–50

Chen C-Y, Lin Y-H, Huang Y-C, Liu M-Y (2006) Crystal structure of rubredoxin from *Desulfovibrio gigas* to ultra-high 0.68 Å resolution. Biochem Biophys Res Commun 349:79–90

Chen L, Liu MY, LeGall J, Fareleira P, Santos H, Xavier AV (1993a) Rubredoxin oxidase, a new flavo-hemo-protein, is the site of oxygen reduction to water by the "strict anaerobe" *Desulfovibrio gigas*. Biochem Biophys Res Commun 193:100–110

Chen L, Liu MY, LeGall J, Fareleira P, Santos H, Xavier AV (1993b) Purification and characterization of an NADH-rubredoxin oxidoreductase involved in the utilization of oxygen by *Desulfovibrio gigas*. Eur J Biochem 216:443–448

Chen L, Liu M-Y, LeGall J (1995) Characterization of electron transfer proteins. In: Barton LL (ed) Sulfate-reducing bacteria. Plenum Press, New York, pp 113–149

Chen L, Pereira MM, Teixeira M, Xavier AV, Le Gall J (1994a) Isolation and characterization of a high molecular weight cytochrome from the sulfate reducing bacterium *Desulfovibrio gigas*. FEBS Lett 347:295–299

Chen L, Sharma P, Le Gall J, Mariano AM, Teixeira M, Xavier AV (1994b) A blue non-heme iron protein from *Desulfovibrio gigas*. Eur J Biochem 226:613–618

Chongdar N, Birrell JA, Pawlak K, Sommer C, Reijerse EJ, Rüdiger O, Lubitz W, Ogata H (2018) Unique spectroscopic properties of the H-cluster in a putative sensory [FeFe] hydrogenase. J Am Chem Soc 140:1057–1068

Clark ME, He Z, Redding AM, Joachimiak MP, Keasling JD, Zhou JZ et al (2012) Transcriptomic and proteomic analyses of *Desulfovibrio vulgaris* biofilms: carbon and energy flow contribute to the distinct biofilm growth state. BMC Genomics 13:138. https://doi.org/10.1186/1471-2164-13-138

Coelho AV, Matias P, Fulop V, Thompson A, Gonzalez A, Carrondo MA (1997) Desulfoferrodoxin structure determined by MAD phasing and refinement to 1.9 Angstroms resolution reveals a unique combination of a tetrahedral FeS_4 centre with a square pyramidal FeS_4 centre. J Biol Inorg Chem 2:680–689

Costa C, Moura JJG, Moura I, Liu MY, Peck HD Jr, LeGall J, Wang Y, Huynh BH (1990) Hexaheme nitrite reductase from *Desulfovibrio desulfuricans* – Mössbauer and EPR characterization of the heme groups. J Biol Chem 265:143132–114387

Costa C, Moura JJG, Moura I, Wang Y, Huynh BH (1996) Redox properties of cytochrome *c* nitrite reductase from *Desulfovibrio desulfuricans* ATCC 27774. J Biol Chem 271:23191–23196

Costa C, Teixeira M, LeGall J, Moura JJG, Moura I (1997) Formate dehydrogenase from *Desulfovibrio desulfuricans* ATCC 27774: isolation and spectroscopic characterization of the active sites (heme, iron-sulfur centers and molybdenum). J Biol Inorg Chem 2:198–208

Coulter ED, Kurtz DM Jr (2001) A role for rubredoxin in oxidative stress protection in *Desulfovibrio vulgaris:* catalytic electron transfer to rubrerythrin and two-iron superoxide reductase. Arch Biochem Biophys 394:76–86

Coutinho IB, Xavier AV (1994) Tetraheme cytochromes. Meth Enzymol 243:119–155

Cunha CA, Macieira S, Dias JM, Almeida G, Goncalves LL, Costa C, Lampreia J, Huber R, Moura JJ, Moura I, Romão MJ (2003) Cytochrome *c* nitrite reductase from *Desulfovibrio desulfuricans* ATCC 27774. The relevance of the two calcium sites in the structure of the catalytic subunit (NrfA). Biol Chem 278:17455–17465

Curley GP, Carr MC, Mayhew SC, Voordouw G (1991) Redox and flavin-binding properties of recombinant flavodoxin from *Desulfovibrio vulgaris* (Hildenborough). Eur J Biochem 202: 1091–1100

Curley GP, Voordouw G (1988) Cloning and sequencing of the gene encoding flavodoxin from *Desulfovibrio vulgaris* Hildenborough. FEMS Microbiol Lett 49:295–299

Cusanovich MA, Bartsch RG (1969) A high potential cytochrome *c* from *Chromatium* chromatophores. Biochim Biophys Acta 189:245–255

Czjzek M, Arnoux P, Haser R, Shepard W (2001) Structure of cytochrome c_7 from *Desulfuromonas acetoxidans* at 1.9 Å resolution. Acta Crystallogr D Biol Crystallogr 57:670–678

Czjzek M, El Antak L, Zamboni V, Morelli X, Dolla A, Guerlesquin F, Bruschi M (2002) The crystal structure of the hexadeca-heme cytochrome Hmc and a structural model of its complex with cytochrome c_3. Structure 10:1677–1686

Czjzek M, Guerlesquin F, Bruschi M, Haser R (1996) Crystal structure of a dimeric octaheme cytochrome c_3 (M_r 26000) from *Desulfovibrio desulfuricans* Norway. Structure 4:395–404

Czaja C, Litwiller R, Tomlinson AJ, Naylor S, Tavares P, LeGall J, Moura JJG, Moura I, Rusnak F (1995) Expression of *Desulfovibrio gigas* desulforedoxin in *Escherichia coli.* Purification and characterization of mixed metal isoforms. J Biol Chem 270:20273–20277

Czechowski M, Fauque G, Galliano N, Dimon B, Moura I, Moura JJG, Xavier AV, Barato BAS, Lino AR, LeGall J (1986) Purification and characterization of three proteins from a halophilic sulfate-reducing bacterium, *Desulfovibrio salexigens*. J Ind Microbiol Biotechnol 1:139–147. https://doi.org/10.1007/BF01569265

Darwin A, Hussain H, Griffiths L, Grove J, Sambongi Y, Busby S, Cole J (1993) Regulation and sequence of the structural gene for cytochrome c_{552} from *Escherichia coli*: not a hexahaem but a 50 kDa tetrahaem nitrite reductase. Mol Microbiol 9:1255–1265 da Costa PN, Marujo PE, van Dongen W, Arraiano CM, Saraiva LM (2000) Cloning, sequencing and expression of the tetraheme cytochrome c_3 from *Desulfovibrio gigas*. Biochim Bipphys Acta 1492:271–275

da Costa PN, Romão CV, LeGall J, Xavier AV, Melo E, Teixeira M, Saraiva LM (2001) The genetic organization of *Desulfovibrio desulphuricans* ATCC 27774 bacterioferritin and rubredoxin-2 genes: involvement of rubredoxin in iron metabolism. Mol Microbiol 41:217–227

da Silva SM, Pimentel C, Valente FMA, Rodrigues-Pousada C, Pereira IAC (2011) Tungsten and molybdenum regulation of formate dehydrogenase expression in *Desulfovibrio vulgaris* Hildenborough. J Bacteriol 193:2909–2917

da Silva SM, Voordouw J, Leitão C, Martins M, Voordouw G, Pereira IA (2013) Function of formate dehydrogenases in *Desulfovibrio vulgaris* Hildenborough energy metabolism. Microbiol. 159:1760–1769

de Bok FA, Hagedoorn PL, Silva PJ, Hagen WR, Schiltz E, Fritsche K, Stams AJ (2003) Two W-containing formate dehydrogenases (CO_2-reductases) involved in syntrophic propionate oxidation by *Syntrophobacter fumaroxidans*. Eur J Biochem 270:2476–2485

deMaré F, Kurtz D, Nordlund P (1996) The structure of *Desulfovibrio vulgaris* rubrerythrin reveals a unique combination of rubredoxin-like FeS_4 and ferritin-like diiron domains. Nat Struct Mol Biol 3:539–546

Dalsgaard T, Bak F (1994) Nitrate reduction in a sulphate-reducing bacterium, *Desulfovibrio desulfuricans,* isolated from rice paddy soil: sulfide inhibition, kinetics and regulation. Appl Environ Microbiol 66:291–297

Dementin S, Burlat B, Fourmond V, Leroux F, Liebgott P-P, Hamdan AA et al (2011) Rates of intra- and intermolecular electron transfers in hydrogenase deduced from steady-state activity measurements. J Am Chem Soc 133:10211–10221

Dervartanian DV, Xavier AV, Le Gall J (1978) EPR determination of the oxidation-reduction potentials of the hemes in cytochrome c_3 from *Desulfovibrio vulgaris*. Biochimie 60:321–325

Devreese B, Costa C, Demol H, Papaefthymiou V, Moura, JJG M, Van Beeumen J (1997) The primary structure of the split-Soret cytochrome *c* from *Desulfovibrio desulfuricans* ATCC 27774 reveals an unusual type of diheme cytochrome *c*. J Eur J Biochem 248:445–451

Devreese B, Tavares P, Lampreia J, Van Damme N, Le Gall J, Moura JJ, Van Beeumen J, Moura I (1996) Primary structure of desulfoferrodoxin from *Desulfovibrio desulfuricans* ATCC 27774, a new class of non-heme iron proteins. FEBS Lett 385:138–142

Dias JM, Than ME, Humm A, Huber R, Bourenkov GP, Bartunik HD et al (1999) Crystal structure of the first dissimilatory nitrate reductase at 1.9 Å solved by MAD methods. Structure 7:65–79

Di Paolo RE, Pereira PM, Gomes I, Valente FM, Pereira IA, Franco R (2006) Resonance Raman fingerprinting of multiheme cytochromes from the cytochrome c_3 family. J Biol Inorg Chem 11: 217–224

Dolla A, Pohorelic BK, Voordouw JK, Voordouw G (2000) Deletion of the *hmc* operon of *Desulfovibrio vulgaris* subsp. *vulgaris* Hildenborough hampers hydrogen metabolism and low-redox-potential niche establishment. Arch Microbiol 174:143–151

Dong X, Dröge J, von Toerne C, Marozava S, McHardy AC, Meckenstock RU (2017) Reconstructing metabolic pathways of a member of the genus *Pelotomaculum* suggesting its potential to oxidize benzene to carbon dioxide with direct reduction of sulfate. FEMS Microbiol Ecol 93(3):fiw254. https://doi.org/10.1093/femsec/fiw254

Dorosh LS, Peretyatk TB, Hudz SP (2016) Nitrate reductase activity of sulphate-reducing bacteria *Desulfomicrobium sp.* CrR3 at the different conditions of the cultivation. Biotechnologia Acta 9(1):2016. https://doi.org/10.15407/biotech9.01.097

Dos Santos WG, Pacheco I, Liu MY, Teixeira M, Xavier AV, LeGall J (2000) Purification and characterization of an iron superoxide dismutase and a catalase from the sulfate-reducing bacterium *Desulfovibrio gigas*. J Bacteriol 182:796–804

Dubourdieu M, Fox JM (1977) Amino acid sequence of *Desulfovibrio vulgaris* flavodoxin. J Biol Chem 252:1453–1462

Dubourdieu M, Le Gall J (1970) Chemical study of two flavodoxins extracted from sulfate-reducing bacteria. Biochem Biophys Res Commun 38:965–972
Dubourdieu M, Le Gall J, Favaudon V (1975) Physicochemical properties of flavodoxin from *Desulfovibrio vulgaris*. Biochem Biophys Acta. 376:519–532
Dubourdieu M, LeGall J, Fox JL (1973) The amino acid sequence of *Desulfovibrio vulgaris* flavodoxin. Biochem Biophys Res Commun 52:1418–1425
Efremov RG, Sazanov LA (2012) The coupling mechanism of respiratory complex I - a structural and evolutionary perspective. Biochim Biophys Acta 1817:1785–1795
Eidsness MK, Scott RA, Prickril BC, DerVartanian DV, LeGall J, Moura I, Moura JJ, Peck HD Jr (1989) Evidence for selenocysteine coordination to the active site nickel in the [NiFeSe] hydrogenases from Desulfovibrio baculatus. Proc Natl Acad Sci U S A 86:147–151
Eng LH, Neujahr HY (1989) Purification and characterization of periplasmic C-type cytochromes from *Desulfovibrio desulfuricans* (NCIMB 8372). Arch Microbiol 153:60–63
Einsle O, Messerschmidt A, Stach P, Bourenkov GP, Bartunik HD, Huber R, Kroneck PM (1999) Structure of cytochrome *c* nitrite reductase. Nature 400:476–480
Einsle O, Stach P, Messerschmidt A, Simoni J, Krögeri A, Huber R, Kroneck PMH (2000) Cytochrome *c* nitrite reductase from *Wolinella succinogenes* structure at 1.6 Å resolution, inhibitor binding, and heme-packing motifs. J Biol Chem 275:39608–39616
Fauque G, Barton LL (2012) Hemoproteins in dissimilatory sulfate- and sulfur-reducing prokaryotes. Adv Microb Physiol 30:2–92
Fauque G, Bruschi M, LeGall J (1979a) Purification and some properties of cytochrome $c_{553(550)}$ isolated from *Desulfovibrio desulfuricans* Norway. Biochem Biophys Res Commun 86:1020–1029
Fauque G, Herve D, LeGall J (1979b) Structure-function relationship in hemoproteins: the role of cytochrome c_3 in the reduction of colloidal sulfur by sulfate-reducing bacteria. Arch Microbiol 121:261–264
Fauque GD, Klimmer O, Kröger A (1994) Sulfur reductases from spirilloid mesophilic sulfur-reducing eubacteria. Meth Enzymol. 243:367–386
Fauque G, LeGall J (1988) Dissimilatory reduction of sulfur compounds. In: Zehnder AJ (ed) Biology of anaerobic microorganisms. Wiley, New York, pp 587–639
Fauque G, LeGall J, Barton LL (1991) Sulfate-reducing and sulfur-reducing bacteria. In: Shively JM, Barton LL (eds) Variations in autotrophic life. Academic Press, New York, pp 271–337
Fauque G, Lino AR, Czechowski M, Kang L, DerVartanian DV, Moura JJG, LeGall J, Moura I (1990) Purification and characterization of bisulfite reductase (desulfofuscidin) from *Desulfovibrio thermophilus* and its complexes with exogenous ligands. Biochim Biophys Acta 1040:112–118
Fauque GD, Moura I, Moura JJG, Xavier AV, Galliano N, LeGall J (1987) Isolation and characterization of a rubredoxin and flavodoxin from *Desulfovibrio desulfuricans* Berre-Eau. FEBS Lett 215:63–67
Fauque G, Peck HD Jr, Moura JJG, Huynh BH, Berlier Y, DerVartanian DV et al (1988) The three classes of hydrogenases from sulfate-reducing bacteria of the genus *Desulfovibrio*. FEMS Microbiol Rev 4:299–344
Frank YA, Kadnikov V, Anastasia P, Lukina AP, David Banks D, Beletsky AV, Mardanov AV et al (2016) Characterization and genome analysis of the first facultatively alkaliphilic *Thermodesulfovibrio* isolated from the Deep Terrestrial Subsurface. Front Microbiol 7(8365): 2000. https://doi.org/10.3389/fmicb.2016.02000
Frazão C, Morais J, Matias PM, Carrondo MA (1994) Crystallization and preliminary X-ray diffraction analysis of tetra-heme cytochrome c_3 from sulfate- and nitrate-reducing *Desulfovibrio desulfuricans* ATCC 27774. Acta Cryst D50:233–236
Frey M (2002) Hydrogenases: hydrogen-activating enzymes. Chembiochem 3:153–160
Frey M, Sieker L, Payan F, Haser R, Bruschi M, Pepe G, LeGall J (1987) Rubredoxin from *Desulfovibrio gigas*. A molecular model of the oxidized form at 1.4 A resolution. J Mol Biol 197:525–541

Friedrich B, Buhrke T, Burgdorf T, Lenz O (2005) A hydrogen-sensing multiprotein complex controls aerobic hydrogen metabolism in *Ralstonia eutropha*. Biochem Soc Trans 33:99–101

Fritz G, Griesshaber D, Seth O, Kroneck PM (2001) Nonaheme cytochrome *c*, a new physiological electron acceptor for [Ni,Fe] hydrogenase in the sulfate-reducing bacterium *Desulfovibrio desulfuricans* Essex: primary sequence, molecular parameters, and redox properties. Biochemist 40:1317–1324

Fu R, Wall JD, Voordouw G (1994) DcrA, a heme-containing methyl-accepting protein from *Desulfovibrio vulgaris* Hildenborough, senses the oxygen concentration or redox potential of the environment. J Bacteriol 176:344–350

Garcin E, Vernede X, Hatchikian EC, Volbeda A, Frey M, Fontecilla-Camps JC (1999) The crystal structure of a reduced [NiFeSe] hydrogenase provides an image of the activated catalytic center. Structure Fold Des 7:557–566

Giuffrè A, Borisov VB, Arese M, Sarti P, Forte E (2014) Cytochrome *bd* oxidase and bacterial tolerance to oxidative and nitrosative stress. Biochim Biophys Acta 1837:1178–1187

Goldet G, Brandmayr C, Stripp ST, Happe T, Cavazza C, Fontecilla-Camps JC, Armstrong FA (2009) Electrochemical kinetic investigations of the reactions of [FeFe]-hydrogenases with carbon monoxide and oxygen: comparing the importance of gas tunnels and active-site electronic/redox effects. J Am Chem Soc 131:14979–14989

Gomes CM, Vicente JB, Wasserfallen A, Teixeira M (2000) Spectroscopic studies and characterization of a novel electron-transfer chain from *Escherichia coli* involving a flavorubredoxin and its flavoprotein reductase partner. Biochemist 39:16230–16237

Grahame DA, DeMoll E (1995) Substrate and accessory protein requirements and thermodynamics of acetyl-CoA synthesis and cleavage in *Methanosarcina barkeri*. Biochemist 34:4617–4624

Greening C, Biswas A, Carere CR, Jackson CJ, Taylor MC, Stott MB, Cook GM, Morales SE (2016) Genomic and metagenomic surveys of hydrogenase distribution indicate H_2 is a widely utilized energy source for microbial growth and survival. ISME J 10:761–777

Guan H-H, Hsieh Y-C, Lin P-J, Huang Y-C, Yoshimura M, Chen L-Y, Chen S-K et al (2018) Structural insights into the electron/proton transfer pathways in the quinol:fumarate reductase from *Desulfovibrio gigas*. Sci Reports 8:14935. https://doi.org/10.1038/s41598-018-33193

Guerlesquin F, Bovier-Lapierre G, Bruschi M (1982) Purification and characterization of cytochrome c_3 (Mr 26,000) isolated from *Desulfovibrio desulfuricans* Norway strain. Biochem Biophys Res Commun 105:530–538

Guerlesquin F, Bruschi M, Bovier-Lapierre G, Fauque G (1980) Comparative study of two ferredoxins from *Desulfovibrio desulfuricans* Norway. Biochem Biophys Acta 626:127–135

Guerlesquin F, Moura JJG, Cammack R (1985) Iron-sulphur cluster composition and redox properties of two ferredoxins from *Desulfovibrio desulfuricans* Norway. Biochem Biophys Acta. 679:422–427

Gupta N, Bonomi F, Kurtz DM Jr, Ravi N, Wang DL, Huynh BH (1995) Recombinant *Desulfovibrio vulgaris* rubrerythrin. Isolation and characterization of the diiron domain. Biochemist 34:3310–3318

Hagen WR, van Berkel-Arts A, Wolters KM, Voordouw G, Veeger C (1986) The iron-sulfur composition of the active site of hydrogenase from *Desulfovibrio vulgaris* (Hildenborough) deduced from its subunit structure and total iron-sulfur content. FEBS 203:59–63

Haser R, Pierrot M, Frey M, Payan F, Astier JP, Bruschi M, LeGall J (1979) Structure and sequence of the multihaem cytochrome c_3. Nature (London) 282:806–810

Hatchikian EC (1975) Purification and properties of thiosulfate reductase from *Desulfovibrio gigas*. Arch Microbiol 105:249–256

Hatchikian EC, Bruschi M, Le Gall J (1978) Characterization of the periplasmic hydrogenase from *Desulfovibrio gigas*. Biochem Biophy Res Commun 82:451–461

Hatchikian EC, Forget N, Fernandez VM, Williams R, Cammack R (1992) Further characterization of the [Fe]-hydrogenase from *Desulfovibrio desulfuricans* ATCC 7757. Eur J Biochem 209: 357–365

Hatchikian CE, LeGall J, Bell GR (1977) Significance of superoxide dismutase and catalase activities in the strict anaerobes, sulfate reducing bacteria. In: Michelson AM, McCord JN, Fridovich I (eds) Superoxide and superoxide dismutases. Academic Press, New York, pp 159–172

Hatchikian EC, Papavassiliou P, Bianco P, Haladjian J (1984) Characterization of cytochrome c_3 from the thermophilic sulfate reducer *Thermodesulfobacterium commune*. J Bacteriol 159: 1040–1046

Hatchikian CE, Traore AS, Fernandez VM, Cammack R (1990) Characterization of the nickel-iron periplasmic hydrogenase from *Desulfovibrio fructosovorans*. Eur J Biochem 187:635–643

Hatchikian EC, Zeikus JG (1983) Characterization of a new type of dissimilatory sulfite reductase present in *Thermodesulfobacterium commune*. J Bacteriol 153:1211–1220

Hatfield DL, Gladyshev VN (2002) How selenium has altered our understanding of the genetic code. Mol Cell Biol 22:3565–3576

He SH, Dervartanian DV, LeGall J (1986) Isolation of fumarate reductase from *Desulfovibrio multispirans*, a sulfate-reducing bacterium. Biochem Biophys Res Commun 135:1000–1007

He SH, Teixeira M, LeGall J, Patil DS, Moura I, Moura JJG et al (1989) EPR studies with ^{77}Se-enriched [NiFeSe] hydrogenase of *Desulfovibrio baculatus*. J Biol Chem 264:2678–2682

Hedderich R (2004) Energy-converting [NiFe] hydrogenases from archaea and extremophiles: ancestors of complex I. J Bioenerg Biomembr 36:65–75

Hedderich R, Forzi L (2005) Energy-converting [NiFe] hydrogenases: more than just H_2 activation. J Mol Microbiol Biotechnol 10:92–104

Helms LR, Krey GD, Swenson RP (1990) Identification, sequence determination, and expression of the flavodoxin gene from *Desulfovibrio salexigens*. Biochem Biophys Res Commun 168:809–817

Helms LR, Swenson RP (1991) Cloning and characterization of the flavodoxin gene from *Desulfovibrio desulfuricans*. Biochim Biophys Acta 1089:417–419

Helms LR, Swenson RP (1992) The primary structures of the flavodoxins from two strains of *Desulfovibrio gigas*. Cloning and nucleotide sequence of the structure genes. Biochim Biophys Acta 1131:325–328

Higuchi Y, Inaka K, Yasuoka N, Yagi T (1987) Isolation and crystallization of high molecular weight cytochrome from *Desulfovibrio vulgaris* Hildenborough. Biochim Biophys Acta 911: 341–348

Higuchi Y, Kusunoki M, Matsuura Y, Yasuoka N, Kakudo M (1984) Refined structure of cytochrome c_3 at 1.8 Å resolution. J Mol Biol 172:109–139

Higuchi Y, Ogata H, Miki K, Yasuoka N, Yagi T (1999) Removal of the bridging ligand atom at the Ni-Fe active site of [NiFe] hydrogenase upon reduction with H2, as revealed by X-ray structure analysis at 1.4 Å resolution. Structure 7:549–556

Higuchi Y, Yagi T, Voordoow G (1994) Hexadecane cytochrome *c*. Meth Enzymol 243:155–165

Higuchi Y, Yagi T, Yasuoka N (1997) Unusual ligand structure in Ni-Fe active center and an additional Mg site in hydrogenase revealed by high resolution X-ray structure analysis. Structure 5:1671–1680

Hocking WP, Stokke R, Roalkvam I, Steen IH (2014) Identification of key components in the energy metabolism of the hyperthermophilic sulfate-reducing archaeon *Archaeoglobus fulgidus* by transcriptome analyses. Front Microbiol 5:95. 1. https://doi.org/10.3389/fmicb.2014.00095

Hsieh Y-C, Chia TS, Fun H-K, Chen C-J (2013) Crystal structure of dimeric flavodoxin from *Desulfovibrio gigas* suggests a potential binding region for the electron-transferring partner. Int J Mol Sci 14:1667–1683

Hube M, Blokesch M, Böck A (2002) Network of hydrogenase maturation in *Escherichia coli*: role of accessory proteins HypA and HybF. J Bacteriol 184:3879–3885

Huynh BH (1995) Mossbauer spectroscopy in study of cytochrome cd_1 from *Thiobacillus denitrificans*, desulfoviridin, and iron hydrogenase. Meth Enzymol 243:523–543

Huynh BH, Czechowski MH, Kruger H-J, DerVartanian DV, Peck HD Jr, LeGall J (1984) *Desulfovibrio vulgaris* hydrogenase: a nonheme iron enzyme lacking nickel that exhibits anomalous EPR and Mossbauer spectra. Proc Natl Acad Sci U S A 81:3728–3732

Huynh BH, Moura JJG, Moura I, Kent TA, Le Gall J, Xavier AV, Munck E (1980) Evidence of a three-iron center in a ferredoxin from *Desulfovibrio gigas*. J Biol Chem 255:3242–3244

Igarashi N, Moriyama H, Fujiwara T, Fukumori Y, Tanaka N (1997) The 2.8 Å structure of hydroxylamine oxidoreductase from a nitrifying chemoautotrophic bacterium, *Nitrosomonas europaea*. Nat Struct Mol Biol 4:276–284

Irie K, Kobayashi K, Kobayashi M, Ishimoto M (1973) Biochemical studies on sulfate-reducing Bacteria XII. Some properties of flavodoxin from *Desulfovibrio vulgaris*. J Biochem 73:353–366

Ishida T, Yu L, Akutu H, Ozawa K, Kawanishi S, Seto A, Inubushi T, Sano S (1998) A primitive pathway of porphyrin biosynthesis and enzymology in *Desulfovibrio vulgaris*. Proc Natl Acad Sci U S A 95:4853–4858

Ishimoto M, Koyama J, Nagi Y (1954a) Role of a cytochrome in thiosulfate reduction by a sulfate-reducing bacterium. Seikagaku Zasshi 26:303

Ishimoto M, Koyama J, Nagi Y (1954b) Biochemical studies on the sulfate-reducing bacteria. IV, the cytochrome system of sulfate-reducing bacteria. J Biochem (Tokyo) 41:763–77O

Iyer RB, Silaghi-Dumitrescu R, Kurtz DM, Lanzilotta WN (2005) High-resolution crystal structures of *Desulfovibrio vulgaris* (Hildenborough) nigerythrin: facile, redox-dependent iron movement, domain interface variability, and peroxidase activity in the rubrerythrins. J Biol Inorg Chem 10:407–416

Jones HE (1971) A re-examination of *Desulfovibrio africanus*. Arch Mikrobiol 80:78–86

Junier P, Junier T, Podell S, Sims DR, Detter JC, Lykidis A et al (2010) The genome of the gram-positive metal- and sulfate-reducing bacterium *Desulfotomaculum reducens* strain MI-1. Environ Microbiol 12:2738–2754

Keith SM, Herbert RA (1983) Dissimilatory nitrate reduction by a strain of *Desulfovibrio desulfuricans*. FEMS Microbiol Lett 18:55–59

Keller K, Wall JD (2011) Genetics and molecular biology of the electron flow for sulfate respiration in *Desulfovibrio*. Front Microbiol 2:135. https://doi.org/10.3389/fmicb.2011.00135

Keon RG, Voordouw G (1996) Identification of the HmcF and topology of the HmcB subunit of the Hmc complex of *Desulfovibrio vulgaris*. Anaerobe 2:231–238

Kent TA, Moura I, JJG M, Lipscomb JD, Huynh BH, LeGall J, Xavier AV, Münck E (1982) Conversion of [3Fe-3S] into [4Fe-4S] clusters in a *Desulfovibrio gigas* ferredoxin and isotopic labeling of iron-sulfur cluster subsites. FEBS Lett 138:55–58

Kim JH, Akagi JM (1985) Characterization of a trithionate reductase system from *Desulfovibrio vulgaris*. J Bacteriol 163:472–475

King NE, Winfield ME (1955) The assay of soluble hydrogenase. Biochim Biophys Acta 18:431–432

Kissinger CR, Adman ET, Sieker LC, Jensen LH, LeGall L (1989) The crystal structure of the three-iron ferredoxin II from *Desulfovibrio gigas*. FEBS Lett 244:47–450

Kissinger CR, Sieker LC, Adman ET, Jensen LH (1991) Refined crystal structure of ferredoxin II from *Desulfovibrio gigas* at 1.7 Å. J Mol Biol 219:693–715

Kitamura M, Koshino Y, Kamikawa Y, Kohno K, Kojima S, Miura K-i et al (1997) Cloning and expression of the rubredoxin gene from *Desulfovibrio vulgaris* (Miyazaki F) – comparison of the primary structure of desulfoferrodoxin. Biochim Biophys Acta (BBA) 1351:239–247

Kitamura M, Mizugai K, Taniguchi M, Akutsu H, Kumagai I, Nakaya T (1995) A gene encoding a cytochrome *c* oxidase-like protein is located closely to the cytochrome *c*-553 gene in the anaerobic bacterium, *Desulfovibrio vulgaris* (Miyazaki F). Microbiol Immunol 39:75–80

Kitamura M, Nakanishi T, Kojima S, Kumagai I, Inoue H (2001) Cloning and expression of the catalase gene from the anaerobic bacterium *Desulfovibrio vulgaris* (Miyazaki F). J Biochem 129:357–364

Kitamura M, Sagara T, Taniguchi M, Ashida M, Ezoe K, Kohno K, Kojim S (1998) Cloning and expression of the gene encoding flavodoxin from *Desulfovibrio vulgaris* (Miyazaki F). J Biochem 123:891–898

Klausner RD, Harford JB (1989) Cis-trans models for posttranscriptional gene regulation. Science 246:870–872

Klemm DJ, Barton LL (1985) Oxidation of protoporphyrinogen in the obligate anaerobe *Desulfovibrio gigas*. J Bacteriol 164:316–320

Klemm DJ, Barton LL (1987) Purification and properties of protoporphyrinogen oxidase from an anaerobic bacterium, *Desulfovibrio gigas*. J Bacteriol 169:5209–5215

Knight E Jr, Hardy RWF (1966) Isolation and characteristics of flavodoxin from nitrogen-fixing *Clostridium pasteurianum*. J Biol Chem 241:2752–2756

Koller KB, Hawkridge FM, Fague G, Le Gall J (1987) Direct electron transfer reactions of cytochrome-c_{553} from *Desulfovibrio vulgaris* Hildenborough at indium oxide electrodes. Biochem Biophys Res Commun 145:619–624

Kpebe A, Benvenuti M, Guendon C, Rebai A, Fernandez V, Le Laz S et al (2018) A new mechanistic model for an O_2-protected electron-bifurcating hydrogenase, Hnd from *Desulfovibrio fructosovorans*. Biochim Biophys Acta Bioenerg 1859:1302–1312

Krekeler D, Teske A, Cypionka H (1998) Strategies of sulfate-reducing bacteria to escape oxygen stress in a cyanobacterial mat. FEMS Microbiol Ecol 25:89–96

Krey GD, Vanin EF, Swenson RP (1988) Cloning, nucleotide sequence and expression of the flavodoxin gene from *Desulfovibrio vulgaris* (Hildenborough). J Biol Chem 263:15436–15443

Kuchenreuther JM, Britt RD, Swartz JR (2012) New insights into [FeFe] hydrogenase activation and maturase function. PLoS One 7(9):e45850. https://doi.org/10.1371/journal.pone.0045850

Kuever J, Visser M, Loeffler C, Boll M, Worm P, Sousa DZ et al (2014) Genome analysis of *Desulfotomaculum gibsoniae* strain GrollT a highly versatile gram-positive sulfate-reducing bacterium. Stand Genomic Sci 9:821–839

Kusche WH, Trüper HG (1984) Cytochromes of the purple sulfur bacterium *Ectothiorhodospira shaposhnikovii*. Z Naturforsch 39c:894–901

Kushkevych IV, Fafula RV, Antonyak HL (2014) Catalase activity of sulfate-reducing bacteria *Desulfovibrio piger* Vib-7 and *Desulfomicrobium* sp. Rod-9 isolated from human large intestine. Microbes Health 3:15–20

Laishley EJ, Travis J, Peck HD, JR. (1969) Amino acid composition of ferredoxin and rubredoxin isolated from *Desulfovibrio gigas*. J Bacteriol 98:302–333

Lamrabet O, Pieulle L, Aubert C, Mouhamar F, Stocker P, Dolla A, Brasseur G (2011) Oxygen reduction in the strict anaerobe *Desulfovibrio vulgaris* Hildenborough: characterization of two membrane-bound oxygen reductases. Microbiol 157:2720–2732

Lee J-P, Peck HD Jr (1971) Purification of the enzyme reducing bisulfite to trithionate from *Desulfovibrio gigas* and its identification as desulfoviridin. Biochem Biophys Res Commun 45:583–589

Lee J-P, LeGall J, Peck HD Jr (1973a) Isolation of assimilatory- and dissimilatory-type sulfite reductases from *Desulfovibrio vulgaris*. J Bacteriol 115:529–542

Lee J-P, Yi C-S, LeGall J, Peck HDJr. (1973b) Isolation of a new pigment, desulforubidin, from *Desulfovibrio desulfuricans* (Norway strain) and its role in sulfite reduction. J Bacteriol 115: 453–455

Le Gall J (1968) Partial purification and study of NAD:rubredoxin oxidoreductase from *D. gigas*. Ann Inst Pasteur (Paris) 114:109–115

LeGall J, DerVartanian DV, Peck HD Jr (1979) Flavoproteins, iron proteins and hemoproteins as electron-transfer components of the sulfate-reducing bacteria. Curr Top Bioenerg 9:2137–2265

LeGall J, Dragoni N (1966) Dependence of sulfite reduction on a crystallized ferredoxin from *Desulfovibrio gigas*. Biochem Biophys Res Commun 23:145–149

LeGall J, Fauque G (1988) Dissimilatory reduction of sulfur compounds. In: Zehnder AJB (ed) Biology of anaerobic microorganisms. Wiley, New York, pp 587–639

LeGall J, Hatchikian EC (1967) Purification et propriétés d'une flavoprotein intervenant dans la reduction du sulfite par *D. gigas*. C R Acad Sci Paris 264:2580–2583

LeGall J, Liu MY, Gomes CM, Braga V, Pacheco I, Regalla M, Xavier AV, Teixeira M (1998) Characterization of a new rubredoxin isolated from *Desulfovibrio desulfuricans* 27774: definition of a new family of rubredoxins. FEBS Lett 429:295–298

LeGall J, Moura JJG, Peck HD Jr, Xavier AV (1982) Hydrogenase and other iron-sulfur proteins from sulfate-reducing and methane-forming bacteria. In: Spiro TG (ed) Iron-sulfur proteins. Wiley Interscience Publications, New York, pp 177–248

Le Gall J, Payne WJ, Chen L, Liu MY, Xavier AV (1994) Localization and specificity of cytochromes and other electron transfer proteins from sulfate-reducing bacteria. Biochimie 76:655–665

LeGall J, Peck HD Jr (1987) NH_2-terminal amino acid sequences of electron transfer proteins from gram-negative bacteria as indicators of their cellular localization: the sulfate-reducing bacteria. FEMS Microbiol Rev 46:35–40

LeGall J, Prickril BC, Moura I, Xavier AV, Moura JJ, Huynh BH (1988) Isolation and characterization of rubrerythrin, a non-heme iron protein from *Desulfovibrio vulgaris* that contains rubredoxin centers and a hemerythrin-like binuclear iron cluster. Biochemist 27:1636–1642

Lemos RS, Gomes CM, LeGall J, Xavier AV, Teixeira M (2002) The quinol:fumarate oxidoreductase from the sulphate-reducing bacterium *Desulfovibrio gigas*: spectroscopic and redox studies. J Bioenerg Biomembr 34:21–30

Lemos RS, Gomes CM, Santana M, LeGall J, Xavier AV, Teixeira M (2001) The 'strict' anaerobe *Desulfovibrio gigas* contains a membrane-bound oxygen-reducing respiratory chain. FEBS Lett 496:40–43

Lenz O, Zebger I, Hamann J, Hildebrandt P, Friedrich B (2007) Carbamoyl phosphate serves as the source of CN^-, but not of the intrinsic CO in the active site of the regulatory [NiFe]-hydrogenase from *Ralstonia eutropha*. FEBS Lett 581:3322–3326

Li X, Luo Q, Wofford NQ, Keller KL, McInerney MJ, Wall JD, Krumholz LP (2009) A molybdopterin oxidoreductase is involved in H_2 oxidation in *Desulfovibrio desulfuricans* G20. J Bacteriol 191:2675–2682

Lie TJ, Clawson MI, Godchaux W, Leadbetter ER (1999) Sulfidogenesis from 2-aminoethanesulfonate (taurine) fermentation by a morphologically unusual sulfate-reducing bacterium, *Desulforhopalus singaporensis* sp. nov. Appl Environ Microbiol 65:3328–3334

Littlewood D, Postgate JR (1956) Substrate inhibition of hydrogenase enhanced by sodium chloride. Biochim Biophys Acta 20:399–400

Liu MC, Peck HD Jr (1981) The isolation of a hexaheme cytochrome from *Desulfovibrio desulfuricans* and its identification as a new type of nitrite reductase. J Biol Chem 256: 13159–13164

Liu MY, LeGall J (1990) Purification and characterization of two proteins with inorganic pyrophosphatase activity from *Desulfovibrio vulgaris*: Rubrerythrin and a new, highly active enzyme. Biochem Biophys Res Commun 171:313–318

Liu MC, Costa C, Coutinho IB, Moura JJG, Moura I, Xavier AV, LeGall J (1988) Studies on the cytochrome components of nitrate-respiring and sulfate-respiring *Desulfovibrio desulfuricans* ATCC 27774. J Bacteriol 170:5545–5551

Lobo SAL, Almeida CC, Carita JN, Teixeira M, Saraiva LM (2008a) The haem–copper oxygen reductase of *Desulfovibrio vulgaris* contains a dihaem cytochrome *c* in subunit II. Biochim Biophys Acta (BBA) 1777:1528–1534

Lobo SA, Brindley AA, Romano CV, Leech HK, Warren MJ, Saraiva LM (2008b) Two distinct roles for two functional cobaltochelatases (CbiK) in *Desulfovibrio vulgaris* Hildenborough. Biochemist 47:5851–5857

Lobo SAL, Brindley A, Warren MJ, Saraiva LM (2009) Functional characterization of the early steps of tetrapyrrole biosynthesis and modification in *Desulfovibrio vulgaris* Hildenborough. Biochem J 420:317–325

Lobo SAL, Warren W, Saraiva LM (2012) Sulfate-reducing bacteria reveal a new branch of tetrapyrrole metabolism. Adv Microbial Physiol 61:267–295

López-Cortés A, Fardeau M-L, Fauque G, Joulian C, Ollivier B (2006) Reclassification of the sulfate- and nitrate-reducing bacterium *Desulfovibrio vulgaris* subsp. *oxamicus* as *Desulfovibrio oxamicus* sp. nov., comb. nov. Int J Sys Evol Microbiol 56:1495–1499

Louro RO (2006) Proton thrusters: overview of the structural and functional features of soluble tetrahaem cytochromes c_3. J Biol Inorg Chem 12:1–10

Louro RO, Bento I, Matias PM, Catarino T, Baptista AM, Soares CM, Carrondo MA, Turner DL, Xavier AV (2001) Conformational component in the coupled transfer of multiple electrons and protons in a monomeric tetraheme cytochrome. J Biol Chem 276:44044–44051

Louro RO, Catarino T, Turner DL, Piçarra-Pereira MA, Pacheco I, LeGall J, Xavier AV (1998) Functional and mechanistic studies of cytochrome c_3 from *Desulfovibrio gigas*: thermodynamics of a "proton thruster". Biochemist 37:15808–15815

Lumppio HL, Shenvi NV, Garg RP, Summers AO, Kurtz DM Jr (1997) A rubrerythrin operon and nigerythrin gene in *Desulfovibrio vulgaris* (Hildenborough). J Bacteriol 179:4607–4615

Lumppio HL, Shenvi NV, Summers AO, Voordouw G, Kurtz DM Jr (2000) Rubrerythrin and rubredoxin oxidoreductase in *Desulfovibrio vulgaris*: a novel oxidative stress protection system. J Bacteriol 183:101–108

Lyon EJ, Shima S, Buurman G, Chowdhuri S, Batschauer A, Steinbach K, Thauer RK (2004) UV-A/blue-light inactivation of the 'metal-free' hydrogenase (Hmd) from methanogenic archaea. Eur J Biochem 271:195–204

Machado P, Félix R, Rodrigues R, Oliveira S, Rodrigues-Pousada C (2006) Characterization and expression analysis of the cytochrome *bd* oxidase operon from *Desulfovibrio gigas*. Curr Microbiol 52:274–281

Magalon A, Böck A (2000) Analysis of the HypC-HycE complex, a key intermediate in the assembly of the metal center of the *Escherichia coli h*ydrogenase 3. J Biol Chem 275:21114–21220

Maier T, Lottspeich F, Böck A (1995) GTP hydrolysis by HypB is essential for nickel insertion into hydrogenases of *Escherichia coli*. Eur J Biochem 230:133–138

Marietou A, Richardson D, Cole J, Mohan S (2005) Nitrate reduction by *Desulfovibrio desulfuricans*: a periplasmic nitrate reductase system that lacks NapB, but includes a unique tetraheme *c*-type cytochrome, NapM. FEMS Microbiol Lett 248:217–225

Marques MC, Coelho R, De Lacey AL, Pereira IA, Matias PM (2010) The three-dimensional structure of [NiFeSe] hydrogenase from *Desulfovibrio vulgaris* Hildenborough: a hydrogenase without a bridging ligand in the active site in its oxidized, "as-isolated" state. J Mol Biol 396: 893–907

Marques MC, Tapia C, Gutiérrez-Sanz O, Ramos AR, Keller KL, Wall JD et al (2017) The direct role of selenocysteine in [NiFeSe] hydrogenase maturation and catalysis. Nat Chem Biol 13: 544–550

Martins M, Mourato C, Pereira IAC (2015) *Desulfovibrio vulgaris* growth coupled to formate-driven H_2 production. Environ Sci Technol 49:14655–14662

Martins M, Mourato C, Morais-Silva FO, Rodrigues-Pousada C, Voordouw G, Wall JD, Pereira IA (2016) Electron transfer pathways of formate-driven H_2 production in *Desulfovibrio*. Appl Microbiol Biotechnol 100:8135–8146

Matthews JC, Timkovich R, Liu M-Y, LeGall J (1995) Siroamide: a prosthetic group isolated from sulfite reductases in the genus *Desulfovibrio*. Biochemist 34:5248–5251

Matias PM, Coelho R, Pereira IA, Coelho AV, Thompson AW, Sieker LC, LeGall J, Carrondo MA (1999a) The primary and three-dimensional structures of a nine-haem cytochrome *c* from *Desulfovibrio desulfuricans* ATCC 27774 reveal a new member of the Hmc family. Structure 7:119–130

Matias PM, Coelho AY, Valente FM, Plácido D, LeGall J, Xavier AY, Pereira A, Carrondo MA (2002) Sulfate respiration in *Desulfovibrio vulgaris* Hildenborough – structure of the 16-heme cytochrome *c* HmcA at 2.5-Å resolution and a view of its role in transmembrane electron transfer. J Biol Chem 277:47907–47916

Matias PM, Morais J, Coelho R, Carrondo MA, Wilson K, Dauter Z, Sieker L (1996) Cytochrome c_3 from *Desulfovibrio gigas*: crystal structure at 1.8 Å resolution and evidence for a specific calcium-binding site. Protein Sci 5:1342–1354

Matias PM, Morais J, Coelho AV, Meijers R, Gonzalez A, Thompson AW et al (1997) A preliminary analysis of the three-dimensional structure of dimeric di-haem split-Soret cytochrome *c* from *Desulfovibrio desulfuricans* ATCC 27774 at 2.5-Å resolution using the MAD phasing method: a novel cytochrome fold with a stacked-haem arrangement. J Biol Inorg Chem 2:507–514

Matias PM, Pereira IAC, Soares CM, Carrondo MA (2005) Sulphate respiration from hydrogen in *Desulfovibrio* bacteria: a structural biology review. Prog Biophys Mol Biol 89:292–329

Matias PM, Saraiva LM, Soares CM, Coelho AV, LeGall J, Carrondo MA (1999b) Nine-haem cytochrome *c* from *Desulfovibrio desulfuricans* ATCC 27774. J Biol Inorg Chem 4:478–494

Matias PM, Soares CM, Saraiva LM, Coelho R, Morais J, Le Gall J, Carrondo MA (2001) [NiFe] hydrogenase from *Desulfovibrio desulfuricans* ATCC 27774: gene sequencing, three-dimensional structures determination and refinement at 1.8 Å and modeling studies of its interaction with the tetrahaem cytochrome c_3. J Biol Inorg Chem 6:63–81

Mayhew SG, van Dijk C, van der Westen HM (1978) Properties of hydrogenases from the anaerobic bacteria *Megasphaera elsdenii* and *Desulfovibrio vulgaris* (Hildenborough). In: Schlegel HG, Schneider K (eds) Hydrogenase: their catalytic activity, structure, and function. Erich Göltze, KG, Göttingen, Germany, pp 125–140

McDowall JS, Murphy BJ, Haumann M, Palmer T, Armstrong FA, Sargent F (2014) Bacterial formate hydrogenlyase complex. PNAS, USA 111:E3948–E3956. https://doi.org/10.1073/pnas.1407927111

McCready RGL, Gould WD, Cook ED (1983) Respiratory nitrate reduction by *Desulfovibrio* sp. Arch Microbiol 135:182–185

Meuer J, Bartoschek S, Koch J, Kunkel A, Hedderich R (1999) Purification and catalytic properties of Ech hydrogenase from *Methanosarcina barkeri*. Eur J Biochem 265:325–335

Meuer J, Kuettner HC, Zhang JK, Hedderich R, Metcalf WW (2002) Genetic analysis of the archaeon *Methanosarcina barkeri* Fusaro reveals a central role for Ech hydrogenase and ferredoxin in methanogenesis and carbon fixation. Proc Natl Acad Sci U S A 99:5632–5637

Meyer J (2001) Ferredoxins of the third kind. FEBS Lett 509:1–5

Meyer TE, Kennel SJ, Tedro SM, Kamen MD (1973) Iron protein content of *Thiocapsa pfennigii*, a purple sulfur bacterium of atypical chlorophyll composition. Biochim Biophys Acta 292:634–643

Misaki S, Morimoto Y, Ogata M, Yagi T, Higuchi Y, Yasuoka N (1999) Structure determination of rubredoxin from *Desulfovibrio vulgaris* Miyazaki F in two crystal forms. Acta Cryst D55:408–413

Mitchell GJ, Jones JG, Cole JA (1986) Distribution and regulation of nitrate and nitrite reduction by *Desulfovibrio* and *Desulfotomaculum* species. Arch Microbiol 144:35–40

Montet Y, Amara P, Volbeda A, Vernede X, Hatchikian EC, Field MJ, Frey M, Fontecilla-Camps JC (1997) Gas access to the active site of Ni-Fe hydrogenases probed by X-ray crystallography and molecular dynamics. Nat Struct Biol 4:523–526

Morais J, Palma PN, Frazao C, Caldeira J, LeGall J, Moura I, Moura JJG, Carrondo MA (1995) Structure of the tetraheme cytochrome from *Desulfovibrio desulfuricans* ATCC 27774: X-ray diffraction and electron paramagnetic resonance studies. Biochemist 34:12830–12841

Morais-Silva FO, Rezende AM, Pimentel C, Santos CI, Clemente C, Varela-Raposo A et al (2014) Genome sequence of the model sulfate reducer *Desulfovibrio gigas*: a comparative analysis within the *Desulfovibrio* genus. Microbiol Open 3:513–530

Morais-Silva FO, Santos CI, Rodrigues R, Pereira IAC, Rodrigues-Pousada C (2013) Roles of HynAB and Ech, the only two hydrogenases found in the model sulfate reducer *Desulfovibrio gigas*. J Bacteriol 195:4753–4760

Morelli X, Dolla A, Czjzek M, Palma PN, Blasco F, Krippahl L, Moura JJG, Guerlesquin F (2000) Heteronuclear NMR and soft docking: an experimental approach for a structural model of the cytochrome c_{553}–ferredoxin complex. Biochemist 39:2530–2537
Morelli X, Guerlesquin F (1999) Mapping the cytochrome c_{553} interacting site using ^{1}H and ^{15}N NMR. FEBS Lett 460:77–80
Mori K, Kim H, Kakegawa TH (2003) A novel lineage of sulfate-reducing microorganisms: *Thermodesulfobiaceae* fam. nov., Thermodesulfobium narugense, gen. nov., sp. nov., a new thermophilic isolate from a hot spring. Extremophiles 7:283–290
Mortenson LE, Valentine RC, Carnahan JE (1962) An electron transport factor from *Clostridium pasteurianum*. Biochem Biophys Res Commun 7:448–452
Moser CC, Anderson JL, Dutton PL (2010) Guidelines for tunneling in enzymes. Biochim Biophys Acta 1797:1573–1586
Moura I, Bruschi M, Le Gall J, Moura JJ, Xavier AV (1977) Isolation and characterization of desulforedoxin, a new type of non-heme iron protein from *Desulfovibrio gigas*. Biochem Biophys Res Commun 75:1037–1044
Moura I, Bursakov S, Costa C, JJG M (1997) Nitrate and nitrite utilization in sulfate-reducing bacteria. Anaerobe 3:279–290
Moura I, Fauque G, LeGall J, Xavier AV, Moura JJG (1987) Characterization of the cytochrome system of a nitrogen-fixing strain of a sulfate-reducing bacterium: *Desulfovibrio desulfuricans* strain Berre-Eau. Eur J Biochem 162:547–554
Moura I, Huynh BH, Hausinger RP, Le Gall J, Xavier AV, Munck E (1980b) Mossbauer and EPR studies of desulforedoxin from *Desulfovibrio gigas*. J Biol Chem 255:2493–2498
Moura I, Moura JJ, Bruschi M, LeGall J (1980a) Flavodoxin and rubredoxin from *Desulphovibrio salexigens*. Biochim Biophys Acta 591:1–8
Moura I, Moura JJG, Santos MH, Xavier AV, Le Gall J (1979) Redox studies on rubredoxins from sulphate and sulphur reducing bacteria. FEBS Lett 107:419–421
Moura I, Tavares P, Moura JJ, Ravi N, Huynh BH, Liu MY, LeGall J (1990) Purification and characterization of desulfoferrodoxin. A novel protein from *Desulfovibrio desulfuricans* (ATCC 27774) and from *Desulfovibrio vulgaris* (strain Hildenborough) that contains a distorted rubredoxin center and a mononuclear ferrous center. J Biol Chem 265:21596–21602
Moura I, Tavares P, Ravi N (1994a) Characterization of three proteins containing multiple iron sites: rubrerythrin, desulfoferrodoxin, and a protein containing a six-iron cluster. Meth Enzymol 243:216–240
Moura I, Teixeira M, LeGall J, Moura JJ (1991a) Spectroscopic studies of cobalt and nickel substituted rubredoxin and desulforedoxin. J Inorg Biochem 44:127–139
Moura I, Xavier AV, Cammack RC, Bruschi M, LeGall J (1978a) A comparative spectroscopic study of two non-haem iron proteins lacking labile sulphide from *Desulphovibrio gigas*. Biochim Biophys Acta 533:156–162
Moura JJG, Costa C, Liu MY, Moura, LeGall J (1991b) Structural and functional approach toward a classification of the complex cytochrome *c* system found in sulfate-reducing bacteria. Biochim Biophys Acta 1058:61–66
Moura JJG, Gonzales P, Moura I, Fauque G (2007) Dissimilatory nitrate and nitrite ammonification by sulphate-reducing eubacteria. In: Barton LL, Hamilton WA (eds) Sulphate-reducing bacteria environmental and engineered systems. Cambridge University Press, Cambridge, UK, pp 241–264
Moura JJG, LeGall J, Xavier AV (1984) Interconversion from 3Fe into 4Fe clusters in the presence of *Desulfovibrio gigas* cell extracts. European J Biochem 41:319–322
Moura JJG, Macedo AL, Palma PN (1994b) Ferredoxins. Meth Enzymol 243:165–203
Moura JJG, Xavier AV, Hatchikian CE, LeGall J (1978b) Structural control of the redox potentials and of the physiological activity by oligomerization of ferredoxin. FEBS Lett 89:177–179
Mulder DW, Shepard EM, Meuser JE, Joshi N, King PW, Posewitz MC, Broderick JB, Peters JW (2011) Insights into [FeFe]-hydrogenase structure, mechanism, and maturation. Structure 19: 1038–1052

Najmudin S, Gonzalez PJ, Trincao J, Coelho C, Mukhopadhyay A, Cerqueira NM et al (2008) Periplasmic nitrate reductase revisited: a sulfur atom completes the sixth coordination of the catalytic molybdenum. J Biol Inorg Chem 13:737–753

Nakagawa A, Higuchi Y, Yasuoka N, Katsube Y, Yagi T (1990) S-class cytochromes *c* have a variety of folding patterns: structure of cytochrome *c*-553 from *Desulfovibrio vulgaris* determined by the multi-wavelength anomalous dispersion method. J Biochem (Tokyo) 108:701–703

Newman D, Postgate JR (1968) Rubredoxin from a nitrogen-fixing variety of *Desulfovibrio desulfuricans*. Eur J Biochem 7:45–50

Nicolet Y, Piras C, Legrand P, Hatchikian CE, Fontecilla-Camps JC (1999) *Desulfovibrio desulfuricans* iron hydrogenase: the structure shows unusual coordination to an active site Fe binuclear center. Structure 7:13–23

Nonaka K, Nguyen NT, Yoon KS, Ogo S (2013) Novel H_2-oxidizing [NiFeSe]-hydrogenase from *Desulfovibrio vulgaris* Miyazaki F. J Biosci Bioeng 115:366–371

Nørager S, Legrand P, Pieulle L, Hatchikian C, Roth M (1999) Crystal structure of the oxidised and reduced cytochrome c_3 from *Desulfovibrio africanus*. J Mol Biol 290:881–902

Odom JM, Peck HD Jr (1981) Localization of dehydrogenases, reductases, and electron transfer components in the sulfate-reducing bacterium *Desulfovibrio gigas*. J Bacteriol 147:161–169

Odom JM, Peck HD Jr (1984) Hydrogenase, electron-transfer proteins and energy coupling in the sulfate-reducing bacteria *Desulfovibrio*. Ann Rev Microbiol 38:551–592

Ogata M, Arihara K, Yagi T (1981) D-lactate dehydrogenase of *Desulfovibrio vulgaris*. J Biochem (Tokyo) 89:1423–1431

Ogata M, Kiuchi N, Yagi T (1993) Characterization and redox properties of high molecular mass cytochrome c_3 (Hmc) isolated from *Desulfovibrio vulgaris* Miyazaki. Biochimie 75:977–983

Ogata M, Kondo S, Okawara N, Yagi T (1988) Purification and characterization of ferredoxin from *Desulfovibrio vulgaris* Miyazaki. J Biochem 103:121–125

Ogata H, Mizoguchi Y, Mizuno N, Miki K, Adachi S-I, Yasuoka N et al (2002) Structural studies of the carbon monoxide complex of [NiFe]hydrogenase from *Desulfovibrio vulgaris* Miyazaki F: suggestion for the initial activation site for dihydrogen. J Am Chem Soc 124:11628–11635

Okawara N, Ogata M, Yagi T, Wakabayashi S, Matsubara H (1988) Amino acid sequence of ferredoxin I from *Desulfovibrio vulgaris* Miyazaki. J Biochem 104:196–199

Olson JW, Mehta NS, Maier RJ (2001) Requirement of nickel metabolism proteins HypA and HypB for full activity of both hydrogenase and urease in *Helicobacter pylori*. Mol Microbiol 39: 176–182

Palma PN, Moura I, LeGall J, Van Beeumen J, Wampler JE, Moura JJG (1994) Evidence for a ternary complex formed between flavodoxin and cytochrome c_3. Biochemist 33:6394–6407

Paquete CM, Pereira PM, Catarino T, Turner DL, Louro RO, Xavier AV (2007) Functional properties of type I and type II cytochromes c_3 from *Desulfovibrio africanus*. Biochim Biophys Acta Bioenerg 1767:178–188

Papavassiliou P, Hatchikian EC (1985) Isolation and characterization of a rubredoxin and a two-[4Fe–4S] ferredoxin *Thermodesulfobacterium commune*. Biochim Biophys Acta Bioenerg 810:1–11

Paschos A, Bauer A, Zimmermann A, Zehelein E, Böck A (2002) HypF, a carbamoyl phosphate-converting enzyme involved in [NiFe] hydrogenase maturation. J Biol Chem 277:49945–49951

Patil DS, Moura JJ, He SH, Teixeira M, Prickril BC, DerVartanian DV, Peck HD Jr, LeGall J, Huynh BH (1988) EPR-detectable redox centers of the periplasmic hydrogenase from *Desulfovibrio vulgaris*. J Biol Chem 263:18732–18738

Pattarkine MV, Tanner JJ, Bottoms CA, Lee Y-H, Wall JD (2006) *Desulfovibrio desulfuricans* G20 tetraheme cytochrome structure at 1.5Å and cytochrome interaction with metal complexes. J Mol Biol 358:1314–1327

Peck HD Jr, LeGall J (1982) Biochemistry of dissimilatory sulphate reduction. Phil Trans R Soc Lond B298:443–466

Pereira IAC (2008) Membrane complexes in *Desulfovibrio*. In Friedrich C, Dahl C (eds), Microbial sulfur metabolism (pp. 24–35). Berlin: Springer-Verlag

Pereira IC, Abreu IA, Xavier AV, LeGall J, Teixeira M (1996a) Nitrite reductase from *Desulfovibrio desulfuricans*(ATCC 27774) – a heterooligomer heme protein with sulfite reductase activity. Biochem Biophys Res Commun 224:611–618

Pereira I, Haveman S, Voordouw G (2007) Biochemical, genetic and genomic characterization of anaerobic electron transport pathways in sulphate-reducing Delta proteobacteria. In: Barton L, Hamilton W (eds) Sulphate-reducing bacteria: environmental and engineered systems. Cambridge University Press, Cambridge, UK, pp 215–240

Pereira IA, LeGall J, Xavier AV, Teixeira M (2000) Characterization of a heme *c* nitrite reductase from a non-ammonifying microorganism, *Desulfovibrio vulgaris* Hildenborough. Biochim Biophys Acta 1481:119–130

Pereira PM, He Q, Valente FM, Xavier AV, Zhou J, Pereira IA, Louro RO (2008) Energy Metabolism in *Desulfovibrio vulgaris* Hildenborough: insights from transcriptome analysis. Antonie van Leeuwenhoek 93:347–362

Pereira PM, Pacheco I, Turner D, Louro L (2002) Structure-function relationship in type II cytochrome c_3 from *Desulfovibrio africanus*: a novel function in a familiar heme core. J Biol Inorg Chem 7:815–822

Pereira PM, Teixeira M, Xavier AV, Louro RO, Pereira IA (2006) The Tmc complex from *Desulfovibrio vulgaris* Hildenborough is involved in transmembrane electron transfer from periplasmic hydrogen oxidation. Biochemist 45:10359–10367

Pereira IC, Abreu IA, Xavier AV, LeGall J, Teixeira M (1996b) Nitrite reductase from *Desulfovibrio desulfuricans* (ATCC 27774)-a heterooligomer heme protein with sulfite reductase activity. Biochem Biophys Res Commun 224:611–618

Pereira IA, Ramos AR, Grein F, Marques MC, da Silva SM, Venceslau SS (2011) A comparative genomic analysis of energy metabolism in sulfate-reducing bacteria and archaea. Front Microbiol 19(2):69. https://doi.org/10.3389/fmicb.2011.00069

Pereira IAC, Romão CV, Xavier AV, LeGall J, Teixeira M (1998) Electron transfer between hydrogenases and mono- and multiheme cytochromes in *Desulfovibrio* sp. J Biol Inorg Chem 3:494–498

Peters JW, Beratan DN, Bothner B, Dyer RB, Harwood CS, Heiden ZM et al (2018a) A new era for electron bifurcation. Current Opinion Chem Biol 47:32–38

Peters JW, Beratan DN, Schut GJ, Adams MWW (2018b) On the nature of organic and inorganic centers that bifurcate electrons, coupling exergonic and endergonic oxidation-reduction reactions. Chem Commun (Camb) 54:4091–4099

Peters JW, Lanzilotta WN, Lemon BJ, Seefeldt LC (1998) X-ray crystal structure of the Fe-only hydrogenase (CpI) from *Clostridium pasteurianum* to 1.8 angstrom resolution. Science 282: 1853–1858

Peters JW, Schut GJ, Boyd ES, Mulder DW, Shepard EM, Broderick JB, King PW, Adams MW (2015) [FeFe]- and [NiFe]-hydrogenase diversity, mechanism, and maturation. Biochim Biophys Acta 1853:1350–1369

Pieiuk AJ, Wolbert RBG, Porter GL, Verhagen MFJM, Hagen WR (1993) Nigerythrin and rubrerythrin from *Desulfovibrio vulgaris* each contain two mononuclear iron centers and two dinuclear iron clusters. Eur J Biochem 212:237–245

Pierrot M, Haser R, Frey M, Payan F, Astier JP (1982) Crystal structure and electron transfer properties of cytochrome c_3. J Biol Chem 257:14341–14348

Pieulle L, Morelli X, Gallice P, Lojou E, Barbier P, Czjzek M, Bianco P, Guerlesquin F, Hatchikian EC (2005) The Type I/Type II cytochrome c_3 complex: an electron transfer link in the hydrogen-sulfate reduction pathway. J Mol Biol 354:73–90

Pires RH, Lourenço AI, Morais F, Teixeira M, Xavier AV, Saraiva LM, Pereira IA (2003) A novel membrane-bound respiratory complex from *Desulfovibrio desulfuricans* ATCC 27774. Biochim Biophys Acta 1605:67–82

Plugge CM, Balk M, Stams AJM (2002) *Desulfotomaculum thermobenzoicum* subsp *thermosyntrophicum* subsp nov., a thermophilic, syntrophic, propionate-oxidizing, spore-forming bacterium. Int J Syst Evol Microbiol 52:391–399

Pollock WBR, Loutfi M, Bruschi M, Rapp-Giles BJ, Wall JD, Voordouw G (1991) Cloning, sequencing, and expression of the gene encoding the high-molecular-weight cytochrome *c* from *Desulfovibrio vulgaris* Hildenborough. J Bacteriol 173:220–228

Postgate JR (1951a) On the nutrition of *Desulphovibrio desulphuricans*. J Gen Microbiol 5:714–724

Postgate JR (1951b) The reduction of sulphur compounds by *Desulphovibrio desulphuricans*. J Gen Microbiol 5:725–538

Postgate JR (1954) Presence of cytochrome in an obligate anaerobe. Biochem J 65:xi

Postgate JR (1979) The sulphate-reducing bacteria. Cambridge University Press, Cambridge, UK

Postgate JR (1984) Genus *Desulfovibrio*. In: King NR, Holt JG (eds) Bergey's manuel of systematic bacteriology. Williams and Wilkins, Baltimore, MD, pp 666–673

Potter L, Angove H, Richardson D, Cole JA (2001) Nitrate reduction in the periplasm of gram-negative bacteria. Adv Microbiol Physiol 45:51–112

Price MN, Ray J, Wetmore KM, Kuehl JV, Bauer S, Deutschbauer AM, Arkin AP (2014) The genetic basis of energy conservation in the sulfate-reducing bacterium *Desulfovibrio alaskensis* G20. Front Microbiol 5:577. https://doi.org/10.3389/fmicb.2014.00577

Prickril BC, Kurtz DM Jr, LeGall J, Voordouw G (1991) Cloning and sequencing of the gene for rubrerythrin from *Desulfovibrio vulgaris* (Hildenborough). Biochemist 30:11118–11123

Rabus R, Venceslau SS, Wöhlbrand L, Voordouw G, Judy D, Wall JD, Inês AC, Pereira IAC (2015) A post-genomic view of the ecophysiology, catabolism and biotechnological relevance of sulphate-reducing prokaryotes. Adv Microbial Physiol 66:58–321

Rajeev L, Chen A, Kazakov AE, Luning EG, Zane GM, Novichkov PS, Wall JD, Mukhopadhyay A (2015) Regulation of nitrite stress response in *Desulfovibrio vulgaris* Hildenborough, a model sulfate-reducing bacterium. J Bacteriol 197:3400–3408

Ramel F, Amrani A, Pieulle L, Lamrabet O, Voordouw G, Seddiki N, Brèthes D, Company M, Dolla A, Brasseur G (2013) Membrane-bound oxygen reductases of the anaerobic sulfate-reducing *Desulfovibrio vulgaris* Hildenborough: roles in oxygen defence and electron link with periplasmic hydrogen oxidation. Microbiol 159:2663–2673

Riederer-Henderson MA, Peck HD Jr (1986a) In vitro requirements for formate dehydrogenase activity from *Desulfovibrio gigas*. Can J Microbiol 32:425–429

Riederer-Henderson MA, Peck HD Jr (1986b) Properties of formate dehydrogenase from *Desulfovibrio gigas*. Can J Microbiol 32:430–435

Rodrigues JV, Abreu IA, Saraiva LM, Teixeira M (2005) Rubredoxin acts as an electron donor for neelaredoxin in *Archaeoglobus fulgidus*. Biochem Biophys Res Commun 329:1300–1305

Rodrigues ML, Oliveira TF, Pereira IAC, Archer M (2006) X-ray structure of the membrane-bound cytochrome *c* quinol dehydrogenase NrfH reveals novel haem coordination. EMBO J 25:5951–5960

Rodriguez JA, Abreu IA (2005) Chemical activity of iron in [2Fe-2S]-protein centers and FeS2 (100) surfaces. J Phys Chem B 109:2754–2762

Romão CV, Archer M, Lobo SA, Louro RO, Pereira IAC, Saraiva LM, Teixeira M, Matias PM (2012) Diversity of heme proteins in sulfate-reducing bacteria. In: Kadish KM, Smith KM, Guilard R (eds) Handbook of porphyrin science – with applications to chemistry, physics, materials science, engineering, biology and medicine. World Scientific Publishing Co., Pte. Ltd., Singapore, pp 139–228

Romão CV, Ladakis D, Lobo SAL, Carrondo MA, Brindley AA, Deery S et al (2011) Evolution in a family of chelatases facilitated by the introduction of active site asymmetry and protein oligomerization. PNAS 108:97–102

Romão CV, Louro R, Timkovich R, Lubben M, Liu MY, LeGall J, Xavier AV, Teixeira M (2000) Iron-coproporphyrin III is a natural cofactor in bacterioferritin from the anaerobic bacterium, *Desulfovibrio desulfuricans*. FEBS Lett 480:213–216

Romero A, Caldeira J, LeGall J, Moura I, Moura JJG, Romao MJ (1996) Crystal structure of flavodoxin from *Desulfovibrio desulfuricans* ATCC 27774 in two oxidation states. Eur J Biochem 239:190–196

Rossi M, Pollock WB, Reij MW, Keon RG, Fu R, R, Voordouw G. (1993) The *hmc* operon of *Desulfovibrio vulgaris* sub sp. *vulgaris* Hildenborough encodes a potential transmembrane redox protein complex. J Bacteriol 175:4699–4711

Rossmann R, Maier T, Lottspeich F, Böck A (1995) Characterisation of a protease from *Escherichia coli* involved in hydrogenase maturation. Eur J Biochem 227:545–550

Rouault TA, Stout CD, Kaptain S, Harford JB, Klausner RD (1991) Structural relationship between an iron-regulated RNA-binding protein (IRE-BP) and aconitase: functional implications. Cell 64:881–883

Rousset M, Montet Y, Guigliarelli B, Forget N, Asso M, Bertrand P, Fontecilla-Camps JC, Hatchikian EC (1998) [3Fe-4S] to [4Fe-4S] cluster conversion in *Desulfovibrio fructosovorans* [NiFe] hydrogenase by site-directed mutagenesis. Proc Natl Acad Sci U S A 95:11625–11630

Rozanova EP, Pivovarova TA (1988) Reclassification of *Desulfovibrio thermophilus* (Rozanova, Khudyakova, 1974). Microbiology (Engl Tr) 57:102–106

Sadana JC, Rittenberg D (1963) Some observations on the enzyme hydrogenase of *Desulfovibrio desulfuricans*. Proc Natl Acad Sci U S A 50:900–904

Sadana JC, Jagannathan V (1956) Purification and properties of the hydrogenase of *Desulphovibrio desulphuricans*. Biochim Biophys Acta 19:440–452

Saint-Martin P, Lespinat PA, Fauque G, Berlier Y, Legall J, Moura I et al (1988) Hydrogen production and deuterium-proton exchange reactions catalyzed by *Desulfovibrio* nickel(II)-substituted rubredoxins (hydrogenase). Proc Natl Acad Sci U S A 85:9378–9380

Samain E, Albagnac G, LeGall J (1986) Redox studies of the tetraheme cytochrome c_3 isolated from the propionate- oxidizing, sulfate-reducing bacterium. FEBS Lett 204:247–250

Sancho J (2006) Flavodoxins: sequence, folding, binding, function and beyond. Cell Mol Life Sci 63:855–864

Santos-Silva T, Dias JM, Dolla A, Durand M-C, Gonçalves LL, Lampreia J, Moura I, Romão MJ (2007) Crystal structure of the 16 heme cytochrome from *Desulfovibrio gigas*: a glycosylated protein in a sulphate-reducing bacterium. J Mol Biol 370:659–673

Saraiva LM, da Costa PN, Conte C, Xavier AV, LeGall J (2001) In the facultative sulphate/nitrate reducer *Desulfovibrio desulfuricans* ATCC 27774, the nine-haem cytochrome c is part of a membrane-bound redox complex mainly expressed in sulphate grown cells. Biochim Biophys Acta 1520:63–70

Saraiva LM, da Costa PN, LeGall J (1999) Sequencing the gene encoding *Desulfovibrio desulfuricans* ATCC 27774 nine-heme cytochrome *c*. Biochem Biophs Res Commun 262: 629–634

Sato M, Shibata N, Morimoto Y, Takayama Y, Ozawa K, Akutsu H, Higuchi Y, Yasuoka N (2004) X-ray induced reduction of the crystal of high-molecular-weight cytochrome *c* revealed by microspectrophotometry. Synchrotron Radiat 11:113–116

Schuchmann K, Chowdhury NP, Müller V (2018) Complex multimeric [FeFe] hydrogenases: biochemistry, physiology and new opportunities for the hydrogen economy. Front Microbiol 9:2911. https://doi.org/10.3389/fmicb.2018.02911

Schuchmann K, Müller V (2012) A bacterial electron-bifurcating hydrogenase. J Biol Chem 287: 31165–31171

Schumacher W, Hole UH, Kroneck PMH (1994) Ammonia-forming cytochrome *c* nitrite reductase from *Sulfurospirillum deleyianum* is a tetraheme protein: new aspects of the molecular composition and spectroscopic properties. Biochem Biophys Res Commun 205:911–916

Schut GJ, Adams MW (2009) The iron-hydrogenase of *Thermotoga maritima* utilizes ferredoxin and NADH synergistically: a new perspective on anaerobic hydrogen production. J Bacteriol 191:4451–4457

Sebban C, Blanchard L, Bruschi M, Guerlesquin F (1995) Purification and characterization of the formate dehydrogenase from *Desulfovibrio vulgaris* Hildenborough. FEMS Microbiol Lett 133: 143–149

Sebban-Kreuzer C, Blackledge M, Dolla A, Marion D, Guerlesquin F (1998a) Tyrosine 64 of cytochrome c_{553} is required for electron exchange with formate dehydrogenase in *Desulfovibrio vulgaris* Hildenborough. Biochemist 37:8331–8340

Sebban-Kreuzer C, Dolla A, Guerlesquin F (1998b) The formate dehydrogenase-cytochrome c_{553} complex from *Desulfovibrio vulgaris* Hildenborough. Eur J Biochem 253:645–652

Seitz H, Cypionka H (1986) Chemolithotrophic growth of *Desulfovibrio desulfuricans* with hydrogen coupled to ammonification of nitrate or nitrite. Arch Microbiol 146:63–67

Senez JC, Pichinoty F (1958) Réduction de l'hydroxylamine liée a l'activité de l'hydrogénase de *Desulfovibrio desulfuricans*. II Nature du système enzymatique et du transporteur d'électrons intervenant dans la réaction. Biochim Biophys Acta 28:355–369

Shafaat HS, Rüdiger O, Ogata H, Lubitz W (2013) [NiFe] hydrogenases: a common active site for hydrogen metabolism under diverse conditions. Biochim Biophys Acta 1827:986–1002

Shepard EM, Mus F, Betz JN, Byer AS, Duffus BR, Peters JW, Broderick JB (2014) [FeFe]-hydrogenase maturation. Biochemist 53:4090–4104

Shima S, Ermler U (2011) Structure and function of [Fe] hydrogenase structure and function of [Fe]-hydrogenase and its iron–guanylylpyridinol (FeGP) cofactor. Eur J Inorg Chem 2011:963–972

Shima S, Thauer RK (2006) A third type of hydrogenase catalyzing H_2 activation. Chem Rec 7:37–46

Shimizu F, Ogata M, Yagi T, Wakabayashi S, Matsubara H (1989) Amino acid sequence and function of rubredoxin from *Desulfovibrio vulgaris* Miyazaki. Biochimie 71:1171–1177

Sieker LC, Jensen LH, LeGall J (1986) Preliminary X-ray studies of the tetra-heme cytochrome *c*, and the octa-heme cytochrome *c*, from *Desulfovibrio gigas*. FEBS Lett 209:261–264

Sieker LC, Jensen LH, Prickril B, LeGall J (1983) Crystallographic study of rubredoxin from *Desulfovibrio desulfuricans* strain 27774. J Mol Biol 171:101–103

Sieker LC, Stenkamp RE, LeGall J (1994) Rubredoxin in crystalline state. Meth Enzymol 243:203–216

Sieker LC, Turley S, Prickril BC, LeGall J (1988) Crystallization and preliminary X-ray diffraction study of a protein with a high potential rubredoxin center and a hemerythrin-type Fe center. Proteins 3:184–186

Silva G, LeGall J, Xavier AV, Teixeira M, Rodrigues-Pousada C (2001) Molecular characterization of *Desulfovibrio gigas* neelaredoxin, a Protein involved in oxygen detoxification in anaerobes. J Bacteriol 183:4413–4420

Silva G, Oliveira S, Gomes CM, Pacheco I, Liu MY, Xavier AV, Teixeira M, Legall J, Rodrigues-pousada C (1999) *Desulfovibrio gigas* neelaredoxin. A novel superoxide dismutase integrated in a putative oxygen sensory operon of an anaerobe. Eur J Biochem 259:235–243

Silva JJRF, Williams RJP (1991) The biological chemistry of the elements – the inorganic chemistry of life. Oxford University Press, New York

Sonne-Hansen J, Ahring BK (1999) *Thermodesulfobacterium hveragerdense* sp. nov., and *Thermodesulfovibrio islandicus* sp. nov., two thermophilic sulfate-reducing bacteria isolated from a Icelandic hot spring. Syst Appl Microbiol 22:559–564

Sousa DZ, Visser M, van Gelder AH, Boeren S, Pieterse MM, Pinkse MWH et al (2018) The deep-subsurface sulfate reducer *Desulfotomaculum kuznetsovii* employs two methanol-degrading pathways. Nature Commun 9:239. https://doi.org/10.1038/s41467-017-02518-9

Spring S, Visser M, Lu M, Copeland AC, Lapidus A, Lucas S et al (2012) Complete genome sequence of the sulfate-reducing Firmicute *Desulfotomaculum ruminis* type strain (DLT). Stand Genomic Sci 7:304–319. https://doi.org/10.4056/sigs.3226659

Stephenson M, Strickland LH (1931) Hydrogenase: a bacterial enzyme activating molecular hydrogen: the properties of the enzyme. Biochem J 25:205–214

Stewart DE, LeGall J, Moura I, Moura JJ, Peck HD Jr, Xavier AV, Weiner PK, Wampler JE (1988) A hypothetical model of the flavodoxin-tetraheme cytochrome c_3 complex of sulfate-reducing bacteria. Biochemist 27:2444–2450

Stewart DE, LeGall J, Moura I, JJG M, Peck HD Jr, Xavier AV, Weiner PK, Wampler JE (1989) Electron transport in sulfate-reducing bacteria. Molecular modeling and NMR studies of the rubredoxin - tetraheme-cytochrome-c_3 complex. Eur J Biochem 185:695–700

Stewart DE, Wampler JE (1991) Molecular dynamics simulations of the cytochrome c_3–rubredoxin complex from *Desulfovibrio vulgaris*. Proteins 11:142–152

Tamura T, Tsunekawa N, Nemoto M, Inagaki K, Hirano T, Sato F (2016) Molecular evolution of gas cavity in [NiFeSe] hydrogenases resurrected *in silico*. Sci Rep 6(19742):2016. https://doi.org/10.1038/srep19742

Tavares P, Ravi N, Moura JJ, LeGall J, Huang YH, Crouse BR et al (1994) Spectroscopic properties of desulfoferrodoxin from *Desulfovibrio desulfuricans* (ATCC 27774). J Biol Chem 269: 10504–10510

Teixeira VH, Baptista AM, Soares CM (2004) Modeling electron transfer thermodynamics in protein complexes: interaction between two cytochromes c_3. Biophys J 86:2773–2785

Teixeira M, Fauque G, Moura I, Lespinat PA, Berlier Y, Prickril B et al (1987) Nickel-[iron-sulfur]-selenium-containing hydrogenases from *Desulfovibrio baculatus* (DSM 1743). Redox centres and catalytic properties. Eur J Biochem 167:47–58

Teixeira M, Moura I, Fauque G, Czechowski M, Berlier Y, Lespinat PA et al (1986) Redox properties and activity studies on a nickel-containing hydrogenase isolated from a halophilic sulfate reducer *Desulfovibrio salexigens*. Biochimie 68:75–84

Thomson AJ, Robinson AE, Johnson MK, Cammack R, Rao KK, Hall DO (1981) Low-temperature magnetic circular dichroism evidence for the conversion of four iron-sulphur clusters in a ferredoxin from *Clostridium pasteurianum* into three iron-sulphur clusters. Biochim Biophys Acta 637:423–427

Tikhonova TV, Trofimov AA, Popov VO (2012) Octaheme nitrite reductase: structure and properties. Biochem Mosc 77:1129–1138

Timkovich R, Burkhalter RS, Xavier AV, Chen L, LeGall J (1994) Iron uroporphyrin I and a heme c - derivative are prosthetic groups in *Desulfovibrio gigas* rubredoxin oxidase. Bioorg Chem 22: 284–293

Trinkerl M, Breunig A, Schauder R, König H (1990) *Desulfovibrio termitidis* sp. nov., a carbohydrate-degrading sulfate-reducing bacterium from the hindgut of a termite. Syst Appl Microbiol 13:372–377

Tucker MD, Barton LL, Thomson BM (1997) Reduction and immobilization of molybdenum by *Desulfovibrio desulfuricans*. J Environ Qual 26:1146–1152

Trudinger PA (1970) Carbon monoxide-reacting pigment from *Desulfotomaculum nigrificans* and its possible relevance to sulfite reduction. J Bacteriol 104:158–170

Umhau S, Fritz G, Diederichs K, Breed J, Welte W, Kroneck PMH (2001) Three-dimensional structure of the nonaheme cytochrome *c* from *Desulfovibrio desulfuricans* Essex in the Fe(III) state at 1.89 Å resolution. Biochemist 40:1308–1316

Valente FMA, Almeida CC, Pacheco I, Carita J, Saraiva LM, Pereira IAC (2006) Selenium is involved in regulation of periplasmic hydrogenase gene expression in *Desulfovibrio vulgaris* Hildenborough. J Bacteriol 188:3228–3235

Valente FM, Oliveira AS, Gnadt N, Pacheco I, Coelho AV, Xavier AV et al (2005) Hydrogenases in *Desulfovibrio vulgaris* Hildenborough: structural and physiologic characterisation of the membrane-bound [NiFeSe] hydrogenase. J Biol Inorg Chem 10:667–682

Valente FMA, Pereira PM, Venceslau SS, Regalla M, Coelho AV, Pereira IAV (2007) The [NiFeSe] hydrogenase from *Desulfovibrio vulgaris* Hildenborough is a bacterial lipoprotein lacking a typical lipoprotein signal peptide. FEBS Lett 581:3341–3344

Valente FM, Saraiva LM, LeGall J, Xavier AV, Teixeira M, Pereira IA (2001) A membrane-bound cytochrome c_3: a type II cytochrome c_3 from *Desulfovibrio vulgaris* Hildenborough. Chembiochem 2:895–905

Valentine RC (1964) Bacterial ferredoxin. Bacteriol Rev 28:497–517

Van Beeumen JJ, Van Driessche G, Liu MY, LeGall J (1991) The primary structure of rubrerythrin, a protein with inorganic pyrophosphatase activity from *Desulfovibrio vulgaris*. Comparison with hemerythrin and rubredoxin. J Biol Chem 266:20645–20653

van der Spek TM, Arendsen AF, Happe RP, Yun S, Bagley KA, Stufkens DJ, Hagen WR, Albracht SP (1996) Similarities in the architecture of the active sites of Ni-hydrogenases and Fe-hydrogenases detected by means of infrared spectroscopy. European J Biochem. 237:629–634

van der Westen HM, Mayhew SG, Veeger C (1978) Separation of hydrogenase form intact cells of *Desulfovibrio vulgaris*. FEBS Lett 86:122–126

Van Rooijen GJH, Bruschi M, Voordouw G (1989) Cloning and sequencing of the gene encoding cytochrome c_{553} from *Desulfovibrio vulgaris* Hildenborough. J Bacteriol 171:3575–3578

Venceslau SS, Lino RR, Pereira IA (2010) The Qrc membrane complex, related to the alternative complex III, is a menaquinone reductase involved in sulfate respiration. J Biol Chem 285: 22774–22783

Verhagen MF, O'Rourke T, Adams MW (1999) The hyperthermophilic bacterium, *Thermotoga maritima*, contains an unusually complex iron-hydrogenase: amino acid sequence analyses versus biochemical characterization. Biochim Biophys Acta 1412:212–229

Verhagen MF, Voorhorst WG, Kolkman JA, Wolbert RB, Hagen WR (1993) On the two iron centers of desulfoferrodoxin. FEBS Lett 336:13–18

VerVoot J, Heering D, Peelen S, van Berkel W (1994) Flavodoxins. Meth Enzymol 241:188–203

Vignais PM, Billoud B (2007) Occurrence, classification and biological function of hydrogenases: an overview. Chem Rev 107:4206–4272

Vignais PM, Billoud B, Meyer J (2001) Classification and phylogeny of hydrogenases. FEMS Microbiol Rev 25:455–501

Vignais PM, Colbeau A (2004) Molecular biology of microbial hydrogenases. Curr Issues Mol Biol 6:159–188

Visser M, Parshina SN, Alves JI, Sousa DZ, Pereira IAC, Muyzer G et al (2014) Genome analyses of the carboxydotrophic sulfate-reducers *Desulfotomaculum nigrificans* and *Desulfotomaculum carboxydivorans* and reclassification of *Desulfotomaculum caboxydivorans* as a later synonym of *Desulfotomaculum nigrificans*. Stand Genomic Sci 9(3):655–675. https://doi.org/10.4056/sigs.4718645

Visser M, Worm P, Muyzer G, Pereira IAC, Schaap PJ, Plugge CM et al (2013) Genome analysis of *Desulfotomaculum kuznetsovii* strain 17^{T} reveals a physiological similarity with *Pelotomaculum thermopropionicum* strain SI^{T}. Stand Genomic Sci 8:69–87

Volbeda A, Charon MH, Piras C, Hatchikian EC, Frey M, Fontecilla-Camps JC (1995) Crystal structure of the nickel-iron hydrogenase from *Desulfovibrio gigas*. Nature 373:580–587

Volbeda A, Garcin E, Piras C, de Lacey AL, Fernandez VM, Hatchikian EC, Frey M, Fontecilla-Camps JC (1996) Structure of the [NiFe] hydrogenase active site: evidence for biologically uncommon Fe ligands. J Am Chem Soc 118:12989–12996

Volbeda A, Martin L, Cavazza C, Matho M, Faber BW et al (2005) Structural differences between the ready and unready oxidized states of [NiFe] hydrogenases. J Biol Inorg Chem 10:239–249

Volbeda A, Montet Y, Vernède X, Hatchikian EC, Fontecilla-Camps JC (2002) High-resolution crystallographic analysis of *Desulfovibrio fructosovorans* [NiFe] hydrogenase. Intern J Hydrogen Energy 27:1449–1461

Vogel H, Bruschi M, Le Gall J (1977) Phylogenetic studies of two rubredoxins from sulfate-reducing bacteria. J Mol Evol 9:111–119

Voordouw G (1988) Cloning of genes encoding redox proteins of known amino acid sequence from a library of the *Desulfovibrio vulgaris* (Hildenborough) genome. Gene (Amst.) 69:75–83

Voordouw G, Brenner S (1985) Nucleotide sequence of the gene encoding the hydrogenase from *Desulfovibrio vulgaris* (Hildenborough). Eur J Biochem 148:515–520

Voordouw G, Brenner S (1986) Cloning and sequencing of the gene encoding cytochrome c_3 from *Desulfovibrio vulgaris* (Hildenborough). Eur J Biochem 159:347–351

Voordouw G, Menon NK, LeGall J, Chol E-S, Peck HD Jr, Przybyla AE (1989) Analysis and comparison of nucleotide sequences encoding the genes for [NiFe] and [NiFeSe] hydrogenases from *Desulfovibrio gigas* and *Desulfovibrio baculatus*. J Bacteriol 171:2894–2899

Voordouw G, Walker JE, Brenner S (1985) Cloning of the gene encoding the hydrogenase from *Desulfovibrio vulgaris* (Hildenborough) and determination of the NH_2-terminal sequence. Eur J Biochem 148:509–514

Walker CB, He Z, Yang ZK, Ringbauer JA Jr, He Q, Zhou J et al (2009) The electron transfer system of syntrophically grown *Desulfovibrio vulgaris*. J Bacteriol 191:5793–5801

Watenpaugh KD, Sieker LC, Jensen LH, LeGall J, Doubourdieu M (1972) Flavodoxin from the sulfate-reducing bacterium *Desulfovibrio vulgaris*, its structure at 2.5 Å resolution. Z Naturforsch 27b:1094–1095

Wang S, Huang H, Kahnt J, Thauer RK (2013a) A reversible electron-bifurcating ferredoxin- and NAD-dependent [FeFe]-hydrogenase (HydABC) in *Moorella thermoacetica*. J Bacteriol 195: 1267–11275

Wang S, Huang H, Kahnt J, Mueller AP, Köpke M, Thauer RK (2013b) NADP-specific electron-bifurcating [FeFe]-hydrogenase in a functional complex with formate dehydrogenase in *Clostridium autoethanogenum* grown on CO. J Bacteriol 195:4373–4386

Warren YA, Citron DM, Merriam CV, Goldstein EJC (2005) Biochemical differentiation and comparison of *Desulfovibrio* species and other phenotypically similar genera. J Clin Microbiol 43:4041–4045

Welte C, Kratzer C, Deppenmeier U (2010) Involvement of Ech hydrogenase in energy conservation of *Methanosarcina mazei*. FEBS J 277:3396–3403

Winkler M, Esselborn J, Happe T (2013) Molecular basis of [FeFe]-hydrogenase function and insight into the complex interplay between protein and catalytic factor. Biochim Biophys Acta 1827:974–985

Wombwell C, Caputo CA, Reisner E (2015) [NiFeSe]-Hydrogenase chemistry. Acc Chem Res 48: 2858–2865

Worm P, Stams AJM, Cheng X, Plugge CM (2011) Growth- and substrate-dependent transcription of formate dehydrogenase and hydrogenase coding genes in *Syntrophobacter fumaroxidans* and *Methanospirillum hungatei*. Microbiol 157:280–289

Xavier AV, Moura JJG, Moura I (1981) Novel structures in iron-sulfur proteins. In: Goodenough B, Hemmerich P, Ibers JA, Jørgensen CK, Neilands JB, Reinen D, Williams RJP (eds) Structure and bonding, vol 43. Springer-Verlag, Berlin, pp 187–213

Yagi T (1979) Purification and properties of cytochrome *c*-553, an electron acceptor for formate dehydrogenase of *Desulfovibrio vulgaris*. Miyazaki Biochim Biophys Acta 548:96–105

Yagi T (1994) Monoheme cytochromes. Meth Enzymol 243:104–118

Yagi T, Kimura K, Daidoji H, Sakai F, Tamura S (1976) Properties of purified hydrogenase from the particulate fraction of *Desulfovibrio vulgaris* Miyazaki. J Biochem 79:661–671

Yagi T, Kimura K, Inokuchi H (1985) Analysis of the active center of hydrogenase from *Desulfovibrio vulgaris* Miyazaki by magnetic measurements. J Biochem 97:181–187

Yoch DC, Carithers RP (1979) Bacterial iron-sulfur proteins. Microbiol Rev 43:384–421

Yoch DC, Valentine RC (1972) Ferredoxins and flavodoxins of bacteria. Ann Rev Microbiol 26: 139–162

Zacarias S, Vélez M, Pita M, De Lacey AL, Matias PM, Pereira IAC (2018) Characterization of the [NiFeSe] hydrogenase from *Desulfovibrio vulgaris* Hildenborough. Meth Enzymol 613:169–201

Zaunmüller T, Kelly DJ, Glöckner FO, Unden G (2006) Succinate dehydrogenase functioning by a reverse redox loop mechanism and fumarate reductase in sulphate-reducing bacteria. Microbiol 152:2443–2453

Zeikus JG, Dawson MA, Thompson TE, Ingvorsen K, Hatchikian EC (1983) Microbial ecology of volcanic sulphidogenesis: isolation and characterization of *Thermodesulfobacterium commune* gen. nov. and sp. nov. J Gen Microbiol 129:1159–1169

Zellner G, Messner P, Kneifel H, Winter J (1989) *Desulfovibrio simplex* spec. nov., a new sulfate-reducing bacterium from a sour whey digester. Arch Microbiol 152:329–334

Zubieta JA, Mason R, Postgate JR (1973) A four-iron ferredoxin from *Desulfovibrio desulfuricans*. Biochem J 133:851–854

Chapter 5
Systems Contributing to the Energetics of SRBP

5.1 Introduction

As with all biological cells, energy is required for growth with ATP referred to as the energy currency since it is critical for cell growth. The estimated amount of energy required by *Escherichia coli* for biosynthesis of protein, lipopolysaccharide, RNA, lipid, and DNA for a cell has been estimated to be 2.9×10^9 ATP molecules (Neidhardt and Umbarger 1966; Stouthamer 1979); and it may be considered that a Gram-negative sulfate-reducing bacterium would have a similar requirement. However, sulfate-reducing bacteria must activate sulfate using ATP before reduction can proceed. As discussed in Chap. 3, there is the requirement of ATP for sulfate to be converted to bisulfite before sulfite can be used as the electron acceptor. Thus, SRP are unique because of the requirement to expend energy before electrons can pass to the final electron acceptor.

Using growth coefficients based on ATP yield for known metabolic pathways, growth of SRB provided information to support the observations of anaerobic oxidative phosphorylation coupled to sulfate respiration. Additionally, SRP adjust to the changing chemical environment by using several alternate electron acceptors, and in some SRB, supplemental energy is generated in the internal cycling of H_2 and CO. As is the case with aerobic heterotrophic bacteria, energy produced from anaerobic respiration is not fixed but often varies with different SRP species. For most of the major steps in energetics, there appears to be a major route of activity and electron flow with alternate possibilities and the use of these alternate possibilities may be highly regulated by environmental and chemical activities.

The energetics of SRP is a broad topic, and often the approach is to focus on suspended or planktonic bacteria with little mention of symbiotic and biofilm associations. SRP cultures growing with lactate-sulfate are one of the most common conditions studied, and the energetic processes found in sulfate reducers have been reviewed (Barton et al. 2014; Fauque and Barton 2012; Matias et al. 2005; Peck and LeGall 1982; Rabus et al. 2015; Thauer et al. 2007). Additionally, the energetics of

L. L. Barton, G. D. Fauque, *Sulfate-Reducing Bacteria and Archaea*,
https://doi.org/10.1007/978-3-030-96703-1_5

specific SOP have been discussed in papers concerning monocultures of *D. vulgaris* Hildenborough (Cypionka 1995; Heidelberg et al. 2004; Keller and Wall 2011; Pereira et al. 2008; Pereira et al. 2011; Vita et al. 2015; Zhang et al. 2006a, b), *D. gigas* (Morais-Silva et al. 2014), *D. alaskensis* G20 (Price et al. 2014); *A. fulgidus* (Hocking et al. 2014; Kunow et al. 1995), *Dst. kuznetsovii* (Sousa et al. 2018), *Dst. reducens* (Otwell et al. 2016), *D. nigrificans*, and *D. carboxydivorans* (Visser et al. 2014). The interactions between SRB and bacteria in biofilms have resulted in specific adjustments concerning energetics which are addressed in reports focused on *D. vulgaris* Hildenborough (Clark et al. 2007); *D. vulgaris* Hildenborough and *Desulfobacterium corrodens* (Sivakumar et al. 2019); *Archaeoglobus fulgidus* (LaPaglia and Hartzell 1997). Symbiotic growth requires a specific set of energy-based reactions, and these are discussed in papers concerning *D. vulgaris* Hildenborough and *Methanococcus maripaludis* (Brileya et al. 2014; Walker et al. 2009), *D. vulgaris* Hildenborough, *Dehalococcoides ethenogenes* and *Methanobacterium congolense* (Men et al. 2012); *D. vulgaris* Hildenborough and *Methanosarcina barkeri* (Plugge et al. 2010) and a triculture of *D. alaskensis* G20 and *Methanococcus maripaludis* (Meyer et al. 2013, 2014). Discussions concerning SRP activities focused on symbiosis, co-cultures, and mat formations are in Chap. 2.

5.2 Growth and Yield Coefficients

An early microbial physiology review stated that a unique characteristic of autotrophic and chemolithotrophic bacteria oxidizing inorganic molecules was the mechanisms of coupling energy released to growth and not the activity of substrate oxidation (Koffler and Wilson 1951). While that statement was intended for aerobic mineral oxidizing bacteria, it would also be relevant for anaerobic bacteria oxidizing H_2 with sulfate as the electron acceptor. This interest in energy production provided a common link in several research laboratories where H_2 was used as the inorganic substrate to energize growth of SRB (Mechalas and Rittenberg 1960; Senez 1953; Senez and Volcani 1951; Sisler and ZoBell 1951; Sorokin 1954). The enzymology of H_2 oxidation became known following the initial reports (Yagi 1970; Yagi et al. 1968), and the coupling of H_2 oxidation to cell energetics in SRP is a topic of considerable interest. Lactate and pyruvate are commonly used as electron sources for growth of SRP, and their metabolic pathway is presented in Eq. (5.1):

$$\text{Lactate} \rightarrow \text{pyruvate} + 2e^- + 2H^+ \rightarrow \text{acetate} + CO_2 + 2e^- + 2H^+ \tag{5.1}$$

Reduction of sulfate to sulfide requires two moles of ATP for each mole of sulfate as indicated from the following reactions:

$$SO_4^{2-} + 2\,H^+ + ATP \rightarrow APS + PP_i \qquad \Delta G^{0'} = +46\ kJ/mol \tag{5.2}$$

$$PP_i + H_2O \rightarrow 2\,P_i \qquad \Delta G^{0'} = -21.9\ kJ/mol \tag{5.3}$$

$$APS + H_2 \rightarrow HSO_3^- + AMP + H^+ \qquad \Delta G^{0'} = -68.6\ kJ/mol \tag{5.4}$$

$$HSO_3^- + 3H_2 \rightarrow HS^- + 3\,H_2O \qquad \Delta G^{0'} = -171.7\ kJ/mol \tag{5.5}$$

The free energy from oxidation of lactate, pyruvate, formate, and H_2 takes into consideration the requirement of eight electrons for bisulfite reduction. With ATP used to activate and AMP released, two moles of ATP are required of each mole of sulfate activated to bisulfite. Considerations to address energy yields for SRP include substrate-level phosphorylation and phosphorylation coupled to electron transport. To understand the energy balance for SRP growing with different electron donor/acceptor combinations, the demand for ATP and the production of ATP is presented in Table 5.1.

With the establishment of catabolic pathways in biology and their presence in bacteria, growth of anaerobic bacteria was correlated with ATP production from various substrates. The pioneering work of Bauchop and Elsden (1960) established the yield coefficient where Y_{ATP} was the amount of growth produced from one mol of ATP generated from an energy source. Using heterotrophs growing anaerobically, Bauchop and Elsden established the value of Y_{ATP} to be 10.5 ± 0.5. Another evaluation of metabolic energetics is to calculate the amount of cell dry weight per mole of substrate used and this is expressed as $Y_{substrate}$. The following $Y_{pyruvate}$ values were reported for the following SRB: *D. vulgaris* Hildenborough $= 9.67 \pm 1.45$

Table 5.1 Energy balance for *Desulfovibrio* with different electron donors[a]

Reactions	Moles ATP from substrate phosphorylation	Net moles ATP available for growth	ATP required from phosphorylation coupled to electron transport to account for growth
$2Lactate^- + Sulfate^{2-} \rightarrow 2Acetate^- + H_2S + 2HCO_3^-$ $\Delta G^{o\prime} = -169.03$ kJ reaction^{-1}	2	0	Yes
$4Pyruvate^- + Sulfate^{2-} + 2H^+ \rightarrow 4Acetate^- + H_2S + 4HCO_3^-$ $\Delta G^{o\prime} = -356.5$ kJ reaction^{-1}	4	2	No
$4Formate^- + Sulfate^{2-} + 2H^+ \rightarrow H_2S + 4HCO_3^-$ $\Delta G^{o\prime} = -120.5$ kJ reaction^{-1}	0	0	Yes
$4H_2 + Sulfate^{2-} \rightarrow H_2S + 4HCO_3^- + 2H^+$ $\Delta G^{o\prime} = -152.2$ kJ reaction^{-1}	0	0	Yes
$4H_2 + Thiosulfate^{2-} \rightarrow 2HS^- + 3H_2O$ $\Delta G^{o\prime} = -173.8$ kJ reaction^{-1}	0	0	Yes

[a]Data for table taken from Fauque et al. (1991)

(Magee et al. 1978), *D. desulfiricans* canet 41 = 9.41 ± 1.58 (Senez 1962) and 9.9 ± 1.8 (Khosrovi and Miller 1975), an undesignated *D. desulfuricans* = 6.7 ± 0.6 (Vosjan 1970), and *D. gigas* = 7.56 ± 1.05 (Magee et al. 1978). The $Y_{lactate}$ values were generally less than the $Y_{pyruvate}$ values with $Y_{lactate}$ for *Dv*H = 5.99 ± 0.19 (Magee et al. 1978), undesignated *D. desulfuricans* = 5.3 ± 0.6 (Vosjan 1970), and *D. gigas* = 1.89 ± 1.05 (Magee et al. 1978) while the $Y_{lactate}$ for *D. desulfuricans* canet 41was 9.90 ± 1.80 (Senez 1962). Since large cells often rupture with forceful centrifugation, it would be interesting to reexamine the growth yield coefficients for *D. gigas* using measurement for growth which avoided the potential for cell lysis. With SRB, the values for and $Y_{substrate}$ and Y_{ATP} vary with the electron donor and this no doubt reflects a varied efficiency in the coupling of electron transport with concomitant phosphorylation. Molar growth yields for Y_{ATP} for *D. vulgaris* Hildenborough were reported to be 10.1 ± 1.7 and 6.7 ± 1.3 g/mol for pyruvate and lactate, respectively (Traore et al. 1981). This difference in Y_{ATP} between lactate and pyruvate is not due to changes in cell composition because with lactate as the energy source, cell carbon content was 48.92% with 48.39% cell carbon in pyruvate energized cells. Additionally, the cell formula for *D. vulgaris* Hildenborough was similar for both electron donors with lactate having an expression of $CH_{1.64}O_{0.33}N_{0.23}S_{0.01}P_{0.014}$ and for pyruvate the expression was $CH_{1.64}O_{0.31}N_{0.25}S_{0.007}P_{0.012}$ (Traore et al. 1981).

With *D. vulgaris* Marburg, the molar growth yield for H_2 plus sulfate medium was optimal at pH 6.5 to pH 6.8 with values of 11.4–14.6 g dry cells/mol ATP. The H_2 plus thiosulfate medium had almost three times the amount of dry cell mass/mol ATP (Badziong and Thauer 1978). These results suggest that with H_2 oxidized, 3 mol of ATP were produced for each mol thiosulfate reduced, and with sulfate, a net of 1 mol ATP produced for each mol of sulfate reduced.

5.3 Energetic Considerations with Organic Acids and Ethanol

Using electron transport proteins distributed in the periplasm, plasma membrane, and cytoplasm, the SRP employ an electron transport system which not only conducts electrons to sulfate but also energizes a system for ATP synthesis. The SRP display a capacity to use several different electron donors which range from H_2 to organic acids as well as several electron acceptors in addition to sulfate. Figure 5.1 provides an overview of energy metabolism for a specific SRB, *D. alaskensis* G20 (Keller et al. 2014), and this illustrates the energy-related metabolism that is found in sulfate reducers.

A The sulfate reduction pathway (cytoplasmic)

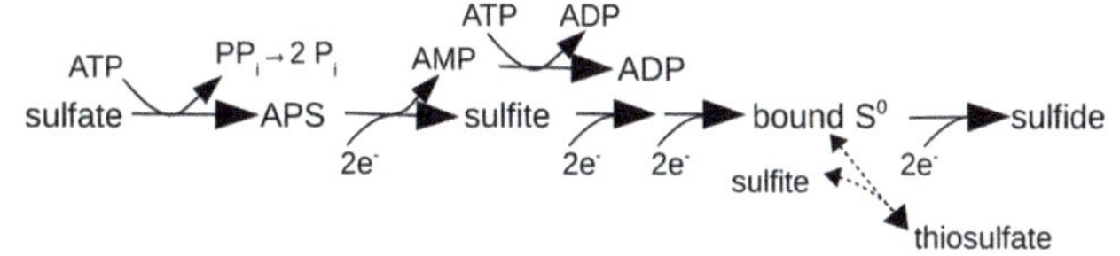

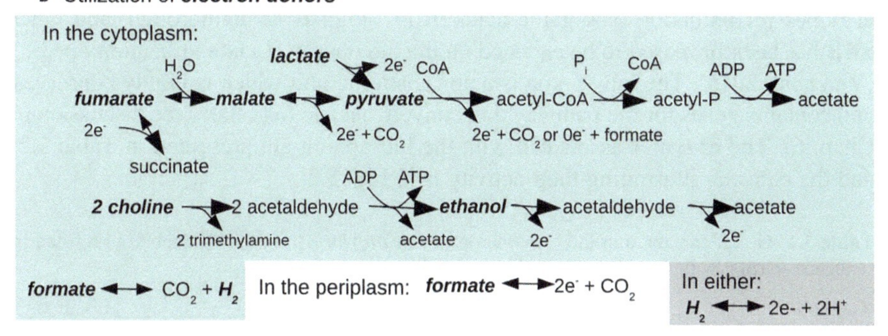

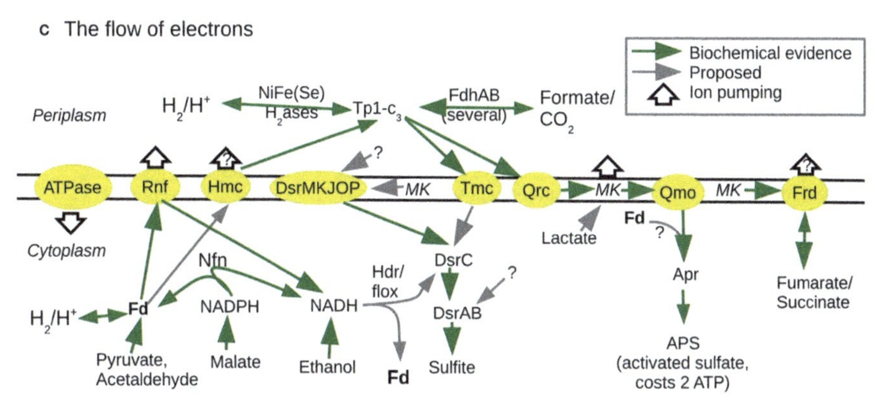

Fig. 5.1 Overview of energy metabolism of *D. alaskensis* G20. (**a**) Sulfate reduction. (**b**) Utilization of electron donors (which are inbold). (**c**) Overview of electron flow. X → Y indicates that X is oxidized while Y is reduced. Fd is ferredoxin; MK is menaquinone; Tp1-c_3 is type1 cytochrome c_3; APS is adenosine5-phosphosulfate; Apr is APS reductase; Dsr is dissimilatory sulfite reductase; H_2ase is hydrogenase; Fdh is formate dehydrogenase; Frd is fumarate reductase; Qrc is menaquinone: Tp1-c_3 oxidoreductase. Steps for which the electron donor is uncertain are marked with "?." Similarly, if the ion pumping is uncertain, it is marked with"?." (Price et al. 2014). (Copyright Information: Copyright©2014 Price, Ray, Wetmore, Kuehl, Bauer, Deutschbauer and Arkin. This is an open-access article distributed under the terms of the Creative Commons Attribution License (CC BY). The use, distribution or reproduction in other forums is permitted, provided the original author(s) or licensor are credited and that the original publication in this journal is cited, in accordance with accepted academic practice. No use, distribution or reproduction is permitted which does not comply with these terms)

5.3.1 Lactate Oxidation

From biochemical evidence, there were two principle groups of SRP, those that oxidize lactate completely with CO_2 as the carbon end product and a second group which incompletely oxidizes lactate with acetate as an end product. A coordinated expression of lactate oxidizing enzymes would be advantageous for SRBs and enzymes for oxidation of organic acids in *D. vulgaris* Hildenborough, and other SRB has been proposed to be encoded on the luo operon (*l*actate *u*tilization *o*peron) (Vita et al. 2015). The lou operon is a nonacistronic unit which is highly conserved and contains genes for the pathway that converts lactate to acetate (see discussion in Chap. 6). The enzymes associated with the luo operon are presented in Table 5.2, and the pathway illustrating their activity is in Fig. 5.2.

Table 5.2 Genes contained in the Luo operon of *Desulfovibrio vulgaris* Hildenborough listed in sequence starting with the first gene in this operon[a]

Gene	Proposed function	Locus tag
por	Pyruvate:ferredoxin oxidoreductase	DVU3025
ltp	D.L-lactate permease	DVU3026
dld-llA	D-lactate dehydrogenase subunit A	DVU3027
dld-llB	D-lactate dehydrogenase subunit B	DVU3028
pta	Phosphate acetyltransferase	DVU3029
ack	Acetate kinase	DVU3030
Undesignated	Hypothetical protein	DVU3031
lldG	L-lactate dehydrogenase subunit A	DVU3032
lldH	L-lactate dehydrogenase subunit B	DVU3033

[a]Reference: Vita et al. (2015)

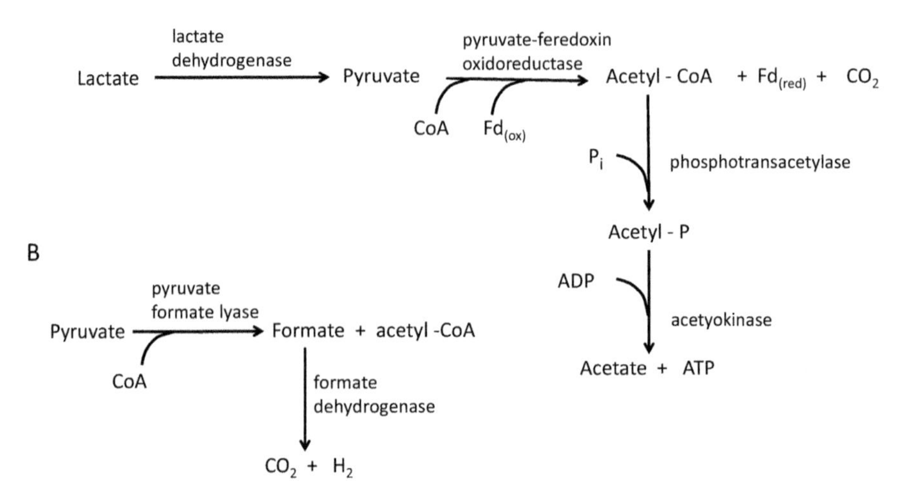

Fig. 5.2 Pathways for metabolism of lactate and pyruvate by *Desulfovibrio*. (**a**) Pathway indication use of enzymes for oxidation of lactate to acetate. (**b**) Pathway metabolizing pyruvate using formate as an intermediate

Table 5.3 Characteristics of membrane-bound respiratory coupled $NAD(P)^+$-independent lactate dehydrogenases reported in sulfate-reducing microorganisms

Microorganisms	Enzyme	Characteristics	References
Desulfovibrio vulagris Miyazaki	D-Ldh	Electron acceptors are DCIP methylene blue, tetrazolium salts, ferricyanide. Does not reduce FAD, FMN, or NAD $(P)^+$. Reduces Cytochrome c_{553}. Alternate substrate is D-2-hydroxybutyrate.	Ogata et al. (1981)
Desulfovibrio desulfuricans ATCC 27774	D-Ldh	Requires Zn^{2+} Stabilized by iodoacetamide.	Steenkamp and Peck (1981)
Desulfovibrio desulfuricans HL 21	D-Ldh	Not sensitive to O_2.	Stams and Hansen (1982)
Desulfovibrio desulfuricans ATCC 7757	D-Ldh	Molecular weight of 440,000 Da. 4 protein bands.	Czechowski and Rossmore (1990)
Archaeoglobus fulgidus	D-Ldh	Molecular weight of 50 kDa. Stable in O_2. A Zn^{2+} flavoprotein. Reduces MTT, DCIP, PMS.	Reed and Hartzell (1999)
Desulfovibrio desulfuricans HL 21	L+Ldh	Electron acceptors are DCIP, MTT, PMS ferricyanide. Inactivated by O_2. Does not reduce $NAD(P)^+$, FAD, orFMN. Alternate substrates: 2-hydroxybutyrate. malate, 3-hydroxybutyrate.	Stams and Hansen (1982) and Hansen (1994)

A lactate-sulfate medium is frequently used to support the growth of SRB with sodium lactate as a 60% solution containing a mixture of D(−)- and L(+)-lactate stereoisomers and hereafter these isomers are designated as D-lactate and L-lactate. Relatively little is known about the molecular structure and activity of Ldhs purified from SRP (Table 5.3). Following the initial report by Czechowski and Rossmore (1980), the D-lactate dehydrogenase was partially purified from *D. vulgaris* Miyazaki and found to donate electrons to C_{553} cytochrome but not to $NAD(P)^+$ (Ogata et al. 1981). Subsequently, a membrane-bound $NAD(P)^+$-independent *D*-lactate dehydrogenase was purified from *D. desulfuricans* ATCC 7757 (Czechowski and Rossmore 1990). The $NAD(P)^+$-independent L-lactate dehydrogenase was partially purified from *D. baculatus* HL21 DSM2555 (Hansen 1994; Stams and Hansen 1982). Future studies will provide the needed information concerning molecular structure and transfer of electrons into the membrane. Since the fraction of solubilized plasma membrane from *D. desulfuricans* ATCC 7757 had a mol mass of about 440 kDa with four proteins (Czechowski and Rossmore 1990), it may be important to reexamine the Ldhs isolated by this method to determine if this is another example of a membrane complex.

The overall reaction of lactate oxidation with sulfate reduction indicates a highly exothermic reaction, see Eq. (5.6):

$$\begin{aligned} 2\text{Lactate}^- + \text{SO}_4{}^{2-} + \text{H}^+ = 2\text{acetate}^- + \text{HS}^- + 2\text{CO}_2 + 2\text{H}_2\text{O} \\ \Delta\text{G}^{\circ\prime} = -196.4\text{kJ/mol} \end{aligned} \tag{5.6}$$

However, examination reveals that several reactions contribute to the transfer of electrons from lactate to sulfate. As described by Thauer et al. (2007), this reaction can be divided into a series of seven steps given as follows:

$$\text{Lactate}^- + 2\text{cyto}c_{3(\text{ox})} + \Delta\mu\text{H}^+ = \text{pyruvate}^- + 2\text{cyto}c_{3(\text{red})} \tag{5.7}$$

$$\text{Pyruvate}^- + \text{CoA} + 2\text{Fd}_{\text{ox}} = \text{acetyl} - \text{CoA} + \text{CO}_2 + 2\text{Fd}_{\text{red}} + 2\text{H}^+ \tag{5.8}$$

$$2\text{Fd}_{\text{red}} + 2\text{H}^+ = 2\text{Fd}_{\text{ox}} + \text{H}_2 + \Delta\mu\text{H}^+ \tag{5.9}$$

$$\text{H}_2 + 2\text{cyto}c_{3(\text{ox})} = 2\text{cyto}c_{3(\text{red})}{}^{-1} + 2\text{H}^+ \tag{5.10}$$

$$4\text{Cyto}c_{3(\text{red})}{}^{1-} + 0.5\text{SO}_4{}^{2-} = 0.5\text{HS}^- + 0.5\text{H}^+ + 4\text{cyto}c_{3(\text{ox})} + 2\text{H}_2\text{O} \tag{5.11}$$

$$\text{Acetyl} - \text{CoA} + \text{P}_\text{i} = \text{acetyl} - \text{P} + \text{CoA} \tag{5.12}$$

$$\text{Acetyl} - \text{P} + \text{ADP} = \text{ATP} + \text{acetate}^- \tag{5.13}$$

The energy release from the lactate/pyruvate half reaction is $E^{0\prime} = -270$ mV is not sufficient for H_2 formation because the $2e^-$ + $2H^+/H_2$ is −420 mV even when considering environmental concentrations the corrected value would be $E^{0\prime} = -270$ mV (Keller and Wall 2011). The flow of electrons from LdH to menaquinone (MK) is readily achieved since the MK_{ox}/MK_{red} is −75 mV (Keller and Wall 2011; Reed and Hartzell 1999; Thauer et al. 2007). The generation of membrane-associated energy coupled to lactate oxidation was demonstrated by the use of protonophores to inhibit H_2 reduction (Pankhania et al. 1988). With *Desulfovibrio* HL21, evidence has been presented that suggests membrane-bound lactate dehydrogenase produces chemiosmotic energy (Stams and Hansen 1982). It is of course possible that there is electron bifurcation involving electrons from reduced ferredoxin and MK to produce H_2 production (Herrmann et al. 2008; Keller and Wall 2011). This problem of requiring membrane-associated energy production from conversion from lactate oxidation may account for the inability for some SRB to grow on pyruvate-sulfate but not on lactate-sulfate. *Clostridium nigrificans* strain 55 (currently *Dst. nigrificans*) grows in pyruvate-sulfate but not in lactate-sulfate medium (Postgate 1963). Also, the lack of tight coupling for electron movement from Ldh into the plasma membrane may be reflected in the lower molar growth yield for lactate than for pyruvate. A model proposed by Keller and Wall (2011) for *D. vulgaris* Hildenborough illustrates the role of plasma membrane in transfer of electrons from lactate to sulfate (Fig. 5.3).

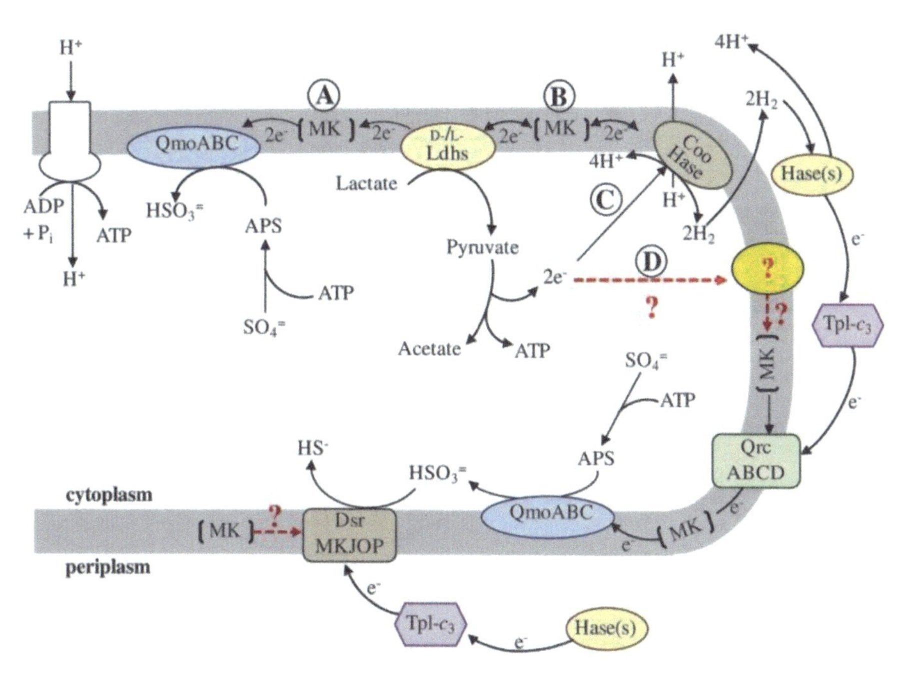

Fig. 5.3 Proposed model for the flow of electrons during sulfate respiration in *Desulfovibrio vulgaris* Hildenborough. Abbreviations: QmoABC, Quinone-interacting membrane-bound oxidoreductase (DVU0848–0850); Ldhs, lactate dehydrogenases (nine annotated); Coo Hase, CO-induced membrane-bound hydrogenase (DVU2286–2293); Hase(s), periplasmic hydrogenases (four annotated); Tpl-c_3, Type-1 tetraheme cytochrome c_3 (DVU3171); QrcABCD, Type-1 cytochrome c_3:menaquinone oxidoreductase, formerly molybdopterin oxidoreductase (DVU0692–0695); DsrMKJOP, (DVU1290–1286); and MK, Menaquinone pool. Red, dashed lines and (?) indicate metabolic pathways for which less evidence is available. The reaction arrows were drawn as unidirectional for clarity of the model and electron flow. (Keller and Wall 2011). (Copyright © 2011 Keller and Wall. This is an open-access article subject to a non-exclusive license between the authors and Frontiers Media SA, which permits use, distribution and reproduction in other forums, provided the original authors and source are credited and other Frontiers conditions are complied with)

With *D. alaskensis* G20, it has been proposed that different routes for electron flow occur with lactate and pyruvate as electron donors for sulfate reduction (Keller et al. 2014). Using mutants for cytochrome c_3 (Tplc_3), they reported that cytoplasmic Tplc$_3$ is required with H_2 or formate as the electron donor for sulfate reduction but that Tplc$_3$ is not essential for lactate-driven reduction of sulfate. Electrons from oxidation of lactate to pyruvate are directed via the membrane to reduce menaquinone and reduced menaquinone donates electrons to the QmoABC complex for the reduction of APS (Fig. 5.4). Pyruvate released from lactate reduces ferredoxin which provides a second set of electrons directed to the QmoABC complex. With pyruvate in the absence of lactate, the mutant *D. alaskensis* G20 lacking Tplc$_3$ did not reduce sulfate to sulfide even though all genes for bisulfite reduction and the QmoABC complex were present. Thus, it was proposed that reduced ferredoxin from oxidation of pyruvate interacted with an unidentified transmembrane complex

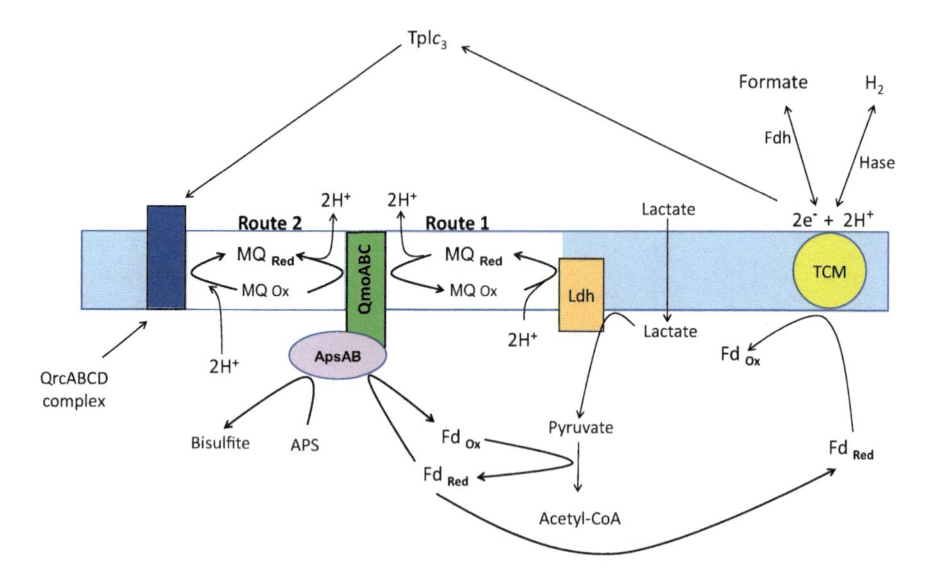

Fig. 5.4 Electron flow from organic acids to sulfate in *Desulfovibrio alaskensis* G20. With cells growing in a lactate-sulfate medium, electron flow is through membrane-associated route 1 and route 2 while with a pyruvate-sulfate medium, electron flow in the membrane is through route 2. This model indicated the role of Tplc_3 in pyruvate oxidation. Abbreviations are as follows: Fdh, formate dehydrogenase; Hase, hydrogenase; MQ, menaquinone; Fd, ferredoxin; APS adenosine phosphosulfate; Ldh, lactate dehydrogenase; APSAB, APS reductase and TCM, unidentified transmembrane complex. Source of information for this figure is in the text

(TMC) to reduce Tplc_3 which in turn reduced the QmoABC complex by way of the membrane Qrc ABC complex. Presumedly, reduced ferredoxin participates in overcoming the membrane potential and enables the menaquinone ($E^{o\prime}$ for $MK_{ox}/MK_{red} = -75$ mV) to donate electrons to APS ($E^{o\prime}$ for $APS/HSO_3^- = -75$ mV) (Keller et al. 2014).

In *Dst. reducens* MI-1, lactate oxidation proceeds by a process which results in the production of pyruvate and NADH (Otwell et al. 2016). This set of coupled reactions involves reduced ferredoxin to facilitate the reduction of NAD^+ to NADH. While some organisms such as *Acetobacterium woodii* produce reduced ferredoxin from NADH using the Rnf complex, this complex is not in the genome of *Dst. reducens* and reduced ferredoxin would be from another other metabolic system.

5.3.2 *Pyruvate Oxidation*

In SRB, pyruvate can be metabolized by pyruvate: ferredoxin oxidoreductase, pyruvate formate-lyase, pyruvate synthase, and an oxo-organic acid oxidoreductase (Tang et al. 2007). The *por* gene in the Luo operon encodes pyruvate:ferredoxin oxidoreductase (PFOR), and this enzyme is accounts for production of acetyl~CoA which leads into an energy-producing path. In many instances, pyruvate:formate lyase (PFL) is an alternate pyruvate catalyzing enzyme which produces formate which can be metabolized with the production of NADH (Fig. 5.1). Another alternate enzyme is pyruvate synthase (PS) which is commonly associated with the production of pyruvate but in appropriate chemical environments is readily reversed with the catalysis of pyruvate. While a gene in *D. vulgaris* Hildenborough has been annotated as oxo-organic acid oxidoreductase, the biochemical characteristics of this enzyme have not been established.

In the scheme for lactate oxidation with the production of acetate as an end product, the oxidation of pyruvate is by pyruvate:ferredoxin oxidoreductase (PFOR) encoded by *por* and the reaction is given in Eq. (5.9). While the lactate/pyruvate couple ($E^{0\prime} = -190$ mV) is insufficient to reduce sulfate since the SO_4^{2-}/HS^- couple is $E^{0\prime} = -200$ mV with HS^- at 0.1 mM (Thauer et al. 2007). The acetyl~CoA + CO_2/pyruvate couple is −498 mM and could drive the reduction of sulfate (Keller and Wall 2011).

5.3.3 Formate Dehydrogenases

Multiple forms of formate dehydrogenase (isozymes) are found in SRP, and the localization of these enzyme suggests distinct functions for each isozyme of formate dehydrogenase. *Desulfobacterium autotrophicum* HRM2 has seven formate dehydrogenases: four in the cytoplasm (FdhA1, FdhA3, FdhA5, FdhA6) and three are periplasmic (FdhA2, FdhA7, FdhA8) (Strittmatter et al. 2009). *D. vulgaris* Hildenborough has three formate dehydrogenase enzymes with FdhAB and $FdhABC_3$ as soluble proteins in the periplasm and FdhM (also referred to as FdhABD) (*fdnG*, DVU2482) in the periplasm but associated with the plasma membrane. The $FdhABC_3$ (DVU2809-11) enzyme contains molybdenum while the FdhAB (DVU0587-88) binds either molybdate or tungstate (da Silva et al. 2011). In the presence of molybdate, *D. vulgaris* Hildenborough produces primarily $FdhABC_3$, and in the presence of tungstate, FdhAB is the major isozyme of formate dehydrogenase produced (da Silva et al. 2013). Formate dehydrogenase, $FdhABC_3$, from *D. vulgaris* Hildenborough consists of three subunits (Sebban et al. 1995). The largest subunit (83.5 kDa) contains a molybdenum cofactor, the27 kDa subunit contains a [Fe-S] center, and the smallest subunit (14 kDa) contains a *c*-type heme. The physiological electron acceptor is considered to be a monoheme cytochrome c_3 (Sebban et al. 1995). *D. alaskensis* has three periplasmic formate dehydrogenases which bind either molybdate or tungstate. The formate dehydrogenase from *D. alaskensis* NCIMB 13491 is a heterodimer composed of 93 kDa and 32 kDa protein subunits which binds 6 Fe atoms/molecule, 0.4 Mo atoms/molecule, and 0.3 W atoms/molecule and 1.3 guanine monophosphate nucleotides /molecule (Brondino et al. 2004).

The partially purified formate dehydrogenase from *D. gigas* can be reactivated with reduced thiols following exposure to air, and cytochrome c_3 is preferred over ferredoxin or flavodoxin as the electron acceptor (Riederer-Henderson and Peck 1986a, b). Biochemical characterization and crystallization data of this formate dehydrogenase from *D. gigas* reveals that the enzyme occurs as a heterodimer (110 kDa and 24 kDa subunits) which contains tungsten, two [4Fe-4S] centers, and the essential residue of SeCys 158 at the active site (Almendra et al. 1999; Raaijmakers et al. 2002).

A formate dehydrogenase complex (Desku_0187–0192) occurs in *Dst. kuznetsovii* (Sousa et al. 2018), and a [4Fe-4S] dicluster for formate dehydrogenase has been suggested from Desku_0190. Formate dehydrogenase from *D. desulfuricans* ATCC 27774 contains molybdenum and selenium (Costa et al. 1997), and the physiological electron acceptor of formate dehydrogenase partially purified from *D. vulgaris* Miyazaki is suggested to be cytochrome *c* (Yagi 1979). Additional SRB reported to contain genes for formate dehydrogenase include several species of *Desulfovibrio* (*aespoeensis, magneticus* RS-1, *piger, salexigens*), *Desulfomicrobium baculatum, Desulfobacterium autotrophicum* HRM2, *Desulfococcus oleovorans* Hxd3, *Desulfohalobium retbaense* DSM 5692, *Desulfonatronospira thiodismutans* ASO3-1, *Syntrophobacter fumaroxidans* MPOB, *Ammonifex degensii* KC4, and *Thermodesulfovibrio yellostonii* (Pereira et al. 2011).

From initial studies, it appeared that the formate dehydrogenase from *D. vulgaris* Miyazaki (Yagi 1979) and from *D. vulgaris* Hildenborough transferred electrons to monoheme c-type cytochromes (Sebban et al. 1995). In the presence of sulfate, electrons from formate dehydrogenase could be directed by transmembrane carriers to the cytoplasm for sulfate reduction. Formate as the energy source is oxidized by SRB according to the following reaction:

$$\begin{aligned} 4HCOO^- + SO_4{}^{2-} + H^+ \rightarrow HS^- + 4HCO_3{}^- \\ \Delta G^{\circ\prime} = -46.9\ \text{kJ/mol} \end{aligned} \quad (5.14)$$

In the absence of sulfate, electrons and protons released from formate dehydrogenase could be converted by hydrogenases in the periplasm to H_2 (Martins et al. 2015).

Considerable attention has been given to formate dehydrogenases in SRP, and as reviewed by Pereira et al. (2011), several generalization concerning periplasmic formate dehydrogenases can be made:

- The formate dehydrogenase may be of three types and one microorganism may have one or several types: FdhAB, FdhABC3 (which has a dedicated cytochrome c_3), and FdhABD (which has a subunit for quinone reduction).
- The electron acceptor for FdhAB appears to be the soluble Tp$\text{I}c_3$ (periplasmic type I cytochrome c_3).

- Sulfate-reducing archaea and several sulfate-reducing bacteria do not have formate dehydrogenases, and this indicated that formate dehydrogenase is not an essential component for sulfate reduction.
- Gram-positive sulfate reducers do not have a periplasm but have FdhABD whith FdhB subunit bound to the outer side of the plasma membrane. The bound FdhABD is presumed to interface with cytochrome c_3.
- Selenocysteine is found in some formate dehydrogenases of SRB with some bacteria having only selenocyseine-containing formate dehydrogenase while some bacteria have both selenocysteine and cysteine-containing enzymes.

5.3.4 Alcohol Dehydrogenase

Several strains of SRB oxidize ethanol to support growth on ethanol-sulfate medium using alcohol dehydrogenase, and in *D. gigas* this enzyme is a homo-octomer with subunits of 43 kDa (Hensgens et al. 1993). This enzyme oxidizes several alcohols and the reaction with ethanol proceeds as follows:

$$\text{Ethanol} + NAD^+ \rightarrow \text{acetaldehyde} + NADH + H^+ \quad (5.15)$$

Acetaldehyde is catalyzed by acetaldehyde oxidoreductase, and this enzyme was purified from *D. gigas* as a homodimer (Hensgens et al. 1995) to acetic acid according to the following reaction:

$$CH_3CHO + H_2O \rightarrow CH_3CO_2H + 2H^+ + 2\,e^- \quad (5.16)$$

The energy for the acetaldehyde/acetate couple is $E^{\circ\prime} = -0.62$ V, and for ethanol/acetaldehyde, $E^{\circ\prime} = -0.197$ V (Thauer et al. 1977). The gene for ethanol dehydrogenase (ORF2977) is expressed during the growth of *D. vulgaris* with lactate, pyruvate, and formate with sulfate but is markedly increased with H_2-supported growth where acetate is required (Haveman et al. 2003). With lactate oxidation by *Desulfovibrio* spp., acetate is a metabolic end product with acetate being converted to ethanol and in late stages of growth ethanol is oxidized by an unknown pathway. With the growth ofSRB on ethanol, the metabolic conversion of ethanol to acetate does not produce substrate-level oxidative phosphorylation to energize sulfate reduction. The oxidation of 2 mol of ethanol to 2 mol of acetate through alcohol dehydrogenase and aldehyde dehydrogenase provides 8 electrons for sulfate reduction. Presumedly 2.7 mol of ATP are produced with the transfer of electrons from the two dehydrogenases to sulfate by a mechanism not fully understood (Haveman et al. 2003). A proteomic study of metabolism by *D. vulgaris* DSM644 supports the suggestion that alcohol dehydrogenase produces a proton gradient across the plasma membrane and may function in ATP synthesis (Zhang et al. 2006a).

5.4 Location of Soluble Hydrogenases, Formate Dehydrogenases and Cytochromes

5.4.1 Periplasmic Activity

Gram-negative SRB have a cell envelope distinct from Gram-positive SRB in that Gram-negatives have an outer membrane and a periplasmic region. With respect to cell energetics, the proteins of interest included uptake hydrogenases, formate dehydrogenases, and cytochromes demonstrated to be located in the periplasm (Odom and Peck 1981a). Most of the SRB have one or two periplasmic hydrogenases; however, *Dv*H and *D. desulfuricans* G20 have four hydrogenases in the periplasm. These periplasmic hydrogenases function to cleave H_2 to $2H^+$ and $2e^-$ where the electrons are directed to periplasmic cytochrome c_3 before they are relayed to the membrane. This action serves to separate protons and electrons across the plasma membrane and contribute to proton motive force. With DvH, three periplasmic hydrogenases are [NiFe] hydrogenases which donate electrons to TpI-c_3 and the fourth is [NiFe] hydrogenases 2 which donates electrons to another periplasmic poly-heme cytochrome, TpII-c_3 (Matias et al. 2005). When *Dv*H is growing with H_2, all four of the hydrogenases are capable of replacing the other as long as H_2 is provided at a high concentration (Thauer et al. 2007). *A. fulgidus* produces only two hydrogenases: a periplasmic hydrogenase (Vht) and a cytoplasmic hydrogenase (MvH:Hdl); and the reaction associated with Vht may contribute to proton motive force (Hocking et al. 2014). Since *A. fulgidus* does not produce Tplc$_3$, electrons from the VhtABC hydrogenase presumedly reduce menaquinone and the resulting menaquinol oxidation would be attributed to a membrane-bound Hdr complex which is facing the periplasm.

Periplasmic formate dehydrogenases would be similar to hydrogenases in that electrons are fist transferred to periplasmic cytochrome c_3 and subsequently to the membrane. *D. vulgaris* Hildenborough has three periplasmic formate dehydrogenases (see Sect. 5.3.4) and the two major enzymes are FdhAB$_3$ and FdhAB with similar function in transferring electrons to the cytochrome pool in the periplasm. If Mo is available in the growth medium, FdhAB$_3$ is produced, and if W is available, FdhAB is the main formate dehydrogenase (da Silva et al. 2013). *D. alaskensis* NCIMB 13491 responds similarly to the presence of Mo and W in the medium (Mota et al. 2011).

5.4.2 Cytoplasmic Activity

D. vulgaris Hildenborough produces two hydrogenases (Ech hydrogenase and Coo hydrogenase) that are associated with the plasma membrane and orientated toward the cytoplasm. With growth on H_2 or formate, genes for the *ech*ABCDEF hydrogenase are markedly up-regulated while genes for the *coo*MKLXUHF hydrogenase are markedly down-regulated (Pereira et al. 2008; Zhang et al. 2006b). This information suggests that with growth on H_2, an important role for the Ech hydrogenase would be to reduce ferredoxin for CO_2 fixation; however, additional research is needed to confirm this (Pereira et al. 2008). The Coo hydrogenase would function with lactate as the electron donor. *D. vulgaris* Hildenborough lacks the genes for a cytoplasmic formate dehydrogenase (da Silva et al. 2013), but high levels of cytoplasmic pyruvate:formate lyase (pflA) occur with H_2 and formate as electron donors (Zhang et al. 2006a).

Cytoplasmic hydrogenases have been reported for several other SRP. In *A. fulgidus*, the reduction of ferredoxin is suggested to be by the Mvh:Hdr hydrogenase (Hocking et al. 2014). Unlike *D. vulgaris* Hildenborough, *D. fructosovorans* produces a unique cytoplasmic hydrogenase which reduces $NADP^+$ for biosynthetic activities and not for energizing the plasma membrane (Malki et al. 1997). A cytoplasmic pyruvate:formate lyase is not present in *D. gigas* (Morais-Silva et al. 2014) but a membrane-bound Ech hydrogenase containing [NiFe] cluster is exposed to the cytoplasm (Rodrigues et al. 2003).

5.5 Protons and Energetic Considerations

5.5.1 *Proton Motive Force*

The chemiosmotic theory for oxidative phosphorylation (Mitchell 1966) takes into consideration the pH difference across the plasma membrane (expressed as $\Delta pH = pH_{out} - pH_{in}$) and the transmembrane electrical difference ($\Delta\psi$ internal negative in mV). The proton motive force (pmf) was calculated as follows: $pmf = -58\ \Delta pH + \Delta\psi$. At neutral pH, the charge ($\Delta\psi$) on the membrane would be primarily attributed to protons pumped from the cytoplasm outward across the plasma membrane. The chemiosmotic theory of coupling ATP production with electron transport was proposed for anaerobic SRB, and it was determined that the plasma membranes of SRB met some of the characteristics required for bacterial oxidative phosphorylation in that SRB membranes contained menaquinones, cytochromes of the *b*-type, ATP synthases sensitive to uncouplers and the enzyme fumarate reductase (Wood 1978).

The pH difference across the plasma membrane of *D. desulfuricans* has been examined using ^{31}P-NMR methods (Kroder et al. 1991) and distribution of ^{14}C-benzoate across the membrane (Kreke and Cypionka 1992). In a neutral environment, the ΔpH across the plasma membrane is about 0.5 units, and at pH 5.9, the ΔpH is 1.2 pH units. Several freshwater strains of SBR were examined and found to have a charge across the plasma membrane of −80 to −140 mV and the pmf of these strains was −110 to −155 mV (Kreke and Cypionka 1992). With a marine strain of *D. salexigens*, the membrane potential was −140 mV and the ΔpH at pH 7 was −0.3 mV which produced a pmf of −158 mV (Kreke and Cypionka 1994). In a Na^+ gradient, the Na^+ motive force was −194 mV which enabled *D. salexigens* to accumulate sulfate by Na^+ symport at a ratio of three Na^+ for each sulfate. The Na^+ gradient was established in the cell by electrogenic Na^+/H^+ antiport. It was proposed that in the marine *D. salexigens*, the pmf and proton translocating ATP synthesis was important in cell energetics (Kreke and Cypionka 1994).

5.5.2 Proton Translocation Experiments Using Cells

Using *D. vulgaris* strain MK, the translocation of protons from the cytoplasm resulting in acidification of the extracellular environment was demonstrated with sulfite reduction with H_2. The ratio of H^+ extrusion was about 3 pair of protons extruded for each sulfite reduced according to the following reaction: $SO_3^{2-} + 6H^+ + 6e- = S^{2-} + 3H_20$ (Kobayashi et al. 1982). After conducting rigorous tests with cells of *D. desulfuricans,* it was determined that proton extrusion across the plasma membrane occurred with H_2 or lactate as the electron donor and nitrite as the electron acceptor (Steenkamp and Peck 1981). Similarly, proton extrusion from cells of *D. gigas* occurred with H_2 oxidation and nitrite reduction (Barton et al. 1983). Respiratory-driven proton translocation was reported for cells of *D. desulfuricans* strain Essex exposed to H_2 and sulfate with a H^+/H_2 ratio approaching 4.4 (Fitz and Cypionka 1989).

D. vulgaris strain Marburg which lacks a periplasmic hydrogenase displayed proton translocation with a H^+/H_2 ratio of 3.9 for H_2, 1.6 for lactate and 2.4 for pyruvate. However, it was concluded that with lactate and pyruvate, protons were not pumped outward across the plasma membrane but protons generated in the periplasm from periplasmic hydrogenases attacking H_2 released from the cytoplasm (Fitz and Cypionka 1991). Protons released by hydrogenases in the periplasm could be distinguished from protons pumped into the periplasm by using copper to inhibit periplasmic hydrogenases.

5.5.3 Proton-Translocating Pyrophosphatase: HppA Complex

Pyrophosphate produced from the initial activation of sulfate produces pyrophosphate (PPi), see reaction as follows:

$$SO_4^{2-} + ATP \rightarrow APS + PP_i \qquad (5.17)$$

As discussed in Chap. 2, hydrolysis of PP_i to 2 P_i is considered to enable the reaction to proceed with the formation of APS (adenosine phosphosulfate). The free energy of PP_i hydrolysis at pH 7.4 in an environment of low ionic strength is calculated to be $\Delta G^{\circ\prime} = -23.56$ kJ/mole and with K^+ and Mg^{2+} at typical cytoplasmic concentrations, the energy of PP_i hydrolysis at typical in the cytoplasm is calculated to be $\Delta G^{\circ\prime} = -16.74$ kJ/mole pyrophosphate (Flodgaard and Fleron 1974).

As discussed by Serrano et al. (2007), the Nobel laureates Melvin Calvin and Fritz Lipmann were the first to propose a possible role for inorganic pyrophosphate as an energy donor in metabolic reactions and thereby function as an alternative to ATP. Over the years, a membrane-associated inorganic pyrophosphatase (PPase) was demonstrated to be a proton pump and sequences of H^+-PPase genes have been demonstrated in hundreds of biological species including bacteria and archaea. The crystal structure of the H(+)-PPase from mung bean indicates the presence of 16 transmembrane helices with a core of six transmembrane helices accounting for the transport of protons (Lin et al. 2012).

A few SRP have developed a mechanism whereby they can couple the hydrolysis of PP_i to proton extrusion from the cytoplasm and thereby conserve energy in pyrophosphate. In a survey of 25 genomes of SRP, genes for H^+-PPase were associated with only a few microorganisms (Pereira et al. 2011). Single genes for H^+-PPase were identified in genomes of *Candidatus* Desulforudis audaxviator MP104C (Chivian et al. 2008), *Desulfotomaculum reducens* strain MI-1 (Vecchia et al. 2014), *Desulfatibacillum alkenivorans* AK-01 (Callaghan et al. 2012), *Desulfococcus oleovorans* Hxd3 (Pereira et al. 2011), and *Dst. acetoxidans* DSM 77 (Spring et al. 2009) while two genes for H^+-PPase were detected in *Syntrophobacter fumaroxidans* strain $MPOB^T$ (Plugge et al. 2012) and *Caldivirga maquilingensis* (Itoh et al. 1999).

5.6 Transmembrane Electron Transport Complexes

In bacteria, the separation of protons and electrons across the plasma membrane results in a charge on the membrane and this energizes ATP synthase. In mitochondria, where the electron transport system is fixed, the sites for proton pumping across the membrane are well established. However, with aerobic bacteria, electron transfer varies with the chemical environment and proton pumps are not fixed as in the case of mitochondria. Most of the periplasmic hydrogenases and formate dehydrogenases in SRP do not interface with the plasma membrane but rely on cytochromes to direct electrons to transmembrane respiratory complexes. Examples of transmembrane electron transfer complexes found in SRP are listed in Table 5.4. Using genome analysis, the composition and topography of electron transport complexes on the plasma membrane of *D. gigas* have been established (Morais-Silva et al. 2014) and are presented in Fig. 5.5. The characteristics and distribution of various electron transport complexes in several SRB are discussed in recent reviews (Grein et al. 2013; Keller and Wall 2011; Morais-Silva et al. 2014; Pereira et al. 2008; Pereira et al. 2011; Price et al. 2014; Rabus et al. 2015). The composition and orientation of several electron transport complexes, many of which are discussed here, are provided in Fig. 5.6.

Table 5.4 Examples and distribution of transmembrane and soluble electron transfer complexes in SRB[a]

Complex	ΔE^a (mV)	Physiological activity: Electron donor/electron acceptors	Distribution in taxonomic groups: Found in several species	Found in only a few species
Transmembrane complexes				
QrcABCD	*+340*	Tplc_3/MQ	+	
QmoABC	*+14*	MQH_2, APS reductase	+	
DsrMKJOP		MQH_2/DsrC	+	
Hmc		Tplc_3/DsrC	+	
Tmc	*+264*	Tplc_3/DsrC	+	
Rnf	+180	Fd/NAD(H)	+	
Nuo		NADH/MQ, Fd?	+	
Nhc	*+340*	Tplc_3/MQ		+
Ohc	*+340*	Tplc_3/MQ		+
Na^+-Nqr	*+433*	NADH/MQ		+
bd oxidase	*+705*	MQ/O_2		+
Periplasmic proteins				
Tplc_3		Hydrogenases, Formate dehydrogenase, MQ	+ (Gram-negative bacteria)	
Cytoplasmic complexes				
Hdr–Flx		Fd, NADH	+	
NfnAB		Fd_{red}, NADH, $NADP^+$	+	

[a]Source for information: Rabus et al. (2015); Calisto et al. (2021)

The transmembrane complexes provide for an organization of electron flow and ion distribution with an emphasis on proton expulsion to energize the plasma membrane. As demonstrated in Fig. 5.6, the pumping of ion and especially protons with is by way of specific protein/MK complexes. As described by Keller et al. (2014), these ion pumping complexes are active with sulfate reducers growing with various electron sources including malate, formate, and pyruvate as electron sources (Fig. 5.7). Although there is tremendous metabolic variation in SRP, the sites for pumping protons are limited. In *Dst. acetoxidans*, the oxidation of NADH is attributed to a proton-translocating NADH dehydrogenase complex that requires menaquinones and this complex is strikingly similar to Complex I in *Escherichia coli* (Spring et al. 2009).

5.6.1 *Quinone Oxidoreductase Complex: Qmo*

A quinone-interacting membrane-bound oxidoreductase complex (Qmo) was isolated from *D. desulfuricans* ATCC 27774, and it consists of three subunits which contain hemes *b*, FAD, and several FeS centers (Pires et al. 2003). Based on the

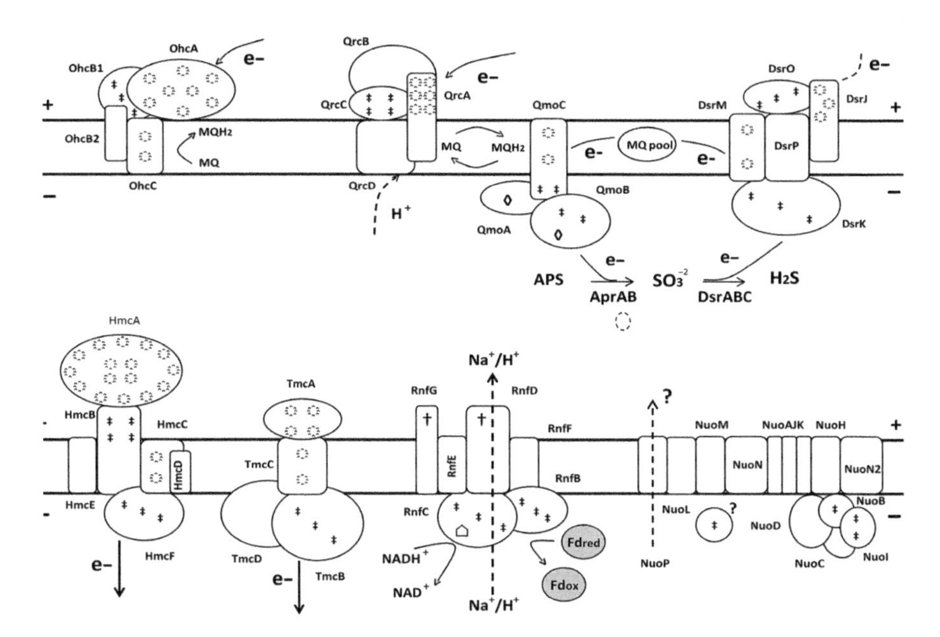

Fig. 5.5 Schematic representation of membrane-bound electron-transfer complexes present in *Desulfovibrio gigas* genome. The complexes were identified in the genome according to their predicted function: quinone reduction, Ohc, and Qrc; quinol oxidation, Qmo; transmembrane electron transfer/sulfite reduction DsrMKJOP, Hmc, and Tmc; and NADH/Fd oxidation, Rnf, and Nuo. Symbols represent: circle with dash boundary, heme; ‡, iron sulfur center; †, FMNcofactor; ⌂, flavin cofactor, and ◇, FAD cofactor. Dashed lines represent hypothetical pathways for electron/proton flow (Morais-Silva et al. 2014)

reduction potentials for the two Qmo hemes b, Pires et al. (2003) proposed that Qmo could be involved in passage of electrons from menaquinone to reduction of APS or sulfite. The QmoABC complex is not required for sulfite reduction but appears to be essential for sulfate reduction (Zane et al. 2010). This complex is annotated in 22 of the first genomes examined and is present in two of the genomes of the sulfate-reducing archaea (Pereira et al. 2011). Genes for the QmoABC complex are not present in *Caldivirga maquilingensis, Desulfotomaculum acetoxidans*, and *C.* Desulforudis audaxviator; and in these microorganisms, genes for heterodisulfide reductase (*hdrBC* genes) are present suggesting that the products of these genes replace the activity of the QmoABC complex (Junier et al. 2010). The initial report of a QmoABC complex followed examination of *D. desulfuricans* ATCC 27774 (Pires et al. 2003). Two hemes *b,* two FAD moieties, and several Fe-S centers are bound to the three subunits of QmoABC. Both QmoA and QmoB are soluble proteins that are similar to a subunit of the soluble heterodisulfide reductases (HdrA) (Hedderich et al. 2005). The QmoC subunit contains a transmembrane

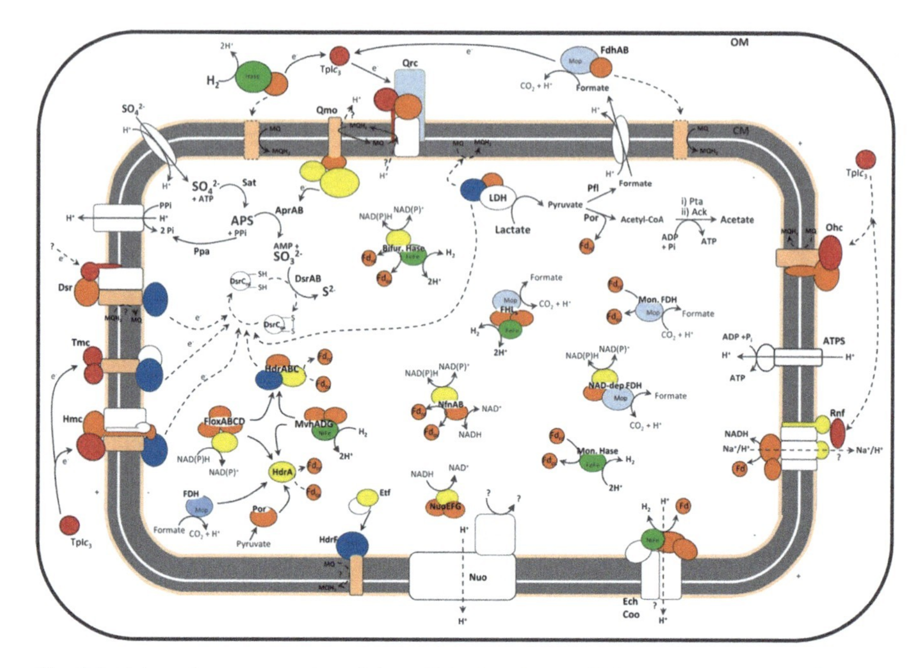

Fig. 5.6 Schematic representation of the cellular location of SRO main energy metabolism proteins. No single organism is represented. For the exact distribution of proteins in each organism, please refer to the tables. The dashed lines represent hypothetical pathways, or (in the case of periplasmic Hases and FDHs) pathways present in only a few organisms. For the sake of clarity, a few proteins discussed are not represented. Color code is red for cytochromes *c*, pale orange for cytochromes *b*, yellow for flavoproteins, dark orange for FeS proteins, light blue for proteins of molybdopterin family, dark blue for CCG proteins, and green for catalytic subunits of Hases. (Pereira et al. 2011). (Copyright © 2011 Pereira, Ramos, Grein, Marques, Marques da Silva and Venceslau. This is an open-access article subject to a non-exclusive license between the authors and Frontiers Media SA, which permits use, distribution and reproduction in other forums, provided the original authors and source are credited and other Frontiers conditions are complied with)

protein segment fused to a hydrophilic Fe-S moiety. It is considered that the Qmo complex transfers electrons from the quinone pool to APS (adenosine phosphosulfate) reductase and that this is an example of energy conservation (Pires et al. 2003; Venceslau et al. 2010).

5.6.2 *Sulfite Reduction Complex: Dsr*

Several protein subunits are associated with the Dsr complex. The DsrMKJOP complex was found in *A. fulgidus* (Mander et al. 2002) and *D. desulfuricans* ATCC 27774 (Pires et al. 2003) and was initially reported as Hme in *A. fulgidus*. The Dsr complex contains the following protein subuits: DsrM, a membrane subunit containing cytochrome *b*; DsrK, protein subunit that binds a special [4Fe-4S] cluster;

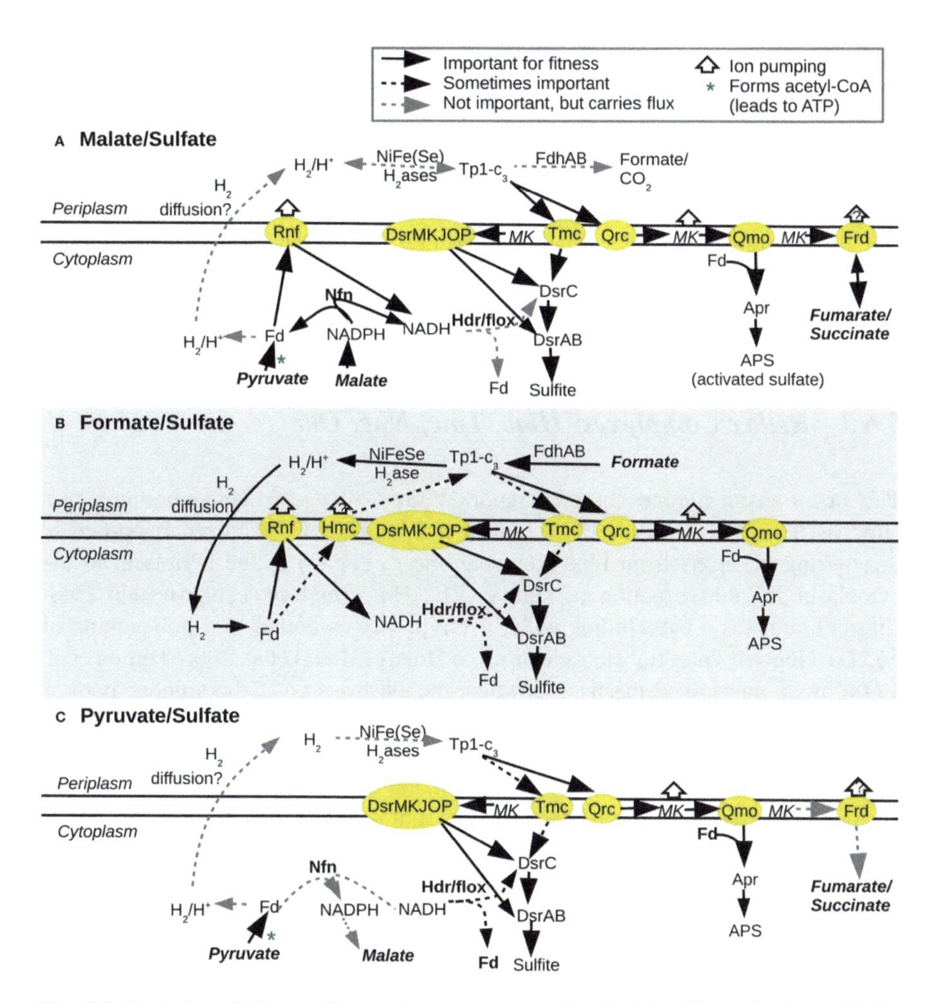

Fig. 5.7 Scenarios of electron flow and energy conservation during sulfate reduction with the electron donors (**a**) malate, (**b**) formate, or (**c**) pyruvate. In each panel, the electron donor and related metabolites are in bold italics. Only electron paths that are important for fitness (or essential) or that we have other reason to believe are carrying flux are shown. Genes are "sometimes important" if they were important for fitness in some experiments with this electron donor but not others. In (**a**, **c**), the path for electrons from the cytoplasm to Tp1-c_3 is uncertain and could be via formate diffusion or Hmc (not shown) instead of by the diffusion of hydrogen. (Price et al. 2014). (Copyright©2014 Price, Ray, Wetmore, Kuehl, Bauer, Deutschbauer and Arkin. This is an open-access article distributed under the terms of the Creative Commons Attribution License (CC BY). The use, distribution or reproduction in other forums is permitted, provided the original author(s) or licensor are credited and that the original publication in this journal is cited, in accordance with accepted academic practice. No use, distribution or reproduction is permitted which does not comply with these terms)

DsrJ, binding a triheme cytochrome *c*, DsrO, an iron-sulfur protein; and DsrP, a protein of the Nrfl family. The DsrJOP subunits are important in bacteria with an outer membrane, and these three subunit proteins serve to direct electrons from periplasmic hydrogenases and formate dehydrogenases not associated with the plasma membrane to the menaquinone pool; however, it is not clear which direction the electrons travel (Pereira et al. 2011).

In Gram-positive sulfate reducers and the archaeon *Caldivirga maquilingensis*, the DsrMK complex is present but not the protein subunits of DsrJOP (Junier et al. 2010). Also, the MK protein subunits are found in all Gram-negative sulfate reducers and these proteins are important components for bisulfite reduction. In Gram-negative sulfate reducers, the presence of a periplasm with hydrogenase or formate dehydrogenase requires the presence of DsrJOP.

5.6.3 Redox Complexes: Hmc, Tmc, Nhc, Ohc

The first transmembrane complex reported for *D. vulgaris* Hildenborough was HmcABCDEF (Rossi et al. 1993), and it has long been considered important in transferring electrons from type I cytochrome c_3 (Tplc_3) in the periplasm to the cytoplasm for the reduction of sulfate. The Hmc (*h*igh-*m*olecular-weight *c*ytochrome) contains a cytochrome with 16 *c*-type hemes bound to a polypeptide of 56 kDa, HcmA). Tplc$_3$ transfers electrons to HmcA (Matias et al. 2005; Pereira et al. 1998) by an unresolved mechanism where the electrons go to the quinone pool or directly to the cytoplasm. The Hmc complex is considered to be involved in symbiotic growth of *D. vulgaris* Hildenborough in the transfer of electrons from the cytoplasm to the periplasm (Walker et al. 2009).

Another complex that transfers electrons from the periplasm to the cytoplasm is TmcABCD (Valente et al. 2001). TmcA is an acidic tetraheme cytochrome c_3 (Type IIc_3), TmcB is a protein homologous to HmcF, TmcC is a cytochrome *b*, and TmcD is a tryptophan rich protein. The reduced Tplc$_3$ readily donates electrons to TmcA and efficiently transfers electrons from the periplasm to the cytoplasm (Pereira et al. 2006, 2011). The Tmc complex is up-regulated with growth on H_2 as compared to lactate, and this complex is important with H_2 as the electron donor for growth (Pereira et al. 2011). The SRB strains that have Hmc also have Tmc; however, neither Hmc or Tmc have been reported for archaeal sulfate reducers and have not been established for all the SRB genomes (Pereira et al. 2011).

Two redox active complexes similar to Hmc and Tmc are Nhc and Ohc. The *n*ine *h*eme *c*ytochrome complex (Nhc) is in the NhcA subunit and it is readily reduced by Tplc_3 (Saraiva et al. 2001). The octa-heme cytochrome complex (Ohc) is found in some SRB and the OhcA contains an eight heme cytochrome which is also reduced by Tplc_3. Both the Ohc and Nhc lack the protein to transfer electrons to the cytoplasm and as a result their electrons are transferred to the quinone pool (Matias et al. 2005). The Nhc is present in only a few SRB including *D. aespoeensis, D. africanus* DSM 2603, *D. desulfuricans* ATCC 27774, *D. desulfuricans* DSM 642, *D. piger, D. putealis* DSM 16056, *Desulfatibacillum alkenivorans*, and *Desulfonatronospira thiodismutans* (Pereira et al. 2011; Rabus et al. 2015). Several SRB have the Ohc complex and these include: *D. magneticus, D. longus* DSM 6739,

D. hydrothermalis DSM 14728, *D. gigas* DSM 1382, *D. bastinii* DSM 16055, *D. aminophilus* DSM 12254, *D. alcoholivorans* DSM 5433, *Desulfotignum phosphitoxidans* FiPS-3, DSM 13687, *Desulfosarcina variabilis* Montpellier, *Desulfonema limico*la Jadebusen, DSM 2076, *Desulfococcus biacutus* KMRActS, *Desulfobacula toluolica* Tol2, *Desulfobacterium anilini* DSM 4660, *Desulfobacter vibrioformis* DSM 8776, *Desulfobacter curvatus* DSM 3379, *Desulfatirhabdium butyrativorans* DSM 18734, *Desulfatibacillum aliphaticivorans* DSM 15576, *Desulfomicrobium baculatum, Desulfatibacillum alkenivorans, Desulfococcus oleovorans*, *Desulfohalobium rethaense*, and *Syntrophobacter fumaroxidans* (Pereira et al. 2011; Rabus et al. 2015).

5.6.4 Quinone Reductase Complex: Qrc

The *q*uinone *r*eductase *c*omplex (Qrc) was isolated from *D. vulgaris* Hildenborough, and it consists of several subunits, QrcABCD (Venceslau et al. 2010). The composition of the protein subunits are as follows: QrcA is a hexaheme cytochrome *c*, QrcB is a protein of the molybdopterin family, QrcC is a periplasmic Fe-S protein, and QrcD is an integral membrane protein. Electrons from periplasmic formate dehydrogenase or hydrogenase reduce Tplc_3, and this cytochrome interfaces with QrcA to reduce the menaquinone. The Qrc complex and the Qmo complex have been proposed to have important roles in promoting proton motive force with SRB using periplasmic hydrogenases and/or formate dehydrogenases with sulfate as the electron acceptor (Simon et al. 2008; Venceslau et al. 2010). While the Qrc complex is found in most SRB that have an outer membrane and periplasm, there are some exceptions. Soluble periplasmic hydrogenases and formate dehydrogenases are found in *D. piger* and *Desulfonatronospira thiodismutans* and *Thermodesulfovibrio yellowstonii,* but these bacteria lack the Qrc complex (Pereira et al. 2011) and would use alternate systems for electron transfer.

5.6.5 Ion-Translocating NADH Dehydrogenase Complexes: Rnf, Nqr, Nuo

The Rnf complex is found in almost all prokaryotes, and since it was first reported for *Rhodobacter capsulatus,* it carries the abbreviation for *Rhodobacter n*itrogen *f*ixation (Rnf) (Schmehl et al. 1993). The Rnf complex contains six to eight subunits and is characteristically found in all bacteria except *Clostridia* and *Archaea.* The role of Rnf is proposed to transport electrons from reduced ferredoxin to NAD^+ (reverse electron flow) which is coupled to electrogenic H+ or Na+ translocation (Müller et al. 2008). Additionally, *Acetobacter woodii* appears to have the capability of reducing NAD^+ from ferredoxin which is coupled to electrogenic Na^+ transport

(Biegel and Müller 2010). Since many sulfate reducers do not encode the Rnf but do produce proton-pumping hydrogenase (Ech or Coo), the role of the Rnf complex in SRB is not fully understood (Price et al. 2014).

Nqr (*N*ADH:*q*uinone oxido*r*eductse) is a Na^+-translocating complex that has been reported for a few SRB; and these include *Desulfobulbus japonicus* DSM 18378, *Desulfobulbus mediterraneus* DSM 13871, *Desulfotalea psychrophila* LSv54, *Desulfobacterium autotrophicum* HRM2, *Desulfococcus oleovorans* Hxd3, *Desulfurivibrio alkaliphilus* AHT2, *D. bastinii* DSM 16055, *D. inopinatus* DSM 10711, *D. salexigens* DSM 2638, *D. zosterae* DSM 11974, and *Desulfonatronovibrio hydrogenovorans* DSM 9292 (Pereira et al. 2011; Rabus et al. 2015).

The Nuo complex (NADH;oxidoreductase) constitutes Complex I which links proton expulsion to electron transfer from NADH to quinone (Efremov et al. 2010). Nuo has more subunits and a greater mass than Nqr or Rnf. In terms of distribution throughout the SRB, Nuo is commonly found in *Desulfovibrio* spp. However, the only redox complexes found in *Desulfotomaculum* spp. are Nuo, Qmo, and Dsr (Pereira et al. 2011; Rabus et al. 2015). In *Escherichia coli*, each of the three subunits (NuoL, NuoM and NuoN) contains 14 transmembrane helices. The architecture of the Nuo complex of sulfate reducers has not been addressed at this time.

5.7 Cytoplasmic Electron Transport Complexes

5.7.1 NADH-Ferredoxin Complex: Nfn

Broadly distributed in most SRB is the cytoplasmic Nfn (*N*ADH-dependent reduced *f*erredoxin:*N*ADP$^+$ oxidoreductase) protein complex. The physiological role of NfnAB flavoprotein is to use NADH and reduced ferredoxin to reduce NADP$^+$ or to function as a bifurcating enzyme to use NADPH to reduce ferredoxin and NAD$^+$ (Huang et al. 2012; Wang et al. 2010). The NfnA subunit contains a [2Fe-2S] cluster and a FAD cofactor while the NfnB contains two [4Fe-4S] clusters and a FAD (Buckel and Thauer 2013). In some instances, the NfnAB is encoded on two genes and in other instances it is encoded on a fused gene.

5.7.2 Heterodisulfide Reductase: Hdr/flox

Heterodisulfide reductase (Hdr) catalyzes a bifurcating reaction that converts the reversible NAD$^+$-dependent reduction of NADP$^+$ by reduced ferredoxin or the NAD$^+$-dependent reduction of ferredoxin by NADPH. This transhydrogenase consists of two subunits, and HdrAB is found in the genomes of many SRP (Pereira et al. 2011; Price et al. 2014). The Hdr complex consists of HdrABC where HdrA is a flavoprotein, HdrB subunits contain a [4Fe-4S] cluster, and HdrC is an iron-sulfur protein. Methanogens are known to form a complex consisting of HdrABC and MvhADG (*m*ethyl-*v*iologen-reducing [NiFe]-*h*ydrogenase) which links H_2 reduction of CoM-S-S-CoB with ferredoxin reduction using a mechanism involving flavin such as contained by HrdA (Kaster et al. 2011). This is an example of electron bifurcation involving a flavin. The cytoplasmic HdrABC enzyme is proposed to complex with flavin oxidoreductase (flox) to produce the Hrd/flox complex (Pereira et al. 2011). This Hrd/flox complex would carry electrons from NADH to ferredoxin and to a heterodisulfide molecule such as DsrC. *D. alaskensis* G20 produces two complexes named Hrd/flox-1 and Hrd/flox-2 (Price et al. 2014). Mutation in the Hrd/flox-1 complex resulted in diminished growth of *D. alaskensis* growing on formate/acetate/sulfate medium and H_2/acetate/sulfate medium (Price et al. 2014).

In *D. vulgaris* Hildenborough growing on ethanol-sulfate, the expression of hdr-flx (heterodisulfide reductase-flavin oxidoreductase genes) is at a higher level than when *D. vulgaris* Hildenborough is growing on lactate-sulfate (Ramos et al. 2014). The proteins designated as FlxABCD-HdrABC participate in NADH oxidation with ethanol employing flavin-based electron bifurcation resulting in the reduction of ferredoxin, and when sulfate is limiting and ethanol fermentation occurs, the reaction catalyzed by FlxABCD-HdrABC proteins would function in reverse with the production of ethanol and NAD^+. In *D. alaskensis* G20, the Hdr/flox-1 complex appears to be involved in electron bifurcation to produce reduced ferredoxin from NADH and contribute to ion pumping across the membrane (Price et al. 2014).

In *Dst. acetoxidans*, some of the genes for heterodisulfide reductase are located near genes associated with electron transport and this location on the chromosome is used to support the proposal that heterodisulfide reductases is used to generate a proton gradient (Spring et al. 2009). Heterodisulfide reductase has been proposed to establish a proton gradient in *Moorella thermoacetica* (Pierce et al. 2008). This flavin-based electron bifurcation of heterodisulfide reductase in bacteria has been suggested to be an energy-conserving process in anaerobic microorganisms, and it would supplement substrate-level phosphorylation and electron coupled oxidative phosphorylation (Buckel and Thauer 2013).

5.8 ATPase (F-Type and V-Type)

Membrane-associated ATPase was considered to be a coupling factor for oxidative phosphorylation in mammalian systems, and a characteristic of this coupling-factor activity was the stimulation of ATPase by 2,4-dinitrophenol (DNP) (Pullman and Schatz 1967). Membrane fragments of *D. gigas* containing ATPase activity was inhibited by DNP, and this was taken to suggest that the respiratory-coupled phosphorylation in *Desulfovibrio* was characteristic of anaerobic oxidative phosphorylation (Guarraia and Peck 1971). While it has been assumed that the ATP synthase/ATPase protein unit in *Desulfovibrio* was the F_0F_1-type, positive evidence for an F-type unit was provided for *D. vulgaris* strain MF (Ozawa et al. 2000). The presence of genes for the production of F_0 protein subunits (F_0A and F_0B) along with the alpha, beta, gamma, delta, and epsilon subunits of F_1 is readily observed in the genomes of *Desulfovibrio* (see supplemental information in Morais-Silva et al. 2014). It is interesting that *D. gigas* (Morais-Silva et al. 2014) and *Dst. reducens* (Otwell et al. 2016) not only have genes for complete production of the F_0F_1 type ATPase, but they also have several genes for a secondary V (vacuolar)-type ATPase. Genes for this vacuolar-type ATPase (V_0V_1) are found in *D. gigas* and include V-type ATPase synthase subunits A, B, D, E, K (see supplemental information in Morais-Silva et al. 2014). V-type ATPase is not found in other *Desulfovibrio* spp. but is found in other anaerobes including *Enterococcus hirae* where it is a sodium pump (Kakinuma et al. 1999) and in *Desulfohalovibrio reitneri* DSM 26903^T where it has been proposed to regulate cytoplasmic pH homeostasis (Spring et al. 2019). The activity of V-type ATPase remains to be established in *D. gigas*. Bacterial ATPases require Mg^{2+} for activity and recently it has been demonstrated that bismuth as Bi^{3+} inhibits the oxidation of H_2 with sulfate as the electron acceptor in *D. desulfuricans* 27774 and it was proposed that Bi^{3+} inhibits the F_1 subunit of ATP synthase (Barton et al. 2019).

5.9 Direct Measurement of ATP Production via Anaerobic Oxidative Phosphorylation

The impetus to study energy production coupled to electron transport in *Desulfovibrio* can be traced to the compelling reviews of Peck (1962a, b) and Senez (1962). It had long been held that oxidative phosphorylation was a characteristic of aerobic respiration and the scientific community was slow to accept the possibility of oxidative phosphorylation to occur in anaerobic bacteria. The chemistry of inorganic sulfur compounds is complex, and biochemical activities of sulfur compounds always included divergent chemical reactions. The establishment of sulfate-reduction pathway by SRB required the mastery of sulfur chemistry. Additionally, the relevance of H_2 oxidation to the energetics of a chemoheterotrophic bacterium oxidizing lactate or pyruvate was being addressed at this time.

5.9.1 Phosphorylation Coupled to Dissimilatory Sulfate Reduction: Bisulfite and Associated Reactions

The demonstration of ATP requirement for sulfate reduction by cells of *D. desulfuricans* came from two laboratories that had been working independently (Ishimoto 1959; Peck 1959). The reaction of sulfite reduction with molecular hydrogen is as follows with energetic calculated according to Thauer et al. (1977):

$$\begin{aligned} S0_3{}^{2-} + 2H^+ + 3H_2 &\rightarrow H_2S + 3H_20 \\ \Delta G^{0\prime} &= -172.8\ \text{kJ/reaction} \end{aligned} \tag{5.18}$$

An early publication reported a diminished level of H_2 oxidized by intact cells when 2,4-dinitrophenol, a known uncoupled of mitochondrial oxidative phosphorylation, was added to the reaction (Peck 1960). To become an electron acceptor, ATP was required for sulfate activation (Ishimoto and Fujimoto 1961; Peck 1961) and it was logical to propose that phosphorylation was coupled to electron transport since there was no mechanism known to produce substrate-level phosphorylation from metabolism of H_2 or sulfate.

While cells of *D. desulfuricans* would oxidize H_2 with sulfate reduction (Postgate 1951), the reduction of sulfate by H_2 was not observed with cell-free extracts unless ATP was added at substrate levels (Ishimoto 1959; Peck 1959). Using many of the procedures developed by scientists working with mitochondrial respiratory systems, Peck (1966) demonstrated phosphorylation to be coupled to electron transport from H_2 to bisulfite using a membrane fraction and soluble protein from *D. gigas.* Inhibition of phosphorylation was observed with the addition of pentachlorophenol or gramicidin while oligomycin had an effect on phosphorylation. Only slight uncoupling activity was observed with 2,4-dinitrophenol because *D. desulfuricans* reduces 2,4-dinitrophenol to 2,4-diaminophenol (Peck 1961). Additions of uncoupling agents (i.e. pentachlorophenol or gramicidin) did not inhibit respiration which indicates that phosphorylation and respiration were not tightly coupled. Calculation of esterification coupled to respiratory reduction of bisulfite in *D. gigas* was expressed as the amount of phosphorylation for each 2 electrons (or $P/2e^-$ or P/H_2) of 0.13–0.17 was reported and P/H_2 approaching 0.4 had been obtained (Peck 1966). In comparison, the reconstituted cell-free fraction of *Micrococcus lysodeikticus* gave $P/2e^-$ values ranging from 0.13 to 0.66 and phosphorylation coupled to respiration was not inhibited by oligomycin and phosphorylation was not tightly coupled to respiration (Ishikawa and Lehninger 1962). Unlike mitochondrial systems but similar to other bacterial systems, respiration by *D. gigas* was not tightly coupled to phosphorylation and ATP synthesis could be uncoupled from respiration. The electron transport with concomitant phosphorylation between H_2 and bisulfite required ferredoxin, and esterification of ADP was observed with the reaction containing H_2 and thiosulfate; however, the $P/2e^-$ values with thiosulfate were lower than with bisulfite and this may reflect that bisulfite was a product of thiosulfate and phosphorylation was attributed to bisulfite reduction (Barton et al. 1972). The requirement of electron transport for phosphorylation and reduction of phosphorylation by known uncoupling agents in the *D. gigas* reconstituted reactions met the standard for the demonstration of oxidative phosphorylation (Chance and Williams 1956). It is assumed that the disruption of the cells of *D. gigas* by French pressure process leads to the formation of both inverted membrane vesicles and linear membrane fragments. Since both the membrane vesicles and linear membrane fragments participated in electron flow, the presence of not closed membrane vesicles could have contributed to low P/2electron ratio.

Using *D. gigas*, hydrogen oxidation and concomitant phosphorylation were not observed when soluble proteins treated for the removal of ferredoxin and flavodoxin; however, when ferredoxin was added to the reactions, both electron transport and phosphorylation occurred (Barton et al. 1972; Peck and LeGall 1982). The addition of flavodoxin in place of ferredoxin to the treated cell-free preparation resulted in electron transport but not phosphorylation, and this suggests that ferredoxin and not flavodoxin energizes the membrane with H_2 oxidation.

5.9.2 Elemental Sulfur Reduction

With examination of anaerobes growing on elemental sulfur, *D. desulfuricans* Norway 4 and *D. gigas* were found to grow with ethanol as the electron donor coupled to S^0 reduction. However, *D. desulfuricans* Essex 6, *D. vulgaris* Hildenborough, *Dst. nigrificans* Delft 74, and *Desulfomonas pigra* 11112 did produce H_2S when placed in media containing ethanol and S^0; however, no growth occurred with these SRB (Biebl and Pfennig 1977). Using a membrane fraction from an extract of *D. gigas*, phosphorylation of ADP was coupled to the oxidation of H_2 and reduction of S^0 to H_2S (Fauque 1994; Fauque et al. 1980). As in the phosphorylation reactions concomitant with hydrogen oxidation with bisulfite, phosphorylation with the H_2-S^0 reaction was uncoupled with pentachlorophenol and methyl viologen. The P/2e$^-$ ratio was 0.1 with the H_2-S^0 reaction and this low value may reflect the random attachment of hydrogenase as a result of cell breakage in the French Press. It was proposed by Fauque et al. (1980) that colloidal sulfur was reduced by electrons from H_2 according to the following reaction:

$$H_2 + S^0 \rightarrow H^+ + HS^- \qquad \Delta G^{o'} = 3\tilde{0}.6\ \text{kJ/mol}\ H_2 \qquad (5.19)$$

However, it is most likely that the electron acceptor is polysulfide (Hedderich et al. 1998) and reduction would be according to the following reaction:

$$H_2 + {S_n}^{2-} \rightarrow H^+ + HS^- + {S_{(n-1)}}^{2-} \qquad \Delta G^{o'} = 3\tilde{1}.0\ \text{kJ/mol}\ H_2 \qquad (5.20)$$

Unlike colloidal sulfur, polysulfide as ${S_4}^{2-}$ or ${S_5}^{2-}$ would be able to traverse the membrane. The redox potential for the S^0/HS^- and ${S_4}^{2-}/HS^-$ coupled reactions are −275 mV and −260 mV, respectively (Hedderich et al. 1998). *Wolinella succinogenes* uses the same hydrogenase in electron transport with polysulfide reduction as with fumarate reduction (Schröder et al. 1988) and it would be interesting to learn if this is the same with *D. gigas*.

5.9.3 Fumarate as an Electron Acceptor

Several strains of SRB were reported to grow in the absence of sulfate by fumarate dismutation; and these include *D. gigas* NCIB 9332 and strains of *D. desulfiricans* including El Agheila A, Norway 4, California 29:137:5, Australia, California 48:68, El Agheila 4, Teddington R and Byron (Miller and Wakerley 1966). Using a cell-free extract of *D. gigas* growing on lactate-sulfate medium, fumarate reduction was observed with pyruvate as the electron donor (Hatchikian and LeGall 1970). A membrane fraction from *D. gigas* grown in lactate-sulfate medium was shown to couple oxidative phosphorylation to electron transfer from H_2 to fumarate (Barton et al. 1970, 1972). The reaction of H_2 oxidation with fumarate reduction would be as follows (Thauer et al. 1977):

$$H_2 + \text{fumarate}^{2-} \rightarrow \text{succinate}^{2-} \qquad \Delta G^{o'} = -86.2\ \text{kJ/reaction} \qquad (5.21)$$

The P/2e$^-$ for this H_2-fumarate electron transport coupled phosphorylation was ~0.4; and this system was sensitive to gramicidin, pentachlorophenol, dinitrophenol, and methyl viologen as uncouplers of phosphorylation. In the *D. gigas* system, H_2 oxidation and coupled phosphorylation were observed with malate (P/2e$^-$ = 0.35), and this oxidation of hydrogen with addition of malate could be attributed to fumarase which produces fumarate from malate. It has been proposed that anaerobic phosphorylation could be functioning in *Selenomonas ruminantium* (Hobson 1965), *Bacteroides amylophilus* (Hobson and Summers 1967), and *Ruminococcus albus* (Hungate (1963) which have higher molar growth yields than explained by substrate phosphorylation.

Subsequently, several reports have been published that support the proposal of phosphorylation coupled to electron transport from H_2 to fumarate in *D. gigas.* In *D. multispirans,* 2-heptyl-4-hydroxy-quinoline-N-oxide inhibits fumarate reduction and this suggests a role for a menaqunone in the electron transport scheme (He et al. 1986). The fumarate reductase of *D. gigas* contains two B-type hemes with reduction potentials dependent on the pH which is an agreement that fumarate reduction participates in membrane proton gradient (Lemos et al. 2002). Subsequently, the coupling of phosphorylation to fumarate reduction has been shown in *Proteus rettgeri* (Kröger 1974) and in *Wolinella* (formerly *Vibrio*) *succinogenes* (Kröger and Winkler 1981; Reddy and Peck 1978).

5.9.4 Nitrite as an Electron Acceptor

The use of molecular hydrogen as an electron donor for the reduction of nitrite and hydroxylamine in *D. desulfuricans* was initially examined by Senez and Pichinoty (1958a, b, c). However, the demonstration of nitrite reductase in the plasma membrane of *D. gigas* (Odom and Peck 1981a) and the report of proton translocation across the plasma membrane with H_2, formate, or lactate (Steenkamp and Peck 1981) provided a background for the examination of phosphorylation coupled to nitrite reduction by H_2. The publication by Barton et al. (1983) provides evidence for oxidative phosphorylation using the plasma membrane of *D. gigas* with H_2 as the electron donor and nitrite as the electron acceptor. This phosphorylation in *D. gigas* produced a P/2e$^-$ of 0.32 with the H_2-nitrite reaction and a P/2e$^-$ of 0.08 with the H_2-hydroxylamine reaction. As in other oxidative phosphorylation systems in *D. gigas*, phosphorylation in the H_2-nitrite reaction was uncoupled by gramicidin, pentachlorophenol, 2,4-dinitrophenol, and methyl viologen. A thermodynamic evaluation of the H_2-nitrite and H_2-hydroxylamine reactions reveals sufficient energy for esterification of orthophosphate to ATP (Thauer et al. 1977):

$$\begin{aligned} 3H_2 + NO_2{}^- + 2H^+ &\rightarrow NH_4{}^+ + 2H_2O \\ \Delta G^{o'} &= -436.4\ \mathrm{kJ/reaction} \end{aligned} \tag{5.22}$$

$$\begin{aligned} H_2 + NH_2OH + H^+ &\rightarrow NH_4{}^+ + 2H_2O \\ \Delta G^{o'} &= -248.3\ \mathrm{kJ/reaction} \end{aligned} \tag{5.23}$$

5.9.5 Energy Generated by Heme Biosynthesis

SRB have several heme proteins which include cytochromes and respiratory-coupled enzymes. In the penultimate step of heme biosynthesis, there is the oxidation of protoporphyrinogen IX to protoporphyrin IX with the release of six electrons. It has been reported that these electrons from protoporphyrinogen IX oxidation are directed to the electron transport system with the reduction of O_2 in aerobic bacteria (Dailey et al. 1994; Dailey and Dailey 1996) and to nitrite or fumarate in anaerobically grown bacteria (Jacobs and Jacobs 1975, 1976). Using a membrane fraction of *D. gigas*, the oxidation of protoporphyrinogen IX with nitrite reduction, the esterification of orthophosphate was observed with a P/2e$-$ ration of 0.9 (Klemm and Barton 1989). With anaerobically grown *Escherichia coli*, the oxidation of protoporphyrinogen IX with fumarate and nitrate was reported to direct electrons into the membrane where a proton-generated force would support ATP synthesis by way of proton driven ATP synthase (Möbius et al. 2010). As suggested by Dailey and Dailey (1996), aerobic bacteria may use an oxygen-dependent synthesis for protoporphyrinogen oxidase while anaerobes and facultative anaerobes would use a multisubunit enzyme.

5.10 Energy Conservation by Metabolite Cycling

Since SRP require energy to activate sulfate before respiration can proceed, there has been an interest in cellular energy production and energy conservation employing alternate pathways. Some of the proposals have focused on mechanisms that would enhance the establishment of a charge on the plasma membrane and thereby provide additional energy to the growing cell. Other suggestions have addressed alternate metabolic pathways that would enhance cellular energetics.

5.10.1 Hydrogen Cycling

Considerable discussion has resulted from the hypothesis presented by Odom and Peck (1981b) concerning the possibility of obtaining energy from hydrogen cycling within the Gram-negative cell. A critical evaluation of the hydrogen cycling model in *Desulfovibrio* and anaerobic bacteria has been provided by Peck and LeGall (1982), Keller and Wall (2011), Price et al. (2014), and Kulkarni et al. (2018). H_2 bursts from cells of *D. vulgaris* Miyazaki had been reported to promote substrate-level phosphorylation as lactate was oxidized to acetate and CO_2, and in later stages of growth, H_2 oxidation was suggested to supply electrons for the reduction of sulfate and for electron-coupled phosphorylation (Tsuji and Yagi 1980). As reviewed for *Desulfovibrio* by Odom and Peck (1984a, b), hydrogen cycling employed activity of a cytoplasmic hydrogenase and a periplasmic hydrogenase and based on this requirement genomic information for *D. vulgaris* H supported this hypothesis (Heidelberg et al. 2004). A model (see Fig. 5.8) for hydrogen cycling in *D. vulgaris* H has been constructed by Sim et al. (2013).

Protons and electrons from the oxidation of pyruvate and lactate would be combined by a cytoplasmic membrane-bound Ech to produce H_2 in the cytoplasm. H_2 would diffuse across the plasma membrane where periplasmic hydrogenase would release protons and electrons. The electrons would be shuttled to the transmembrane complexes by cytochrome c_3 for the reduction of APS and bisulfite while the protons would remain at the exterior of the plasma membrane to generate a proton gradient. The result of this would be to pump one proton across the membrane for each electron from H_2 which would increase the proton motive force and support electron-coupled phosphorylation. While the hydrogen cycle was proposed for species of *Desulfovibrio*, it would not be an absolute requirement for all species of this genus because *D. sapovorans* lacks hydrogenase (Postgate 1984). In support of the hypothesis of hydrogen cycling, H_2 released from cells of *D. vulgaris* growing on pyruvate-sulfate medium was collected and measured by membrane-inlet mass spectrometry (Peck et al. 1987). A proteomic analysis of *D. vulgaris* metabolism revealed that the hydrogenase HynAB-1 was present in considerable quantity which would support the presence of hydrogen cycling (Zhang et al. 2006b). A most convincing argument for hydrogen cycling for *D. vulgaris* growing on lactate-

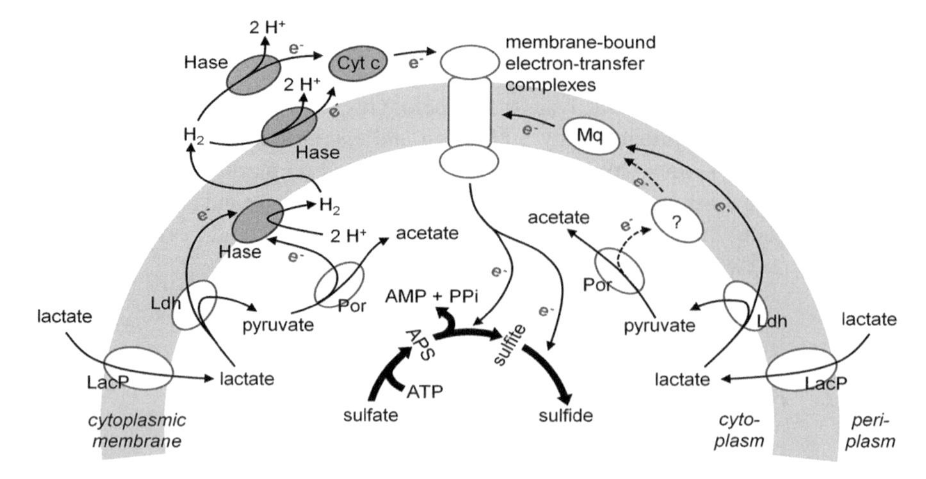

Fig. 5.8 Schematic representation of two proposed pathways for electron transport during sulfate reduction in *Desulfovibrio vulgaris* Hildenborough (Keller and Wall 2011). The hydrogen cycling model (Odom and Peck 1981a) describes the flow of reducing equivalents from the electron donor to oxidized sulfur species through hydrogen metabolism. This flow can be mediated by hydrogenases and other electron carriers, including cytochromes (Heidelberg et al. 2004). The second pathway can transfer electrons to the membrane-associated menaquinone pool (Keller and Wall 2011). Dark gray ovals indicate enzymes and electron carriers deleted in the mutant strains used in this study. Abbreviations: LacP, lactate permease; Ldh, lactate dehydrogenase; Por, pyruvate-ferredoxin oxidoreductase; Hase, hydrogenase; Cyt. *c*, cytochrome *c*; Mq, menaquinone pool; APS, adenosine 5-phosphosulfate. Dashed lines and the question mark indicate currently hypothetical pathways and components (Sim et al. 2013).

sulfate medium considered that electron flow used multiple pathways simultaneously and that free energy available influenced the direction of electron movement (Noguera et al. 1998). They proposed a model predicted that 52% of electrons from lactate moved through H_2 to sulfate and 48% of the electrons moved from lactate to sulfate without H_2 formation. Experimental analysis of electron movement associated with respiration revealed that 52% of electrons directed to sulfate reduction involved hydrogen cycling while 47% of electrons contribution to sulfate reduction were direct and not involving H_2 production (Noguera et al. 1998).

Some publications have been critical of a hydrogen cycle in *D. vulgaris.* It was reported that H_2 was produced and ultimately consumed by *D. vulgaris* growing on organic acids and sulfate but H_2 production was also observed when thiosulfate was the electron acceptor and this negated the requirement for hydrogen cycle to energize cells for sulfate reduction (Lupton et al. 1984). Furthermore, elevated concentrations of H_2 did not inhibit the growth of *D. vulgaris* on lactate (Pankhania et al. 1986) and hydrogenase or TpI-c_3 mutants of *D. desulfuricans* ATCC 27774, *D. desulfuricans* G20, *D. vulgaris* Hildenborough, and *D. gigas* remained capable of growing on lactate-sulfate even with the mutation (Keller et al. 2014; Li et al. 2009; Morais-Silva et al. 2014; Odom and Wall 1987; Rapp-Giles et al. 2000; Sim et al. 2013). Additionally, it was proposed that hydrogen cycles did not occur in SRP but growth was attributed to constitutive hydrogenase production or hydrogenase activity was a reflection of redox status within the bacterial cell (Rabus et al. 2006; Widdel and Hansen 1991). Since neither *D. alaskensis* G20 nor *Desulfotalea psycrophila* have both cytoplasmic and periplasmic hydrogenases (Pereira et al. 2008), it has been suggested that these two *SRB* do not employ hydrogen cycling.

However, an argument to support the presence of a hydrogen cycle in SRP was the suggestion that the electron transport chains in *Desulfovibrio* are branched involving several alternative or supplemental electron transport processes (Noguera et al. 1998). Hydrogen cycling has been suggested for several anaerobic acetogenic bacteria where growth is greater than the quantity of ATP known to be produced (Odom and Peck 1984a, b). Anaerobic bacteria which may use hydrogen cycling include *Acetobacterium woodii* (Winter and Wolfe 1980) and *Clostridium thermoaceticum* (Andressen et al. 1973), *Selenomonas ruminantium* (Hobson and Summers 1972), *Methanosarcina* spp. (Lovley and Ferry 1985), *Geobacter sulfurreducens* (Coppi 2005), *Methanococcus maripaludis* (Lupa et al. 2008), and *Methanosarcina barkeri* (Kulkarni et al. 2018). The energy coupling attributed to hydrogen cycling in sulfate-reducing bacteria and other anaerobic bacteria awaits additional research results.

5.10.2 CO Cycling

CO is an appropriate electron donor for SRB. For the purpose of ATP production, oxidation of CO to CO_2 is coupled to H_2 oxidation with sulfate reduction (Lupton et al. 1984) as indicated in the following reactions:

$$4CO + 4H_20 = 4CO_2 + 4H_2 \qquad \Delta G^{o'} = -80\ \mathrm{kJ/mol\ CO} \quad (5.24)$$

$$4H_2 + SO_4^{2-} + 2H^+ = H_2S + 4H_20 \qquad \Delta G^{o'} = -152.2\ \mathrm{kJ/mol\ SO_4^{2-}} \quad (5.25)$$

CO is produced in the cytoplasm from pyruvate and in pyruvate-sulfate medium, *D. vulgaris* Hildenborough was observed to produce bursts of CO in addition to H_2 bursts and CO would be used to support membrane charging (Voordouw 2002). It is considered that CO cycling is important with *D. vulgaris* Hildenborough growing on lactate and pyruvate but not with hydrogen as the electron donor with sulfate (Pereira et al. 2008). To facilitate CO cycling, a CO dehydrogenase is required along with a hydrogenase resistant to inhibition by CO, and the genome of *D. vulgaris* Hildenborough contains genes for a membrane-associated CO-dependent hydrogenase (Voordouw 2002). CO is an inhibitor of most hydrogenases of the *Desulfovibrio* and there are only about a dozen sulfate-reducing bacteria/archaea that are carboxydotrophic (Parshina et al. 2010).

5.10.3 Formate Cycling

Formate cycling in SRP has the following characteristic. Formate which is produced in the cytoplasm by pyruvate:formate lyase is transported across the plasma membrane to the periplasm where it is oxidized by formate dehydrogenase and electrons are moved via cytochrome to the membrane for sulfate reduction. Coupled to this electron movement across the plasma membrane is the formation of proton motive force (Badziong et al. 1979; Tang et al. 2007). Several reports have suggested the functioning of formate cycling in *D. vulgaris* Hildenborough (da Silva et al. 2013; Heidelberg et al. 2004; Voordouw 2002). With *D. vulgaris* Hildenborough growing on hydrogen or formate with sulfate, the level of pyruvate:formate lyase (DVU2271, *pfl*A) was greater relative to growth with lactate (Zhang et al. 2006a), and this observation was interpreted to suggest that formate cycling occurred (Pereira et al. 2008). Using *D. vulgaris* ATCC29579 growing on lactate-sulfate as a biofilm as compared to planktonic growth, there was a down-expression of genes for pyruvate: ferredoxin oxidoreductases with *por* genes (DVU1569-70) and an upshift in the expression of *oor* genes (oxo-organic acid oxidoreductase, DVU1944-47) as well as an increase in the expression of the gene for cytochrome c_{553} (DVU1817), the electron transfer partner for formate dehydrogenase (Clark et al. 2012). With the absence of a pyruvate:formate lyase (*pfl*) gene in *D. gigas*, formate cycling would not occur with this bacterium (Morais-Silva et al. 2014). It has been reported that in *D. alaskensis* G20, molecular cycling of formate is not involved in growth on lactate-sulfate medium and neither is hydrogen cycling or carbon monoxide cycling (Price et al. 2014).

5.11 Substrate-Level Phosphorylation

5.11.1 Fermentation

While sulfate respiration is the hallmark feature of SRP, there are numerous instances where SRP grow on organic compounds in the absence of sulfate by fermentation processes. In the case of *A. fulgidus*, metabolic activity is strain specific. *A. fulgidus* strain 7324 grows on starch and ferments glucose to acetate and CO_2 by a modified Embden-Meyerhoff pathway (Labes and Schönheit 2001). The genome of *A. fulgidus* strain VC 16 lacks hexokinase, pyruvate kinase, and 6-phosphofructokinase of the classical Embden-Meyerhoff pathway (Cordwell 1999) and genes encoding for ADP-dependent hexokinase and ADP-dependent phosphofructokinase in the modified Embden-Meyerhoff pathway are absent (Labes and Schönheit 2001). The metabolism of pyruvate with the formation of acetate is attributed to the action of pyruvate:ferredoxin oxidoreducxtase and NAD-forming acetyl-CoA synthetase.

A few SRB have the capability of growing using glucose or fructose as an energy source with sulfate as the electron acceptor. Growth with fructose and sulfate has been reported for *D. fructosovorans* (Cord-Ruwisch et al. 1986; Ollivier et al. 1988), *D. inopinatus* HHQ 20 (Reichenbecher and Schink 1997), *D. termitidis* (Trinkerl et al. 1990), *Dst. nigrificans* DSM575 and *Dst. geothermicum* (Klemps et al. 1985), *Dst. carboxydivorans* DSM14880 (Visser et al. 2014), *Dst. defluvii* DSM23699 (Krishnamurthi et al. 2013), *Dst. copahuensis* CINDEFII (Poratti et al. 2016), *Dst. varum* (Ogg and Patel 2011), and *Dst. ferrireducens* GSS09 (Guo et al. 2016; Yang et al. 2016). Growth in glucose–sulfate medium has been reported for *Dst. nigrificans* (Akagi and Jackson 1967), *D. simplex* (Zellner et al. 1989), *D. termitidis* (Trinkerl et al. 1990), *Dst. defluvii* DSM23699 (Krishnamurthi et al. 2013), and *Dst. carboxydivorans* DSM14880 (Visser et al. 2014). Growth in the absence of sulfate by fructose fermentation has been reported for *Dst. ferrireducens* GSS09 (Yang et al. 2016) and *D. fructosovorans* (Ollivier et al. 1988) while fermentation of glucose was reported for *Desulfothermobacter acidiphilus* DSM 105356^T (Frolov et al. 2018).

Internal granules of polyglucose are found in several *Desulfovibrio* species, and utilization of glucose from this carbon reserve appears to be attributed to the classical Embden-Meyerhoff pathway (Fareleira et al. 1997; Stams et al. 1983). The inability for *D. gigas* and *D. vulgaris* Hildenborough to grow on glucose is attributed to the absence of appropriate transport systems (Wall et al. 2003).

While various strains of *Desulfovibrio* are capable of fermenting pyruvate, fumarate, and malate, choline fermentation is attributed only to *D. desulfuricans* (Postgate 1984). Additional examples of fermentation include methanol + CO_2, ethanol + CO_2, and lactate by *Dst. orientis* (Klemps et al. 1985). Ethanol + CO_2 is fermented by *Desulfobulbus propionicus* (Laanbroek et al. 1982), lactate and pyruvate are fermented by *Desulfobulbus,* lactate and pyruvate are fermented by *Desulfosarcina* (Widdel and Pfennig 1984), and glycerol is fermented by *D. carbinolicus* (Nanninga and Gottschal 1987). Usually, acetate is an end product of these examples of fermentation. Fermentation of bisulfite and thiosulfate by *D. sulfodismutans* (Bak and Cypionka 1987) is covered in Chap. 3. It should be noted that sulfate respiration and fermentation should not be considered as one excluding the other since simultaneous dissimilatory sulfate reduction and fermentation have been observed with *D. vulgaris* Hildenborough (Sim et al. 2013) and *Desulfobulbus mediterraneus* (Sass et al. 2002) and has been proposed to occur in certain marine environments (Sim et al. 2011).

5.11.2 Pyruvate Phosphoroclastic Reaction

Pyruvate is an important source of energy for sulfate-reducing microorganisms, and in general, the pyruvate phosphoroclastic reaction refers to the conversion of pyruvate to acetate + CO_2 with the production of ATP (Barton 1994). Acetyl phosphate is an intermediate in the metabolism of pyruvate to acetate, and the reactions associated with this activity are indicated as follows:

$$\text{Pyruvate} + P_i = \text{acetyl phosphate} + CO_2 + H_2 \tag{5.26}$$

$$\text{Pyruvate} + P_i = \text{acetyl phosphate} + \text{formate} \tag{5.27}$$

Reaction Eq. (5.26) is catalyzed by pyruvate:ferredoxin 2-oxidoreductase, and reaction Eq. (5.27) requires pyruvate formate lyase. Initially Eq. (5.26) was designated as a clostridial-type reaction, and Eq. (5.27) was called the *Enterobacteriaceae*-type reaction (Doelle 1975). In general use, Eq. (5.26) is now designated as the pyruvate phosphoroclastic reaction (Gottschalk 1986) with the conversion of acetyl-phosphate to acetate and ATP (Eq. (5.31)):

$$\text{Pyruvate} + \text{CoA} + \text{ferredoxin}_{(ox)} = \text{acetyl} - \text{CoA} + CO_2 + \text{ferredoxin}_{(red)} \tag{5.28}$$

$$\text{Ferredoxin}_{(red)} + 2\,H^+ = H_2 + \text{ferredoxin}_{(ox)} \tag{5.29}$$

$$\text{Acetyl} - \text{CoA} + P_i = \text{acetyl} - \text{phosphate} + \text{CoA} \tag{5.30}$$

$$\text{Acetyl} - \text{phosphate} + \text{ADP} = \text{acetate} + \text{ATP} \tag{5.31}$$

Based on activity using proteins purified from *D. vulgaris* Hildenborough, the rate of electron transfer from ferredoxin to H^+ by hydrogenase in Eq. (5.29) is greatly enhanced if cytochrome c_3 is present (Akagi 1967). The enzyme catalyzing reaction Eq. (5.31) is commonly referred to as acetokinase but is also known as acetate kinase, acetylkinase, or ATP:acetate phosphotransferase). Acetokinase isolated from *D. vulgaris* Hildenborough required Mg^{2+} which could be replaced by Mn^{2+}, Co^{2+}, and Zn^{2+} (Brown and Akagi 1966). The physiological electron carrier for acetokinase in *D. vulgaris* strain Miyazaki is reported to be flavodoxin (Ogata and Yagi 1986).

5.11.3 Succinate-Fumarate Reactions

D. desulfuricans strain Essex 6 displays fumarate respiration with formate or H_2 as electron donors. Metabolic activity with these two electron donors indicated that about 70–80% of fumarate (see reactions below) was dedicated to electron accepting activity while a small amount of fumarate is oxidized to acetate (Zaunmüller et al. 2006). The growth yield for this bacterium oxidizing formate or H_2 was about 5 g dry wt/mol fumarate with about 0.5 ATP/fumarate:

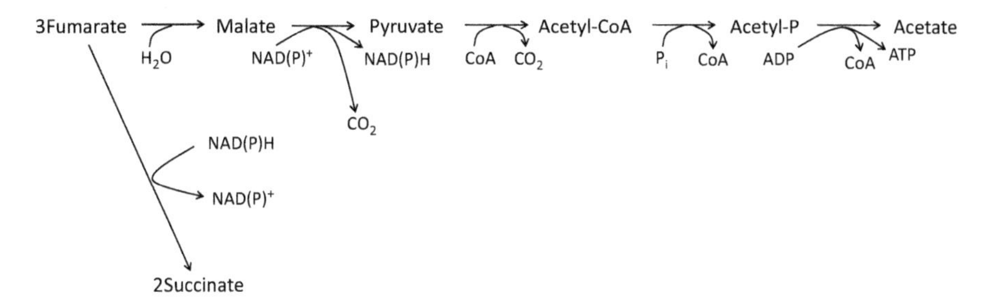

Fig. 5.9 Pathway for fumarate disproportionation by SRB. Energy is obtained from substrate-level phosphorylation with acetyl-P converted to acetate + ATP (Zaunmüller et al. 2006)

$$1.0\ \text{Fumarate} + 0.8\ \text{formate} \rightarrow 0.8\ \text{succinate} + 0.2\ \text{acetate} + 1.2\ CO_2 \quad (5.32)$$

$$1.0\ \text{Fumarate} + 0.8\ H_2 \rightarrow 0.8\ \text{succinate} + 0.2\ \text{acetate} + 0.4\ CO_2 \quad (5.33)$$

Disproportionation of fumarate (fermentation) occurs in *D. desulfuricans* strain Essex 6 with 1 mol of fumarate oxidized to acetate with 2 mol fumarate reduced to succinate (Zaunmüller et al. 2006) according to the following reaction:

$$3.0\ \text{Fumarate} \rightarrow 2.0\ \text{succinate} + 2.0\ CO_2 + 1.0\ \text{acetate} \quad (5.34)$$

While malate was initially suggested to be a product of fumarate disproportionation (Miller and Wakerley 1966), malate is an intermediate in the reaction. The metabolic pathway for fumarate dismutation is given in Fig. 5.9. In bacteria, the activity attributed to succinate:quinone oxidoreductase may be as succinate:quinone reductase (succinate dehydrogenase, Sdh) or quinol:fumarate reductase (fumarate reductase, Frd). These two enzymes display structural similarities but are distinguished by their Frd/Sdh activity. For example, the Frd/Sdh activity for *D. vulgaris* Hildenborough with formate + fumarate is 2.0 while with *D. desulfuricans* Essex 6 is 17.0. Additionally, succinate:menaquinone reductase requires proton potential while menaquinone:fumarate reductase is independent of membrane charge (Zaunmüller et al. 2006).

SRB employ menaquinones (Collins and Widdel 1986) which are associated with respiratory quinone activity. The redox potential of the fumarate/succinate reaction is electropositive ($E^{0\prime} = +33$ mV) while the menaquinone (MK) ox/red is electronegative ($E^{0\prime} = -74$ mV) (Thauer et al. 1977). It is considered that the oxidation of succinate occurs on the cytoplasmic side of the membrane while the reduction of menaquinone occurs on the outer side of the membrane, and based on the unfavorable energetics of the reaction, a proton potential on the membrane is required to facilitate the transfer of electrons from succinate to menaquinone which enables reverse electron transport to proceed (Hägerhäll and Hederstedt 1996; Körtner et al. 1990; Lancaster 2002; Schirawski and Unden 1998). In general, it appears that bacteria with menaquinone-dependent succinate dehydrogenases (such as in the SRB), a proton potential is required for succinate dehydrogenase to be functional. With *D. vulgaris* Hildenborough and *D. desulfuricans* Essex 6, the proton-associated membrane potential can be disrupted when the cells are disrupted when carbonyl cyanide *m*-chlorophenylhydrazone (CCCP) is added as an uncoupling agent to disrupt the proton charge (Zaunmüller et al. 2006).

5.12 Summary and Perspective

While energy production in SRP is an area of interest, an understanding of the mechanisms accounting for energy production proceeds slowly due to the diversity of processes involved. The amount of energy produced from oxidation of lactate and pyruvate is readily predicted from growth yield values. Also, the minimum level of energy produced from H_2 and formate oxidation can be estimated from the biochemical demands for sulfate activation. For each mol of sulfate reduced by four mol H_2, three mol of ATP is produced by electron coupled phosphorylation which is also known as anaerobic oxidative phosphorylation. The mechanism accounting for oxidative phosphorylation in SRP follows the principles of aerobic oxidative phosphorylation (i.e., a charge on the plasma membrane of -130 ± 20 mV, proton pumps in the membrane, lipid soluble quinones to carry electrons and protons, *b*-type cytochromes, F-type ATP synthase, and a sensitivity to uncoupling agents). As demonstrated from physiological and biochemical studies, concomitant phosphorylation associated with electron transport in SRB is loosely coupled as is the case for aerobic respiratory bacterial systems. With membranes from *D. gigas*, H_2 oxidation provides for ATP production with nitrite, fumarate, and elemental sulfur as well as with bisulfite. SRB have numerous transmembrane electron transport systems which interface with periplasmic and cytoplasmic electron transport complexes which contribute to cell energetic. Energy-producing systems in several species of SRB are associated with the pyruvate phosphoroclastic reaction, alcohol dehydrogenase and heterodisulfide reduction. Additionally, in some species of SRB, energy is provided from fermentation of organic and inorganic compounds. Finally, energy-conserving systems associated with cycling of H_2, CO, and formate have been proposed for several SRB.

References

Akagi JM (1967) Electron carriers for the phosphoroclastic reaction of *Desulfovibrio desulfuricans*. J Biol Chem 242:2478–2483

Akagi JM, Jackson G (1967) Degradation of glucose by proliferating cells of *Desulfotomaculum nigrificans*. Appl Microbiol 15:1427–1430

Almendra MJ, Brondino CD, Gavel O, Pereira AS, Tavares P, Bursakov S et al (1999) Purification and characterization of a tungsten-containing formate dehydrogenase from *Desulfovibrio gigas*. Biochemist 38:16366–16372

Andressen JR, Schaupp A, Neurauter C, Brown A, Ljungdahl LG (1973) Fermentation of glucose, fructose, and xylose by *Clostridium thermoaceticum*: effect of metal on growth yield, enzymes, and the synthesis of acetate from CO_2. J Bacteriol 114:743–751

Badziong W, Bernhard D, Thauer RK (1979) Acetate and carbon dioxide assimilation by *Desulfovibrio vulgaris* (Marburg), growing on hydrogen and sulfate as sole energy source. Arch Microbiol 123:301–305

Badziong W, Thauer RK (1978) Growth yields and growth rates of *Desulfovibrio vulgaris* (Marburg) growing on hydrogen plus sulfate and hydrogen plus thiosulfate as the sole energy sources. Arch Microbiol 117:209–214

Bak F, Cypionka H (1987) A novel type of energy metabolism involving fermentation of inorganic sulphur compounds. Nature 326:891–892

Barton LL (1994) Pyruvic acid phosphoroclastic system. Methods Enzymol 243:94–104

Barton LL, Fardeau M-L, Fauque GD (2014) Hydrogen sulfide. A toxic gas produced by dissimilatory sulfate and sulfur reduction and consumed by microbial oxidation. In: Sigel A, Sigel H, Sigel RKO (eds) Metal ions in life. Springer Science & Business Media B.V, Dordrecht, pp 237–278

Barton LL, Granat AS, Lee S, Xu H, Ritz NL, Hider R, Lin HC (2019) Bismuth(III) interactions with *Desulfovibrio desulfuricans*: inhibition of cell energetics and nanocrystal formation of Bi_2S_3 and Bi^0. Biometals 32:803–811

Barton LL, LeGall J, Odom JM, Peck HD (1983) Energy coupling to nitrite respiration in the sulfate-reducing bacterium *Desulfovibrio gigas*. J Bacteriol 153:867–871

Barton LL, Le Gall J, Peck HD (1970) Phosphorylation coupled to oxidation of hydrogen with fumarate in extracts of the sulfate reducing bacterium, *Desulfovibrio gigas*. Biochem Biophys Res Commun 41:1036–1042

Barton LL, LeGall J, Peck HD (1972) Oxidative phosphorylation in the obligate anaerobe, *Desulfovibrio gigas*. In: Pietro AS, Gest H (eds) Horizons of bioenergetics. Academic Press, New York, pp 33–51

Bauchop T, Elsden SR (1960) The growth of micro-organisms in relation to their energy supply. J Gen Microbiol 23:457–469

Biebl H, Pfennig NT (1977) Growth of sulfate-reducing bacteria with sulfur as electron acceptor. Arch Microbiol 112:115–117

Biegel E, Müller V (2010) Bacterial Na+ translocating ferredoxin: NAD^+ oxidoreductase. Proc Natl Acad Sci U S A 107:18138–18142

Brileya KA, Camilleri LB, Zane GM, Wall JD, Fields MW (2014) Biofilm growth mode promotes maximum carrying capacity and community stability during product inhibition syntrophy. Front Microbiol 2014:693. https://doi.org/10.3389/fmicb.2014.00693

Brondino CD, Passeggi MC, Caldeira J, Almendra MJ, Feio MJ, Moura JJ, Moura I (2004) Incorporation of either molybdenum or tungsten into formate dehydrogenase from *Desulfovibrio alaskensis* NCIMB 13491: EPR assignment of the proximal iron-sulfur cluster to the pterin cofactor in formate dehydrogenases from sulfate-reducing bacteria. J Biol Inorg Chem 9:145–151

Brown MS, Akagi JM (1966) Purification of acetokinase from *Desulfovibrio desulfuricans*. J Bacteriol 92:1273–1274

Buckel W, Thauer RK (2013) Energy conservation via electron bifurcating ferredoxin reduction and proton/Na^+ translocating ferredoxin oxidation. Biochim Biophys Acta 1827:94–113

Calisto F, Sousa FM, Sena FV, Refojo RN, Pereira MM (2021) Mechanisms of energy transduction by charge translocating membrane proteins. Chem Rev 121:1804–1844

Callaghan AV, Morris BEL, Pereira IAC, McInerney MJ, Austin RN, Groves JY et al (2012) The genome sequence of *Desulfatibacillum alkenivorans* AK-01: a blueprint for anaerobic alkane oxidation. Environ Microbiol 14:101–113

Chance B, Williams GR (1956) The respiratory chain and oxidative phosphorylation. Adv Enzymol Relat Subj Biochem 17:65–134

Chivian D, Brodie EL, Alm EJ, Culley DE, Dehal PS, DeSantis TZ et al (2008) Environmental genomics reveals a single-species ecosystem deep within the Earth. Science 322:275–278

Clark ME, Edelmann RE, Duley ML, Wall JD, Fields MW (2007) Biofilm formation in *Desulfovibrio vulgaris* Hildenborough is dependent upon protein filaments. Environ Microbiol 9:2844–2854

Clark ME, He Z, Redding AM, Joachimiak MP, Keasling JD, Zhou JZ, Arkin AP, Mukhopadhyay A, Fields MW (2012) Transcriptomic and proteomic analyses of *Desulfovibrio vulgaris* biofilms: carbon and energy flow contribute to the distinct biofilm growth state. BMC Genomics 13:138. https://doi.org/10.1186/1471-2164-13-138. PMID: 22507456; PMCID: PMC3431258.

Collins MD, Widdel F (1986) Respiratory quinones of sulphate-reducing and sulphur-reducing bacteria: a systematic investigation. Syst Appl Microbiol 8:8–18

Coppi MV (2005) The hydrogenases of *Geobacter sulfurreducens*: a comparative genomic perspective. Microbiology (Reading) 151:1239–1254

Cordwell SJ (1999) Microbial genomes and "missing enzymes": redefining biochemical pathways. Arch Microbiol 172:269–279

Cord-Ruwisch R, Ollivier B, Garcia J (1986) Fructose degradation by *Desulfovibrio* sp. in pure culture and in coculture with *Methanospirillum hungatei*. Curr Microbiol 13:285–289

Costa C, Teixeira M, LeGall J, Moura JJG, Moura I (1997) Formate dehydrogenase from *Desulfovibrio desulfuricans* ATCC 27774: isolation and spectroscopic characterization of the active sites (heme, iron-sulfur centers and molybdenum). J Biol Inorg Chem 2:198–208

Cypionka H (1995) Solute transport and cell energetics. In: Barton LL (ed) Sulfate-reducing bacteria: biotechnology handbook. Plenum Publishing, London, pp 151–184

Czechowski MH, Rossmore HW (1980) Factors affecting *Desulfovibrio desulfuricans* lactate dehydrogenase. Dev Ind Microbiol 21:349–356

Czechowski MH, Rossmore HW (1990) Purification and partial characterization of aD(−)-lactate dehydrogenase from*Desulfovibrio desulfuricans* (ATCC 7757). J Indust Microbiol 6:117–122

Dailey HA, Dailey TA (1996) Protoporphyrinogen oxidase of *Myxococcus xanthus*. Expression, purification, and characterization of the cloned enzyme. J Biol Chem 271:8714–8718

Dailey LA, Meissner P, Dailey HA (1994) Expression of a cloned protoporphyrinogen oxidase. J Biol Chem 269:813–815

Doelle HA (1975) Bacterial metabolism. Academic Press, New York

da Silva SM, Pimentel C, Valente FMA, Rodrigues-Pousada C, Pereira IAC (2011) Tungsten and molybdenum regulation of formate dehydrogenase expression in *Desulfovibrio vulgaris* Hildenborough. J Bacteriol 193:2909–2916

da Silva SM, Voordouw J, Leitão C, Martins M, Voordouw G, Pereira IAC (2013) Function of formate dehydrogenases in *Desulfovibrio vulgaris* Hildenborough energy metabolism. Microbiology (Reading) 159:1760–1769

Efremov RG, Baradaran R, Sazanov LA (2010) The architecture of respiratory complex I. Nature 465:441–445

Fareleira P, LeGall J, Xavier AV, Santos H (1997) Pathways for utilization of carbon reserves in *Desulfovibrio gigas* under fermentative and respiratory conditions. J Bacteriol 179:3972–3980

Fauque GD (1994) Sulfur reductase from thiophilic sulfate-reducing bacteria. Methods Enzymol 243:353–367

Fauque GD, Barton LL (2012) Hemoproteins in dissimilatory sulfate- and sulfur-reducing prokaryotes. Adv Microb Physiol 60:1–263

Fauque GD, Barton LL, LeGall J (1980) Oxidative phosphorylation linked to the dissimilatory reduction of elemental sulfur by *Desulfovibrio*. In: Ciba foundation symposium 72. Sulphur in biology. Excerpta Medica, New York, pp 71–86

Fauque GD, LeGall J, Barton LL (1991) Sulfate-reducing and sulfur-reducing bacteria. In: Shively JM, Barton LL (eds) Variations in autotrophic life. Academic Press, London, pp 271–337

Fitz RM, Cypionka H (1989) A study on electron transport-driven proton translocation in *Desulfovibrio desulfuricans*. Arch Microbiol 152:369–376

Fitz RM, Cypionka H (1991) Generation of a proton gradient in *Desulfovibrio vulgaris*. Arch Microbiol 155:444–448

Flodgaard H, Fleron P (1974) Thermodynamic parameters for the hydrolysis of inorganic pyrophosphate at pH 7.4 as a function of [Mg+], [K+], and ionic strength determined from equilibrium studies of the reaction. J Biol Chem 249:3465–3474

Frolov EN, Zayulina KS, Kopitsyn DS, Kublanov IV, Bonch-Osmolovskaya EA, Chernyh NA. 2018. *Desulfothermobacter acidiphilus* gen. nov., sp. nov., a thermoacidophilic sulfate-reducing bacterium isolated from a terrestrial hot spring. Int J Syst Evol Microbiol, 68:871–875

Gottschalk G (1986) Bacterial metabolism. Springer, New York

Grein F, Ramos AR, Venceslau SS, Pereira IAC (2013) Unifying concepts in anaerobic respiration: insights from dissimilatory sulfur metabolism. Biochim Biophys Acta 1827:145–160

Guarraia LJ, Peck HD (1971) Dinitrophenol stimulated adenosine triphosphatase activity in extracts of *Desulfovibrio gigas*. J Bacteriol 106:890–895

Guo J, Zhuang L, Yuan Y, Zhou S (2016) *Desulfotomaculum ferrireducens* sp. nov., a moderately thermophilic sulfate-reducing and dissimilatory Fe(III)-reducing bacterium isolated from compost. Int J Syst Evol Microbiol 66:3022–3028

Hägerhäll C, Hederstedt L (1996) A structural model for the membrane-integral domain of succinate: quinone oxidoreductases. FEBS Lett 389:25–31

Hansen TL (1994) NAD(P)-independent lactate dehydrogenase from sulfate-reducing prokaryotes. Meth Enzymol 243:21–43

Hatchikian EC, LeGall J (1970) Étude du métabolisme des acides dicarboxyliqueset du pyruvate chez les bactéries sulfato-réductases. I. Étude de l'oxidation enzymatique du fumarate en acetate. Ann Inst Pasteur 118:125–142

Haveman SA, Brunelle V, Voordouw JK, Voordouw G, Heidelberg JF, Rabus R (2003) Gene expression analysis of energy metabolism mutants of *Desulfovibrio vulgaris* Hildenborough indicates an important role for alcohol dehydrogenase. J Bacteriol 185:4345–4353

He SH, DerVartanian DV, LeGall J (1986) Isolation of fumarate reductase from *Desulfovibrio multispirans*, a sulfate-reducing bacterium. Biochem Biophys Res Commun 135:1000–1007

Hedderich R, Hamann N, Bennati M (2005) Heterodisulfide reductase from methanogenic archaea: a new catalytic role for an iron-sulfur cluster. Biol Chem 386:961–970

Hedderich R, Klimmek O, Kröger A, Dirmeier R, Keller M, Stetter KO (1998) Anaerobic respiration with elemental sulfur and with disulfides. FEMS Microbiol Rev 22:353–381

Heidelberg JF, Seshadri R, Haveman SA, Hemme CL, Paulsen IT, Kolonay JF et al (2004) The genome sequence of the anaerobic, sulfate-reducing bacterium *Desulfovibrio vulgaris* Hildenborough. Nat Biotechnol 22:554–559

Hensgens CMH, Vonck J, Van Beeumen J, Ernst FJ, Van Bruggen EFJ, Hansen TA (1993) Purification and characterization of an oxygen-labile, NAD-dependent alcohol dehydrogenase from *Desulfovibrio gigas*. J Bacteriol 175:2859–2863

Hensgens CMH, Hagen WR, Hansen TA (1995) Purification and characterization of a benzyl viologen-linked, tungsten-containing aldehyde oxidoreductase from *Desulfovibrio gigas*. J Bacteriol 177:6195–6200

Hobson PN (1965) Continuous culture of some anaerobic and facultatively anaerobic rumen bacteria. J Gen Microbiol 38:167–180

Hobson PN, Summers R (1967) The continuous culture of anaerobic bacteria. J Gen Microbiol 47: 53–65

Hobson PN, Summers R (1972) ATP pool and growth yield in *Selenomonas ruminantiurn*. J Gen Microbiol 70:351–360

Herrmann G, Jayamani E, Mai G, Buckel W (2008) Energy conservation via electron-transferring flavoprotein in anaerobic bacteria. J Bacteriol 190:784–791

Hocking WP, Stokke R, Roalkvam I, Steen IJ (2014) Identification of key components in the energy metabolism of the hyperthermophilic sulfate-reducing archaeon *Archaeoglobus fulgidus* by transcriptome analyses. Front Microbiol 2014(5):95. https://doi.org/10.3389/fmicb.2014.00095

Huang H, Wang S, Moll J, Thauer RK (2012) Electron bifurcation involved in the energy metabolism of the acetogenic bacterium *Moorella thermoacetica* growing on glucose or H_2 plus CO_2. J Bacteriol 194:3689–3699

Hungate RE (1963) Polysaccharide storage and growth efficiency in *Ruminococcus albus*. J Bacteriol 86:848–854

Ishikawa S, Lehninger AL (1962) Reconstitution of oxidative phosphorylation in preparations from *Micrococcus lysodeikticus*. J Biol Chem 237:2401–2408

Ishimoto M (1959) Sulfate reduction in cell-free extracts of *Desulfovibrio*. J Biochem (Tokyo) 46: 105–106

Ishimoto M, Fujimoto D (1961) Biochemical studies on sulfate-reducing bacteria. X. Adenosine-5′-phosphosulfate reductase. J Biochem (Tokyo) 50:299–304

Itoh T, Suzuki K, Sanchez PC, Nakase T (1999) *Caldivirga maquilingensis* gen. nov., sp. nov., a new genus of rod-shaped crenarchaeote isolated from a hot spring in the Philippines. Int J Syst Bacteriol 49:1157–1163

Jacobs NJ, Jacobs JM (1976) Nitrate, fumarate and oxygen as electron acceptors for a late step in microbial heme synthesis. Biochim Biophys Acta 449:1–9

Jacobs NJ, Jacobs JM (1975) Fumarate as an alternate electron acceptor for the late steps of anaerobic heme synthesis in *Escherichia coli*. Biochem Biophys Res Commun 65:435–441

Junier P, Junier T, Podell S, Sims DR, Detter JC, Lykidis A et al (2010) The genome of the Gram-positive metal- and sulfate-reducing bacterium *Desulfotomaculum reducens* strain MI-1. Environ Microbiol 12:2738–2754

Kakinuma Y, Yamato I, Murata T (1999) Structure and function of vacuolar Na^+-translocating ATPase in *Enterococcus hirae*. J Bioenerg Biomembr 31:7–14

Kaster AK, Moll J, Parey K, Thauer RK (2011) Coupling of ferredoxin and heterodisulfide reduction via electron bifurcation in hydrogenotrophic methanogenic archaea. Proc Natl Acad Sci U S A 108:2981–2986

Keller KL, Rapp-Giles BJ, Semkiw ES, Porat I, Brown SD, Wall JD (2014) New model for electron flow for sulfate reduction in *Desulfovibrio alaskensis* G20. Appl Environ Microbiol 80:855–868

Keller KL, Wall JD (2011) Genetics and molecular biology of the electron flow for sulfate respiration in *Desulfovibrio*. Front Microbiol 2:135. https://doi.org/10.3389/fmicb.2011.00135

Khosrovi R, Miller JDA (1975) A comparison of the growth of *Desulfovibrio vulgaris* under a hydrogen and under an inert atmosphere. Plant Soil 43:171–187

Klemm DJ, Barton LL (1989) Protoporphyrinogen oxidation coupled to nitrite reduction with membranes from *Desulfovibrio gigas*. FEMS Microbiol Lett 61:61–64

Klemps R, Cypionka H, Widdel F, Pfennig N (1985) Growth with hydrogen, and further physiological characteristics of *Desulfotomaculum* species. Arch Microbiol 143:203–208

Kobayashi K, Hasegawa H, Takagi M, Ishimoto M (1982) Proton translocation associated with sulfite reduction in a sulfate-reducing bacterium, *Desulfovibrio vulgaris*. FEBS Lett 142:235–237

Koffler H, Wilson PW (1951) The comparative biochemistry of molecular hydrogen. In: Werkman CW, Wilson PW (eds) Bacterial physiology. Academic Press, New York, pp 517–530

Körtner C, Lauterbach F, Tripier D, Unden G, Kröger A (1990) *Wolinella succinogenes* fumarate reductase contains a dihaem cytochrome *b*. Mol Microbiol 4:855–860

Kreke B, Cypionka H (1992) Proton motive force in freshwater sulfate-reducing bacteria, and its role in sulfate accumulation in *Desulfobulbus propionicus*. Arch Microbiol 158:183–187

Kreke B, Cypionka H (1994) Role of sodium ions for sulfate transport and energy metabolism in *Desulfovibrio salexigens*. Arch Microbiol 161:55–61

Krishnamurthi S, Spring S, Kumar PA, Mayilraj S, Klenk H-P, Suresh K (2013) *Desulfotomaculum defluvii* sp. nov., a sulfate-reducing bacterium isolated from the subsurface environment of a landfill. Int J Syst Evol Microbiol 63:2290–2295

Kroder M, Kroneck PMH, Cypionka H (1991) Determination of the transmembrane proton gradient in the anaerobic bacterium, *Desulfovibrio desulfuricans* by ^{31}P nuclear magnetic resonance. Arch Microbiol 156:145–147
Kröger A, Winkler E (1981) Phosphorylative fumarate reduction in *Vibrio succinogenes:* stoichiometry of ATP synthesis. Arch Microbiol 129:100–104
Kröger A (1974) Electron-transport phosphorylation coupled to fumarate reduction in anaerobically grown *Proteus rettgeri*. Biochim Biophys Acta 347:273–289
Kulkarni G, Mand TD, Metcalf WW (2018) Energy conservation via hydrogen cycling in the methanogenic archaeon *Methanosarcina barkeri*. MBio 9:e01256–e01218
Kunow J, Linder D, Thauer RK (1995) Pyruvate:ferredoxin oxidoreductase from the sulfate-reducing *Archaeoglobus fulgidus*: molecular composition, catalytic properties, and sequence alignments. Arch Microbiol 163:21–28
Laanbroek HJ, Abee T, Voogd IL (1982) Alcohol conversion by *Desulfobulbus propionicus* Lindhorst in the presence and absence of sulfate and hydrogen. Arch Microbiol 133:178–184
Labes A, Schönheit P (2001) Sugar utilization in the hyperthermophilic, sulfate-reducing archaeon *Archaeoglobus fulgidus* strain 7324: starch degradation to acetate and CO_2 via a modified Embden-Meyerhof pathway and acetyl-CoA synthetase (ADP-forming). Arch Microbiol 176: 329–338
Lancaster CRD (2002) Succinate:quinone oxidoreductases: an overview. Biochim Biophys Acta 1553:1–6
LaPaglia C, Hartzell PL (1997) Stress-induced production of biofilms in the hyperthermophile *Archaeoglobus fulgidus*. Appl Environ Microbiol 63:3158–3163
Lemos RS, Gomes CM, LeGall J, Xavier AV, Teixeira M (2002) The quinol:fumarate oxidoreductase from the sulphate reducing bacterium *Desulfovibrio gigas*: spectroscopic and redox studies. J Bioenerg Biomembr 34:21–30
Li X, Luo Q, Wofford NQ, Keller KL, McInerney MJ, Wall JD, Krumholz LR (2009) A molybdopterin oxidoreductase is involved in H_2 oxidation in *Desulfovibrio desulfuricans* G20. J Bacteriol 191:2675–2682
Lin S-M, Tsai J-Y, Hsiao C-D, Huang Y-T, Chiu C-L, Liu M-H et al (2012) Crystal structure of a membrane-embedded H+-translocating pyrophosphatase. Nature 484:399–403
Lovley DR, Ferry JG (1985) Production and consumption of H_2 during growth of *Methanosarcina* spp. on acetate. Appl Environ Microbiol 49:247–249
Lupa B, Hendrickson EL, Leigh JA, Whitman WB (2008) Formate dependent H_2 production by the mesophilic methanogen *Methanococcus maripaludis*. Appl Environ Microbiol 74:6584–6590
Lupton FS, Conrad R, Zeikus JG (1984) Physiological function of hydrogen metabolism during growth of sulfidogenic bacteria on organic substrates. J Bacteriol 159:843–849
Magee EL, Ensley BD, Barton LL (1978) An assessment of growth yields and energy coupling in *Desulfovibrio*. Arch Microbiol 117:21–26
Malki S, De Luca G, Fardeau ML, Rousset M, Bélaïch JP, Dermoun Z (1997) Physiological characteristics and growth behavior of single and double hydrogenase mutants of *Desulfovibrio fructosovorans*. Arch Microbiol 167:38–45
Mander GJ, Duin EC, Linder D, Stetter KO, Hedderich R (2002) Purification and characterization of a membrane-bound enzyme complex from the sulfate-reducing archaeon *Archaeoglobus fulgidus* related to heterodisulfide reductase from methanogenic archaea. Eur J Biochem 269: 1895–1904
Martins M, Mourato C, Pereira IAC (2015) *Desulfovibrio vulgaris* growth coupled to formate-driven H_2 production. Environ Sci Technol 49:14655–14662
Matias PM, Pereira IA, Soares CM, Carrondo MA (2005) Sulphate respiration from hydrogen in *Desulfovibrio* bacteria: a structural biology overview. Prog Biophys Mol Biol 89:292–329
Mechalas BJ, Rittenberg SC (1960) Energy coupling in *Desulfovibrio desulfuricans*. J Bacteriol 80: 501–507
Men Y, Feil H, VerBerkmoes NC, Shah MB, Johnson DR, Lee PKH et al (2012) Sustainable syntrophic growth of *Dehalococcoides ethenogenes* strain 195 with *Desulfovibrio vulgaris*

Hildenborough and *Methanobacterium congolense*: global transcriptomic and proteomic analyses. ISME J 6:410–421
Meyer B, Kuehl J, Deutschbauer AM, Price MN, Arkin AP, Stahl DA (2013) Variation among *Desulfovibrio* species in electron transfer systems used for syntrophic growth. J Bacteriol 195: 990–1004
Meyer B, Kuehl JV, Deutschbauer AM, Arkin AP, Stahl DA (2014) Flexibility of syntrophic enzyme systems in *Desulfovibrio* species ensures their adaptation capability to environmental changes. J Bacteriol 195:4900–4914
Miller JDA, Wakerley DS (1966) Growth of sulphate-reducing bacteria by fumarate dismutation. J Gen Microbiol 43:101–107
Mitchell P (1966) Chemiosmotic coupling in oxidative and photosynthetic phosphorylation. Biol Rev 41:445–502
Möbius K, Arias-Cartin R, Breckau D, Hännig A-L, Riedma K, Biedendieck R et al (2010) Heme biosynthesis is coupled to electron transport chains for energy generation. Proc Natl Acad Sci U S A 107:10436–10441
Morais-Silva FO, Rezende AM, Pimentel C, Santos CI, Clemente C, Varela-Raposo A et al (2014) Genome sequence of the model sulfate reducer *Desulfovibrio gigas*: a comparative analysis within the *Desulfovibrio* genus. Microbiologyopen 3:513–530
Mota CS, Valette O, González PJ, Brondino CD, Moura JJG, Moura I, Dolla A, Rivas MG (2011) Effects of molybdate and tungstate on expression levels and biochemical characteristics of formate dehydrogenases produced by *Desulfovibrio alaskensis* NCIMB 13491. J Bacteriol 193: 2917–2923
Müller V, Imkamp F, Biegel E, Schmidt S, Dilling S (2008) Discovery of a ferredoxin:NAD+-oxidoreductase (Rnf) in *Acetobacterium woodii*: a novel potential coupling site in acetogens. Ann N Y Acad Sci 1125:137–146. https://doi.org/10.1196/annals.1419.011
Nanninga HJ, Gottschal JC (1987) Properties of *Desulfovibrio carbinolicus* sp. nov. and other sulfate-reducing bacteria isolated from an anaerobic-purification plant. Appl Environ Microbiol 53:802–809
Neidhardt EC, Umbarger HE (1966) Chemical composition of *Escherichia coli*. In: Neidhardt FC, Curtis R III, Ingraham JL, Lin ECC, Brooks Low K, Magasanik B, Reznikoff WS, Riley M, Schaechter M, Umbarger HE (eds) Escherichia coli and salmonella cellular and molecular biology. American Society for Microbiology Press, Washington, DC, pp 13–16
Noguera DR, Brusseau GA, Rittmann BE, Stahl DA (1998) A unified model describing the role of hydrogen in the growth of *Desulfovibrio vulgaris* under different environmental conditions. Biotechnol Bioeng 59:732–746
Odom JM, Peck HD (1981a) Hydrogen cycling as a general mechanism for energy coupling in the sulfate-reducing bacteria, *Desulfovibrio* sp. FEMS Microbiol Lett 12:47–50
Odom JM, Peck HD (1981b) Localization of dehydrogenases, reductases, and electron transfer components in the sulfate-reducing bacterium *Desulfovibrio gigas*. J Bacteriol 147:161–169
Odom JM, Peck HD (1984a) Hydrogen cycling in *Desulfovibrio*: a new mechanism for energy coupling in anaerobic microorganisms. In: Halvorsen H, Cohen Y (eds) Microbial mats: stromatolites. Alan R. Liss, Inc, New York, pp 215–243
Odom JM, Peck HD (1984b) Hydrogenase, electron-transfer proteins, and energy coupling in the sulfate-reducing bacteria *Desulfovibrio*. Ann Rev Microbiol 38:551–592
Odom JM, Wall JD (1987) Properties of a hydrogen-inhibited mutant of *Desulfovibrio desulfuricans* ATCC 27774. J Bacteriol 169:1335–1337
Ogata M, Yagi T (1986) Pyruvate dehydrogenase and the path of lactate degradation in *Desulfovibrio vulgaris* Miyazaki F. J Biochem 100:311–318
Ogata M, Arihara K, Yagi T (1981) D-lactate dehydrogenase of *Desulfovibrio vulgaris*. J Biochem 89:1423–1431
Ogg CD, Patel BKC (2011) *Desulfotomaculum varum* sp. nov., a moderately thermophilic sulfate-reducing bacterium isolated from a microbial mat colonizing a Great Artesian Basin bore well runoff channel. 3 Biotech 1:139–149

Ollivier B, Cord-Ruwisch R, Hatchikian EC, Garcia JL (1988) Characterization of *Desulfovibrio fructosovorans* sp. nov. Arch Microbiol 149:447–450

Otwell AE, Callister SJ, Zink EM, Smith RD, Richardson RE (2016) Comparative proteomic analysis of *Desulfotomaculum reducens* MI-1: insights into the metabolic versatility of a Gram-positive sulfate- and metal-reducing bacterium. Front Microbiol 7:191. https://doi.org/10.3389/fmicb.2016.00191

Ozawa K, Meikari T, Motohashi K, Yoshida M, Akutsu H (2000) Evidence for the presence of an F-Type ATP synthase involved in sulfate respiration in *Desulfovibrio vulgaris*. J Bacteriol 182: 2200–2206

Pankhania IP, Gow LA, Hamilton WA (1986) The effect of hydrogen on the growth of *Desulfovibrio vulgaris* (Hildenborough) on lactate. Microbiology 132:3349–3356

Pankhania IP, Spormann AM, Hamilton WA, Thauer RK (1988) Lactate conversion to acetate, CO_2 and H_2 in cell suspensions of *Desulfovibrio vulgaris* (Marburg): indications for the involvement of an energy driven reaction. Arch Microbiol 150:26–31

Parshina SN, Sipma J, Henstra AM, Stams AJ (2010) Carbon monoxide as an electron donor for the biological reduction of sulphate. Int J Microbiol 2010:319527. https://doi.org/10.1155/2010/319527

Peck HD (1959) The ATP-dependent reduction of sulfate with hydrogen in extracts of *Desulfovibrio desulfuricans*. Proc Natl Acad Sci U S A 45:701–708

Peck HD (1960) Evidence for oxidative phosphorylation during the reduction of sulfate with hydrogen by *Desulfovibrio desulfuricans*. J Biol Chem 235:2734–2738

Peck HD (1961) Evidence for the reversibility of adenosine-5′-phosphosulfate reductase. Biochim Biophys Acta 49:621–624

Peck HD (1962a) Symposium on metabolism of inorganic compounds V. Comparative metabolism of inorganic sulfur compounds in microorganisms. Bacteriol Rev 26:67–94

Peck HD (1962b) The role of adenosine-5′-phosphosulfate in the reduction of sulfate to sulfite by *Desulfovibrio desulfuricans*. J Biol Chem 237:198–203

Peck HD (1966) Phosphorylation coupled with electron transfer in extracts of the sulfate-reducing bacterium, *Desulfovibrio gigas*. Biochem Biophys Res Commun 22:112–118

Peck HD, LeGall J (1982) Biochemistry of dissimilatory sulphate reduction. Philos Trans R Soc Lond B Biol Sci 298:443–466

Peck HD, LeGall J, Lespinat PA, Berlier Y, Fauque G (1987) A direct demonstration of hydrogen cycling by *Desulfovibrio vulgaris* employing membrane-inlet mass spectrometry. FEMS Microbiol Lett 40:295–299

Pereira PM, He Q, Filipa MA, Valente FM, Xavier AV, Zhou J, IAC P, Louro RO (2008) Energy metabolism in *Desulfovibrio vulgaris* Hildenborough: insights from transcriptome analysis. Antonie Van Leeuwenhoek 93:347–362

Pereira IAC, Romão CV, Xavier AV, LeGall J, Teixeira M (1998) Electron transfer between hydrogenases and mono- and multiheme cytochromes in *Desulfovibrio*. Eur J Biochem 3: 494–498

Pereira IAC, Ramos AR, Grein F, Marques MC, da Silva SM, Venceslau SS (2011) A comparative genomic analysis of energy metabolism in sulfate reducing bacteria and archaea. Front Microbiol 2:69. https://doi.org/10.3389/fmicb.2011.00069

Pereira PM, Teixeira M, Xavier AV, Louro RO, Pereira IAC (2006) The Tmc complex from *Desulfovibrio vulgaris* Hildenborough is involved in transmembrane electron transfer from periplasmic hydrogen oxidation. Biochemist 45:10359–10367

Pierce E, Xie G, Barabote RD, Saunders E, Han CS, Detter JC et al (2008) The complete genome sequence of *Moorella thermoacetica* (f. *Clostridium thermoaceticum*). Environ Microbiol 10: 2550–2573

Pires RH, Lourenço AI, Morais F, Teixeira M, Xavier AV, Saraiva LM, Pereira IA (2003) A novel membrane-bound respiratory complex from *Desulfovibrio desulfuricans* ATCC 27774. Biochim Biophys Acta 1605:67–82

Plugge CM, Henstra AM, Worm P, Swarts DC, Paulitsch-Fuchs AH, Scholten JCM et al (2012) Complete genome sequence of *Syntrophobacter fumaroxidans* strain ($MPOB^T$). Stand Genomic Sci 7:91–106

Plugge CM, Scholten JCM, Culley DE, Nie L, Brockman FJ, Zhang W (2010) Global transcriptomics analysis of *Desulfovibrio vulgaris* lifestyle change from syntrophic growth with *Methanosarcina barkeri* to sulfate reducer. Microbiology 156:2746–2756

Poratti GW, Yaakop AS, Chan CS, Urbieta MS, Chan K-G, Ee R et al (2016) Draft genome sequence of the sulfate-reducing bacterium *Desulfotomaculum copahuensis* strain CINDEFI1 isolated from the geothermal Copahue System, Neuquén, Argentina. Genome Announc 4(4): e00870–e00816. https://doi.org/10.1128/genomeA.00870-16

Postgate JR (1951) The reduction of sulfur compounds by *Desulfovibrio desulfuricans*. J Gen Microbiol 5:725–738

Postgate JR (1963) Sulfate-free growth of *Clostridium nigrificans*. J Bacteriol 85:1450–1451

Postgate JR (1984) Genus *Desulfovibrio*. In: King NR, Holt JG (eds) Bergey's manuel of systematic bacteriology. Williams & Wilkins, Baltimore, pp 666–673

Price MN, Ray J, Wetmore KM, Kuehl JV, Bauer S, Deutschbauer AM, Arkin AP (2014) The genetic basis of energy conservation in the sulfate-reducing bacterium *Desulfovibrio alaskensis* G20. Front Microbiol 2014(5):577. https://doi.org/10.3389/fmicb.2014.00577

Pullman ME, Schatz G (1967) Mitochondrial oxidations and energy coupling. Annu Rev Biochem 36:539–610

Raaijmakers H, Macieira S, Dias J, Teixeira S, Bursakov S, Huber R et al (2002) Gene sequence and the 1.8 Å crystal structure of the tungsten-containing formate dehydrogenase from *Desulfovibrio gigas*. Structure 10:1261–1272

Rabus R, Venceslau SS, Wöhlbrand L, Voordouw G, Wall JD, Pereira IAC (2015) A post-genomic view of the ecophysiology, catabolism and biotechnological relevance of sulphate-reducing prokaryotes. Adv Microbial Physiol 66:56–321

Rabus R, Hansen TA, Widdel F (2006) Dissimilatory sulfate and sulfur-reducing prokaryotes. In: Dworkin M, Falkow S, Rosenberg E, Schleifer K-H, Stackebrandt E (eds) The prokaryotes. Springer-Verlag, New York, pp 659–768

Ramos AR, Grein F, Oliveira GP, Venceslau SS, Keller KL, Wall JD, Pereira IAC (2014) The FlxABCD-HdrABC proteins correspond to a novel NADH dehydrogenase/heterodisulfide reductase widespread in anaerobic bacteria and involved in ethanol metabolism in *Desulfovibrio vulgaris* Hildenborough. Environ Microbiol 17:2288–2305

Rapp-Giles BJ, Casalot L, English RS, Ringbauer JA, Dolla A, Wall JD (2000) Cytochrome c_3 mutants of *Desulfovibrio desulfuricans*. Appl Environ Microbiol 66:671–677

Reddy CA, Peck HD (1978) Electron transport phosphorylation coupled to fumarate reduction by H_2- and Mg^{2+}-dependent adenosine triphosphatase activity in extracts of the rumen anaerobe *Vibrio succinogenes*. J Bacteriol 78:982–991

Reed DW, Hartzell PL (1999) The *Archaeoglobus fulgidus* D-lactate dehydrogenase is a Zn^{2+} flavoprotein. J Bacteriol 181:7580–7587

Reichenbecher W, Schink B (1997) *Desulfovibrio inopinatus*, sp. nov., a new sulfate-reducing bacterium that degrades hydroxyhydroquinone (1,2,4-trihydroxybenzene). Arch Microbiol 168: 338–344

Riederer-Henderson MA, Peck HD (1986a) *In vitro* requirements for formate dehydrogenase activity from *Desulfovibrio*. Can J Microbiol 32:425–429

Riederer-Henderson MA, Peck HD (1986b) Properties of formate dehydrogenase from *Desulfovibrio gigas*. Can J Microbiol 32:430–435

Rodrigues R, Valente FM, Pereira IA, Oliveira S, Rodrigues-Pousada C (2003) A novel membrane-bound Ech [NiFe] hydrogenase in *Desulfovibrio gigas*. Biochim Biophys Res Commun 306: 366–375

Rossi M, Pollock WB, Reij MW, Keon RG, Fu R, Voordouw G (1993) The hmc operon of *Desulfovibrio vulgaris* subsp. *vulgaris* Hildenborough encodes a potential transmembrane redox protein complex. J Bacteriol 175:4699–4711

Saraiva LM, da Costa PN, Conte C, Xavier AV, LeGall J (2001) In the facultative sulphate/nitrate reducer *Desulfovibrio desulfuricans* ATCC 27774, the nine-haem cytochrome c is part of a membrane-bound redox complex mainly expressed in sulphate-grown cells. Biochim Biophys Acta 1520:63–70

Sass A, Rütters H, Cypionka H (2002) *Desulfobulbus mediterraneus* sp. nov., a sulfate-reducing bacterium growing on mono- and disaccharides. Arch Microbiol 177:468–474

Schirawski J, Unden G (1998) Menaquinone-dependent succinate dehydrogenase of bacteria catalyzes reversed electron transport driven by the proton potential. Eur J Biochem 257:210–215

Schmehl M, Jahn A, Meyer zu Vilsendorf A, Hennecke S, Masepohl B, Schuppler M, Marxer M, Oelze J, Klipp W (1993) Identification of a new class of nitrogen fixation genes in *Rhodobacter capsulatus*: a putative membrane complex involved in electron transport to nitrogenase. Mol Gen Genet 241:602–615

Schröder I, Kröger A, Macy JM (1988) Isolation of the sulphur reductase and reconstitution of the sulphur respiration of *Wolinella succinogenes*. Arch Microbiol 149:572–579

Sebban C, Blanchard L, Bruschi M, Guerlesquin F (1995) Purification and characterization of the formate dehydrogenase from *Desulfovibrio vulgaris* Hildenborough. FEMS Microbiol Lett 133: 143–149

Senez J (1953) Sur l'activité et la croissance des bacteries anaérobies sulfato-réductrices en cultures semi-autotrophes. Ann Inst Pasteur 84:595–604

Senez JC (1962) Some considerations on the energetics of bacterial growth. Bacteriol Rev 26:95–105

Senez JC, Pichinoty F (1958a) Sur la réduction du nitrite aux dépens de l'hydrogène moléculaire par *Desulfovibrio desulfuricans* et d'autres espèces bactériennes. Bull Soc Chim Biol 40:2099–2117

Senez JC, Pichinoty F (1958b) Réduction de l'hydroxylamine de liée à l'activité de l'hydrogénase de *Desulfovibrio desulfuricans*. I. Activité des cellules et des extraits. Biochim Biophys Acta 27: 569–580

Senez JC, Pichinoty F (1958c) Réduction de l'hydroxylamine liée a l'activité de l'hydrogénase de *Desulfovibrio desulfuricans*. II. Nature du système enzymatique et du transporteur d'électrons intervenant dans la réaction. Biochim Biophys Acta 28:355–369

Senez J, Volcani B (1951) Utilization de l'hydrogène moleculaire par des souches pures de bactéries sulfato-réductrices d'origine marine. Compt Rend 232:1035–1036

Serrano A, Pérez-Castiñeira JR, Baltscheffsky M, Baltscheffsky H (2007) H^+-PPases: yesterday, today and tomorrow. UBMB Life 59:76–83

Sim MS, Bosak T, Ono S (2011) Large sulfur isotope fractionation does not require disproportionation. Science 333:74–77

Sim MS, Wang DT, Zane GM, Wall JD, Bosak T, Ono S (2013) Fractionation of sulfur isotopes by *Desulfovibrio vulgaris* mutants lacking hydrogenases or type I tetraheme cytochrome c_3. Front Microbiol 4:171. https://doi.org/10.3389/fmicb.2013.00171

Simon J, van Spanning RJ, Richardson DJ (2008) The organization of proton motive and non-proton motive redox loops in prokaryotic respiratory systems. Biochim Biophys Acta 1777:1480–1490

Sisler FD, ZoBell CE (1951) Hydrogen utilization by some marine sulfate-reducing bacteria. J Bacteriol 62:117–127

Sivakumar K, Scarascia G, Zaouri N, Wang T, Kaksonen AH, Hong P-Y (2019) Salinity-mediated increment in sulfate reduction, biofilm formation, and quorum sensing: a potential connection between quorum sensing and sulfate reduction? Front Microbiol 10:188. https://doi.org/10.3389/fmicb.2019.00188

Sorokin YI (1954) Chemistry of the process of hydrogen reduction of sulfates. Trudy Inst Microbiol Akad Nauk S S S R 3:21–34

Sousa DZ, Visser M, van Gelder AH, Boeren S, Pieterse MM, Pinkse MWH et al (2018) The deep-subsurface sulfate reducer *Desulfotomaculum kuznetsovii* employs two methanol-degrading pathways. Nat Commun 9:239. https://doi.org/10.1038/s41467-017-02518-9

Spring S, Lapidus A, Schröder M, Gleim D, Sims D, Meincke L et al (2009) Complete genome sequence of *Desulfotomaculum acetoxidans* type strain (5575^T). Stand Genomic Sci 1:242–253

Spring S, Sorokin DY, Verbarg S, Rohde M, Woyke T, Kyrpides NC (2019) Sulfate-reducing bacteria that produce exopolymers thrive in the calcifying zone of a hypersaline cyanobacterial mat. Front Microbiol 24. https://doi.org/10.3389/fmicb.2019.00862

Stams AJM, Hansen TA (1982) Oxygen-labile L(+)-lactate dehydrogenase activity in *Desulfovibrio* HL21. FEMS Microbiol Lett 13:389–394

Stams FJM, Veenhuis M, Weenk GH, Hansen TA (1983) Occurrence of polyglucose as a storage polymer in *Desulfovibrio* species and *Desulfobulbus propionicus*. Arch Microbiol 136:54–59

Steenkamp DJ, Peck HD (1981) Proton translocation associated with nitrite respiration in *Desulfovibrio desulfuricans*. J Biol Chem 256:5450–5458

Stouthamer AH (1979) The search for correlation between theoretical and experimental growth yields. Int Rev Biochem Microb Biochem 21:1–47

Strittmatter AW, Liesegang H, Rabus R, Decker I, Amann J, Andres S et al (2009) Genome sequence of *Desulfobacterium autotrophicum* HRM2, a marine sulfate reducer oxidizing organic carbon completely to carbon dioxide. Environ Microbiol 11:1038–1055

Tang Y, Pingitore F, Mukhopadhyay A, Phan R, Hazen TC, Keasling JD (2007) Pathway confirmation and flux analysis of central metabolic pathways in *Desulfovibrio vulgaris* Hildenborough using gas chromatography-mass spectrometry and Fourier transform-ion cyclotron resonance mass spectrometry. J Bacteriol 189:940–949

Thauer RK, Jungermann K, Decker K (1977) Energy conservation in chemotrophic anaerobic bacteria. Bacteriol Rev 41:100–180

Thauer RK, Stackebrandt E, Hamilton W (2007) Energy metabolism and phylogenetic diversity of sulphate-reducing bacteria. In: Barton LL, Hamilton WA (eds) Sulphate-reducing bacteria: environmental and engineered systems. Cambridge University Press, Cambridge, pp 1–38

Traore AS, Hatchikian CE, Belaich J-P, Le Gall J (1981) Microcalorimetric studies of the growth of sulfate-reducing bacteria:energetics of *Desulfovibrio vulgaris* growth. J Bacteriol 145:191–199

Trinkerl M, Breunig A, Schauder R, König H (1990) *Desulfovibrio termitidis* sp. nov., a carbohydrate-degrading sulfate-reducing bacterium from the hindgut of a termite. Syst Appl Microbiol 13:372–377

Tsuji K, Yagi T (1980) Significance of hydrogen burst from growing cultures of *Desulfovibrio vulgaris* Miyazaki, and the role of hydrogenase, and cytochrome c_3 in energy production system. Arch Microbiol 125:35–42

Valente FM, Saraiva LM, LeGall J, Xavier AV, Teixeira M, Pereira IA (2001) A membrane-bound cytochrome c_3: a type II cytochrome c_3 from *Desulfovibrio vulgaris* Hildenborough. Chembiochem 2:895–905

Vecchia ED, Shao PP, Suvorova E, Chiappe D, Hamelin R, Bernier-Latmani R (2014) Characterization of the surfaceome of the metal-reducing bacterium *Desulfotomaculum reducens*. Front Microbiol 5:432. https://doi.org/10.3389/fmicb.2014.00432

Venceslau SS, Lino RR, Pereira IAC (2010) The Qrc membrane complex, related to the alternative complex III, is a menaquinone reductase involved in sulfate respiration. J Biol Chem 285: 22774–22783

Visser M, Parshina SN, Alves JI, Sousa DZ, Pereira IAC, Muyzer G et al (2014) Genome analyses of the carboxydotrophic sulfate-reducers *Desulfotomaculum nigrificans and Desulfotomaculum carboxydivorans* and reclassification of *Desulfotomaculum caboxydivorans* as a later synonym of *Desulfotomaculum nigrificans*. Stand Genomic Sci 9(3):655–675. PMC4149029. https://doi.org/10.3389/fmicb.2014.00095

Vita N, Valette O, Brasseur G, Lignon S, Denis Y, Ansaldi M, Dolla A, Pieulle L (2015) The primary pathway for lactate oxidation in *Desulfovibrio vulgaris*. Front Microbiol 6:606. https://doi.org/10.3389/fmicb.2015.00606

Voordouw G (2002) Carbon monoxide cycling by *Desulfovibrio vulgaris* Hildenborough. J Bacteriol 184:5903–5911

Vosjan JH (1970) ATP generation by electron transport in *Desulfovibrio desulfuricans*. Antoine von Leeuwenhoek 36:585–586

Walker CB, He Z, Yang ZK, Ringbauer JA, He Q, Zhou J et al (2009) The electron transfer system of syntrophically grown *Desulfovibrio vulgaris*. J Bacteriol 191:5793–5801

Wall JD, Hemme CL, Rapp-Giles B, Ringbauer JA, Casalot L, Giblin T (2003) Genes and genetic manipulations of *Desulfovibrio*. In: Ljungdahl LG, Adams MW, Barton LL, Ferry JG, Johnson MK (eds) Biochemistry and physiology of anaerobic bacteria. Springer-Verlag, New York, pp 85–98

Wang S, Huang H, Moll J, Thauer RK (2010) $NADP^+$ reduction with reduced ferredoxin and $NADP^+$ reduction with NADH are coupled via an electron-bifurcating enzyme complex in *Clostridium kluyveri*. J Bacteriol 192:5115–5123

Widdel F, Hansen TA (1991) The dissimilatory sulfate- and sulfur-reducing bacteria. In: Balows A, Truper HG, Dworkin M, Harder W, Schleifer K-H (eds) The prokaryotes. Springer-Verlag, New York, pp 659–768

Widdel F, Pfennig N (1984) Dissimilatory sulfate-and sulfur-reducing bacteria. In: Krieg NR, Holt JG (eds) Bergey's manual of systematic bacteriology, vol 1. Williams & Wilkins, Baltimore, pp 663–679

Winter JU, Wolfe RS (1980) Methane formation from fructose by syntrophic associations of *Acetobacterium woodii* and different strains of methanogens. Arch Microbiol 124:73–79

Wood PM (1978) A chemiosmotic model for sulfate respiration. FEBS Lett 95:12–18

Yagi T (1970) Solubilization, purification and properties of particulate hydrogenase from *Desulfovibrio vulgaris*. J Biochem (Tokyo) 68:649–657

Yagi T (1979) Purification and properties of cytochrome *c*-553, an electron acceptor for formate dehydrogenase of *Desulfovibrio vulgaris,* Miyazaki. Biochim Biophys Acta 548:96–105

Yagi T, Honya M, Tamiya N (1968) Purification and properties of hydrogenase of different origins. Biochim Biophys Acta 153:699–705

Yang G, Guo J, Zhuang L, Yuan Y, Zhou S (2016) *Desulfotomaculum ferrireducens* sp. nov., a moderately thermophilic sulfate-reducing and dissimilatory Fe(III)-reducing bacterium isolated from compost. Int J Syst Evol Microbiol 66:3022–3028

Zane GM, Yen HC, Wall JD (2010) Effect of the deletion of qmoABC and the promoter-distal gene encoding a hypothetical protein on sulfate reduction in *Desulfovibrio vulgaris* Hildenborough. Appl Environ Microbiol 76:5500–5509

Zaunmüller T, Kelly DJ, Glöckner FO, Unden G (2006) Succinate dehydrogenase functioning by a reverse redox loop mechanism and fumarate reductase in sulphate-reducing bacteria. Microbiology 152:2443–2453

Zellner G, Messner P, Kneifel H, Winter J (1989) *Desulfovibrio simplex* spec. nov., a new sulfate-reducing bacterium from a sour whey digester. Arch Microbiol 152:329–334

Zhang W, Culley DE, Scholten JC, Hogan M, Vitiritti L, Brockman FJ (2006a) Global transcriptomic analysis of *Desulfovibrio vulgaris* on different electron donors. Antonie Van Leeuwenhoek 89:221–237

Zhang W, Gritsenko MA, Moore RJ, Culley DE, Nie L, Petritis K et al (2006b) A proteomic view of *Desulfovibrio vulgaris* metabolism as determined by liquid chromatography coupled with tandem mass spectrometry. Proteomics 6:4286–4299

Chapter 6
Cell Biology and Metabolism

6.1 Introduction

For many years, the bacteria that grew with sulfate respiration with the production of H_2S were regarded as biological curiosities. In one of the first reviews of sulfate-reducing bacteria (Postgate 1979), the isolation of SRB from sites around the world, their activity in corrosion of iron popes and conditions supporting growth was discussed. In the past few decades, an understanding of the mechanisms accounting for adjustments in cell biology of SRB has been developed using biochemistry and genome analysis. Identification of genes present in *Desulfovibrio vulgaris*, the model organism of the sulfate reducers, has provided evidence that SRB share many characteristics of genetic expression with other bacteria. The unraveling of intermediary metabolism and the biosynthetic processes has benefited greatly from the development of proteonomics and transcriptomics. This chapter focuses on the response of SRB and especially *D. vulgaris* Hildenborough to chemical perturbations from the environment, and when appropriate, comparisons to other bacteria are provided.

6.2 Using Genomic, Proteomic, and Biochemical Analysis

Genome analysis of *D. vulgaris* Hildenborough DSM 644/ATCC29579/NCIMB 8303 (Heidelberg et al. 2004) has been revised (Price et al. 2011), and annotation of genes is available for comparison with other bacteria using Microbeson Line (http://www.microbesonline.org/). A database for the genome of *D. vulgaris* Hildenborough and other sulfate-reducing microorganisms is available at several websites: http://www.tigr.org; www.jgi.doe.gov; www.genome.jp/kegg/. At this time, there is an interest in enhancing the identification of proteins encoded in the

L. L. Barton, G. D. Fauque, *Sulfate-Reducing Bacteria and Archaea*,
https://doi.org/10.1007/978-3-030-96703-1_6

D. vulgaris Hildenborough genome and a method for integrating proteomic analysis with transcriptomic analysis is being pursued (Torres-García et al. 2009).

6.2.1 Cell Surface

6.2.1.1 Outer Membrane, Porins, and Vesicles

Outer membranes of *D.vulgaris* Hildenborough have been partially characterized. Following the generalization that 2–3% of all genes coding for proteins in a Gram-negative bacterium would be associated with the outer membrane (Casadio et al. 2003), the outer membrane proteome of *D. vulgaris* Hildenborough would be predicted to be 85 proteins. Mass spectrometry analysis of proteins isolated from the outer membrane of *D. vulgaris* Hildenborough revealed the presence of 70 proteins with over half of these as lipoproteins, and several were associated with stress response (Walian et al. 2012). A series of proteins were proposed to be associated with efflux systems of *D. vulgaris* Hildenborough with putative proteins (DVO1013, 2815, 3097, and 0062) which were channel-forming proteins which spanned the periplasm and outer membrane in type I secretion systems. Also present was secretin (DVO1273) of type II secretory system and DVUA0117 associated with type III secretory system. The periplasmic [NiFeSe] (DVU1719,1718) hydrogenase and [NiFe] (DVU1921,1922) hydrogenase were isolated with the outer membrane and may reflect the strong lipophilic character of these hydrogenases (Walian et al. 2012).

The moving of small molecular weight molecules through the outer membrane of Gram-negative bacteria employs porins, and these structures are also present in the sulfate-reducing bacteria. The outer membrane of *D. vulgaris* Hildenborough contains the protein (DVU0799) with 466 amino acids which forms a transmembrane 18-strand β-barrel trimer structure with an 8 nm pore (Zeng et al. 2017). Using proteoliposomes constructed with protein (DVU0799), the porin channel was determined to facilitate the passage of uncharged sugar molecules such as sucrose and raffinose which are 342 Da and 504 Da, respectively. This trimeric porin was more permeable to sugar acids as compared to uncharged sugars, and this may indicate a preference for negatively charges molecules be important for SRB because of the quantity of anionic sulfate required for anaerobic respiration. In comparison, the outer membrane porin isolated from *D. piger* contains 480 amino acids and has an exclusion of 300 Da (Avidan et al. 2008).

Vesicles associated with the surface of SRB were initially observed in electron micrographs of *D. gigas* (Thomas 1972). Using electron microscopy, the formation of extracellular vesicles that appeared to originate from the cell surface was observed with *D. carbinolicus* (Hanninga and Gottschal 1987). Outer membrane vesicles (80–800 Å diameter) were isolated from *D. gigas* using differential ultracentrifugation. These vesicles were found to contain about 1% of total cellular formate dehydrogenase, and this represented formate dehydrogenase captured from the periplasm to the closed vesicles (Haynes et al. 1995). Outer membrane vesicles produced by Gram-negative bacteria are proposed to have several different toles including horizontal gene transfer, and gene transfer has been suggested from research involving *D. alaskensis* DSM16109 (Crispim et al. 2019).

6.2.1.2 Cell Size of *Desulfovibrio Gigas*

While most *Desulfovibrio* spp. are 3–5 μm in length and 0.5–1.0 μm in diameter (Postgate and Campbell 1966), a few *Desulfovibrio* strains have much larger cells. The first large SRB reported was *D. gigas* DSM 1382 ATCC 19364 NCIMB 9332 isolated from the Etang de Berre near Marseilles, France, by Jean LeGall (Le Gall 1963). *D. gigas* DSM 1382 is 5–10 μm long and 1.2–1.5 μm wide. Another strain of *D. gigas* DSM 490 ATCCDSM 490 ATCC 9494 was isolated from sewage mud in Göttingen, Germany. Subsequently, *D. giganteus* was also isolated from the Berre Lagoon ("Etang de Berre") near Marseille and it is 5.0–10.0 μm long and 1.0 μm wide (Esnault et al. 1988). Another bacterium, *D. longus*, was isolated from an oil well environment and it is 0.4–0.5 μm wide and 5–10 μm long (Magot et al. 1992). Bacterial cell elongation is regulated by FtsZ and MreB proteins (Marshall et al. 2012), and evaluation of the *D. gigas* genome reveals the presence of the *minCDE* system which is an inhibitor of the FtsZ and not observed in other genomes of *Desulfovibrio* that have been sequenced (Morais-Silva et al. 2014). While other *Desulfovibrio* spp have two MreB proteins, *D. gigas* has three. Future research is required to see if the regulation resulting from the presence of the *minCDE* system and MreB proteins contributes to the large size of *D. gigas*. The role of cell division proteins in *D. giganteus* (Esnault et al. 1988) and *D. longus* (Magot et al. 1992) is not resolved at this time.

6.2.2 Metabolism of Carbon Compounds

From a perspective of biochemistry and enzymology, the pathways of lactate or pyruvate oxidation with sulfate as electron acceptor have been addressed in early summations: Peck Jr. (1962, 1993), Postgate (1965), Senez (1962), Thauer et al. (1977), Peck Jr. and LeGall (1982), Peck Jr. and Lissolo (1988), Thauer and Kunow (1995) and Cypionka (1995). And, more recent reviews: Thauer et al. (2007), Barton and Fauque (2009), Fauque and Barton (2012), Barton et al. (2014) and Rabus et al. (2015).

6.2.2.1 Lactate Oxidation

Lactate is an important substrate for *Desulfovibrio* with about 95% of lactate used for cellular energy and the remaining 5% for biosynthesis of carbon structures (Noguera

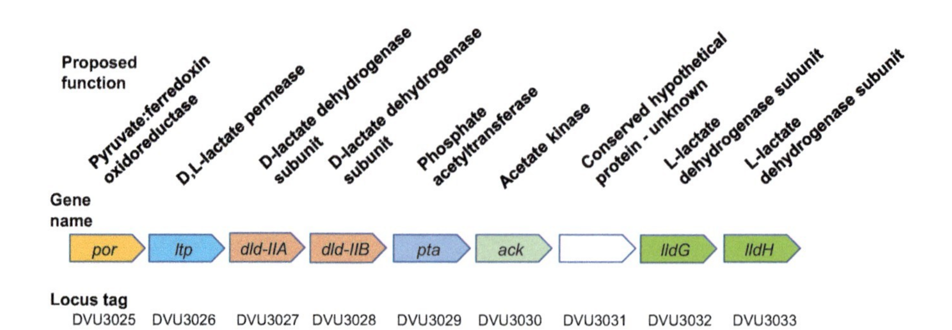

Fig. 6.1 The operon luo in *D. vulgaris* Hildenborough. Image bearing the gene name does not reflect gene length

et al. 1998). A coordinated expression of lactate-oxidizing enzymes would be advantageous for SRBs, and enzymes for oxidation of organic acids in *D. vulgaris* Hildenborough have been proposed to be encoded on an operon designated as luo (*l*actate *u*tilization *o*peron) which is highly conserved throughout *Desulfovibrio* (see Fig. 6.1) and also present in *Desulfohalobium* and *Desulfocurvus*. Transcription of the operon luo appears to be under the control of four regulators (DVU3023, DVU0539, DVU0621, and DVU1083), and upstream of the transcription start site, there are two sigma (σ) factor-dependent promoters (σ^{70} and σ^{54}) (Vita et al. 2015).

The growth medium commonly used for SRB uses lactate as the electron donor with lactate supplied as mixture of D(−)- and L(+)-lactate stereoisomers hereafter designated as D-lactate and L-lactate. Although reports indicate that these lactate dehydrogenase (Ldh) for D-lactate (Czechowski and Rossmoore 1980; Ogata et al. 1981; Czechowski and Rossmoore 1990) or L-lactate (Stams and Hansen 1982; Hansen 1995) displayed $NAD(P)^+$ independence, the physiological electron acceptor for these Ldhs has not been established. In addition to the D-ildh ($NAD(P)^+$-independent lactate dehydrogenase) and D-ildh in SRB, there is a fermentative LDH in the cytoplasm which interfaces with NAD^+.

Since the lactate dehydrogenase proteins reside on the cytoplasmic side of the plasma membrane, a permease is required to shuttle lactate into the cytoplasm from the periplasm. *D. vulgaris* Hildenborough has three putative permease genes for L-lactate (DVU2110, DVU2683, and DVU3284), permeases specific for D-lactate (DVU2285, DVU2451), a permease for both D- and L-lactate (DVU3026). In terms of moving sulfate into the cytoplasm where reduction occurs, three putative genes for sulfate permease (DVU0053, DVU1999, and DVU0279) are present in *D. vulgaris* Hildenborough (Vita et al. 2015; Clark et al. 2006). There is considerable redundancy in oxidation of lactate by *D. vulgaris* Hildenborough with three sets of genes that are proposed to participate. One group contains D-Ldh A/B which is encoded in genes DVU3027-28 and this is the major enzyme accounting for D-lactate oxidation. Paralogs of DVU3027-28 are DVU 3026-27, DVU3071, DVU0390, DVU0253 which are annotated as oxidoreductases or glycolate oxidases (Vita et al. 2015). The second group contains L-Ldh G/H which is encoded in

DVU3032-33, and this is the principle enzyme for oxidation of L-lactate. Paralogs of DVU3032-33 are DVU1781-82-83. Another gene proposed for L-lactate oxidation is DVU2784 which is an ortholog of L-Ldh from *Escherichia coli* (Vita et al. 2015). The third group contains DVU0600 and DVU1412 which are orthologs of L-Ldh of *Escherichia coli* with 40–50% sequence similarity to the fermentative Ldhs of *Clostridium cellulolyticum* and *Shewanella oneidensis* (Vita et al. 2015). Since about 95% of lactate used by *Desulfovibrio* is for cellular energy (Noguera et al. 1998), it can be assumed that the membrane-bound D-ldh and L-ldh are the primary respiratory-coupled enzymes for lactate metabolism.

Lactate-sulfate metabolism With respect to oxidation of lactate, there are three possibiliyies to consider: (i) sulfate as the electron acceptor, (ii) fermentation in the absence of sulfate, and (iii) fumarate as the electron acceptor. In the presence of sulfate, lactate is oxidized to acetate plus CO_2 by the following reaction (Pankhania et al. 1988):

$$\begin{gathered} 2\,\mathrm{Lactate}^- + \mathrm{SO_4}^{2-} + \mathrm{H}^+ = 2\,\mathrm{acetate}^- + \mathrm{HS}^- + 2\mathrm{CO_2} + 2\mathrm{H_2O} \\ \Delta G^{\circ\prime} = -196.4\,\mathrm{kJ/mol} \end{gathered} \tag{6.1}$$

Using *D. vulgaris* Hildenborough as the model SRB, the carbon pathway for lactate oxidation is provided in Fig. 6.2 and it provides information on the formation

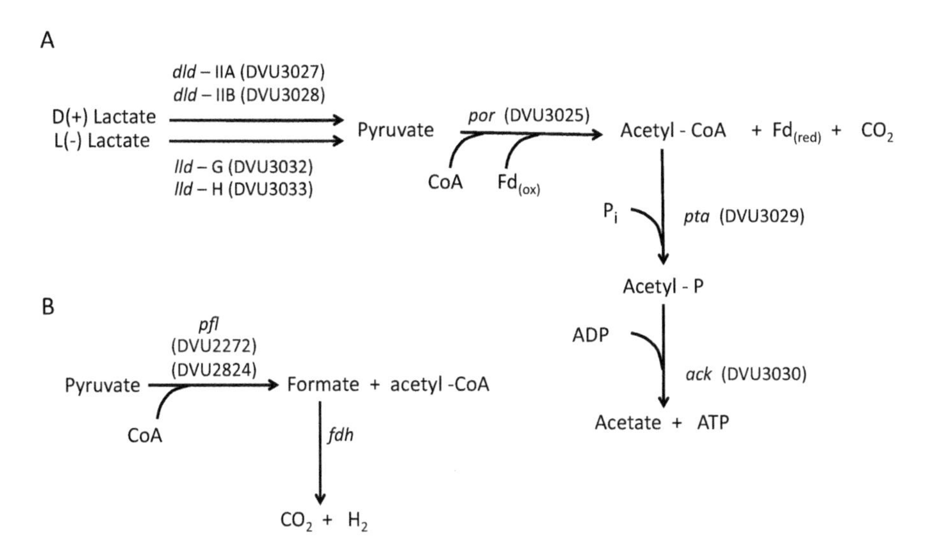

Fig. 6.2 Pyruvate oxidation pathway in *D. vulgaris*. (**A**) The use of genes encoded on the Luo operon accounts for oxidation of lactate to acetate. (**B**) Metabolism of pyruvate with the production of acetate, CO_2, and H_2 involves formate as an intermediate. The function of genes are as follows: *dld*-llA/llB = D-; lactate dehydrogenase subunits; *lld*-G/H, L-lactate dehydrogenase subunits; *por*, pyruvate:ferredoxin oxidoreductase; *pta*, phosphotransacetylase; *ack*, acetate kinase; *pfl*, pyruvate: formate lyase; *fdh*, formate dehydrogenase. Locus tag for genes is as (DVUXXXX). Fd = ferredoxin

of acetate, formate, and CO_2. Oxidative phosphorylation coupled to electron transport requires the plasma membrane, and these activities are discussed in Chap. 5.

Lactate fermentation If *D. vulgaris* Hildenborough grows on lactate in the absence of sulfate using fermentation or syntrophic relationships, the energetics of lactate oxidation is markedly reduced as compared to growth by sulfate respiration. The overall reaction follows:

$$\text{Lactate}^- + H_2O = \text{acetate}^- + CO_2 + 2H_2 \qquad \Delta G^{\circ\prime} = -8.8\,\text{kJ/mol} \tag{6.2}$$

The reaction steps include:

$$\text{Lactate}^- = \text{pyruvate}^- + 2H_2 \qquad \Delta G^{\circ\prime} = +43.2\,\text{kJ/mol} \tag{6.3}$$

$$\text{Pyruvate}^- + H_2O = \text{acetate}^- + CO_2 + 2H_2 \qquad \Delta G^{\circ\prime} = -47.3\,\text{kJ/mol} \tag{6.4}$$

The energy release from the lactate/pyruvate half reaction is $E^{\circ\prime} = -190$ mV, which is not sufficient for H_2 formation because the $2e^- + 2H^+/H_2$ is $E^{\circ\prime} = -420$ mV. Special processes may be required to energize the reaction of lactate fermentation and this results in very little growth.

6.2.2.2 Pyruvate Oxidation

Pyruvate as an electron donor Whether supplied externally or as a product of lactate oxidation, pyruvate is the hub of intermediary metabolism and is important for cell energetics as well as for biosynthesis. Since carbohydrates are not used as an energy source in *D. vulgaris* Hildenborough and many SRB, the intermediary carbon pathways are limited for biosynthesis. About 84% of lactate entering *D. vulgaris* Hildenborough cells in a lactate-sulfate medium are directed via pyruvate to acetyl-CoA for energy and about 5% of lactate is oxidized via CO dehydrogenase (Tang et al. 2007). The biosynthetic activities in *D. vulgaris* Hildenborough are prioritized with the greatest flow of carbon from pyruvate directed towards the bifurcated TCA cycle, the second greatest flow is for generation of compounds for reverse glycolysis and the phosphate pentose pathway and the third flow of carbon is directed to amino acid, nucleic acid and related biosynthetic activities for biomass production. While a pyruvate dehydrogenase is not present in *D. vulgaris* Hildenborough, the production of acetyl-CoA from pyruvate is a key reaction in *D. vulgaris* Hildenborough and there are several avenues that have been suggested to account for this reaction (see Fig. 6.2). In one case, acetyl-CoA + CO_2 can be produced from pyruvate by pyruvate:ferredoxin oxidoreductase (PFOR, Garczarek et al. 2007). Another enzyme, pyruvate:formate lyase (*pfl*, DVU2272 and DVU2824), converts pyruvate to CO_2 + $2H^+$ + $2e^-$ with hydrogenase converting $2H^+$ + $2e^-$ to H_2.

$$C_3H_3O_3^- + 2H_2O \rightarrow C_2H_3O_2^- + HCO_3^- + H_2 + H^+ \qquad \Delta G^{\circ\prime} = -47.1\,\text{kJ/mol} \tag{6.5}$$

The reactions of pyruvate = acetyl-CoA + CO_2 or formate are reversed by pyruvate synthase, pyruvate formate-lyase, pyruvate:ferredoxin oxidoreductase and oxo-organic acid oxidoreductase (Tang et al. 2007).

PFOR has been purified from *D. africanus* (Pieulle et al. 1995) and the homodimer has a molecular mass of 267 kDa with one thiamin pyrophosphate (TPP) cofactor and three $[4Fe\pm4S]^{2+/1+}$ clusters with two as [4Fe±4S] ferredoxin. This PFOR is stable in air and donates electrons to ferredoxin (Fd) I, Fd II and Fd III which are also produced by *D. africanus* (Pieulle et al. 1999a, b). The PFOR from *D. africans* has been crystalized and the arrangement of the TPP cofactor along with the three [4Fe-4S] clusters accounts for an electron transfer pathway (Chabrière et al. 1999). Purification of PFOR from *D. vulgaris* Hildenborough indicates that the enzyme is homo-octomer with a molecular mass of 1056 kDa (Garczarek et al. 2007). In reality, this homo-octomer of *D. vulgaris* Hildenborough is a tetramer of the dimeric enzyme of *D. africanus* and the difference in macromolecular organization is the presence of an addition valine in a surface loop involved in the dimer-dimer interaction site (Val 383) in the *D. vulgaris* Hildenborough enzyme (Garczarek et al. 2007). The PFOR from *D. desulfuricans* was also found to have the inserted valine residue. The PFOR from *A. fulgidus* is distinct from PFOR from *Desulfovibrio* in that the enzyme from *A. fulgidus* has a molecular mass of 120 kDA and consists of four dissimilar subunits which have a molecular mass of 45, 33, 25, and 13 kDa (Kunow et al. 1995). The physiological electron acceptor for PFOR in SRP is considered to be ferredoxin. A gene encoding pyruvate-flavodoxin oxidoreductase (DMR_20070) has been reported for *D. magneticus* strain RS-1 and this enzyme awaits characterization (Nakazawa et al. 2009). Pyruvate oxidation in *Dst. reducens* MI-1 is considered to proceed by several pathways: pyruvate:ferredoxin oxidoreductase pathway, pyruvate-formate lyase and a third pathway which does not appear to include a pyruvate dehydrogenase (Otwell et al. 2016).

Another enzyme involved in pyruvate oxidation is pyruvate synthase which acts in reverse with the degradation of pyruvate and sometimes may be referred to as 'pyruvate dehydrogenase'. Pyruvate synthase was demonstrated in extracts of *D. vulgaris* Miyazaki F with FMN, FAD or flavodoxin as electron acceptors. Reduction of cytochrome c_3 is very slow but when flavodoxin is added, the rate of reduction is greater than when either flavodoxin of cytochrome were used alone (Ogata and Yagi 1986). Pyruvate dehydrogenase activity with methyl viologen as the electron acceptor has been reported for *Dst. thermobenzoicum* (Plugge et al. 2002). Putative pyruvate dehydrogenase has been reported for *Dst. nigrificans* (DesniDRAFT_1250, 2504, 1245) and *Dst. carboxydivorans* (Desca_0770, 0146, 0775) (Visser et al. 2014).

$$\text{Pyruvate} + \text{CoA} + \text{FMN} = \text{acetyl-CoA} + CO_2 + \text{FMNH}_2 \tag{6.6}$$

$$\text{Pyruvate} + \text{CoA} = \text{acetyl-CoA} + \text{formate} \quad (6.7)$$

The cleavage of pyruvate by formate acetyltransferase, EC 2.3.1.54 (also known as (pyruvate:formate lyase, PFL) to acetyl~CoA and formate is a nonoxidative system which does not generate reducing equivalents such as NADH. It has been proposed that the production of formate by *D. vulgaris* Hildenborough may be important to support symbiotic microorganisms (da Silva et al. 2013). Production of PFL in D. vulgaris Hildenborough involves expression of DVU2272-DVU2269 and DVU2820, DVU2822-DVU2825 which are regulated by DVU2275 and DVU2827. Enhancer binding proteins involved in influencing DVU2272-DVU2269 include the regulator gene DVU2275 and influencing DVU2820, DVU2822-DVU2825 is the regulator gene DVU2827. The PFL is inactive as it is translated and must be converted to an active form to be functional. This activation activity follows a report for *Escherichia coli*. The PFL of *Escherichia coli* is subjected to post-translation by the Fe^2-activating enzyme which requires S-adenosylmethionine and reduced flavodoxin (Knappe et al. 1984; Crain and Broderick 2013). The PFL activating enzyme in *D. vulgaris* Hildenborough is encoded on DVU2271 and this enzyme is also found in the genomes of several *Desulfovibrio* and *Desulfotomaculum* spp. Formate produced from PFL action could be oxidized by formate dehydrogenase with the generation of NADH. It has been suggested that PFL is preferred over PFOR in *Desulfovibrio* because PFL could provide energy for growth in the cases of formate cycling (Pereira et al. 2008a, b). *Dst. reducens* produces three of the pyruvate-transforming systems: (i) pyruvate dehydrogenase (dred_1893) yields acetate, CO2 and NADH; (ii) decarboxylation by the pyruvate-ferredoxin oxidoreductase (PFOR) (dred_0047-50), pyruvate-formate lyase (dred_2750-53) and (iii) pyruvate-ferredoxin oxidoreductase (dred_0047-50) (Junier et al. 2010). Pyruvate decarboxylase, a non-oxidative enzyme forming acetaldehyde and CO_2 from pyruvate, has not been reported in SRP.

Pyruvate fermentation From the early publications of Postgate (1952, 1963), it became apparent that some SRB were capable of growing on pyruvate in sulfate-free media. Growth yield of SRB growing by pyruvate fermentation were poor. Using continuous culture systems, the cell mass of *D. desulfuricans* was markedly greater in pyruvate respiration than pyruvate fermentation (Vosjan 1975). There is no fixed pathway for acetogenesis with pyruvate fermentation. Products of pyruvate metabolism varies with the species with *D. vulgaris* Marburg produces acetate, CO_2 and H_2: *Desulfobulbus propionicus* MUD converts 3 mol pyruvate to 2 mol acetate and 1 mol propionate; *Dst. thermobenzoicum* TSB utilizes 4 mol pyruvate to produce 5 mol acetate (Tasaki et al. 1993). Although the growth is relatively poor, *Desulfomicrobium norvegicum, Desulfomicrobium escambiense* and *Desulfomicrobium apsheronum* produce acetate, lactate and succinate from pyruvate while *Desulfomicrobium baculatum* produce only acetate and succinate (Sham Genthner et al. 1997).

There is greater cell density with *D. vulgaris* Hildenborough growing with pyruvate as compared to growing with lactate in sulfate-free medium (Walker et al. 2009). As evidenced by the low energy production by pyruvate fermentation, the doubling time for growth of *D. vulgaris* Hildenborough is 100–130 h and the maximum OD of the culture is 0.15 (Voordouw 2002). In pyruvate fermentation there is a "burst" of H_2 and CO which occurs early in the growth. The reaction for pyruvate fermentation is considered to be as follows (Voordouw 2002):

$$\text{Pyruvate}^- + 2\text{H}_2\text{O} = \text{acetate}^- + \text{HCO}_3{}^- + \text{H}_2 + \text{H}^+ \qquad \Delta G^{\circ\prime} = -47.1\,\text{kJ/mol} \tag{6.8}$$

$$\text{Pyruvate}^- = \text{acetate}^- + \text{CO} \qquad \Delta G^{\circ\prime} = -32.0\,\text{kJ/mol} \tag{6.9}$$

$$\text{CO} + 2\text{H}_2\text{O} = \text{HCO}_3{}^- + \text{H}_2 + \text{H}^+ \qquad \Delta G^{\circ\prime} = -15.1\,\text{kJ/mol} \tag{6.10}$$

$$\text{Pyruvate}^- + \text{H}_2\text{O} = \text{acetate}^- + \text{formate}^- + \text{H}^+ \qquad \Delta G^{\circ\prime} = -48.5\,\text{kJ/mol} \tag{6.11}$$

$$\text{Formate}^- + \text{H}_2\text{O} = \text{HCO}_3{}^- + \text{H}_2 + \text{H}^+ \qquad \Delta G^{\circ\prime} = +1.4\,\text{kJ/mol} \tag{6.12}$$

6.2.2.3 Phosphotransacetylase

Phosphotransacetylase (Pta, EC 2.3.1.8) catalyzes the reaction of acetyl~CoA conversion to acetyl~pyruvate. This is a highly reversible reaction and it has been demonstrated in several SRB including *D. inopinatus* (Reichenbecher and Schink 1997), *Dst. acetoxidans* (Spormann and Thauer 1989), *Dst. thermobenzoicum* (Plugge et al. 2002) and *Dst. acetoxidans* (Spormann and Thauer 1989).

6.2.2.4 Acetate Kinase

Acetate kinase (ATP:acetate phosphotransferase, EC 2.7.2.1) is a cytoplasmic enzyme that was purified from *Desulfovibrio desulfuricans* strain 8303 (reclassified as *D. vulgaris* 8303) with a requirement for Mg^{2+} which can be replaced by Mn^{2+}, Co^{2+}, or Zn^{2+} (Brown and Akagi 1966). Acetate kinase has been reported for several SRP including *D. inopinatus* (Reichenbecher and Schink 1997), *D. piger* Vib-7 and *Desulfomicrobium* sp. rod-9 (Kushkevych 2014), by *Dst. acetoxidans* (Spormann and Thauer 1989) and *Dst. thermobenzoicum* (Plugge et al. 2002). Acetate kinase is encoded on DVU3030 (*D. vulgaris* Hildenborough), Dde3242 (*D. alaskensis* strain G20), dred_2094 (*Dst. reducens*). *D. vulgaris* Miyazaki F produces two forms of acetate kinase with molecular masses of 49.3 and 47.8 kDa. While these two enzymes from *D. vulgaris* Miyazaki F have similar properties, it is proposed that these two enzymes result from post-translational modifications and are used to provide fine regulation of enzyme function (Yu et al. 2001). Although a pyrophosphate:acetate kinase has been demonstrated in a protista (Reeves and Guthrie 1975), this enzyme does not appear to be present in *Desulfovibrio* or *Desulfotomaculum* (Thebrath et al. 1989).

6.2.2.5 Fumarate Respiration

Fumarate-sulfate system A characteristic used to classify SRB at one time was the growth on small organic acids in the presence of sulfate. Species of *Desulfovibrio* that grew on malate in the presence of sulfate included *desulfuricans, salexigens, africanus* but not *gigas* or *vulgaris* (Postgate and Campbell 1966). *D. desulfuricans* grows in sulfate-free media by choline fermentation (Baker et al. 1962). *Desulfomicrobium apsheronum* and *Desulfomicrobium baculatum* grow in fumarate-sulfate medium with the production of acetate and succinate while *Desulfomicrobium escanbiense* and *Desulfomicrobium norvegicum* have acetate, lactate and succinate as end products. However, *Desulfomicrobium baculatum* and *Desulfomicrobium norvegicum* grow in the fumarate-sulfate medium without reducing sulfate by fumarate disproportionation (Sham Genthner et al. 1997). An unresolved issue is the transporters for release of carboxylic acids and especially C-4 dicarboxylic acids into the culture medium.

Fumarate disproportionation An early report evaluating growth of sulfate-reducing bacteria reveled that *D. gigas* and 10 strains of *D. desulfuricans* were capable of growing on fumarate in sulfate-free medium by fumarate disproportionation (Miller and Wakerley 1966). *D. desulfuricans* DSM642 (formerly known as Essex 6) has a fumarate reductase which is distinct from succinate reductase that enables it to grow on fumarate without sulfate with secretion of succinate and acetate into the medium (Eq. 6.13) (Zaunmüller et al. 2006).

$$3\,\text{Fumarate}^- \rightarrow 2\,\text{succinate}^- + 2CO_2 + \text{acetate}^- \qquad (6.13)$$

There is a difference in the fumarate dismutation reactions with *Desulfomicrobium apsheronum* and *Desulfomicrobium baculatum* which have acetate and succinate as major end products while *Desulfomicrobium escanbiense* and *Desulfomicrobium norvegicum* produce acetate, lactate and succinate (Sham Genthner et al. 1997).

Fumarate reduction Fumarate reductase is an important electron transport component in some SRB but is not found in *D. vulgaris Hildenborough*. Fumarate reductase is not the reverse of succinate dehydrogenase and as observed from genome analysis, both of these enzymes are encoded on separate genes. Fumarate reductase has been isolated from *D. gigas* (Lemos et al. 2002) and *D. multispirans*, (He et al. 1986). In *D. desulfuricans* DSM642, fumarate may also serve as a terminal electron acceptor with formate or H_2 as the electron donor (Zaunmüller et al. 2006). Sufficient energy is available from the transfer of electrons from H_2 to fumarate (see Eq. 6.14) to support growth and phosphorylation coupled to this reaction has been demonstrated (Barton et al. 1970). The reaction with lactate and fumarate accounts for sufficient energy to support growth (see Eq. 6.15); however, there is little information available on this topic.

$$\text{Fumarate}^- + H_2 \rightarrow \text{succinate}^- \qquad \Delta G^{\circ\prime} = -85.8\,\text{kJ/mol} \tag{6.14}$$

$$\text{Lactate}^- + 2\text{fumarate} \rightarrow \text{acetate}^- + HCO_3{}^- + 2\text{succinate}^- \quad \Delta G^{\circ\prime} = -179.1\,\text{kJ/mol} \tag{6.15}$$

6.2.2.6 Intermediary Metabolism

Central metabolism of SRP is an interaction of metabolic pathways and cycles where carbon units from catabolism are used for biosynthesis of cellular structures and biopolymers. While the electron donors and acceptors supporting cell energetics of SRP are distinct from heterotrophic microorganisms, there is considerable similarity between these chemoheterotrophs and heterotrophs in the metabolic pathways employed. In both groups, pyruvate and acetyl-CoA serve as the hub of intermediary metabolism. Carbon compounds enter the SRP cells by specific transport systems (see lactate uptake in a previous section) or as alternate substrates in uptake processes. Transport systems for low-molecular-weight organic acids have not been addressed at this time, and research is needed to understand the metabolic versatility displayed by certain SRP in utilizing different substrates for growth. It is likely that a monocarboxylic acid transporter for acetate, pyruvate, and propionate is used by SRP as has been reported for bacteria that do not respire sulfate (Jolkver et al. 2009). Similarly, the uptake of fumarate, succinate, and malate by SRB may be attributed to a C_4-dicarboxylic acid transporter as has been demonstrated in numerous bacteria (Rhie et al. 2014).

Cycles and pathways The TCA cycle is a key component in aerobic metabolism where it functions to oxidize acetyl-CoA completely to CO_2 and produce NAD(P)H along with appropriate carbon structures for carbon units such as amino acids. With the SRP being anaerobes, the TCA cycle is incomplete although segments present are used to produce essential units for biosynthesis. In D. vulgaris Hildenborough and many other SRB, the TCA cycle does not function because the genes encoding for malate $\leftrightarrow$ oxaloacetate and 2-oxoglutarate $\leftrightarrow$ are not present. In many SRP, the pentose phosphate pathway is incomplete due to the lack of genes for the production of glucose-6-phosphate dehydrogenase and phosphogluconate dehydrogenase. The presence of metabolic pathways in a few of the SRP is given in Table 6.1. In many of the SRP, including *D. vulgaris* Hildenborough, enzymes for glycolysis (Embden-Meyerhof-Parnas pathway) are present even though most SRP do not use glucose as an energy source (Table 6.2). Under nutrient-limited conditions, SRP may use glycolysis for the utilization of internal nutrient reserves of polyglucose, and in many SRP, enzymes of the glycolytic pathway may be used in gluconeogenesis. The Entner-Douderoff pathway does not appear to be important for SRP. Oxidation of

Table 6.1 Status of metabolic pathways in a selection of sulfate-reducing microorganisms

Organism	Entner-Doudoroff pathway	Embden-Meyerhof- Parnas pathway	TCA (tricarboxylic acid) cycle	Wood-Ljungdahl pathway
Desulfovibrio gigas[a]	*Lacks* gene for hexokinase and 2-keto-3-deoxy-gluconate-3-P (KDGP) aldolase	All genes are present to make pathway functional and enzymes were demonstrated	Oxidative and reductive TCA cycle not fully functional. Genes absent: 2-oxoglutarate dehydrogenase 2-oxoglutarate synthase Succinyl-CoA ligase	*Lacks* bifunctional CO dehydrogenase/acetyl-CoA synthase (CODH/ACS) in energy conservation.
Desulfovibrio vulgaris Hildenborough[b]		All genes present to make the pathway functional	Enzymes *not* annotated include: 2-oxoglutarate dehydrogenase Malate dehydrogenase	A *codh*/ *acs* gene is annotated *Lacks* formyltetrahydrofolate synthase
Desulfotomaculum nigrificans[c]	Enzymes present to make pathway functional, genes not annotated: 6-P-gluconate dehydrogenase 2-keto-3-deoxy-6-P-gluconate aldolase	Pathway appears to function;	*Lacks* genes for Citrate synthase Isocitrate dehydrogenase Aconitase	
Desulfotomaculum reducens MI-1[d]	*Lacks* gene for 2-keto-3-deoxygluconate-3-P aldolase	All genes for pathway are annotated[e]	*Lacks* gene for Succinyl-CoA synthase	
Archaeoglobus fulgidus VC-16[f]	Most genes are absent	Most genes are absent	*Lacks* several genes of TCA cycle	A *codh*/ *acs* gene is present

[a]Morais-Silva et al. (2014) and Fareleira et al. (1997)
[b]Santana and Crasnier-Mednansky (2006), Tang et al. (2007) and da Silva et al. (2013)
[c]Akagi and Jackson (1967) and Visser et al. (2014)
[d]Junier et al. (2010)
[e]Reducing pentose phosphate pathway present
[f]Klenk et al. (1997)

Table 6.2 Glycolytic capability of *Desulfovibrio vulgaris* Hildenborough

Enzymes	Gene	Locus
Hexokinase	Not identified	
Glucose phosphate isomerase	*pgi*	DVU3222
Phosphofructokinase	*pfkA*	DVU2061
Fructose-1,6- bisphosphate aldolase	*fba*	DVU2143
Triose phosphate isomerase	*tpiA*	DVU1677
Glyceraldehyde-3-phosphate dehydrogenase	*gap-1*	DVU0565
	gap-2	DVU2144
3-phosphoglycerate kinase	*pgk*	DVU2529
Phosphoglycerate mutase	*gpm*	DVU2935
Phosphoglyceromutase	*gpmA*	DVU1619
Enolase	*eno*	DVU0322
Pyruvate kinase	*pyk*	DVU2514

Source of data is Santana and Crasnier-Mednansky, 2006

acetyl-CoA to CO_2 in many SRP is by a modified TCA (citric acid) cycle. Numerous SRP including *Desulfobacterium autotrophicum* and *Dst. acetoxidans* 5575^T use the Wood-Ljungdahl pathway to oxidize acetyl-CoA to CO_2 and the presence of this pathway provides a pathway for CO_2 assimilation and autotrophic growth (Strittmatter et al. 2009; Spring et al. 2009).

Fueling substrates Many SRP have unique enzymes which enable the sulfate reducer to use complex organic compounds as a carbon and electron source to energize the sulfate reduction process (Table 6.3). Often these organic compounds occur as environmental toxins and SRP would potentially have a role in bioremediation. Not only is there a large number of organic compounds utilized by SRP but many different species of SRP aslo participate in these oxidation reactions with the result being acetyl-CoA. Amino acids are commonly used as carbon sources for SRP and the entry of the carbon unit from amino acid catabolism into central metabolism is indicated in Fig. 6.3.

6.2.2.7 Sugars and Amino Acids as Energy Substrates

In the early 1900s, growth on sugars was often considered to be attributed to contaminants in SRB cultures. Akagi and Jackson (1967) demonstrated that *Dst. nigrificans* used glucose for growth with the production of acetate, ethanol, and CO_2. The phosphorotransferase system is important for transport of sugars in sulfate reducers (Santana and Crasnier-Mednansky 2006). Functional sugar transport systems are not present in many SRB including *D. vulgaris* Hildenborough; however, several genes for the phosphotransferase system for uptake of mannose have been reported for *D. vulgaris* Hildenborough even though this bacterium is unable to grow on mannose (Santana and Crasnier-Mednansky 2006). The putative phosphotransferase genes for *D. vulgaris* Hildenborough are homologues for the *Escherichia coli* genes for the phosphotransferase system and are found at three

Table 6.3 A selection of complex organic compounds oxidized by sulfate-reducing microorganisms

Organic compound	Microorganism	Reference
Aniline, dihydroxybenzenes	*Desulfobacterium anilini*	Schnell et al. (1989)
Aniline, dihydroxybenzenes	*Desulfococcus multivorans*	Widdel (1988) and Schnell et al. (1989)
Benzoate	*Desulfobacterium phenolicum*[a]	Bak and Widdel (1986)
Benzoates	*Desulfotomaculum sapomandens*	Widdel (1988) and Cord-Ruwisch and Garcia (1985)
Benzoates	*Desulfotomaculum australicum*	Love et al. (1993)
Benzoates	*Desulfosarcina variabilis*	Widdel (1988)
Benzoates	*Desulfonema magnum*	Widdel (1988)
Benzoates	*Desulfoarculus* sp. Strain SAX[a]	Drzyzga et al. (1993)
Catechol	*Desulfobacterium catecholicum*	Szewzyk and Pfenning (1987)
Fatty acids, ketones	*Desulfobacterium cetonicum*	Galusko and Rozanova (1991)
Furfural	*Desulfovibrio* sp. Strain B	Boopathy and Daniels (1991)
	Desulfovibrio furfuralis	Folkerts et al. (1989)
	Desulfovibrio gigas	Brune et al. (1983)
Kraft lignin	*Desulfovibrio desulfuricans* ATCC7757	Ziomek and Williams (1989)
Toluene	*Desulfobacula toluolica*	Rabus et al. (1993)
2,4,6 Trinitrotoluene	*Desulfovibrio gigas, Desulfovibrio vulgaris, Desulfovibrio desulfuricans* ATCC43938	Boopathy (2007)
n-Do-, Tetra, Penta-, Hepta-, Octa-decane	*Desulfobacterium oleovorans*	Aeckersberg et al. (1991)

[a]*Desulfobacterium phenolicum* has been reclassified as *Desulfobacula phenolica* comb.nov. and *Desulfoarculus* sp. Strain SAX as *Desulfotignum balticum* (Kuever et al. 2001)

distinct locations on the *D. vulgaris* Hildenborough chromosome with Enzyme I, HPr, and Enzyme IID (mannose) in one group, Enzyme IIA (mannose), Enzyme IIA (fructose), Enzyme IIB (mannose), and Enzyme IIC (mannose) in another and the multiphosphoryl transfer protein (mannose) in a third site. Sequence alignment of the Enzyme IIA (mannose) component is suggested to be mannose/fructose specific while the Enzyme IIA (fructose) appears to be specific for mannitol/fructose. Enzyme IIB, Enzyme IIC, and Enzyme D are proposed to be mannose/fructose/N-acetylglactosamine specific. These Enzyme II genes are also present in *D. alaskensis* G20 (Santana and Crasnier-Mednansky 2006). *Dst. nigrificans* and *Dst. carboxydivorans* utilize fructose, and both have an operon encoding for the fructose-specific phosphotransferase system. However, *Dst. ruminis* and *Dst. kuznetsovii* do not use fructose, and even though they have the genes for the Embden-Meyerhof-Parnas pathway, they apparently lack the specific transporter in

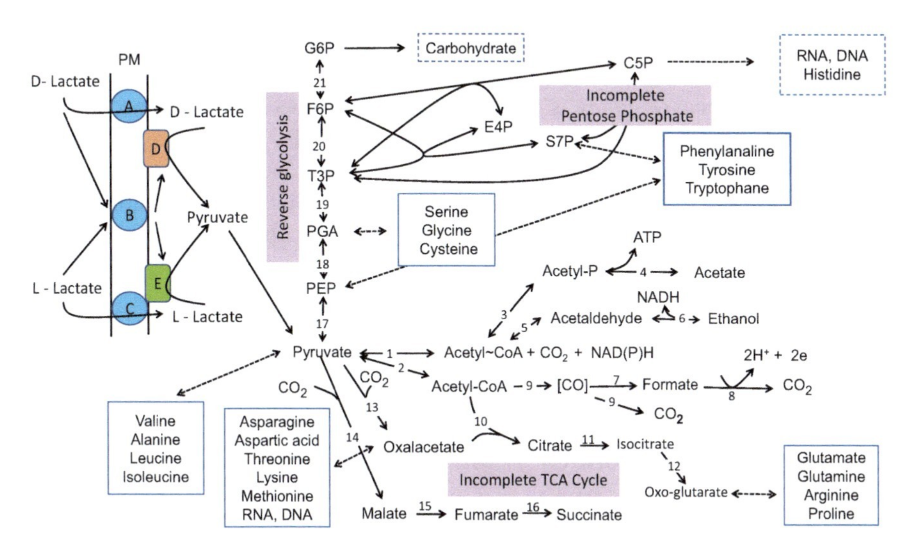

Fig. 6.3 Intermediary carbon metabolism in *D. vulgaris* Hildeborourgh. Key for abbreviations of carbon compounds and enzymes. In the plasma membrane (PM) are A = D-lactate permease, B = D, L-lactate permease, C = L-lactate permease, D = D-lactate dehydrogenase, E = L-lactate dehydrogenase. Enzymes in the cytoplasm include 1 = pyruvate:ferredoxin oxidoreductase, 2 = pyruvate formate-lyase, 3 = phosphate acetyltransferase, 4 = acetate kinase, 5 = aldehyde dehydrogenase, 6 = alcohol dehydrogenase, 7 = methylenetetrahdrofolate dehydrogenase/methenyltetrahydrofolate cyclohydrolase, 8 = formate dehydrogenase, 9 = CO dehydrogenase, 10 = citrate sunthase, 11 = cis-aconitase, 12 = isocitrate dehydrogenase, 13 = pyruvate carboxylase, 14 = malic enzyme, 15 = fumarate hydratase, 16 = fumarate reductase. The following enzymes are associated with the glycolytic pathway, and while key compounds are relevant to the pentose pathway, all enzymes associated with each conversion are given without listing all the intermediate compounds. The glycolytic enzymes are as follows: 17 = pyruvate kinase, 18 = enolase plus phosphoglycerate mutase, 19 = 3-phosphoglycerate kinase plus glyceraldehyde-3-phosphate dehydrogenase plus triose phosphate isomerase, 20 = fructose bisphosphate aldolase plus phosphofructokinase, 21 = glucose phosphate isomerase

the phosphotransferase system for fructose (Visser et al. 2014). *Dst. reducens* grows on glucose and very poorly on fructose with genes for glucose-specific phosphotransferase tentatively identified (Visser et al. 2016). It has been suggested that fructose may be recognized as an alternate substrate by the glucose permease and moved slowly into the cytoplasm but specific details of this activity await further research.

Although rare, some SRP grow on carbohydrates as the energy source and some of these are indicated as follows: *D. fructosovorans* DSM 3604 (Ollivier et al. 1988), *Dst. antarcticum* (Iizuka et al. 1969), *Dst. geothermicum* (Daumas et al. 1988), and *Dst. nigrificans* Delft 74 (Klemps et al. 1985) grow with fructose; *Desulfoconvexum algidum* DSM 21856 grows on sorbitol and mannitol (Könneke et al. 2013); *Desulfovibrio* strain D uses glucose, fructose, galactose, ribose, and mannose (Huisingh et al. 1974); *Desulfobulbus mediterraneus* 86FS1 uses glucose, fructose, galactose, ribose, sucrose, and lactose as energy substrates (Sass et al. 2002);

D. inopinatus grows on fructose and ribose (Reichenbecher and Schink 1997); and *Dst. carboxydivorans* grows on fructose and glucose (Visser et al. 2014; Parshina et al. 2005). While *D. gigas* is unable to grow on glucose, it does accumulate polyglucose and has enzymes of the Embden-Meyerhof-Parnas pathway to support the cytoplasmic utilization of glucose under fermentative and respiratory conditions (Santos et al. 1993; Fareleira et al. 1997). *A. fulgidus* VC-16 is capable of growing on glucose; however, metabolic pathways for glucose are unresolved (Klenk et al. 1997). *A. fulgidus* strain 7324 degrades starch to acetate a + CO_2 using a modified Embden-Meyerhof pathway and acetyl-CoA synthetase (Labes and Schönheit 2001). According to Cao et al. (2012), greater levels of energy are derived by bacteria using complete oxidation (Eq. 6.16) while acetogenic SRB receive a lower level of energy (Eq. 6.17):

$$C_6H_{12}O_6 + 3SO_4{}^{2-} \rightarrow 6HCO_3{}^- + 3H^+ + 3HS^- \qquad \Delta G = -4542.5\,\text{kJ/rxn} \tag{6.16}$$

$$C_6H_{12}O_6 + SO_4{}^{2-} \rightarrow 2HCO_3{}^- + 3H^+ + HS^- + 2CH_3COO^- \qquad \Delta G = -358.2\,\text{kJ/rxn} \tag{6.17}$$

A few SRB use amino acids as energy sources for the reduction of sulfate and some will grow in the absence of sulfate with energy obtained from fermentation of amino acids. Following the transport of the amino acid into the cytoplasm, the amino acid is deaminated and the carbon structure of the amino acid would interface with appropriate enzymes in intermediary metabolism using the standard metabolic pathways. Alanine utilization is one of the more common substrates since deamination of alanine produces pyruvate which is widely metabolized by SRB. Some of the reported instances of amino acids as energy and carbon sources are presented in Table 6.4. Members of the *Desulfovibrio* genus utilize the greatest number of amino acids and only a few species of *Desulfotomaculum* can grow on amino acids with sulfate.

6.2.2.8 Acetate, Butyrate, and Propionate Oxidation

Acetate oxidation A characteristic that distinguishes the taxonomic groups of sulfate reducers is the complete oxidation of an organic substrate to CO_2 or the incomplete oxidation with the production of acetate as the end product. The oxidation of acetate by *Dst. acetoxidans* was the first bacterium demonstrated to use acetate and while it can also use butyrate to reduce sulfate but is unable to oxidize lactate or pyruvate (Widdel and Pfennig 1977). *Desulfobacter postgatei* was also observed to use acetate as the sole electron source for sulfate respiration, and some strains of this bacterium are also able to oxidize ethanol and lactate (Widdel and Pfennig 1981). The energetics of the acetate/sulfate reaction is given in Eq. 6.18. The oxidation of acetate to CO_2 by *Desulfobacter postgatei* and *Desulfobacter*

Table 6.4 Amino acid used by SRB to support sulfate respiration and growth

Organism	*L*-amino acids used with sulfate	Amino acids used without sulfate	Reference
Desulfovibrio acrylicus	gly, ser, ala, cys	Not reported	Van der Maarel et al. (1996)
Desulfovibrio aminophilus DSM 12257	ala, val leu, lys, asp met	ser, gly, cys, thr	Baena et al. (1998)
Desulfovibrio sp. strain 20020	ala, ser, gly, val, leu, lys, cys, thr, asp	ser	Baena et al. (1998)
Desulfovibrio sp. strain HL21	ala, ser, asp, glutamate	Not reported	Stams et al. (1986)
Desulfovibrio mexicanus DSM 13116	ser, cys	Not reported	Hernandez-Eugenio et al. (2000)
Desulfovibrio marinizediminis DSM 17456	gly, ala, ser, asp	Not reported	Takii et al. (2008)
Desulfovibrio salexigens DSM 2638	ala	Not reported	Van Niel et al. (1996)
Desulfovibrio zosterae	ala	Not reported	Nielsen et al. (1999)
Desulfobacterium vacuolatum	pro, glu, gly, ser, ala	Not reported	Rees et al. (1998)
Desulfobulbus mediterraneus 86FS1	ala	Not reported	Sass et al. (2002)
Desulfococcus niacini	glu	Not reported	Imhoff-Stuckle and Pfennig (1983)
Desulfotomaculum carboxydivorans	ala	Not reported	Visser et al. (2014)
Desulfotomaculum nigrificans DSM 574	ala	Not reported	Visser et al. (2014)
Desulfotomaculum reducens DSM 100696	ala	Not reported	Visser et al. (2016)
Desulfotomaculum ruminis DSM 2154	ala	Not reported	Visser et al. (2016)

hydrogenophilus is attributed to the use of a modified citric acid cycle (Brandis-Heep et al. 1983; Schauder et al. 1986, 1987; Möller et al. 1987). As indicated in Fig. 6.3, acetate is activated by succinyl-CoA to produce acetyl-CoA and succinate and the loss of CO_2 occurs as isocitrate is converted to oxoglutarate and oxoglutarate is converted to succinyl-CoA. It is interesting to note that this modified citric acid cycle has a membrane-bound malate dehydrogenase and the electron acceptor with the decarboxylation by oxoglutarate dehydrogenase is ferredoxin. Many of the SRP (e.g. *Dst. acetoxidans, Desulfobacterium* spp., *Desulfosarcina* spp., *Desulfococcus* spp., *D. baarsii*, and *A. fulgidus*) use a reductive acetyl-CoA/CO dehydrogenase pathway to completely oxidize acetate to CO_2. In the reductive acetyl-CoA/CO dehydrogenase pathway, the carboxyl group is converted to CO_2 by a carbon monoxide dehydrogenase while the methyl group of acetyl-CoA is bound to tetrahydropterin, as in the case of many of the bacteria, and to tetrahydromethanopterin in *A. fulgidus* (Thauer et al. 1989; Klenk et al. 1997).

$$CH_3COO^- + SO_4{}^{2-} \rightarrow HS^- + 2HCO_3{}^- \qquad \Delta G^{\circ\prime} = -47.36\,\text{kJ/mol} \tag{6.18}$$

Butyric and propionic acid oxidation Several SRB use butanol and propionate as electron sources for dissimilatory sulfate reduction, and these bacteria can either complete oxidizers that convert organic acids to CO_2 or incomplete oxidizers where the final product of metabolism is acetate. Equations 6.19 and 6.20 reflect incomplete oxidation by dissimilatory sulfate reducers. Incomplete oxidation of butyrate is characteristic of *D. butyratiphilus* (Suzuki et al. 2010), *Desulfoluna butyratoxydans* (Suzuki et al. 2008), and *D. portus* (Suzuki et al. 2009); incomplete oxidation of propionate occurs with *Desulforegula conservatrix* (Rees and Patel 2001) while incomplete oxidation of propionate and butyrate occurs with *Desulfofaba gelida* strain PSv29 (Knoblauch et al. 1999) and *Desulfobotulus alkaliphilus* (Sorokin et al. 2010). Incomplete oxidation of propionate is characteristic of members of the genus *Desulfobulbus* which are represented by *Desulfobulbus propionicus* (Widdel and Pfennig 1982; Pagani et al. 2011), *Desulfobulbus elongatus* (Samain et al. 1984), *Desulfobulbus rhabdoformis* (Lien et al. 1998), *Desulfobulbus mediterraneus* (Sass et al. 2002), *Desulfobulbus japonicus* (Suzuki et al. 2007), *Desulfobulbus marinus* (previously *Desulfobulbus* 3pr10 (El Houari et al. 2017), *Desulfobulbus alkaliphilus* (Sorokin et al. 2012), and *Desulfobulbus oligotrophicus* (El Houari et al. 2017):

$$\text{Butyrate}^- + 0.5\,SO_4{}^- \rightarrow 2\,\text{acetate}^- + 0.5\,HS^- + 0.5\,H^+ \quad \Delta G^{\circ\prime} = -27.8\,\text{kJ/rxn} \tag{6.19}$$

$$\begin{gathered}\text{Propionate}^- + 0.75\,SO_4{}^{2-} \rightarrow \text{acetate}^- + HCO_3{}^- + 0.75\,HS^- + 0.25\,H^+ \\ \Delta G^{\circ\prime} = -37.7\,\text{kJ/rxn}\end{gathered} \tag{6.20}$$

Some SRB grow with complete oxidation of butyrate and include *Desulfatirhabdium butyrativorans* (Balk et al. 2008), *Dst. geothermicum* (Daumas et al. 1988), *Dst. 47thermosapovorans* (Fardeau et al. 1995), and *Dst. sapomandens* (Cord-Ruwisch and Garcia 1985). With *Dst. gibsoniae*, complete oxidation occurs if the concentration of low-molecular-weight organic acid is low but it displays incomplete oxidation with acetate accumulation when the substrate concentration is elevated (Kuever et al. 1999). *Dst.geothermicum* uses butyrate as electron and carbon sources with part of butyrate being completely oxidized and the other part incompletely oxidized with the released as acetate (Daumas et al. 1988).

A group of propionate-oxidizing bacteria are better known for syntrophic relationships and not their capability of growing with sulfate as the electron acceptor. Members of the genus *Syntrophobacter* oxidize propionate and some oxidize other fatty acids including butyrate with the reduction of sulfate. *Syntrophobacter* are best known for symbiotic growth with the coupling of propionate oxidation with a partner organism with growth rates and cell yields being lower than when sulfate serves as the electron acceptor (Van Kuijk and Stams 1995). Examples include *Syntrophobacter pfennigii* (Wallrabenstein et al. 1995), *Syntrophobacter sulfatireducens* (Chen et al. 2005), and *Syntrophobacter fumaroxidans* (Harmsen et al. 1998). The genomic organization of genes for dissimilatory sulfite reduction in *Syntrophobacter fumaroxidans* has been described in the literature and has been compared to other SRB (Grein et al. 2013). The syntrophic relationship between a *Syntrophobacter* sp. and a sulfate-reducing bacterium enables the growth of both organisms by efficient electron transfer. As shown in Eqs. 6.21 and 6.22, there is no energy released when *Syntrophobacter wolinii* metabolizes either propionate or butyrate (Boone and Bryant 1980); however, when co-cultured with a *Desulfovibrio* sp., sufficient energy is released (see Eqs. 6.19 and 6.20):

$$\text{Butyrate}^- + 2H_2O \rightarrow 2\,\text{acetate}^- + 2H_2 + H^+ \qquad \Delta G^{\circ\prime} = +48.1\,\text{kJ/rxn} \quad (6.21)$$

$$\text{Propionate}^- + 3H_2O \rightarrow \text{acetate}^- + HCO_3^- + 3H_2 + H^+ \qquad \Delta G^{\circ\prime} = +76.1\,\text{kJ/rxn} \quad (6.22)$$

6.2.2.9 Oxidation of Alcohols

Ethanol Several of the SRB such as *D. gigas* produce ethanol in small quantities as an end product of metabolism (Fareleira et al. 1997). The purified alcohol dehydrogenase from *D. gigas* oxidizes several low–molecular-weight alcohols (C1 to C5) in addition to ethanol (Hensgrens et al. 1993). The growth of *D. vulgaris* Hildenborough in ethanol-sulfate medium (see Eqs. 6.23 and 6.24) is attributed to the oxidation of ethanol to acetate by alcohol dehydrogenase (adh) with the release of NADH to provide energy for growth (Haveman et al. 2003). Ethanol oxidation appears to be an accessory energy pathway since the *adh* gene is up-regulated during the growth of *D. vulgaris* Hildenborough on lactate; pyruvate is also important in the growth of *D. vulgaris* Hildenborough in sulfate medium and lactate, pyruvate, formate, and H_2 as the electron donor (Haveman et al. 2003). The use ethanol as an energy source is reported for some SRB and includes: *D. aminophilus* DSM12254 (Baena et al. 1998):

$$2CH_3CH_2OH + SO_4^{2-} = 2CH_3COO^- + HS^- + H^+ + 2H_2O \qquad \Delta G^{\circ\prime} = -22\,\text{kJ/rxn} \quad (6.23)$$

For SRB that are not acetogens on ethanol, the following reaction would be appropriate:

$$CH_3CH_2OH + SO_4^{2-} = 2HCO_3^- + HS^- \qquad \Delta G^{\circ\prime} = -48\,\text{kJ/rxn} \quad (6.24)$$

Table 6.5 Sulfate-reducing microorganisms that are energized by alcohol additions to sulfate media

Microorganisms	Alcohol added to media along with sulfate for growth	Reference
Archaeoglobus lithotrophicus	Ethanol	Stetter et al. (1993)
Desulfatiglans parachlorophenolica	Ethanol, methanol	Suzuki et al. (2014)
Desulfobacterium catecholicum	Methanol	Szewzyk and Pfennig (1987)
Desulfobacterium anilini	Methanol	Schnell et al. (1989)
Desulfobulbus propionicus strain DKpr12[a]	Ethanol, propanol	Nanninga and Gottschal (1987)
Desulfosporosinus orientis[b]	Methanol	Klemps et al. (1985)
Desulfotomaculum alkaliphilum	Ethanol	Pikuta et al. (2000)
Desulfotomaculum australicum	Ethanol	Love et al. (1993)
Desulfotomaculum geothermicum	Ethanol	Daumas et al. (1988)
Desulfotomaculum kuznetsovii	Ethanol, methanol	Nazina et al. (1988)
Desulfotomaculum luciae	Ethanol	Liu et al. (1997)
Desulfotomaculum nigrificans	Ethanol	Klemps et al. (1985)
Desulfotomaculum putei	Ethanol	Liu et al. (1997)
Desulfotomaculum thermocisternum	Ethanol	Nilsen et al. (1996)
Desulfotomaculum thermosyntrophicum	Ethanol	Nilsen et al. (1996)
Desulfotomaculum thermosyntrophicum Strain T93B	Ethanol, methanol	Rosnes et al. (1991)
Desulfovibrio alcoholivorans	Ethanol, methanol Butanol, propanol pentanol	Qatibi et al. (1991)
Desulfovibrio carbinolicus strain EDK82	Ethanol, methanol, propanol, butanol	Nanninga and Gottschal (1987)
Desulfovibrio desulfuricans strain DK81	Ethanol, propanol	Nanninga and Gottschal (1987)
Thermodesulforhabdus norvegicus	Ethanol	Beeder et al. (1995)

[a]Grows by ethanol fermentation
[b]Grows slowly by methanol fermentation

Methanol The anaerobic degradation of methanol is accomplished by sulfate reducers along with acetogenic bacteria and methanogenic archaea. Several SRP can grow with methanol as the electron donor with sulfate and some of these are listed in Table 6.5. The methanol-sulfate couple is highly exergonic as seen in the following reaction:

$$4\,CH_3OH + 3SO_4{}^{2-} = 4HCO_3{}^{-} + 3\,HS^{-} + H^{+} + 4H_2O \qquad \Delta G^{\circ\prime} = -364\,kJ/rxn \tag{6.25}$$

Examination of *Dst. kuznetsovii* revealed the presence of two pathways for the degradation of methanol which is unique for bacteria (Sousa et al. 2018). In one pathway designated as the alcohol dehydrogenase system, methanol is oxidized to formaldehyde by an alcohol dehydrogenase, the formaldehyde is oxidized to formate by an aldehyde ferredoxin oxidoreductase, and formate is oxidized to CO_2 by formate dehydrogenase. In an alternate pathway referred to as the methyl transferase system, methanol is initially methylated to form methyl-tetrahydrofolate which is subsequently oxidized to CO_2. Expression of the methyltransferase system requires the presence of cobalt, and this pathway may be important when the methanol concentration is low. Genes have been identified for both pathways for the oxidation of methanol, and an evaluation of gene products suggests that the methyl transferase system is evolutionary closer to the methanogens while the alcohol dehydrogenase system is closer to the acetogens (Sousa et al. 2018).

6.2.2.10 CO as an Electron Donor

There are several metabolic activities in SRP that involve CO and these include: (i) oxidation of CO to CO_2 by CO dehydrogenase with CO as an electron donor for sulfate respiration, (ii) production of CO as growth occurs on lactate or pyruvate, (iii) CO cycle in cells producing CO, and (iv) inhibition of metabolism due to CO. The first series of reactions concerns CO dehydrogenase. Many SRP produce a cytoplasmic reaction that oxidizes carbon monoxide (CO) to carbon dioxide using CO dehydrogenase.

Growth with CO Enzymatic oxidation of carbon monoxide was first reported using cell extracts of *D. desulfuricans* (Yagi 1958, 1959); however, the use of CO to grow bacteria with sulfate respiration was first reported for *Desilfotomaculum* sp. (Klemps et al. 1985). Subsequently, several SRB (e.g. *D. vulgaris* Madison, *D. desulfuricans, D. baculatus, D. africanus, Dst. Nigrificans*, and *Desullfosporosinus orientis*) have been demonstrated to grow in 2–20% CO in sulfate media containing acetate and end products of H_2, CO_2, and H_2S were reported (Parshina et al. 2010). At $\leq$50% CO with a sulfate medium, *Dst. kuznetsovii* DSM6115 and *Dst. thermobenoicum* subsp. *thermosyntrophicum* DSM14055 produce acetate, CO_2, and H_2 as end products and the same end products are produced by *Dst. carboxydivorans* CO-1-SRB as it grows on 100% CO (Parshina et al. 2010). In co-culture consisting of *Dst. kuznetsovii* and *Dst. thermobenoicum* with H_2 production by *Carboxydothermus hydrogenoformans*, sulfate was reduced when CO was supplied at 100% (Parshina et al. 2005).

Several SRB have a single gene encoding for CO dehydrogenase and these include: *D. vulgaris* Hildenborough, *D. vulgaris* PD4, *D. vulgaris* Miyazaki, *D. alaskensis* G20, *D. desulfuricans* ATCC27774, *D. salexingens* DSM2638, *Dst. reducens* MI-1, and *Thermodesulfovibrio yellowstonii* DSM11347. The SRB that have two genes in the genome for CO dehydrogenase include *Desulfococcus oleovorans* Hxd3, *Desulforudis audaxviator* MP104C (with one archaeal type and one bacterial type enzyme) and SRP with three genes of CO dehydrogenase per genome are *Desulfobacterium autotrophicum* HRM2 and *A. fulgidus* DSM4304 (which has two archaeal type and one bacterial type enzyme). It is noteworthy that *A. fulgidus* VC16 oxidizes CO with sulfate reduction and acetate formation and that there is a transient production of formate in the media (Henstra et al. 2007). It is assumed that those SRP with multiple genes encoding CO dehydrogenase have specific activities for each of these enzymes.

A soluble CO dehydrogenase has been isolated from *Desulfovibrio vulgaris*, and it has three [4Fe- 4 S] clusters and two [NiFeS] clusters in the active site (Hadj-Saïd et al. 2015). The CO dehydrogenase is encoded on *cooS* (coo refers to CO oxidation), and the insertion of nickel into the active center is presumed to be attributed to a chaperone protein (*cooC*). The overall reaction and energy yield for CO as the electron donor for sulfate reduction is as follows:

$$4CO + SO_4^{2-} + 4H_2O^- \rightarrow 4HCO_3^- + HS^- + 3H^+ \qquad \Delta G^{\circ\prime} = -37.1\,kJ/mol\,CO \tag{6.26}$$

The reaction was found to be completed by three enzymes: CODH, an electron transfer protein, and an energy-converting hydrogenase (EcH). CO is oxidized via the CODH complex, and electrons are transferred to a "ferredoxin-like" electron carrier. Oxidation of this electron carrier can be coupled to proton reduction via an EcH complex, producing hydrogen and simultaneously generating an ion motive force (Hedderich and Forzi 2005). The following reactions are important for the growth of SRB on CO (Parshina et al. 2010):

$$4CO + SO_4^- + 4H_2O \rightarrow 4HCO_3^- + HS^- + 3H^+ \qquad \Delta G^{\circ} = -37.1\,kJ/mol\,CO \tag{6.27}$$

$$4CO + 4H_2O \rightarrow acetate^- + 2HCO_3^- + 3H^+ \qquad \Delta G^{\circ} = -28.2\,kJ/mol\,CO \tag{6.28}$$

$$CO + 2H_2O \rightarrow HCO_3^- + H_2 + H^+ \qquad \Delta G^{\circ} = -15\,kJ/mol\,CO \tag{6.29}$$

$$4H_2 + SO_4^{2-} + H^+ \rightarrow HS^- + 4H_2O \qquad \Delta G^{\circ} = -45.2\,kJ/mol\,H_2 \tag{6.30}$$

$$4H_2 + 2HCO_3^- + H^+ \rightarrow acetate^- + 4H_2O \qquad \Delta G^{\circ} = -36.4\,kJ/mol\,H_2 \tag{6.31}$$

CO as a metabolite A fermentation pulse has often been associated with the growth of sulfate reducers and it would include the transient production of H_2, CO, and formate early in the growth phase. The growth of *D. vulgaris Hildenborough* (Voordouw 2002) and *D. vulgaris* Madison (Lupton et al. 1984) on lactate-sulfate and pyruvate-sulfate media produces CO as an end product of metabolism. CO consumption by *D. vulgaris* is coupled to hydrogenase activity since H_2 (Eq. 6.29) contributes to production of acetate (Eq. 6.31). Trace levels of H_2

production have been reported with *D, vulgaris* (Badziong and Thauer 1980; Traore et al. 1981) and the oxidation of CO appears linked to H_2 oxidation.

CO cycle The endogenous generation of CO from pyruvate and subsequent oxidation of CO to CO_2 is part of the CO cycle proposed by Voordouw (2002). With the cytoplasmic oxidation of pyruvate to acetyl-CoA + H_2O + CO, CO would be oxidized by a carbon monoxide dehydrogenase by a transmembrane hydrogenase with the production of CO_2 + H_2 in the periplasm. Periplasmic hydrogenases and cytochromes would direct the protons and electrons to a transmembrane electron transport (e.g. Hmc) with the release of H_2 into the cytoplasm. Energy production would be associated with cycling of H_2.

Inhibition by CO Progress is continuing to understand the mechanisms of CO inhibition of SRB growth. Cultivation of *D. desulfuricans* B-1388 in the presence of 5% CO resulted in an increase of superoxide radicals as compared to cultivation in media without CO added and suggests that CO toxicity is attributed to increased oxidative stress (Davydova et al. 2004; Davydova and Tarasova 2005). While CO is known to inhibit enzymes with metal centers, the focus at this time is on CO as an inhibitor of metalloenzymes and specifically hydrogenases. As reviewed by Parshina et al. (2010), CO inhibits [Fe-Fe]-hydrogenases by binding to iron in the active site. Hydrogenases that are not inhibited by CO could have an active site with H_2 accessing the active site through a narrow channel which would restrict CO entry. The growth of *D. vulgaris* Hildenborough on lactate-sulfate is more sensitive to exogenous CO than growth on pyruvate sulfate, and this is attributed to the uptake hydrogenase, HydAB, which is important in energetics of lactate oxidation (Rajeev et al. 2012). Activity of CO dehydrogenase in SRP may be, in part, detoxification as well as for cell energetics and synthesis.

6.2.2.11 Formate Oxidation

Some SRB couple formate oxidation to drive sulfate reduction with concomitant growth. The overall reaction is as follows:

$$4\,HCOO^- + SO_4{}^{2-} + H^+ \rightarrow HS^- + 4HCO_3{}^- \qquad -46.9\,kJ/mol \tag{6.32}$$

However, since *D. vulgaris Hildenborough* does not have the appropriate enzymes to synthesize carbon compounds from CO_2 with energy sources of H_2 and formate, acetate is usually added to the medium as a carbon source. Acetate is taken up by the SRB cell, converted to acetyl-CoA (see Fig. 6.3) and CO_2 is added to acetyl-CoA to produce pyruvate. Additional CO_2 fixing reactions contribute to synthesis of oxaloacetate and malate (see Fig. 6.3). In *D. vulgaris* Marburg, only about 30% of cell carbon comes from CO_2 with the remainder from acetate (Badziong et al. 1979).

In the cytoplasm of SRB growing on lactate, the pyruvate:formate lyase reaction generates formate according to the following reaction:

$$\text{Pyruvate}^- + \text{CoA} = \text{acetyl-CoA} + \text{formate}^- \qquad \Delta G^{\circ\prime} = -16.3\,\text{kJ/mol} \quad (6.33)$$

Formate would be transported from the cytoplasm to the periplasm where formate dehydrogenases are localized in *D. vulgaris* Hildenborough and *D. gigas* or exported from the cell for intraspecies formate transfer (Thauer et al. 2007).

Formate dehydrogenase has been characterized from *D. gigas* NCIB 9332 (Almendra et al. 1999; Riederer-Henderson and Peck Jr. 1986a, b), *D. vulgaris* Hildenborough (Sebban et al. 1995), *D. vulgaris* Miyazaki (Yagi 1979), *D. desulfuricans* ATCC 27774 (Costa et al. 1997), and *D. alaskensis* NCIMB 13491 (Brondino et al. 2004). *D. vulgaris* Hildenborough has three formate dehydrogenase enzymes (Pereira et al. 2007) with FdhAB and $FdhABC_3$ as soluble proteins in the periplasm and FdhM (also referred to as FdhABD) (*fdnG*, DVU2482) in the periplasm but associated with the plasma membrane. The $FdhABC_3$ (DVU2809-11) enzyme contains molybdenum while the FdhAB (DVU0587-88) binds either molybdate or tungstate. *D. alaskensis* has three periplasmic formate dehydrogenases with one enzyme binding molybdate and the other binding tungstate (Martins et al. 2015, 2016; Brondino et al. 2004). Sulfide produced from SRB respiration will precipitate molybdenum (Tucker et al. 1997); however, molybdenum sulfide is markedly more stable than tungsten sulfide and this may reflect the necessity of having multiple forms of formate dehydrogenases (da Silva et al. 2011). In the presence of molybdate, *D. vulgaris* Hildenborough produces primarily $FdhABC_3$ and in the presence of tungstate, the primary formate dehydrogenase produced is FdhAB (da Silva et al. 2013). Formate as the energy source is oxidized by the following reaction in the periplasm:

$$\text{Formate}^- + H^+ = CO_2 + 2\,H^+ + 2e^- \qquad (6.34)$$

Under appropriate environmental conditions, it has been proposed that formate can be produced according to the following reaction:

$$H_2 + CO_2 = HCOO^- + H^+ \qquad (6.35)$$

With 1 Pa H_2 and 10 μM formate, the $\Delta G^{\circ\prime}$ is −33.5 kJ/mol; and with 10 Pa H_2 and 1 μM formate, the $\Delta G^{\circ\prime}$ is −44.9 kJ/mol. These free energy values would be sufficient to support bacterial growth since the minimum free energy required for growth has been suggested to be −19 to −2 kJ/mol (Hoehler et al. 2001; Schink 1997). It has been proposed that formate cycling in SRB could be functioning under certain conditions as evidenced by increased production of pyruvate:formate lyase with growth on formate (da Silva et al. 2013). Clearly, this utilization and production of formate by SRB indicates a unique energy versatility in anaerobes.

6.2.3 Transition to Stationary Phase

A change in biochemical and physiological activities occurs as bacterial cell transition to the stationary phase from the exponential growth phase. Transcriptomic analysis of *Desulfovibrio vulgaris Hildenborough* progressing through phases of growth revealed differential expression of several metabolic genes (Clark et al. 2006). At the end of log phase when lactate concentration was greatly reduced, the expression of one of the lactate permease genes was increased, reading of two of the permease genes was decreased, and three of the lactate permeases had no change in transcription. While expression of genes (DVU0253, DVU0600) for lactate dehydrogenase was enhanced in the log phase, greatest levels of lactate dehydrogenase were attributed to reading of DVU0253. The expression of both of these genes was diminished as cells entered the stationary phase. Expression of DVU2784 gene was the lowest of the three genes for lactate dehydrogenase and the production of this enzyme remained constant as cells entered the stationary phase. The genes encoding for sulfate permease were differentially expressed. The expression of gene DVU0053 increased in the log phase and declined in the stationary phase. The reading of DVU0279 was reduced throughout the log phase and remained at a low level in the stationary phase. DVU1999 was expressed at a low level growing logarithmically but slowly increased throughout the stationary phase.

Like many bacteria, it appears that metabolic systems in SRB may be subject to global regulators. While a transition of gene expression as cells enter stationary phase in many bacteria is attributed to *rpoS* (*R*NA *po*lymerase sigma *S*) which encodes for a unique sigma factor (σ^S), there was no predicted *rpoS* gene in *Desulfovibrio vulgaris Hildenborough* jmi8 (Clark et al. 2006) and the identity of a possible alternative to a RpoS regulon was not made. A partial review of SRB genomes reveals that an annotated *rpoS* gene is not found in *D. alaskensis* G20, *Desulfotalea psychrophile*, and *Desulfuromonas acetoxidans*. Based on transcriptomic analysis, the following generalization concerning transition from exponential to log phase growth of *D. vulgaris* was made (Clark et al. 2006).

- Cells in the stationary phase were experiencing a limitation of electron donor and 130–250 presumptive OPFs (open reading frames) were up-expressed while about 90–13 OPFs were down-expressed.
- About 110 genes up-expressed were unique for transition into stationary phase while many others up-expressed were also associated with stress responses to heat, nitrite, pH, and salt. Of the up-expressed genes, 13% reside on the megaplasmid.
- Four categories containing a total of 33 up-regulated genes included lipoproteins, metal binding, amino acid metabolism, and phage proteins.
- Also, up-expressed genes included those for DNA repair, carbohydrate metabolism, and nucleic acid metabolism.
- Down-expressed genes could be clustered into three groups: amino acid transport and amino acid metabolism, translation, and energy production and energy conversion.

The up-expressed carbohydrate-related genes for *D. vulgaris* Hildenborough (Clark et al. 2006) included DVUA0037 (a glucosyl transferase), DVU1128 (lysozyme), DVUA0043 (a polysaccharide deacetylase), DVU0351 (glycosyl transferase), DVU0448 (manno-dehydrase), DVU2455 (an epimerase), DVU1035 (glucokinase), and DVUA0096 (a sugar facilitator). Two genes up-expressed in stationary phase were DVU0103 and DVU3170 which are proposed to be an ABC-type cobalamin/Fe(III)-siderophore transporter and a precorrin-3B C_{17}-methyltransferase, respectively (Clark et al. 2006). Also, up-regulated genes that are associated with peptide utilization during carbon starvation include DVU2336 (a protease located in the periplasm) along with DVU0555 and DVU0598 (membrane metabolic proteins).

Several genes were up-expressed in the log phase but down-expressed in the stationary phase. As is characteristic of numerous SRB (Santos et al. 1993), carbohydrate levels associated with cells of *D. vulgaris* Hildenborough in the late log phase maybe attributed to polyglucose (glycogen) accumulation and genes associated with this carbon reserve would be expected to be down-expressed. With the transition of *D. vulgaris* Hildenborough into stationary phase, genes associated with ATP synthase, ribosomal proteins and elongation factors are down-expressed (Clark et al. 2006). Acquisition of Fe(II) in the active phase of growth but decline in the stationary phase may involve a Fe(II) transport system (DVU2571, DVU2572, and DVU2574) and nigerythrin production (DVU0019).

Proteomic and transcriptomic analysis of *D. vulgaris* Hildenborough growing in lactate-sulfate and formate-sulfate medium was conducted with cultures in the exponential and stationary phase of growth (Zhang et al. 2006a, b, c). With both lactate and formate, the expression of genes encoding for ribosomal proteins was down-regulated as the cells entered stationary phase; however, there was no change in the production of translation elongation factors (G, P, Tu and Ts), translation initiation factors (InfB, and InfC), and release of peptide chain by factor 1 (PrfA) (Zhang et al. 2006b). Several genes associated with stress response were up-regulated in stationary phase and these included: catalase, KatA, (DVUA0091); alkyl hydroperoxide reductase, AhpC (DVU2247); and universal stress proteins that may be important for cell aging (DVU0261, DVU0423, DVU0893, DVU1030, DVU2100). Genes down-regulated in stationary phase of cells growing in lactate-sulfate medium included those for the production of lactate dehydrogenase, Ldh; pyruvate oxidoreductase, Por; phosphate acetyltransferase, Pta; and acetate kinase, AckA. An interesting result is the up-regulated alcohol dehydrogenase, *adh* gene (DVU2405), and H_2:heterodisulfide oxidoreductase complex, HdrABC (DVU2402, DVU2403 and DVU2404), in the stationary phase with both lactate and formate as electron donors (Zhang et al. 2006b). To place these proteomic and transcriptomic results into perspective, the end products of lactate-sulfate and pyruvate-sulfate by *D. vulgaris* Hildenborough both involve minimal production of ethanol and H_2 (Traore et al. 1981) as shown below:

$$\text{Lactate} + 0.37\,SO_4^{2-} + 0.56\,H^+ \rightarrow CO_2 + 0.98\,\text{acetate} + 0.02\,\text{ethanol} + 0.16\,H_2S + 0.215\,HS^- + 0.5H_2O + 0.48\,H_2 \quad (6.36)$$

$$\text{Pyruvate} + 0.2\,SO_4^{2-} + 0.33H^+ + 0.15H_2O \rightarrow CO_2 + 0.95\,\text{acetate} + 0.05\,\text{ethanol} + 0.087H_2S + 0.113HS^- + 0.1\,H_2 \quad (6.37)$$

Transcription of *D. vulgaris* Hildenborough at mid-exponential grow phase growing in various electron donor-electron acceptors was compared to mRNA production in lactate-sulfate grown cells. Transcription profiles indicated that the number of up- and down-regulated genes for growth in H_2-sulfate, lactate-thiosulfate, pyruvate with limiting sulfate, and pyruvate were 761, 272, 73, and 96, respectively (Pereira et al. 2008a). The greatest change was in the category of energy and central metabolism. With growth on H_2-sulfate, the greatest increase was seen in genes for periplasmic [NiFeSe] hydrogenase and Ech hydrogenase production. Also, with H_2-sulfate growth, formate cycling and the ethanol pathway were predicted from the transcriptome. With growth on lactate and pyruvate, CO cycling was confirmed while thiosulfate as the electron acceptor; numerous genes associated with sulfate reduction were down-regulated.

Examination of *A. fulgidus* cultures growing on lactate-sulfate or H_2/CO_2-thiosulfate revealed an up-regulation of genes in late log cultures for cell wall/membrane/envelope processing, histone modification, and inhibitor of cell division (Hocking et al. 2014). With cultures of *A. fulgidus* in late log growth, the response of genes to change in growth phase was specific and reflected a change in cellular metabolism and not similar to a change in electron flow attributed to changes in electron donors or acceptors.

To maintain the organization of metabolism in sulfate-reducing microorganisms, regulatory systems to control transcription must be functioning and it has been suggested that *D. vulgaris* Hildenborough may have at least 150 transcriptional regulators (Christensen et al. 2015). *D. vulgaris* Hildenborough has four genes (DVU0379, DVU2097, DVU2547, DVU3111) that are homologs of the Crp/Fnr (*c*yclic AMP *r*egulator *p*rotein/*f*umarate and *n*itrate *r*eductase regulator protein) regulators, and using knockout mutants, it was determined that these genes function as global transcriptional regulators for stress response to chromate, NaCl, and nitrite (Zhou et al. 2013). An examination of the Crp/Fnr, ArsR, and GntR families in *Desulfovibrionales* revealed that 65% of the regulators are highly conserved while 35% are species specific (Kazakov et al. 2013a). *D. desulfuricans* 27774 has the capability of using nitrate as well as sulfate as a final electron acceptor and this metabolism of nitrate is regulated by HcpR1(*h*ybrid *c*luster *p*rotein) and HcpR2 which are members of the Crp/Fnr family of transcription factors (Cadby et al. 2016).

From an evaluation of the Crp/Fnr, ArsR and GntR families in *Desulfovibrionales* revealed that 65% of the regulators are highly conserved while 35% are species specific (Kazakov et al. 2013a). Rajeev et al. (2012, 2015) identified 200 genes representing 84 operons that were influenced by response regulators which appear unique to *D. vulgaris* Hildenborough or closely related species and reflect the environment where these sulfate reducers are found. Similar to heterotrophic bacteria, SRB respond to the changing environment through two component signal transduction systems containing a response regulator and sensory histidine kinases. *D. vulgaris* Hildenborough has 72 response regulators and 64 histidine kinases. Two component systems predicted for regulation of metabolic activities include: pili assembly, flagella regulation, exopolysaccharide and biofilm synthesis, lipid synthesis, nitrogen and phosphate metabolism, low potassium stress, stress response (nitrite, pH, heat shock, and stationary phase), and carbon starvation (Rajeev et al. 2011). Multiple response regulators are associated with lactate utilization (including lactate permease, lactate dehydrogenase pyruvate oxidation, and acetate kinase), regulation of alcohol dehydrogenase, phosphate starvation (including DNA replication, nitrogen metabolism and cyclic-di-GMP), regulation of acetyl-CoA levels, and potassium uptake. A very high percentage of response regulators are of the NtrC family of σ^{54}-dependent response regulators. It is interesting to note that a significant number of hypothetical proteins of unknown function are associated with many of these regulations.

6.2.3.1 Sigma Factors

An efficient process involving regulation of transcription in bacteria is the control of promoters by the σ^{54} subunit. The gene encoding for σ^{54} is *npoN* and this gene is found in a large number of bacterial genomes. The σ^{54}-containing RNA polymerase requires enhancer-binding proteins (EBP) to interact with a conserved upstream activating sequence (UAS) which contains an activating transcription factor (TF) site. The σ^{54}-dependent regulome in *D. vulgaris* Hildenborough has been examined and consists of 37 regulons containing 201 coding genes and 4 non-coding RNAs (Kazakov et al. 2015).

Environmental signals frequently modulate transcription activation using signal response systems across the plasma membrane involving two-component or one-component regulatory systems and *D. vulgaris* Hildenborough has 22 EBPs for two-component systems and 15 EBPs for one-component systems. The number of EBPs varies within the SRB with 37 in *D. vulgaris* Hildenborough, 41 in *Desulfomicrobium baculatum*, 33 in *D. vulgaris* Miyazaki, 24 in *D. alaskensis* G20, 15 in *Desulfohalobium retaensae*, 9 in *D. desulfuricans*, and <8 in *D. piger* (Kazakov et al. 2015). This variation in σ^{54}-dependent regulation may reflect adjustments to environmental exposure since a large percentage of regulons in *D. vulgaris* Hildenborough are associated with life style issues such as stress response with adaptation to diverse environments. In *D. vulgaris* Hildenborough, a small number of regulons control genes associated with nitrogen metabolism, flagellar synthesis, and stress response. Four expressions in *D. vulgaris* Hildenborough that are σ^{54}-dependent have a relevance to SRB physiology (Kazakov et al. 2015): (i) Type III secretion is attributed to a multiprotein complex across the cell membranes and is regulated by DVUA0100. (ii) Pyruvate: formate lyase produces formate and acetyl-CoA from formate and this gene is under the control of DVU2275 and DVU2827. There is some suggestion that sites next to these genes are activated and promote C4 dicarboxylate transport system and pyruvate biosynthesis when CO_2 and acetate are the carbon sources. (iii) The regulatory gene DVU3142 for *ohcBAC* genes encode for the cytoplasmic octaheme c-type cytochrome (OhcA), the iron-sulfur membrane protein (OhcB), and the membrane protein with a *b*-containing heme are under control of the σ^{54}-dependent promoters. (vi) An σ^{54}-dependent regulatory gene (DVU0569) is located upstream of the alanine dehydrogenase (*ald*) gene (DVU0571) which encodes for the reversible conversion of *L*-alanine to pyruvate and ammonia. When the σ^{54}-dependent regulator DVU2956 is repressed, biofilm development occurs and H_2S production is regulated in *D. vulgaris* Hildenborough (Zhu et al. 2019). The σ^{54}-dependent transcriptional regulator (DVU2894) is found in the genomic island which enables *D. vulgaris* Hildenborough to respond to oxygen and nitrite stress (Johnston et al. 2009). *D. vulgaris* Hildenborough has more than 91 response regulators, and of these, 37 are σ^{54}-dependent regulators (Kazakov et al. 2015). A copper and molybdenum containing cluster $[S_2MoS_2CuS_2MoS_2]^{3-}$ held by non-covalent bonds into a small protein (George et al. 2000) was isolated from *D. gigas* and this orange colored protein binds a [2Fe-2S] cluster (Maiti et al. 2016). Research with *D. vulgaris* Hildenborough reveals that expression of this orange protein complex is regulated by the σ^{54} factor and a proposed activity is appropriate positioning of the Z ring for cell division (Fiévet et al. 2021). Thus, it is apparent that the σ^{54}-dependent regulome serves an important role in *D. vulgaris* Hildenborough and other SRB as they adjust to the environment. The sigma factor profile for some SRB is remarkable complete with *Desulfotignum phosphitoxidans* containing σ^{70} (RpoD), σ^{54} (RpoN), $\sigma 3^2$ (RpoH), and σ^{24} (RpoE) but no genes for σ^{38} or sigma σ^{28} (Poehlein et al. 2013). Additional information concerning σ^{54} and σ^{32} sigma factors is covered in Sect. 6.3.

Extracytoplasmic function (ECF) σ factors are a set of regulatory proteins distinct from the primary and alternate sigma factors. ECF σ factors are characteristically found in complex bacterial genomes, and the number of ECF σ factors ranges from 7 in *Bacillus subtilis* to ~19 in *Pseudomonas aeruginosa* (Helmann 2002). In *Desulfobacterium autotrophicum* HRM2, five alternative sigma factors have been identified which included two ECF σ factors. Additionally, six anti-σ factors and two anti-anti-σ factors were identified in *Desulfobacterium autotrophicum* (Strittimatter et al. 2009).

6.2.3.2 Rex Regulon

The regulation of gene expression in SRB is attracting considerable attention, and new sysgv tems of control are being presented. The Rex regulon has been described as an important transcription control mechanism for regulating energy metabolism in many bacteria, and perhaps as many as 50 genes are under the control of Rex by members of the *Desulfovibrionales* (Ravcheev et al. 2012). From a taxonomic survey, Rex is predicted to regulate the following genes in 9 of 10 SRB species: *apsBA* (adenylyl-sulfate reductase), *sat* (dissimilatory sulfate adenylyltransferase), *adk* (adenylate kinase), and *ppaC* (manganese-dependent inorganic pyrophosphatase). Also, *dsrABD* (sulfite reductase) is under Rex control in 8 of 10 SRB species with 7 of 10 species having *qrcABCD* (cytochrome c3: menaquinone oxidoreductase) and *qmoABC* (Quinone membrane-bound oxidoreductase) under the influence of Rex. It has been reported that the gene *rex* in *D. vulgaris* Hildenborough (DVU0916) functions as a repressor of *sat* (gene for *sulfate adenylyltransferase), is regulated by the re-dox state inside the cell, and may function in regulating genes in other sulfate reduction activities.* (Christensen et al. 2015).

6.2.3.3 SahR Regulon

Cells of *D. vulgaris* Hildenborough and *D. alaskensis* G20 regulate the amount of methionine, S-adenosylmethionine (SAM), and S-adenosylhomocysteine (SAH) by SahR which is a negative transcriptional regulator (Novichkov et al. 2014). SAM is the coenzyme for methyltransferases. Methionine is directly involved in the production of SAM and SAH, which is the demethylated product of SAM (see Fig. 6.4). In *D. vulgaris* Hildenborough, s*ahR* (DvU0606) regulates m*etF* (DVU0997), m*etE* (DVU3371), *metK* (DVU2449), and *ahcY* (DVU0607). Pantothenic acid is used for the production of CoA and in some *Desulfovibrio jmi8* sp. *srhR* also regulates pantothenic acid production, which is in *D. vulgaris* Hildenborough encoded by *panC* (DVU2448).

6.2.3.4 Fur Regulon

The *f*erric *u*ptake *r*egulator (*fur,* DVU0942) is a powerful global regulator in bacterial transcription and influences ferrous uptake in *D. vulgaris* Hildenborough (Bender et al. 2007).The Fur operon contains genes for ferrous transport (*feoA,B*;

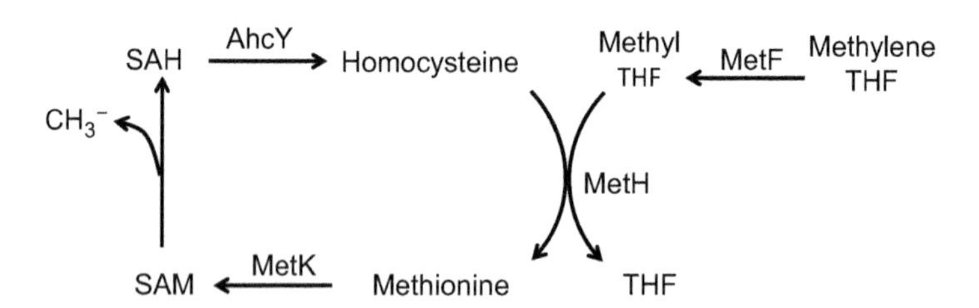

Fig. 6.4 Regulation of S-adenosylmethionine production. Abbreviations are as follows: SAM = S-adenosylmethionine, SAH = S- adenosylhomocysteine, THF = tetrahydrofolate, MetK = S-adenosylmethionine synthetase (DVU2449), MetF = 5,10-methylenetetrahydrofolate reductase (DVU0997), MetH = Methionine synthase (DVU1585), AhcY = S-adenosylhomocysteine hydrolase (DVU0607)

DVU2572, DVU2571), transporter component for FeoA (*feoA*, DVU2574), a gene for flavodoxin (*fld*, DVU2680), ATP transporter proteins (DVU2380,DCU2384, DVU2384, DVU2385, DVU2386, DVU2387), iron regulated P-type ATPase (DVU3330), and heavy metal translocating P-type ATPase (DVU3332), Fur is also suggested to participate in osmotic and nitrite stress response (Bender et al. 2007). Genes for siderophore production have not been documented in *D. vulgaris* Hildenborough, but a putative transporter for the enterobactin siderophore (*fepC*, DVU0648) has been identified. A 12 gene operon in *D. vulgaris* Hildenborough includes orthologs of TonB (COG-*tonB*, DVU2390; DVU2383) and TolQR (*tolG1*, DVU2388; *tolR*, DVU2389) which is interesting since these are characteristic of aerobic bacteria. Many SRB show microaerophilic growth and it will be interesting to see if *D. vulgaris* Hildenborough is capable of utilizing enterobactin or other siderophores of bacteria under an aerobic environment.

6.2.3.5 TunR Regulon

The tungstate-responsive regulator (TunR) is involved in maintaining homeostasis of molybdate and tungstate in SRB. The TunR proteins are classified into two groups: (i) activate the *modABC* genes for molybdate and tungstate transport and (ii) repress the TSUP genes for toluene sulfonate permease. The *modABC* system contains a periplasmic protein-binding molybdate (ModA), a transmembrane permease (ModB), and a protein accessible to the cytoplasm which functions to energize transport (ModC). In *D. vulgaris* Hildenborough, the regulatory protein for the molybdate/tungstate transport is encoded on the gene DVU0179, with DVU0177 for *modA*, DVU0181 for *modB*, and DVU0180 for modC. The TunR system would regulate the intracellular concentration of these molybdate and tungstate at a level which would not inhibit sulfate metabolism. This TunR system of regulation has been proposed to be restricted to *Deltaproteobacteria* SRB (Kazakov et al. 2013a). A high affinity tungstate transport system (TupABC) has been reported for *D. desulfuricans* ATCC 27774, *D. vulgaris* Miyazaki (Kazakov et al. 2013a), and *D. alaskensis* G20 (Otrelo-Cardoso et al. 2014, 2017). The regulator protein for the TupABC transport system is activated only when molybdate is absent in the environment. In addition to TunR proteins present in SRB, there is a TunR2 protein which is not sensitive to molybdate or tungstate and encodes transporters from the TSUP and SulP (sulfate permease MFS superfamily). The impact of the TunP2 regulator is proposed to promote sulfur uptake and thereby avoid inhibition of sulfate metabolism by molybdate or tungstate (Kazakov et al. 2013b).

6.2.4 Genomic Islands

Mobile genetic elements moved between bacteria by horizontal gene transfer are referred to as genomic islands. There has been a long-standing speculation that SRB have a metabolic island for dissimilatory sulfate reduction (Klein et al. 2001; Friedrich 2002); however, the enzymes of dissimilatory sulfate reduction (ATP sulfurylase), APS reductase, and dissimilatory sulfite reductase are distributed around the chromosomes in genomes of cultivated SRB. Gene analysis of uncultured SRB reveals a gene cluster for these key enzymes of dissimilatory sulfate reduction and would represent a genomic metabolic island for sulfate reduction (Mussmann et al. 2005). The distribution of these metabolic genes around the chromosome would result from gene rearrangement by the recipient cells. An array of 52 genes makeup a 47 kb region on the genome of *D. vulgaris* Hildenborough (DVU2000 -DVU2051) is designated as a genomic island (GEI^H) which contributes to stress response (Johnston et al. 2009). Included in this array of genes are rubredoxin: oxygen oxidoreductase-1 (Roo1) and hybrid cluster protein-1 (Hcp-1) which confer resistance to oxygen and nitrite stress. Also, present are genes which encode for specific recombinases. About 0.3% of the cells spontaneously delete the GEI^H when cells are growing under anaerobic conditions in the laboratory. The *D. vulgaris* Hildenborough cells that are GEI^- grow on lactate-sulfate at a rate faster than GEI^+ cells (Johnston et al. 2009). A review of GEIs in *D. vulgaris* Dupre (indicated as GEI^D) revealed five large GEIs and four smaller units (Walker et al. 2009). The genomic island of *D. vulgaris* Dupre occupies the same region of the chromosome as in *D. vulgaris* Hildenborough and consists of a 57 kb region with 33 open reading frames. A cluster of genes for *D. vulgaris* Dupre in this region (Dvul_2580-Dvul_2594) encode for exopolysaccharide biosynthesis and export which account for the production of extracellular sialic acid. Since *D. vulgaris* reduces chromate to Cr(III), the acidic sialic acid polymer produced by *D. vulgaris* Dupre would bind Cr (III) and enable this bacterium to tolerate high levels of chromium and enable this organism to persist in metal contaminated sites (Walker et al. 2009). Thus, GEIs in different SRB strains would be expected to have unique GEIs that reflect the environment where they were isolated.

6.3 Stress Response

The exposure of bacteria to an environmental condition suboptimal for growth would be a stressful condition and several of these stresses are presented in Table 6.6. Following exposure to a stress, genes are either up-regulated or down-regulated in expression. There is a focus on the 46 "signature" genes which are characteristic of SRB; and a few of these signature genes include the following: *dsrAB* (dissimilatory sulfite reductases), *apsB-qmoC* (adenylylsulfate reductase and quinone-interaction oxidoreductase), and *rbO-roO* (desulforrodoxin and rubredoxin-oxygen oxidoreductase) (Chhabra et al. 2006). Expression of genes in stress response is attributed, in part, to σ^{54} and σ^{32} sigma factors which are alternates to the housekeeping sigma factor σ^{70}. In the review by Wall et al. (2007), the genes in *D. vulgaris* Hildenborough associated with various stress responses are reviewed and the number of genes responding to the stresses is presented in Table 6.6.

Table 6.6 Number of genes changed (up-regulated and down-regulated as a result of stress response in *Desulfovibrio vulgaris* Hildenborough (Wall et al. 2007)

Stress	Number of genes	Stressed condition or concentration	Unstressed culture
Cold	68	8 °C	30 °C
Heat	474	50 °C	37 °C
Oxygen	70	0.1%	0 °C
Nitrate	298	105 mM	None
Nitrite	305	2.5 mM	None
Na^+	428	250 mM	No Na added
K^+	399	250 mM	No K added
Chromate	337	0.55 mM	None
Alkaline (pH 10)	280	pH 10	pH 7.0
Acid (pH 5.5)	471	pH 5.5	pH 7.0

In *Escherichia coli*, the master regulator, RpoS or σ^s, induces expression of numerous genes following stress (Hengge-Aronis 2000) and would account for a universal stress response. However, in *D. vulgaris* Hildenborough and numerous other non-γ-proteobacteria, there is no identified RpoS (Nies 2004) and stress response appears uncoordinated and not consistent. *D. vulgaris* Hildenborough does contain nine open reading frames annotated as members of the universal stress family, but the expression of these genes does not fit the description of a universal stress response (Wall et al. 2007). Using a system-level analysis, several stress responses in *Desulfovibrio vulgaris* have been reviewed (Zhou et al. 2011).

6.3.1 Oxidative Stress

Historically, SRB were considered to be obligate anaerobes; however, many of the strains are oxygen tolerant or capable of microaerophilic growth (see discussions in Chap. 2). The exposure of SRB to oxygen has a pronounced effect on gene expression, and this has been best studied in *D. vulgaris* Hildenborough following exposure to 100% oxygen, atmospheric oxygen level, and 0.1% oxygen (Mukhopadhyay et al. 2007; Fournier et al. 2006; Zhang et al. 2006b; Pereira et al. 2008b). With 0.1% oxygen bubbled through the *D. vulgaris* Hildenborough culture, only 12 genes were up-regulated which included genes of the predicted PerR regulon, *ahpC* (alkyl hydroperoxide reductase) and *tmc* (transmembrane cytochrome c_3) complex (Mukhopadhyay et al. 2007). In contrast, exposure to air initiated a strong stress response with 393 genes up-regulated and 454 genes down-regulated (Mukhopadhyay et al. 2007). It has been speculated by Pereira et al. (2008b) that oxygen stress accounts for degradation of the metal clusters of the metalloproteins and several genes up-regulated are for repair of these essential metalloproteins.

Hydrogen peroxide has been used to promote oxidative stress in *D. vulgaris* Hildenborough; and at low concentrations (~0.1 mM H_2O_2), cell viability is not diminished, but with elevated concentrations, hydrogen peroxide is highly toxic. At sublethal concentration of hydrogen peroxide, a transcriptional analysis revealed that genes for superoxide dismutase (*sod*), superoxide reductase (*sor*), nigerythrin (*ngr*), thiolperoxidase (*tpx*), and the PerR regulor were up-regulated and were considered to be part of the hydrogen peroxide stimulon (Brioukhanov et al. 2010). At elevated concentrations of hydrogen peroxide, hundreds of *D. vulgaris* Hildenborough genes displayed modified expression (485 genes up-regulated and 527 genes down-regulated) (Zhou et al. 2010). In addition to the increased production of detoxification enzymes, sulfur metabolizing enzymes were down-regulated as well as the Fur regulon. The overlap of the PerR regulon and Fur regulon expression as a defense against hydrogen oxide stress distinguishes this response from the hydrogen peroxide stress response observed in *Escherichia coli* and *Bacillus subtilis* (Zhou et al. 2010).

6.3.2 Starvation Response and CO_2 Stress

Nutrients required for cell growth may become limiting under certain conditions, and these deficiencies stimulate adjustments in cells. As SRB such as *D. vulgaris* Hildenborough grow in batch liquid culture and transition through the various phases of growth, changes in gene expression occur as the cell adjusts to the environment. Several adjustments occur as a growing culture in defined medium adjusts to stationary phase (Clark et al. 2006). With the depletion of lactate as a carbon source in late log/early stationary phase, glycogen stored as an internal granule becomes an important source of carbon material for SRB (Santos et al. 1993). Transcriptional analysis of *D. vulgaris* Hildenborough in stationary phase (see Sect. 6.2.3) where electron donor is limiting reveals that 130–250 open reading frames are up-regulated while 90–130 open reading frames were down-regulated (He et al. 2006). Genes up-regulated in stationary phase included lipoproteins, iron binding proteins, amino acid metabolism, and phage-related proteins. Two genes up-regulated were similar to *cstA,* a membrane protein expressed during carbon starvation in *E. coli.* Down-regulated genes were those encoding for amino acid transport and metabolism, energy production and conversion and transcription. A group (~110) of the up-expressed genes were associated with conversion from log to stationary phase, and the remaining genes up-expressed in stationary phase were associated with other stress responses (i.e., heat, pH temperature and nitrite) (He et al. 2006).

In response to the concern of increasing levels of greenhouse emissions in the atmosphere, a technology is being developed to pump CO_2 into various geological environments. In a study where CO_2 was injected under pressure into subsurface environments, most of the *D. vulgaris* Hildenborough cells were viable after 4 h at 50 bar CO_2 and 2 h at 80 bar CO_2 (Wilkins et al. 2014). The number of up-regulated genes at 25 bar, 50 bar, and 80 bar CO_2 were 71, 84, and 57, respectively. The number of down-regulated genes at the same three pressures was 40, 38, and 38.

Production of leucine and isoleucine as a CO_2 stress response was observed and was considered to be part of the general stress response (Wilkins et al. 2014).

6.3.3 *Extreme Temperatures*

6.3.3.1 Heat Shock

Heat shock response by *D. vulgaris* Hildenborough was reviewed by Wall et al. (2007). In heat shock studies conducted with *D. vulgaris* Hildenborough, the culture was grown at 37 °C and when the stationary phase was reached the culture was subjected to 50 °C (Chhabra et al. 2006; Zhang et al. 2006b). A large number of genes appear regulated by heat shock; and the reports by Chhabra et al. (2006), Zhang et al. (2006b), and Wall et al. (2007) reflect the different criteria used for assessing transcription. In general, 58.9% of the genes are up-regulated and 41.1% are down-regulated (Chhabra et al. 2006). Regulation of heat shock involves the use of alternate sigma factors of σ^{54} (RpoN) and σ^{32} (RpoH) sigma factors and the controlling inverted repeat of chaperone expression (CIRCE) element. Predicted from the genome of *D. vulgaris* Hildenborough are four σ^{32}-dependent promoters, one CIRCE site, and <98 genes with σ^{54} binding sites (Chhabra et al. 2006). The up-regulated genes included those in the categories of protein turnover, chaperones, and posttranslational modification while the down-regulated genes were for energy production, nucleotide transport, metabolism, translation, ribosomal structure, and biogenesis.

A. fulgidus is a thermophilic sulfate-reducing archaea which grows from 60 to 95 °C, and it has been examined for heat shock response by growing the microorganism at 78 °C and shifting the temperature to 89 °C. The first genes in *A. fulgidus* reported to be up-shifted in response to heat shock were the *Cpn-α* and *Cpn-β* subunits of chaperonin (Emmerhoff et al. 1998). Using whole-genome microarrays, after 5 min at 89 °C about 10% of the open reading frames (ORF) were influenced by heat shock with 118 ORF with increased mRNA production and 120 ORF with a decrease (Rohlin et al. 2005). Five genes highly up-regulated by heat shock had been annotated as heat shock genes. Heat shock proteins produced by *A. fulgidus* included HSP60s (often referred to as thermosomes), the small protein sHSP20 and a subtilisin-like peptidase, an enzyme in the mevalonate pathway important for isoprenoid synthesis, and increased production of myoinositol phosphate and diglycerol phosphate. Little is known about the regulation of genes associated with heat shock in these thermophilic archaeal organisms.

6.3.3.2 Cold Shock

To evaluate the response of *D. vulgaris* Hildenborough to low temperature shock, cells grown at 20 °C were transferred to 8 °C and transcriptional responses were noted in 68 genes (Wall et al. 2007). Relying on a report concerning cold shock response in *Escherichia coli* (Phadtare 2012), cold shock has a greatest impact on ribosome structure, plasma membrane fluidity, and nucleic acid structure. The complement of cold shock proteins of *E. coli* was not found in *D. vulgaris* Hildenborough. Even though a homologue to *yfiA* (a gene encoding for protein Y binding to ribosomes of cold shock *E. coli*) was present in the *D. vulgaris* Hildenborough genome, expression of *yfiA* in *D. vulgaris* Hildenborough was unchanged following exposure to cold temperature. Up-regulated in *D. vulgaris* Hildenborough was a putative homolog gene for an RNA helicase which effects a change in translation and nucleic acid structure. Of the 68 genes responding to cold stress in *D. vulgaris* Hildenborough, 15 genes addressing cell membrane fluidity and transport were reported (Wall et al. 2007). It will be important to address the response that other sulfate reducers have upon exposure to cold environments. It has been determined that the genome of *Desulfotalea psychrophile,* a psychrophilic bacterium that grows in Artic sediments, contains several genes which encode for cold shock proteins and several inducible cold shock proteins (Rabus et al. 2004).

6.3.4 Salt Adaptation

The cultivation of SRB in high levels of NaCl accounts for phenotypic changes and differences in transcription as compared to cultures growing in low salt environments (Mukhopadhyay et al. 2006). An examination of *D. vulgaris* Hildenborough growing in 250 mM NaCl or KCl revealed that the cells became elongated with diminished flagellar movement. Additional examination revealed that in the presence of salt stress, cells of *D. vulgaris* Hildenborough displayed an induction of ABC transporters for uptake of ectoine, glycine betaine, and proline (known osmoregulators in other bacteria). Also up-regulated were genes for efflux systems, and this may suggest that elevated salt levels affected the stability of DNA and RNA base pairing. With high salt, there was an increase in branched fatty acids which would contribute to changes in membrane fluidity and gene expression for chemotaxis were up-regulated. Down-regulated were genes for lactate uptake and ABC transport systems.

Following cultivation in 250 mM NaCl for 5000 generations, phenotypic changes of *D, vulgaris* Hildenborough cells indicated that glutamate was the key osmolyte for salt tolerance (Zhou et al. 2017). The transcriptional changes of *D. vulgaris* Hildenborough under high salt conditions resulted in 71 genes with an increased expression and 128 genes had a diminished expression. Genetic adaptations of *D. vulgaris* Hildenborough indicated mutations in 36 genes that may be important in salt tolerance. Other changes observed in adaptive evolution of *D. vulgaris* Hildenborough after 5000 generations under salt stress (300 mM) included the following:

- In the initial adjustment following salt stress, the cells displayed characteristics associated with general stress response such as phage shock protein and heat shock protein genes.
- A deletion of 49 genes in the chromosome which included prophage-related genes. A loss of plasmid DNA accounting for about 52% of the plasmid including genes for type III secretion associated with horizontal gene transfer.
- Increased abundance of glutamate and glutamine in cells subjected to salt stress. In early phases of adjustment to high salt environments, alanine and branched amino acids are considered to be important.
- Enhanced abundance of branched phospholipid fatty acids (PLFA) and unsaturated PLFA. A greater contribution to salt tolerance may be attributed to branched PLFA than unsaturated PLFA.
- After long periods of low flagella motility following exposure to salt, high flagellar activity and high cell yields may indicate a more conserved energy management than unstressed cells.
- Increased transcription of genes expressing the TonB domain associated with iron and cobalamin transport, Hmc production, and a sigma-54 transcriptional regulator gene.
- Decreased expression of genes for cytochrome *c* and periplasmic [Ni-Fe] hydrogenase.
- Decreased expression of genes involved in carbohydrate transport and metabolism.

6.3.5 Nitrogen Stress

The environment where dissimilatory sulfate reducers are found commonly contains nitrate which is assimilated by sulfate reducers for the production of amino acids and other nitrogen-containing compounds. Some anaerobic bacteria use nitrate as the electron acceptor with the production of N_2 with the accumulation of nitrite and nitric oxide ion the environment as intermediates in the nitrate reduction process. While most of the sulfate-reducing bacteria do not couple nitrate reduction to anaerobic respiration, some of the sulfate-reducing bacteria use nitrate as an alternate electron acceptor. The accumulation of nitrate, nitrite, and nitric oxide in the environment induces stress on sulfate-reducing bacteria including those that have dissimilatory nitrate reduction. *D. vulgaris* Hildenborough, a sulfate-reducing bacterium unable to use nitrate in respiration, and *D. desulfuricans* 27774, a bacterium with both sulfate and nitrate dissimilatory respiration, are the model organisms used to study nitrate and nitrite stress response.

6.3.5.1 Nitrate Stress

The presence of nitrate may pose a specific stress to SRB as nitrate has been observed to suppress sulfate reduction activity in situ (Jenneman et al. 1986; Davidova et al. 2001). In SRB, nitrate stress appears to be independent form nitrite stress and may be a result of nitrate entering the cell by a leaky thiosulfate transporter (Korte et al. 2014). While nitrate at 70 mM accounts for a reduction of growth, nitrate at 105 mM inhibits growth by about 50%. Nitrate inhibition may be attributed to nitrite inhibition of sulfite reductase with nitrite produced from nitrate by metabolic nitrate reductase. Specific to nitrate stress in *D. vulgaris* Hildenborough is the up-regulation of numerous ABC transport systems, [Fe-S] proteins, and a series of hypothetical proteins (Redding et al. 2006). In *D. vulgaris* Hildenborough and *D. alaskensis* 20, nitrate resistance is not associated with nitrate consumption but with changes in the Rex transcriptional regulator and a series of genes found in both organisms (DVU0251-DVU0245; Dde_0597- Dde_0605). From a global perspective, nitrate stress in *D. vulgaris* Hildenborough accounts for a differential expression of 298 genes and this response appears to overlap with nitrite stress, oxidative stress, and osmotic stress response (Redding et al. 2006; He et al. 2010).

6.3.5.2 Nitrite Stress

Nitrite may be found in the environment in considerable abundance especially if the nitrate level is considerable. It has often been speculated that nitrite inhibition of SRB was attributed to interference with the process of dissimilatory sulfate reduction as a competitive inhibitor in sulfite reductase (Haveman et al. 2003; Haveman et al. 2004); however, it now appears that nitrite has a marked effect on energy metabolism, iron uptake, oxidative stress, and nitrogen metabolism (He et al. 2006). Global transcription studies revealed that *D. vulgaris* Hildenborough responds to nitrite concentrations greater than 2.5 mM by having a down-regulation of genes which function in ATP-mediated transport of amino acid, protein biosynthesis, and genes associated with oxidative phosphorylation. Also, down-regulated genes in the presence of nitrite were for the periplasmic Ech hydrogenase (*echABCD*), the decaheme electron transport membrane complex (*rnfABCDEFG*), and the triheme membrane cytochrome (*dsrMKJOP*) (He et al. 2006). Up-regulated in the presence of nitrite included genes for enhanced nitrogen metabolism, and these genes included glutamine synthesis (DVU3392), aspartate ammonia-lyase (DVU1766), and asparaginase (DVU2242) (He et al. 2006). Genes up-regulated for substrate metabolism and oxidation/electron transfer included the [NiFe] periplasmic hydrogenase (hynBA-2), tetraheme cytochrome (DVU2524-2526), lactate dehydrogenase (*ldh*), pyruvate ferredoxin:oxidoreductase (*porAB*), and formate dehydrogenase (DVU2819-2812) (He et al. 2006). An increase in iron for the cytochromes, hydrogenases, and dehydrogenases appears to account for an up-regulation of ferrous ion transport. In general, it appears that *D. vulgaris* Hildenborough in the presence of nitrite shifts from an oxidative phosphorylation to a substrate level phosphorylation process for ATP production.

Oxidative stress response due to nitrite in *D. vulgaris* Hildenborough may be attributed to a specific set of genes in a genomic island and this is discussed in Sect. 6.2.4. *D. vulgaris* Hildenborough and a few other SRB have the capability of using nitrite as an electron acceptor at low nitrite levels. The enzyme accounting for this reduction activity is nitrite reductase (NrfHA) and it is constitutively produced. At elevated concentrations of nitrite, NrfHA production is induced in response to NrfR (DVU0621) which is a sigma-54 response regulator (Kazakov et al. 2015).

6.3.5.3 Nitric Oxide Stress

Nitric oxide, NO, is produced by denitrifying bacteria including SRB as nitrate is reduced to N_2 and as nitrite is reduced to ammonia (Wang et al. 2016). Nitric oxide readily reacts with redox-sensitive thiols and the nitrosylation product is toxic to cells. Nitrosative stress, which is observed in SRB and other bacteria, is distinct from stress due to nitrate and nitrite. The gene encoding for nitric oxide reductase to protect the cytoplasm from nitrosative products has been designated as the hybrid protein cluster, *hpc*, and transcription regulators for this gene are indicated as HpcR. Additionally, the flavodiiron protein (rubredoxin:oxidoreductase, ROO) protects against nitrosative stress in enteric bacteria. Both Hpc and ROO proteins promote the resistance of sulfate-reducing bacteria to NO exposure (Rodrigues et al. 2006; Johnston et al. 2009;Wildschut et al. 2006; Yurkiw et al. 2012; Figueiredo et al. 2013) and the activities of these genes are reviewed by Cadby et al. (2017). *D. vulgaris* contains two copies of the hpc gene and two copies of the flavodiiron protein ROO gene (Johnston et al. 2009), and the *D. desulfuricans* ATCC 27774 genome contains *hprcR1* and *hprcR2* which are induced by NO exposure (Cadby et al. 2016). Since NO is generated by animal metabolism, establishment of SRBs in animals is dependent on the presence of a system that detoxifies nitrosative stress and this subject is discussed in Sect. 6.3.5.3.

6.3.6 pH Extremes

6.3.6.1 Acid Stress

In a report concerning acid pH stress in *D. vulgaris* Hildenborough, there was a change in expression of 471 genes (Wall et al. 2007). Unlike *Escherichia coli* which uses RpoS to regulate response to acid stress, *D. vulgaris* Hildenborough does not have the RpoS system. Several genes in the operon for arginine decarboxylation were up-regulated in *D. vulgaris* Hildenborough but, unlike acid stress in *E. coli*, decarboxylation of glutamate was not affected. Acidic treatment results in the up-regulation of 14 genes for flagella synthesis and DNA repair. Cells of

D. vulgaris Hildenborough exposed to acid readily bind to glass surfaces, and this may be due to the stimulation of biofilm formation which is enhanced by polyamine formation associated with acid exposure. Acid stress response in Gram-negative enteric bacteria has focused on periplasmic proteins designated as HdeA and HdeB (52); however, SRB do not have these chaperone proteins. Under acid stress, *D. vulgaris* ATCC 7757 has 57 genes proposed to facilitate acid tolerance and an upshift in of proteins for tolerance to low pH included rplC, rpsQ, rpsO, rpsJ, and yfiA which constitute a segment of the "GO:0005840-ribosome" unit (Yu et al. 2020).

6.3.6.2 Alkaline Stress

The exposure of *D. vulgaris* Hildenborough to pH 10 resulted in specific transcriptional expression (Stolyar et al. 2007a, b; Wall et al. 2007). The number of genes regulated by *D. vulgaris* Hildenborough alkaline stress at 240 min following a shift from pH 7.5 to pH 10 was 367 with an up-regulation of 184 genes and a down-regulation of 183 genes. While the regulatory agents in *D. vulgaris* Hildenborough for this stress response are not the same as for *Escherichia coli,* several genes up-regulated are the same for both cultures and include genes for ATPases, amino acid synthesis, lipopolysaccharide synthesis, outer membrane protein, peptidoglycan synthesis, proteases, and chaperones. Genes down-regulated following alkaline stress were associated with signal transduction, transcription, and phage-associated activities.

6.4 Biofilm

Biofilm formation by SRB has been studied in pure culture and in coculture. Coculture of *D. vulgaris* and *Methanococcus maripaludis* S2 is shown in Fig. 6.5. Biofilm continuous reactor growth of *D. vulgaris* Hildenborough was compared to planktonic cells using transcriptomic and proteomic analysis. While the biofilm transcriptome has a cluster of about 70 genes with similar expressions, a significant number of these genes are annotated as hypothetical proteins (Clark et al. 2012). Biofilm cells, as compared to batch grown planktonic cells, contained 231 transcripts up-regulated and 358 genes down-regulated. Altered abundance of proteins and expression of genes in biofilm occur, and additional summaries are listed below according to some of the COG categories (Clark et al. 2012):

- Energy conversion – 44 genes were up-regulated and 17 down-regulated in biofilm cells. Two different genes for conversion of pyruvate to acetyl-CoA were differentially regulated with the *oor* gene up-regulated and the *por* gene down-regulated. An increase in the quantity of formate dehydrogenases (HybA

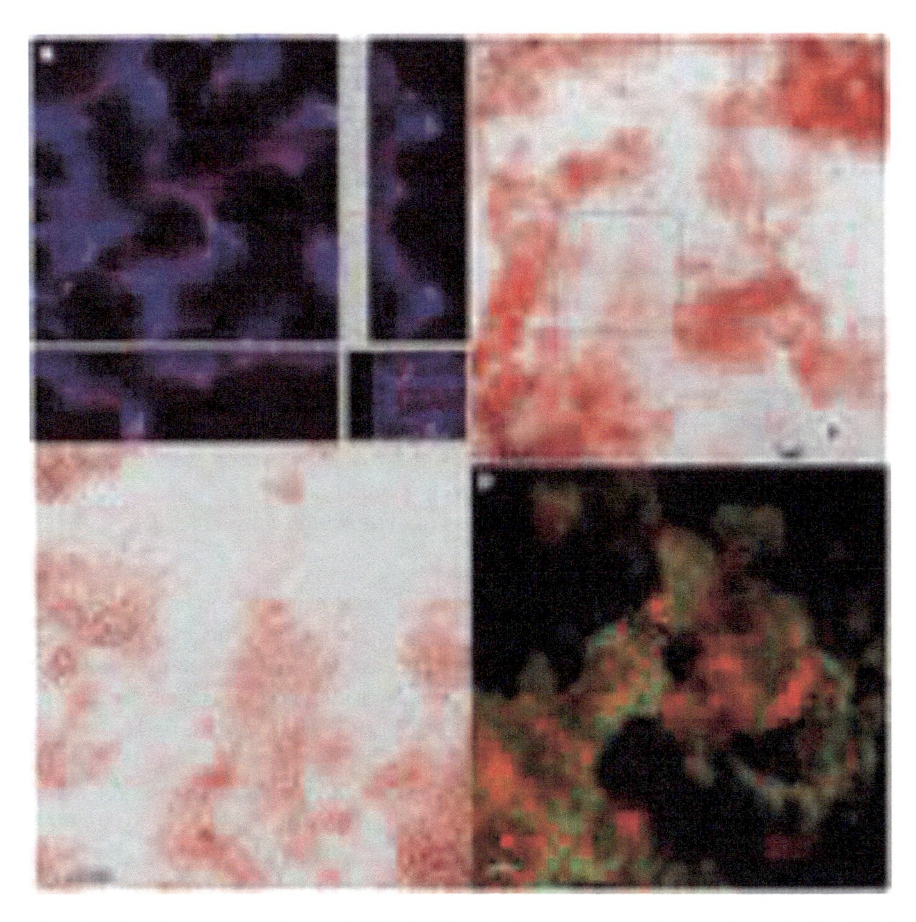

Fig. 6.5 Coculture of *D. vulgaris* ATCC 29579 and *Methanococcus* (*M.*) *maripaludis* S2 (DSM 14266). Coculture biofilm (**a**) stained with CTC while intact and hydrated and showing all biomass stained with DAPI in blue, and CTC in red, or purple where both DAPI and CTC are present (**b**) scraped from the slide after CTC staining. (**c**) Zoomed in from the inset in (**b**) showing individual grains of red fluorescent CT-formazan in each cell. (**d**) Coculture biofilm scraped from the substratum, fixed and hybridized with domain-specific probes for *D. vulgaris* (green) and *M. maripaludis* (red) (Brileya et al. 2014).

and FdnG1) along with the increase in cytochrome c_{553} supported the suggestion that formate cycling was important within the biofilm.

- Carbohydrate metabolism/fatty acid and lipid synthesis – Glycerol from lysed bacterial cells may be an important carbon source for amino acid biosynthesis in biofilms. Enzymes associated with gluconeogenesis via glycolytic pathway were not enhanced; fructose-1,6-bisphosphatase was down-regulated. This may indicate a slower rate of cell wall synthesis in biofilm cells.

- Amino acid and nucleotide production – A reduced level of transcripts for amino acids and nucleotides in the biofilms indicates a reduced rate of protein synthesis. Only genes for glutamate synthesis (*gltA*) and cysteine synthesis (*cysK*) were increased in biofilm cells.
- Transporters – With biofilm cells, several ABC transporter genes were up-regulated with the corresponding increased production of some transport proteins. There was a down-regulation of several genes encoding for subunits of the Sec transport system while components of the type I and type III secretion system were enhanced.
- Stress response – Within the biofilm, cells had an up-regulation of *uspA* and *hsp20* genes which encoded for chaperone proteins and the up-regulation of stress response proteins of superoxide dismutase (*sodB*), alkyl hydroperoxide reductase (*ahpC*), *msrB* and *ahpC*/TCA (thiol specific antioxidant).
- Nitrogen metabolism – An up-regulation of cytochrome c nitrite reductase (*nrfA*) gene in biofilm cells was observed.

A report using single-cell-based RT-qPCR examined eight genes that are of potential importance in biocorrosion and biofilm formation under planktonic and biofilm growth. In comparison to planktonic *D. vulgaris* Hildenborough cultures, biofilm cells displayed up-regulation of genes encoding for exopolysaccharide protein (DVU0281), ferric uptake repressor (DVU1340), and iron storage (DVU1397) while genes down-regulated included energy metabolism (DVU0434 and DVU0588), stress responses (DVU2410), histidine kinase/response regulator (DVU3062), and iron transport (DVU2571) (Qi et al. 2016). The interpretation of some of these results of transcription regulation is as follows. Exopolysaccharide is an important component for biofilm matrix and up-regulation of DVU0281 would enhance exopolysaccharide protein to be produced. Excess iron may be detrimental to biofilms and up-regulation of genes for ferric uptake repressor (Fur) regulators (DVU1340) and bacterioferritin for iron storage (DVU1397) would manage iron metabolism in *D. vulgaris* Hildenborough. Further limitation of iron to the biofilm would be the down-regulation of the ferrous iron transport protein (DVU2571).

Since biocorrosion is often occurring in sea and brackish water, evaluation of biofilm formation in saline environments is important. Biofilm formation by *D. vulgaris* Hildenborough and *Desulfobacterium corrodens* grown in saline (salinity = 25.9 g/L) and fresh water (salinity = 4.17 g/L) media was examined by Sivakumar et al. (2019). In high saline media, both cultures exhibited an increase in sulfate respiration, carbon utilization, histidine kinase signal transduction, exopolysaccharide synthesis, and biofilm formation. Specific genes up-regulated in high saline encoded for lactate dehydrogenase, pyruvate:ferredoxin oxidoreductase, formate dehydrogenase, and pyruvate formate lyase dissimilatory sulfite reductase. These upshifts of gene expression support previous observations that the electron flow and sulfate respiration are dependent on carbon metabolism genes (Pereira et al. 2011; Keller and Wall 2011).

Laboratory-driven evolution of *D. vulgaris* Hildenborough has been observed with differences in biofilm formation. When comparing the *D. vulgaris* Hildenborough culture from the University of Montana (*D. vulgaris* HildenboroughMT) with that from the University of Missouri (*D. vulgaris* HildenboroughMO), one culture had lost the capability of producing a biofilm (De León et al. 2017.) The loss of biofilm production was attributed to one nucleotide change in the ABC transporter (DVU1017) of the type I secretion system (TISS). Without the transport of biofilm proteins, no biofilm is produced. While the strain *D. vulgaris* HildenboroughMT produced biofilm within 24 h in an appropriate biofilm reactor, *D. vulgaris* HildenboroughMO did not produce biofilm until after 100 h of cultivation which may reflect that a very small number of bacteria in the population that did not carry the mutation.

The biofilm produced by *A. fulgidus* is a response to a variety of environmental stresses including extreme temperatures, high pH, presence of metal ions, O_2, antibiotics, and xenobiotics (LaPaglia and Hartzell 1997). This biofilm has a variable structure and contains proteins, polysaccharides, and metals. The function of the biofilm appears to protect the cells and serve as a nutrient reserve.

6.5 CRISPR-Cas Systems, Proviruses, and Viruses

An adaptive immunity system found in many bacteria and archaea is CRISPR-Cas. Crisper refers to a segment of DNA that contains "clustered regularly interspersed short palindromic repeats," and Cas is the "CRISPR-associated proteins." As genomes of SRB are being sequenced, many have CRISPR-Cas which is to provide immunity to invading plasmids and viruses. The complexity of the CRISPR-Cas system in microorganisms is being addressed, and three types of CRISPR-Cas systems have been described with the *cas1* and *cas2* genes as major features in each type (Rath et al. 2015; Makarova et al. 2011). While the CRISPR-Cas system is absent in the genomes of *Desulfovibrio aepoeensis* Apso-2, *D. africanus* Wavis Bay, *D. piezophilus* CITI.V30, *D. salexigens* DSM 2638 and *D. piger* F111049, several strains of *Desulfovibrio* spp. have defined regions containing the CRISPR-Cas system (Crispim et al. 2018; Morais-Silva et al. 2014). The presence of CRISPR-Cas systems in *Desulfovibrio* spp. is presented in Fig. 6.5. *D. gigas* has two CRISPR-Cas systems with one classified as Type I-F with six CRISPR repeat genes and two of these genes are flanked by Cas operons. *D. vulgaris* has CRISPR-Cas system of the Type I-C and the CRISPR genes reside on the megaplasmid in the Hildenborough strain and on the chromosome in the Miyazaki strain (Morais-Silva et al. 2014) (Fig. 6.6).

In a recent survey of 47 genomes of *Desulfovibrio*, 53 proviruses were identified as complete and 75 segments were characterized as degenerate prophage or related to gene transfer agents (Crispim et al. 2018). As listed in Table 6.7, 17 strains of *Desulfovibrio* spp. contain one to eight complete prophages and from genetic characterization are classified into the order of *Caudovirales*. The distribution and

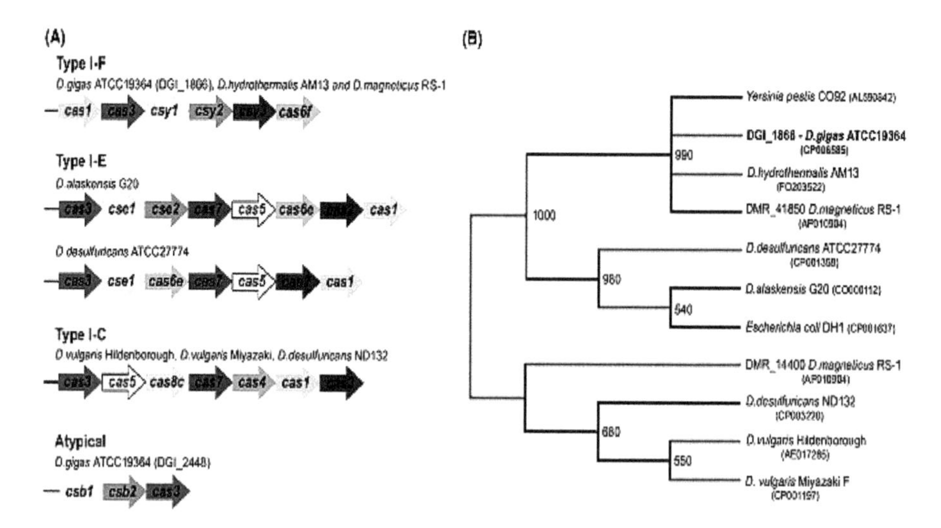

Fig. 6.6 Distribution of different types of CRISPR/Cas systems among *Desulfovibrio* spp. (**a**) Operon structure of cas genes from the indicated *Desulfovibrio* spp. The operon organization was assessed using the DOE Joint Genome Institute (JGI) website (http://www.jgi.doe.gov/). Classification into the distinct Type I subtypes is according to Makarova et al. (2011). (**b**) The evolutionary history of Cas1 proteins was inferred by using the Maximum Likelihood method. The bootstrap consensus tree inferred from 1000 replicates was taken to represent the evolutionary history of the taxa analyzed. Branches corresponding to partitions reproduced in less than 50% bootstrap replicates were collapsed. The percentage of replicate trees in which the associated taxa clustered together in the bootstrap test (1000 replicates) is shown next to the branches. Accession numbers are indicated after species names. Gene sequence in cas operons listed as left to right in (**a**) are as follows: Type I-F *D. gigas* ATCC 19364, *D. hydrothermalis* AM13 and *D. magneticus* S-1: *cas1, cas3, csy1, csy2, csy3, cas6f*; Type 1-E *D. alasekensis* G20: *cas3, cscf, csc2, cas7, cas5, cas6c, cas2, casf*; *D. desulfuricans* ATCC 27774: *cas3, cse7,cas6e, cas7, cas5, cas2,cas1*; Type I-C *D. vulgaris* Hildenborough. *D. vulgaris* Miyazaki and *D. desulfuricans* ND 132: *cas3, cas5, cas8c, cas7, cas4, cas1, cas2*; Atypical *D. gigas* ATCC 19364: *csb1, csb2, cas3*.

abundance of proviruses belonging to the *Myoviridae, Siphoviridae*, and *Podoviridae* families are 69.82%, 22.64%, and 7.7.54%, respectively (Crispim et al. 2018). A couple of the proviruses of *D. vulgaris* Hildenborough have been induced by Mitomycin C or UV light, and the resulting viruses have a 50 nm icosahedral head with a contractile tail (Walker et al. 2006; Seyedirashti et al. 1991; Handley et al. 1973). This *D. vulgaris* Hildenborough phage fails to replicate in *D. desulfuricans* ATCC 13541 (Kamimura and Araki 1989).

A nonplaque-forming virus was isolated from culture supernatant of *D. desulfuricans* ATCC 27774 and this phage contained dsDNA with 13.6 kb (Rapp and Wall 1987). One of the few plaque-forming phages of the SRB is the one isolated from *D. salexigens* (Kamimura and Araki 1989). Viable phages with icosahedral heads and tails were isolated from deep groundwater, and these viruses specifically targeted *D. aespoeensis* (Eydal et al. 2009).

Table 6.7 Distribution of proviruses and CRISPR arrays in *Desulfovibrio* spp

Microorganism	Number of prophage elements in genome[a]	CRISPR arrays (number of spacers)[a]
D. aepoeensis Aspo 2	4	UK[b]
D. africanus Walvis Bay	4	UK
D. alaskensis G20	4	1 (19)
D. desulfuricans ND132	3	1 (98)
D. desulfuricans ATCC 27774	1	1 (29)
D. fairfeldensis CCUG45958	5	UK
D. gigas ATCC19364	3	1 (4)
D. hydrothermalis DSM14728	3	5 (35)
D. indicus J2	2	1 (19)
D. magneticus RS-1	5	1 (30)
D. piezophilus C1TLV30	1	UK
D. piger FI11049	5	UK
D. salexigens DSM2638	1	UK
D. vulgaris DP4	6	1 (44)
D. vulgaris Hildenborough	8	1 (27)
D. vulgaris Miyazaki	4	1 (54)
D. vulgaris RCH1	7	1 (27)

[a]Source is Crispim et al. (2018)
[b]*UK* unknown

6.6 Perspective

Great advances in the understanding of the cellular and metabolic processes of SRP have occurred in the last few years with the application of genomics, transcriptomics, proteomics, and biochemical analysis. While *D. vulgaris* Hildenborough has been the model organism for numerous studies, more recent studies involving a diversity of sulfate reducers are providing an understanding of those microorganisms that have the phenotypic characteristic of using sulfate as the electron acceptor. While lactate is the hallmark energy and carbon source for the SRP, each species has the metabolic capability of using a few different carbon compounds and collectively; the SRP use a broad range of substrates which results in their wide distribution in the environment. Many of the metabolic pathways and cycles associated with heterotrophic microorganism are also found in the SRP; however, based on genomic and transcriptomics these pathways and cycles are used for biosynthesis and are often incomplete. Survival of SRP cells under conditions where energy production is low has resulted in obtaining supplemental energy for reactions associated with CO, formate, and H_2 cycling. Numerous regulators for transcription are being found in the SRP and the mechanism of transcription regulation appears similar to that found throughout biology. The details of metabolism and transcription are unique to the SRP and this accounts for their persistence in changing environments. Stress response from exposure to pH extremes, O_2, temperature changes, and inorganic nitrogen compounds involves a relatively large number of genes; and these responses allow the SRP to persist in the environmental changes that continuously occur. Cellular activities in biofilm formation and syntrophic partnerships are attributed to expression of specific genes. While horizontal gene transfer in SRP is considered to be important to increase cell vitality, the presence of numerous proviruses in SRP genomes indicates a presence of potential vehicles for DNA exchange while CRISPR genes would provide a record of previous provirus encounters. Further studies are needed to fully develop the cellular activities of SRP.

References

Aeckersberg F, Bak F, Widdel F (1991) Anaerobic oxidation of saturated hydrocarbons to CO_2 by a new type of sulfate-reducing bacteria. Arch Microbiol 156:5–14

Akagi JM, Jackson G (1967) Degradation of glucose by proliferating cells of *Desulfotomaculum nigrificans*. Appl Microbiol 15:1427–1430

Almendra MJ, Brondino CD, Gavel O, Pereira AS, Tavares P, Bursakov S et al (1999) Purification and characterization of a tungsten-containing formate dehydrogenase from *Desulfovibrio gigas*. Biochem 38:16366–16372

Avidan O, Kaltageser E, Pechatnikov I, Wexler HM, Shainskaya A, Nitzan Y (2008) Isolation and characterization of porins from *Desulfovibrio piger* and *Bilophila wadsworthia*: structure and gene sequencing. Arch Microbiol 190:641–650

Badziong W, Thauer RK (1980) Vectorial electron transport in *Desulfovibrio vulgaris* (Marburg) growing on hydrogen plus sulfate as sole energy source. Arch Microbiol 125:167–174

Badziong W, Ditter B, Thauer RK (1979) Acetate and carbon dioxide assimilation by *Desulfovibrio vulgaris* (Marburg), growing on hydrogen and sulfate as sole energy source. Arch Microbiol 123:301–305

Baena S, Fardeau M-L, Labat M, Ollivier B, Garcia J-L, Patel BKC (1998) *Desulfovibrio aminophilus* sp. nov., a novel amino acid degrading and sulfate reducing bacterium from an anaerobic dairy wastewater lagoon. Syst Appl Microbiol 21:498–504

Bak F, Widdel F (1986) Anaerobic degradation of phenol and phenol derivatives by *Desulfobacterium phenolicum* sp. nov. Arch Microbiol 146:177–180

Baker FD, Papiska HR, Campbell LL (1962) Choline fermentation by *Desulfovibrio desulfuricans*. J Bacteriol 84:973–978

Balk M, Altinbas M, Rijpstra WIC, Damste JSS, Stams AJM (2008) *Desulfatirhabdium butyrativorans* gen. nov., sp. nov., a butyrate-oxidizing, sulfate-reducing bacterium isolated from an anaerobic bioreactor. Int J Syst Evol Microbiol 58:110–115

Barton LL, Fauque GD (2009) Biochemistry, physiology and biotechnology of sulfate-reducing bacteria. Adv Appl Microbiol 68:41–98

Barton LL, LeGall J, Peck HD Jr (1970) Phosphorylation coupled to oxidation of hydrogen with fumarate in extracts of the sulfate reducing bacterium, Desulfovibrio gigas. Biochem Biophys Res Commun 41:1036–1042

Barton LL, Fardeau M-L, Fauque GD (2014) Hydrogen sulfide: a toxic gas produced by dissimilatory sulfate and sulfur reduction and consumed by microbial oxidation. In: Sigel A, Sigel H, Sigel RKO (eds) Metal ions in life sciences. Springer Science & Business Media B.V., Dordrecht, pp 237–278

Beeder J, Torsvik T, Lien T (1995) *Thermodesulforhabdus norvegicus* gen. nov., sp. nov., a novel thermophilic sulfate-reducing bacterium from oil field water. Arch Microbiol 164:331–336

Bender KS, Yen HCB, Hemme CL, Yang Z, He Z, He Q et al (2007) Analysis of a ferric uptake regulator (Fur) mutant of *Desulfovibrio vulgaris* Hildenborough. Appl Environ Microbiol 73: 5389–5400

Boone DR, Bryant MP (1980) Propionate-degrading bacterium, *Syntrophobacter wolinii* sp. nov. gen. nov., from methanogenic ecosystems. Appl Environ Microbiol 40:626–632

Boopathy R (2007) Anaerobic metabolism of nitroaromatic compounds and bioremediation of explosives by sulphate-reducing bacteria. In: Barton LL, Hamilton WA (eds) Sulphate-reducing bacteria. Cambridge University Press, Cambridge, pp 483–502

Boopathy R, Daniels L (1991) Isolation and characterization of a furfural degrading sulfate-reducing bacterium from an anaerobic digester. Curr Microbiol 23:327–332

Brandis-Heep A, Gebhardt NA, Thauer RK, Widdel F, Pfennig N (1983) Anaerobic acetate oxidation to CO2 by *Desulfobacter postgatei*. 1. Demonstration of all enzymes required for the operation of the citric acid cycle. Arch Microbiol 136:222–229

Brileya KA, Camilleri LB, Zane GM, Wall JD, Fields MW (2014) Biofilm growth mode promotes maximum carrying capacity and community stability during product inhibition syntrophy. Front Microbiol 5:693. https://doi.org/10.3389/fmicb.2014.00693

Brioukhanov AL, Durand M-C, Dolla A, Aubert C (2010) Response of *Desulfovibrio vulgaris* Hildenborough to hydrogen peroxide: enzymatic and transcriptional analyses. FEMS Microbiol Lett 310:175–181

Brondino CD, Passeggi MC, Caldeira J, Almendra MJ, Feio MJ, Moura JJG, Moura I (2004) Incorporation of either molybdenum or tungsten into formate dehydrogenase from *Desulfovibrio alaskensis* NCIMB 13491: EPR assignment of the proximal iron-sulfur cluster to the pterin cofactor in formate dehydrogenases from sulfate-reducing bacteria. J Biol Inorg Chem 9:145–151

Brown MS, Akagi JM (1966) Purification of acetokinase from *Desulfovibrio desulfuricans*. J Bacteriol 92:1273–1274

Brune G, Schoberth SM, Sahm H (1983) Growth of a strictly anaerobic bacterium on furfural (2-furaldehyde). Appl Environ Microbiol 46:1187–1192

Cadby I, Ibrahim SA, Faulkner M, Lee DJ, Browning D, Busby SJ et al (2016) Regulation, sensory domains and roles of the two *Desulfovibrio desulfuricans* ATCC 27774 Crp, family transcription factors, Hpc1 and HpcR2, in response to nitrosative stress. Mol Microbiol 102:1120–1137

Cadby I, Faulkner M, Chèneby J, Long J, Van Helden J, Dolla A, Cole JA (2017) Coordinated response of the *Desulfovibrio desulfuricans* 27774 transcriptome to nitrate, nitrite and nitric oxide. Sci Rep 7:16228. https://doi.org/10.1038/s41598-017-16403-4

Cao J, Zhang G, Mao Z-S, Li Y, Fang Z, Yang C (2012) Influence of electron donors on the growth and activity of sulfate-reducing bacteria. Int J Miner Process 106–109:58–64

Casadio R, Fariselli P, Finocchiaro G, Martelli PL (2003) Fishing new proteins in the twilight zone of genomes: the test case of outer membrane proteins in Escherichia coli K12, Escherichia coli O157:H7, and other Gram-negative bacteria. Protein Sci 12:1158–1168

Chabrière E, Charon M-H, Volbeda A, Pieulle L, Hatchikian EC, Fontecilla-Camps J-C (1999) Crystal structure of pyruvate: ferredoxin oxidoreductase, a central enzyme in anaerobic metabolism, and its complex with pyruvate. Nat Struct Biol 6:182–190

Chen S, Liu X, Dong X (2005) *Syntrophobacter sulfatireducens* sp. nov., a novel syntrophic, propionate-oxidizing bacterium isolated from UASB reactors. Int J Syst Evol Microbiol 55: 1319–1324

Chhabra SR, He Q, Huang KH, Gaucher SP, Alm EJ, He Z et al (2006) Global analysis of heat shock response in *Desulfovibrio vulgaris* Hildenborough. J Bacteriol 188:1817–1828

Christensen GA, Zane GM, Kazakov AE, Li X, Rodionov DA, Novichkov PS et al (2015) Rex (encoded by DVU_0916) in *Desulfovibrio vulgaris* Hildenborough is a repressor of sulfate adenylyl transferase and is regulated by NADH. J Bacteriol 197:29–39

Clark ME, He Q, He Z, Huang KH, Alm EJ, Wan XF et al (2006) Temporal transcriptomic analysis as Desulfovibrio vulgaris Hildenborough transitions into stationary phase during electron donor depletion. Appl Environ Microbiol 72:5578–5588

Clark ME, He Z, Redding AM, Joachimiak MP, Keasling JD, Zhou JZ, Arkin AP, Mukhopadhyay A, Fields MW (2012) Transcriptomic and proteomic analyses of *Desulfovibrio vulgaris* biofilms: carbon and energy flow contribute to the distinct biofilm growth state. BMC Genomics 13:138. https://doi.org/10.1186/1471-2164-13-138

Cord-Ruwisch R, Garcia JL (1985) Isolation and characterization of an anaerobic benzoate-degrading spore-forming sulfate-reducing bacterium *Desulfotomaculum sapomandens* sp. nov. FEMS Microbiol Lett 29:325–330

Costa C, Teixeira M, LeGall J, Moura JJG, Moura I (1997) Formate dehydrogenase from *Desulfovibrio desulfuricans* ATCC 27774: isolation and spectroscopic characterization of the active sites (heme, iron-sulfur centers and molybdenum). J Biol Inorg Chem 2:198–208

Crain AV, Broderick BJ (2013) Flavodoxin cofactor binding induces structural changes that are required for protein-protein interactions with $NADP^+$ oxidoreductase and pyruvate formate-lyase activating enzyme. Biochim Biophys Acta 1834:2512–2519

Crispim JS, Dias RS, Vidigal PMP, de Sousa MP, da Silva CC, Santana MF, de Paula SO (2018) Screening and characterization of prophages in *Desulfovibrio* genomes. Sci Rep 8:9273. https://doi.org/10.1038/s41598-018-27423-z

Crispim JS, Dias RS, Laguardia CN, Araújo LC, da Silva JD, Vidigal PMP et al (2019) *Desulfovibrio alaskensis* prophages and their possible involvement in the horizontal transfer of genes by outer membrane vesicles. Gene 703:50–57

Cypionka H (1995) Solute transport and cell energetics. In: Barton LL (ed) Sulfate-reducing bacteroa: biotechnology handbook. Plenum Publishing, London, pp 151–184

Czechowski MH, Rossmoore HW (1980) Factors affecting Desulfovibrio desulfuricans lactate dehydrogenase. Dev Ind Microbiol 21:404–410

Czechowski MH, Rossmoore MH (1990) Purification and partial characterization of a D(-)-lactate dehydrogenase from *Desulfovibrio desulfuricans* (ATCC 7757). J Ind Microbiol 6:117–122

da Silva SM, Pimentel C, Valente FMA, Rodrigues-Pousada C, Pereira IAC (2011) Tungsten and molybdenum regulation of formate dehydrogenase expression in *Desulfovibrio vulgaris* Hildenborough. J Bacteriol 193:2909–2917

da Silva SM, Voordouw J, Leitão C, Martins M, Voordouw G, Pereira IA (2013) Function of formate dehydrogenases in *Desulfovibrio vulgaris* Hildenborough energy metabolism. Microbiology 159:1760–1769

Daumas S, Cordruwisch R, Garcia JL (1988) *Desulfotomaculum geothermicum* sp. nov., a thermophilic, fatty acid-degrading, sulfate-reducing bacterium isolated with H2 from geothermal ground-water. Antonie Van Leeuwenhoek J Microb *54*:165–178

Davidova I, Hicks MS, Fedorak PM, Suflita JM (2001) The influence of nitrate on microbial processes in oil industry production waters. J Ind Microbiol Biotechnol 27:80–86

Davydova MN, Tarasova NB (2005) Carbon monoxide inhibits superoxide dismutase and stimulates reactive oxygen species production by *Desulfovibrio desulfuricans* 1388. Anaerobe 11: 335–338

Davydova M, Sabirova R, Vylegzhanina N, Tarasova N (2004) Carbon monoxide and oxidative stress in *Desulfovibrio desulfuricans* B-1388. J Biochem Mol Toxicol 18:61–114

De León KB, Zane GM, Trotter VV, Krantz GP, Arkin AP, Butland GP et al (2017) Unintended laboratory-driven evolution reveals genetic requirements for biofilm formation by *Desulfovibrio vulgaris* Hildenborough. mBio 8:e01696–e01617. https://doi.org/10.1128/mBio.01696-17

Drzyzga O, Kuever J, Blotevogel K-H (1993) Complete oxidation of benzoate and 4-hydroxybenzoate by a new sulfate-reducing bacterium resembling *Desulfoarculus*. Arch Microbiol 159:109–113

El Houari A, Ranchou-Peyruse M, Ranchou-Peyruse A, Dakdaki A, Guignard M, Idouhammou L et al (2017) *Desulfobulbus oligotrophicus* sp. nov., a sulfate-reducing and propionate-oxidizing bacterium isolated from a municipal anaerobic sewage sludge digester. Int J Syst Evol Microbiol 67:275–281

Emmerhoff OJ, Klenk H-P, Birkelanda N-K (1998) Characterization and sequence comparison of temperature-regulated chaperonins from the hyperthermophilic archaeon *Archaeoglobus fulgidus*. Gene 215:431–438

Esnault G, Caumette P, Garcia J-L (1988) Characterization of *Desulfovibrio giganteus* sp. nov., a sulfate-reducing bacterium isolated from a brackish coastal lagoon. Syst Appl Microbiol 10: 147–151

Eydal HSC, Jägevall S, Hermansson M, Pedersen K (2009) Bacteriophage lytic to *Desulfovibrio aespoeensis* isolated from deep groundwater. ISEM J 3:1139–1147

Fardeau M-L, Ollivier B, Patel BKC, Dwivedi P, Ragot M, Garcia J-L (1995) Isolation and characterization of a thermophilic sulfate-reducing bacterium, *Desulfotomaculum thermosapovorans* sp. nov. Int J Syst Bacteriol 45:218–221

Fareleira P, LeGall J, Xavier AV, Santos H (1997) Pathways for utilization of carbon reserves in *Desulfovibrio gigas* under fermentative and respiratory conditions. J Bacteriol 179:3972–3980

Fauque GD, Barton LL (2012) Hemoproteins in dissimilatory sulfate- and sulfur- reducing prokaryotes. Adv Microb Physiol 60:1–90

Fiévet A, My L, Cascales E, Ansaldi M, Pauleta SR, Moura I et al (2021) The anaerobe-specific orange protein pomplex of *Desulfovibrio vulgaris* Hildenborough is encoded by two divergent operons coregulated by σ^{54} and a cognate transcriptional regulator. J Bacteriol 193:3207–3219

Figueiredo MCO, Lobo SAL, Sousa SH, Pereira FP, Wall JD, Nobre LS, Saraiva LM (2013) Hybrid cluster proteins and flavodiiron proteins afford protection to *Desulfovibrio vulgaris* upon macrophage infection. J Bacteriol 195:2684–2690

Folkerts M, Ney U, Kneifel H, Stackebrandt E, Witte EG, Förstel H, Schoberth SM, Sahm H (1989) *Desulfovibrio furfuralis* sp. nov., a furfural degrading strictly anaerobic bacterium. Syst Appl Microbiol 11:161–169

Fournier M, Aubert C, Dermoun Z, Durand M-C, Moinier D, Dolla A (2006) Response of the anaerobe *Desulfovibrio vulgaris* Hildenborough to oxidative conditions: proteome and transcript analysis. Biochimie 88:85–94

Friedrich MW (2002) Phylogenetic analysis reveals multiple lateral transfers for adenosine-5-′-phosphosulfate reductase genes among sulfate-reducing microorganisms. J Bacteriol 184: 278–289

Galushko AS, Rozanova EP (1991) *Desulfobacterium cetonicum* sp. nov.: a sulfate-reducing bacterium which oxidizes fatty acids and ketones. Mikrobiologiya 60:102–107

Garczarek F, Dong M, Typke D, Witkowska HE, Hazen TC, Nogales E, Biggin MD, Glaeser RM (2007) Octomeric pyruvate-ferredoxin oxidoreductase from *Desulfovibrio vulgaris*. J Struct Biol 159:9–18

George GN, Pickering IJ, Yu EY, Prince RC, Bursakov SA, Gavel OY, Moura I, Moura JJG (2000) A novel protein-bound copper – molybdenum cluster. J Am Chem Soc 122:8321–8322

Grein F, Ramos AR, Venceslau SS, Pereira IAC (2013) Unifying concepts in anaerobic respiration: insights from dissimilatory sulfur metabolism. Biochem Biophys Acta 1827:145–160

Hadj-Saïd J, Pandelia ME, Léger C, Fourmond V, Dementin S (2015) The carbon monoxide dehydrogenase from *Desulfovibrio vulgaris*. Biochim Biophys Acta 1847:1574–1583

Handley J, Adams V, Akagi JM (1973) Morphology of bacteriophage-like particles from *Desulfovibrio vulgaris* Hildenborough. J Bacteriol 115:1205–1207

Hanninga HJ, Gottschal JC (1987) Properties of *Desulfovibrio carbinolicus* sp. n ov. and other sulfate-reducing bacteria isolated from an anaerobic-purification plant. Appl Environ Microbiol 53:802–809

Hansen TA (1995) NAD(P)-independent lactate dehydrogenase from sulfate-reducing prokaryotes. Meth Enzymol 243:21–23

Harmsen HJM, Kuijk BLM, Plugge CM, Akkermans ADL, De Vos WM, Stams AJM (1998) *Syntrophobacter fumaroxidans* sp. nov., a syntrophic propionate- degrading sulfate-reducing bacterium. Int J Syst Bacteriol 48:1383–1387

Haveman SA, Brunelle V, Voordouw JK, Voordouw G, Heidelberg JF, Rabus R (2003) Gene expression analysis of energy metabolism mutants of *Desulfovibrio vulgaris* Hildenborough indicates an important role for alcohol dehydrogenase. J Bacteriol 185:4345–4353
Haveman SA, Greene EA, Stilwell CP, Voordouw JK, Voordouw G (2004) Physiological and gene expression analysis of inhibition of *Desulfovibrio vulgaris* Hildenborough in nitrite. J Bacteriol 186:7944–7950
Haynes TS, Klemm DJ, Ruocco JJ, Barton LL (1995) Formate dehydrogenase activity in cells and outer membrane blebs of Desulfovibrio gigas. Anaerobe 1:175–182
He SH, DerVartanian DV, LeGall J (1986) Isolation of fumarate reductase from *Desulfovibrio multispirans*, a sulfate-reducing bacterium. Biochem Biophys Res Commun 135:1000–1007
He Q, Huang KH, He Z, Alm EJ, Fields MW, Hazen TC, Arkin AP, Wall JD, Zhou J (2006) Energetic consequences of nitrite stress in Desulfovibrio vulgaris Hildenborough inferred from global transcriptional analysis. Appl Environ Microbiol *72*:4370–4381
He Q, He Z, Joyner DC, Joachimiak M, Price MN, Yang ZK et al (2010) Impact of elevated nitrate on sulfate-reducing bacteria: a comparative study of *Desulfovibrio vulgaris*. ISME J 4:1386–1397
Hedderich R, Forzi L (2005) Energy-converting [NiFe] hydrogenases: more than just H_2 activation. J Mol Microbiol Biotechnol 10:92–104
Heidelberg JF, Seshadri R, Haveman SA, Hemme CL, Paulsen IT, Kolonay JF et al (2004) The genome sequence of the anaerobic, sulfate-reducing bacterium *Desulfovibrio vulgaris* Hildenborough. Nat Biotechnol 22:554–559
Helmann JD (2002) The extracytoplasmic function (ECF) sigma factors. Adv Microbial Physiol 46: 47–110
Hengge-Aronis R (2000) The general stress response in *Escherichia coli*. In: Storz G, Hengge-Aronis R (eds) Bacterial stress responses. ASM Press, Washington, DC, pp 161–178
Hensgrens CMH, Vonck J, van Beeumen J, van Bruggen EFJ, Hansen TA (1993) Purification and characterization of an oxygen-labile NAD-dependent alcohol dehydrogenase from *Desulfovibrio gigas*. J Bacteriol 175:2859–2863
Henstra AM, Dijkema C, Stams AJM (2007) *Archaeoglobus fulgidus* couples CO oxidation to sulfate reduction and acetogenesis with transient formate accumulation. Environ Microbiol 9: 1836–1841
Hernandez-Eugenio G, Fardeau M-L, Patel BKC, Macarie H, Garcia J-L, Ollivier B (2000) *Desulfovibrio mexicanus* sp. nov., a sulfate-reducing bacterium isolated from an upf low anaerobic sludge blanket (UASB) reactor treating cheese wastewaters. Anaerobe 6:305–312
Hocking WP, Stokke R, Roalkvam I, Steen IH (2014) Identification of key components in the energy metabolism of the hyperthermophilic sulfate-reducing archaeon *Archaeoglobus fulgidus* by transcriptome analyses. Front Microbiol 5:95. https://doi.org/10.3389/fmicb.2014.00095
Hoehler TM, Alperin MJ, Albert DB, Martens CS (2001) Apparent minimum free energy requirements for methanogenic archaea and sulfate-reducing bacteria in an anoxic marine sediment. FEMS Microbiol Ecol 38:33–41
Huisingh J, McNeill JJ, Matrone G (1974) Sulfate reduction by a *Desulfovibrio* species isolated from sheep rumen. Appl Microbiol 28:489–497
Iizuka H, Okazaki H, Seto N (1969) A new sulfate-reducing bacterium isolated from Antarctica. J Gen Appl Microbiol 15:11–18
Imhoff-Stuckle D, Pfennig N (1983) Isolation and characterization of a nicotinic acid-degrading sulfate-reducing bacterium, *Desulfococcus niacini* sp nov. Arch Microbiol 136:194–198
Jenneman GE, McInerney MJ, Knapp RM (1986) Effect of nitrate on biogenic sulfide production. Appl Environ Microbiol 51:1205–1211
Johnston S, Lin S, Lee P, Caffrey SM, Wildschut J, Voordouw JK et al (2009) A genomic island of the sulfate-reducing bacterium *Desulfovibrio vulgaris* Hildenborough promotes survival under stress conditions while decreasing the efficiency of anaerobic growth. Environ Microbiol 11: 981–991

Jolkver E, Emer D, Ballan S, Krämer R, Eikmanns BJ, Marin K (2009) Identification and characterization of a bacterial transport system for the uptake of pyruvate, propionate and acetate in *Corynebacterium glutamicum*. J Bacteriol 191:940–948

Junier P, Junier T, Podell S, Sims DR, Detter JC, Lykidis A et al (2010) The genome of the Gram-positive metal- and sulfate-reducing bacterium *Desulfotomaculum reducens* strain MI-1. Environ Microbiol 12:2738–2754

Kamimura K, Araki M (1989) Isolation and characterization of a bacteriophage lytic for *Desulfovibrio salexigens*, a salt-requiring, sulfate-reducing bacterium. Appl Environ Microbiol 55:645–648

Kazakov AE, Rodionov DA, Price MN, Arkin AP, Dubchak I, Novichkov PS (2013a) Transcription factor family-based reconstruction of singleton regulons and study of the Crp/Fnr, ArsR, and GntR families in *Desulfovibrionales* genomes. J Bacteriol 195:29–38

Kazakov AE, Rajeev L, Luning EG, Zane GM, Siddartha K, Rodionov DA et al (2013b) New family of tungstate-responsive transcriptional regulators in sulfate-reducing bacteria. J Bacteriol 195:4466–4475

Kazakov AE, Rajeev L, Chen A, Luning EG, Dubchak I, Mukhopadhyay A, Novichkov PS (2015) σ^{54}-dependent regulome in *Desulfovibrio vulgaris* Hildenborough. BMC Genomics 16:919. https://doi.org/10.1186/s12864-015-2176-y

Keller KL, Wall JD (2011) Genetics and molecular biology of the electron flow for sulfate respiration in *Desulfovibrio*. Front Microbiol 2:135. https://doi.org/10.3389/fmicb.2011.00135

Klein M, Friedrich RAJ, Hugenholtz P, Fishbain S, Abicht H et al (2001) Multiple lateral transfers of dissimilatory sulfite reductase genes between major lineages of sulfate-reducing prokaryotes. J Bacteriol 183:6028–6035

Klemps R, Cypionka H, Widdel F, Pfennig N (1985) Growth with hydrogen, and further physiological characteristics of *Desulfotomaculum* species. Arch Microbiol 143:203–208

Klenk H-P, Clayton RA, Tomb J-F, White O, Nelson KE, Ketchum KA et al (1997) The complete genome sequence of the hyperthermophilic, sulphate-reducing archaeon *Archaeoglobus fulgidus*. Nature 390:364–370

Knappe J, Neugebauer FA, Blaschkowski HP, Gänzler M (1984) Post-translational activation introduces a free radical into pyruvate formate-lyase. Proc Natl Acad Sci U S A 81:1332–1335

Knoblauch C, Sahm K, Jørgensen BB (1999) Psychrophilic sulfate-reducing bacteria isolated from permanently cold Arctic marine sediments: description of *Desulfofrigus oceanense* gen. nov., sp. nov., *Desulfofrigus fragile* sp. nov., *Desulfofaba gelida* gen. nov., sp. nov., *Desulfotalea psychrophila* gen. nov., sp. nov., and *Desulfotalea arctica* sp. nov. Int J Syst Bacteriol 49:1631–1643

Könneke M, Kuever J, Galushko A, Jørgensen BB (2013) *Desulfoconvexum algidum* gen. nov., sp. nov., a psychrophilic sulfate-reducing bacterium isolated from a permanently cold marine sediment. Int J Syst Bacteriol 63:959–964

Korte HL, Fels SR, Christensen GA, Price MN, Kuehl JV, Zane GM et al (2014) Genetic basis for nitrate resistance in *Desulfovibrio* strains. Front Microbiol. https://doi.org/10.3389/fmicb.2014.00153

Kuever J, Rainey FA, Hippe H (1999) Description of *Desulfotomaculum* sp. Groll as *Desulfotomaculum gibsoniae* sp. nov. Int J Syst Bacteriol 49:1801–1808

Kuever J, Konneke M, Galushko A, Drzyzga O (2001) Reclassification of *Desulfobacterium phenolicum* as *Desulfobacula phenolica* comb. nov. and description of strain SaxT as *Desulfotignum balticum* gen. nov., sp. nov. Int J Syst Evol Microbiol 51:171–177

Kunow J, Linder D, Thauer RK (1995) Pyruvate: ferredoxin oxidoreductase from the sulfate-reducing Archaeoglobus fulgidus: molecular composition, catalytic properties, and sequence alignments. Arch Microbiol 163:21–28

Kushkevych IV (2014) Acetate kinase activity and kinetic properties of the enzyme in *Desulfovibrio piger* Vib-7 and *Desulfomicrobium* sp. rod-9 intestinal bacterial strains. Open Microbiol J 8:138–143. https://doi.org/10.2174/1874285801408010138

Labes A, Schönheit P (2001) Sugar utilization in the hyperthermophilic, sulfate-reducing archaeon *Archaeoglobus fulgidus* strain 7324: starch degradation to acetate and CO_2 via a modified Embden-Meyerhof pathway and acetyl-CoA synthetase (ADP-forming). Arch Microbiol 176: 329–338

LaPaglia C, Hartzell PL (1997) Stress-induced production of biofilms in the hyperthermophile *Archaeoglobus fulgidus*. Appl Environ Microbiol 63:3158–3163

Le Gall J (1963) A new species of *Desulfovibrio*. J Bacteriol 86:1120

Lemos RS, Gomes CM, LeGall J, Xavier AV, Teixeira M (2002) The quinol: fumarate oxidoreductase from the sulphate reducing bacterium *Desulfovibrio gigas:* spectroscopic and redox studies. J Bioenerg Biomembr 34:21–30

Lien T, Madsen M, Steen IH, Gjerdevik K (1998) *Desulfobulbus rhabdoformis* sp. nov., a sulfate reducer from a water-oil separation system. Int J Syst Bacteriol 48:469–474

Liu Y, Karnauchow TM, Jarrell KF, Balkwill DL, Drake GR, Ringelberg D, Clarno R, Boone DR (1997) Description of two new thermophilic *Desulfotomaculum* spp., *Desulfotomaculum putei* sp. nov., from a deep terrestrial subsurface, and *Desulfotomaculum luciae* sp. nov., from a hot spring. Int J Syst Bacteriol 47:615–621

Love CA, Patel BKC, Nichols PD, Stackebrandt E (1993) *Desulfotomaculum australicum*, sp. nov., a thermophilic sulfate-reducing bacterium isolated from the Great Artesian Basin of Australia. Syst Appl Microbiol 16:244–251

Lupton FS, Conrad R, Zeikus JG (1984) CO metabolism of *Desulfovibrio vulgaris* Madison physiological function in the absence or presence of exogenous substrates. FEMS Microbiol Lett 23:263–268

Magot M, Caumette P, Desperrier JM, Matheron R, Dauga C, Grimont F, Carreau L (1992) *Desulfovibrio longus* sp. nov., a sulfate-reducing bacterium isolated from an oil-producing well. Int J Syst Bacteriol 42:398–403

Maiti BK, Moura I, Moura JJG, Pauleta SR (2016) The small iron-sulfur protein from the ORP operon binds a [2Fe-2S] cluster. Biochim Biophys Acta 1857:1422–1429

Makarova KS, Haft DH, Barrangou R, Brouns SJ, Charpentier E, Horvath P et al (2011) Evolution and classification of the CRISPR-Cas systems. Nat Rev Microbiol 9:467–477

Marshall WF, Young KD, Swaffer M, Wood E, Nurse P, Kimura A et al (2012) What determines cell size? BMC Biol 10:101. https://doi.org/10.1186/1741-7007-10-101

Martins M, Mourato C, Pereira IAC (2015) *Desulfovibrio vulgaris* growth coupled to formate-driven H_2 production. Environ Sci Technol 49:14655–14662

Martins M, Mourato C, Morais-Silva FO, Rodrigues-Pousada C, Voordouw G, Wall JD, Pereira IA (2016) Electron transfer pathways of formate-driven H_2 production in *Desulfovibrio*. Appl Microbiol Biotechnol 100:8135–8146

Miller JDA, Wakerley DS (1966) Growth of sulphate-reducing bacteria by fumarate dismutation. J Gen Microbiol 43:101–107

Möller D, Schauder R, Fuchs G, Thauer RK (1987) Acetate oxidation to CO_2 via a citric acid cycle involving an ATP-citrate lyase: a mechanism for the synthesis of ATP via substrate level phosphorylation in Desulfobacter postgatei growing on acetate and sulfate. Arch Microbiol 148:202–207

Morais-Silva FO, Rezende AM, Pimentel C, Santos CI, Clemente C, Ana Varela–Raposo A et al (2014) Genome sequence of the model sulfate reducer *Desulfovibrio gigas*: a comparative analysis within the *Desulfovibrio* genus. Microbiologyopen 3:513–530

Mukhopadhyay A, He Z, Alm EJ, Arkin AP, Baidoo EE, Borglin SC et al (2006) Salt stress in Desulfovibrio vulgaris Hildenborough: an integrated genomics approach. J Bacteriol *188*:4068–4078

Mukhopadhyay A, Redding AM, Joachimiak MP, Arkin AP, Borglin SE, Dehal PS et al (2007) Cell-wide responses to low-oxygen exposure in *Desulfovibrio vulgaris* Hildenborough. J Bacteriol 189:5996–6010

Mussmann M, Richter M, Lombardot T, Meyerdierks A, Kuever J, Kube M et al (2005) Clustered genes related to sulfate respiration in uncultured prokaryotes support the theory of their concomitant horizontal transfer. J Bacteriol 187:7126–7137

Nakazawa H, Arakaki A, Narita-Yamada S, Yashiro I, Jinno K, Aoki N et al (2009) Whole genome sequence of Desulfovibrio magneticus strain RS-1 revealed common gene clusters in magnetotactic bacteria. Genome Res 19:1801–1808

Nanninga HJ, Gottschal JC (1987) Properties of *Desulfovibrio carbinolicus* sp. nov. and other sulfate-reducing bacteria Isolated from an anaerobic-purification plant. Appl Environ Microbiol 53:802–809

Nazina TN, Ivanova AE, Kanchaveli LP, Rozanova EP (1988) A new sporeforming thermophilic methylotrophic sulfate-reducing bacterium, *Desulfotomaculum kuznetsovii* sp. nov. Mikrobiologia (Russian) 57:823–827

Nielsen JT, Liesack W, Finster K (1999) Desulfovibrio zosterae sp. nov., a new sulfate reducer isolated from surface-sterilized roots of the seagrass Zostera marina. Int J Syst Bacteriol 49:859–865

Nies DH (2004) Incidence and function of sigma factors in *Ralstonia metallidurans* and other bacteria. Arch Microbiol 181:255–268

Nilsen RK, Torsvik T, Lien T (1996) *Desulfotomaculum thermocisternum* sp. nov., a sulfate reducer isolated from a hot North Sea oil reservoir. Int J Syst Bacteriol 46:397–402

Noguera DR, Brusseau GA, Rittmann BE, Stahl DA (1998) A unified model describing the role of hydrogen in the growth of *Desulfovibrio vulgaris* under different environmental conditions. Biotechnol Bioeng 59:732–746

Novichkov PS, Li X, Kuehl JV, Deutschbauer AM, Arkin AP, Price MN, Rodionov DA (2014) Control of methionine metabolism by the SahR transcriptional regulator in *Proteobacteria*. Environ Microbiol 16:1–8

Ogata M, Yagi T (1986) Pyruvate dehydrogenase and the path of lactate degradation in *Desulfovibrio vulgaris* Miyazaki F. J Biochem 100:311–318

Ogata M, Arihara K, Yagi T (1981) D-lactate dehydrogenase of *Desulfovibrio vulgaris*. J Biochem 89:1423–1431

Ollivier B, Cord-Ruwisch R, Hatchikian EC, Garcia LL (1988) Characterization of *Desulfovibrìo fructosovorans* sp. nov. Arch Microbiol 149:447–450

Otrelo-Cardoso AR, Nair R, Correia MAS, Rivas MG, Santos-Silva T (2014) TupA: a tungstate binding protein in the periplasm of *Desulfovibrio alaskensis* G20. Int J Mol Sci 15:11783–11798

Otrelo-Cardoso AR, Nair RR, Correia MAS, Correia Cordeiro RS, Panjkovich A, Svergun DI, Santos-Silva T, Rivas MG (2017) Highly selective tungstate transporter protein TupA from *Desulfovibrio alaskensis* G20. Sci Rep 7:5798. https://doi.org/10.1038/s41598-017-06133-y

Otwell AE, Callister SJ, Zink EM, Smith RD, Richardson RE (2016) Comparative proteomic analysis of *Desulfotomaculum reducens* MI-1: insights into the metabolic versatility of a Gram-positive sulfate- and metal-reducing bacterium. Front Microbiol 7:191. https://doi.org/10.3389/fmicb.2016.00191

Pagani I, Lapidus A, Nolan M, Lucas S, Hammon N, Deshpande S et al (2011) Complete genome sequence of *Desulfobulbus propionicus* type strain (1pr3^T). Stand Genomic Sci 4:100–110

Pankhania IP, Spormann AM, Hamilton WA, Thauer RK (1988) Lactate conversion to acetate, CO_2 and H_2 in cell suspensions of *Desulfovibrio vulgaris (*Marburg): indications for the involvement of an energy driven reaction. Arch Microbiol 150:26–31

Parshina SN, Sipma J, Nakashimada Y, Henstra AM, Smidt H, Lysenko AM et al (2005) *Desulfotomaculum carboxydivorans* sp. nov., a novel sulfate-reducing bacterium capable of growth at 100% CO. Int J Syst Evol Microbiol 55:2159–2165

Parshina SN, Sipma J, Henstra AM, Stams AJM (2010) Carbon monoxide as an electron donor for the biological reduction of sulphate. Int J Microbiol 2010:319527, 9 pages. https://doi.org/10.1155/2010/319527

Peck HD Jr (1962) Comparative metabolism of inorganic sulfur compounds in microorganisms. Bacteriol Rev 26:67–94

Peck HD Jr (1993) Bioenergetic strategies of the sulphate-reducing bacteria. In: Odom JM, Rivers Singleton J (eds) The sulphate-reducing bacteria: contemporary perspectives. Springer-Verlag, New York, pp 41–74

Peck HD Jr, LeGall J (1982) Biochemistry of dissimilatory sulfate reduction. Philos Trans R Soc Lond B Biol Sci 298:443–466

Peck HD Jr, Lissolo T (1988) Assimilatory and dissimilatory sulphate-reduction: enzymology and bioenergetics. In: Cole JA, Ferguson SJ (eds) The nitrogen and sulfur cycles. 42nd Symposium of the Society for General Microbiology. Cambridge University Press, Cambridge, pp 99–132

Pereira IAC, Haveman SA, Voordouw G (2007) Biochemical, genetic and genomic characterization of anaerobic electron transport pathways in sulphate-reducing *Delta proteobacteria*. In: Barton LL, Hamilton WA (eds) Sulphate-reducing bacteria – environmental and engineered systems. Cambridge University Press, Cambridge, pp 215–240

Pereira PM, He Q, Valente FMA, Xavier AV, Zhou J, Pereira IAC, Louro RO (2008a) Energy metabolism in *Desulfovibrio vulgaris* Hildenborough: insights from transcriptome analysis. Antonie Van Leeuwenhoek *93*:347–362

Pereira P, He Q, Xavier A, Zhou J, Pereira I, Louro R (2008b) Transcriptional response of *Desulfovibrio vulgaris* Hildenborough to oxidative stress mimicking environmental conditions. Arch Microbiol 189:451–461

Pereira IAC, Raquel Ramos A, Grein F, Marques MC, da Silva SM, Venceslau SS (2011) A comparative genomic analysis of energy metabolism in sulfate-reducing bacteria and archaea. Front Microbiol 2:69. https://doi.org/10.3389/fmicb.2011.00069

Phadtare S (2012) *Escherichia coli* cold-shock gene profiles in response to overexpression/deletion of CsdA, RNase R and PNPase and relevance to low-temperature RNA metabolism. Genes Cells 17:850–874

Pieulle LB, Guigliarelli M, Asso F, Dole BA, Hatchikian EC (1995) Isolation and characterization of the pyruvate-ferredoxin oxidoreductase from the sulfate-reducing bacterium *Desulfovibrio africanus*. Biochim Biophys Acta 1250:49–59

Pieulle L, Chabrière E, Hatchikian EC, Fontecilla-Camps JC, Charon M-H (1999a) Crystallization and preliminary crystallographic analysis of the pyruvate: ferredoxin oxidoreductase from *Desulfovibrio africanus*. Acta Crystallogr D Biol Crystallogr 55:329–331

Pieulle L, Charon M-H, Bianco P, Bonicel J, Pétillot Y, Hatchikian EC (1999b) Structural and kinetic studies of the pyruvate:ferredoxin oxidoreductase/ferredoxin complex from *Desulfovibrio africanus*. Eur J Biochem 264:500–508

Pikuta E, Lysenko A, Suzina N, Osipov G, Kuznetsov B, Tourova T, Akimenko V, Laurinavichius K (2000) *Desulfotomaculum alkaliphilum* sp. nov., a new alkaliphilic, moderately thermophilic, sulfate-reducing bacterium. Int J Syst Evol Microbiol 50:25–33

Plugge CM, Balk M, Stams AJM (2002) *Desulfotomaculum thermobenzoicum* subsp. *thermosyntrophicum* subsp. nov., a thermophilic, syntrophic, propionate-oxidizing, spore-forming bacterium. Int J Syst Evol Microbiol 52:391–399

Poehlein A, Daniel R, Schink B, Simeonova DD (2013) Life based on phosphite: a genome-guided analysis of *Desulfotignum phosphitoxidans*. BMC Genomics 14:753. http://www.biomedcentral.com/1471-2164/14/753*BMC*

Postgate JR (1952) Growth of sulphate reducing bacteria in sulphate-free media. Research (London) 5:189–190

Postgate JR (1963) Sulfate-free growth of *Clostridium nigrificans*. J Bacteriol 85:1450–1451

Postgate JR (1965) Recent advances in the study of the sulfate-reducing bacteria. Bacteriol Rev 29: 425–441

Postgate JR (1979) The sulphate-reducing bacteria. Cambridge University Press, Cambridge

Postgate JR, Campbell LL (1966) Classification of *Desulfovibrio* species, the nonsporulating sulfate-reducing bacteria. Bacteriol Rev 30:732–737

Price MN, Deutschbauer AM, Kuehl JV, Liu H, Witkowska HE, Arkin AP (2011) Evidence-based annotation of transcripts and proteins in the sulfate-reducing bacterium *Desulfovibrio vulgaris* Hildenborough. J Bacteriol 193:5716–5727

Qatibi AI, Nivière V, Garcia JL (1991) *Desulfovibrio akoholovorarzs* sp. nov., a sulfate-reducing bacterium able to grow on glycerol, 1,2- and 1,3-propanediol. Arch Microbiol 155:143–148

Qi Z, Chen L, Zhang W (2016) Comparison of transcriptional heterogeneity of eight genes between batch *Desulfovibrio vulgaris* biofilm and planktonic culture at a single-cell level. Front Microbiol 7:597. https://doi.org/10.3389/fmicb.2016.00597

Rabus R, Nordhaus R, Ludwig W, Widdel F (1993) Complete oxidation of toluene under strictly anoxic conditions by a new sulfate-reducing bacterium. Appl Environ Microbiol 59:1444–1451

Rabus R, Ruepp A, Frickey T, Rattei T, Fartmann B, Stark M et al (2004) The genome of *Desulfotalea psychrophila*, a sulfate-reducing bacterium from permanently cold Arctic sediments. Environ Microbiol 6:887–902

Rabus R, Venceslau SS, Wöhlbrand L, Voordouw G, Wall JD, Pereira IAC (2015) A post-genomic view of the ecophysiology, catabolism and biotechnological relevance of sulphate-reducing prokaryotes. Adv Microb Physiol 66:56–321

Rajeev L, Luning EG, Dehal PS, Price MN, Arkin AP, Mukhopadhyay A (2011) Systematic mapping of two component response regulators to gene targets in a model sulfate-reducing bacterium. Genome Biol 12:R99

Rajeev L, Hillesland KL, Zane GM, Zhou A, Joachimiak MP, He Z et al (2012) Delection of the *Desulfovibrio vulgaris* carbon monoxide sensor invokes global changes in transcription. J Bacteriol 194:5783–5793

Rajeev L, Chen A, Kazakov AE, Luning EG, Zane GM, Novichkov PS, Wall JD, Mukhopadhyay A (2015) Regulation of nitrite stress response in *Desulfovibrio ulgaris* Hildenborough, a model sulfate-reducing bacterium. J Bacteriol 197:3400–3408

Rapp BJ, Wall JD (1987) Genetic transfer in *Desulfovibrio desulfuricans*. Proc Natl Acad Sci U S A 84:9128–9130

Rath D, Amlinger L, Rath A, Lundgren M (2015) The CRISPR-Cas immune system: biology, mechanisms and applications. Biochimie 117:119–128

Ravcheev DA, Li X, Latif H, Zengler K, Leyn SA, Korostelev YD et al (2012) Transcriptional regulation of central carbon and energy metabolism in bacteria by redox-responsive repressor Rex. J Bacteriol 94:1145–1157

Redding AM, Mukhopadhyay A, Joyner DC, Hazen TC, Keasling JD (2006) Study of nitrate stress in *Desulfovibrio vulgaris* Hildenborough using iTRAQ proteomics. Brief Funct Genomic Proteomic 5:133–143

Rees GN, Patel BK (2001) *Desulforegula conservatrix* gen. nov., sp. nov., a long-chain fatty acid-oxidizing, sulfate-reducing bacterium isolated from sediments of a freshwater lake. Int J Syst Evol Microbiol 51:1911–1916

Rees GN, Harfoot CG, Sheehy AJ (1998) Amino acid degradation by the mesophilic sulfate-reducing bacterium *Desulfobacterium vacuolatum*. Arch Microbiol 169:76–80

Reeves RE, Guthrie JD (1975) Acetate kinase (pyrophosphate). A fourth pyrophosphate-dependent kinase from *Entamoeba histolytica*. Biochem Biophys Res Commun 66:1389–1395

Reichenbecher W, Schink B (1997) *Desulfovibrio inopinatus*, sp. nov., a new sulfate-reducing bacterium that degrades hydroxyhydroquinone (1,2,4-trihydroxybenzene). Arch Microbiol 168: 338–344

Rhie MN, Yoon HE, Oh HY, Zedler S, Unden G, Kim OB (2014) A Na^+-coupled C_4-dicarboxylate transporter (Asuc_0304) and aerobic growth of *Actinobacillus succinogenes* on C_4-dicarboxylates. Microbiology 160:1533–1544

Riederer-Henderson MA, Peck HD Jr (1986a) In vitro requirements for formate dehydrogenase activity from *Desulfovibrio gigas*. Can J Microbiol 32:425–429

Riederer-Henderson MA, Peck HD Jr (1986b) Properties of formate dehydrogenase from *Desulfovibrio gigas*. Can J Microbiol 32:430–435

Rodrigues R, Vicente JB, Félix R, Oliveira S, Teixeira M, Rodrigues-Pousada C (2006) *Desulfovibrio gigas* flavodiiron protein affords protection against nitrosative stress in vivo. J Bacteriol 188:2745–2751

Rohlin L, Trent JD, Salmon K, Kim U, Gunsalus RP, Liao JC (2005) Heat shock response of *Archaeoglobus fulgidus*. J Bacteriol 187:6046–6057

Rosnes JT, Torsvik T, Lien T (1991) Spore-forming thermophilic sulfate-reducing bacteria isolated from North Sea oil field waters. Appl Environ Microbiol 57:2302–2307

Samain E, Dubourguier HC, Albagnac G (1984) Isolation and characterization of *Desuljobulbus elongatus* sp. nov. from a mesophilic industrial digester. Syst Appl Microbiol 5:391–401

Santana M, Crasnier-Mednansky M (2006) The adaptive genome of *Desulfovibrio vulgaris* Hildenborough. FEMS Microbiol Lett 260:127–133

Santos H, Fareleira P, Xavier AV, Chen L, Liu MY, LeGall J (1993) Aerobic metabolism of carbon reserves by the obligate anaerobe *Desulfovibrio gigas*. Biochem Biophys Res Commun 195: 551–557

Sass A, Rutters H, Cypionka H, Sass H (2002) *Desulfobulbus mediterraneus* sp. nov., a sulphate-reducing bacterium growing on mono- and disaccharides. Arch Microbiol 177:468–474

Schauder R, Eikmanns B, Thauer RK, Widdel F, Fuchs G (1986) Acetate oxidation to CO_2 in anaerobic bacteria via a novel pathway not involving reactions of the citric acid cycle. Arch Microbiol 145:162–172

Schauder R, Widdel F, Fuchs G (1987) Carbon assimilation pathways in sulfate-reducing bacteria. II. Enzymes of a reductive citric acid cycle in the autotrophic *Desulfobacter hydrogenophilus*. Arch Microbiol 148:218–225

Schink B (1997) Energetics of syntrophic cooperation in methanogenic degradation. Microbiol Mol Biol Rev 61:262–280

Schnell S, Bak F, Pfennig N (1989) Anaerobic degradation of aniline and dihydroxybenzenes by newly isolated sulfate-reducing bacteria and description of *Desulfobacterium anilini*. Arch Microbiol 152:556–563

Sebban C, Blanchard L, Bruschi M, Guerlesquin F (1995) Purification and characterization of the formate dehydrogenase from Desulfovibrio vulgaris Hildenborough. FEMS Microbiol Lett 133: 143–149

Senez JC (1962) Some considerations on the energetics of bacterial growth. Bacteriol Rev 26:95–105

Seyedirashti S, Wood C, Akagi JM (1991) Induction and partial purification of bacteriophages from *Desulfovibrio vulgaris* (Hildenborough) and *Desulfovibrio desulfuricans* ATCC 13541. J Gen Microbiol 137:1545–1549

Sham Genthner BR, Friedman SD, Devereux R (1997) Reclassification of *Desulfovibrio desulfuricans* Norway 4 as *Desulfomicrobium nowegicum* comb. nov. and confirmation of *Desulfomicrobium escambiense* (corrig., formerly "escambium") as a new species in the genus *Desulfomicrobium*. Int J Syst Bacteriol 47:889–892

Sivakumar K, Scarascia G, Zaouri N, Wang T, Kaksonen AH, Hong P-Y (2019) Salinity-mediated increment in sulfate reduction, biofilm formation, and quorum sensing: a potential connection between quorum sensing and sulfate reduction? Front Microbiol 10:188. https://doi.org/10.3389/fmicb.2019.00188

Sorokin DY, Detkova EN, Muyzer G (2010) Propionate and butyrate dependent bacterial sulfate reduction at extremely haloalkaline conditions and description of Desulfobotulus alkaliphilus sp. nov. Extremophiles 14:71–77

Sorokin DY, Tourova TP, Panteleeva AN, Muyzer G (2012) *Desulfonatronobacter acidivorans* gen. nov., sp. nov. and *Desulfobulbus alkaliphilus* sp. nov., haloalkaliphilic heterotrophic sulfate-reducing bacteria from soda lakes. Int J Syst Evol Microbiol 62:2107–2113

Sousa DZ, Visser M, van Gelder AH, Boeren S, Pieterse MM, Pinkse MWH et al (2018) The deep-subsurface sulfate reducer *Desulfotomaculum kuznetsovii* employs two methanol-degrading pathways. Nat Commun 9:239. https://doi.org/10.1038/s41467-017-02518-9

Spormann AM, Thauer RK (1989) Anaerobic acetate oxidation to CO_2 by *Desulfotomaculum acetoxidans*. Arch Microbiol 150:374–380

Spring S, Lapidus A, Schröder M, Gleim D, Sims D, Meincke L et al (2009) Complete genome sequence of *Desulfotomaculum acetoxidans* type strain (5575). Stand Genomic Sci 1:242–253

Stams AJM, Hansen TA (1982) Oxygen-labile L(+) lactate dehydrogenase activity in *Desulfovibrio desulfuricans*. FEMS Microbiol Lett 13:389–394

Stams AJM, Hoekstra LG, Hansen TA (1986) Utilization of L-alanine as carbon and nitrogen source by *Deusulfovibrio* HL21. Arch Microbiol 145:272–276

Stetter KO, Huber R, Blöchl E, Kurr M, Eden RD, Fielder M, Cash H, Vance I (1993) Hyperthermophilic archaea are thriving in deep North Sea and Alaskan oil reservoirs. Nature 365:743–745

Stolyar S, He Q, Joachimiak MP, He Z, Yang ZK, Borglin SE, Joyner DC et al (2007a) Response of *Desulfovibrio vulgaris* to alkaline stress. J Bacteriol 189:8944–8952

Stolyar S, Van Dien S, Hillesland KL, Pinel N, Lie TJ, Leigh JA, Stahl DA (2007b) Metabolic modeling of a mutualistic microbial community. Mol Syst Biol 3:92. https://doi.org/10.1038/msb4100131

Strittimatter AW, Liesegang H, Rabus R, Decker I, Amann J, Andres S et al (2009) Genome sequence of *Desulfobacterium autotrophicum* HRM2, a marine sulfate reducer oxidizing organic carbon completely to carbon dioxide. Environ Microbiol 11:1038–1055

Suzuki D, Ueki A, Amaishi A, Ueki K (2007) *Desulfobulbus japonicus* sp. nov., a novel Gram-negative propionate-oxidizing, sulfate-reducing bacterium isolated from an estuarine sediment in Japan. Int J Syst Evol Microbiol 57:849–855

Suzuki D, Ueki A, Amaishi A, Ueki K (2008) *Desulfoluna butyratoxydans* gen. nov., sp. nov., a novel, Gram-negative, butyrate oxidizing, sulfate-reducing bacterium isolated from an estuarine sediment in Japan. Int J Syst Evol Microbiol 58:826–832

Suzuki D, Ueki A, Amaishi A, Ueki K (2009) *Desulfovibrio portus* sp. nov., a novel sulfate-reducing bacterium in the class Deltaproteobacteria isolated from an estuarine sediment. J Gen Appl Microbiol 55:125–133

Suzuki D, Ueki A, Shizuku T, Ohtaki Y, Ueki K (2010) *Desulfovibrio butyratiphilus* sp. nov., a Gram negative, butyrate-oxidizing, sulfate-reducing bacterium isolated from an anaerobic municipal sewage sludge digester. Int J Syst Evol Microbiol 60:595–602

Suzuki D, Cui X, Li Z, Zhang C, Katayama A (2014) Reclassification of *Desulfobacterium anilini* as *Desulfatiglans anilini* comb. nov. within *Desulfatiglans* gen. nov., and description of a 4-chlorophenol-degrading sulfate-reducing bacterium, *Desulfatiglans parachlorophenolica* sp. nov. Int J Syst Evol Microbiol 64:3081–3086

Szewzyk R, Pfennig N (1987) Complete oxidation of catechol by the strictly anaerobic sulfate-reducing *Desulfobacterium catecholicum* sp. nov. Arch Microbiol 147:163–168

Takii S, Hanada S, Hase Y, Tamaki H, Uyeno Y, Sekiguchi Y, Matsuura K (2008) *Desulfovibrio marinisediminis* sp. nov., a novel sulfate-reducing bacterium isolated from coastal marine sediment via enrichment with casamino acids. Int J Syst Evol Microbiol 58:2433–2438. https://doi.org/10.1099/ijs.0.65750-0. Erratum in: Int J Syst Evol Microbiol. 2008;58(Pt 11): 2673. PMID: 18842870

Tang Y, Pingitore F, Mukhopadhyay A, Phan R, Hazen TC, Keasling JD (2007) Pathway confirmation and flux analysis of central metabolic pathways in *Desulfovibrio vulgaris* Hildenborough using gas chromatography-mass spectrometry and Fourier transform-ion cyclotron resonance mass spectrometry. J Bacteriol *189*:940–949

Tasaki M, Kamagata Y, Nakamura K, Okamura K, Minami K (1993) Acetogenesis from pyruvate by *Desulfotomaculum thermobenzoicum* and differences in pyruvate metabolism among three sulfate-reducing bacteria in the absence of sulfate. FEMS Microbiol Lett 106:259–263

Thauer RK, Kunow J (1995) Sulphate-reducing Archaea. In: Barton LL (ed) Sulfate-reducing bacteroa: biotechnology handbook. Plenum Publishing, London, pp 33–48

Thauer RK, Jungermann K, Decker K (1977) Energy conservation in chemotrophic anaerobic bacteria. Bacteriol Rev 41:100–180

Thauer RK, Möller-Zinkhan D, Spormann AM (1989) Biochemistry of acetate catabolism in anaerobic chemotrophic bacteria. Annu Rev Microbiol 43:43–67

Thauer RK, Stackebrandt E, Hamilton W (2007) Energy metabolism and phylogenetic diversity of sulphate-reducing bacteria. In: Barton LL, Hamilton WA (eds) Sulphate-reducing bacteria: environmental and engineered systems. Cambridge University Press, Cambridge, pp 1–38

Thebrath B, Dilling W, Cypionka H (1989) Sulfate activation in *Desulfotomaculum*. Arch Microbiol 152:296–301

Thomas P (1972) Ultrastructure de *Desulfovibrio gigas* LeGall et de D*esulfovibrio vulgaris* Hildenborough. J Microsc (Paris) 13:349–360

Torres-García W, Zhang W, Runger GC, Johnson RH, Meldrum DR (2009) Integrative analysis of transcriptomic and proteomic data of *Desulfovibrio vulgaris*: a non-linear model to predict abundance of undetected proteins. Bioinformatics 25:1905–1914

Traore AS, Hatchikian CE, Belaich J-P, Le Gall J (1981) Microcalorimetric studies of the growth of sulfate-reducing bacteria: energetics of *Desulfovibrio vulgaris* growth. J Bacteriol 145:191–199

Tucker MD, Barton LL, Thomson BM (1997) Reduction and immobilization of molybdenum by *Desulfovibrio desulfuricans*. J Environ Qual 26:1146–1152

van der Maarel MJEC, van Bergeijk S, van Werkhoven AF, Laverman AM, Meijer WG, Stam WT, Hansen TA (1996) Cleavage of dimethylsulfoniopropionate and reduction of acrylate by *Desulfovibrio acrylicus* sp. nov. Arch Microbiol 166:109–115

Van Kuijk BLM, Stams AJM (1995) Sulfate reduction by a syntrophic propionate- oxidizing bacterium. Antonie Van Leeuwenhoek 68:293–296

Van Niel EWJ, Gomes TMP, Willems A, Collins MD, Prins RA, Gottschal JC (1996) The role of polyglucose in oxygen-dependent respiration by a new strain of Desulfovibrio salexigens. FEMS Microbiol Ecol 21:243–253

Visser M, Parshina SN, Alves JI, Sousa DZ, Pereira IAC, Muyzer G et al (2014) Genome analyses of the carboxydotrophic sulfate-reducers *Desulfotomaculum nigrificans* and *Desulfotomaculum carboxydivorans* and reclassification of *Desulfotomaculum caboxydivorans* as a later synonym of *Desulfotomaculum nigrificans*. Stand Genomic Sci 9(3):655–675. https://doi.org/10.3389/fmicb.2014.00095. PMC4149029

Visser M, Stams AJM, Frutschi M, Bernier-Latmani R (2016) Phylogenetic comparison of *Desulfotomaculum* species of subgroup 1a and description of *Desulfotomaculum reducens* sp. nov. Int J Syst Evol Microbiol 66:762–767

Vita N, Valette O, Brasseur G, Lignon S, Denis Y, Ansaldi M, Dolla A, Pieulle L (2015) The primary pathway for lactate oxidation in *Desulfovibrio vulgaris*. Front Microbiol 6:606. https://doi.org/10.3389/fmicb.2015.00606

Voordouw G (2002) Carbon monoxide cycling by *Desulfovibrio vulgaris* Hildenborough. J Bacteriol 184:5903–5911

Vosjan JH (1975) Respiration and fermentation of the sulphate-reducing bacterium *Desulfovibrio desulfuricans* in a continuous culture. Plant Soil 43:141–152

Walian PJ, Allen S, Shatsky M, Zeng L, Szakal ED, Liu H et al (2012) High-throughput isolation and characterization of untagged membrane protein complexes: outer membrane complexes of *Desulfovibrio vulgaris*. J Proteome Res 11:5720–5735. https://doi.org/10.1021/pr300548d

Walker CB, Stolyar SS, Pinel N, Yen HC, He Z, Zhou J, Wall JD, Stahl DA (2006) Recovery of temperate *Desulfovibrio vulgaris* bacteriophage using a novel host strain. Environ Microbiol 8: 1950–1959

Walker CB, Stolyar S, Chivian D, Pinel N, Gabster JA, Dehal PS et al (2009) Contribution of mobile genetic elements to *Desulfovibrio vulgaris* genome plasticity. Environ Microbiol 11: 2244–2252

Wall JD, Bill Yen HC, Drury EC (2007) Evaluation of stress response in sulphate-reducing bacteria through genome analysis. In: Barton LL, Hamilton WA (eds) Sulphate-reducing bacteria – environmental and engineered systems. Cambridge University Press, Cambridge, pp 141–165

Wallrabenstein C, Hauschild E, Schink B (1995) *Syntrophobacter pfennigii* sp. nov., new syntrophically propionate-oxidizing anaerobe growing in pure culture with propionate and sulfate. Arch Microbiol 164:346–352

Wang J, Vine CE, Balasiny BK, Rizk J, Bradley CL, Tinajero-Trejo M et al (2016) The roles of the hybrid cluster protein, Hcp and its reductase, Hcr, in high affinity nitric oxide reduction that protects anaerobic cultures of *Escherichia coli* against nitrosative stress. Mol Microbiol 100: 877–892

Widdel F (1988) Microbiology and ecology of sulfate-and sulfur-reducing bacteria. In: Zehnder AJB (ed) Biology of anaerobic microorganisms. John Wiley and Sons, Inc., New York, pp 469–585

Widdel F, Pfennig N (1977) A new anaerobic, sporing, acetate-oxidizing, sulfate reducing bacterium, *Desulfotomaculum* (emend.) *acetoxidans*. Arch Microbiol 112:119–122

Widdel F, Pfennig N (1981) Studies on dissimilatory sulfate-reducing bacteria that decompose fatty acids. I. Isolation of new sulfate-reducing bacteria enriched with acetate from saline environments. Description of *Desulfobacter postgatei* gen. nov., sp. nov. Arch Microbiol 129:395–400

Widdel F, Pfennig N (1982) Studies on dissimilatory sulfate-reducing bacteria that decompose fatty acids II. Incomplete oxidation of propionate by *Desulfobulbus propionicus* gen. nov., sp. nov. Arch Microbiol 131:360–365

Wildschut JD, Lang RM, Voordouw JK, Voordouw G (2006) Rubredoxin: oxygen oxidoreductase enhances survival of *Desulfovibrio vulgaris* Hildenborough under microaerophilic conditions. J Bacteriol 188:6253–6260

Wilkins MJ, Hoyt DW, Marshall MJ, Alderson PA, Plymate AE, Markillie LM et al (2014) CO_2 exposure at pressure impacts metabolism and stress responses in the model sulfate-reducing bacterium *Desulfovibrio vulgaris* strain Hildenborough. Front Microbiol 5:507. https://doi.org/10.3389/fmicb.2014.00507

Yagi T (1958) Enzymic oxidation of carbon monoxide. Biochim Biophys Acta 30:194–195

Yagi T (1959) Enzymic oxidation of carbon monoxide II. J Biochem (Tokyo) 46:949–955

Yagi T (1979) Purification and properties of cytochrome c-553, an electron acceptor for formate dehydrogenase of *Desulfovibrio vulgaris*, Miyazaki. Biochim Biophys Acta 548:96–105

Yu L, Ishida T, Ozawa K, Akutsu H, Horiike K (2001) Purification and characterization of homo- and hetero-dimeric acetate kinases from the sulfate-reducing bacterium *Desulfovibrio vulgaris*. J Biochem 129:411–421

Yu H, Jiang Z, Lu Y, Yao X, Han C, Ouyang WH et al (2020) Transcriptome analysis of the acid stress response of *Desulfovibrio vulgaris* ATCC 7757. Curr Microbiol 77:2702–2712

Yurkiw MA, Voordouw J, Voordouw G (2012) Contribution of rubredoxin: oxygen oxidoreductases and hybrid cluster proteins of *Desulfovibrio vulgaris* Hildenborough to survival under oxygen and nitrite stress. Environ Microbiol 14:2711–2725

Zaunmüller T, Kelly DJ, Glöckner FO, Unden G (2006) Succinate dehydrogenase functioning by a reverse redox loop mechanism and fumarate reductase in sulphate-reducing bacteria. Microbiology 152:2443–2453

Zeng L, Wooton E, Stahl DA, Walian PJ (2017) Identification and characterization of the major porin of *Desulfovibrio vulgaris* Hildenborough. J Bacteriol 199:e00286–e00217. https://doi.org/10.1128/JB.00286-17

Zhang W, Culley DE, Scholten JCM, Hogan M, Vitiritti L, Brockman FJ (2006a) Global transcriptomic analysis of *Desulfovibrio vulgaris* on different electron donors. Antonie Van Leeuwenhoek 89:221–237

Zhang W, Culley DE, Hogan M, Vitiritti L, Brockman FJ (2006b) Oxidative stress and heat-shock responses in *Desulfovibrio vulgaris* by genome-wide transcriptomic analysis. Antonie Van Leeuwenhoek 90:41–55

Zhang W, Gritsenko MA, Moore RJ, Culley DE, Nie L, Petritis K et al (2006c) A proteomic view of *Desulfovibrio vulgaris* metabolism as determined by liquid chromatography coupled with tandem mass spectrometry. Proteomics 6:4286–4299

Zhou A, He Z, Redding-Johanson AM, Mukhopadhyay A, Hemme CL, Joachimiak MP et al (2010) Hydrogen peroxide-induced oxidative stress responses *in Desulfovibrio vulgaris* Hildenborough. Environ Microbiol 12:2645–2657

Zhou J, He Q, Hemme CL, Mukhopadhyay A, Hillesland K, Zhou A et al (2011) How sulphate-reducing microorganisms cope with stress: lessons from systems biology. Nat Rev Microbiol 9: 452–466

Zhou A, Chen YI, Zane GM, He Z, Hemme CL, Joachimiak MP et al (2013) Functional characterization of Crp/Fnr-type global transcriptional regulators in *Desulfovibrio vulgaris* Hildenborough. J Bacteriol 195:29–38

Zhou A, Lau R, Baran R, Ma J, von Netzer F, Shi W et al (2017) Key metabolites and mechanistic changes for salt tolerance in an experimentally evolved sulfate-reducing bacterium, *Desulfovibrio vulgaris*. mBio 8(6):pii: e01780–17. https://doi.org/10.1128/mBio.01780-17. PMID: 29138306

Zhu L, Gong T, Wood TL, Yamasaki R, Wood TK (2019) σ_{54}-Dependent regulator DVU2956 switches *Desulfovibrio vulgaris* from biofilm formation to planktonic growth and regulates hydrogen sulfide production. Environ Microbiol 21:3564–3576

Ziomek E, Williams RE (1989) Modification of lignins by growing cells of the sulfate-reducing anaerobe *Desulfovibrio desulfuricans*. Appl Environ Microbiol 55:2262–2266

Chapter 7
Geomicrobiology, Biotechnology, and Industrial Applications

7.1 Introduction

Early research on sulfate-reducing bacteria (SRB) was to assess which bacteria were to be credited with the copious production of hydrogen sulfide. With the isolation of bacteria by the Delph group led by Beijerinck, sulfate-reducing bacteria may have been considered biological curiosities by scientists; however, gradually, associations between these bacteria and environmental activities were recognized because of economic importance or for their potential contribution to scientific concerns of the day. Although the number of sulfate-reducing archaea is less numerous than sulfate-reducing bacteria, the sulfate-respiring archaea occupy an important niche in the biosphere. The SRB are significant members of the microbial community associated with nutrient cycles by producing end products used by other bacteria and consumption of end products of other bacteria. This chapter focuses on the metabolism of organic carbon compounds including the possible use of SRB in bioremediation of polluted environments containing hydrocarbons. Also discussed here are activities of SRB in the production of minerals, the transformation of radionuclides, and the production of metallic nanoparticles. In addition to bioremediation of hydrocarbon environments and acid mine waters by SRB, the "green" industrial process for production of metallic nanoparticles is discussed.

7.2 Contributions of SRB to Major Nutrient Cycles

In an open environment, SRB use oxidized inorganic sulfur compounds for respiration, and in the absence of readily available inorganic sulfur compounds, SRB will use oxidized minerals and compounds from other nutrient cycles as electron acceptors. These alternate electron acceptors enable the SRB to persist and grow in the environment with significant contributions to the global cycles involving carbon,

L. L. Barton, G. D. Fauque, *Sulfate-Reducing Bacteria and Archaea*,
https://doi.org/10.1007/978-3-030-96703-1_7

nitrogen, and sulfur. SRB may be found in extreme environments such as alkaline soda lakes, and their participation in carbon, sulfur, and nitrogen cycles have been documented (Sorokin et al. 2014). The diversity of SRB throughout the environment provides an important contribution to nutrient cycling.

7.2.1 Sulfur and Nitrogen Cycling

The topics of sulfur cycling and nitrogen cycling are covered in Chap. 3.

7.2.2 Carbon Cycling

Sulfate-reducing microorganisms have an important role in the global carbon cycle because these microorganisms are broadly distributed in the anaerobic terrestrial ecosystems and their capability of using a large number of compounds as sources of energy (Fauque et al. 1991; Hansen 1993). Substrates found in the sediment environments that support growth of SRB include H_2 (Widdel and Hansen 1992), sugars (Ollivier et al. 1988); methanol (Klemps et al. 1985; Nanninga and Gottschal 1987; Nazina et al. 1988), amino acids (Stams et al. 1985; Baena et al. 1998), alkanes (Aeckersberg et al. 1998; So and Young 1999; Cravo-Laureau et al. 2004; Davidova et al. 2006), alkenes (Grossi et al. 2007), CO (Parshina et al. 2005a, b; Henstra et al. 2007; Parshina et al. 2010), methanethiol (Tanimoto and Bak 1994), and aromatic hydrocarbons (Rabus et al. 1993; Ensley and Suflita 1995; Harms et al. 1999; Morasch et al. 2004). SRB participate in the oxidation of acetate, propionate, and butyrate, which are commonly found in the decomposition of petroleum hydrocarbon (PHC) (Rozanova et al. 1991; Rabus et al. 1993; Cozzarelli et al. 1994; Ensley and Suflita 1995; Zhang and Young 1997; Harms et al. 1999; Morasch et al. 2004). Acetate, propionate, and butyrate are commonly present as fermentation products in salt marsh (Balba and Nedwell 1982) and in marine, and estuary sediments (Sørensen et al. 1981; Parkes et al. 1989). Lactate is commonly found as a fermentation product in fresh water sediments (Cappenberg and Prins 1974; Hordijk and Cappenberg 1983). Other environments containing SRB are wetland sediments (Martins et al. 2017) and aquifers (Kleikemper et al. 2002a, b). SRB have two metabolic strategies for the decomposition of organic material with the production of CO_2 and these are sulfidogenic respiration and syntrophic activities. While the hallmark characteristic of SRB is the dissimilatory reduction of sulfate with the production of hydrogen sulfide, several strains will use nitrate and nitrite as alternate electron acceptors when sulfate is unavailable. Anaerobic syntrophy involving several different strains of bacteria is important in the global carbon cycle (McInerney et al. 2009), and syntrophic activities involving SRB are discussed by Muyzer and Stams (2008) and Plugge et al. (2011). A summary of interactions of SRB with the global carbon cycle is presented in Fig. 7.1.

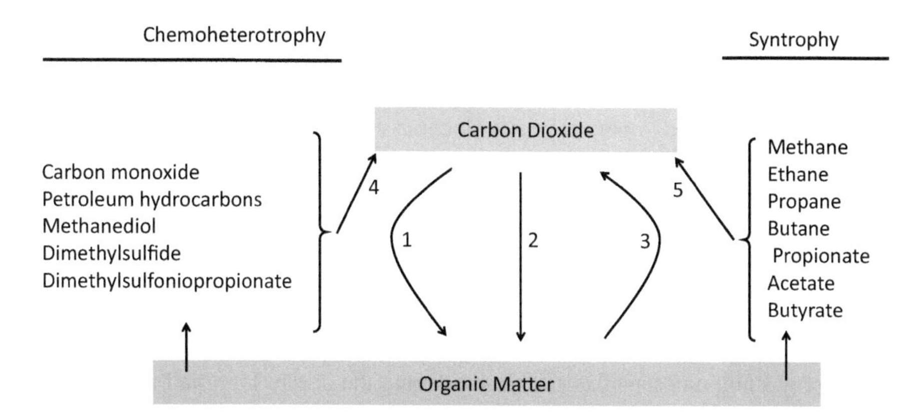

Fig. 7.1 Interaction of SRB with the global carbon cycle. Stages as follows: (1) assimilation, (2) autotrophic CO_2 fixation, (3) fermentation, (4) an array of carbon material mineralized that may not be in organic matter or are minor compounds in the sediment, (5) organic compounds oxidized by syntrophic metabolism involving SRB

7.2.2.1 Decomposition of Organic Matter

Following the initial reports of SRB in marine waters, attention was directed to SRB in marine sediments and associated microbial decomposition. Reports by Issatschenko (1924, 1929) and Ravich-Sherbo (1930) indicated that SRB were present in the Black Sea waters and this supported early suggestions that hydrogen sulfide in the Black Sea with a stench of the "blue mud" sediment was attributed to biological activity (Zelinski 1893; Murray and Irvine 1895). SRB were found in the Salton Sea (Brannon 1914), the Mediterranean Sea (Bertel 1935), and the Russian lagoons in the Odessa region (Rubentschik and Goicherman 1936). Sediment near the Bahama Islands (Bavendamm 1932) contained SRB, and in a broad survey, Zobell (1938) recognized that SRB were more abundant in marine sediments than in water above the sediment. As reported by Zobell and Rittenberg (1948), depth profiles of marine sediment indicated the number of SRB in the sediment decreased with an increase in depth. The vertical zones used to describe costal marine sediment decomposition generally used are designated by Chen et al. (2017) as follows: (1) surface zone, which is subject to bioturbation; (2) sulfate-rich zone, which is nonbioturbated and where most dissimilatory sulfate reduction occurs; (3) sulfate–methane transition zone where methane oxidation dependent on sulfate reduction occurs; and (4) methane zone where methanogenesis occurs and sulfate is limiting. In a report on abundance of bacteria in costal marine sediment in Aarhus Bay (Denmark), SRB accounted for 13% of the bacteria present in the surface zone, 22% of the bacteria in the sulfate–methane zone and 8% of the bacteria in the methane zone (Leloup et al. 2009).

Using qPCR to quantify *dsrB* and *aprA* gene copies, the surface layer of the sediments of Aarhus Bay contained about 10^8 cells/g sediment (Jochum et al. 2017), and this was comparable to the abundance of SRB measured in different regions using different methods (Sahm et al. 1999a; Schippers and Neretin 2006; Leloup et al. 2007, 2009; Blazejak and Schippers 2011). In the Aarhus Bay sediments, 12–17% of the surface bacterial community were SRB and 2% of the microbial cells in the sulfate-depleted zone were SRB (Jochum et al. 2017). It has been estimated that 2.9×10^{29} microbial cells occur in the marine sediment, and these microorganisms would have a considerable metabolic impact on the degradation of organic carbon matter in the subseafloor (Singh et al. 2017). Organic matter deposited is not uniform throughout the oceans with plant-derived material associated with lagoons and deposits near shore regions and algal material as a source of organic material in open waters (Westrich and Berner 1984). Additionally, algal distribution in open seas is irregular, and in some areas, algal growth is nutrient limited. In some regions, there are gas and oil seeps that release fluids into the ocean floor, and this input of fluids has an impact on carbon cycling in marine sediments (Chakraborty et al. 2020). With an abundant population of SRB in marine sediment, it was evident that SRB were involved in decomposition activities of marine sediment with the production of hydrogen sulfide. The role of SRB in coastal including estuaries and buried marine sediments has been reviewed by Jørgensen (1982), Ferdelman et al. (1997, 1999), Knoblauch et al. (1999a, b), Sahm et al. (1999b), Thamdrup et al. (2000), Jørgensen et al. (2001), D'Hondt et al. (2004), Parkes et al. (2005), Schippers et al. (2005), and Schippers et al. (2010). Decomposition of organic matter in the seafloor involved several different metabolic types of bacteria and archaea with fermenting, sulfate-reducing, denitrifying, and methane-producing groups of microorganisms being the most dominant (Jørgensen 1982). While SRB use small organic substrates to support growth (Fauque et al. 1991; Hansen 1993), the substrates commonly used by SRB (lactate, acetate, and propionate) supported growth of SRB in marine sediments of the Gulf of Gdańsk (Mudryk et al. 2000). The number of SRB present in the Gulf of Gdańsk varied from 0.76×10^3 to 1.27×10^4 cells/g wet sediment, and these bacteria accounted for sulfate reduction at 1.9–31 nM sulfate/g sediment/day (Mudryk et al. 2000).

Anaerobic heterotrophic bacteria are required for the decomposition of natural polymers, and the steps of anaerobic decomposition of marine sediment is outlined in Fig. 7.2. When considering the degradation of organic matter in marine and estuary sediments, decomposition attributed to SRB is about 12 times greater than the combined activities of aerobic and denitrifying bacteria (Jørgensen 1977; Howarth 1979). It has been estimated that over 50% of the organic matter in marine shelf sediments is decomposed by SRB (Jørgensen 1982; Canfield 1993). Estuary sediments account for only 0.7% of marine sediments, but estuary sediments produce 7–10% of carbon released into the atmosphere (Bange et al. 1984; Abril and Iversen 2002; Sela-Adler et al. 2017). A survey of marine sediments at various geological locations reveals that the *Desulfosarcina*/*Desulfococcus* clade of Desulfobacteraceae are the most abundant types of microorganisms in the SRB community (Ravenschlag et al. 2000; Llobet-Brossa et al. 2002) and the complete oxidizer *Desulfococcus multivorans* is important in mineralization of organic matter in marine sediments (Robador et al. 2016). The in situ temperature of the marine sediment is the major factor and organic concentration or C–N ratio in the sediment

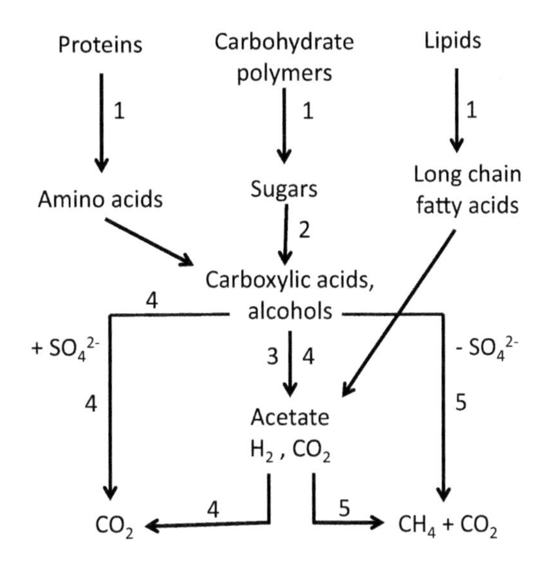

Fig. 7.2 Processes associated with anaerobic mineralization of organic matter. Abbreviations are as follows: (1) hydrolysis, (2) fermentation, (3) acetogenesis, (4) sulfate reduction, (5) methanogenesis. (Information for this figure is from Fauque (1995) and Visser (1995))

in establishing the SRB community (Robador et al. 2016). However, in Arctic sediments, the supply of organic materials has a greater control on organic decomposition that does temperature (Arnosti et al. 1998). In marine sediments near coastal region, sulfate reduction by SRB correlates with the production of "humic-like" fluorescence compounds and the role of these compounds in the environmental processes is unknown (Luek et al. 2017). From wastewater treatment studies, it is recognized that low-molecular-weight compounds released by fermentation of organic matter are consumed by SRB, acetogens, and methanogens. In terms of catabolism of organic molecules, there are the complete oxidizers, which result in complete mineralization of organic matter with the production of CO_2, and incomplete oxidizers, which produce acetate and CO_2 as end products of metabolism. As indicated in the following reactions conducted by SRB, complete oxidation of lactate is more energetic than incomplete oxidization (Dar et al. 2008).

Incomplete oxidation:

$$C_3H_5O_3^- + 1.5SO_4^{2-} \rightarrow C_2H_3O_2^- + HCO_3^- + 0.5HS^- + 0.5H^+ \\ - 80.8\ \Delta G^{0'}\ \text{kJ/reaction} \tag{7.1}$$

Complete oxidation:

$$C_3H_5O_3^- + 0.5SO_4^{2-} \rightarrow 3HCO_3^- + 1.5HS^- + 0.5H^+ \\ - 128.5\ \Delta G^{0'}\ \text{kJ/reaction} \tag{7.2}$$

A competition for substrates occurs between the microbial groups with SRB outcompeting acetogens and methanogens if the sulfate concentration is in excess and H_2 is available; SRB will outcompete methanogens (Lupton and Zeikus 1984; Uberoi and Bhattacharya 1995) because SRB have a greater V_{max} for H_2 and a shorter generation time. However, several mechanisms including syntrophic partnerships have evolved to enable SRB and archaea members to coexist in marine environments (Dar et al. 2008; Ozuolmez et al. 2015; Sela-Adler et al. 2017). It has been suggested that oxidation of short carbon chain, volatile hydrocarbons such as butane may be attributed to a syntrophic partnership consisting of archaea members oxidizing the hydrocarbon and members of SRB that serve as the electron acceptors (Laso-Perez et al. 2016). In this regard, oxidation of butane and similar hydrocarbons by relevant archaea would be similar to ANME-1 archaea, which oxidize methane. The net effect is that these archaea would recycle volatile hydrocarbons from the organic sediment and reduce the amount of carbon emission from the marine environment (Laso-Perez et al. 2016; Seitz et al. 2019). From a deep subseafloor off Costa Rica margin containing a pile of organic sediment estimated to be 890 m thick, several new strains of Helarchaeota of the Asgard superphylum of archaea and sulfate-reducing bacteria were isolated and were assumed to have the capability of developing syntrophic relationships (Zhao and Biddle 2021). Sediments from the Yangon (Israel) estuary supported production of hydrogen sulfide and methane at the same time with hydrogen sulfide reaching 680 μmol/L/day while methane production was two orders of magnitude lower (Sela-Adler et al. 2017).

7.2.2.2 Anaerobic Methane Oxidation

Marine sediments contain 50,000–10,000,000 teragrams (Tg) methane (Reeburgh 2007) and anaerobic methane oxidation (AMO) consumes 70–77,300 Tg methane each year (Conrad 2009). Anaerobic methane oxidation in marine sediments has attracted the attention of scientists with the proposal that methane oxidation in the Cariaco Trench of the Caribbean Sea was attributed to co-metabolism of sulfate-reducing bacteria (Reeburgh 1976). The observation that methane oxidization in marine sediments involved sulfate reducers was supported by analysis of sediments from Skan Bay, Alaska (Reeburgh 1980); Saanich Inlet, British Columbia (Devol 1983); and Kattegat and Skagerrak, Denmark (Iversen and Jørgensen 1985). This metabolic activity was considered to occur at the sulfate–methane transition zone, which was just above the marine sediment. Reports for anaerobic consumption of methane were extended to include salt marsh and coastal marine sediments (Howarth 1984); reefs in the Black Sea (Michaelis et al. 2002); sediment surface of Kysing Fjord, Denmark, where the temperature ranged from 4 to 21 °C (Iversen and Blackburn 1981); laboratory experiments (Hoehler et al. 1994); and Norwegian Sea (Lazar et al. 2011).

For anaerobic methane oxidation, the oxidation of methane coupled to sulfate reduction was proposed to involve a consortium consisting of methanogenic archaea and sulfate-reducing bacteria (Zehnder and Brock 1980; Hoehler et al. 1994). The demonstration of nine laboratory strains of methanogenic archaea to oxidize methane by reverse methanogenesis (Zehnder and Brock 1979) and the demonstration of genes for methane oxidation by methanogenic archaea from deep sea environments (Hallam et al. 2004) were also reviewed. Based on phylogenic (16S rRNA gene sequence analysis) and biochemical analysis, a consortium of microorganisms consisting of sulfate-reducing bacteria and anaerobic methanotrophic archaea (ANME) was determined to be responsible for anaerobic methane oxidation (Hoehler et al. 1994; Orphan et al. 2002). With the hypothesis that anaerobic oxidation of methane (AOM) was a reversal of methanogenesis, the key enzyme in methane formation methyl-coenzyme M reductase (mcrA; Thauer 1998; Wolfe 1991) would be required to be reversed. The biochemical considerations involved with transfer of the methyl group in this oxidative process are discussed by Widdel et al. (2007).

Over the years, there have been several reviews addressing the co-metabolism by methylotrophic archaea and sulfate-reducing bacteria associated with the oxidation of methane (Reeburgh et al. 1993; Hinrichs and Boetius 2002; Valentine 2002; Jørgensen and Kasten 2006; Widdel et al. 2006, 2007; Conrad 2009; Knittel and Boetius 2009). Because there are numerous unresolved issues in this open environment, this remains as an active research area. It has been reported that in regions where methane production occurs, there is an abundance of non-sulfate-reducing, syntrophic bacteria, which are affiliated with *Desulfotomaculum* (Imachi et al. 2006). Members of the sulfate-reducing bacteria and ANME associated with methane oxidation have never been grown in pure culture. Environmental samples containing anaerobic methane-oxidizing microorganisms have been used to obtain information about syntrophic organization (Boetius et al. 2000; Wilms et al. 2007; Pernthaler et al. 2008; Vigneron et al. 2014), and active stable enrichment cultures have provided useful information concerning biochemical pathway (Hinrichs et al. 1999; Nauhaus et al. 2007; Webster et al. 2011; Milucka et al. 2012). An organism Ca. Desulfofervidus auxilii is the first ANME-free enrichment of an AOM partner bacterium (Krukenberg et al. 2016), and physiological analysis remains to be reported for this SRB.

Based on 16S rRNA gene, 16S rRNA, methyl-coenzyme M reductase (*mcrA*), and d*srA*B clone libraries, the consortium of methane-oxidizing microorganisms is placed into three phylogenic groups of archaea (ANME-1a/b, ANME-2a/b/c, and ANME-3), and the syntrophic sulfate reducers are either members of the Desulfobacteraceae (*Desulfosarcina/Desulfococcus* cluster) or of the Desulfobulbaceae (Lloyd et al. 2006; Cui et al. 2015; Milucka et al. 2013; Kurth et al. 2019). The ANME-1 and ANME-2 groups are the most commonly detected, and members of the three ANME groups can be distinguished by the type of membrane lipid, ionic strength of media for growth, and substrates used (Cui et al. 2015; Timmers et al. 2015; Kurth et al. 2019). *Desulfosarcina-/Desulfococcus*-related cells are examples of sulfate-reducing bacteria commonly associated with ANME-1 and ANME-2, while *Desulfobulbus*-related cells are found with ANME-3 (Cui et al. 2015). ANME-1 and ANME-2 groups are the most commonly encountered in the environment. While sulfate reduction is associated with ANME (also called sulfate-dependent anaerobic methane oxidation [S-DAMO]), methane oxidation coupled to reduction of nitrate or nitrite is designated as nitrogen-DAMO

(N-DAMO) and methane oxidation coupled to Fe^{3+} or Mn^{4+} is identified as metal-DAMO (M-DAMO; Green-Saxena et al. 2014; Cui et al. 2015). N-DAMO organisms using these alternate electron acceptors could couple energy production from methane oxidation to electron acceptor for sufficient energy to support growth (Raghoebarsing et al. 2006; Beal et al. 2009); see Eqs. (7.3)–(7.6):

$$5CH_4 + 8NO_3^- + 8H^+ \rightarrow 5CO_2 + 4N_2 + 14H_2O \qquad \Delta G'^{\circ} = -765\,\mathrm{kJ/mol\,CH_4} \tag{7.3}$$

$$3CH_4 + 8NO_2^- + 8H^+ \rightarrow 3CO_2 + 4N_2 + 10H_2O \qquad \Delta G'^{\circ} = -928\,\mathrm{kJ/mol\,CH_4} \tag{7.4}$$

$$CH_4 + 4MnO_2 + 7H^+ \rightarrow HCO_3^- + 4Mn^{2+} + 5H_2O \qquad \Delta G'^{\circ} = -556\,\mathrm{kJ/mol\,CH_4} \tag{7.5}$$

$$CH_4 + 8Fe(OH)_3 + 15H^+ \rightarrow HCO_3^- + 8Fe^{2+} + 2H_2O \qquad \Delta G'^{\circ} = -270.3\,\mathrm{kJ/mol\,CH_4} \tag{7.6}$$

However, organisms associated with S-DAMO are encountered in greater numbers in the environment than N-DAMO or M-DAMO organisms. Of concern over the years was the thermodynamics of the reaction using anaerobic methane oxidation with sulfate. When using in situ concentrations (Martens and Berner 1977; Hoehler et al. 1994), the amount of energy released would not appear to be sufficient for ATP production; see Eq. (7.7):

$$CH_4 + SO_4^{2-} + 2H^+ \rightarrow CO_2 + 2H_2O + H_2S \qquad \Delta G'^{\circ} = -24.1\,\mathrm{kJ/mol\,CH_4} \tag{7.7}$$

Equation (7.7) is an overall reaction resulting from metabolism of the archaeal organisms plus the reactions by sulfate reducers. The products of archaeal oxidation of methane may result in the formation of H_2, CO_2, acetate, or methanol, which would be readily oxidized by exergonic reactions employed by sulfate-reducing bacteria. As reviewed by Hansen et al. (1998) and Cui et al. (2015), there may be interspecies carbon transfer from methane oxidation to the sulfate-reducing bacteria. Clearly, anaerobic methane oxidation is an important area for future investigations.

7.3 H_2S Pollution: Agricultural and Commercial Impact

While sulfate-reducing bacteria were associated with corrosion and known to be detrimental to the oil industry, production of hydrogen sulfide contributes to environmental pollution and with an undesirable impact on commercial processes. As reviewed by Postgate (1979), toxic levels of hydrogen sulfide have resulted in death of sewage workers, aquatic birds, and fish. Periodically, elevated levels of hydrogen sulfide in Walvis Bay off of South West Africa have resulted in death to aquatic plants and fish (Gilchrist 1914; Copenhagen 1934, 1953, 1954; Currie 1953; Butlin 1949). Aquatic regions containing high levels of hydrogen sulfide may become bright red due to the growth of anaerobic photosynthetic bacteria (*Chromatium* and *Thiopedia*), which use sulfide as an electron source (Postgate 1965, 1979). Lake Faro in Messina, Sicily, was reported (Genovese 1961, 1963; Genovese and Macri 1964) to contain anoxic photosynthetic bacteria, which use sulfide generated by sulfate-reducing bacteria as the electron donor and producing a red phototrophic bloom. This photosynthetic layer was at the interface of the oxic–anoxic zone and contained a red layer of photosynthetic bacteria (Thiorhodaceae) located 8–13 m below the surface (Trüper and Genovese 1968). A sessional hydrogen sulfide pollution occurs on the Namibian shelf and a unique ecological cycle has developed (Currie et al. 2018; Ohde and Dadou 2018). The upwelling activity of the Namibian shelf has been occurring for thousands of years (Diester-Haass et al. 2002) and results in levels of hydrogen sulfide that may exceed 0.1 mM (Currie 1953). A high organic load in the sediments of the Namibian continental shelf provides nutrients for the sulfate-reducing bacteria. As the hydrogen sulfide reaches the aerobic surface, the sulfide is chemically oxidized to elemental sulfur by O_2, and large plumes of sulfur float on the marine surface forming a distinct mass, which has been observed from space (Ohde et al. 2007). In deep wells where anaerobic conditions exist, corrosion of iron pipes due to sulfate-reducing bacteria contributes to rusty water in drinking water (Wood 1961).

Rice plants are susceptible to elevated levels of sulfide (Tanaka et al. 1968; Freney et al. 1982), and the physiological response to sulfide toxicity in rice has been described (Armstrong and Armstrong 2005). Additionally, the soil chemistry of rice paddies may contribute to the toxicity of hydrogen sulfide to rice plants (Fryer 2018). High levels of hydrogen sulfide in soil lead to iron stress and various phytotoxic responses (Lamers et al. 2013). However, at sublethal levels, hydrogen sulfide is proposed to signal plant hormones in response to various types of environmental stress (Lisjak et al. 2013; Jin and Pei 2015; Li et al. 2016).

In the paper industry, SRB have been reported to account for some pollution problems (Starkey 1961). Sulfite liquor from a paper mill entering the Ottawa River in Canada contained nutrients that enhanced the population of sulfate-reducing bacteria several miles downstream (Desrochers and Fredette 1960). Wood rot resulted from sulfate-reducing bacteria converting waterlogged pulp wood to thiolignin, and the thiolignin reacted with the preservative to enable microbial wood rotting (Russell 1961). Modification of lignin has also been reported by Ziomek and Williams (1989).

One of the microbial activities associated with food spoilage is the production of hydrogen sulfide, which has the characteristic "rotten egg" odor, and this spoilage was attributed to *Desulfotomaculum* (formerly *Clostridium*) *nigrificans* (Speck 1981). Spoilage of canned corn, mushrooms, and vegetables has been attributed to *Dst. nigrificans* (Werkman and Weaver 1927; Werkman 1929; Lin and Lin 1970). Sulfate-reducing bacteria growing in the brines for preservation of olives have been implicated in spoilage of olives (Levin et al. 1959). The spores of this sulfate-reducing bacterium are markedly heat resistant, and this presents special problems in food processing (Donnelly and Busta 1982; Doores 1983).

7.4 Oil Technology and SRB

Oil-well waters in Illinois and California were found to have a low content of sulfate, and this was attributed to the action of sulfate-reducing bacteria (Bastin 1926a; Gahl and Anderson 1928). Water from several oil-producing wells in Illinois was reported to contain appreciable levels of hydrogen sulfide (Bastin and Greer 1930), and water associated with petroleum deposits contained sulfide and carbonate, which was presumed to result from sulfate-reducing bacteria (Ginter 1930). Also, sulfate-reducing bacteria were present in the oil sands from the Grozny oil field formation (Ginsburg-Karagitscheva and Rodionova 1935). Core samples taken at a depth of 10–37 m in the Pleistocene sediment near Amsterdam were reported to contain microorganisms capable of sulfate reduction (von Wolzogen Kühr 1922). The demonstration of sulfate-reducing bacteria in sulfur–limestone samples collected at a depth of 1500 ft indicated that sulfate-reducing bacteria could survive in deep subsurface marine or petroleum deposits (Zobell 1946).

With the reports of sulfate-reducing bacteria in marine sediments containing petroleum, the suggestion was raised that sulfate-reducing bacteria may be important in formation or modification of petroleum (Zobell and Rittenberg 1948). Several reports indicate the role of sulfate-reducing bacteria in metabolism of petroleum hydrocarbons (Tausson and Vesselov 1934; Novelli and ZoBell 1944; Rosenfeld 1947). It was proposed that sulfate-reducing bacteria may have a role in recovery of petroleum from oil-bearing environments (Zobell and Rittenberg 1948). Enhanced release of oil could be attributed to dissolving of carbonate and sulfate rocks following acidification of the environment by sulfate-reducing bacteria or their production of carbon dioxide, which decreases the viscosity of crude oil (Zobell 1947a, b).

A common practice in the oil drilling industry is to inject water from an environment adjacent into a production well, and the emulsion of oil and water is separated at the surface. These pioneering activities related to the oil and gas industry have led to a great contribution to the industry. Oil souring may be associated with this injection process where the sweet oil (no hydrogen sulfide) becomes laden with hydrogen sulfide (souring) as a result of high levels of sulfate in the injection oil. The souring of oil fields and control of this deleterious activity are of ongoing concern in the oil industry (Cord-Ruwisch et al. 1987; Youssef et al. 2009; Gieg et al. 2011; Basafa and Hawboldt 2019). The use of biocides in industrial settings to inhibit activities and control growth of bacterial communities including those with SRB has been discussed in previous reviews (Jack and Westlake 1995; Greene et al. 2006).

In marine oil fields, the standard practice is to inject seawater, which contains sulfate into the oil reservoir. There are numerous reports indicating that bacteria in oil fields and reservoirs are highly diverse (Jeanthon et al. 2005; Berdugo-Clavijo and Gieg 2014; Nazina et al. 2017; Pannekens et al. 2019). The dominant species of SRB in oil wells is relatively limited. For example, thermophilic SRB (*Desulfotomaculum* spp.) and thermophilic sulfate-reducing archaea (*Archaeoglobus* spp.) are found in the Terra Nova field east of Newfoundland. The Terra Nova oil field is at a depth of 3200–3700 m below the seafloor where the temperature is 95 °C (Okpala et al. 2017). Sulfide produced from these microorganisms diffused through the reservoir where it reacted with reservoir rocks, and after 8 years, hydrogen sulfide appeared in the produced water. In comparison to the Terra Nova field, sulfide production in North Sea oil reservoirs was primarily the activity of *Thermodesulforhabdus norvegicus* and *A. fulgidus* strains (Nilsen et al. 1996; Beeder et al. 1994, 1995). A listing of 14 strains of mesophilic and thermophilic bacteria and archaea isolated from several different oil producing sites is provided by (Magot et al. 2000). In the review by Youssef et al. (2009), the SRB isolated from oil fields are listed as members of the Deltaproteobacteria class and include Desulfovibrionales (*Desulfovibrio* and *Desulfomicrobium*), Desulfobacterales (*Desulfobacter*, *Desulfobulbus*, *Desulfotignum*, and *Desulfobacterium*), and Syntrophobacterales (*Desulfacinum* and *Thermodesulforhabdus*). SRB isolated from oil environments with the Firmicutes phylum were members of the Clostridiales order and included species belonging to the *Desulfotomaculum*, *Desulfurispora*, *Desulfovirgula*, *Desulfosporosinus*, and *Thermodesulfobium* genera. Additionally, some SRB isolated from oil reservoirs were with the phylum Thermodesulfobacteria and species were members of the *Thermodesulfobacterium* and *Thermodesulfatator* genera (Youssef et al. 2009).

The control of sulfidogenic microorganisms by stimulating the growth of other microorganisms present has been pursued with some success. Additions of nitrate or nitrite have been used as a bioprocess to control the growth and sulfide production by SRB (Voordouw 2008). Nitrate supports the growth of sulfite-oxidizing bacteria and heterotrophic nitrate-reducing bacteria in the water associated with the oil wells. Organic compounds in the water are consumed with the growth of nitrate-respiring bacteria, and organic compounds as electron donors become limiting for SRB (Hubert and Voordouw 2007; Okpala et al. 2017; Suri et al. 2017). This competition for nutrients to control biogenic sulfide production was first proposed by Hitzman and Dennis (1997). Incomplete bacterial reduction of nitrate with the accumulation of nitrite is desirable over complete reduction of nitrate to N_2 because nitrite inhibits sulfite reduction (Greene et al. 2006). However, in a complex microbial community, there may be SRB that couple nitrite respiration to cell growth (Greene et al. 2006), and sulfidogenesis persists in nitrate additions to oil fields (Hulecki et al. 2009). Control of biogenic sulfide formation in field environments is subject to site characteristics including oil chemistry, chemicals available as microbial nutrients, and composition of the microbial community.

7.5 Metabolism of Hydrocarbons

7.5.1 *Oxidation of Environmentally Relevant Organic Compounds*

There are a large number of compounds derived from the coal, gas, and oil industries that enter the environment and are metabolized by SRB. Anaerobic oxidation of methane is considered to be attributed to a syntrophic process involving anaerobic methanotrophic archaea (ANME) and SRB (see Sect. 1.3 in Chap. 1). The net energy yield of the reaction of methane oxidation with sulfate reduction as shown in Eq. (7.8) is unresolved with −16.6 kJ/mol reported by Schink (1997) and −22 to −36 kJ/mol listed by Valentine and Reeburgh (2000). It appears that the energy released by methane oxidation may be close to the amount of energy needed for ATP synthesis (Caldwell et al. 2008; Müller and Hess 2017).

$$CH_4 + SO_4{}^{2-} \rightarrow HS^- + HCO_3{}^- + H_2O \tag{7.8}$$

The first report of alkane oxidation attributed to SRB was by Novelli and Zobell (1944) and by Davis and Yarbrough (1966), but these cultures were lost. Subsequently, *Desulfatibacillum aliphaticivorans* strain Hxd3 (Aeckersberg et al. 1991) and sulfate-reducing bacteria strain TD3 (Rueter et al. 1994), Pnd3 (Aeckersberg et al. 1998), and AK-01 (So and Young 1999) were reported to oxidize alkanes. *Desulfatibacillum aliphaticivorans* oxidizes alkanes (C13–C18) and alkenes (C7–C23) (Cravo-Laureau et al. 2004), while *Desulfoglaeba alkanexedens* degrades n-alkanes (C6–C12) (Davidova et al. 2006).

There are numerous reports of decomposition of hydrocarbons in the open environment by SRB and several of these are presented in Table 7.1. Soil contamination with gasoline leaking from tanks and pumps has been a serious environmental problem (Weelink et al. 2010) due to benzene, toluene, ethylbenzene, and xylene (BTEX) in gasoline. Additionally, BTEX compounds have been widely employed in industry for an array of activities. The oxidation of benzene by sulfate-reducing bacteria has been demonstrated in sediment from the San Diego Bay, California (Lovley et al. 1995), where benzene was mineralized completely to CO_2 without any intermediates. Several reports highlight the decomposition of benzene and using enrichment cultures growing under sulfate-reducing conditions (Musat and Widdel 2008; Kleinsteuber et al. 2008). A Gram-positive bacterium *Pelotomaculum propionicicum* grows syntrophically on propionate with methanogens, and while it has not been reported to reduce sulfate, it has *dsrAB* genes for dissimilatory sulfate reduction (Laban et al. 2009). Benzene is a difficult substrate for the anaerobic SRB to degrade, while aromatic substrates such as phenol and benzene, which have a functional group or indole with a non-carbon atom in the ring, are more readily oxidized (Rabus et al. 1993). The energy yield for benzene oxidation and sulfate

Table 7.1 Examples of environmental sites with hydrocarbon decomposition attributed to sulfate-reducing bacteria

Substrates	Comment	Reference
Methane	A hydrogenic bacterium *Candidatus* Desulfofervidus auxilii is involved in oxidation of methane	Krukenberg et al. (2016)
Alkanes	Alkanes are degraded in marine seeps by sulfate-reducing bacteria of the *Desulfosarcina/Desulfococcus* clade	Kleindienst et al. (2014)
Alkenes	Sulfate-reducing bacteria oxidize ethylene in a fresh water canal	Fullerton et al. (2013)
Cyclohexane	Degradation of cyclohexane occurs in marine sediments by sulfate-reducing bacteria	Jaekel et al. (2015)
BTEX	Degradation of benzene, toluene, ethyl, benzene, and xylenes in oil sands tailings ponds is attributed to sulfate-reducing bacteria	Stasik et al. (2015) Meckenstock et al. (2016) Huang et al. (2017) Beller et al. (1996) Kolhatkar et al. (2008) Kleindienst et al. (2008) Kolhatkar and Schnobrich (2017)
Benzenes	*Desulfotomaculum* spp. degraded benzene, toluene, and ethylbenzene in an extended deep gas storage subterranean aquifer	Aüllo et al. (2016) Anderson and Lovley (2000)
Naphthalene	In oil sands process waters, greater decomposition of 2-methyl naphthalene under sulfate-reducing conditions as compared to methanogenic conditions	Folwell et al. (2015)
	Degradation of naphthalene in anaerobic aquifers is attributed to Desulfobacteraceae members	Kümmel et al. (2015)
2,4,6-trinitrotoluene	Degradation is faster under sulfate addition as compared to nitrate addition	Boopathy (2007) Boopathy (2014)
Polyaromatic hydrocarbons	Coal tar–contaminated sediments in Boston Harbor were oxidized under conditions supporting growth of sulfate-reducing bacteria	Rothermich et al. (2002)
	Decomposition of high-molecular-weight hydrocarbons occurred under sulfate-reducing conditions	Kuwano and Shimizu (2006)

reduction is not as favorable as reactions where nitrate or Fe(III) is the electron acceptor in place of sulfate (Table 7.2). Many different hydrocarbons are derived from oil, and some of the compounds are used as electron donors by SRB (see Table 7.3), and some SRBs are generalist in that they can attack more than one organic substrate.

Table 7.2 Energetics of reactions where benzene is oxidized and nitrate, Fe(III), and sulfate are the electron acceptors

Reactions	Equations	$\Delta G^{0'}$ (kJ/mol)
Denitrification	$C_6H_6 + 6NO_3^- \rightarrow 6HCO_3^- + 3N_2$	−2978
	$C_6H_6 + 15NO_3^- + 3H_2O \rightarrow 6HCO_3^- + 15NO_2^- + 6H^+$	−2061
Iron reduction	$C_6H_6 + 30Fe^{3+} + 18H_2O \rightarrow 6HCO_3^- + 30Fe^{2+} + 36H^+$	−3040
Sulfate reduction	$C_6H_6 + 3.75SO_4^{2-} + 3H_2O \rightarrow 6HCO_3^- + 1.875H2S + 1.875HS^- + 0.375H^+$	−186

Adopted from Weelink et al. (2010)

Table 7.3 Organic substrates oxidized by sulfate-reducing bacteria[a]

Bacteria	Compounds	References
Desulfovibrio (*D.*) *vulgaris* Marburg	2-Methoxyethanol	Tanaka (1992)
D. desulfuricans ATCC 7757	Kraft lignin, lignosulfonate	Ziomek and Williams (1989)
D. desulfuricans M6	Phenylsulfide, benzylsulfide, benzothiophene, dibenzothiophene, crude oil	Kim et al. (1990a) Kim et al. (1990b)
D. furfuralis	Furfural	Folkerts et al. (1989)
Desulfovibrio sp.	Dibenzothiophene, dibenzylsulfide, benzothiophene, dibenzyldisulfide, butylsulfide, octylsulfide	Köhler et al. (1984)
Desulfovibrio sp. B	Furfural, 2,4,6-trinitrotoluene, 2,4-dinitrophenol, 2,6-dinitrophenol, aniline	Boopathy and Daniels (1991) Boopathy and Kulpa (1993) Boopathy et al. (1993)
Desulfovibrio sp. F-1	Furfural	Brune et al. (1983)
Desulfobacterium (*Dsfb.*) *phenolicum*[b]	Benzoate, phenol, *p*-cresol, 2-hydroxybenzoate, 4-hydroxybenzoate, phenylacetate, indole, toluene, 4-hydroxyphenylacetate	Bak and Widdel (1986b) Rabus et al. (1993)
Dsfb. catecholicum	Benzoate, 4-hydroxybenzoate, protocatechuate, 2-aminobenzoate, 3,4-dihydroxybenzoate, catechol, resorcinol, hydroquinone, pyrogallol, phloroglucinol	Szewzyk and Pfennig (1987)
Dsfb. anilini	Benzoate, 3,4-dihydroxybenzoate, quinoline, 2,5-dihydroxybenzoate, 2,4-dihydroxybenzoate, aniline, 2-aminobenzoate, 4-hydroxybenzoate, phenol, *p*-cresol, pyrogallol, phenylacetate, phenylpropionate, catechol, 2-hydrtoxtbenzoate, 3-hydroxybenzoate	Schnell et al. (1989)
Dsfb. sp. Cat 2	Benzoate, catechol, *m*-cresol, *p*-cresol, indole, pyrogallol, phenylacetate, phenylpropionate, hippurate, 2-aminobenzoate, quinoline, 2-hydroxybenzoate, 3-hydroxybenzoate, 4-hydroxybendoate, phenol, 3,4-dihydroxybenzoate, 3,5-dihydroxybenzoate	Schnell et al. (1989)
Dsfb. oleovorans Hxd3	*n*-Dodecane, *n*-tetradecane, *n*-pentadecane, *n*-heptadecane, *n*-octadecane, *n*-eicosane, *l*-hexadecane	Aeckersberg et al. (1991)
Dsfb. vacuolatum	Phenylpropionate	Widdel (1988)
Dsfb. cetonicum	Benzoate	Galushko and Rozanova et al. (1991)
Dsfb. indolicum	Indole, quinoline, 2-aminobenzoate	Bak and Widdel (1986a)

(continued)

Table 7.3 (continued)

Bacteria	Compounds	References
Desulfotomaculum (Dsm.) sapomandens	Benzoate, phenylacetate, 3-phenylpropionate, 4-hydroxybenzoate	Widdel (1988); Cord-Ruwisch and Garcia (1985)
Dsm. orientis	3,4,5-Trimethoxybenzoate	Klemps et al. (1985)
Dsm. sp. strain TEP	3,4,5-Trimethoxybenzoate	Klemps et al. (1985)
Dsm. sp. strain TWC	3,4,5-Trimethoxybenzoate	Klemps et al. (1985)
Dsm. sp. strain TWC	Benzoate	Klemps et al. (1985)
Dsm. sp. strain Groll	Benzoate, 3-hydroxybenzoate, phenol, catechol, 4-hydroxybenzoate, 2-aminobenzoate, vanillate, 3,4-dihydroxybenzoate, phenylacetate, ferulate, phenylpropionate, protocatechuate, cinnamate, m-cresol, p-cresol, syringate, 2,6 dimethoxyphenol	Kuever et al. (1993)
Dsm. thermobenzoicum	Benzene	Tasaki et al. (1991)
Dsm. australicum	Benzoate	Love et al. (1993)
Desulfococcus (Dc.) multivorans	Benzoate, phenylacetate, phenylpropionate, 2-hydroxybenzene, hydroquinone, 2,5-dihydroxybenzoate	Widdel (1988) Schnell et al. (1989)
Dc. niacini	Nicotinic acid, phenylpropionate	Imhoff-Stuckle and Pfennig (1983)
Desulfosarcina variabilis	Benzoate, phenylacetate, 3-phenylpropionate	Widdel (1988) Widdel and Pfennig (1984)
Desulfomonile tiedjei	Benzoate, 3-anisate, 4-anisate	DeWeerd et al. (1990)
Desulfonema magnum	Benzoate, phenylacetate, 3-phenylpropionate, hippurate, 4-hydroxybenzoate	Widdel et al. (1983)
Desulfoarculus sp. SAX	Benzoate, 4-hydroxybenzoate, phenol, phenylacetate	Drzyzga et al. (1995)
Desulfobacula toluolica	Toluene, benzoate, p-cresol, benzaldehyde, phenylacetate, p-hydroxybenzaldehyde, 4-hydroxybenzoate	Rabus et al. (1993)
Desulfoprunum benzoelyticum	Benzoate	Junghare and Schink (2015)
SRB strain oXyS1	*o*-Xylene, toluene, *o*-ethyltoluene, benzoate, *o*-methylbenzoate, benzylsuccinate	Harms et al. (1999)
SRB strain mXyS1	*m*-Xylene, toluene, *m*-ethyltoluene, benzoate, *m*-isopropyltoluene, *m*-methylbenzoate	Harms et al. (1999)
Desulfosarcina sp. strain PP31	*p*-Xylene, *o*-xylene, *m*-xylene, *n*-hexane, benzene, toluene, ethylbenzene	Higashioka et al. (2012)

(continued)

Table 7.3 (continued)

Bacteria	Compounds	References
Desulfotomaculum AR1	Xylene, toluene	Verkholiak and Peretyatko (2019)
Desulfovibrio desulfuricans Ya-11	Xylene, toluene	Verkholiak and Peretyatko (2019)

[a]Modified from Ensley and Suflita (1995); Barton and Fauque (2009)
[b]*Desulfobacterium phenolicum* has been reclassified as *Desulfobacula phenolica* (Kuever et al. 2001)

7.5.2 *Reductive Dehalogenation*

Shelton and Tiedje (1984) isolated a unique SRB capable of using the organohalides as a final electron acceptor. While *Desulfomonile tiedjei* DCB-1 uses sulfate, sulfite, and thiosulfate as a final electron acceptor, it is also capable of dehalogenating tetrachloroethene (TCE), 3-chlorobenzoate (Dolfing and Tiedje 1987; Dolfing 1990; Mohn and Tiedje 1990; DeWeerd et al. 1990), and pentachlorophenol (Mohn and Kennedy 1992) with electrons derived from oxidation of formate. *Desulfomonile tiedjei* DCB-1 removes bromine and iodine from all positions on the benzoate ring, while chlorine was removed only from the *meta* position (Mohn and Tiedje 1992). With the transfer of electrons from formate to 3-chlorobenzoate, there is the production of ATP coupled to generation of proton motive force (Mohn and Tiedje 1991).

Members of the *Desulfitobacterium* genus are known for their ability to degrade halogenated organic compounds. Several *Desulfitobacterium* strains including TCE1 were initially described as *Desulfitobacterium hafniense*, but taxonomic analysis prompted the reclassification of these strains as *Desulfitobacterium frappieri* (Villemur et al. 2006). A syntrophic relationship between *D. fructosivorans* and *Desulfitobacterium* strain TCE1 contributed to reductive dehalogenation of tetrachloroethene (TCE; Drzyzga et al. 2001; Drzyzga and Gottschal 2002). When *D. fructosivorans* was grown in sulfate-limiting concentrations, H_2 produced from fructose fermentation was consumed by *Desulfitobacterium* strain TCE1 to drive dehalogenation reactions. Additional information about co-culture of bacteria is in Sect. 1.3.

7.6 Magnetosomes and Iron Mineralization

A novel dissimilatory sulfate-reducing bacterium that contained magnetosomes was isolated from a sulfide-rich sediment of the Kameno River, Japan (Sakaguchi et al. 1993, 1996), and was designated *D. magneticus* RS-1 DSM 13731 (hereafter referred to as RS-1) (Sakaguchi et al. 2002). This is the only magnetotactic bacterium of the δ-Proteobacteria phylum with the other magnetotactic bacteria with the α-Proteobacteria phylum. Intracellular magnetite ($Fe^{2+}Fe_2^{3+}O_4^{2-}$) or greigite (Fe_3S_4) bullet-shaped crystals are produced when this bacterium is cultured on lactate–fumarate medium and capable of producing extracellular magnetic iron sulfide. Only a few of the cells of RS-1 contained magnetite crystals, which were ~40 nm, and when present, the range of internal magnetite particles per cell was 1–18 (Pósfai et al. 2006). These nanomagnetite particles inside RS-1 are not organized into chains, and the magnetic moments of individual cells are not sufficient to arrange the cells in response to geomagnetic field lines. It has been speculated that the magnetite crystals inside the cells of this organism are not for magnetotaxis but rather for iron storage or detoxification (Pósfai et al. 2006). Extracellular formation of hematite was observed, and this observation is interesting in that while the culture environment is anaerobic, iron is in the oxidized state. However, intracellular magnetite granules also have the same environmental exposure where the cytosol is anaerobic. Since small quantities of magnetite are associated with the core of animal ferritin (Quintana et al. 2004), perhaps magnetite in RS-1 functions in maintaining appropriate redox status inside the cell in a manner similar to ferritin in anaerobic bacteria. A diffuse organic layer was reported to surround isolated magnetosomes of PS-1 (Matsunaga et al. 2009); however, after an extensive examination of magnetosomes from RS-1, Byrne et al. (2010) concluded that these magnetosomes were not surrounded by bilayer membranes. In addition to the magnetosome, cells of the bacterial strain RS-1 also contain an organelle that contains iron and phosphorus, and this amorphous iron–phosphorus granule is surrounded by a membrane (Byrne et al. 2010). The synthesis of the magnetite granule and the iron–phosphorus granule in RS-1 is suggested to be independently controlled, and the role of this iron–phosphorus granule remains unresolved.

Genome analysis of RS-1 revealed the presence of numerous genes found in magnetotactic bacteria (i.e., *Magnetospirillum magneticum* strain AMB-1, *Candidatus* Magnetococcus sp. strain MC-1, *Magnetospirillum gryphiswaldense* strain MSR-1, and *Magnetospirillum magnetotacticum* strain MS-1) and is considered important for magnetite particle formation (Matsunaga et al. 2009). As with other magnetotactic bacteria, RS-1 contained three distinct gene regions, which include *nuo* genes, *mamAB*-like genes, and several genes in a cryptic plasmid. The *nuo* (NADH–quinone oxidoreductase [complex I]) genes are found as a single operon *nuoABCDEFGHIJKLM* (DMR_13310-420) and as another separate gene set (DMR_02470, DMR_27770-27880). Other *Desulfovibrio* strains have only a single copy of *nuo* genes. The genome of PS-1 was reported to contain orthologous genes for *mam* (magnetosome membrane proteins) that in magnetotactic α-Proteobacteria constitute a *mamAB* gene cluster (Grünberg et al. 2001). The orthologs of *MamA*, *MamB*, *MamE*, *MamK*, *MamM*, *MamO*, *MamQ*, *MomP*, and *MamT* were present in PS-1 and were proposed to consist of a single operon, and the cryptic plasmid pDMC1 was reported to contain three genes and their function was unresolved (Nakazawa et al. 2009). Using proteomic analysis, three magnetosome membrane proteins (MamA, MamK, and MamM) found in RS-1 were also present in magnetotactic bacteria (*Magnetospirillum* [*M.*] *magneticum* AMB-1 and

M. gryphiswaldense MSR-1), and the cryptic plasmid of RS-1 (pDMC1) contained two genes encoding proteins homologenous to *Magnetococcus* MC-1 (Matsunaga et al. 2009). A comparative genomic analysis for PS-1 with known magnetotactic bacteria including the δ-Proteobacteria was prepared by Lefèvre et al. (2013), and the *mad* genes for production of the bullet-shaped magnetite crystals in PS-1 were reviewed. McCausland and Komeili (2020) suggest that different types of magnetotactic bacteria such as PS-1 have specific genes for magnetosome formation, and while suicide vectors for specific gene deletions are not useful for PS-1, the approach by Grant et al. (2018) using replicative plasmids to replace genes of interest has considerable promise.

While the magnetotactic bacteria are reported to be broadly distributed around the world, the geographic information of the magnetotactic SRB requires additional study (Lin et al. 2014). The biosphere appears to be rich with magnetotactic sulfate-reducing bacteria that either remain to be isolated or to be further characterized. Bacteria with spiral cells containing bullet-shaped magnetite granules were isolated from alkaline (pH 9.0–9.5) mud environments and were closely related to *Desulfonatronum thiodismutans*, a non-magnetotactic sulfate reducer (Lefèvre et al. 2011a). A magnetotactic sulfate-reducing bacterium, strain BW-1, of the Deltaproteobacteria group was isolated from Death Valley National Park, California, USA. The strain BW-1 biomineralizes both greigite and magnetite, and it has distinct gene clusters for production of each type of magnetosome (Lefèvre et al. 2011b). Sulfate-reducing bacteria of the *Desulfosarcina/Desulfococcus* group in sediments of the Black Sea contained iron sulfide aggregates characteristic of ferrimagnetic greigite in magnetotactic bacteria (Reitner et al. 2005).

"*Candidatus* Magnetoglobus multicellularis" consists of 10–40 individual cells, and this multicellular prokaryote was isolated from a hypersaline lagoon in Brazil with each cell containing about 60–100 pleomorphic magnetosomes. With identification based on 16S rRNA gene, this multicellular magnetotactic organism belongs to the family Desulfobacteraceae and is closely related to *Desulfonema*, *Desulfosarcina*, and *Desulfococcus* (Abreu et al. 2007). Another multicellular magnetotactic prokaryote was isolated from the North Sea sediment, and the cells were reported to contain genes for dissimilatory adenosine-5′-phosphate reductase and dissimilatory sulfite reductase. The 16S rRNA gene analysis indicated that this multicellular organism from the North Sea was closely related to *Desulfosarcina variabilis* (Wenter et al. 2009).

7.7 Mercury Methylation

In sediments, hydrogen sulfide immobilizes Hg^{2+}, and insoluble HgS is converted to methyl mercury at a rate markedly less than with soluble Hg^{2+}. At high levels of H_2S, it was proposed that volatile dimethylmercury production is favored over methylmercury formation (Craig and Bartlett 1978). Initial papers (Blum and Bartha 1980; Compeau and Bartha 1983) revealed that salinity had a role in bacterial methylation of mercury. Compeau and Bartha (1984) demonstrated that demethylation of CH_3HgCl was inhibited at low levels of salinity, but at high salinity, this inhibition was reversed. Methylation of mercury was greater if the environment had a redox potential of −220 mV as compared to methylation activity at +110 mV (Compeau and Bartha 1984). The formation of methylmercury by sulfate-reducing bacteria was established by Compeau and Bartha (1985) as they demonstrated that 95% of methylmercury synthesis was inhibited by molybdate, while 2-bromoethane sulfonate (inhibitor of methanogens) had no effect on methylmercury formation. Two strains of *D. desulfuricans*, which were capable of methylmercury formation, were isolated, and one required 0.5% NaCl while the other grew without the addition of salt (Compeau and Bartha 1985, 1987). In addition to *D. desulfuricans* LS, *D. desulfuricans* Norway 4 displayed mercury methylation activity, while *D. desulfuricans* ATCC 27774, *D. gigas*, *Dst. orientis*, and *Desulfobulbus propionicus* FP did not have this capability (Choi and Bartha 1993). Cobalamin (vitamin B_{12}) was proposed as the methyl carrier for the mercury methylation reaction in *D. desulfuricans* LS (Choi and Bartha 1993), and the cobalamin protein (40 kDa) received the methyl group from CH_3-tetrahydrofolate (Choi et al. 1994a, b). Each gram of cell protein of *D. desulfuricans* LS was found to contain 4–35 μg tetrahydrofolate and 58–161 ng of cobalamin, which supported the proposed involvement of these components in mercury methylation reactions (Berman et al. 1990). The origin of the methyl group was proposed to be from formate or C-3 of serine employing the acetyl-coenzyme A synthase pathway (Berman et al. 1990; Choi et al. 1994a, b). However, not all bacteria displaying mercury methylation have a functioning reductive acetyl-coenzyme A pathway, which is also known as the Wood–Ljungdahl pathway. Mercury methylation has been reported for *D. africanus* (DSMZ 2603), *Desulfobacter* strain BG-8, *Desulfobulbus propionicus* 1pr3 (DSMZ 2302), and *Desulfobulbus propionicus* MUD (DSMZ 6523), which lack the acetyl-coenzyme A pathway and are incomplete oxidizers (Ekstrom et al. 2003). *Desulfococcus multivorans* 1be1 (DSMZ 2302), *Desulfosarcina variabilis* 3be13, *Desulfobacterium autotrophicum*, and *D. desulfuricans* LS methylate mercury and also have the acetyl-coenzyme A pathway. Thus, mercury methylation may be independent of the reductive acetyl-coenzyme A pathway (Ekstrom et al. 2003), and cobalt limitation in the growth medium for *D. africanus* DSM 2603 does not inhibit mercury methylation (Ekstrom and Morel 2008). This suggests that vitamin B_{12} may not be required for mercury methylation in all sulfate-reducing bacteria. The rate of mercury methylation was found to vary considerably with different species of sulfate-reducing bacteria, and with the mercury methylation activity normalized to cell number, the following order was observed with laboratory experiments: *Desulfobacterium* sp. strain BG-33 $\gg$ *Desulfobacter* strain BG-8 $\approx$ *Desulfococcus multivorans* ATCC 33890 $\gg$ *D. desulfuricans* ATCC 1354 $\approx$ *Desulfobulbus propionicus* ATCC 33891 (King et al. 2000). Some strains of *Desulfovibrio* (i.e., *D. desulfuricans* Essex 6, *D. desulfuricans* El Agheila Z, and *D. alaskensis* G20) were not observed to methylate mercury, while methylation of mercury by *D. aestuarii* Sylt3 and *D. vulgaris* Hildenborough was only twice the level of the control (Gilmour et al. 2011).

Model organisms are used to further understand microbial processes, and since the *D. desulfuricans* LS culture is no longer available, a new bacterium was isolated. Recently, *D. desulfuricans* ND132 has become the model SRB for methylmercury research due to its high methylation activity and biochemical similarity to *D. desulfuricans* LS (Gilmour et al. 2011). Analysis of the genome of *D. desulfuricans* ND132 has been released (Brown et al. 2011). Production of methylmercury by *D. desulfuricans* ND132 is constitutive and does not account for mercury resistance, and a greater production occurs with fumarate as the electron acceptor as compared to sulfate. Following active transport of mercury into the cell of *D. desulfuricans* ND132, mercury methylation occurs in the cytoplasm and methylmercury is transported from the cell into the culture medium (Schaefer et al. 2011). The mechanisms for mercury uptake may vary with the mercury-methylating bacteria (Benoit et al. 1999; Schaefer and Morel 2009). Methylmercury was slowly degraded by *D. desulfuricans* ND132, and the rates of degradation were higher in sulfidogenic cultures (Gilmour et al. 2011). While some degradation of methylmercury can be attributed to chemical disproportionation, the presence of Hg^0 has never been reported for cultures of sulfate-reducing bacteria methylating mercury (Craig and Moreton 1984; Wallschläger et al. 1995); Gilmour et al. (2011) considered that bacterial metabolism would account for some degradation of methylmercury. Upon examination of lake water, the presence of SRB and mercury methylation in the water column was reported (Achá et al. 2012).

Methylation of mercury varies between the strains of sulfate-reducing bacteria. While four species of SRB (*D. alcoholivorans* DSM 5233, *D. tunisiensis* DSM 19275, *D. carbinoliphilus* DSM 17524, and *D. piger* DSM 749) were unable to methylate mercury, *D. aespoeensis* DSM 10631, *D. alkalitolerans* DSM 16529, *D. psychrotolerans* DSM 19430, and *D. sulfodismutans* DSM 3696 methylated mercury but at rates lower than *D. desulfuricans* ND132, isolate 12, or *Desulfovibrio* sp. strain X2, which were isolated from the Chesapeake Bay sediments (Graham et al. 2012). With the addition of Hg(II) to strains of SRB capable of methylation and non-methylation, Hg(II) was rapidly associated with the SRB cells, which could perhaps be attributed to binding of Hg(II) onto the cell surface. To maintain inorganic mercury in solution, the presence of *l*-cysteine was required. Mercury methylation is a cytoplasmic activity, and once produced, methylmercury quickly exits the cell. Neither methylating nor non-methylating strains of SRB were found to degrade methylmercury (Graham et al. 2012).

D. desulfuricans DN132 has the capability of transforming gaseous Hg^0 to methylmercury at a rate about one-third slower than conversation of soluble Hg^{2+} to methylmercury (Colombo et al. 2013; Hu et al. 2013). Mercury methylation for *D. desulfuricans* DN132 and *Geobacter sulfurreducens* PCA requires the genes *hgcA* and *hgcB*, and if these genes are deleted, mercury methylation does not occur (Parks et al. 2013). HgcA is a putative corrinoid protein and HgcB is a 2 [4Fe-4S] ferredoxin. This two-gene cluster is also found in the genomes of other bacteria (*D. aespoeensis* Aspo-2, *D. africanus* Walvis Bay, and *Desulfobulbus propionicus* DSM 2032) capable of mercury methylation. It has been reported that the transfer of *hgcA* and *hgcB* genes into a non-methylating *Desulfovibrio* does not confer mercury methylation to the recipient cells, and this suggests that additional gene functions are required (Smith et al. 2015). Additional information is needed to understand the HgcA/HgcB structures and the mechanisms of methylation of mercury by sulfate reducers.

Methylmercury in the environment is being assessed using the *hgcAB* genes. Using cloning/sequencing *hgcAB* gene products and 16S rRNA gene pyrosequencing, the presence of mercury-methylating genes and Deltaproteobacteria was observed in soil where methylmercury was observed (Christensen et al. 2019). Another study used the abundance of the mercury-methylating gene *hgcA* in ocean samples obtained from the Ocean Gene Atlas, which were clustered into three groups (Villar et al. 2020): Cluster 1 contained Desulfobacterales, Clostridiales, and Desulfovibrionales. Cluster 2 gathered together sequences closely related to *Smithella*, *Desulfomonile tiedjei*, and Chloroflexi, while Cluster 3 consisted to be members of *Nitrospina*. The global distribution of *hgcA* genes indicates that members of cluster 3 are broadly distributed throughout the world, while species of cluster 1 and 2 are markedly restricted (Villar et al. 2020).

Mercury resistance and detoxification are associated with the *mer operon*, which is found in many bacteria. A few sulfate-reducing prokaryotes have putative genes that are relevant to the *mer operon* (Bruschi et al. 2007). The putative gene for *merA*, which encodes for the reduction of Hg^{2+} to Hg^0 by mercuric reductase, has been identified in *D. vulgaris* Hildenborough with a gene locus of DVU1037 on the chromosome and DVUA0093 on the plasmid. In *D. alaskensis* G20, the gene locus for putative gene for *merA* is Dde1463. The gene locus in *Desulfotalea psychrophia* for a putative gene related to *merA* is DP0504. The putative gene for *merR* (the mercuric resistance regulatory protein) in *A. fulgidus* is AF00675. The locus for the putative mercuric transport gene in *D. vulgaris* Hildenborough is DVU2325, in *D. alaskensis* G20 is De1312, and in *A. fulgidus* is AF0346. The putative gene for Hg^{2+} binding protein in *Desulfotalea psychrophila* is DP1460. It remains to be demonstrated that these presumptive *mer* genes are expressed and are functional in sulfate-reducing bacteria.

An important application of mercury-methylating and mercury-reducing bacteria is their interaction with mercury in the environment. The toxicity of mercury and its distribution in the environment are well established (National Academy of Sciences 1978). Due to human activities, 2500 tons of mercury are released into the environment every year (Outridge et al. 2018) and 3.7 tons of mercury are present in dental amalgam waste released annually into municipal wastewater treatment systems. Since silver, zinc, and copper are commonly associated with mercury in the dental waste, it is important that bacteria in the wastewater are resistant to these heavy metal ions. Members of the Desulfovibrionaceae and Desulfobulbaceae were present in significant numbers in dental amalgam waste containing high levels of mercury, and it is inferred that these sulfate-reducing bacteria would be important in detoxification of mercury (Rani et al. 2015).

Methylmercury may be degraded by sulfate-reducing bacteria. With iodomethylmercury (CH_3HgI) as a substrate, cells of *D. gigas*, *D. africanus*, or anaerobic sediment from Mono Lake, California, were found to release trace levels of CO_2, while an obligately methylotrophic methanogen strain GS-16 released high levels of CO_2 and CH_4 (Oremland et al. 1991). With *D. desulfuncans* strains LS and API grown in media with sulfate as the electron acceptor with the release of hydrogen sulfide, the addition of chloromethylmercury (CH_3HgCl) to the culture resulted in the immediate formation of a white precipitate, which was dimethylmercury sulfide [$(CH_3Hg)_2S$] (Baldi et al. 1993). In comparison, the addition of HgCl to cultures of LS and API resulted in a black precipitate, which was mercury sulfide. Dimethylmercury sulfide slowly decomposes under anaerobic conditions to metacinnabar, dimethylmercury, and methane (Baldi et al. 1993).

7.8 Biologically Induced Minerals

When the formation of biogenic metals by *Desulfovibrio* is discussed, this generally refers to hydrogen sulfide as the end product of reacting with soluble metals to produce metal sulfides (Barton and Tomei 1995). The formation of metal sulfides attributed to the activities of sulfate reducers was first reported by Miller (1950). Subsequently, the impact of sulfate reducers "breathing" hydrogen sulfide became recognized. As reported by Trudinger et al. (1985; Rickard 2012), sulfate-reducing bacteria and archaea are the predominant microorganisms responsible for 97% of sulfide produced on Earth. The remaining 3% of sulfide is attributed to volcanoes and vents in the ocean (Elderfield and Schultz 1996; Andres and Kasgnoc 1998). It has been estimated that seven megatons of H_2S is produced each day in marine sediments (Jørgensen and Kasten 2006). The release of sulfide into metal-laden environments results in the precipitation of metals as metal sulfides, and the order for removal of heavy metals may be difficult to predict due to the phases of the metals involved, the pH, and other environmental conditions. Additionally, there is an array of metal sulfide formations ranging from disordered precipitations to defined crystalline minerals. The association of microorganisms and sulfide minerals has been recently reviewed (Picard et al. 2016), and the following narrative discusses some of these sulfide mineral formations.

7.8.1 Iron Sulfide Mineral Precipitation

Hydrogen sulfide of microbial origin was considered to be responsible for enrichments of pyrite formations in Mediterranean and Caspian Seas (Doss 1912), Bay of Fundy (Kindle 1926), the Sea of Azov (Issatschenko 1929), Norwegian fjords (Strøm 1939), Walvis Bay (Copenhagen 1934), Cu–Zn Kidd Creek Mine (Fortin and Beveridge 1997), and salt marshes (Howarth 1979). While pyrite formation is rarely produced by sulfate-reducing bacteria (Rickard 1969), pyrite is suggested to be driven by abiotic processes (diagenesis) and reduction of iron and sulfur by microorganisms (Archer and Vance 2006; Guilbaud et al. 2011). After several months of incubation, *D. desulfuricans* forms pyrite and marcasite from goethite (Rickard 1969). Laboratory experiments describing the formation of extracellular iron sulfide by several strains of sulfate-reducing bacteria provide conditions for bioformation of iron sulfide minerals (Picard et al. 2016). After incubation for 16 days on pyruvate–sulfate medium, *D. vulgaris* produced mackinawite, and with accumulation of sulfide, greigite is formed from mackinawite (Zhou et al. 2014). With a consortium containing *Desulfovibrio* and *Sulfurospirillum* as the dominant bacteria, pyrite was formed around the *Desulfovibrio* cells, and this was attributed to the presence of polysulfide around these cells (Berg et al. 2020). In other reports, bacterial cells serve as mineral nucleation sites for the formation of amorphous FeS, greigite, pyrrhotite, or mackinawite (Neal et al. 2001; Stanley and Southam 2018; Picard et al. 2016, 2018). In media rich with phosphate and *Desulfovibrio*, the precipitation of iron and phosphate as vivianite [$Fe_3(PO_4)_2 \cdot 8H_2O$] occurs (Kraal et al. 2012, 2017; März et al. 2018). It was proposed that under appropriate circumstances, sulfate-reducing bacteria could participate in the formation of magnetic iron sulfide (Freke and Tate 1961). Internal granules of magnetite and greigite are found in magnetotactic sulfate-reducing bacteria, and this topic is discussed in the section of magnetosomes.

7.8.2 Cu Sulfide Deposits

There had been suggestion that copper sulfide deposits in marine settings may have been established by sulfate-reducing bacteria (Schneiderholm 1923; Trask 1925; Thiel 1926; Bastin 1933). Deposits of organic matter would support the growth of sulfate-reducing bacteria, and the sulfide generated could contribute to ores of lead sulfide and zinc sulfide (Bastin 1926b). Some researchers (Davidson 1962; Booth and Mercer 1963) were critical of the possibility that copper was precipitated by sulfide from sulfate-reducing bacteria because copper is highly toxic to sulfate-reducing bacteria. Cheney and Jensen (1962) supported the biogenic formation of copper sulfide ore because soil copper lacks bioavailability and copper readily forms copper sulfide in the presence of hydrogen sulfide, which would, in effect, reduce copper toxicity (Karnachuk et al. 2008). With an actively metabolizing *Desulfovibrio* culture, sufficient hydrogen sulfide is produced to prevent toxicity until the concentration of copper sulfate exceeds 0.25–0.29% (Temple and Le Roux 1964). Three copper-tolerant cultures of *Desulfovibrio* were isolated from wastewater effluents from a zinc smelter, and these cultures precipitated copper as covellite (CuS), chalcocite (Cu_2S), and chalcopyrite ($CuFeS_2$) (Karnachuk et al. 2008). An interesting bacterial isolate, *Citrobacter* sp. strain DBM, produces hydrogen sulfide from sulfate with the precipitation of copper (Qiu et al. 2009).

7.8.3 Zn Sulfide Deposits

Low-temperature formation of minerals has been a consideration of some time. The formation of strata-bound deposits of zinc sulfide (Siebenthal 1915) and the formation of Mississippi Valley–type Pb–Zn deposits (Bastin 1926a) have been proposed to involve sulfate-reducing bacteria. When the environmental impact of sulfide production and rates of hydrogen sulfide production were examined, Trudinger et al. (1972) concluded that insufficient information was available to reach a decision about the role of sulfate-reducing bacteria on mineral ore formation. The minerals in the Mississippi Valley–type deposits have been extensively characterized (Sverjensky 1986), and a scenario involving respiration of bacteria supported by organic and thiosulfate has been proposed to be a common theme for the formation of the Mississippi Valley–type deposits (Spirakis and Heyl 1995). While the role of sulfate-reducing bacteria in mineral formation has been highly controversial, several reports consider the conditions where bacterial participation may be useful (Sangameshwar and Barnes 1983; Vasconcelos and McKenzie 2000).

Now, there are several reports that sulfate reducers are capable of producing insoluble zinc sulfide even though zinc is toxic to sulfate-reducing bacteria. About 20% of the biofilm from the Piquette mine in Wisconsin contained particles that were 2–5 nm in diameter. These metal-containing particles consisted of ZnS (sphalerite) with less than 0.3% contamination attributed to Mn, Fe, As, and Se. Present in the biofilm containing the ZnS nanoparticles were aerotolerant members of the Desulfobacteraceae (Labrenz et al. 2000; Labrenz and Banfield 2004). A model has been proposed that involves the production of ZnS crystalline particles into micron-scale aggregates by sulfate-producing bacteria converting a few mg/L of Zn^{2+} to ZnS (Druschel et al. 2002). This model is considered to be useful in understanding the metal sulfide mineral paragenesis applied to low-temperature ore deposits involving Cu, Pb, and Zn. An extensive evaluation of crystal growth of biogenic sphalerite and wurtzite of biogenic was reported by Moreau et al. (2004). A soil enrichment that contained Zn^{2+}-resistant sulfate-reducing bacteria (*Desulfovibrio* spp., *Desulfobulbus* spp., and *Desulfotomaculum* spp.) produces sphalerite in a laboratory setting (Wolicka et al. 2015).

7.8.4 Ni Sulfide Formations

The concentrations of Ni in marine water are in the range of 2–12 nM, while in drainage streams from industrial or mining wastes, they can approach 2 mM (Mansor et al. 2019). While nickel sulfides rarely are found in nature (Huang et al. 2010), however, in laboratory experiments using nickel concentration greater than found in nature, Ni sulfides are produced (Fortin et al. 1994; Lewis and Swartbooi 2006; Gramp et al. 2007; Karbanee et al. 2008; Cao et al. 2009; Sampaio et al. 2010; Reis et al. 2013; Kiran et al. 2015). In lake sediments, bacterial surfaces are nucleation sites for NiS formation (Ferris et al. 1987). In a study evaluating the precipitation of various Ni-hosting phases, polyphasic Ni sulfide of <20 nm was produced under abiotic conditions, and after a week, these precipitates formed hexagonal α-NiS. In the presence of *D. vulgaris* and in the absence of Fe, the crystalline phases of polydymite (Ni_3S_4) and vaesite (NiS_2) appear along with monosulfide phases (Mansor et al. 2019). With biotic and abiotic activity in the presence of Fe^{2+}, the formation of crystalline Ni sulfides is low, and Ni–Fe phases appear resulting in the formation of Ni-rich mackinawite (FeS).

7.8.5 Mo Sulfide Minerals

Molybdenum occurs in various ore deposits as molybdenite (MoS_2), and Mo sulfides in black shale may be a result of metabolism attributed to sulfate-reducing bacteria (Li and Gao 2000; Han et al. 2014). Mo sulfide minerals and Mo sulfides closely associated with organic matter are found in the black shale in the Guizhou and Hunan areas of southern China (Fan 1983). Molybdate is an analogue of sulfate and was reported to inhibit the first enzyme of sulfate activation (ATP sulfurylase) of *Desulfovibrio* (Peck Jr. 1959). Molybdate inhibits the growth of sulfate-reducing bacteria by inhibiting sulfate transport and other metabolic systems in sulfate-reducing bacteria (Newport and Nedwell 1988; Jesus et al. 2015). H_2S produced by *D. desulfuricans* was found to promote dissolution of Mo powder introduced into the culture and MoS_2 was one of the products (Chen et al. 1998). When using washed cells of *D. desulfuricans* DSM 642 and H_2 as the electron donor, Mo^{6+} was reduced to Mo^{4+} with the formation of crystalline MoS_2 (Tucker et al. 1997). In column experiments containing immobilized cells of *D. desulfuricans*, ~90% of a 1 mM solution of sodium molybdate was removed as a solid phase of MoS_2 (Tucker et al. 1998). The addition of a sublethal concentration of molybdate to *D. gigas* ATCC 19364, *D. vulgaris* Hildenborough, *D. desulfuricans* DSM 642, and *D. desulfuricans* DSM 27774 reduced Mo^{6+} to Mo^{4+} with the formation of MoS_2 (Biswas et al. 2009). The impact of sulfate-reducing bacteria on formation of Mo ores and metabolism of molybdate-contaminated acid mine waters requires additional study.

7.8.6 Co Sulfides

In laboratory experiments, *D. vulgaris* reduced Co^{III}EDTA (ethylenediaminetetraacetic acid) to Co^{II}EDTA, and this reduction of cobalt was attributed to respiratory production of hydrogen sulfide (Blessing et al. 2001). Sulfide reacts with Co^{III}EDTA to produce Co^{II}EDTA and form polysulfide. As Co^{2+} is released from EDTA, CoS is quickly formed and cobalt is removed from the solution. In another experiment, *D. desulfuricans*, *Dst. gibsoniae*, and

Desulfomicrobium hypogeia were grown in the presence of 1 mM $CoCl_2$, and 99.99% of cobalt was removed from the solution with the formation of CoS (Krumholz et al. 2003). The cultivation of *D. desulfuricans*, *Dst. gibsoniae*, *Desulfomicrobium hypogeia*, and *Desulfoarcula baarsi* in media containing Co^{2+} chelated by nitrilotriacetate resulted in 98–99.94% of the soluble Co(II). Using core and sediment samples from the environment, removal of Co^{2+} was enhanced when sulfate-reducing bacteria were stimulated through addition of electron donors such as ethanol or formate.

7.8.7 Carbonate Minerals and Dolomite

The origin of dolomite [$CaMg(CO_3)_2$], which is associated with Precambrian sedimentary rock, is not a finished science (Warren 2000; McKenzie and Vasconcelos 2009), and its precipitation is considered to proceed through several metastable precursors (Kaczmarek and Thornton 2017). Microorganisms have been proposed to facilitate the formation of Holocene dolomite, which contains elevated levels of Ca and a poorly ordered crystal (McKenzie and Vasconcelos 2009; Petrash et al. 2017). Extracellular polymeric substances and capsules produced by sulfate-reducing bacteria are considered to enhance the formation of protodolomite (Beech and Cheung 1995; Krause et al. 2012; Bontognali et al. 2014). Several strains of sulfate-reducing bacteria have been isolated from microbial mats with lithification zones. Strains H12.1 and H2.3 were obtained from the Highborne Cay mats, and purified exopolymeric substance from these bacterial strains had high calcium-binding capacity (Braissant et al. 2007; Gallagher et al. 2012). Sulfate-reducing bacteria from the Coorong region, South Australia, precipitated dolomite (Wright and Wacey 2005). *Desulfohalovibrio reitneri* was isolated from Lake 21 on the Kiritimati Atoll (Kiribati, Central Pacific) (Spring et al. 2015, 2019). Strain LV representing *D. brasiliensis* was isolated from a coastal lagoon in Brazil, and it was reported to induce dolomite formation (Bontognali et al. 2014). The extracellular polymeric substance from *Desulfotomaculum ruminis* isolated from a protodolomite environment induced the formation of Mg calcite and protodolomite (Liu et al. 2020). The metabolism of sulfate-reducing bacteria contributes to the alkalinity of the mats due to the use of H_2 and formate as electron donors (Gallagher et al. 2012).

7.9 Reduction of Redox Active Metals and Metalloids Including U and Radionucleotides

7.9.1 *Reduction of Metal(loid)s: Cr, Mo, Se, and As*

Heavy metal ions are toxic to biological systems including SRB. The greatest concentration of metal ions tolerated by *D. vulgaris* has been reported to be as follows: 15 ppm Cr(III), 4 ppm Cu(II), 10 ppm Mn(II), 8.5 ppm Ni(I), and 30 ppm Zn(II) (Cabrera et al. 2006). One mechanism of heavy metal resistance used by DRB is the reduction of the toxic ion to a less toxic form, which is frequently of lower solubility. The reduction of heavy metals has been the subject of several reviews that include activities of SRB (Lloyd et al. 2001; Lloyd 2003; Barton et al. 2003, 2015; Hocking and Gadd 2007; Bruschi et al. 2007; Barton and Fauque 2009; Newsome et al. 2014). Metal-reducing bacteria (MRB) found around the world and the thermodynamics of dissimilatory metal reductions are provided in Chap. 1, Table 1.4. This discussion of SRB will be focused on their metabolism of (i) transition metals, (ii) metalloids, and (iii) actinides and fission products. With respect to transition metals, iron- and manganese-reducing bacteria have received considerable attention, and their impact on the environment has been reviewed (Lovley 1991; Lovley et al. 2004; Lovley 2006). Only a few SRB have the capability to reduce Fe(III) or Mn(IV) as the final electron acceptor (Table 7.4), and the mechanism accounting for this reduction has not been explained. Growth with dissimilatory Fe(III) or Mn(IV) reduction is limited to a few SRB; see Table 7.5. Additionally, some recent SRB isolates have the capability of using iron and manganese as electron acceptors; however, details of this reduction are not provided in the characterization of the isolate.

7.9.1.1 Chromium

Hexavalent chromium (chromate) is reduced to trivalent chromium by numerous heterotrophic aerobic bacteria (Wang and Shen 1995) and a few SRB. While chromium reduction may occur by biogenic production of hydrogen sulfide, it is reported that an enzymatic activity is involved. *Pseudomonas* sp. uses a soluble chromate reductase flavoprotein (ChrR) to reduce Cr(VI) to Cr(III) (Gonzalez et al. 2003; Ackerley et al. 2004), but this enzyme has not been reported to be present in SRB. Reduction of chromium by SRB appears to be attributed to interaction with low-potential polyheme cytochrome c_3 from *D. vulgaris* and *Desulfomicrobium norvegicum* and the triheme cytochrome c_7 from *Desulfuromonas acetoxidans* (a sulfur reducer) (Table 7.6). The monoheme cytochrome c_{553} from *D. vulgaris* has a redox potential of about +50 mV and did not reduce Cr(VI) (Michel et al. 2001). Using mutations, it was determined that heme 1 but not heme 2 of cytochrome c_3 from *D. vulgaris* was needed for reduction of Cr(VI) (Michel et al. 2001). Other proteins that reduced Cr(VI) included [Fe] hydrogenase, [FeNi] hydrogenase,

Table 7.4 Reduction of metals, metalloids, and radionuclides by SRB cells or cell-free extracts

Redox couple	Bacteria	Reference
Precious metals		
Ag(I)/Ag(0)	*Desulfovibrio alaskensis* G20	Capeness et al. (2019a)
Au(III)/Au(0)	*Desulfovibrio desulfuricans*	Deplanche and Macaskie (2008)
Au(I)/Au(0)	*Desulfovibrio* sp.	Lengke and Southam (2006, 2007)
Pd(II)/Pd(0)	*Desulfovibrio desulfuricans*	Lloyd et al. (1998) Omajali et al. (2015)
	Desulfovibrio alaskensis G20	Capeness et al. (2019a)
Pt(II)/Pt(0)	*Desulfovibrio desulfuricans*	de Vargas et al. (2005)
Pt(IV)/Pt(0)	*Desulfovibrio alaskensis* G20	Capeness et al. (2019b)
Transition metals		
Cr(VI)/Cr(III)	*Desulfovibrio vulgaris*	Lovley and Phillips (1994) Goulhen et al. (2006) Franco et al. (2018)
	Desulfovibrio desulfuricans	Tucker et al. (1998) Chovanec et al. (2012)
	Desulfovibrio alaskensis G20	Li and Krumholz (2009)
	Desulfomicrobium norvegicum	Chardin et al. (2002)
Fe(III)/Fe(II)	*Desulfobacter postgatei*	Lovley et al. (1993b)
	Desulfobacterium autotrophicum	
	Desulfobulbus propionicus	
	Desulfovibrio baarsii	
	Desulfovibrio baculatus	
	Desulfovibrio desulfuricans	
	Desulfovibrio sulfodismutans	
	Desulfovibrio vulgaris	
	Desulfotomaculum acetoxidans	
Mn(IV)/Mn(II)	*Desulfobacterium autotrophicum*	Lovley (1995)
	Desulfotomaculum reducens	Tebo and Obraztsova (1998)
	Desulfotomaculum acetoxidans	Nealson and Saffarini (1994)
	Desulfomicrobium baculatum	Lovley (1995)
Metalloids		
As(V)/As(III)	*Desulfotomaculum auripigmentum*	Newman et al. (1997)
	Desulfovibrio strain Ben-RA	Macy et al. (2000)
Mo(VI)/Mo(IV)	*Desulfovibrio desulfuricans*	Chen et al. (1998) Tucker et al. (1997, 1998) Biswas et al. (2009)
Rh(VII)/Rh(0)	*Desulfovibrio desulfuricans*	Xu et al. (1999)
Se(VI)/Se(0)	*Desulfovibrio desulfuricans*	Woolfolk and Whiteley (1962); Tomei-Torres et al. (1995)
Se(IV)/Se(0)	*Desulfovibrio desulfuricans*	Tomei-Torres et al. (1995)
Tc(VII)/Tc(IV)	*Desulfovibrio desulfuricans*	Lloyd et al. (1999a, b)
	Desulfovibrio fructosovorans	De Luca et al. (2001)
Te(IV)/Te(0)	*Desulfovibrio desulfuricans*	Woolfolk and Whiteley (1962)

(continued)

Table 7.4 (continued)

Redox couple	Bacteria	Reference
V(V)/V(IV)	*Desulfovibrio desulfuricans*	Woolfolk and Whiteley (1962)
Heavy metal		
Bi(III)/Bi(0)	*Desulfovibrio desulfuricns*	Barton et al. (2019)
Radionuclide		
U(VI)/U(IV)	*Desulfovibrio alaskensis* G20	Payne et al. (2002) Li and Krumholz (2009)
	Desulfovibrio baarsii	Lovley et al. (1993b)
	Desulfovibrio baculatus	Lovley et al. (1993b)
	Desulfovibrio desulfuricans	Tucker et al. (1996, 1998)
	Desulfovibrio gigas	Barton et al. (1996)
	Desulfovibrio sulfodismutans	Lovley et al. (1993b)
	Desulfovibrio vulgaris	Lovley et al. (1993a) Franco et al. (2018)
	Desulfovibrio strain UFZ/B490	Pietzsch et al. (1999)
	Desulfotomaculum reducens	Tebo and Obraztsova (1998)

Table 7.5 Growth of SRB cultures with dissimilatory reduction of metals or metalloids

Electron acceptor	Bacteria	References
Cr(VI)	*Desulfotomaculum reducens*	Tebo and Obraztsova (1998)
Mn(IV)	*Desulfotomaculum reducens*	Tebo and Obraztsova (1998)
Fe(III)	*Desulfotomaculum reducens*, *Desulfofrigus oceanense*, *Desulfofrigus fragile*, *Desulfotalea psychrophila*, *Desulfotalea arctica*	Tebo and Obraztsova (1998) Knoblauch et al. (1999b)
U(VI)	*Desulfovibrio* strain UFZ B-490	Pietzsch et al. (1999)
	Desulfotomaculum reducens	Tebo and Obraztsova (1998)
Arsenate	*Desulfotomaculum auripigmentum*	Newman et al. (1997)
	Desulfovibrio strain Ben-RA	Macy et al. (2000)

[FeNiSe] hydrogenase, and thioredoxin (Michel et al. 2001; Chardin et al. 2003; Li and Krumholz 2009). The mechanism for chromate reduction in SRB appears to involve low-potential cytochromes and proteins with [Fe-S] clusters, which interact with the chromate ion if the redox potential is sufficiently negative. *Dst. reducens* M1-1 is capable of growing with Cr(VI) as alternate electron acceptors to sulfate (Tebo and Obraztsova 1998). The growth and physiological response of SRB in a microbial consortium have been shown to respond to polylactate compounds to stimulate chromate reduction (Brodie et al. 2011). With a pure culture and in the absence of sulfate, *D. desulfuricans* reduced chromate when immobilized in acrylamide and energized by lactate or H_2 (Tucker et al. 1998). *D. vulgaris* Hildenborough reduced chromium when immobilized in agar and agarose but not

Table 7.6 Reduction of metals with proteins from SRB

Proteins	Bacteria	Metal ions reduced	References
Cytochromes			
Tetraheme Cytochrome c_3	*Desulfovibrio vulgaris*	Cr(VI)	Lovley and Phillips (1992, 1994)
		Se(VI)	Abdelouas et al. (2000)
		U(VI)	Lovley et al. (1993b)
		Fe(III)	Lojou et al. (1998)
	Desulfovibrio desulfuricans	Fe(III)	Lojou et al. (1998)
	Desulfovibrio gigas	Fe(III)	Lojou et al. (1998)
	Desulfomicrobium norvegicum	Cr(VI) Fe(III)	Michel et al. (2001, 2003)
Triheme Cytochrome c_7	*Desulfuromonas acetoxidans*[a]	Cr(VI) Fe(III) Mn(IV)	Michel et al. (2001, 2003)
Fe-S proteins			
Hydrogenase	*Desulfovibrio alaskensis* G20	Pd(II) Tc(III)	Lloyd et al. (1998) Lloyd et al. (1999a)
[Fe] hydrogenase	*Desulfovibrio vulgaris*	Cr(VI)	Michel et al. (2003)
[NiFe] hydrogenase	*Desulfovibrio fructosivorans*	Cr(VI)	Chardin et al. (2003)
	Desulfovibrio fructosivorans	Tc(VII)	De Luca et al. (2001)
[NiFeSe] hydrogenase	*Desulfomicrobium norvegicum*	Cr(VI)	Michel et al. (2001)
Thioredoxin	*Desulfovibrio alaskensis* G20	U(VI) Cr(VI)	Li and Krumholz (2009)

[a]Cytochrome also reduces S^0

when immobilized in chitosan or PVA-borate beads (Humphries et al. 2005, 2006). The rate of chromium reduction by *D. vulgaris* Hildenborough was enhanced when chelators were used to complex Cr(III) (Mabbett et al. 2002). Chromate has an impact on the metabolic activity and especially sulfate metabolism as observed with experiments using *D. vulgaris* (Klonowska et al. 2008; Franco et al. 2018). It was reported by Bruschi et al. (2007) that there is an energy cost to SRB with Cr (VI) presence in the medium since 25 μM Cr(VI) accounts for a redox potential of the medium to be Eh = +135 mV and reduction to Cr(III) produces an optimal redox value for growth of −150 mV. Thus, it was suggested that chromate reduction was a protective mechanism to enable SRB to grow. With *D. vulgaris* Hildenborough growth in 250 μM chromate, Cr particles were identified on the inner and outer membranes of the cell, which supported the suggestion that hydrogenases and cytochromes accounted for Cr(VI) reduction (Goulhen et al. 2006). It has also been suggested that chromate may enter the SRB cell by way of a sulfate transporter with Cr(III) accumulating near the plasma membrane or outside the cell (Bruschi et al. 2007). The chromosome of *D. vulgaris* Hildenborough contains a gene (locus DVU0426) that has been identified as a member of the chromate transport family, and the role this has on chromium metabolism is unknown. A similar gene identified as a member of the chromate transport family has been identified in the plasmid of *D. vulgaris* Hildenborough, and the role of this gene has not been established. These chromate genes have not been identified in the genomes of *D. alaskensis* G20, A. fulgidus, or *Desulfotalea psychrophila* (Bruschi et al. 2007). Chromate stress influences proteins associated with redox regulation and is associated with upregulation of superoxide dismutans (Bruschi et al. 2007). While the ingested chromium may be toxic and kill the cells, subsequent generations would be provided with a toxin-free environment. This sequestering of toxic ions by dead bacteria could also explain the resistance of SRB to Cr(VI), Tc(VII), Pd(II), and V(V) (Goulhen et al. 2006).

7.9.1.2 Molybdenum

Molybdate has long been known to inhibit the initial enzyme in activation of sulfate (sulfate adenylyl transferase) (Peck Jr. 1959), which is a futile energy cycle where ADP-molybdate is transiently formed before it spontaneously degrades to ADP plus molybdate and thereby depletes cytoplasmic ATP concentrations (Taylor and Oremland 1979). It has been suggested that molybdate competitively inhibits sulfate transport in SRB (Newport and Nedwell 1988; Cypionka 1989). Using mutagenized cells, the resistance to molybdate by *D. vulgaris* Hildenborough was reported to be attributed to involve YcaO-like proteins, which have an unknown function (Zane et al. 2020). At sublethal concentrations where the molybdate–sulfate ratio is $<1:10$, molybdate is reduced from Mo(VI) to Mo(IV) by *Desulfovibrio desulfuricans* DSM 642 in a sulfide environment. The reduced molybdenum ion combines with sulfide to produce MoS_2 (molybdenite), which accumulates in the periplasm, on the cell surface, or in the extracellular medium (Tucker et al. 1997, 1998; Biswas et al. 2009). When *D. gigas*, *D. vulgaris* Hildenborough, *D. desulfuricans* DSM 642, and *D. desulfuricans* DSM 27774 were cultivated in sulfate medium containing elevated levels of Mo(VI), the culture fluid developed a red-brown color. *D. desulfuricans* is also reported to convert elemental selenium to MoS_2 by an unknown process (Chen et al. 1998). Southern China contains black shale deposits of the Lower Cambrian (Fan 1983), and this shale contains unusually elevated concentrations of Mo (Li and Gao 2000). It is possible that SRB in the paleo-seawater could account for this enrichment of molybdenum in the shale.

Molybdate has been used to control growth of SRB populations in many different environments. The addition of 0.095–0.47 mM molybdate to a consortium of bacteria obtained from a Canadian oil field revealed that inhibition of SRB was dependent on the composition of the microbial population and the physiological state of the bacteria (Nemati et al. 2001). Biosulfide production by SRB is inhibited by adding molybdate to sulfate-rich wastewater (Tanaka and Lee 1997). Using a microbial consortium of bacteria containing *Desulfovibrio vulgaris* and other SRB found in water from oil wells of the Reconcavo Basin, Brazil, the addition of molybdate at a molar ratio of molybdate–sulfate of 0.004 inhibits hydrogen sulfide production (de Jesus et al. 2015). With the addition of 2.5 mM molybdate to anaerobic reactors, a differential impact was noted with molybdate being bacteriostatic for SRB bactericidal for methanogens (Isa and Anderson 2005).

7.9.1.3 Selenium

Selenium as selenate [Se(VI)] and selenite [Se(IV)] interacts with bacteria using either of the three processes: assimilation with synthesis of selenocysteine, detoxification with the reduction of soluble Se(VI) and Se(IV) to insoluble Se(0), and dissimilatory reduction with selenite respiratory activity coupled to cellular growth (Staicu and Barton 2017, 2021). Selenate was reported by Peck Jr. (1959) to inhibit the activation of sulfate by the enzyme adenylyl transferase, but at sublethal concentrations, *D. desulfuricans* reduces selenate to elemental Se^0, which is present as an extracellular red precipitate (Tomei-Torres et al. 1995). Selenate has been shown to inhibit sulfate transport in SRB (Postgate 1949; Newport and Nedwell 1988), and this would be another mechanism for selenate inhibition of SRB. Selenite reduction by *D. desulfuricans* produces cytoplasmic and extracellular particles, which are readily observed by electron microscopy. Selenate and selenite may enter SRB cells using sulfate transporter systems. The enzymology of selenate or selenite reduction by SRB is unresolved and may follow the pathway proposed for *Escherichia (E.) coli*, which used reduced glutathione (Turner et al. 1998). However, production of Se^0 by SRB does not use glutathione but may use thioredoxin (Bruschi et al. 2007). In *E. coli*, production of Se^0 from selenocysteine is attributed to cysteine desulfurase and selenocysteine lyase, but most SRB lack cysteine desulfurase. Another possible mechanism of selenate and selenite reduction would be by enzymes associated with denitrification. Selenate is reduced by bacterial nitrate reductases (Sabaty et al. 2001) and selenite by bacterial nitrite reductase (Demoll-Decker and Macy 1993). Of the genes for nitrate reductase (*narGHIJ*), only *narI* was present in *A. fulgidus* (AF0546), but none of the *nar* genes were present in *D. vulgaris* Hildenborough, *D. alaskensis* G20, or *Desulfotalea psychrophila* (Bruschi et al. 2007). However, the gene for the γ-subunit of nitrate reductase was identified in *D. vulgaris* Hildenborough (DVU1290), *D. alaskensis* G20 (Dde2271), and *Desulfotalea psychrophila* (Dp3075) (Bruschi et al. 2007). While there are several potential mechanisms for selenate and selenite reduction by SRB, definitive experiments are need to establish this enzymology in SRB.

7.9.1.4 Arsenic

The mechanisms used by SRB to provide resistance to toxic arsenic in the environment have been reviewed by Bruschi et al. (2007). Ecosystems containing arsenic benefit from the presence of SRB, which detoxify the environment by reducing arsenate and precipitating arsenite with the formation of arsenic sulfide (Kirk et al. 2004). The mechanisms of enzymatic transformation of As(V) to As(III) by bacteria have been reviewed by Silver and Phung (2005), and several of the genes associated with this process have been identified in SRB. The first of a two-step process is the cytoplasmic reduction of As(V) by a putative arsenate reductase (arsC) with arsenate imported into the cell by a phosphate transport system. The putative arsC gene has been identified in *D. vulgaris* Hildenborough (DVU1646), *D. alaskensis* G20 (Dde2792-2793), *A. fulgidus* (AF1361), and *Desulfotalea psychrophila* (Dp1829) (Bruschi et al. 2007). The second step is the removal of As(III) from the cytoplasm by an ATP-independent export system encoded on an ars operon or by an ABC transporter. A putative arsenite transporter is encoded at the AF2308 locus of *A. fulgidus* genome, and putative genes for the arsenite efflux pump (acr3) are on the genomes of *D. alaskensis* G20 and *Desulfotalea psychrophila* at the Dde2791 and DP1779 loci, respectively. Regulation of these arsenate genes would be by the transcriptional regulator gene (*arsR*), which is found in *D. vulgaris* Hildenborough (DVU0606), *D. alaskensis* G20 (Dde0747), *A. fulgidus* (AF1270), and *Desulfotalea psychrophila* (Dp1300) (Bruschi et al. 2007). It should be noted that the four SRB used for arsenate resistance in silico study were the only sequenced genomes available by 2007. With many genomes of SRB now published, this search for arsenate resistance genes should be expanded to include other SRB genomes.

7.10 Pollutants and Bioremediation Processes

One of the important components in devising a bioremediation process is the selection of nutrients to support SRB in the environment, and a second requirement is the presence of appropriate indigenous bacteria to support the targeted bioremediation. Liamleam and Annachhatre (2007) have provided an approach to select the appropriate electron donors for a bioremediation process involving SRB. Cultivation of SRB provides only partial information of the capability of the microbial community in a contaminated site, and a more complete assessment is to use geochip as has been established in the literature (Kang et al. 2013) or another molecular technique to more accurately establish the breadth of the SRB species present.

7.10.1 Biogenic Hydrogen Sulfide Production

Bioproduction of hydrogen sulfide is important for removal of toxic metal ions in acid rock drainage (ARD) and acid mine drainage (AMD) (Hocking and Gadd 2007), and the rate of metal sulfide formation is influenced by environmental chemical and physical conditions. Inhibition of growth occurs if the concentrations of metal ions are high and lethal concentrations have been reported for Cu (2–20 mg/L), Zn (13–40 mg/L), Pb (75–125 mg/L), Cd (4–54 mg/L), Ni (10–20 mg/L), and Hg

(74 mg/L) (Jamil and Clarke 2013). Also, the bulk water pH influences the amount of sulfide in the water. With bulk water at pH 6.5, 50% of the hydrogen sulfide species is HS^- and 50% is H_2S (Neto et al. 2018) with a maximum dissolved sulfide concentration of 100–140 mg/L (Magowo et al. 2020). There are commercial programs that use biogenic sulfide production to remediate leach solutions.

In 1992, the first process to be commercialized uses THIOPAQ® technology and is marketed by PAQUES (https://www.paquues.nl.paques) in Balk, Netherlands. As reviewed by Hocking and Gadd (2007), in Budel-Dorplein, Netherlands, the Budelco zinc refinery uses the THIOPAQ® technology 400 m^3/h of process water that is treated using H_2 as the electron donor to produce several tons of hydrogen sulfide per day. As summarized by Jamil and Clarke (2013), the THIOPAQ® technology was used to recover about 99% of copper from leach water streams from the Kennecott Utah Copper mining facility in Brigham Canyon, Utah. The BiotecQ's BioSulphide® process has been applied to leach solution from Copper Queen Mine in Bisbee, Arizona (Jamil and Clarke 2013). This plant uses elemental sulfur to produce up to 3.7 tons of sulfides per day and recovers copper at a rate of about 63.500 kg/month.

7.10.2 *Acid Mine and Acid Rock Drainage Bioremediation*

Underground water draining through abandoned mines, water in unworked pits, or diffuse water streams running over mine waste piles are an environmental problem in many areas throughout the world. Acid rock drainage (ARD) from coal or gold fields and acid mine drainage (AMD) from metal or coal mines characteristically have a low pH and elevated concentrations of toxic metals (Johnson and Hallberg 2005). A strategy to control ARD (McCarthy 2011) and AMD was to use SRB in a bioremediation process (Kim et al. 1999; Kolmert and Johnson 2001; Lens et al. 2008; Bai et al. 2013; Giloteaux et al. 2013; Church et al. 2007; Nancucheo et al. 2017; Ayangbenro et al. 2018). Indigenous SRB grow on a diversity of electron donors producing end products of CO_2 and hydrogen sulfide and couple growth to dissimilatory reduction of sulfate present in the ARD and AMD. Bicarbonate would neutralize acid in the drainage water and sulfide would precipitate the toxic heavy metals. Detoxification of Cd(II), Ni(II), and Cr(VI) in marine sediments has been attributed to biogenic hydrogen sulfide produced by SRB (Joo et al. 2015). The dissimilatory SRB are important for bioremediation of acidic waters and, based on end products of metabolism, are divided into two different groups. The complete oxidizers produce CO_2 from a substrate, while the incomplete oxidizers produce acetate (see Chap. 2). It is desirable to employ complete oxidizing SRB (*Desulfosarcina*, *Desulfobacter*, *Desulfarculus*, *Desulfomonas*, and others indicated in Table 2.1) because carbonate produced from carbon dioxide will neutralize acidic waters. While selection of complete oxidizing SRB may be difficult in an open environment, this would be a consideration when bioreactors are a component of the bioremediation process (Kolmert and Johnson 2001; Lens et al. 2002; Zagury and

Neculita 2007; Hessler et al. 2018). Recently, there has been an interest in applying permeable barriers to detoxify waters from acid mines and acid rock piles (Benner et al. 1999; Gibert et al. 2002; Ludwig et al. 2002). Because organic carbon levels may limit the growth of SRB in the ARD and AMD environments, based on local commercial availability, various organic additions have been made to enhance the bioremediation process (Boshoff et al. 2004; Zagury et al. 2006; Liamleam and Annachhatre 2007; Gonçalves et al. 2008; Martins et al. 2009; Choudhary and Sheoran 2011; Jamil and Clarke 2013; Magowo et al. 2020). A dynamic community structure is required in the acidic waters because SRB depend on microbial decomposition of complex organic materials to provide appropriate substrates for SRB. With excess sulfate, SRB would outcompete acetogens and methanogens for nutrients.

7.10.3 Uranium Remediation

The demand for uranium required extraction of ores with sulfuric acid, and as a result, various toxic metals were solubilized along with uranium. Groundwater at the milling site had become contaminated, and due to environmental concerns, various bioremediation processes were pursued to detoxify the site. Bioremediation of metal-contaminated sites relies on converting the soluble highly toxic ion to an insoluble form, which is less toxic. In aerobic groundwaters, uranyl carbonate is the predominant uranium form present, and with respect to bioremediation, the soluble oxidized uranyl ion (UO_2^{2+}) would be converted to the insoluble reduced uraninite ion (UO_2). Attention given to remediation of uranium mill tailings could be attributed to the demonstration of uranium reduction by SRB (Lovley et al. 1991, 1993b) and funding through the Uranium Mill Tailings Remedial Action (UMTRA) program of the US Department of Energy (US DOE). Reviews by Abdelouas (2006), Wall and Krumholz (2006), and Newsome et al. (2014) provide information on microorganisms and uranium remediation sites. A comprehensive review of field studies discussing the DOE Rifle site (Colorado), US DOE Oak Ridge site (Tennessee), and US DOE Hanford site (Washington) is provided by Newsome et al. (2014). Bench and pilot studies provided parameters for uranium reduction and an indication of enzymology of uranium bioreduction. Cytochrome c_3 from *D. vulgaris* Hildenborough was reported to reduce U(VI) to U(IV) (Lovley et al. 1993a), but it was not the only mechanism that would reduce U(VI) because a mutant of *D. alaskensis* G20 lacking cytochrome c_3 was able to reduce uranium (Payne et al. 2002). Uranium reduction may not be to a specific protein, but reduction may be attributed to redox proteins where the uranyl ion has access to the redox center. As summarized by Barton et al. (2003), cytochromes and enzymes from several different bacteria are capable of metal(loid) reduction. Laboratory studies indicated that numerous SRB were capable of reducing U(VI) to U(IV), and these cultures included several species of *Desulfovibrio* (*gigas*, *desulfuricans*, *vulgaris* strain Hildenborough, *baculatus*, *baarsii*, *sulfodismutans*, *alaskensis* G20, strain UFZ B

490; Lovley et al. 1993b; Barton et al. 1996; Tucker et al. 1996, 1998; Pietzsch et al. 1999; Payne et al. 2002) and *Dst. reducens* (Tebo and Obraztsova 1998). However, only *Desulfovibrio* sp. strain UFZ B 490 (Pietzsch et al. 1999) and *Dst. reducens* (Tebo and Obraztsova 1998) are capable of coupling growth to reduction of uranium. In situ bioremediation studies demonstrate the presence of SRB at uranium sites where reduction of uranium is occurring. Sulfate-reducing bacteria were demonstrated in U(VI) treatment ponds near Oak Ridge, TN (Wu et al. 2005). In groundwater at a uranium mill tailings site at Shiprock, NM, the abundance of *Desulfotomaculum* spp. suggested these SRB were involved in uranium reduction (Chang et al. 2001), and in uranium at the Midnite Mine in eastern Washington, *Desulfosporosinus* spp. were abundant (Suzuki et al. 2003). Bacterial activity is highly dynamic in remediation sites as seen by changes in the SRB population following injections of nutrients (Hwang et al. 2009; Marsh et al. 2010; Gihring et al. 2011). Prior to injection of emulsified vegetable oil (EVO) into subsurface wells, DNA sequences did not correspond to known strains of SRBs, but by 4–1 days after EVO amendment, *Desulfovibrio* spp. and *Desulfococcus* spp. were the dominate SRB, and at 80–140 days, species of *Desulfobacterium* became abundant (Zhang et al. 2017).

Long-term stability of reduced uranium as U(IV) in bioremediation sites is of some concern. U(IV) as solid UO_2 is subject to oxidation when exposed to the atmosphere (Van Nostrand et al. 2009), or when organic substrates are no longer added to the uranium site, reoxidation of U(IV) to U(VI) occurs (Wu et al. 2007; Hwang et al. 2009). A study conducted at an inactive uranium mining district of Ronneburg (Thuringia, Germany) indicated the presence of bacteria that were members of the Desulfobacterales, Desulfovibrionales, Syntrophobacteraceae, and Clostridiales in addition to numerous soil bacteria (Sitte et al. 2010). Reduced uranium was present in the soil environment and waters draining from the site contained U(IV). The processes contributing to the leaching of U(IV) from the site were unresolved. It is known that in an environment where a diversity of bacterial types is present, siderophores could be produced and the mobility of U(IV) could be attributed to siderophores by various bacteria. Siderophore chelation of U(VI) is well documented in the literature (Frazier et al. 2005; Schalk et al. 2011; Ahmed and Holmström 2014) and is a factor to be considered when long-term bioremediation of U(VI) is being implemented.

7.10.4 Perchlorate Reduction and Use as an Inhibitor

Several species of bacteria reduce perchlorate, and recently some strains are capable of growing by perchlorate respiration. Two bacterial isolates, *Desulfotomaculum* AR1 and *Desulfovibrio desulfuricans* Ya-11, grew with lactate as the electron donor and perchlorate ($ClO4^-$) as the final electron acceptor (Verkholiak et al. 2020). These two SRB isolates readily reduced perchlorate when cells were immobilized on wood chips or agar. Research is required to assess the presence of perchlorate reductase (EC 1.97.1.1), which is present in *Dechloromonas aromatica* RCB and

Pseudomonas PK (Bender et al. 2002). In SRB that do not reduce perchlorate, inhibition of sulfate respiration occurs. Addition of perchlorate to *D. alaskensis* G20 results in inhibiting the central sulfate reduction pathway, and since this inhibition is more effective than nitrate, Carlson et al. (2015) suggest the use of perchlorate to control sulfidogenesis.

7.10.5 Bioremediation of Petroleum Hydrocarbons

Treatment of soils contaminated with hydrocarbons has often used chemical systems or aerobic bioremediation processes. If unconfined hydrocarbon contamination is in fractured bedrock or assessment to the site is prevented by aboveground utilities, anaerobic bioremediation is a viable option. Sulfate-reducing bacteria are important in petroleum hydrocarbon (PHC) bioremediation and have been implicated in 25 of 38 PHC-contaminated aquifer sites (Wiedemeier et al. 1999). While anaerobic bioremediation by indigenous bacteria proceeds faster with additions of nitrate or ferric ion as compared to sulfate (Cunningham et al. 2001), environmental restrictions limit the introduction of nitrate into a contaminated site, and the low solubility of iron at near-neutral pH limits its use. Acetate, butyrate, and propionate are generated as PHCs are degraded (Cozzarelli et al. 1994), and these organic acids are considered to be used as carbon sources for SRB in the PHC sites because organic acids produced in marine sediments are readily used by SRB (Purdy et al. 1997; Kniemeyer et al. 2007). Some SRB associated with bioremediation have a highly versatile metabolism as seen with *Desulforhabdus amnigenus* (Elferink et al. 1995), which uses acetate, lactate, butyrate, and propionate. SRB have been used to remediate several PHC-contaminated environments (Kleikemper et al. 2002a, b; Sublette et al. 2006; van Stempvoort et al. 2007; Morasch and Meckenstock 2005; Kniemeyer et al. 2007; Chin et al. 2008; Wei et al. 2018), and in most instances, organic electron donors or sulfate are added to the site to selectively support growth of the sulfate reducers (Morasch & Meckenstock 2005).

7.11 Industrial Applications

Throughout time, there has been the interest in understanding the impact of bioproduction of H_2S from SRB respiration on the environment. Gradually, there developed the use of H_2S to detoxify metals in the environment using SRB, and this application was initially the reactions associated with H_2S, and later on, the oxidative capability of SRB became an activity of interest in bioremediation. Several applications of SRB in bioremediation processes have already been discussed in detail in the previous section, and in this section, additional applications are presented. Several additional biotechnical applications using SRB are listed in Table 7.7 and not discussed further in this chapter. In this section, applications concerning electrochemistry, metallic nanoparticles, and energy development are discussed.

Table 7.7 Application of SRB to detoxify or remediate polluted environments

Treatments	References
Gypsum waste	Maree and Hill (1989); Deswaef et al. (1996); Kijjanapanich et al. (2014)
SO_2 from flue gases	Gasiorek (1994); Selvaraj et al. (1997); Lens et al. (2003)
Tannery wastewater	Genschow et al. (1998); Durai and Rajasimman (2011); Mannucci et al. (2013); Igiri et al. (2018)
Sulfate containing wastewater	Lens et al. (1998); Lens et al. (2002); Geets et al. (2006); Miao et al. (2012); van den Brand et al. (2018)
Metal-contaminated soil	Maree et al. (1989, 1991); White et al. (1998); Groudev et al. (2001); Jiang and Fan (2008); Hussain et al. (2016)
Transformation of explosives	Boopathy et al. (1998a, b, c); Boopathy (2007)
Degradation of organic pollutants	Beller et al. (1996); Anderson and Lovley (2000); Kleikemper et al. (2002a, b); Roychoudhury and McCormick (2006); Kolhatkar et al. (2008); Abu Laban et al. (2010); Wei et al. (2018)

7.11.1 Production of Metallic Nanoparticles

Nanoparticles of metal(loid)s are produced by several different species of SRB as a result of reduction of oxidized metals or production of metallic sulfides (Table 7.8). Biosulfide produced by SRB will also reduce metals with the production of nanoparticles, which are a few microns in diameter, but these nanoparticles can aggregate to become several hundred micrometers in diameter, and depending on the bacterium and the chemical conditions, the particles are found in the cytoplasm, in the periplasm, or on the exterior of the cell. With a growing use and specific design required of metallic nanoparticles in electrical and medical application, this "green fabrication" approach in production of nanoparticles has stimulated nanoparticle production by microorganisms including SRB. With biosulfide produced by SRB, nanocrystals of CdS and ZnS semiconductors have been formed (Qi et al. 2013; Rangel-Chavez et al. 2015).

The interaction of an element with reduced cytochrome was first reported with cytochrome c_3 reducing elemental selenium to hydrogen sulfide. The growth of SRB on elemental sulfur (Biebl and Pfennig 1977) led to the discovery that the sulfur reductase was in fact the tetraheme cytochrome c_3 (Fauque et al. 1979). It is interesting to note that reduction of S^0 by cytochrome c_3 was first reported by Ishimoto et al. (1958); however, this observation was dismissed as unimportant in physiological activity. As reviewed by Fauque (1994), cytochrome c_3 from *D. gigas*, *D. multispirans*, *D. africanus* Benghazi, *D. desulfuricans* Berre-Eau, *D. desulfuricans* British Guiana, *Desulfomicrobium baculatum* DSM 1743, and *Desulfomicrobium baculatum* Norway 4 reduced S^0. With this background in cytochrome reduction of S^0, the examination of other elements interacting with cytochrome c_3 was a natural extension. As indicated in Table 7.6, several of these cytochromes were also found to reduce metals including Fe(III) as final electron acceptors. Additionally, redox active proteins such as hydrogenase and thioredoxin contain iron–sulfur clusters, which also reduce metals.

Table 7.8 Production of inorganic nanoparticles by SRB

Nanoparticle	Substrate	Bacteria	Reference
Se^0	Selenate	*Desulfovibrio desulfuricans*	Tomei-Torres et al. (1995)
Se^0	Selenite	*Desulfovibrio desulfuricans*	Tomei-Torres et al. (1995)
Re	Potassium perrhenate	*Desulfovibrio desulfuricans*	Xu et al. (2000)
Au^0	Au(I)-thiosulfate	*Desulfovibrio* sp.	Lengke and Southam (2006, 2007)
Pd^0	Sodium palladium chloride	*Desulfovibrio alaskensis* G20	Capeness et al. (2019b)
		Desulfovibrio desulfuricans	Omajali et al. (2015)
Pt	Platinum chloride	*Desulfovibrio alaskensis* G20	Capeness et al. (2019b)
UO_2	Uranyl acetate		
ZnS	Zn(II)	*Desulfobacteraceae* member	Labrenz et al. (2000) Qi et al. (2013) Gong et al. (2018)
NiS	Ni chloride & Fe sulfate	*D. vulgaris* Hildenborough	Mansor et al. (2019)
		D. desulfuricans 8307	Capeness et al. (2015)
		D. alaskensis G20	Capeness et al. (2015)
		Desulfotomaculum sp.	Fortin et al. (1994)
FeNiS	Ni chloride & Fe sulfate	*D. vulgaris* Hildenborough	Mansor et al. (2019)
Pd/Pt	Sodium palladium chloride	*Desulfovibrio alaskensis* G20	Capeness et al. (2019b)
Magnetite	Fe(II)	*Desulfovibrio magneticus*	Pósfai et al. (2006)

Gold as Au(III) or Au(I) is reduced by numerous bacteria including SRB to elemental gold [Au(0)], and this activity has been reviewed by Reith et al. (2007). *Desulfovibrio* sp. obtained from a bacterial consortium of SRB obtained from the Driefontein gold mine in South Africa reduced gold(I)-thiosulfate to gold nanoparticles (<10 nm) inside the cell and was released to collect on the cell surface or in the extracellular fluid (Lengke and Southam 2006). These nanoparticles of gold of bacterial origin contributed to the formation of octahedral gold crystals, framboid-like structures, and gold foil (Lengke and Southam 2007). With H_2 as the electron donor, cells of *D. vulgaris* Hildenborough did reduce Au(III) as gold chloride to elemental gold although no details of the reduction were provided (Kashefi et al. 2001), and *D. desulfuricans* was determined to contribute to the reduction of Au(III) with the formation of Au(0) by an unresolved mechanism (Creamer et al. 2006). Using 2 mM $HAuCl_4$ and a H_2 atmosphere, *D. desulfuricans* produced gold particles of 20–50 nm diameter in the periplasm and cell surface with smaller nanoparticles distributed throughout the cytoplasm (Deplanche and Macaskie 2008). A suspension containing the nanoparticles of *D. desulfuricans* was red at pH 2.0, purple at pH 6.0–7.0, and blue at pH 9.0.

Palladium and platinum nanoparticles were reported to be produced by *D. desulfuricans* and *D. alaskensis* (Yong et al. 2002a, b; Omajali et al. 2015; Capeness et al. 2019b). With *D. desulfuricans* NCIMB 8307 and *D. desulfuricans* NCIMB 8326 in a 2 mM Na_2PdCl_4 solution, intracellular accumulation of palladium nanoparticles (Pd-NP) with diameters of 0.2–8.0 nm was detected (Omajali et al. 2015). Using an electron microscopy to locate distribution of palladium in *D. fructosivorans*, palladium was localized in the periplasm in wild-type cells but was distributed along the plasma membrane in mutant cells lacking hydrogenase (Mikheenko et al. 2008). Exposure of *D. alaskensis* to platinum and palladium revealed an increased production of the small subunit (Dde_2137) for [NiFe] hydrogenase, and this subunit is one of the 13 proteins upregulated following platinum and palladium stress (Capeness et al. 2019a, b). Evidence is fairly strong that *D. desulfuricans* and *D. alaskensis* produce Pd-NP and Pt-NP consisting of metal sulfides. For production of zero valent metals, palladium is proposed to be transported to the periplasm where it is reduced to Pd(0), which is exported across the outer membrane where it collects on the cell surface. Further studies are required to establish the mechanism of Pd(0) and Pt(0) nanoparticle formation.

The production of NiS nanoparticles by *Desulfotomaculum* sp. was proposed as a mechanism for resistance to nickel salts (Fortin et al. 1994). An enrichment culture containing *Desulfosporosinus auripigmenti* was reported to produce nanocrystalline α-NiS and nanocrystalline cobalt pentlandite (Sitte et al. 2013). Bismuth as bismuth subsalicylate is commonly administered to correct gastrointestinal discomfort, and *D. desulfuricans* ATCC 27774 produces Bi_2S_3 when cultivated in media with sulfate, but with nitrate as the final electron acceptor, these SRB reduce Bi(III) to nanoparticles of Bi(0) (Barton et al. 2019). *Desulfovibrio* spp. reduce Tc(VII) to Tc (IV) (Lloyd et al. 1999a, b), and the reduction of Tc(VII) *D. desulfuricans* is reported to involve a [NiFe] hydrogenase (De Luca et al. 2001).

7.11.2 Energy Technology

From research with SRB, several developments may be important for energy technology. The use of H_2 as an alternate fuel source has been considered for several decades, and biological systems have been considered as a possible source of H_2. Almost 50 years ago, an approach to obtain H_2 involved biophotolysis of water using spinach chloroplast to capture light and bacterial ferredoxin to transfer electrons from chloroplasts to bacterial hydrogenase, which releases H_2 (Barton and Tomei 1995). A considerable research was devoted to maximize the rate and quantity of H_2 that could be generated by this biophotolysis system (Rao et al. 1976). Variations to this system employed hydrogenase from *D. desulfuricans* with cytochrome c_3 with proflavine, 5-deazariboflavin, or Zn-tetraphenylporphyrin sulfonate to transfer electrons (Eng et al. 1993a, b). Also, reports have focused on enhancing the production of H_2 (Velázquez-Sánchez et al. 2016; Martins et al. 2021) and on using immobilized cells or enzymes (Plasterk et al. 1981; Yu et al. 2011; Miller et al.

2019; Moore et al. 2020). With the production of H2 as a biofuel, storage of H_2 becomes important, and it has been suggested that SRB can be used to hydrogenate CO_2 with the formation of formic acid/formate (Schuchmann et al. 2018). The storage of formic acid/formate becomes a "H_2 sink" where formic acid can be used in a fuel cell or formate dehydrogenase could be used to release H_2 to power cars. Reactions are being studied for solar-driven reduction of carbon dioxide (Alvarez-Malmagro et al. 2021), which could be applied in a future system for the reduction of atmospheric CO_2. Several reports consider the use of pure or mixed cultures containing SRB in biofuel cells (Cooney et al. 1996; Miran et al. 2017; Murugan et al. 2018). An interesting report is the application of SRB in electrochemistry with *Desulfovibrio ferrophilus* generating an extracellular current (Deng et al. 2015; Murugan et al. 2018). A new physiological activity for SRB is the fixation of CO_2 energized by a bioelectrochemical system without H_2, and these bacteria would be referred to as electroautotrophs (Agostino and Rosenbaum 2018).

7.11.3 Dye Decolorization

Wastewaters from the textile industry carry dyes that are detrimental to the aquatic environment. Removal of dyes that are polluting the environment can be achieved by the use of SRB, and dyes with an azo bond (R^1–N=N–R^2) are the most difficult for bacteria to metabolize. The decolorization of the azo dyes Reactive Orange 96 and Reactive Red 120 by *D. desulfuricans* occurs under sulfate-limiting conditions using electrons from fermentation of pyruvate (Yoo et al. 2000). The removal of the azo dye Reactive Black 5 using sludge is enhanced when *D. desulfuricans* is added to the digestion mixture (Kim et al. 2007). Azo dye solutions containing Orange II, Reactive Black 5, Reactive Red 120, Reactive Brilliant Violet 5R, and the anthraquinone dye Reactive Blue 2 are reduced under sulfidogenic conditions (Togo et al. 2008), and the sulfonated azo dye Congo Red is reduced by H_2S (Diniz et al. 2002). H_2S attacks the azo bond to convert the azo dye to a colorless aromatic amine. Hydrogenase has been suggested to reduce textile dyes (Mutambanengwe et al. 2007) with anthraquinone 2,6-disulfonate and riboflavin functioning as shuttle molecules to transfer electrons from H_2 to the dye molecule (Van der Zee et al. 2001; dos Santos et al. 2007).

7.12 Perspective

The SRB are chemolithotrophs with a metabolism that impacts several global nutrient cycles with considerable effect on the carbon, nitrogen, and sulfur cycles. With respect to the carbon cycle, which is emphasized in this chapter, SRB not only consume carbon compounds that range from inorganic CO_2 to complex heterocyclic organic molecules, but due to the incomplete oxidation of some of the SRB, they also generate numerous small organic compounds. The environmental metabolism of the syntrophic SRB is being appreciated, and this activity is especially relevant in explaining the disappearance of methane (or other small organic compounds) and sulfate simultaneously in marine environments. SRB have the hallmark characteristic of using sulfate as the terminal electron acceptor with the production of H_2S. This bioproduction of H_2S is chemically reactive with various elements in the environment, and there is a strong interest in assessing the formation of mineral sulfides by SRB. The applications of SRB to remediate water streams polluted with heavy metals using biosulfide production from SRB have been used with appreciable success. The formation of heavy metal sulfide precipitates can be readily demonstrated, and SRB have the potential for formation of metal sulfide mineral deposits. Additionally, with an alkaline shift in the environment due to SRB reduction of sulfate, carbonate minerals including dolomite may result. For SRB to induce mineral formation, specific geochemical characteristics would be required, and establishing the appropriate chemical environment to study bioformation of mineral sulfides or dolomite in the laboratory may be difficult to achieve. The SRB have a considerable array of metalloproteins with a low redox potential. It appears that redox proteins (cytochromes and dehydrogenases) that have a metallocenter that can be accessed by redox-active metal(loid)s are nonspecifically reduced. Additionally, if shuttle molecules are reduced in the environment, this reduction of metallic ions and actinides can be accomplished extracellularly. Formation of metallic nanoparticles is a relatively new application of SRB for use in "green" applications to detoxify the environment, and as the nanotechnology develops, additional interest may be directed to activities of the SRB. The interaction of SRB with electrodes in electrochemistry is a new research avenue, and exciting developments in energy production await applications that have not yet matured.

References

Abdelouas A (2006) Uranium mill tailings. Elements 2:335–341

Abdelouas A, Gong W, Lutze WF, Shelnutt JA, Franco R, Moura I (2000) Using cytochrome c_3 to make selenium nanowires. Chem Mater 12:1510–1512

Abreu F, Martins JL, Silveira TS, Keim CN, Lins de Barros HGP, Filho FJ, Lins U (2007) "Candidatus Magnetoglobus multicellularis", a multicellular, magnetotactic prokaryote from a hypersaline environment. Int J Syst Evol Microbiol 57:1318–1322

Abril G, Iversen N (2002) Methane dynamics in a shallow non-tidal estuary (Randers Fjord, Denmark). Mar Ecol Prog Ser 230:171–181. https://doi.org/10.3354/meps230171

Abu Laban N, Selesi D, Rattei T, Tischler P, Meckenstock RU (2010) Identification of enzymes involved in anaerobic benzene degradation by a strictly anaerobic iron-reducing enrichment culture. Environ Microbiol 12:2783–2796

Achá D, Hintelmann H, Pabón CA (2012) Sulfate-reducing bacteria and mercury methylation in the water column of the Lake 658 of the experimental lake area. Geomicrobiol J 29:667–674

Ackerley DF, Gonzalez CF, Park CH, Blake R, Keyhan A, Matin A (2004) Chromate-reducing properties of soluble flavoproteins from *Pseudomonas putida* and *Escherichia coli*. Appl Environ Microbiol 70:873–882

Aeckersberg F, Bak F, Widdel F (1991) Anaerobic oxidation of saturated hydrocarbons to CO_2 by a new type of sulfate-reducing bacterium. Arch Microbiol 156:5–14

Aeckersberg F, Rainey FA, Widdel F (1998) Growth, natural relationships, cellular fatty acids and metabolic adaptation of sulfate-reducing bacteria that utilize long-chain alkanes under anoxic conditions. Arch Microbiol 170:361–369

Agostino V, Rosenbaum MA (2018) Sulfate-reducing electroautotrophs and their applications in bioelectrochemical systems. Front Energy Res 6:55. https://doi.org/10.3389/fenrg.2018.00055

Ahmed E, Holmström SJM (2014) Siderophores in environmental research: roles and applications. Microb Biotechnol 7:196–208

Alvarez-Malmagro J, Oliveira AR, Gutiérrez-Sánchez C, Villajos B, Pereira IAC, Vélez M, Pita M, De Lacey AL (2021) Bioelectrocatalytic activity of W-formate dehydrogenase covalently immobilized on functionalized gold and graphite electrodes. ACS Appl Mater Interfaces 13: 11891–11900

Anderson R, Lovley D (2000) Anaerobic bioremediation of benzene under sulfate-reducing conditions in a petroleum-contaminated aquifer. Environ Sci Technol 34:2261–2266

Andres R, Kasgnoc A (1998) A time-averaged inventory of subaerial volcanic sulfur emissions. J Geophys Res 103:25251–25261

Archer C, Vance D (2006) Coupled Fe and S isotope evidence for Archean microbial Fe(III) and sulfate reduction. Geology 34:153–156

Armstrong J, Armstrong W (2005) Rice: Sulfide-induced barriers to root radial oxygen loss, Fe^{2+} and water uptake, and lateral root emergence. Ann Bot 96:625–638

Arnosti C, Jørgensen BB, Sagemann J, Thamdrup B (1998) Temperature dependence of microbial degradation of organic matter in marine sediments: polysaccharide hydrolysis, oxygen consumption, and sulfate reduction. Mar Ecol Prog Ser 165:59–70

Aüllo T, Berlendis S, Lascourrèges JF, Dessort D, Duclerc D, SaintLaurent S et al (2016) New bio-indicators for long term natural attenuation of monoaromatic compounds in deep terrestrial aquifers. Front Microbiol 7:122. https://doi.org/10.3389/fmicb.2016.00122

Ayangbenro AS, Olanrewaju OS, Babalola OO (2018) Sulfate-reducing bacteria as an effective tool for sustainable acid-mine bioremediation. Front Microbiol 9:1986. https://doi.org/10.3389/fmicb.2018.01986

Baena S, Fardeau M-L, Labat M, Ollivier B, Garcia J-L, Patel BKC (1998) *Desulfovibrio aminophilus* sp. nov., a novel amino acid degrading and sulfate reducing bacterium from an anaerobic dairy wastewater lagoon. Syst Appl Microbiol 21:498–504

Bai H, Kang Y, Quan H, Han Y, Sun J, Feng Y (2013) Treatment of acid mine drainage by sulfate-reducing bacteria with iron in bench scale runs. Bioresour Technol 128:818–822

Bak F, Widdel F (1986a) Anaerobic degradation of indolic compounds by sulfate-reducing enrichment cultures, and description of *Desulfobacterium indolicum* gen. nov., sp. nov. Arch Microbiol 146:170–176

Bak F, Widdel F (1986b) Anaerobic degradation of phenol and phenol derivatives by *Desulfobacterium phenolicum* sp. nov. Arch Microbiol 146:177–180

Balba MT, Nedwell DB (1982) Microbial metabolism of acetate, propionate and butyrate in anoxic sediment from the Colne Point Saltmarsh, Essex, UK. J Gen Microbiol 128:1415–1422

Baldi F, Pepi M, Filippelli M (1993) Methylmercury resistance in *Desulfovibrio desulfuricans* strains in relation to methylmercury degradation. Appl Environ Microbiol 59:2479–2485

Bange HW, Bartell UH, Rapsomanikis S, Andreae MO (1984) Methane in the Baltic and North Seas and a reassessment of the marine emissions of methane. Global Biogeochem Cycles 8:465–480

Barton LL, Fauque GD (2009) Biochemistry, physiology and biotechnology of sulfate-reducing bacteria. Adv Appl Microbiol 65:43–98

Barton LL, Tomei FA (1995) Characteristics and activities of sulfate-reducing bacteria. In: Barton LL (ed) Sulfate-reducing bacteria: biotechnology handbooks, vol 8. Plenum Press, Inc., New York, pp 1–32

Barton LL, Choudhury K, Thompson BM, Steehhoudt K, Groffman AR (1996) Bacterial reduction of soluble uranium: the first step of in situ immobilization of uranium. Radioact Waste Manage Environ Restor 20:141–151

Barton LL, Thomson BM, Plunkett RM (2003) Reduction of metals and non-essential elements. In: Adams M, Barton LL, Ferry J, Ljungdahl L, Johnson M (eds) Biochemistry and physiology of anaerobes. Springer, New York, pp 220–234

Barton LL, Tomei-Torres FA, Xu H, Zucco T (2015) Metabolism of metals and metalloids by the sulfate-reducing bacteria. In: Saffarini D (ed) Bacteria-metal interactions. Springer International Publishing, Switzerland, pp 57–83

Barton LL, Granat AS, Lee S, Xu H, Ritz NL, Hider R, Lin HC (2019) Bismuth(III) interactions with *Desulfovibrio desulfuricans*: inhibition of cell energetics and nanocrystal formation of Bi_2S_3 and Bi^0. Biometals 32:803–811

Basafa M, Hawboldt K (2019) Reservoir souring: sulfur chemistry in offshore oil and gas reservoir fluids. J Pet Explor Prod Technol 9:1105–1118. https://doi.org/10.1007/s13202-018-0528-2. OPEN ACCESS use figure

Bastin ES (1926a) The problem of the natural reduction of sulphates. Bull Am Assoc Pet Geol 10: 1270–1299

Bastin ES (1926b) A hypothesis of bacterial influence in the genesis of certain sulfide ores. J Geol 34:773–792

Bastin ES (1933) The chalcocite and native copper types of ore deposits. Econ Geol 28:407–446

Bastin ES, Greer FE (1930) Additional data on sulphate-reducing bacteria in soils and waters of Illinois oil fields. Bull Am Assoc Pet Geol 14:153–159

Bavendamm W (1932) Die mikrobiologische Kalkfiillung in der tropischen See. Arch Mikrobiol 3: 305–276

Beal EJ, House CH, Orohan VJ (2009) Manganese- and iron-dependent marine methane oxidation. Science 325:184–187

Beech IB, Cheung CWS (1995) Interactions of exopolymers produced by sulfate-reducing bacteria with metal ions. Int Biodeterior Biodegrad 35:59–72

Beeder J, Nilsen RK, Rosnes JT, Torsvik T, Lien T (1994) *Archaeoglobus fulgidus* isolated from hot North Sea oil field waters. Appl Environ Microbiol 60:1227–1231

Beeder J, Torsvik T, Lien T (1995) *Thermodesulforhabdus norvegicus* gen. nov., sp. nov., a novel thermophilic sulfate-reducing bacterium from oil field water. Arch Microbiol 164:331–336

Beller HR, Spormann AM, Sharma PK, Cole JR, Reinhardt M (1996) Isolation and characterization of a novel toluene-degrading, sulfate-reducing bacterium. Appl Environ Microbiol 62:1188–1196

Bender KS, O'Connor SM, Chakraborty R, Coates JD, Achenbach LA (2002) Sequencing and transcriptional analysis of the chlorite dismutase gene of *Dechloromonas agitata* and its use as a metabolic probe. Appl Environ Microbiol 68:4820–4826

Benner SG, Blowes DW, Gould WD, Herbert RB, Ptacek CJ (1999) Geochemistry of a permeable reactive barrier for metals and acid mine drainage. Environ Sci Technol 33:2793–2799

Benoit JM, Gilmour CC, Mason RP, Hayes A (1999) Sulfide controls on mercury speciation and bioavailability to methylating bacteria in sediment pore waters. Environ Sci Technol 33:951–957

Berdugo-Clavijo C, Gieg LM (2014) Conversion of crude oil to methane by a microbial consortium enriched from oil reservoir production waters. Front Microbiol 5:197. https://doi.org/10.3389/fmicb.2014.00197

Berg JS, Duverger A, Cordier L, Laberty-Robert C, Guyot F, Miot J (2020) Rapid pyritization in the presence of a sulfur/sulfate-reducing bacterial consortium. Sci Rep 10:8264. https://doi.org/10.1038/s41598-020-64990-6

Berman M, Chase T Jr, Bartha R (1990) Carbon flow in mercury biomethylation by *Desulfovibrio desulfuricans*. Appl Environ Microbiol 56:298–300

Bertel R (1935) Les bacteries marines et leur influence sur la circulation de la matiere dans la mer. Bull Inst Oceanogr Monaco 672:1–12

Biebl H, Pfennig N (1977) Growth of sulfate-reducing bacteria with sulfur as electron acceptor. Arch Microbiol 112:115–117

Biswas KC, Woodards NA, Xu H, Barton LL (2009) Reduction of molybdate by sulfate-reducing bacteria. Biometals 22:131–139

Blazejak A, Schippers A (2011) Real-time PCR quantification and diversity analysis of the functional genes aprA and dsrA of sulfate-reducing prokaryotes in marine sediments of the Peru continental margin and the Black Sea. Front Microbiol 2:253–264. https://doi.org/10.3389/fmicb.2011.00253

Blessing TC, Wielinga BW, Morra MJ, Fendorf S (2001) Co^{III}EDTA- reduction by *Desulfovibrio vulgaris* and propagation of reactions involving dissolved sulfide and polysulfides. Environ Sci Technol 35:1599–1603

Blum JE, Bartha R (1980) Effect of salinity on methylation of mercury. Bull Environ Contam Toxicol 25:404–408

Boetius A, Ravenschlag K, Schubert CJ, Rickert D, Widdel F, Gieseke A et al (2000) A marine microbial consortium apparently mediating anaerobic oxidation of methane. Nature 407:623–626

Bontognali TRR, McKenzie JA, Warthmann RJ, Vasconcelos C (2014) Microbially influenced formation of Mg-calcite and Ca-dolomite in the presence of exopolymeric substances produced by sulphate-reducing bacteria. Terra Nova 26:72–77

Boopathy R (2007) Anaerobic metabolism of nitroaromatic compounds and bioremediation of explosives by sulphate-reducing bacteria. In: Barton LL, Hamilton WA (eds) Sulphate-reducing bacteria – environmental and engineered systems. Cambridge University Press, Cambridge, pp 483–502

Boopathy R (2014) Biodegradation of 2,4,6-trinitrotoluene (TNT) under sulfate and nitrate reducing conditions. Biologia 69:1264–1270

Boopathy R, Daniels L (1991) Isolation and characterization of a furfural degrading sulfate-reducing bacterium from an anaerobic digester. Curr Microbiol 23:327–332

Boopathy R, Kulpa CF (1993) Nitroaromatic compounds serve as nitrogen source for *Desulfovibrio* sp. (B strain). Can J Microbiol 39:430–433

Boopathy R, Kulpa CF, Wilson M (1993) Metabolism of 2,4,6-trinitrotoluene (TNT) by *Desulfovibrio* sp. (B strain). Appl Microbiol Biotechnol 39:270–275

Boopathy R, Gurgas M, Ullian J, Manning JF (1998a) Metabolism of explosive compounds by sulfate-reducing bacteria. Curr Microbiol 37:127–131

Boopathy R, Kulpa CF, Manning J (1998b) Anaerobic biodegradation of explosives and related compounds by sulfate-reducing and methanogenic bacteria: a review. Bioresour Technol 63:81–89

Boopathy R, Manning J, Kulpa CF (1998c) Biotransformation of explosives by anaerobic consortia in liquid culture and in soil slurry. Int Biodeterior Biodegrad 41:167–174

Booth GH, Mercer SJ (1963) Resistance to copper of some oxidizing and reducing bacteria. Nature (London) 199:622

Boshoff G, Duncan J, Rose PD (2004) Tannery effluent as a carbon source for biological sulphate reduction. Water Res 38:2651–2658

Braissant O, Decho AW, Dupraz C, Glunk C, Przekop KM, Visscher PT (2007) Exopolymeric substances of sulfate-reducing bacteria: interactions with calcium at alkaline pH and implication for formation of carbonate minerals. Geomicrobiology 5:401–411

Brannon MA (1914) The action of Salton Sea water on vegetable tissues. Puhl Carnegie Inst 193: 71–78

Brodie EL, Joyner DC, Faybishenko B, Conrad ME, Rios-Velazquez C, Malave J et al (2011) Microbial community response to addition of polylactate compounds to stimulate hexavalent chromium reduction in groundwater. Chemosphere 85:660–665

Brown SD, Gilmour CC, Kucken AM, Wall JD, Elias DA, Brandt CC et al (2011) Genome sequence of the mercury-methylating strain *Desulfovibrio desulfuricans* ND132. J Bacteriol 193:2078–2079

Brune G, Schoberth SM, Sahm M (1983) Growth of a strictly anaerobic bacterium on furfural (2-formaldehyde). Appl Environ Microbiol 46:1187–1192

Bruschi M, Barton LL, Goulhen F, Plunkett RM (2007) Enzymatic and genomic studies on the reduction of mercury and selected metallic oxyanions by sulphate-reducing bacteria. In: Barton LL, Hamilton WA (eds) Sulphate-reducing bacteria: environmental and engineered systems. Cambridge University Press, Cambridge, pp 435–457

Butlin KR (1949) Some malodorous activities of sulphate-reducing bacteria. Proc Soc Appl Bacteriol 2:39–42

Byrne ME, Ball DA, Guerquin-Kern J-L, Rouiller I, Wu T-D, Downing KH, Vali H, Komeili A (2010) *Desulfovibrio magneticus* RS-1 contains an iron- and phosphorus-rich organelle distinct from its bullet-shaped magnetosomes. Proc Natl Acad Sci U S A 107:12263–12268

Cabrera G, Pérez R, Gómez JM, Abalos A, Cantero D (2006) Toxic effects of dissolved heavy metals on *Desulfovibrio vulgaris* and *Desulfovibrio* sp. strains. J Hazard Mater 135:40–46

Caldwell SL, Laidler JR, Brewer EA, Eberly JO, Sandborgh SC, Coldwell FS (2008) Anaerobic oxidation of methane: mechanisms, bioenergetics, and the ecology of associated microorganisms. Environ Sci Technol 42:6781–6799

Canfield DE (1993) Organic matter oxidation in marine sediments. In: Wollast R, Mackenzie FT, Chou L (eds) Interactions of C, N, P and S biogeochemical cycles and global change, NATO ASI Series (Series I: Global environmental change), vol 4. Springer, Berlin, Heidelberg, pp 333–363. https://doi.org/10.1007/978-3-642-76064-8_14

Cao J, Zhang G, Mao Z, Fang Z, Yang C (2009) Precipitation of valuable metals from bioleaching solution by biogenic sulfides. Miner Eng 22:289–295. https://doi.org/10.1016/j.mineng.2008.08.006

Capeness MJ, Edmundson MC, Horsfall LE (2015) Nickel and platinum group metal nanoparticle production by *Desulfovibrio alaskensis* G20. New Biotechnol 6:727–731

Capeness MJ, Echavarri-Bravo V, Horsfall LE (2019a) Production of biogenic nanoparticles for the reduction of 4-nitrophenol and oxidative laccase-like reactions. Front Microbiol 10:997. https://doi.org/10.3389/fmicb.2019.00997. redn of many metals Ag, Pb etc

Capeness MJ, Imrie L, Mühlbauer LF, Le Bihan T, Horsfall LE (2019b) Shotgun proteomic analysis of nanoparticle-synthesizing *Desulfovibrio alaskensis* in response to platinum and palladium. Microbiology (Reading) 165:1282–1294

Cappenberg TE, Prins RA (1974) Interrelations between sulfate-reducing and methane-producing bacteria in bottom deposits of a fresh-water lake. III. Experiments with ^{14}C-labelled substrates. Antonie Van Leeuwenhoek 40:457–469

Carlson HK, Kuehl JV, Hazra AB, Justice NB, Stoeva MK, Sczesnak A et al (2015) Mechanisms of direct inhibition of the respiratory sulfate-reduction pathway by (per)chlorate and nitrate. ISME J 9:1295–1305

Chakraborty A, Ruff SE, Dong X, Ellefson ED, Li C, Brooks JM et al (2020) Hydrocarbon seepage in the deep seabed links subsurface and seafloor biospheres. Proc Natl Acad Sci U S A 117: 11029–11037

Chang Y-J, Peacock AD, Long PE, Stephen JR, McKinley JP, Macnaughton SJ et al (2001) Diversity and characterization of sulfate-reducing bacteria in groundwater at a uranium mill tailings site. Appl Environ Microbiol 67:3149–3160

Chardin B, Dolla A, Chaspoul F, Fardeau ML, Gallice P, Bruschi M (2002) Bioremediation of chromate: thermodynamic analysis of the effects of Cr(VI) on sulfate-reducing bacteria. Appl Microbiol Biotechnol 60:352–360

Chardin B, Giudici-Orticoni MT, De Luca G, Guigliarelli B, Bruschi M (2003) Hydrogenases in sulfate-reducing bacteria function as chromium reductase. Appl Microbiol Biotechnol 63:315–321

Chen G, Ford TE, Clayton CR (1998) Interaction of sulfate-reducing bacteria with molybdenum dissolved from sputter-deposited molybdenum thin films and pure molybdenum powder. J Colloid Interface Sci 204:237–246

Chen X, Andersen TJ, Morono Y, Inagaki F, Jørgensen BB, Lever MA (2017) Bioturbation as a key driver behind the dominance of *Bacteria* over *Archaea* in near-surface sediment. Sci Rep 7: 2400. https://doi.org/10.1038/s41598-017-02295-x

Cheney ES, Jensen ML (1962) Comments on biogenic sulfides. Econ Geol 57:624–627. https://pubs.geoscienceworld.org/segweb/economicgeology/article-abstract/57/4/624/17133/Comments-on-biogenic-sulfides

Chin KJ, MSharma ML, Russell LA, O'Neill KR, Lovley DR (2008) Quantifying expression of a dissimilatory (bi)sulfite reductase gene in petroleum-contaminated marine harbor sediments. Microb Ecol 55:489–499

Choi S-C, Bartha R (1993) Cobalamin-mediated mercury methylation by *Desulfovibrio desulfuricans* LS. Appl Environ Microbiol 59:290–295

Choi S-C, Chase T Jr, Bartha R (1994a) Metabolic pathways leading to mercury methylation in *Desulfovibrio desulfuricans* LS. Appl Environ Microbiol 60:4072–4077

Choi S-C, Chase T Jr, Bartha R (1994b) Enzymatic catalysis of mercury methylation by *Desulfovibrio desulfuricans* LS. Appl Environ Microbiol 60:1342–1346

Choudhary RP, Sheoran AS (2011) Comparative study of cellulose waste versus organic waste as substrate in a sulfate reducing bioreactor. Bioresour Technol 102:4319–4324

Chovanec P, Sparacino-Watkins C, Zhang N, Basu P, Stolz JF (2012) Microbial reduction of chromate in the presence of nitrate by three nitrate respiring organisms. Front Microbiol 3:416. www.frontiers.in.org

Christensen GA, Gionfriddo CM, King AJ, Moberly JG, Miller CL, Somenahally AC et al (2019) Determining the reliability of measuring mercury cycling gene abundance with correlations with mercury and methylmercury concentrations. Environ Sci Technol 53:8649–8663

Church CD, Wilkin RT, Alpers CN, Rye RO, McCleskey RB (2007) Microbial sulfate reduction and metal attenuation in pH 4 acid mine water. Geochem Trans 8:10. https://doi.org/10.1186/1467-4866-8-10

Colombo JH, Reinfelder JR, Barkay T, Yee N (2013) Anaerobic oxidation of Hg(0) and methylmercury formation by *Desulfovibrio desulfuricans* ND132. Geochim Cosmochim Acta 112: 166–177

Compeau GC, Bartha R (1983) Effects of sea salt anions on the formation and stability of methylmercury. Bull Environ Contam Toxicol 31:486–493

Compeau G, Bartha R (1984) Methylation and demethylation of mercury under controlled redox, pH, and salinity conditions. Appl Environ Microbiol 48:1203–1207

Compeau GC, Bartha R (1985) Sulfate-reducing bacteria: principal methylators of mercury in anoxic estuarine sediment. Appl Environ Microbiol 50:498–502

Compeau GC, Bartha R (1987) Effect of salinity on mercury-methylating activity of sulfate-reducing bacteria in estuarine sediments. Appl Environ Microbiol 53:261–265

Conrad R (2009) The global methane cycle: recent advances in understanding the microbial process involved. Environ Microbiol Rep 1:285–292

Cooney MJ, Roschilan E, Marison W, Stockar Ch CU (1996) Physiologic studies with the sulfate-reducing bacterium *Desulfovibrio desulfuricans*: evaluation for use in a biofuel cell. Enzyme Microb Technol 18:358–365

Copenhagen WJ (1934) Occurrence of sulphides in certain areas of the sea bottom on the South African Coast. Union South African Department of Commerce and Industries Fisheries and Marine Biology Survey Division, Investigational Report no. 3

Copenhagen WJ (1953) The periodic mortality of fish in the Walvis region. S Afr J Sci 49:330–344

Copenhagen WJ (1954) The periodic mortality of fish in the Walfis region. S Afr Med J 28:381–391

Cord-Ruwisch R, Garcia JL (1985) Isolation and characterization of an anaerobic benzoate-degrading spore-forming sulfate-reducing bacterium, *Desulfotomaculum sapomandens* sp. nov. FEMS Microbiol Lett 29:325–330

Cord-Ruwisch R, Kleinitz W, Widdel F (1987) Sulfate-reducing bacteria and their activities in oil production. J Pet Technol 39:97–106

Cozzarelli IM, Baedecker MJ, Eganhouse RP, Goerlitz DF (1994) The geochemical evolution of low-molecular-weight organic acids derived from the degradation of petroleum contaminants in groundwater. Geochim Cosmochim Acta 58:863–877

Craig PJ, Bartlett PD (1978) The role of hydrogen sulfide in environmental transport of mercury. Nature (London) 275:635–637

Craig PJ, Moreton PA (1984) The role of sulphide in the formation of dimethyl mercury in river and estuary sediments. Mar Pollut Bull 15:406–408

Cravo-Laureau C, Matheron R, Cayol J-L, Joulian C, Hirschler-Réa A (2004) *Desulfatibacillum aliphaticivorans* gen. nov., sp. nov., an n-alkane- and n-alkene-degrading, sulfate-reducing bacterium. Int J Syst Evol Microbiol 54:77–83

Creamer NJ, Baxter-Plant VS, Henderson J, Potter M, Macaskie LE (2006) Palladium and gold removal and recovery from precious metal solutions and electronic scrap leachates by *Desulfovibrio desulfuricans*. Biotechnol Lett 28:1475–1484

Cui M, Ma A, Qi H, Zhuang X, Zhuang G (2015) Anaerobic oxidation of methane: an "active" microbial process. Microbiology 4:1–11

Cunningham JA, Rahme HA, Hopkins GD, Lebron C, Reinhard M (2001) Enhanced in situ bioremediation of BTEX-contaminated groundwater by combined injection of nitrate and sulfate. Environ Sci Technol 35:1663–1670

Currie R (1953) Upwelling in the Benguela current. Nature 171:497–500. https://doi.org/10.1038/171497a0

Currie B, Utne-Palm AC, Salvanes AGV (2018) Winning ways with hydrogen sulphide on the Namibian Shelf. Front Mar Sci 5:341. https://doi.org/10.3389/fmars.2018.00341

Cypionka H (1989) Characterization of sulfate transport in *Desulfovibrio desulfuricans*. Arch Microbiol 152:237–243

D'Hondt SL, Jørgensen BB, Miller DJ, Batzke A, Blake R, Cragg BA et al (2004) Distributions of microbial activities in deep subseafloor sediments. Science 306:2216–2221

Dar SA, Kleerebezem R, Stams AJM, Kuenen JG, Muyze G (2008) Competition and coexistence of sulfate-reducing bacteria, acetogens and methanogens in a lab-scale anaerobic bioreactor as affected by changing substrate to sulfate ratio. Appl Microbiol Biotechnol 78:1045–1055

Davidova IA, Duncan KE, Choiand OK, Suflita JM (2006) *Desulfoglaeba alkanexedens* nov., sp. nov., an n-alkane-degrading, sulfate-reducing bacterium. Int J Syst Evol Microbiol 56:2737–2742

Davidson CF (1962) On the origin of some strata-bound sulfide ore deposits. Econ Geol 57:265–271

Davis JB, Yarbrough HF (1966) Anaerobic oxidation of hydrocarbons by *Desulfovibrio desulfuricans*. Chem Geol 1:137–144. Used methane, ethane and *n*-octadecane

de Jesus EB, de Andrade Lima LRP, Bernardez LA, Almeida PF (2015) Inhibition of microbial sulfate reduction by molybdate. Braz J Pet Gas 9:95–106

De Luca G, de Philip P, Dermoun Z, Rousset M, Verméglio A (2001) Reduction of technetium (VII) by *Desulfovibrio fructosovorans* is mediated by the nickel-iron hydrogenase. Appl Environ Microbiol 67:4583–4587. https://doi.org/10.1128/aem.67.10.4583-4587.2001

de Vargas I, Sanyahumbi D, Ashworth AM, Hardy CM, Macaskie LE (2005) Use of X-ray photoelectron spectroscopy to elucidate the mechanism of palladium and platinum biosorption by *Desulfovibrio desulfuricans* biomass. In: Harrison STL, Rawlings DE, Petersen J (eds) Proceedings of the 16th International Biohydrometallurgy Symposium, Cape Town, 25th–29th September 2005, pp 605–616, ISBN 1-920051-17-1

Demoll-Decker H, Macy J (1993) The periplasmic nitrite reductase of *Thauera selenatis* may catalyze the reduction of selenite to elemental selenium. Arch Microbiol 160:241–247

Deng X, Nakamura R, Hashimoto K, Okamoto A (2015) Electron extraction from an extracellular electrode by *Desulfovibrio ferrophilus* strain IS5 without using hydrogen as an electron donor. Electrochemistry 83:529–531

Deplanche K, Macaskie LE (2008) Biorecovery of gold by *Escherichia coli* and *Desulfovibrio desulfuricans*. Biotechnol Bioeng 99:1055–1064

Desrochers R, Fredette V (1960) Étude d'une population de bactéries réductrices du soufre. Can J Microbiol 6:349–354. https://doi.org/10.1139/m60-039

Deswaef S, Salmon T, Hiligsmann S, Taillieu X, Milande N, Thonart P, Crine M (1996) Treatment of gypsum waste in a two stage anaerobic reactor. Water Sci Technol 34:367–374

Devol AH (1983) Methane oxidation rates in anaerobic sediments of Sannich Inlet. Limnol Oceanogr 28:783–742

DeWeerd KA, Mandelco L, Tanner RS, Woese CR, Suflita JM (1990) *Desulfomonile tiedjei* gen. nov. and sp. nov., a novel anaerobic, dehalogenating, sulfate-reducing bacterium. Arch Microbiol 154:23–30

Diester-Haass L, Meyers PA, Vidal L (2002) The late Miocene onset of high productivity in the Benguela current upwelling system as part of a global pattern. Mar Geol 180:87–103

Diniz PE, Lopes AT, Lino AR, Serralheiro ML (2002) Anaerobic reduction of a sulfonated azo dye, Congo Red, by sulfate-reducing bacteria. Appl Biochem Biotechnol 97:147–163

Dolfing J (1990) Reductive dechlorination of 3-chlorobenzoate is coupled to ATP production and growth in an anaerobic bacterium, strain DCB-1. Arch Microbiol 153:264–266

Dolfing J, Tiedje JM (1987) Growth yield increase linked to reductive dechlorination in a defined 3-chlorobenzoate degrading methanogenic coculture. Arch Microbiol 149:102–105

Donnelly LS, Busta FF (1982) Characterization of germination of *Desulfotomaculum nigrificans* spores. J Food Prot 45:721–728

Doores S (1983) Bacterial spore resistance – species of emerging importance. Food Technol 37: 127–134

dos Santos AB, Cervantes FJ, van Lier JB (2007) Review paper on current technologies for decolorization of textile wastewaters: perspectives for anaerobic biotechnology. Bioresour Technol 96:2369–2385

Doss B (1912) Melnikowit, ein neues Eisenbisulfid, und seine Bedeutung fiir die Genesis der Kieslagerstiitten. Z prakt Geol 20:453–483

Druschel GK, Labrenz M, Thomsen-Evert T, Fowle DA, Banfield JF (2002) Geochemical modeling of ZnS in biofilms: an example of ore depositional processes. Econ Geol 97:1319–1329

Drzyzga O, Gottschal JC (2002) Tetrachloroethene dehalorespiration and growth of *Desulfitobacterium frappieri* TCE1 in strict dependence on the activity of *Desulfovibrio fructosivorans*. Appl Environ Microbiol 68:642–649

Drzyzga O, Kuver J, Blotevogel K-H (1995) Complete oxidation of benzoate and 40hydroxylbenzoate by a new sulfate-reducing bacterium resembling *Desulfoarculus*. Arch Microbiol 159:109–113

Drzyzga O, Gerritse J, Dijk JA, Elissen H, Gottschal JC (2001) Coexistence of a sulphate-reducing *Desulfovibrio* species and the dehalorespiring *Desulfitobacterium frappieri* TCE1 in defined chemostat cultures grown with various combinations of sulphate and tetrachloroethene. Environ Microbiol 3:92–99

Durai G, Rajasimman M (2011) Biological treatment of tannery wastewater – a review. J Environ Sci Technol 4:1–17

Ekstrom EB, Morel FMM (2008) Cobalt limitation of growth and mercury methylation in sulfate-reducing bacteria. Environ Sci Technol 42:93–99

Ekstrom EB, Morel FM, Benoit JM (2003) Mercury methylation independent of the acetyl-coenzyme A pathway in sulfate-reducing bacteria. Appl Envion Microbiol 69:5414–5422

Elderfield H, Schultz A (1996) Mid-ocean ridge hydrothermal fluxes and the chemical composition of the ocean. Annu Rev Earth Planet Sci 24:191–224

Elferink SJWHO, Maas RN, Harmsen HJM, Stams AJM (1995) *Desulforhabdus amnigenus* gen-nov sp-nov, a sulfate-reducer isolated from anaerobic granular sludge. Arch Microbiol 164:119–124

Eng LH, Lewin MB-M, Neujahr HY (1993a) Kinetic properties of the periplasmic hydrogenase from *Desulfovibrio desulfuricans* NCIMB 8372 and use in photosensitized H_2-production. Chem Technol Biotechnol 56:317–324

Eng LH, Lewin MM-B, Neujahr HY (1993b) Light-driven H_2 oxidation with proflavine and hydrogenase: comparison of cytochrome c_3 and methyl viologen as e- mediators. Photochem Photobiol 58(4):594–599. https://doi.org/10.1111/j.1751-1097.1993.tb04938.x

Ensley BD, Suflita JM (1995) Metabolism of environmental contaminants by mixed and pure cultures of sulfate-reducing bacteria. In: Barton LL (ed) Sulfate-reducing bacteria. Plenum Press, New York, pp 293–332

Fan D (1983) Polyelements in the Lower Cambrian black shale series in southern China. In: Augustitithis SS (ed) The significance of trace elements in solving petrogenetic problems and controversies. Theophrastus Publications S.A, Athens, pp 447–474

Fauque GD (1994) Sulfur reductase from thiophilic sulfate-reducing bacteria. Methods Enzymol 243:353–367

Fauque G (1995) Ecology of sulfate-reducing bacteria. In: Barton LL (ed) Sulfate-reducing bacteria. Biotechnology Handbook 8, Plenum Press, New York, pp 217–241

Fauque G, Herve D, Le Gall J (1979) Structure-function relationship in hemoproteins: the role of cytochrome c_3 in the reduction of colloidal sulfur by sulfate-reducing bacteria. Arch Microbiol 121:261–264

Fauque G, LeGall J, Barton LL (1991) Sulfate-reducing and sulfur-reducing bacteria. In: Shively JM, Batron LL (eds) Variations in autotrophic life. Academic Press, London, pp 271–200

Ferdelman TG, Lee C, Pantoja S, Harder J, Bebout BM, Fossing H (1997) Sulfate reduction and methanogenesis in a *Thioploca*-dominated sediment off the coast of Chile. Geochim Cosmochim Acta 61:3065–3079

Ferdelman TG, Fossing H, Neumann K (1999) Sulfate reduction in surface sediments of the southeast Atlantic continental margin between 158389S and 278579S (Angola and Namibia). Limnol Oceanogr 44(3):650–661

Ferris FG, Fyfe WS, Beveridge TJ (1987) Bacteria as nucleation sites for authigenic minerals in a metal-contaminated lake sediment. Chem Geol 63:225–232

Folkerts M, Ney U, Kneifel H, Stackebrandt E, Witte EG, Förstel H, Schoberth SM, Sahm H (1989) *Desulfovibrio furfuralis* sp. nov., a furfural degrading strictly anaerobic bacterium. Syst Appl Microbiol 11:161–169

Folwell BD, McGenity TJ, Price A, Johnson RJ, Whitby C (2015) Exploring the capacity for anaerobic biodegradation of polycyclic aromatic hydrocarbons and naphthenic acids by microbes from oil-sands-process-affected waters. Int Biodeterior Biodegradation 108:214–221

Fortin D, Beveridge T (1997) Microbial sulfate reduction within sulfidic mine tailings: formation of diagenetic Fe sulfides. Geomicrobiol J 14:1–21

Fortin D, Southam G, Beveridge TJ (1994) Nickel sulfide, iron-nickel sulfide and iron sulfide precipitation by a newly isolated *Desulfotomaculum* species and its relation to nickel resistance. FEMS Microbiol Ecol 14:121–132

Franco LC, Steinbeisser S, Zane GM, Wall JD, Fields MW (2018) Cr(VI) reduction and physiological toxicity are impacted by resource ratio in *Desulfovibrio vulgaris*. Appl Microbiol Biotechnol 102:2839–2850

Frazier SW, Kretzschmar R, Kraemer SM (2005) Bacterial siderophores promote dissolution of UO_2 under reducing conditions. Environ Sci Technol 39:5709–5715

Freke AM, Tate D (1961) Formation of magnetic iron sulphide by bacterial reduction of iron solutions. J Biochem Microbiol Technol Eng 3:29–39

Freney JR, Jacq VA, Baldensperger JF (1982) The significance of the biological sulfur cycle in rice production. In: Dommergues YR, Diem HG (eds) Microbiology of tropical soils and plant productivity. Martinus Nijhoff/Dr W. Junk Publishers, The Hague, pp 271–317

Fryer JM (2018) Soil properties that influence the occurrence of hydrogen sulfide toxicity in rice fields. Theses and Dissertations. 2640. http://scholarworks.uark.edu/etd/2640

Fullerton H, Crawford M, Bakenne A, Freedman DL, Zinder SH (2013) Anaerobic oxidation of ethene coupled to sulfate reduction in microcosms and enrichment cultures. Environ Sci Technol 47:12374–12381

Gahl R, Anderson B (1928) Sulphate-reducing bacteria in California oil waters. Zbl Bakt, II Abt 73: 331–338

Gallagher KL, Kading TJ, Braissant O, Duparaz C, Visscher PT (2012) Inside the alkalinity engine: the role of electron donors in the organomineralization potential of sulfate-reducing bacteria. Geobiology 10:518–530
Galushko A, Rozanova EP (1991) *Desulfobacterium cetonicum* sp. nov.: a sulfate-reducing bacterium which oxidizes fatty acids and ketones. Microbiology 60:742–746
Gasiorek J (1994) Microbial removal of sulfur dioxide from a gas stream. Fuel Process Technol 40: 129–138
Geets J, Borremans B, Diels L, Springael D, Vangronsveld J, Van Der Lelie D, Vanbroekhoven K (2006) DsrB gene-based DGGE for community and diversity surveys of sulfate-reducing bacteria. J Microbiol Methods 66:194–205
Genovese S (1961) Sul fenomeno dell' "aqua rarra" riscontrato nelio stagno salmastro di Faro (Messina). Atti Soc Pelorit Sci Fis Mat Nat 7:269–271
Genovese S (1963) The distribution of H_2S in the Lake of Faro (Messina) with particular regard to the presence of "Red Water". In: Oppenheimer CH (ed) Marine microbiology. Charles C Thomas Publisher, Springfield, pp 194–204
Genovese S, Macri G (1964) Sulle condizioni microbiologiche dello stretto di Messina e di alcum stagni salmastri della costa Tirrenica nord-orientale della Sicilia. Atti Soc Pelorit Sci Fis Mat Nat 7:43–57
Genschow E, Hegemann W, Maschke C (1998) Biological sulfate removal from tannery wastewater in a two-stage anaerobic treatment. Water Res 30:2072–2078
Gibert O, de Pablo J, Cortina JL, Ayora C (2002) Treatment of acid mine drainage by sulphate-reducing bacteria using permeable reactive barriers: a review from laboratory to full-scale experiments. Rev Environ Sci Biotechnol 1:327–333
Gieg LM, Jack TR, Foght JM (2011) Biological souring and mitigation in oil reservoirs. Appl Microbiol Biotechnol 92:263–282
Gihring TM, Zhang GX, Brandt CC, Brooks SC, Campbell JH, Carroll S et al (2011) A limited microbial consortium is responsible for extended bioreduction of uranium in a contaminated aquifer. Appl Environ Microbiol 77:5955–5965
Gilchrist JDF (1914) Marine biological report for the year ending 30th June, 1914. Union of South Africa, Cape Town. Available online at: http://archive.org/details/marinebiological21914cape
Gilmour CC, Elias DA, Kucken AM, Brown SD, Palumbo AV, Schadt CW, Wall JD (2011) Sulfate-reducing bacterium *Desulfovibrio desulfuricans* ND132 as a model for understanding bacterial mercury methylation. Appl Environ Microbiol 77:3938–3951
Giloteaux L, Duran R, Casiot C, Bruneel O, Elbaz-Poulichet F, Goni-Urriza M (2013) Three-year survey of sulfate-reducing bacteria community structure in Carnoulès acid mine drainage (France), highly contaminated by arsenic. FEMS Microbiol Ecol 83:724–737
Ginsburg-Karagitscheva TL, Rodionova K (1935) Beitrag zur Kenntnis der im Tiefseeschlamm stattfindenden biochemischen Prozesse. Biochem Z 275:396–404
Ginter RL (1930) Causative agents of sulphate reduction in oil-well waters. Bull Am Assoc Pet Geol 14:139–152
Gonçalves M, Oliveira Mello L, Costa A (2008) The use of seaweed and sugarcane bagasse for the biological treatment of metal-contaminated waters under sulfate-reducing conditions. Appl Biochem Biotechnol 147:97–105
Gong J, Song XM, Gao Y, Gong SY, Wang YF, Han JX (2018) Microbiological synthesis of zinc sulfide nanoparticles using *Desulfovibrio desulfuricans*. Inorg Nano-Met Chem 48:96–102. https://doi.org/10.1080/15533174.2016.1216451
Gonzalez CF, Ackerley DF, Park CH, Matin A (2003) A soluble flavoprotein contributes to chromate reduction and tolerance by *Pseudomonas putida*. Acta Biotechnol 23:233–239
Goulhen F, Gloter A, Guyot F, Bruschi M (2006) Cr(VI) detoxification by *Desulfovibrio vulgaris* strain Hildenborough: microbe-metal interactions studies. Appl Microbiol Biotechnol 71:892–897

Graham AM, Bullock AL, Maizel AC, Elias DA, Gilmour CC (2012) Detailed assessment of the kinetics of Hg-cell association, Hg methylation, and methylmercury degradation in several *Desulfovibrio* species. Appl Environ Microbiol 78:7337–7346

Gramp JP, Bigham JM, Sasaki K, Tuovinen OH (2007) Formation of Ni- and Zn-sulfides in cultures of sulfate-reducing bacteria. Geomicrobiol J 24:609–614

Grant CR, Rahn-Lee L, LeGault KN, Komeilia A (2018) Genome editing method for the anaerobic magnetotactic bacterium *Desulfovibrio magneticus* RS-1. Appl Environ Microbiol 84(22): e01724–e01718

Greene EA, Brunelle V, Jenneman GF, Voordouw G (2006) Synergistic inhibition of microbial sulfide production by combinations of the metabolic inhibitor nitrite and biocides. Appl Environ Microbiol 72:7897–7901

Green-Saxena A, Dekas AE, Dalleska NF, Orphan VJ (2014) Nitrate-based niche differentiation by distinct sulfate-reducing bacteria involved in the anaerobic oxidation of methane. ISME J 8: 150–163

Grossi V, Cravo-Laureau C, Meóu A, Raphel D, Garzino F, Hirschler-Réa A (2007) Anaerobic 1-alkene metabolism by the alkane- and alkene-degrading sulfate reducer *Desulfatibacillum aliphaticivorans* strain CV2803T. Appl Environ Microbiol 73:7882–7890

Groudev SN, Georgiev PS, Spasova II, Komnitsas K (2001) Bioremediation of a soil contaminated with radioactive elements. Hydrometall 59:311–318. https://doi.org/10.2166/wpt.2018.068

Grünberg K, Wawer C, Tebo BM, Schüler D (2001) A large gene cluster encoding several magnetosome proteins is conserved in different species of magnetotactic bacteria. Appl Environ Microbiol 67:4573–4582

Guilbaud R, Butler IB, Ellam RM (2011) Abiotic pyrite formation produces a large Fe isotope fractionation. Science 332:1548–1551

Hallam SJ, Putnam N, Preston CM, Detter JC, Rokhsar D, Richardson PM, DeLong EF (2004) Reverse methanogenesis: testing the hypothesis with environmental genomics. Science 305: 1457–1462

Han S, Hu K, Cao J, Pan J, Liu Y, Bian L, Shi C (2014) Mineralogy of early Cambrian Ni-Mo polymetallic black shale at the Sancha deposit, South China: implications for ore genesis. Resour Geol 65:1–12

Hansen TA (1993) Carbon metabolism of the sulfate-reducing bacteria. In: Odom JM, Rivers Singleton J (eds) The sulphate-reducing bacteria: contemporary perspectives. Springer-Verlag, New York, pp 21–40

Hansen LB, Finster K, Fossing H, Iversen N (1998) Anaerobic methane oxidation in sulfate depleted sediments: effects of sulfate and molybdate additions. Aquat Microb Ecol 14:195–204

Harms G, Zengler K, Rabus R, Aeckersberg F, Minz D, Rosselló-Mora R, Widdel F (1999) Anaerobic oxidation of *o*-xylene, *m*-xylene, and homologous alkylbenzenes by new types of sulfate-reducing bacteria. Appl Environ Microbiol 65:999–1004

Henstra AM, Dijkema C, Stams AJM (2007) *Archaeglobus fulgidus* couples CO oxidation to sulfate reduction and acetogenesis with transient formate accumulation. Environ Microbiol 9: 1836–1841

Hessler T, Harrison ST, Huddy RJ (2018) Stratification of microbial communities throughout a biological sulphate reducing up-flow anaerobic packed bed reactor, revealed through 16S metagenomics. Res Microbiol 169:543–551

Higashioka Y, Kojima H, Fukui M (2012) Isolation and characterization of novel sulfate-reducing bacterium capable of anaerobic degradation of *p*-xylene. Microbes Environ 27:273–277

Hinrichs K-U, Boetius A (2002) The anaerobic oxidation of methane: new insights in microbial ecology and biogeochemistry. In: Wefer G, Billett D, Hebbeln D, Jørgensen BB, Schlüter M, van Weering T (eds) Ocean margin systems. Springer, Berlin, pp 457–477

Hinrichs KU, Hayes JM, Sylva SP, Brewer PG, DeLong EF (1999) Methane consuming Archaebacteria in marine sediments. Nature 398:802–805

Hitzman DO, Dennis DM (1997) New technology for prevention of sour oil and gas. In: Proceedings SPE/DOE exploration and production environmental conference, Dallas, TX, pp 406–411

Hocking SL, Gadd GM (2007) Bioremediation of metals and metalloids by precipitation and cellular binding. In: Barton LL, Hamilton WA (eds) Sulphate-reducing bacteria: environmental and engineered systems. Cambridge University Press, Cambridge, pp 405–434

Hoehler TM, Alperin MJ, Albert DB, Martens CS (1994) Field and laboratory studies of methane oxidation in an anoxic marine sediment: evidence for a methanogen sulfate reducer consortium. Global Biogeochem Cycles 8:451–463

Hordijk KA, Cappenberg TE (1983) Quantitative high-pressure liquid chromatography-fluorescence determination of some important lower fatty-acids in lake-sediments. Appl Environ Microbiol 46:361–369

Howarth RW (1979) Pyrite: its rapid formation in saltmarsh and its importance to ecosystem metabolism. Science 203:49–51

Howarth RW (1984) The ecological significance of sulfur in the energy dynamics of salt marsh and costal marine sediments. Biogeochemistry 1:5–27

Hu H, Lin H, Zheng W, Tomanicek SJ, Johs A, Feng X et al (2013) Oxidation and methylation of dissolved elemental mercury by anaerobic bacteria. Nat Geosci 6:751–754

Huang S, Lopez-Capel E, Manning DAC, Rickard D (2010) The composition of nanoparticulate nickel sulfide. Chem Geol 277:207–213

Huang W-H, Dong C-D, Chen C-W, Surampalli RY, Kao C-M (2017) Application of sulfate reduction mechanisms for the simultaneous bioremediation of toluene and copper contaminated groundwater. Int Biodeterior Biodegradation 124:215–222

Hubert C, Voordouw G (2007) Oil field souring control by nitrate-reducing *Sulfurospirillum spp.* that outcompete sulfate-reducing bacteria for organic electron donors. Appl Environ Microbiol 73:2644–2652

Hulecki JC, Foght JM, Gray MP, Phillip M Fedorak PM (2009) Sulfide persistence in oil field waters amended with nitrate and acetate. J Indust Microbiol Biotechnol 36(12):1499. https://doi.org/10.1007/s10295-009-0639-3

Humphries AC, Mikheenko IP, Macaskie LE (2006) Chromate reduction by immobilized palladized sulfate-reducing bacteria. Biotechnol Bioengineer 94:81–90

Humphries AC, Nott KP, Hall LD, Macaskie LE (2005) Reduction of Cr(VI) by immobilized cells of *Desulfovibrio vulgaris* NCIMB 8303 and *Microbacterium* sp. NCIMB 13776 Biotechnol Bioengineer 90:589–596

Hussain A, Hasan A, Javid A et al (2016) Exploited application of sulfate-reducing bacteria for concomitant treatment of metallic and non-metallic wastes: a mini review. 3 Biotech 6:119. https://doi.org/10.1007/s13205-016-0437-3

Hwang C, Wu W, Gentry T, Carley J, Corbin GA, Carroll SL and five co-authors (2009) Bacterial community succession during in situ uranium bioremediation: spatial similarities along controlled flow paths. ISME J 3:47–64. https://doi.org/10.1038/ismej.2008.77

Igiri BE, Okoduwa SIR, Idoko GO, Akabuogu EP, Adeyi AO, Ejiogu IK (2018) Toxicity and bioremediation of heavy metals contaminated ecosystem from tannery wastewater: A review. J Toxicol Volume 2018 |Article ID 2568038 | https://doi.org/2018/2018/2568038

Imachi H, Sekiguchi Y, Kamagata Y, Loy A, Qiu Y-L, Hugenholtz P et al (2006) Non-sulfate-reducing, syntrophic bacteria affiliated with *Desulfotomaculum* Cluster I are widely distributed in methanogenic environments. Appl Environ Microbiol 72:2080–2091

Imhoff-Stuckle D, Pfennig N (1983) Isolation and characterization of a nicotinic acid-degrading sulfate-reducing bacterium, *Desulfococcus niacini* sp. nov. Arch Microbiol 136:194–198

Isa MH, Anderson GK (2005) Molybdate inhibition of sulphate reduction in two-phase anaerobic digestion. Process Biochem 40:2079–2089

Ishimoto M, Kondo Y, Kameyama T, Yagi T, Shiraki M (1958) The role of cytochrome in the enzyme system of sulfate-reducing bacteria. In: Science Council of Japan (ed) Proceedings of the International Symposium on Enzyme Chemistry. Tokyo, pp 229–234

Issatschenko BL (1924) Sur la fermentation sulfhydrique dans la mer Noire. C R Acad Sci, Paris 178:2204–2206

Issatschenko BL (1929) Zur Frage der biogenischen Bildung des Pyrits. Int Rev Hydrobiol 22:99–101

Iversen N, Blackburn TH (1981) Seasonal rates of methane oxidation in anoxic marine sediments. Appl Environ Microbiol 41:1295–1300

Iversen N, Jørgensen BB (1985) Anaerobic methane oxidation rates at the sulfate methane transition in marine-sediments from Kattegat and Skagerrak (Denmark). Limnol Oceanogr 30:944–955

Jack TR, Westlake DWS (1995) Control in industrial settings. In: Barton LL (ed) Sulfate-reducing bacteria biotechnology handbook, vol 8. Plenum Press, New York, pp 265–292

Jaekel U, Zedelius J, Wilkes H, Musat F (2015) Anaerobic degradation of cyclohexane by sulfate-reducing bacteria from hydrocarbon-contaminated marine sediments. Front Microbiol 6:116. https://doi.org/10.3389/fmicb.2015.00116

Jamil IN, Clarke WP (2013) Bioremediation for acid mine drainage: organic solid waste as carbon sources for sulfate-reducing bacteria: a review. J Mech Eng Sci 5:569–581. ISSN (Print): 2289-4659; e-ISSN: 2231-8380

Jeanthon C, Nercessian O, Cove E, Grabowski-Lux A (2005) Hyperthermophilic and methanogenic archaea in the oil fields. In: Ollivier B, Magot M (eds) Petroleum microbiology. ASM Press, Washington, DC, pp 55–69

Jesus E, Lima LRP, Bernardez LA, Almeida PF (2015) Inhibition of microbial sulfate reduction by molybdate. Braz J Pet Gas 9:95–106

Jiang W, Fan W (2008) Bioremediation of heavy metal–contaminated soils by sulfate-reducing bacteria. Ann N Y Acad Sci 1140:446–554

Jin Z, Pei Y (2015) Physiological implications of hydrogen sulfide in plants: pleasant exploration behind its unpleasant odour. Oxid Med Cell Longev 2015:397502. 6 p. https://doi.org/2018/2015/397502

Jochum LM, Chen X, Lever MA, Loy A, Jørgensen BB, Schramm A, Kjeldsen KU (2017) Depth distribution and assembly of sulfate-reducing microbial communities in marine sediments of Aarhus Bay. Appl Environ Microbiol 83:e01547-17. https://doi.org/10.1128/AEM.01547-17

Johnson DB, Hallberg KB (2005) Acid mine drainage remediation options: a review. Sci Total Environ 338:3–14

Joo JO, Choi JH, Kim IH, Kim Y-K, Oh B-K (2015) Effective bioremediation of cadmium (II), nickel (II), and chromium (VI) in a marine environment by using *Desulfovibrio desulfuricans*. Biotechnol Bioprocess Eng 20:937–941

Jørgensen BB (1977) The sulfur cycle of a coastal marine sediment (Limfjorden, Denmark). Limnol Oceanogr 22:814–832

Jørgensen BB (1982) Mineralization of organic matter in the sea bed; the role of sulphate reduction. Nature 296:643–645

Jørgensen BB, Kasten S (2006) Sulfur cycling and methane oxidation. In: Schulz HD, Zabel M (eds) Marine geochemistry. Springer, Berlin, pp 271–309

Jørgensen BB, Weber A, Zopfi J (2001) Sulfate reduction and anaerobic methane oxidation in Black Sea sediments. Deep Sea Res 1 Oceanogr Res Pap 48:2097–2120

Junghare M, Schink B (2015) *Desulfoprunum benzoelyticum* gen. nov., sp. nov., a Gram-stain-negative, benzoate-degrading, sulfate-reducing bacterium isolated from a wastewater treatment plant. Int J Syst Evol Microbiol 65:77–84

Kaczmarek SE, Thornton BP (2017) The effect of temperature on stoichiometry, cation ordering, and reaction rate in high-temperature dolomitization experiments. Chem Geol 468:32–41

Kang S, Van Nostrand JD, Gough HD, He Z, Hazen TC, Stahl DA, Zhou J (2013) Functional gene array–based analysis of microbial communities in heavy metals-contaminated lake sediments. FEMS Microbiol Ecol 86(2):200–214

Karbanee N, Van Hille RP, Lewis AE (2008) Controlled nickel sulfide precipitation using gaseous hydrogen sulfide. Ind Eng Chem Res 47:1596–1602

Karnachuk OV, Sasaki K, Gerasimchuk AL, Sukhanova O, Ivasenko DA, Kaksonen AH et al (2008) Precipitation of Cu-sulfides by copper-tolerant *Desulfovibrio* isolates. Geomicrobiol J 25:219–227

Kashefi K, Tor JM, Nevin KP, Lovley DR (2001) Reductive precipitation of gold by dissimilatory Fe(III)-reducing bacteria and archaea. Appl Environ Microbiol 67:3275–3279

Kijjanapanich P, Annachhatre AP, Lens PNL (2014) Biological sulfate reduction for treatment of gypsum contaminated soils, sediments, and solid wastes. Crit Rev Environ Sci Technol 44: 1037–1070

Kim HY, Kim TS, Kim BH (1990a) Degradation of organic sulfur compounds and the reduction of dibenzothiophene to biphenyl and hydrogen sulfide by *Desulfovibrio desulfuricans* M6. Biotechnol Lett 12:761–764

Kim TS, Kim HY, Kim BH (1990b) Petroleum desulfurization by *Desulfovibrio desulfuricans* M6 using electrochemically supplied reducing equivalent. Biotechnol Lett 12:757–760

Kim SD, Kilbaneii JJ, Cha DH (1999) Prevention of acid mine drainage by sulfate-reducing bacteria: organic substrate addition to mine waste piles. Environ Eng Sci 16:139–145

Kim SY, An JY, Kim BW (2007) Improvement of the decolorization of azo dye by anaerobic sludge bioaugmented with *Desulfovibrio desulfuricans*. Biotechnol Bioprocess Eng 12:222. https://doi.org/10.1007/BF02931096

Kindle EM (1926) Notes on the tidal phenomena of Bay of Fundy Rivers. J Geol 34:642–652

King JK, Kostka JE, Frischer ME, Michael F, Sanders M (2000) Sulfate-reducing bacteria methylate mercury at variable rates in pure culture and in marine sediments. Appl Environ Microbiol 66:2430–2437

Kiran MG, Pakshirajan K, Das G (2015) Heavy metal removal using sulfate-reducing biomass obtained from a lab-scale upflow anaerobic packed bed reactor. Environ Eng 142:1–8. https://doi.org/10.1061/(ASCE)EE.1943-7870.0001005

Kirk MF, Holm TR, Park J, Jin Q, Sanford RA, Fouke B, Bethke CM (2004) Bacterial sulfate reduction limits natural arsenic contamination in groundwater. Geology 32:953–956

Kleikemper J, Pelz O, Schroth MH, Zeyer J (2002a) Sulfate-reducing bacterial community response to carbon source amendments in contaminated aquifer microcosms. FEMS Microbiol Ecol 42: 109–118

Kleikemper J, Schroth MH, Sigler WV, Schmucki M, Bernasconi SM, Zeyer JJ (2002b) Activity and diversity of sulfate-reducing bacteria in a petroleum hydrocarbon-contaminated aquifer. Appl Environ Microbiol 68:1516–1523

Kleindienst S, Herbst F-A, Stagars M, von Netzer F, von Bergen M, Seifert J et al (2008) Harnessing sulfate-reducing microbial ecology to enhance attenuation of dissolved BTEX at two petroleum-impacted sites. Ecol Chem Eng 15:535–548

Kleindienst S, Herbst F-A, Stagars MH, von Netzer F, Von Bergen M, Seifert J et al (2014) Diverse sulfate-reducing bacteria of the *Desulfosarcina/Desulfococcus* clade are the key alkane degraders at marine seeps. ISME J 8:2029–2044

Kleinsteuber S, Schleinitz KM, Breitfeld J, Harms H, Richnow HH, Vogt C (2008) Molecular characterization of bacterial communities mineralizing benzene under sulfate-reducing conditions. FEMS Microbiol Ecol 66:143–157

Klemps R, Cypionka H, Widdel F, Pfennig N (1985) Growth with hydrogen, and further physiological characteristics of *Desulfotomaculum* species. Arch Microbiol 143:203–208

Klonowska A, Clark ME, Thieman SB, Giles BJ, Wall JD, Fields MW (2008) Hexavalent chromium reduction in *Desulfovibrio vulgaris* Hildenborough causes transitory inhibition of sulfate reduction and cell growth. Appl Microbiol Biotechnol 78:1007–1016

Kniemeyer O, Musat F, Sievert SM, Knittel K, Wilkes H, Blumenberg M et al (2007) Anaerobic oxidation of short-chain hydrocarbons by marine sulphate-reducing bacteria. Nature 449:898–901

Knittel K, Boetius A (2009) Anaerobic oxidation of methane: progress with an unknown process. Annu Rev Microbiol 63:311–334. https://doi.org/10.1146/annurev.micro.61.080706.093130. PMID: 19575572

Knoblauch C, Jørgensen BB, Harder J (1999a) Community size and metabolic rates of psychrophilic sulfate-reducing bacteria in Arctic marine sediments. Appl Environ Microbiol 65:4230–4233

Knoblauch C, Sahm K, Jørgensen BB (1999b) Psychrophilic sulfate-reducing bacteria isolated from permanently cold Arctic marine sediments: description of *Desulfofrigus oceanense* gen. nov., sp. nov., *Desulfofrigus fragile* sp. nov., *Desulfofaba gelida* gen. nov., sp. nov., *Desulfotalea psychrophila* gen. nov., sp. nov. and *Desulfotalea arctica* sp. nov. Int J Syst Bacteriol 49:1631–1643

Köhler M, Genz IL, Schicht B, Eckart V (1984) Microbial desulfurization of petroleum and heavy petroleum fractions. 4. Communication: anaerobic degradation of organic sulfur compounds of petroleum. Zbl Mikrobiol 139:239–247

Kolhatkar R, Schnobrich M (2017) Land application of sulfate salts for enhanced natural attenuation of benzene in groundwater: a case study. Ground Water Monit Remediat 37:43–57

Kolhatkar A, Bruce L, Kolhatkar R, Flagel S, Beckmann D, Larsen E (2008) Harnessing sulfate-reducing microbial ecology to enhance attenuation of dissolved BTEX at two petroleum-impacted sites. Ecol Chem Eng 15:535–548

Kolmert A, Johnson DB (2001) Remediation of acidic waste waters using immobilised, acidophilic sulfate-reducing bacteria. J Chem Technol Biotechnol 76:836–843

Kraal P, Slomp CP, Reed DC, Reichart G-J, Poulton SW (2012) Sedimentary phosphorus and iron cycling in and below the oxygen minimum zone of the northern Arabian Sea. Biogeosciences 9:2603

Kraal P, Dijkstra N, Behrends T, Slomp CP (2017) Phosphorus burial in sediments of the sulfidic deep Black Sea: key roles for adsorption by calcium carbonate and apatite authigenesis. Geochim Cosmochim Acta 204:140–158

Krause S, Liebetrau V, Gorb S, Sánchez-Román M, McKenzie JA, Treude T (2012) Microbial nucleation of Mg-rich dolomite in exopolymeric substances under anoxic modern seawater salinity: new insight into an old enigma. Geology 40:587–590

Krukenberg V, Harding K, Richter M, Glöckner FO, Gruber-Vodicka HR, Adam B et al (2016) *Candidatus* Desulfofervidus auxilii, a hydrogenotrophic sulfate-reducing bacterium involved in the thermophilic anaerobic oxidation of methane. Environ Microbiol 18(9):3073–3091. https://doi.org/10.1111/1462-2920.13283

Krumholz LR, Elias DA, Suflita JM (2003) Immobilization of cobalt by sulfate-reducing bacteria in subsurface sediments. Geomicrobiol J 20(1):61–72. https://doi.org/10.1080/01490450303892

Kuever J, Kulmer J, Jannsen S, Fischer U, Blotevogel K-H (1993) Isolation and characterization of a new spore-forming sulfate-reducing bacterium growing by complete oxidation of catechol. Arch Microbiol 159:282–288

Kuever J, Könneke M, Galushko A, Drzyzga O (2001) Reclassification of *Desulfobacterium phenolicum* as *Desulfobacula phenolica* comb. nov. and description of strain SaxT as *Desulfotignum balticum* gen. nov., sp. nov. Int J Syst Evol Microbiol 51:171–177

Kümmel S, Herbst FA, Bahr A, Duarte M, Pieper DH, Jehmlich N et al (2015) Anaerobic naphthalene degradation by sulfate-reducing *Desulfobacteraceae* from various anoxic aquifers. FEMS Microbiol Ecol 91(3):fiv006. https://doi.org/10.1093/femsec/fiv006

Kurth JM, Smit NT, Berger S, Schouten S, Jetten MSM, Welte CU (2019) ANME-2d anaerobic methanotrophic archaea differ from other ANME archaea in lipid composition and carbon source. FEMS Microbiol Ecol. https://doi.org/10.1093/femsec/fiz082

Kuwano Y, Shimizu Y (2006) Bioremediation of coal contaminated soil under sulfate-reducing condition. Environ Technol 27:95–102

Laban NA, Selesi D, Jobelius C, Meckenstock RU (2009) Anaerobic benzene degradation by Gram-positive sulfate-reducing bacteria. FEMS Microbiol Ecol 68:300–311

Labrenz M, Banfield JF (2004) Sulfate-reducing bacteria-dominated biofilms that precipitate ZnS in a subsurface circumneutral-pH mine drainage system. Microb Ecol 47:205–217

Labrenz M, Druschel GK, Thomsen-Ebert T, Gilbert B, Welch SA, Kemner KM et al (2000) Formation of sphalerite (ZnS) deposits in natural biofilms of sulfate-reducing bacteria. Science 290:1744–1747

Lamers LPM, Govers LL, Janssen ICJM, Geurts JJM, Van der Welle MEW, Van Katwijk MM et al (2013) Sulfide as a soil phytotoxin—a review. Front Plant Sci 4:268. https://doi.org/10.3389/fpls.2013.00268

Laso-Perez R, Wegener G, Knittel K, Widdel F, Harding KJ, Krukenberg V et al (2016) Thermophilic archaea activate butane via alkyl-coenzyme M formation. Nature 539:396–401

Lazar CS, Dinasquet J, L'Haridon S, Pignet P, Toffin L (2011) Distribution of anaerobic methane-oxidizing and sulfate-reducing communities in the G11 Nyegga pockmark, Norwegian Sea. Antonie Van Leeuwenhoek 100:639–653

Lefèvre CT, Frankel RB, Pósfai M, Prozorov T, Bazylinski DA (2011a) Isolation of obligately alkaliphilic magnetotactic bacteria from extremely alkaline environments. Environ Microbiol 13:2342–2350

Lefèvre CT, Menguy N, Abreu F, Lins U, Pósfai M, Prozorov T et al (2011b) A cultured greigite-producing magnetotactic bacterium in a novel group of sulfate-reducing bacteria. Science 334: 1720–1723

Lefèvre CT, Trubitsyn D, Abreu F, Kolinko S, Jogler C, de Almeida LGP et al (2013) Comparative genomic analysis of magnetotactic bacteria from the *Deltaproteobacteria* provides new insights into magnetite and greigite magnetosome genes required for magnetotaxis. Environ Microbiol 15:2712–2735

Leloup J, Loy A, Knab NJ, Borowski C, Wagner M, Jørgensen BB (2007) Diversity and abundance of sulfate-reducing microorganisms in the sulfate and methane zones of a marine sediment, Black Sea. Environ Microbiol 9:131–142

Leloup J, Fossing H, Kohls K, Holmkvist L, Borowski C, Jørgensen BB (2009) Sulfate-reducing bacteria in marine sediment (Aarhus Bay, Denmark): abundance and diversity related to geochemical zonation. Environ Microbiol 11:1278–1291

Lengke MF, Southam G (2006) Bioaccumulation of gold by sulfate-reducing bacteria cultured in the presence of gold(I)-thiosulfate complex. Geochim Cosmochim Acta 70:3646–3661

Lengke MF, Southam G (2007) The deposition of elemental gold from gold(I)-thiosulfate complex mediated by sulfate-reducing bacterial conditions. Econ Geol 102:109–126

Lens PNL, Visser A, Janssen AJH, Hulshoff Pol LW, Lettinga G (1998) Biotechnological treatment of sulfate-rich wastewaters. Crit Rev Environ Sci Technol 28:41–88. https://doi.org/10.1080/10643389891254160

Lens P, Vallerol M, Esposito G, Zandvoort M (2002) Perspectives of sulfate reducing bioreactors in environmental biotechnology. Rev Environ Sci Biotechnol 1:311–325

Lens P, Gastesi R, Lettinga G (2003) Use of sulfate-reducing cell suspension bioreactors for the treatment of SO_2 rich flue gases. Biodegradation 14:229–240. https://doi.org/10.1023/A:1024222020924

Lens PNL, Meulepas RJW, Sampaio R, Vallero M, Esposito G (2008) Bioprocess engineering of sulfate reduction for environmental technology. In: Dahl C, Friedrich CG (eds) Microbial sulfur metabolism. Springer, Berlin, pp 285–295

Levin RE, Ng H, Nagel CW, Vaughn RH (1959) *Desulfovibrios* associated with hydrogen sulfide formations in olive brines. Bacteriol Proc:7

Lewis A, Swartbooi A (2006) Factors affecting metal removal in mixed sulfide precipitation. Chem Eng Technol 29:277–280

Li S, Gao Z (2000) Tracing the origin of precious metals in Lower Cambrian black shale (in Chinese). Sci China 30D:169–174

Li X, Krumholz LR (2009) Thioredoxin is involved in U(VI) and Cr(VI) reduction in *Desulfovibrio desulfuricans* G20. J Bacteriol 191:4924–4933

Li Z-G, Min X, Zhou Z-H (2016) Hydrogen sulfide: a signal molecule in plant cross-adaptation. Front Plant Sci 7:1621. https://doi.org/10.3389/fpls.2016.01621

Liamleam W, Annachhatre AP (2007) Electron donors for biological sulfate reduction. Biotechnol Adv 25:452–463
Lin C-C, Lin K-C (1970) Spoilage bacteria in canned foods II. Sulfide spoilage bacteria in canned mushrooms and a versatile medium for the enumeration of *Clostridium nigrificans*. Appl Microbiol 19:283–286
Lin W, Bazylinski DA, Xiao T, Wu L-F, Pan Y (2014) Life with compass: diversity and biogeography of magnetotactic bacteria. Environ Microbiol 16:2646–2658
Lisjak M, Teklic T, Wilson ID, Whitman M, Hancock JT (2013) Hydrogen sulfide: environmental factor or signaling molecule? Plant Cell Environ 36:1607–1616
Liu D, Fan Q, Papineau D, Yu N, Chu Y, Wang H, Qiu X, Wang X (2020) Precipitation of protodolomite facilitated by sulfate-reducing bacteria: the role of capsule extracellular polymeric substances. Chem Geol 533:119415. https://doi.org/10.1016/j.chemgeo.2019.119415
Llobet-Brossa E, Rabus R, Böttcher ME, Könneke M, Finke N, Schramm A et al (2002) Community structure and activity of sulfate-reducing bacteria in an intertidal surface sediment: a multimethod approach. Aquat Microb Ecol 29:211–226
Lloyd JR (2003) Microbial reduction of metals and radionuclides. FEMS Microbiol Rev 27:411–425
Lloyd JR, Yong P, Macaskie LE (1998) Enzymatic recovery of elemental palladium by using sulfate-reducing bacteria. Appl Environ Microbiol 64:4607–4609
Lloyd JR, Ridley J, Khizniak T, Lyalikova NN, Macaskie LE (1999a) Reduction of technetium by *Desulfovibrio desulfuricans*: biocatalyst characterization and use in a flowthrough bioreactor. Appl Environ Microbiol 65:2691–2696
Lloyd JR, Thomas GH, Finlay JA, Cole JA, Macaskie LE (1999b) Microbial reduction of technetium by *Escherichia coli* and *Desulfovibrio desulfuricans*: enhancement via the use of high-activity strains and effect of process parameters. Biotechnol Bioeng 66:122–130
Lloyd JR, Mabbett AN, Williams DR, Macaskie LE (2001) Metal reduction by sulphate-reducing bacteria: physiological diversity and metal specificity. Hydrometallurgy 59:327–337
Lloyd KG, Lapham L, Teske A (2006) An anaerobic methane-oxidizing community of ANME-1b archaea in hypersaline Gulf of Mexico sediments. Appl Environ Microbiol 72:7218–7230
Lojou E, Bianco P, Bruschi M (1998) Kinetic studies on the electron transfer between various c-type cytochromes and iron(III) using a voltametric approach. Electrochim Acta 43:205–2013
Love CA, Patel BKC, Nichols PD, Stackebrandt E (1993) *Desulfotomaculum australicum,* sp. nov., a thermophilic sulfate-reducing bacterium isolated from the Great Artesian Basin of Australia. Syst Appl Microbiol 16:244–251
Lovley DR (1991) Dissimilatory Fe(III) and Mn(IV) reduction. Microbiol Rev 55:259–287
Lovley DR (1995) Microbial reduction of iron, manganese, and other metals. Adv Agron 54:175–231
Lovley D (2006) Dissimilatory Fe(III)- and Mn(IV)-reducing prokaryotes. In: Dworkin M, Falkow S, Rosenberg E, Schleifer KH, Stackebrandt E (eds) The Prokaryotes. Springer, New York, NY. pp 635–658 https://doi.org/10.1007/0-387-30742-7_21
Lovley DR, Phillips EJ (1992) Reduction of uranium by *Desulfovibrio desulfuricans*. Appl Environ Microbiol 58:850–856
Lovley DR, Phillips EJP (1994) Reduction of chromate by *Desulfovibrio vulgaris* and its c_3 cytochrome. Appl Environ Microbiol 60:726–728
Lovley DR, Phillips EJP, Gorby YA, Landa ER (1991) Microbial reduction of uranium. Nature 350: 413–416
Lovley DR, Widman PK, Woodward JC, Phillips EJP (1993a) Reduction of uranium by cytochrome c_3 of *Desulfovibrio vulgaris*. Appl Environ Microbiol 59:3572–3576
Lovley DR, Roden EE, Phillips EJP, Woodard JC (1993b) Enzymatic iron and uranium reduction by sulfate-reducing bacteria. Mar Geol 113:41–53
Lovley DR, Coates JD, Woodward JC, Phillips EJP (1995) Benzene oxidation coupled to sulfate reduction. Appl Environ Microbiol 61:953–958

Lovley DR, Holmes DE, Nevin KP (2004) Dissimilatory Fe(III) and Mn(IV) reduction. Adv Microb Physiol 49:219–286. https://doi.org/10.1016/S0065-2911(04)49005-5. PMID: 15518832

Ludwig RD, McGregor RG, Blowes DW, Benner SG, Mountjoy K (2002) A permeable reactive barrier for treatment of heavy metals. Ground Water 40:59–66

Luek JL, Thompson KE, Larsen RK, Heyes A, Gonsior M (2017) Sulfate reduction in sediments produces high levels of chromophoric dissolved organic matter. Sci Rep 7:8829. https://doi.org/10.1038/s41598-017-0922

Lupton FS, Zeikus JG (1984) Physiological basis for sulfate-dependent hydrogen competition between sulfidogens and methanogens. Curr Microbiol 11:7–12

Mabbett AN, Lloyd JR, Macaskie LE (2002) Effect of complexing agents on reduction of Cr(VI) by *Desulfovibrio vulgaris* ATCC 29579. Biotechnol Bioeng 79:389–397

Macy JM, Santini JM, Pauling BV, O'Neill AH, Sly LI (2000) Two new arsenate/sulfate-reducing bacteria: mechanisms of arsenate reduction. Arch Microbiol 173:49–57

Magot M, Ollivier B, Patel BKC (2000) Microbiology of petroleum reservoirs. Antonie Van Leeuwenhoek 77:103–116

Magowo WE, Sheridan C, Rumbold K (2020) Bioremediation of acid mine drainage using Fischer-Tropsch waste water as a feedstock for dissimilatory sulfate reduction. J Water Proc Eng 35: 101229. https://doi.org/10.1016/j.jwpe.2020.101229

Mannucci A, Munz G, Mori G, Lubello C (2013) Factors affecting biological sulphate reduction in tannery wastewater treatment. Environ Eng Manag J 13:1005–1012

Mansor M, Winkler C, Hochella MF Jr, Xu J (2019) Nanoparticulate nickel-hosting phases in sulfidic environments: effects of ferrous iron and bacterial presence on mineral formation mechanism and solid-phase nickel distribution. Front Earth Sci 7:151. https://doi.org/10.3389/feart.2019.00151

Maree JP, Hill E (1989) An integrated process for biological treatment of sulfate-containing industrial effluent. J Water Poll Control Federation 59:1069–1074

Maree JP, Hulse G, Dods D, Schutte CE (1991) Pilot plant studies on biological sulphate removal from industrial effluent. Water Sci Technol 23:1293–1300

Marsh T, Cardenas E, Wu WM, Leigh MB, Carley J, Carroll S et al (2010) Significant association between sulfate-reducing bacteria and uranium-reducing microbial communities as revealed by a combined massively parallel sequencing-indicator species approach. Appl Environ Microbiol 76:6778–6786

Martens CS, Berner RA (1977) Interstitial water chemistry of Long Island Sound sediments. 1. Dissolved gases. Limnol Oceanogr 22:10–24

Martins M, Faleiro M, Barros R, Veríssimo A, Costa M (2009) Biological sulphate reduction using food industry wastes as carbon sources. Biodegradation 20:559–567

Martins PD, Hoyt DW, Bansal S, Mills CT, Tfaily M, Tangen BA et al (2017) Abundant carbon substrates drive extremely high sulfate reduction rates and methane fluxes in Prairie Pothole Wetlands. Glob Chang Biol 23:3107–3120

Martins M, Toste C, Pereira IAC (2021) Enhanced light-driven hydrogen production by self-photosensitized biohybrid systems. Angew Chem 133:9137–9144. https://doi.org/10.1002/ange.202016960

März C, Riedinger N, Sena C, Kasten S (2018) Phosphorus dynamics around the sulphate-methane transition in continental margin sediments: authigenic apatite and Fe(II) phosphates. Mar Geol 404:84–96

Matsunaga T, Nemoto M, Arakaki A, Tanaka M (2009) Proteomic analysis of irregular, bullet-shaped magnetosomes in the sulphate-reducing magnetotactic bacterium *Desulfovibrio magneticus* RS-1. Proteomics 9:3341–3352

McCarthy TS (2011) The impact of acid mine drainage in South Africa. S Afr J Sci 107:1–7

McCausland HC, Komeili A (2020) Magnetic genes: studying the genetics of biomineralization in magnetotactic bacteria. PLoS Genet 16(2):e1008499. https://doi.org/10.1371/journal.pgen.1008499

McInerney MJ, Sieber JR, Gunsalus RP (2009) Syntrophy in anaerobic global carbon cycles. Curr Opin Biotechnol 20:623–632

McKenzie JA, Vasconcelos C (2009) Dolomite Mountains and the origin of the dolomite rock of which they mainly consist: historical developments and new perspectives. Sedimentology 56: 205–219

Meckenstock RU, Boll M, Mouttaki H, Koelschbach JS, Cunha Tarouco P, Weyrauch P, Dong X, Himmelberg AM (2016) Anaerobic degradation of benzene and polycyclic aromatic hydrocarbons. J Mol Microbiol Biotechnol 26:92–118

Miao Z, Brusseau ML, Carroll KC, Carreón-Diazconti C, Johnson B (2012) Sulfate reduction in groundwater: characterization and applications for remediation. Environ Geochem Health 34: 439–450

Michaelis W, Seifert R, Nauhaus K, Treude T, Thie V, Blumenberg M et al (2002) Microbial reefs in the Black Sea fueled by anaerobic oxidation of methane. Science 297:1013–1015

Michel C, Brugna M, Aubert C, Bernadac A, Bruschi M (2001) Enzymatic reduction of chromate: comparative studies using sulfate-reducing bacteria. Key role of polyheme cytochromes *c* and hydrogenases. Appl Microbiol Biotechnol 55:95–100

Michel C, Giudici-Orticoni MT, Baymann F, Bruschi M (2003) Bioremediation of chromate by sulfate-reducing bacteria, cytochromes c_3 and hydrogenases. Water Air Soil Pollut Focus 3: 161–169. https://doi.org/10.1023/A:1023969415656

Mikheenko IP, Rousset M, Dementin S, Macaskie LE (2008) Bioaccumulation of palladium by *Desulfovibrio fructosivorans* wild-type and hydrogenase-deficient strains. Appl Environ Microbiol 74:6144–6146

Miller LP (1950) Formation of metal sulphides through the activities of sulphate-reducing bacteria. Contr Boyce Thomson Inst 16:85–89

Miller M, Robinson WE, Oliveira AR, Heidary N, Kornienko N, Warnan J, Pereira IAE, Reisner E (2019) Interfacing formate dehydrogenase with metal oxides for the reversible electrolysis and solar driven reduction of carbon dioxide. Angew Chem Int Ed 58:4601–4605

Milucka J, Ferdelman TG, Polerecky L, Franzke D, Wegener G, Schmid M et al (2012) Zero-valent sulphur is a key intermediate in marine methane oxidation. Nature 491:541–546

Milucka J, Widdel F, Shima S (2013) Immunological detection of enzymes for sulfate reduction in anaerobic methane-oxidizing consortia. Environ Microbiol 15:1561–1571

Miran W, Jang J, Nawaz M, Shahzad A, Jeong SE, Jeon CO, Lee AS (2017) Mixed sulfate-reducing bacteria-enriched microbial fuel cells for the treatment of wastewater containing copper. Chemosphere 189:134–142

Mohn WW, Kennedy KJ (1992) Reductive dehalogenation of chlorophenols by *Desulfomonile tiedjei* DCB-1. Appl Environ Microbiol 58:1367–1370

Mohn WW, Tiedje JM (1990) Strain DCB-1 conserves energy for growth from reductive dechlorination coupled to formate oxidation. Arch Microbiol 153:267–271

Mohn WW, Tiedje JM (1991) Evidence for chemiosmotic coupling of reductive dechlorination and ATP synthesis in *Desulfomonile tiedjei*. Arch Microbiol 157:1–6

Mohn WW, Tiedje JM (1992) Microbial reductive dehalogenation. Microbiol Rev 56:482–507

Moore EE, Andrei V, Zacarias Z, Pereira IAC, Reisner E (2020) Integration of a hydrogenase in a lead halide perovskite photoelectrode for tandem solar water splitting. ACS Energy Lett 5:232–237

Morasch B, Meckenstock RU (2005) Anaerobic degradation of *p*-xylene by a sulfate-reducing enrichment culture. Curr Microbiol 51:127–130

Morasch B, Schink B, Tebbe CC, Meckenstock RU (2004) Degradation of *o*-xylene and *m*-xylene by a novel sulfate reducer belonging to the genus *Desulfotomaculum*. Arch Microbiol 181:407–417

Moreau JW, Webb RI, Banfield JF (2004) Ultrastructure, aggregation-state, and crystal growth of biogenic nanocrystalline sphalerite and wurtzite. Am Mineral 89:950–960

Mudryk ZJ, Podgórska B, Ameryk A, Bolalek J (2000) The occurrence and activity of sulphate-reducing bacteria in the bottom sediments of the Gulf of Gdańsk. Oceanologia 42:105–117

Müller V, Hess V (2017) The minimum biological energy quantum. Front Microbiol 8:2019. https://doi.org/10.3389/fmicb.2017.02019

Murray J, Irvine R (1895) On the chemical changes which take place in the composition of the sea water associated with blue muds on the floor of the ocean. Earth Environ Sci Trans R Soc Edinb 37:481–507

Murugan M, Miran W, Masuda T, Lee DS, Okamoto A (2018) Biosynthesized iron sulfide nanocluster enhanced anodic current generation by sulfate-reducing bacteria in microbial fuel cells. ChemElectroChem 5:4015–4020. https://doi.org/10.1002/celc.201801086

Musat F, Widdel F (2008) Anaerobic degradation of benzene by a marine sulfate-reducing enrichment culture, and cell hybridization of the dominant phylotype. Environ Microbiol 10: 10–19

Mutambanengwe CCZ, Togo CA, Whiteley CG (2007) Decolorization and degradation of textile dyes with biosulfidogenic hydrogenases. Biotechnol Prog 23:1095–1100

Muyzer G, Stams AJM (2008) The ecology and biotechnology of sulphate-reducing bacteria. Nat Rev Microbiol 6:441–454

Nakazawa H, Arakaki A, Narita-Yamada S, Yashiro I, Jinno K, Aoki N et al (2009) Whole genome sequence of *Desulfovibrio magneticus* strain RS-1 revealed common gene clusters in magnetotactic bacteria. Genome Res 19:1801–1808

Nancucheo I, Bitencourt JAP, Sahoo PK, Oliveira Alves JO, Siqueira JO, Oliveira G (2017) Recent developments for remediating acidic mine waters using sulfidogenic bacteria. BioMed Res Int 2017:7256582, 17 p. https://doi.org/2018/2017/7256582

Nanninga HJ, Gottschal JC (1987) Properties of *Desulfovibrio carbinolicus* sp. nov. and other sulfate-reducing bacteria isolated from an anaerobic-purification plant. Appl Environ Microbiol 53:802–809

National Academy of Sciences (1978) An assessment of mercury in the environment. National Academy of Sciences, Washington, DC

Nauhaus K, Albrecht M, Elvert M, Boetius A, Widdel F (2007) In vitro cell growth of marine archaeal-bacterial consortia during anaerobic oxidation of methane. Environ Microbiol 9:187–196

Nazina TN, Ivanova AE, Kanchaveli LP, Rozanova EP (1988) A new spore-forming thermophilic methylotrophic sulfate-reducing bacterium, *Desulfotomaculum kuznetsovii* sp. nov. Mikrobiologiya (Russian) 57:823–827

Nazina TN, Shestakova NM, Semenova EM, Korshunova AV, Kostrukova NK, Tourova TP et al (2017) Diversity of metabolically active bacteria in water-flooded high-temperature heavy oil reservoir. Front Microbiol 8:707. https://doi.org/10.3389/fmicb.2017.00707

Neal AL, Techkarnjanruk S, Dohnalkova AC, McCready D, Peyton BM, Geesey GG (2001) Iron sulfides and sulfur species produced at hematite surfaces in the presence of sulfate-reducing bacteria. Geochim Cosmochim Acta 65:223–235

Nealson KH, Saffarini D (1994) Iron and manganese in anaerobic respiration: environmental significance, physiology, and regulation. Annu Rev Microbiol 48:311–343

Nemati M, Mazutinec TJ, Jenneman GE, Voordouw G (2001) Control of biogenic H_2S production with nitrite and molybdate. J Ind Microbiol Biotechnol 26:350–355

Neto ESC, Aguiar ABS, Rodriguez RP, Sancineti GP (2018) Acid mine drainage treatment and metal removal based biological sulfate-reducing processes. Braz J Chem Eng 35:543–552

Newman DK, Beveridge TJ, Morel F (1997) Precipitation of arsenic trisulfide by *Desulfotomaculum auripigmentum*. Appl Environ Microbiol 63:2022–2028

Newport PJ, Nedwell DP (1988) The mechanisms of inhibition of *Desulfovibrio* and *Desulfotomaculum* species by selenate and molybdate. J Appl Bacteriol 65:419–423

Newsome L, Morris K, Lloyd JR (2014) The biogeochemical and bioremediation of uranium and other priority radionuclides. Chem Geol 363:164–174

Nilsen RK, Beeder J, Thorstenson T (1996) Distribution of thermophilic marine sulfate reducers in North Sea oil field waters and oil reservoirs. Appl Environ Microbiol 62:1793–1798

Novelli GD, ZoBell CL (1944) Assimilation of petroleum hydrocarbons by sulfate-reducing bacteria. J Bacteriol 47:447–448

Ohde T, Dadou I (2018) Seasonal and annual variability of coastal sulphur plumes in the northern Benguela upwelling system. PLoS One 13:e0192140. https://doi.org/10.1371/journal.pone.0192140

Ohde T, Siegel H, Reißsmann J, Gerth M (2007) Identification and investigation of sulphur plumes along the Namibian coast using the MERIS sensor. Cont Shelf Res 27:744–756

Okpala GN, Chen C, Fida T, Voordouw G (2017) Effect of thermophilic nitrate reduction on sulfide production in high temperature oil reservoir samples. Front Microbiol 8:1573. https://doi.org/10.3389/fmicb.2017.01573

Ollivier B, Cord-Ruwisch R, Hatchikian EC, Garcia JI (1988) Characterization of *Desulfovibrio fructosovorans* sp. nov. Arch Microbiol 149:447–450

Omajali JB, Mikheenko IP, Merroun ML, Wood J, Macaskie LE (2015) Characterization of intracellular palladium nanoparticles synthesized by *Desulfovibrio desulfuricans* and *Bacillus benzeovorans*. J Nanopart Res 17:264. https://doi.org/10.1007/s11051-015-3067-5

Oremland RS, Culbertson CW, Winfrey MR (1991) Methylmercury decomposition in sediments and bacterial cultures: involvement of methanogens and sulfate reducers in oxidative demethylation. Appl Environ Microbiol 57:130–137

Orphan VJ, House CH, Hinrichs KU, McKeegan KD, DeLong EF (2002) Multiple archaeal groups mediate methane oxidation in anoxic cold seep sediments. Proc Natl Acad Sci U S A 99:7663–7668

Outridge PM, Mason RP, Wang F, Guerrero S, Heimbürger-Boavida LE (2018) Updated global and oceanic mercury budgets for the United Nations Global Mercury Assessment. Environ Sci Technol 52:11466–11477

Ozuolmez D, Na H, Lever MA, Kjeldsen KU, Jørgensen BB, Plugge CM (2015) Methanogenic archaea and sulfate-reducing bacteria co-cultured on acetate: teamwork or coexistence? Front Microbiol 6:492. https://doi.org/10.3389/fmicb.2015.00492

Pannekens M, Kroll L, Müller H, Mbow FT, Meckenstock RU (2019) Oil reservoirs, an exceptional habitat for microorganisms. New Biotechnol 49:1–9

Parkes RJ, Gibson GR, Mueller-Harvey I, Buckingham WJ, Herbert RA (1989) Determination of the substrates for sulphate-reducing bacteria within marine and estuarine sediments with different rates of sulphate reduction. J Gen Microbiol 135:175–187

Parkes RJ, Webster G, Cragg BA, Weightman AJ, Newberry CJ, Ferdelman TG et al (2005) Deep sub-seafloor prokaryotes stimulated at interfaces over geological time. Nature 436:390–394

Parks JM, Johs A, Podar M, Bridou R, Hurt RA Jr, Smith SD et al (2013) The genetic basis for bacterial mercury methylation. Science 339:1332–1335

Parshina SN, Kijlstra S, Henstra AM, Sipma J, Plugge CM, Stams AJM (2005a) Carbon monoxide conversion by thermophilic sulfate-reducing bacteria in pure culture and in co-culture with *Carboxydothermus hydrogenoformans*. Appl Microbiol Biotechnol 68:390–396

Parshina SN, Sipma J, Nakashimada Y, Henstra AM, Smidt H, Lysenko AM, Lens PNL, Lettinga G, Stams AJM (2005b) *Desulfotomaculum carboxydivorans* sp. nov., a novel sulfate-reducing bacterium capable of growth at 100% CO. Int J Syst Evol Microbiol 55:2159–2165

Parshina SN, Sipma I, Henstra AM, Stams AJM (2010) Carbon monoxide as an electron donor for the biological reduction of sulphate. Int J Microbiol 2010:319527

Payne RB, Gentry DM, Rapp-Giles BJ, Casalot L, Wall JD (2002) Uranium reduction by *Desulfovibrio desulfuricans* strain G20 and a cytochrome c_3 mutant. Appl Environ Microbiol 68:3129–3132

Peck HD Jr (1959) The ATP-dependent reduction of sulfate hydrogen in extracts of *Desulfovibrio desulfuricans*. Proc Natl Acad Sci U S A 45:701–708

Pernthaler A, Dekas AE, Brown CT, Goffredi SK, Embaye T, Orphan VJ (2008) Diverse syntrophic partnerships from-deep-sea methane vents revealed by direct cell capture and metagenomics. Proc Natl Acad Sci U S A 105:7052–7057

Petrash DA, Bialik OM, Bontognali TRR, Vasconcelos C, Roberts JA, McKenzie JA, Konhauser KO (2017) Microbially catalyzed dolomite formation: from near-surface to burial. Earth-Sci Rev 171:558–582
Picard A, Gartman A, Girguis PR (2016) What do we really know about the role of microorganisms in iron sulfide mineral formation? Front Earth Sci 6:68. https://doi.org/10.3389/feart.2016.00068
Picard A, Gartman A, Clarke DR, Girguis PR (2018) Sulfate-reducing bacteria influence the nucleation and growth of mackinawite and greigite. Geochim Cosmochim Acta 220:367–384
Pietzsch K, Hard BC, Babel W (1999) A *Desulfovibrio* sp. capable of growing by reducing U(VI). J Basic Microbiol 39:365–372
Plasterk RHA, Rao KK, Hall DO (1981) Immobilization of hydrogenases for biophotolytic hydrogen production; stability and kinetics. Biotechnol Lett 3:99–104
Plugge CM, Zhang W, Scholten JCM, Stams AJM (2011) Metabolic flexibility of sulfate reducing bacteria. Front Microbiol 2:81. https://doi.org/10.3389/fmicb.2011.00081
Pósfai M, Moskowitz BM, Arató B, Schüler D, Flies C, Bazylinski DA, Frankel RB (2006) Properties of intracellular magnetite crystals produced by *Desulfovibrio magneticus* strain RS-1. Earth Planet Sci Lett 249:444–455
Postgate JR (1949) Competitive inhibition of sulphate reduction by selenate. Nature 164:67–671
Postgate JR (1965) Recent advances in the study of the sulfate-reducing bacteria. Bacteriol Rev 29: 425–441
Postgate JR (1979) The sulphate-reducing bacteria. Cambridge University Press, Cambridge
Purdy KJ, Nedwell DB, Embley TM, Takii S (1997) Use of 16S rRNA-targeted oligonucleotide probes to investigate the occurrence and selection of sulfate-reducing bacteria in response to nutrient addition to sediment slurry microcosms from a Japanese estuary. FEMS Microbiol Ecol 24:221–234
Qi P, Zhang D, Wan Y (2013) Sulfate-reducing bacteria detection based on the photocatalytic property of microbial synthesized ZnS nanoparticles. Anal Chim Acta 800:65–70
Qiu R, Zhao B, Liu J, Huang X, Li Q, Brewer E, Wang S, Shi N (2009) Sulfate reduction and copper precipitation by a *Citrobacter* sp. isolated from a mining area. J Hazard Mater 164:1310–1315
Quintana C, Cowley JM, Marhic C (2004) Electron nanodiffraction and high-resolution electron microscopy studies of the structure and composition of physiological and pathological ferritin. J Struct Biol 147:166–178
Rabus R, Nordhaus R, Ludwig W, Widdel F (1993) Complete oxidation of toluene under strictly anoxic conditions by a new sulfate-reducing bacterium. Appl Environ Microbiol 59:1444–1451
Raghoebarsing AA, Pol A, de van Pas-Schoonen AKT, Smolders AJ, Ettwig KF, Ripstra T et al (2006) A microbial consortium couples anaerobic methane oxidation to denitrification. Nature 440:918–921
Rangel-Chavez LG, Neria-Gonzalez MI, Márquez-Herrera A, Zapata-Torres M, Campos-González E, Zelaya-Angel O et al (2015) Synthesis of CdS nanocrystals by employing the byproducts of the anaerobic respiratory process of *Desulfovibrio alaskensis* 6SR bacteria. J Nanomater 2015:260397, 7 p. https://doi.org/2018/2015/260397
Rani A, Rockne KJ, Drummond J, Al-Hinai M, Ranjan R (2015) Geochemical influences and mercury methylation of a dental wastewater microbiome. Sci Rep 5:12872. https://doi.org/10.1038/srep12872
Rao KK, Rosa L, Hall DO (1976) Prolonged production of hydrogen gas by a chloroplast biocatalytic system. Biochem Biophys Res Commun 68:21–28
Ravenschlag K, Sahm K, Knoblauch C, Jørgensen BB, Amann R (2000) Community structure, cellular rRNA content, and activity of sulfate-reducing bacteria in marine arctic sediments. Appl Environ Microbiol 66:3592–3602
Ravich-Sherbo J (1930) To the question of bacterial thin layer in the Black Sea according to the hypothesis of Prof. Egounoff. Trav Sta Biol Sébastopol 2:127–141
Reeburgh WS (1976) Methane consumption in Cariaco Trench waters and sediments. Earth Planet Sci Lett 28:337–344

Reeburgh WS (1980) Anaerobic methane oxidation: rate depths distributions in Skan Bay sediments. Earth Planet Sci Lett 47:345–352

Reeburgh WS (2007) Oceanic methane biogeochemistry. Chem Rev 107:486–513

Reeburgh WS, Whalen SC, Alperin MJ (1993) The role of methylotrophy in the global methane budget. In: Murrell JC, Kelly DP (eds) Microbial growth on C1 compounds. Intercept, Andover, pp 1–14

Reis FD, Silva AM, Cunha EC, Leão VA (2013) Application of sodium- and biogenic sulfide to the precipitation of nickel in a continuous reactor. Sep Purif Technol 120:346–353

Reith F, Lengke MF, Falconer D, Craw D, Southam G (2007) The geomicrobiology of gold. ISME J 1:567–584

Reitner J, Peckmann J, Blumenberg M, Michaelis W, Reimer A, Thiel V (2005) Concretionary methane-seep carbonates and associated microbial communities in Black Sea sediments. Palaeogeogr Palaeoclimatol Palaeoecol 227:18–30

Rickard D (1969) The microbiological formation of iron sulphides, Acta Universitatis Stockholmiensis. Almqvist & Wiksell, Stockholm. Contributions in Geology

Rickard DT (2012) Microbial sulfate reduction in sediments. In: Rickard DT (ed) Developments in sedimentology. Elsevier, Oxford, pp 319–351

Robador A, Müller AL, Sawicka JE, Berry D, Hubert CRJ, Loy A, Jørgensen BB, Brüchert V (2016) Activity and community structures of sulfate-reducing microorganisms in polar, temperate and tropical marine sediments. ISME J 10:796–809

Rosenfeld WD (1947) Anaerobic oxidation of hydrocarbons by sulfate-reducing bacteria. J Bacteriol 54:664–665

Rothermich MM, Hayes LA, Lovley DR (2002) Anaerobic, sulfate-dependent degradation of polycyclic aromatic hydrocarbons in petroleum-contaminated harbor sediment. Environ Sci Technol 36:4811–4817

Roychoudhury AN, McCormick DW (2006) Kinetics of sulfate reduction in a coastal aquifer contaminated with petroleum hydrocarbons. Biogeochemistry 81:17–31

Rozanova EP, Galushko AS, Ivanova AE (1991) Distribution of sulfate-reducing bacteria utilizing lactate and fatty-acids in anaerobic ecotopes of flooded petroleum reservoirs. Microbiology 60: 251–256

Rubentschik LI, Goicherman DG (1936) The influence of a decrease in salt-content in limans on the microflora of medicinal muds. Arkh Biol Nauk 43:217–227

Rueter P, Rabus R, Wilkes H, Aeckersberg F, Rainey FA, Jannasch HW, Widdel F (1994) Anaerobic oxidation of hydrocarbons in crude oil by new types of sulphate-reducing bacteria. Nature 372:455–458

Russell P (1961) Microbiological studies in relation to moist groundwood pulp. Chem Ind (London) 20:642–649

Sabaty M, Avazeri C, Pignol D, Vermeglio A (2001) Characterization of the reduction of selenate and tellurite by nitrate reductases. Appl Environ Microbiol 67:5122–5126

Sahm K, Knoblauch C, Amann R (1999a) Phylogenetic affiliation and quantification of psychrophilic sulfate-reducing isolates in marine Arctic sediments. Appl Environ Microbiol 65:3976–3981

Sahm K, MacGregor BJ, Jørgensen BB, Stahl DA (1999b) Sulphate reduction and vertical distribution of sulphate-reducing bacteria quantified by rRNA slot-blot hybridization in a coastal marine sediment. Environ Microbiol 1:65–74. https://doi.org/10.1046/j.1462-2920.1999.00007.x

Sakaguchi T, Burgess JG, Matsunaga T (1993) Magnetite formation by a sulphate-reducing bacterium. Nature 365:47–49

Sakaguchi T, Tsujimura N, Matsunaga T (1996) A novel method for isolation of magnetic bacteria without magnetic collection using magnetotaxis. J Microbiol Methods 26:139–145

Sakaguchi T, Arakaki A, Matsunaga T (2002) *Desulfovibrio magneticus* sp. nov., a novel sulfate-reducing bacterium that produces intracellular single-domain-sized magnetite particles. Int J Syst Evol Microbiol 52:215–221

Sampaio RMM, Timmers RA, Kocks N, André V, Duarte MT, Van Hullebusch ED et al (2010) Zn-Ni sulfide selective precipitation: the role of supersaturation. Sep Purif Technol 74:108–118
Sangameshwar SR, Barnes HL (1983) Supergene processes in zinc lead-silver sulfide ores in carbonates. Econ Geol 78:1379–1397
Schaefer JK, Morel FMM (2009) High methylation rates of mercury bound to cysteine by *Geobacter sulfurreducens*. Nat Geosci 2:123–126
Schaefer JK, Rocks SS, Zheng W, Liang L, Gu B, Morel FMM (2011) Active transport, substrate specificity, and methylation of Hg(II) in anaerobic bacteria. Proc Natl Acad Sci U S A 108: 8714–8719
Schalk IJ, Hannauer M, Braud A (2011) Minireview new roles for bacterial siderophores in metal transport and tolerance. Environ Microbiol 13:2844–2854
Schink B (1997) Energetics of syntrophic cooperation in methanogenic degradation. Microbiol Mol Biol Rev 61:1092–2172
Schippers A, Neretin LN (2006) Quantification of microbial communities in near-surface and deeply buried marine sediments on the Peru continental margin using real-time PCR. Environ Microbiol 8:1251–1260
Schippers A, Neretin LN, Kallmeyer J, Ferdelman TG, Cragg BA, Parkes RJ, Jørgensen BB (2005) Prokaryotic cells of the deep sub-seafloor biosphere identified as living bacteria. Nature 433: 861–864
Schippers A, Köweker G, Höft C, Teichert B (2010) Quantification of microbial communities in three forearc sediment basins off Sumatra. Geomicrobiol J 27:170–182
Schneiderholm H (1923) Chalkographische Untersuchung des Mansfelder Kupferschiefers. N Jb Miner Geol Paliiont Beil 47:1–38
Schnell S, Bak F, Pfennig N (1989) Anaerobic degradation of aniline and dihydroxybenzenes by newly isolated sulfate-reducing bacteria and description of *Desulfobacterium anilini*. Arch Microbiol 152:556–563
Schuchmann K, Chowdhury NP, Müller V (2018) Complex multimeric [FeFe] hydrogenases: biochemistry, physiology and new opportunities for the hydrogen economy. Front Microbiol 9:2911. https://doi.org/10.3389/fmicb.2018.02911
Seitz KW, Dombrowski N, Eme L, Spang A, Lombard J, Sieber JR et al (2019) Asgard archaea capable of anaerobic hydrocarbon cycling. Nat Commun 10:1822. https://doi.org/10.1038/s41467-019-09364-x
Sela-Adler M, Ronen Z, Herut B, Antler G, Vigderovich H, Eckert W, Sivan O (2017) Co-existence of methanogenesis and sulfate reduction with common substrates in sulfate-rich estuarine sediments. Front Microbiol 8:766. https://doi.org/10.3389/fmicb.2017.00766
Selvaraj PT, Little MH, Kaufman EN (1997) Biodesulfurization of flue gases and other sulfate/sulfite waste streams using immobilized mixed sulfate-reducing bacteria. Biotechnol Prog 13: 583–589
Shelton DR, Tiedje JM (1984) Isolation and partial characterization of bacteria in an anaerobic consortium that mineralizes 3-chlorobenzoic acid. Appl Environ Microbiol 48:840–848
Siebenthal CE (1915) Origin of the zinc and lead deposits of the Joplin Region, Missouri, Kansas, and Oklahoma. U.S. Government Publishing Office, Washington, DC
Silver S, Phung LP (2005) Genes and enzymes involved in bacterial oxidation and reduction of inorganic arsenic. Appl Environ Microbiol 71:599–608
Singh P, Singh SM, Singh RN, Naik S, Roy U, Srivastava A, Bölter M (2017) Bacterial communities in ancient permafrost profiles of Svalbard, Arctic. J Basic Microbiol 57(12):1018–1036. https://doi.org/10.1002/jobm.201700061
Sitte J, Akob DM, Kaufmann C, Finster K, Banerjee D, Burkhardt E-M et al (2010) Microbial links between sulfate reduction and metal retention in uranium- and heavy metal-contaminated soil. Appl Environ Microbiol 76:3143–3152
Sitte JM, Pollok K, Langenhorst F, Küsel K (2013) Nanocrystalline nickel and cobalt sulfides formed by a heavy metal-tolerant, sulfate-reducing enrichment culture. Geomicrobiol J 30:36–47. https://doi.org/10.1080/01490451.2011.653082

Smith SD, Bridou R, Johs A, Parks JM, Elias DA, Hurt RA Jr et al (2015) Site-directed mutagenesis of HgcA and HgcB reveals amino acid residues important for mercury methylation. Appl Environ Microbiol 81:3205–3217

So CM, Young LY (1999) Isolation and characterization of a sulfate-reducing bacterium that anaerobically degrades alkanes. Appl Environ Microbiol 65:2969–2976

Sørensen J, Christensen D, Jørgensen BB (1981) Volatile fatty acids and hydrogen as substrates for sulfate-reducing bacteria in anaerobic marine sediment. Appl Environ Microbiol 42:5–11

Sorokin DY, Berben T, Melton ED, Overmars L, Vavourakis CD, Muyzer G (2014) Microbial diversity and biogeochemical cycling in soda lakes. Extremophiles 18:791–809

Speck RV (1981) Thermophilic organisms in food spoilage: sulfide spoilage anaerobes. J Food Prot 44:149–153

Spirakis CS, Heyl AH (1995) Evaluation of proposed precipitation mechanisms for Mississippi Valley-type deposits. Ore Geol Rev 10:1–17

Spring S, Brinkmann N, Murrja M, Spröer C, Reitner J, Klenk H-P (2015) High diversity of culturable prokaryotes in a lithifying hypersaline microbial mat. Geomicrobiol J 32:332–346. https://doi.org/10.1080/01490451.2014.913095

Spring S, Sorokin DY, Verbarg S, Rohde M, Woyke T, Kyrpides NC (2019) Sulfate-reducing bacteria that produce exopolymers thrive in the calcifying zone of a hypersaline cyanobacterial mat. Front Microbiol 10:862. https://doi.org/10.3389/fmicb.2019.00862

Staicu LC, Barton LL (2017) Microbial metabolism of selenium-for survival or profit. In: van Hullebusch ED (ed) Bioremediation of selenium contaminated wastewaters. Springer, Dordrech, pp 11–31

Staicu LC, Barton LL (2021) Selenium respiration in anaerobic bacteria: does energy generation pay off? J Inorg Biochem 222:111509

Stams AJM, Hansen TA, Skyring GW (1985) Utilization of amino acids as energy substrates by two marine *Desulfovibrio* strains. FEMS Microbiol Ecol 31:11–15

Stanley W, Southam G (2018) The effect of gram-positive (*Desulfosporosinus orientis*) and gram-negative (*Desulfovibrio desulfuricans*) sulfate-reducing bacteria on iron sulfide mineral precipitation. Can J Microbiol 64:629–637

Starkey RL (1961) Sulfate-reducing bacteria, their production of sulfide and their economic importance. Tappi 44:493–496

Stasik S, Wick LY, Wendt-Potthoff K (2015) Anaerobic BTEX degradation in oil sands tailings ponds: impact of labile organic carbon and sulfate-reducing bacteria. Chemosphere 138:133–139

Strøm MK (1939) Land-locked waters and the deposition of black muds. In: Trask PD (ed) Recent marine sediments. American Association of Petroleum Geologists, Tulsa, pp 356–372

Sublette K, Peacock A, White D, Davis G, Ogles D, Cook D et al (2006) Monitoring subsurface microbial ecology in a sulfate-amended, gasoline-contaminated aquifer. Ground Water Monit Remediat 26:70–78

Suri N, Voordouw J, Voordouw G (2017) The effectiveness of nitrate-mediated control of the oil field sulfur cycle depends on the toluene content of the oil. Front Microbiol 8:956. https://doi.org/10.3389/fmicb.2017.00956

Suzuki Y, Kelly SD, Kemner KM, Banfield JF (2003) Microbial populations stimulated for hexavalent uranium reduction in uranium mine sediment. Appl Environ Microbiol 69:2337–1346

Sverjensky DA (1986) Genesis of Mississippi Valley-type lead-zinc deposits. Annu Rev Earth Planet Sci 14:177–199

Szewzyk R, Pfennig N (1987) Complete oxidation of catechol by the strictly anaerobic sulfate-reducing *Desulfobacterium catecholicum* sp. nov. Arch Microbiol 147:163–168

Tanaka K (1992) Anaerobic oxidation of isobutyl alcohol, 1-pentanol, and 2-methoxyethanol by *Desulfovibrio vulgaris* strain Marburg. J Ferment Bioeng 73:503–504

Tanaka S, Lee Y-H (1997) Control of sulfate reduction by molybdate in anaerobic digestion. Water Sci Technol 36:143–150

Tanaka A, Mulleriyawa RP, Yasu T (1968) Possibility of hydrogen sulfide induced iron toxicity of the rice plant. Soil Sci Plant Nutr 14:1–6
Tanimoto Y, Bak F (1994) Anaerobic degradation of methylmercaptan and dimethyl sulfide by newly isolated thermophilic sulfate-reducing bacteria. Appl Environ Microbiol 60:2450–2455
Tasaki M, Kamagata Y, Nakamura K, Mikami E (1991) Isolation and characterization of a thermophilic benzoate-degrading, sulfate-reducing bacterium, *Desulfotomaculum thermobenzoicum* sp. nov. Arch Microbiol 155:348–352
Tausson WO, Vesselov IJ (1934) On the bacteriology of the decomposition of cyclical compounds at the reduction of sulphates. Mikrobiologiia 3:360–369
Taylor BF, Oremland RS (1979) Depletion of adenosine triphosphate in *Desulfovibrio* by oxyanions of group VI elements. Curr Microbiol 3:101–103
Tebo BM, Obraztsova AY (1998) Sulfate-reducing bacterium with Cr(VI), U(VI), Mn(IV), Fe(III) as electron acceptors. FEMS Microbiol Lett 162:193–198
Temple KL, Le Roux NW (1964) Syngenesis of sulfide ores: sulfate-reducing bacteria and copper toxicity. Econ Geol 59:271–278
Thamdrup B, Rosselló-Mora R, Amann R (2000) Microbial manganese and sulfate reduction in Black Sea shelf sediments. Appl Environ Microbiol 66:2888–2897
Thauer RK (1998) Biochemistry of methanogenesis: a tribute to Marjory Stephenson. Microbiology 144:2377–2406
Thiel GA (1926) The Mansfeld Kupferschiefer. Econ Geol 21:299–300
Timmers PHA, Widjaja-Greefkes HCA, Ramiro-Garcia J, Plugge CM, Stams AJM (2015) Growth and activity of ANME clades with different sulfate and sulfide concentrations in the presence of methane. Front Microbiol 6:988. https://doi.org/10.3389/fmicb.2015.000988
Togo CA, Mutambanengwe CCZ, Whiteley CG (2008) Decolorization and degradation of textile dyes using sulfate-reducing bacteria (SRB)—biodigestor microflora co-culture. Afr J Biotechnol 7:114–121
Tomei-Torres FA, Barton LL, Lemanski CL, Zocco TG, Fink NH, Sillerud L (1995) Transformation of selenate and selenite to elemental selenium by *Desulfovibrio desulfuricans*. J Ind Microbiol Biotechnol 14:329–336
Trask PD (1925) The origin of the ore of the Mansfeld Kupferschiefer, Germany. A review of the current literature. Econ Geol 20:746–761
Trudinger PA, Lambert IB, Skyring GW (1972) Biogenic sulfide ores: a feasibility study. Econ Geol 67:1114–1127
Trudinger PA, Chamber LA, Smith JW (1985) Low-temperature sulphate reduction: biological versus abiological. Can J Earth Sci 22:1910–1918
Trüper HG, Genovese S (1968) Characterization of photosynthetic sulfur bacteria causing red water in Lake Faro (Messina, Sicily). Limnol Oceanogr 13:225–232
Tucker M, Barton L, Thomson B (1996) Kinetic coefficients for simultaneous reduction of sulfate and uranium by *Desulfovibrio desulfuricans*. Appl Microbiol Biotechnol 46:74–77
Tucker MD, Barton LL, Thomson BM (1997) Reduction and immobilization of molybdenum by *Desulfovibrio desulfuricans*. J Environ Qual 26:1146–1152
Tucker MD, Barton LL, Thomson BM (1998) Reduction of Cr, Mo, Se and U by *Desulfovibrio desulfuricans* immobilized in polyacrylamide gels. J Ind Microbiol Biotechnol 20:13–19
Turner RJ, Weiner JH, Taylor DE (1998) Selenium metabolism in *Escherichia coli*. Biometals 11: 223–227. https://doi.org/10.1023/a:1009290213301. PMID: 9850565
Uberoi V, Bhattacharya SK (1995) Interactions among sulfate reducers, acetogens and methanogens in anaerobic propionate systems. Water Environ Res 67:330–339
Valentine DL (2002) Biogeochemistry and microbial ecology of methane oxidation in anoxic environments: a review. Antonie Van Leeuwenhoek 81:271–282
Valentine DL, Reeburgh WS (2000) New perspectives on anaerobic methane oxidation. Environ Microbiol 2:477–484
van den Brand T, Snip L, Palmen L, Weij P, Jan Sipma J, van Loosdrecht M (2018) Sulfate reducing bacteria applied to domestic wastewater. Water Pract Technol 13:542–554

Van der Zee FP, Bouwman RH, Strik DP, Lettinga G, Field JA (2001) Application of redox mediators to accelerate the transformation of reactive azo dyes to anaerobic bioreactors. Biotechnol Bioeng 75:691–701

Van Nostrand JD, Wu WM, Wu L, Deng Y, Carley J, Carroll S et al (2009) GeoChip-based analysis of functional microbial communities during the reoxidation of a bioreduced uranium-contaminated aquifer. Environ Microbiol 11:2611–2626

van Stempvoort DR, Armstrong J, Mayer BF (2007) Seasonal recharge and replenishment of sulfate associated with biodegradation of a hydrocarbon plume. Ground Water Monit Remediat 27: 110–121

Vasconcelos C, McKenzie JA (2000) Biogeochemistry–sulfate reducers–dominant players in a low-oxygen world? Science 290:1711–1712

Velázquez-Sánchez HI, Puebla-Nuñez HF, Aguilar-López R (2016) Novel feedback control to improve biohydrogen production by *Desulfovibrio alaskensis*. Int J Chem React Eng 14:1255–1264

Verkholiak N, Peretyatko T (2019) Destruction of toluene and xylene by sulfate-reducing bacteria. Ecol Noospherol 30:95–100. https://doi.org/10.15421/031916

Verkholiak NS, Peretyatko TB, Halushka AA (2020) Reduction of perchlorate ions by the sulfate-reducing bacteria *Desulfotomaculum* sp. and *Desulfovibrio desulfuricans*. Regul Mech Biosyst 11:278–282

Vigneron A, Cruaud P, Pignet P, Caprais JC, Gayet N, Cambon-Bonavita MA, Godfroy A, Toffin L (2014) Bacterial communities and syntrophic associations involved in anaerobic oxidation of methane process of the Sonora Margin cold seeps, Guaymas Basin. Environ Microbiol 16: 2777–2790

Villar V, Cabrol L, Heimbürger-Boavida L-E (2020) Widespread microbial mercury methylation genes in the global ocean. Environ Microbiol Rep 12:277–287. https://doi.org/10.1111/1758-2229.12829

Villemur R, Lanthier M, Beaudet R, Lépine F (2006) The *Desulfitobacterium* genus. FEMS Microbiol Rev 30:706–733

Visser A (1995) The anaerobic treatment of sulfate containing wastewater. Doctoral thesis, Wageningen Agricultural University, Wageningen, The Netherlands

von Wolzogen Kühr CAH (1922) Occurrence of sulfate reduction in the deeper layers of the earth. Proc Koninklijke Nederlandse Akademie van Wetenschappen 25:188–198

Voordouw G (2008) Impact of nitrate on the sulfur cycle in oil fields. In: Dahl C, Friedrich CG (eds) Microbial sulfur metabolism. Springer, Berlin, pp 296–303

Wall JD, Krumholz LR (2006) Uranium reduction. Annu Rev Microbiol 60:149–166

Wallschläger D, Hintelmann H, Evans RD, Wilken RD (1995) Volatilization of dimethylmercury and elemental mercury from River Elbe floodplain soils. Water Air Soil Pollut 80:1325–1329

Wang YT, Shen H (1995) Bacterial reduction of hexavalent chromium. J Ind Microbiol 14:159–163

Warren J (2000) Dolomite: occurrence, evolution and economically important associations. Earth Sci Rev 52:1–81

Webster G, Sass H, Cragg BA, Gorra R, Knab NJ, Green CJ et al (2011) Enrichment and cultivation of prokaryotes associated with the sulphate-methane transition zone of diffusion-controlled sediments of Aarhus Bay, Denmark, under heterotrophic conditions. FEMS Microbiol Ecol 77:248–263

Weelink SAB, van Eekert MHA, Stams AJM (2010) Degradation of BTEX by anaerobic bacteria: physiology and application. Rev Environ Sci Biotechnol 9:359–385

Wei Y, Thomson N, Aravena R, Marchesi M, Barker JF, Madsen EL et al (2018) Infiltration of sulfate to enhance sulfate-reducing biodegradation of petroleum hydrocarbons. Ground Water Monit Remediat 38:73–87. https://doi.org/10.1111/gwmr.12298

Wenter R, Wanner G, Schüler D, Overmann J (2009) Ultrastructure, tactic behaviour and potential for sulfate reduction of a novel multicellular magnetotactic prokaryote from North Sea sediments. Environ Microbiol 11:1493–1505

Werkman CH (1929) Bacteriological studies on sulfide spoilage of canned vegetables. Res Bull 9(117):161–180

Werkman CH, Weaver HJ (1927) Studies in bacteriology of sulphur stinker spoilage of canned sweet corn. Iowa State Coll J Sci 2:57–67

Westrich JT, Berner RA (1984) The role of sedimentary organic matter in bacterial sulfate reduction: the G model tested. Limnol Oceanogr 29:236–249

White C, Shaman A, Gadd G (1998) An integrated microbial process for the bioremediation of soil contaminated with toxic metals. Nat Biotechnol 16:572–575

Widdel F (1988) Microbiology and ecology of sulfate- and sulfur-reducing bacteria. In: Zehnder AJB (ed) Biology of anaerobic microorganisms. John Wiley and Sons, New York, pp 469–585

Widdel F, Hansen TA (1992) The dissimilatory sulfate- and sulfur-reducing bacteria. In: Balows A, Trüper HG, Dworkin M, Harder W, Schleifer K-H (eds) The prokaryotes, 2nd edn. Springer, Berlin, pp 583–624

Widdel F, Pfennig N (1984) Dissimilatory sulfate- or sulfur-reducing bacteria. In: Krieg NR, Holt JG (eds) Bergey's manual of systematic bacteriology, vol 1. Williams & Wilkins, Baltimore, pp 663–379

Widdel F, Kohring GW, Maye F (1983) Studies on dissimilatory sulfate-reducing bacteria that decompose fatty acids III. Characterization of the filamentous gliding *Desulfonema limicola* gen. nov. sp. nov., and *Desulfonema magnum* sp. nov. Arch Microbiol 134:286–294

Widdel F, Boetius A, Rabus R (2006) Anaerobic biodegradation of hydrocarbons including methane. In: Dworkin M, Falkow S, Rosenberg E, Schleifer KH, Stackebrandt E (eds) The prokaryotes. Springer Verlag, New York, pp 1028–1049

Widdel F, Musat F, Knittel K, Galushko A (2007) Anaerobic degradation of hydrocarbons with sulfate as electron acceptor. In: Barton LL, Hamilton WA (eds) Sulphate-reducing bacteria: environmental and engineered systems. Cambridge University Press, Cambridge, pp 265–304

Wiedemeier TH, Rifai HS, Newell CJ, Wilson JT (1999) Natural attenuation of fuels and chlorinated solvents in the subsurface. John Wiley & Sons, Inc., New York

Wilms R, Sass H, Köpke B, Cypionka H, Engelen B (2007) Methane and sulfate profiles within the subsurface of a tidal flat are reflected by the distribution of sulfate-reducing bacteria and methanogenic archaea. FEMS Microbiol Ecol 59:611–621

Wolfe RS (1991) My kind of biology. Annu Rev Microbiol 45:1–35

Wolicka D, Borkowski A, Jankiewicz U, Stępień W, Kowalczyk P (2015) Biologically-induced precipitation of minerals in a medium with zinc under sulfate-reducing conditions. Pol J Microbiol 64:149–155

Wood EC (1961) Some chemical and bacteriological aspects of East Anglian waters. Proc Soc Water Treat Exam 10:82–90

Woolfolk CA, Whiteley HR (1962) Reduction of inorganic compounds with molecular hydrogen by *Micrococcus lactilyticus*. I. Stoichiometry with compounds of arsenic, selenium, tellurium, transition and other elements. J Bacteriol 84:647–658

Wright DT, Wacey D (2005) Precipitation of dolomite using sulphate-reducing bacteria from the Coorong Region, South Australia: significance and implications. Sedimentology 52:987–1008

Wu W-M, Gu B, Fields MW, Gentile M, Ku Y-K, Yan H et al (2005) Uranium (VI) reduction by denitrifying biomass. Biorem J 9:49–61

Wu W-M, Carley J, Luo J, Ginder-Vogel M, Cardanans E, Leigh MB et al (2007) In situ bioreduction of uranium(VI) to submicromolar levels and reoxidation by dissolved oxygen. Environ Sci Technol 41:5716–5723

Xu H, Barton LL, Choudhury K, Zhang P, Wang Y (1999) TEM investigation of U6+ and Re7+ reduction by *Desulfovibrio desulfuricans*, a sulfate-reducing bacterium. Sci Basis Nucl Waste Manage 23:299–304

Xu H, Barton LL, Zhang P, Wang Y (2000) TEM Investigation of U^{6+} and Re^{7+} reduction by *Desulfovibrio desulfuricans*, a sulfate-reducing bacterium. Sci Basis Nucl Waste Manage 23:299–304

Yong P, Rowson NA, Farr JPG, Harris IR, Macaskie LE (2002a) Bioreduction and biocrystallization of palladium by *Desulfovibrio desulfuricans* NCIMB 8307. Biotechnol Bioeng 80:369–379
Yong P, Rowson NA, Farr JPG, Harris IR, Macaskie LE (2002b) Bioaccumulation of palladium by *Desulfovibrio desulfuricans*. J Chem Technol Biotechnol 77:593–601
Yoo ES, Libra J, Wiesmann U (2000) Reduction of azo dyes by *Desulfovibrio desulfuricans*. Water Sci Technol 41:15–22
Youssef N, Elshahed MS, McInerney MJ (2009) Microbial processes in oil fields: culprits, problems, and opportunities. Adv Appl Microbiol 66:141–251
Yu L, Duan JZ, Zhao W, Huang YL, Hou BR (2011) Characteristics of hydrogen evolution and oxidation catalyzed by D*esulfovibrio caledoniensis* biofilm on pyrolytic graphite electrode. Electrochim Acta 56:9041–9047
Zagury GJ, Neculita C (2007) Passive treatment of acid mine drainage in bioreactors: short review, applications, and research needs. In: Proceedings of the 60th Canadian geotechnical conference and 8th joint CGS/IAH-CNC specialty groundwater conference, Ottawa, Canada, pp 1439–1446
Zagury GJ, Kulnieks VI, Neculita CM (2006) Characterization and reactivity assessment of organic substrates for sulphate-reducing bacteria in acid mine drainage treatment. Chemosphere 64: 944–954
Zane GM, Wall JD, De León KB (2020) Novel mode of molybdate inhibition of *Desulfovibrio vulgaris* Hildenborough. Front Microbiol 11:610455. https://doi.org/10.3389/fmicb.2020.610455
Zehnder AJB, Brock TD (1979) Methane formation and methane oxidation by methanogenic bacteria. J Bacteriol 137:420–432
Zehnder AJB, Brock TD (1980) Anaerobic methane oxidation: occurrence and ecology. Appl Environ Microbiol 39:194–204
Zelinski ND (1893) On hydrogen sulfide fermentation in the Black Sea and the Odessa estuaries. Proc Russ Phys Chem Soc 25:298–303
Zhang X, Young LY (1997) Carboxylation as an initial reaction in the anaerobic metabolism of naphthalene and phenanthrene by sulfidogenic consortia. Appl Environ Microbiol 63:4759–4764
Zhang P, He Z, Van Nostrand JD, Qin Y, Deng Y, Wum L et al (2017) Dynamic succession of groundwater sulfate-reducing communities during prolonged reduction of uranium in a contaminated aquifer. Environ Sci Technol 51:3609–3620
Zhao R, Biddle JF (2021) *Helarchaeota* and co-occurring sulfate-reducing bacteria in subseafloor sediments from the Costa Rica Margin. bioRxiv preprint. https://doi.org/10.1101/2021.01.19.427333
Zhou C, Vannela R, Hayes KF, Rittmann BE (2014) Effect of growth conditions on microbial activity and iron-sulfide production by *Desulfovibrio vulgaris*. J Hazard Mater 272:28–35
Ziomek E, Williams RE (1989) Modification of lignins by growing cells of the sulfate-reducing anaerobe *Desulfovibrio desulfuricans*. Appl Environ Microbiol 55:2262–2266
Zobell CE (1938) Studies on the bacterial flora of marine bottom sediments. J Sediment Petrol 8:10–18
Zobell CE (1946) Functions of bacteria in the formation and accumulation of petroleum. Oil Wkly 120:30–36
Zobell CE (1947a) Bacterial release of oil from oil-bearing materials. World Oil 126:36–47
Zobell CE (1947b) Bacterial release of oil from oil-bearing materials. World Oil 127:35–41
Zobell CE, Rittenberg SC (1948) Sulfate-reducing bacteria in marine sediments. J Mar Res 7: 602–617

Chapter 8
Biocorrosion

8.1 Destructive Effects of Biocorrosion

As reviewed by Gu (2012), there are several terms used to describe the destruction of materials by microorganisms. The use of "biocorrosion" and "microbial corrosion" suggests that microorganisms are the cause of corrosion. "Microbiologically induced corrosion" (MIC) is used to describe corrosion where microorganisms are present and are assumed to be responsible for corrosion. In many instances, biocorrosion has not been subjected to a rigorous test such as Koch's postulates for pathogenic diseases, and therefore, the agents responsible for corrosion are often speculative (Gu 2012).

8.1.1 Activities of SRB Leading to Corrosion of Non-metallic Surfaces

8.1.1.1 Biocorrosion of Concrete and Stone

Early reports document the biocorrosion of concrete where hydrogen sulfide is involved (Barr and Buchanan 1912; Parker 1945, 1951; Pochon et al. 1951). As reviewed by Wei et al. (2013), microbiologically induced deterioration (MID) of concrete sewer pipes, concrete floors of buildings housing animals, and concrete in wastewater treatment plants involves a succession of bacterial populations.

High alkalinity of concrete attributed to $Ca(OH)_2$ inhibits bacterial growth, but various factors cause the concrete surface to become rough (Ribas-Silva 1995). Bacteria including SRB collect in cracks and grooves on the surface of the concrete, and sulfate from the environment is converted to H_2S by SRB and the pH of the concrete surface approaches neutrality (Lahav et al. 2004; Matos and Aires 1995; Nielsen et al. 2005; Zhang et al. 2008). Volatile H_2S is oxidized by *Acidithiobacillus*

L. L. Barton, G. D. Fauque, *Sulfate-Reducing Bacteria and Archaea*,
https://doi.org/10.1007/978-3-030-96703-1_8

sp. and other sulfide-oxidizing acidophilic bacteria to sulfuric acid, which is highly corrosive and contributes to the deterioration of concrete (Sand and Bock 1984: Satoh et al. 2009; Vollertsen et al. 2008; Zhang et al. 2008). This microbially induced concrete corrosion (MICC) reduces the lifetime of concrete by many years, and the cost for replacement of concrete pies is several billion USD (O'Connell et al. 2010; Gutiérrez-Padilla et al. 2010; Grengg et al. 2015).

SRB have also been associated with corrosion of store artworks; however, since most of these artworks are in a highly aerobic environment, other microorganisms have a greater role in biocorrosion of these stone surfaces. Postgate (1979) suggests that the biocorrosion of stone statues in tropical regions of Cambodia could be attributed to H_2S, which came from the polluted soils where the statues were placed. Recently, it was proposed that deterioration of sandstone in the ancient structures of Angkor, Cambodia, was attributed to sulfur-oxidizing microorganisms (Kusumi et al. 2011).

8.1.1.2 Remediation of Artistic Stoneworks

The pores on surfaces of stone artwork are suitable sites for biocorrosion, and the process to deposit biogenic minerals into the natural patina involves bacterial precipitation of calcite (Rodriguez-Navarro et al. 1997, 2003; De Muynck et al. 2010). Microorganisms have been associated with the formation of dolomite (Vasconcelos et al. 1995), and this precipitation of calcium carbonate by bacteria has been termed "carbonatogenesis" (Le Métayer-Levrel et al. 1999; Castanier et al. 2000). Laboratory experiments indicated that members of *Desulfovibrio* sp. precipitate dolomite (Vasconcelos et al. 1995). Aerotolerant strains of *D. vulgaris* and *D. desulfuricans* have been delivered to stone surfaces using sepiolite, hydrobiogel-97, and carbogel (Ranalli et al. 1997; Cappitelli et al. 2006).

In another stone restoration activity, SRB have been used to remove black crusts on stone artworks. In urban areas, pollution is high and acid rain results in the depositing of sulfuric acid on stone artworks converting insoluble calcium carbonate ($CaCO_3$) into calcium sulfate dihydrate (gypsum, $CaSO_4 \cdot 2H_2O$) (Gauri et al. 1989; Ausset et al. 2000; Bugini et al. 2000). Atmospheric pollutants including carbon-based particles become part of the mineral matrix producing the black crusts on the stone. A review of crust samples collected from marble statues in Milan, Italy, revealed that over a period of 54 years, the amount of gypsum formed was 5–13 mg/cm^2, which was about 0.2 mg/cm^2 per year (Bugini et al. 2000). In restoration of stone artworks, SRB have been used to remove sulfate from calcium sulfate, and when using a medium where iron was removed, the precipitation of black iron sulfide was avoided (Gauri et al. 1992; Cappitelli et al. 2006; Polo et al. 2010). Several strains of SRB have been found to effectively remove the black crusts, and these bacteria include *D. desulfuricans* ATCC 13541, *D. desulfuricans* ATCC 29577, and *D. vulgaris* ATCC 29579 (Ranalli et al. 1997). Subsequent publications support the use of SRB to remove black crusts form stone artworks (Fernandes 2006; Dhami et al. 2014).

8.1.2 Association of SRB with Metal Corrosion

Anaerobic corrosion had been reported in waterlogged soil and sediments of freshwater or marine environments, and early reports addressed this topic (Gaines 1910; Cobb 1911; Stumper 1923; Ginter 1927; Copenhagen 1934; Bunker 1936). Numerous studies addressed the environments and conditions for metallic corrosion by sulfate-reducing bacteria with production of the reports by Starkey and Wright (1945), Butlin et al. (1949), Hunter et al. (1948), Spruit and Wanklyn (1951), Wanklyn and Spruit (1952), Adams and Farrer (1953), Ewing (1955), Updegraff (1955), Starkey (1957), Meyer et al. (1958), Starkey (1958), Harris (1960), Horvath (1960), Booth and Tiller (1962), Tiller and Booth (1962), Churchill (1963), Ward (1963), Booth et al. (1964), Miller et al. (1964), Booth et al. (1966), Iverson (1966, 1968), and Booth et al. (1968). Examples of metals corroded by SRB are listed in Table 8.1. Metal corrosion is a serious economic problem for industries associated with petroleum, water, and cooling towers, and related processes with about 20% of

Table 8.1 Examples of biocorrosion of metals by SRB

Metal/alloy	Species of SRB or source	References
Carbon steel	*Desulfovibrio desulfuricans*	Beech et al. (1994a)
	Desulfovibrio indonesiensis	Beech and Coutinho (2003)
	SRB in soil	Li et al. (2001)
	Desulfovibrio desulfuricans	Javed et al. (2015)
Steel alloys	SRB from water in oil field	Al Abbas et al. (2013)
	SRB from environment	Antony et al. (2008)
	SRB from lake water	Enos and Taylor (1996)
	Desulfovibrio caledoniensis	Guan et al. (2016)
	Desulfovibrio capillatus	Miranda et al. (2006)
	Desulfovibrio desulfuricans	Heggendorn et al. (2015)
	Desulfovibrio fairfieldensis	Heggendorn et al. (2015)
	SRB in seawater	Neville and Hodgkiess (2000)
	Desulfovibrio vulgaris	Newman et al. (1991)
	Desulfopila corrodens	
	Desulfovibrio sp. strain HSE	
	Desulfovibrio vulgaris	Zhang et al. (2015)
Copper/nickel alloys	*Desulfovibrio vulgaris*	Fu et al. (2014)
	SRB in seawater	Huang et al. (2004)
	SRB in soil	Li et al. (2010)
	SRB in saline water	Little et al. (1991)
	SRB in seawater	Little et al. (1992)
Cobalt/titanium alloys	*Desulfotomaculum nigrificans*	Mystkowska (2016)
Copper metal	SRB from seawater	Elmouaden et al. (2016) and Jodeh et al. (2016)
Titanium	*Desulfovibrio vulgaris*	Rao et al. (2005)

all corrosion activities are attributed to biocorrosion. This great interest in corrosion involving SRB continues and reflects the industrial importance of corrosion, which globally has been estimated to account for an annual replacement expense ranging from 20 to 50 billion USD (Guo et al. 2018) to a trillion USD (Li et al. 2018). Several reviews have provided general reviews detailing biocorrosion by SRB (Hamilton 1985, 2003; Odom 1993; Hamilton et al. 1995; Beech and Sunner 2007; Barton and Fauque 2009).

Since the initial report in 1910 suggesting a connection between the bacterial sulfur cycle and iron corrosion (Gaines 1910), sulfate-reducing bacteria (SRB) have been considered to be important in anaerobic iron corrosion. Carbon steel pipes, which are widely used in the gas and oil industry, are subject to biocorrosion, and the cost for replacement of failed iron pipes or structures has been estimated to be 2–3% of the gross domestic product in developed countries (Enning and Garrelfs 2014). The environments and types of metals being impacted by microbially influenced corrosion (MIC) are numerous and have been recently reviewed (Liengen et al. 2014; Little and Lee 2015; Anandkumar et al. 2015). The methods used to characterize MIC are numerous and have been reviewed by Beech and Gaylarde (1999), Beech and Sunner (2004), Graeber et al. (2014), Dall'Agnol and Moura (2014), and Little and Lee (2015). SRB are a physiological group of bacteria having an important role in MIC, and some activities and types of SRB in this corrosion process have been reviewed (Dall'Agnol and Moura 2014; Beech and Sunner 2007; Hamilton 1985). While the environmental parameters required for SRB-related biocorrosion are relatively broad, they generally include the following: organic carbon content of 0.05–1.2%, soil resistivity of 500–30,000 Ω·cm, water content of 5–36%, oxidation reduction potential of −316 to +385 mV, sulfate level of 0.3–200 mg/g soil, and a near-neutral pH (Jack 2002). A *D. vulgaris* biofilm on carbon steel was followed for over 55 days under organic carbon starvation, and the pitting of steel was attributed to *D. vulgaris* extracting electrons from the steel (Chen et al. 2015). An example of biofilm formation and pitting of ferrous material by *D. desulfuricans* is given in Fig. 8.1.

The bulk of the literature concerning ferrous corrosion by sulfate-reducing microorganisms examines the role of agents associated with the bacteria domain. While there are three genera of sulfate reducers in the archaea domain (*Archaeoglobus*, *Thermocladium*, and *Caldivirga*), only a few reports concerning biocorrosion identify *Archaeoglobus* as the important sulfate reducer present. Since *Archaeoglobus* grows anaerobically at 60–95 °C, biocorrosion due to archaea is limited to those sites where the temperature is markedly elevated. Additionally, biocorrosion at elevated temperatures of about 70 °C could be attributed to *Thermodesulfovibrio* sp. of the bacteria domain. Thus, the sulfate-reducing organisms of greatest concern to the industries dealing with ferrous corrosion are the bacteria, and they are the focus of this chapter.

Biocorrosion of copper has been reported, but only a few papers address this process. In an environment containing a mixture of SRB, pitting of copper alloy was reported by Little et al. (1991). There is a strong correlation between the amount of copper corrosion in seawater and the growth of an SRB biofilm on the copper metal,

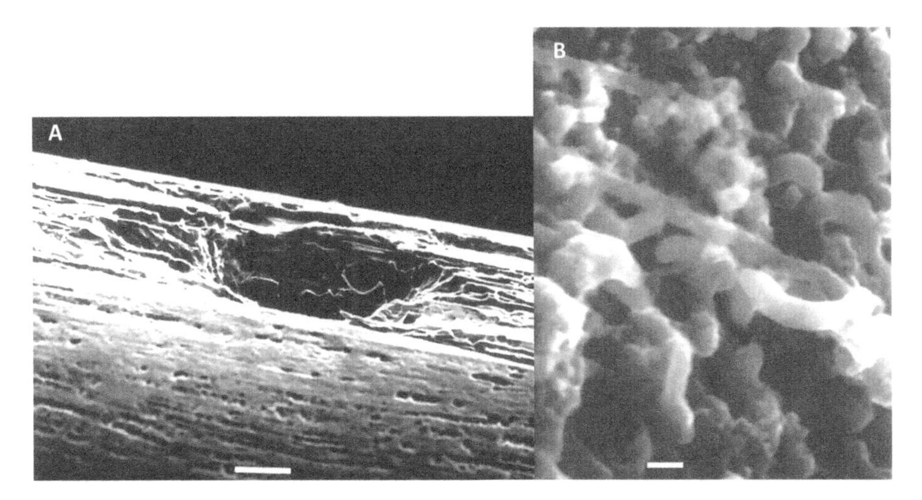

Fig. 8.1 Corrosion of ferrous wire by *D. desulfuricans* as observed with an environmental scanning electron microscope. (**a**) Ferrous wire with a pit attributed to biocorrosion. Bar = 20 μm. (**b**) Biofilm of *D. desulfuricans* on the ferrous wire. Bar = 1 μm. (Reproduced by permission from Barton and Fauque (2009), Copyright 2009, with permission from Elsevier)

production of extracellular polymeric substance (EPS) by the SRB, and the length of time exposed (Elmouaden et al. 2016). Beneath the SRB biofilm, there is the production of cuprous sulfide (Cu_2S), which is a product of hydrogen sulfide reacting with copper oxide on the metal surface (Chen et al. 2014). In comparison to the quantity of ferrous metals used in the environment, the amount of copper employed is relatively small. Thus, the corrosion of iron is of great economic expense, and the focus here is on biocorrosion of ferrous materials.

8.1.3 Mechanisms for Corrosion of Ferrous Metals

The complexity of ferrous biocorrosion has stimulated the suggestions for several different mechanisms to explain activity observed in the field. While the classical theory of cathodic depolarization (discussed in the following paragraph) dominated for many years, other mechanisms proposed included anionic depolarization (Crolet 1992), iron sulfide–induced corrosion (King et al. 1973a), production of a reduced phosphorus compound (Iverson and Olsen 1984), sulfide-induced corrosion (Edyvean et al. 1998), hydrogen-induced blistering (Edyvean et al. 1998), hydrogen-associated weakening of iron (Ford and Mitchell 1990), and metal-binding exopolymer (Beech and Cheung 1995). Hamilton (2003) hypothesized that in metal biocorrosion, the electron flow from the metallic base to oxygen in the atmosphere involved a sequence of biotic and abiotic reactions involving insoluble products of microorganisms.

Various parameters can be controlled by laboratory experiments, but these tests do not include subtle environmental changes observed in open systems. Laboratory experiments provide information on biocorrosion using submersed metal coupons with a specific species of pure cultures of SRB. A variation of chemicals supporting SRB growth has been used, and complex organic materials far greater than found in corrosion environments were present as nutrients in media of Baar or Postgate (Postgate 1979) and Luria-Bertani (LB) broth. Some experiments were with artificial seawater using the Vaatanen nine-salt solution (Jayanaman et al. 1997), while some used natural seawater. Media composition is important as observed from reports by King et al. (1973b) and Booth et al. (1966) where the rate of biocorrosion was dependent on the amount of ferrous iron in the growth medium. In a controlled biocorrosion experiment using *D. desulfuricans*, a decrease in the rate of biocorrosion was observed as the sulfate concentration in the medium was increased from 1.93 to 6.5 g/L, and this was attributed to a decrease in growth of the SRB at elevated sulfate concentrations (Jhobalia et al. 2005). Chloride salts appear to be essential for biocorrosion, and the substitution of non-chloride salts in culture media has been reported to limit biocorrosion (Ringas and Robinson 1988a, b). It has been reported that establishing an anaerobic condition by bubbling N_2 seawater would remove dissolved CO_2 and O_2 with a resulting shift in pH from ~8.0 to >9.0 (Lee et al. 2007). Most of the laboratory experiments involved the closed environment, while field biocorrosion is open with an environments subject to change. It is important to recall that the contribution of one explanation of ferrous corrosion would not exclude other observations of merit, and the overall biocorrosion event may be attributed to the synergism of several different activities and mechanisms. As reported by Mand et al. (2014), acetogenic bacteria are of importance in biocorrosion due to the production of organic acids but also providing acetate as a carbon source for SRB. Acetogenic bacterium such as *Acetobacterium woodii* is important in production of acetate as an energy source for SRB when carbon substrates are limiting.

In 1934, von Wolzogen Kühr and von der Vlugt proposed that SRB were responsible for anaerobic iron corrosion by consumption of H_2 produced in the corrosion event. This "cathodic depolarization theory" relied on the idea that SRB would consume H_2 on the iron surface and this would enable corrosion to proceed. The corrosion of iron pipes and tanks by an electrochemical process is described in the following reactions:

$$Fe^0 \leftrightarrow Fe^{2+} + 2e^- \quad E^o = -0.47\ V\ (\text{anodic reaction}) \tag{8.1}$$

$$2H^+ + 2e^- \leftrightarrow H_2 \quad E^{o'} = -0.41\ V\ (\text{cathodic reaction}) \tag{8.2}$$

A combination of these two reactions produces

$$Fe^0 + 2H^+ \rightarrow Fe^{2+} + H_2 \quad \Delta G^{o'} = -10.6\ kJ \tag{8.3}$$

This proposed mechanism to explain corrosion involved bacterial hydrogenase in depolarization. Several reviews (Starkey 1957; Postgate 1960; Starkey 1960; Booth 1964; Iverson 1972) have examined the extent of anaerobic corrosion and underlying events that could explain this biological process. As reviewed by Enning and Garrelfs (2014), numerous investigators failed to demonstrate the increased rate of corrosion with removal of H_2, and questions were raised pertaining to the experimental procedure used by von Wolzogen Kühr and von der Vlugt. Recently, it was determined that a fresh isolate of SRB from a ferrous corrosion environment with H_2 as the electron donor, *Desulfovibrio* sp. strain HS3, failed to enhance iron corrosion when oxidizing H_2 on the surface of iron placed in water (Venzlaff et al. 2013). When evaluating all the experiments of anaerobic iron corrosion, it has been proposed that oxidation of H_2 is not a major factor contributing to ferrous corrosion (Enning and Garrelfs 2014). However, a role for SRB in ferrous corrosion is strongly suggested since sulfate is present in the anaerobic environment where hydrogen sulfide is present on the corroded iron. The rate of corrosion approaches 0.9 mm Fe^0 per year (Enning and Garrelfs 2014). When the environment is too hot for mesophilic SRB to grow, thermophilic sulfate-reducing archaea (*Archaeoglobus spp.*) may account for biocorrosion of iron (Duncan et al. 2009; Islam and Karr 2013; Jia et al. 2018; Ali et al. 2020), and mechanisms of biocorrosion by these sulfate-reducing archaea have not been detailed.

When testing the cathodic depolarization theory for biocorrosion, it was found that common laboratory cultures of SRB utilizing organic compounds (organotrophs) as the electron donor will not produce anaerobic iron corrosion when using iron as the only electron source. Widdel's group has isolated novel lithotrophic SRB using corroding iron as the electron donor (Dinh et al. 2004). These strains designated as *D. ferrophilus* strain IS5 and *Desulfopila corrodens* strain IS4 acquired electrons directly from iron (Enning et al. 2012; Venzlaff et al. 2013), establishing this as a mechanism for microbial corrosion. It appears that these lithotrophic SRB are widely dispersed in anaerobic mud but are rarely observed because media for isolation favor the faster-growing organotrophic SRB. It appears that the electron-consuming SRB acquire electrons from nonferrous materials present in the environment because they are found in considerable numbers in environments where metallic iron has not been introduced. The lithotrophic cultivation of microorganisms using Fe^0 as the electron donor is not limited to SRB but has been shown also for an archaeal organism known as strain IMI, which is related to *Methanobacterium*, and *Methanococcus maripaludis* strain KA1 (Enning and Garrelfs 2014). With additional studies, the electron-conducting materials on the surface of these ferrous-corroding microorganisms will become established. The mechanism of ferrous corrosion due to anaerobic depletion of electrons was designated "electrical microbially induced corrosion" (EMIC) (Enning et al. 2012). Corrosion models involving EMIC and chemical microbially influenced corrosion (CMIC) are presented in Fig. 8.2.

Several mechanisms have been proposed for SRB-induced biocorrosion of ferrous materials. MIC can be classified into several categories where type I and type II are electrochemical and type III metabolized organic materials (see reviews by

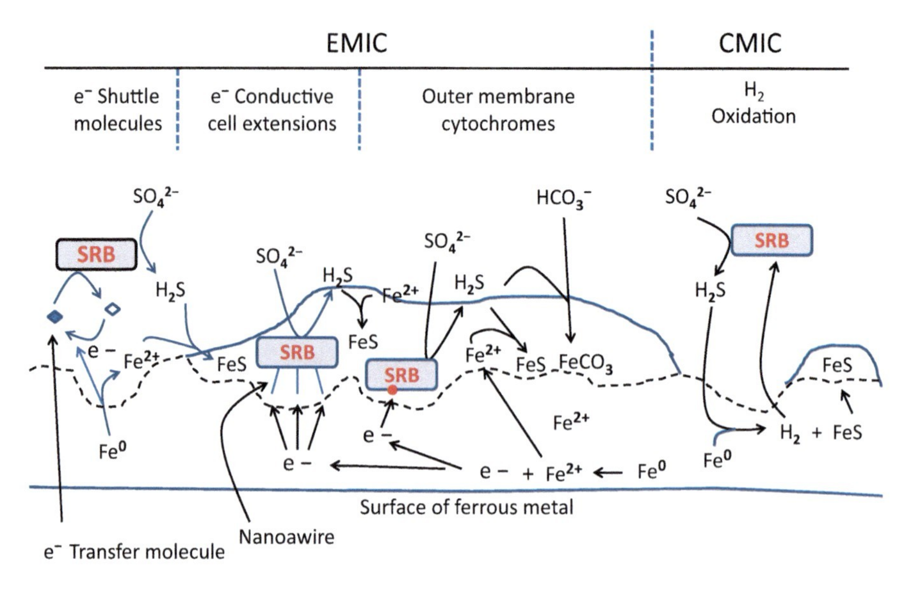

Fig. 8.2 Models depicting the mechanism of corrosion of ferrous metal by SRB

Enning et al. 2012; Li et al. 2018). SRB have been associated with type I and type II biocorrosion. Type I is anaerobic oxidation with electrons extracted anaerobically from the metal (most commonly discussed as iron) by bacteria, and the electrons are used to energize respiration. Type II is acidic corrosion where organic acids such as acetate are produced by SRB and acetogens (Mand et al. 2014; Smith et al. 2019) in the environment and the acid degrades the material. Electrical microbially influenced corrosion (EMIC) would follow the equation listed below:

$$4Fe^0 + SO_4^{2-} + 3HCO_3 + 5H^+ \rightarrow FeS + 3FeCO_3 + 4H_2O \quad (8.4)$$

Chemical microbially influenced corrosion (CMIC) would be as follows (Smith et al. 2019):

$$3[CH_2O] + 2Fe^0 + 2SO_4^{2-} + H^+ \rightarrow 3HCO_{3^-} + 2FeS + 2H_2O \quad (8.5)$$

Type III involves biodegradation of polyurethane and organic plasticizer around pipes or on metallic surfaces, and products of decomposition are used as a carbon and energy source by microorganisms (Gu 2003).

Iron in the steel matrix is insoluble, and with type I MIC, electrons are transported across the bacterial cell wall and membrane into the cytoplasm where sulfate is reduced (Xu et al. 2013; Li et al. 2015, 2018). This extracellular electron transfer (EET) was discussed by Hernandez and Newman (2001), while the types of EET, (i) direct electron transfer (DET) and (ii) mediated electron transfer (MET), have been reviewed by Du et al. (2007), Xu et al. (2014), Li et al. (2015), and Kato (2016).

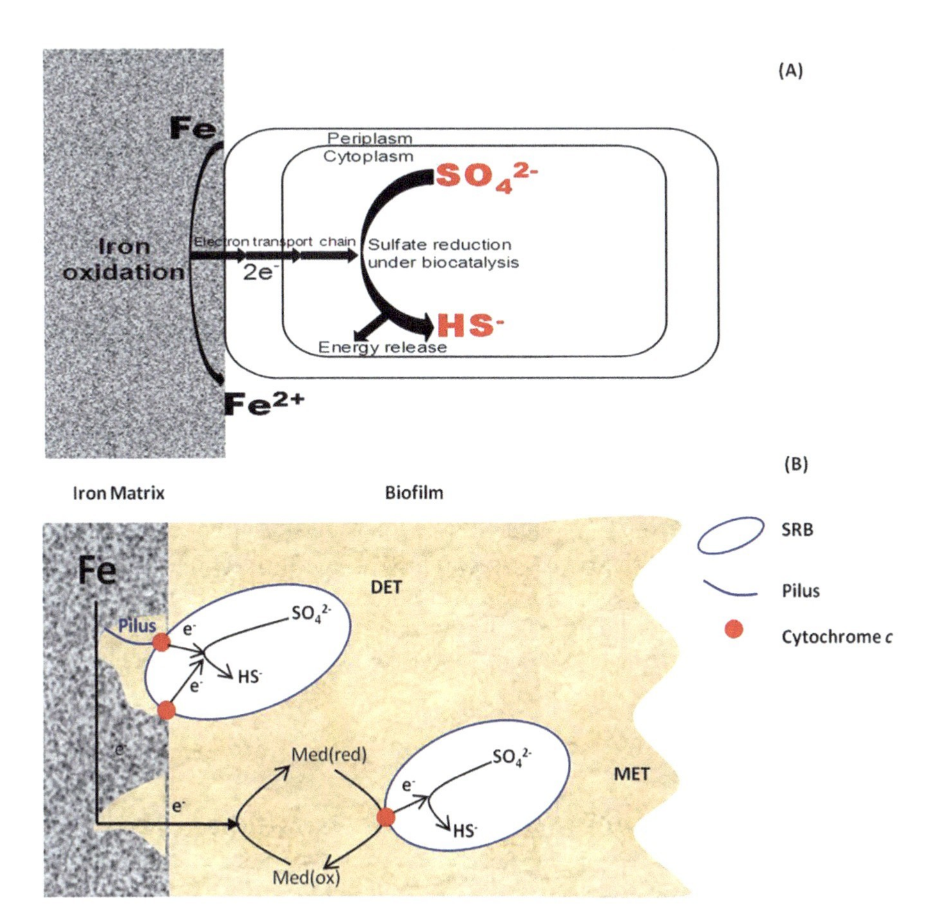

Fig. 8.3 Electron transfer in microbiologically influenced corrosion. (**a**) Mechanism for MIC by SRB due to utilization of extracellular electrons from iron oxidation for intracellular sulfate reduction and (**b**) schematic illustration of direct electron transfer (DET) and mediated electron transfer (MET). (Citation: Li et al. (2015). Copyright: © 2015 Li et al. This is an open access article distributed under the terms of the Creative Commons Attribution License, which permits unrestricted use, distribution, and reproduction in any medium, provided the original author and source are credited)

As indicated in Fig. 8.3, the mechanism of DET is proposed to involve electron conductive pili (nanowires) or cytochrome *c* at the surface of the cell, and the impact of this activity is to produce pits on the surface of the ferrous material (Sherar et al. 2011; Venzlaff et al. 2013). In other DET systems, riboflavin and flavin adenine dinucleotide (FAD) have been found to serve as mediators for transfer of electrons from the steel surface to the SRB resulting in corroding carbon steel (Zhang et al. 2015). Additional information on DET and EET (type I) are in the following sections.

Hydrogen sulfide produced by SRB or thiosulfate-reducing bacteria (Magot et al. 1997) are important in the corrosion process because hydrogen sulfide is a strong reducing agent that chemically reacts with iron. An unusual bacterium that reduces sulfate to sulfide was identified as *Citrobacter amalonaticus*, and it was attributed to account for internal corrosion of sour gas transmission pipelines (Angeles-Ch et al. 2002). Thiosulfate is found in oil fields and is the product of sulfide-oxidizing bacteria. The production of iron sulfide is by the following chemical reaction:

$$H_2S + Fe^0 \rightarrow FeS + H_2 \qquad \Delta G^o = -72.5\ \mathrm{kJ} \tag{8.6}$$

This formation of iron sulfide by biogenic hydrogen sulfide is referred to as chemical microbially influenced corrosion (CMIC) and involves electron transfer of electrons from the galvanic couple of iron to iron sulfide. Metal sulfides are electrically conductive and account for the movement of hydrogen atoms through the ferrous sulfide crust resulting in the formation of H_2 (Enning et al. 2012). Ferrous sulfide also accounts for the formation of H_2, which forms cracks in the metal that accounts for the metal becoming brittle. This process is frequently referred to as sulfide stress cracking. If the iron sulfide forms a film over the metallic iron, the rate of corrosion is reduced because the release of Fe^{2+} is impeded. However, if the iron sulfide produces a fine precipitate and no iron sulfide film, the rate of corrosion is not impeded by formation of iron sulfide. The chemical environment appears to be an important factor in determining the form of iron sulfide produced and, therefore, the rate of iron corrosion.

8.1.4 Biocorrosion of Different Types of Steel Alloys

In addition to iron, various steel alloys are subject to biocorrosion (Javed et al. 2017; Jack 2002). Stainless steel and various metal alloys have been used to reduce the rate of biocorrosion, and some examples of these corrosion events are found in the references of Table 8.1. However, the rate and severity of nonferrous corrosion by SRB are markedly less than with iron. The weld site and heat-affected zones (known as heat tinting) are commonly involved in biocorrosion (Enning and Garrelfs 2014). Welds are more susceptible to corrosion than the base steel and are most pronounced when flow in pipes is intermittent or is stagnant. The corrosion includes an anaerobic zone adjacent to the metal, an aerobic region at the surface of the corrosion, and a microaerophilic zone between the oxic and anoxic regions. To some extent, the corrosion of weldments in steel is a cumulative process involving various physiological types of bacteria (Jack 2002) including the following: (i) manganese-oxidizing bacteria, (ii) sulfate-reducing bacteria that produce iron sulfide, (iii) sulfur-oxidizing bacteria that produce thiosulfate from sulfide, and (iv) iron-oxidizing bacteria in the aerobic zone. The anaerobic zone of biocorrosion is brown to black due to iron sulfide, and the aerobic region is orange due to the presence of iron oxides.

8.1.5 SRB–Metal Interface

Most notable in microbiologically influenced corrosion (MIC) of steel is the formation of a biofilm on the surface of the metal (Beech et al. 2005). Characteristically, the bacterial density in a biofilm is many times greater than in planktonic growth, the metabolic products of the bacteria are concentrated on the metal surface, and the distance for transfer of electrons from the metal surface to the bacterial cell is minimal. With the oxidation of iron occurring outside of the bacterial cells, electrons would need to be transported from the metal surface across the cellular membranes into the bacterial cell. This electrogenic activity occurs in SRB and nitrate-reducing bacteria (NRB) and is referred to as EMIC, but some (Gu and Xu 2013; Xu et al. 2013) have termed it type I MIC, while type II MIC is electrochemical and it involves the anode and cathode reactions listed in Sect. 9.2 (Li et al. 2015).

8.1.5.1 SRB Biofilms

An important feature in biocorrosion is the SRB–metal interface because it is at this interface where electrons are moved from the metal to the bacterial cell. Biofilms occur on the surface of the metal as a diffusible chemical barrier consisting of cellular materials resulting from secretion of biomaterials, cell lysis, and metabolic products (Beech and Sunner 2004). The biofilm influences corrosion by binding metal ions, entrapping microbial produces such as FeS that enhance corrosion by establishing or providing electrochemical pathways (Lewandowski et al. 1997). A component in the corrosion biofilm is the extracellular polymeric substance (EPS), and the EPS of SRB was first recognized by Ochynski and Postgate (1963) using laboratory cultures. The composition of EPS produced by a culture enriched with SRB was determined by Chan et al. (2002) to consist of proteins (60%), polysaccharides (37%), and hydrocarbons (3%). In a similar experiment, EPS from a corrosion biofilm was determined to contain 84–92% proteins and 8–16% polysaccharides (Fang et al. 2002). Biofilm distribution on metal surface is often patchy, and the enhanced biocorrosion with pit formation in the metal is located at the site of the biofilm (Beech et al. 1998).

In an effort to understand the role of biofilms in ferrous corrosion, several laboratory studies have been conducted using pure cultures of SRB. With biocorrosion of carbon steel by *D. alaskensis*, over 150 proteins were found in the biofilm, and it was proposed that some of these proteins were instrumental in surface colonization and development of the biofilm (Wikiel et al. 2014). Fatty acid biosynthesis was upregulated in *D. vulgaris* immobilized in biofilm on steel as compared to planktonic bacteria (Zhang et al. 2016). Interaction of *D. desulfuricans* biofilms with stainless steel surface in a flow cell system reveals a stimulation of metabolic activity of the SRB, and this increased metabolism is suggested to be due to the nickel slowly released from the stainless steel (Lopes et al. 2006). These studies indicate that with establishment of the biofilm, a change in metabolic activity is observed with cultures of SRB.

8.1.5.2 Accelerated Low Water Corrosion

Accelerated low water corrosion (ALWC) is an aggressive form of MIC where steel structures in intertidal zones are corroded by mechanisms involving both sulfate-reducing microorganisms and sulfur-oxidizing bacteria (Smith et al. 2019; Phan et al. 2021). There is considerable background information concerning ALWC including the observation that exposure from the tidal zone into the immersion zone accounted for the greatest iron corrosion (Jeffrey and Melchers 2009). ALWC is associated with the following activities that reveal microbial biogeochemical cycles involving different species of bacteria with specific metabolic capabilities: (i) SRB account for reduction of sulfate, thiosulfate, and S^0 to H_2S. (ii) Fe (II) released from the oxidation of steel reacts with H_2S to produce iron sulfide and pyrite (FeS_2). (iii) H_2S and pyrite are oxidized to iron oxyhydroxides and sulfates. These reactions are summarized in Eqs. (8.4, 8.5, and 8.6). While the most common SRB species associated with ALWC-associated reduction of sulfate is *D. desulfuricans* (Little et al. 2000; Païssé et al. 2013), Malard et al. (2013) examined ALWC tubercles and identified the presence of 15 genera of the order Desulfobacterales. The biofilm on surface of steel undergoing ALWC contains sulfur-oxidizing bacteria in addition to SRB (Beech and Campbell 2008; Smith et al. 2019). In marine and estuary intertidal zones, corrosion of steel structures ranged from 0.3 to 1 mm/y^3 as compared to electrochemical corrosion rates of about 0.05 mm/y^3 (Smith et al. 2019). Thus, ALWC is a process of global significance, and the expense for replacement of ALWC steel in the United Kingdom is estimated to be about 20 million British Pounds (Phan et al. 2021).

An interesting evaluation of ALWC was reported by Smith et al. (2019). Using samples of steel from the Shoreharm Harbour at Harbour, UK, the mechanism accounting for ALWC involves both sulfate-reducing microorganisms and sulfur-oxidizing bacteria (Smith et al. 2019). Geochip technology and 16S rRNA analysis were used to assist in identifying the presence of 32 bacterial phyla and as many as 1958 bacterial species associated with ALWC biofilm in the Harbour, UK, samples. Genes associated with the geochemical cycles were most abundant for the carbon cycle, while genes associated with nitrogen, sulfur, and phosphorus cycles were less abundant. Genes for *aprA*, *aprB*, and *cystH* were found in 51, 30, and 9 species, respectively, and these genes were detected in 34% of the bacteria that are uncultivated. Numerous strains of sulfur-oxidizing bacteria are associated with ALWC (Hansel et al. 2015; Smith et al. 2019). With the combined activities of the reduction of sulfate to sulfide with numerous intermediate sulfur compounds and the oxidation of these reduced sulfur compounds to sulfate reflect the presence of a sulfur redox cycle in ALWC. A summary of reactions associated with steel ALWC is given in Fig. 8.4, and additional details of the sulfur cycle are in Chap. 3.

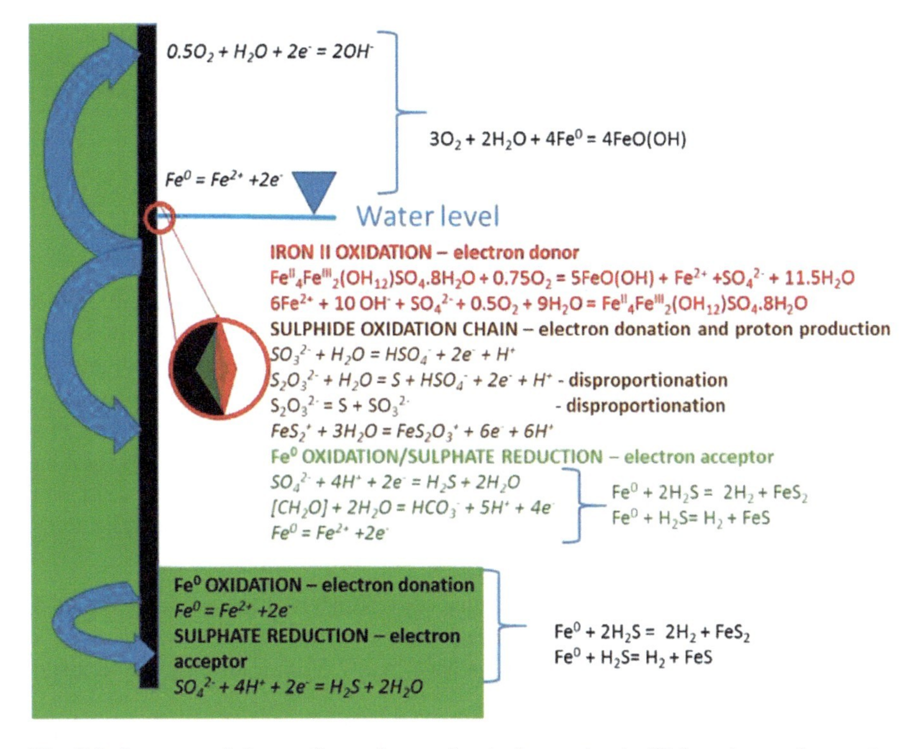

Fig. 8.4 Summary of the reaction pathways for steel corrosion in tidal marine environments, highlighting electron acceptors and electron donors. Half reactions are given in italics and net reactions in plain text. Every step of the sulfide oxidation reaction chain generated electrons and protons that can be consumed by the sulfate reduction process. The development of the full microbial sulfur cycle occurs within corrosion microenvironments and leads ultimately to the development of ALWC blisters with a layered structure. The emergence of blister structures generates an extremely corrosive microenvironment, conducive to the action of sulfate-reducing bacteria, which is protected from the oxic water column via the development of an outer layer of iron oxides and green rusts (Smith et al. 2019).

8.1.5.3 Case Studies

Mixed culture experiments in the environment have been conducted to have a better understanding of the interactions in the biofilm that occur between SRB and bacteria not using sulfate for respiration. Since environmental biocorrosion of ferrous material occurs in the presence of numerous physiological types of bacteria, the microorganisms in corrosion biofilms are highly complementary and instances of synergism occur (Vigneron et al. 2016). In the following narrative, two case studies concerning SRB in biofilms and biocorrosion are discussed.

Oil Facility in the Gulf of Mexico

One study of corroded pipes in a Gulf of Mexico oil facility evaluates the presence and abundance of bacteria and archaea in biofilm using quantitative PCR, ribosomal intergenic spacer analysis, and multigenic DNA next-generation sequencing (Vigneron et al. 2016). While *Desulfovibrio* spp. were the dominant bacteria present in the biofilm community, numerous methanogenic archaea, fermenting bacteria, iron-reducing bacteria, and sulfur-reducing bacteria were also present in the biofilm matrix, which extended from the pipe wall to the flowing oil. The bottom, weld, and leak sites inside the corroded pipe were examined, and the following generalizations were made:

- Based on the number of 16S rRNA gene copy numbers at the weld, bottom, and leak site, ten times more bacteria were present than archaea.
- Microbial abundance (bacteria plus archaea) was 7.5 times lower at the leak site as compared to the pipe bottom.
- At the weld site, 88% of the bacteria were *Desulfovibrio* and *Pelobacter*, while those in low abundance included *Acetobacterium*, *Syntrophomonas*, *Geotoga*, *Pseudomonas*, and *Rhodospirillaceae.*
- At the leak site, 8.1% of the 16S rRNA gene copy numbers were Oceanospirillaceae and *Marinobacter.*
- At the weld and bottom, <10% of the archaeal community contained *Methanosarcina* and Methanomassiliicoccales.
- At the leak site, ~35% of the archaea were *Methanobacterium*, ~35% were Methanomassiliicoccales, and ~20% were related to *Methanosarcina.*

Additionally, several physiological groups of microorganisms could contribute to transformations present in the biofilm of the pipe carrying oil:

- Sulfate reduction would be attributed to *Desulfovibrio* spp.
- Sulfur (S^0) reduction would be attributed to *Geotoga*, *Methanobacterium*, *Methanosarcina*, and *Methanolobus.*
- Iron reduction would be attributed to *Pseudomonas*, *Pelobacter*, and *Desulfovibrio.*
- Fermentation would be associated with *Clostridia*, *Geotoga*, *Acetobacterium*, and *Syntrophomonas*.

Steel Plates Immersed in Seawater

In another case study, steel plates were immersed in seawater near Sanya and Xiamen, China, for 6 months or 8 years (Li et al. 2017). The biofilm was removed from plates, and microorganisms present were identified by extracting DNA and assaying the DNA sequences. While SRB were the dominant group of bacteria present on all plates, a greater diversity of bacterial species was observed on the plates with the 6-month immersion as compared to the 8-year immersion samples. This change in bacterial communities with immersion time was also reported in a previous biocorrosion study (Bermont-Bouis et al. 2007) and is consistent with the concept that the composition of microorganism changes as the biofilm matures (Stoodley et al. 2002). In the 9 samples immersed for 8 years (Li et al. 2017), there was a core group of bacteria that were associated with three phyla: Proteobacteria (63.44%), Firmicutes (19.12%), and Bacteroidetes (6.28%). The abundance of the Proteobacteria reflects the role of these bacteria as pioneers in colonizing surfaces and building of the biofilm, and members of the *Deltaproteobacteria* are sulfate reducers. *Tindallia texcoconensis* and other members of the Firmicutes produce H_2 for the SRB and were most abundant in the rust layer. Members of the Bacteroidetes are abundant in marine environments where they contribute to the survival of colonizing bacteria and hydrolyze various biopolymers that support the growth of other bacteria. Bacteria present in low abundance on the plates in seawater for 8 months included members of the following phyla: Terenicutes (1.57%), Actinobacteria (0.99%), Chloroflexi (0.86%), Thermotogae (0.81%), Cyanobacteria (0.54%), Acidobacteria (0.49%), Planctomycetes (0.36%), Nitrospirae (0.34%), Spirochaetae (0.32%), and Ignavibacteriae (0.23%). These bacteria represented 21 taxonomic classes with the greatest abundance in Deltaproteobacteria, Clostridia, and Gammaproteobacteria. There were 56 genera identified in the immerged steel plate experiment with *Desulfovibrio* being the most abundant and *Desulfobacter* and *Desulfotomaculum* in fewer numbers than *Desulfovibrio*. It is apparent that the biofilm is a dynamic region where various microbial communities are working in a complementary manner, which results in corrosion.

8.1.5.4 Nanowires and Outer Membrane Cytochromes

As reviewed by Sure et al. (2015), many bacteria have electron-conductive extensions from the cell surface, which have been known as nanowires. The size and shape of these nanowires vary with the species of bacteria, and their production is often influenced by nutritional stress. Using nanowires to mediate the transfer of electrons from the oxidizing metal to the bacterial cell would be one possible mechanism associated with biocorrosion. The transfer of electrons along the microbial nanowires has been hypothesized to be by an electron hopping model or by metallic-like conductivity. Aromatic amino acids are proposed to be essential in protein nanowires to achieve electron hopping, and electron-conductive biomolecules such as cytochromes may be used in metallic-like conductivity. Electron-conductive nanowires have been demonstrated for *D. acetoxidans* and *D. desulfuricans* (Alshehri 2017; Eaktasang et al. 2016); however, the specific mechanism for electron transfer for these SRB is unresolved at this time. Nanowires have been demonstrated in *D. ferrophilus* IS5 (Deng et al. 2018), and based on selective staining visualized by electron microscopy, the nanowires have cytochromes along their exterior. *D. ferrophilus* IS5 contains 95 genes for cytochromes (heme proteins), 26 genes producing proteins having four or more heme-binding moieties, 5 periplasmic cytochromes, 2 outer membrane cytochromes, and 2 extracellular cytochromes (Deng et al. 2016, 2018; Deng and Okamoto 2017). The putative genes for extracellular cytochrome production are *DFE_450* and *DFE_464*, and these extracellular cytochromes may be the ones that align with the outer side of the nanowires. A review of the protein database of National Center for Biotechnology Information by Deng and colleagues (Deng et al. 2018) revealed that outer membrane cytochromes homologous to genes *DFE_450* and *DFE_464* were found in Proteobacteria (*D. alkaliphilus*, *Geoalkalibacter subterraneus*, *C. Desulfofervidus auxilii*, and *Desulfuromonas acetoxidans*), Thermodesulfobacteria (*Thermodesulfatator autotrophicus*, *Thermodesulfatator indicus*, and *Thermodesulfatator atlanticus*), and Aquificales (*Thermosulfidibacter takaii*). The expression of DNA to produce cytochromes from genes *DFE_450* and *DFE_464* occurs when electron donor sources are absent. The electron-conductive nanowires in cells of sulfate reducers are of interest not only for biocorrosion of metals but also for functioning in microbial fuel cells (Eaktasang et al. 2016).

Another potential system for the direct transfer of electrons to the bacterial cell would be by electron-conductive proteins such as cytochromes or proteins with electron-carrying centers that are located in the outer membrane. The presence of *c*-type cytochromes in the outer membrane of *Geobacter* and *Shewanella* has been known for some time, and these cytochromes would be important for electron exchange with the environment. Cytochrome in the outer membrane of SRB was inferred to have a role in biocorrosion (Van Ommen Kloeke et al. 1995), and recent genetic and biochemical evidence for cytochrome *c* in outer membranes of *D. ferrophilus* IS5 has been presented (Deng et al. 2016, 2018). If one considers iron reduction for comparison, *Shewanella sulfurreducens* and *Geobacter sulfurreducens* use cytochromes with at least four heme-binding sites to transfer electrons across the outer membrane. An examination of the genome of *D. ferrophilus* IS5 reveals that there are at least 26 genes that produce proteins with four or more heme-binding sites, and the two outer membrane cytochromes have 12 and 14 heme-binding sites each (Deng et al. 2018). These outer membrane cytochromes are products of genes *DFE_449* and *DFE_461*, and cytochrome induction occurs in the absence of lactate as the electron donor. It remains to be established if these outer membrane cytochromes in *D. ferrophilus* IS5 are also associated with nanowires. *Desulfopila corrodens* and *Desulfovibrio* sp. strain IS7 have a mechanism of acquiring electrons directly from an iron surface (Enning and Garrelfs 2014), and direct electron uptake from iron has been observed in field studies (Marty et al. 2014; Smith et al. 2019).

8.1.5.5 Shuttle Molecules

Several molecules have been demonstrated to be important in the extracellular shuttling of electrons from the bacterial cell to electron acceptors such as ferric oxides in the environment. *Desulfovibrio* G11 can donate electrons from H_2 oxidation to humic acids or anthraquinone-2,6-disulfonate (AQDS), the humic analogue (Cervantes et al. 2002). It would be interesting to know if this strain of sulfate-reducing bacteria or other strains could enhance ferrous corrosion in the presence of AQDS. Certain organic compounds can function as electron mediators in biocorrosion of steel by SRB and thereby enhance rate of iron corrosion. The rate of weight loss from stainless steel coupons due to biocorrosion by *D. vulgaris* ATCC 1249 was enhanced with the addition of 10 ppm riboflavin or 10 ppm flavin adenine dinucleotide (FAD) to lactate–sulfate media (Li et al. 2015; Zhang et al. 2015). The source of riboflavin in the environment could be *Shewanella* since bacteria of this genus are known to secrete riboflavin in sufficient quantities to support extracellular electron transport (Marsili et al. 2008; Canstein et al. 2008). In natural environments, *D. vulgaris* and *Shewanella putrefaciens* are found on metal surfaces along with other bacteria (McLeod et al. 2004).

8.1.6 Control of SRB-Based Corrosion

Concrete corrosion in sewage systems is of concern, and approaches to control the problems attributed to the effect of H_2S have been reviewed by Zhang et al. (2008). Biocorrosion of ferrous metals is minimal if stainless steel is used. The most common types of stainless steel are chromium-manganese-nickel (200 series stainless steel) and chromium-nickel-molybdenum (300 series stainless steel) alloys. The presence of nickel in stainless steel accounts for reduced pitting and minimal biocorrosion. Although stainless steel may be desirable to use, the expense in manufacturing makes it economically prohibitive. Thus, a considerable effort is directed to control biocorrosion by physical, chemical, and biological actions.

8.1.6.1 Physical Methods to Control Biocorrosion

An approach in management of biocorrosion is to prevent this activity in pipes and tanks through the use of cathodic protection and coatings. In cathodic protection, a metal is applied to protect iron from corrosion because the metal functions as a sacrificial metal and is corroded instead of iron. A zinc coating may be applied to a variety of pipes, and in this case, zinc in the galvanized steel functions as the anode and prevents corrosion. Alternately, a slight DC external power source is used to provide sufficient current to prevent corrosion. However, cathodic protection is used in a limited number of applications and does not appear to be effective in inhibiting growth of sulfate-reducing bacteria (Guezennec 1994a).

Another protective approach to prevent or reduce the impact of biocorrosion is the use of polymers to coat and provide a physical barrier to ferrous pipes or plates. As recently reviewed, the protective polymers are layered with biocides (e.g., triclosan and quaternary ammonium compounds), and new approaches in applying the coating material to iron use free radical polymerization, chemical oxidative polymerization, or surface-initiated atom transfer, which replace older methods of dipping or spin application (Guo et al. 2018). This area of protective coating to prevent biocorrosion is under considerable attention, and future reports will provide information about inhibitory action of protective coating on SRB.

8.1.6.2 Chemical Control of Iron Biocorrosion

There are no biocides specific for inhibition of biocorrosion. As reviewed by Saleh et al. (1964), Postgate (1979), and Jayaraman et al. (1997), hundreds of chemicals including antibiotics have been reported to inhibit SRB, and few of these compounds are appropriate for use in controlling SRB in an open environmental setting. One chemical, molybdate, is effective in inhibiting growth of sulfate-reducing bacteria by inhibiting ATP sulfurylase; however, the detrimental impact of molybdate on the environment restricts its use in controlling biocorrosion. Examples of chemicals used as biocides for SRB are listed in Table 8.2, and most of these studies are based on laboratory studies and only a few have been applied in the field. Although there has been an extensive use of biocides in an attempt to control biocorrosion, this treatment has not always been successful (Guo et al. 2018) because (i) sessile bacteria beneath the biofilm are more resistant than planktonic bacteria, (ii) bacterial metabolism and release of end products in the biofilm may inactivate the biocide, (iii) the biofilm functions as a passive diffusion barrier to prevent a lethal concentration of the chemical to reach the bacteria attached to the corroding metal, and (IV) over time bacteria develop a resistance to the chemical applied.

Attention has also been directed to the use of chemicals as a general biocide to destroy the biofilm including the SRB. The most commonly employed biocides to control bacterial biofilm containing SRB are glutaraldehyde, formaldehyde, chlorine, isothiazolones, and quaternary ammonium compounds (Franklin et al. 1991; Gardner and Stewart 2002). Due to the size of the environmental setting and the frequency at which the chemical is applied, the amount of biocide used each year could approach 0.1 million liters, which would be a considerable expense. Enhancing agents have been added to ~200 ppm glutaraldehyde to inhibit biofilm growth while using less glutaraldehyde. Ethylenediaminetetraacetic acid (EDTA) was shown to enhance the inhibition of SRB by glutaraldehyde (Zhao et al. 2005; Wen and Gu 2007). However, EDTA is recalcitrant to biodegradation, and its environmental use has considerable risk (European Commission 2004). Ethylenediamine disuccinate (EDDS) is a biodegradable metal chelator similar to EDTA, and the use of EDDS as a "green" enhancer for glutaraldehyde has been proposed (Wen et al.

Table 8.2 Chemicals used to control biocorrosion of ferrous metal

Classes of chemicals	Examples	Advantages for use	Disadvantages for use	References
Oxidizing agents	Chromate	Generally inexpensive	Corrosive, inactivated by H2S; some are carcinogenic or harmful to the environment	Franklin et al. (1991), Jack and Westlake (1995), and Jack (2002)
	Halogens (e.g., chlorine, bromine)			
	Organohalogens (e.g., chloramines, bromamines)			
	Chlorine dioxide and hypochlorite			
	Ozone			
Protein denaturants	Heavy metals	Effective at ~300 ppm	Many react with H2S or are strong disinfectants	Jack and Westlake (1995), Jack (2002), Fang- et al. (2002), Gardner and Stewart (2002), Odom (1993), and Postgate (1979)
Enzyme inhibitors	Aldehydes (formaldehyde, glutaraldehyde, acrolein)			
Membrane disruption	Isothiazolones (benzisothiazolone)	Effective at 30 ppm		
	Thiocyanates			
	2,2-Dibromo-3-nitrilopropionamide	Cationic disinfectant		
	Chlorhexidine (Hibitane)			
Surfactant agents	Quaternary ammonium compounds	High concentration needed	Strong detergents that create foaming	Jack and Westlake (1995), Jack (2002), and Wen et al. (2009)
	Tetrakis (hydroxylmethyl) phosphonium sulfate	Low toxicity		
Miscellaneous	2-methylbenzimidazole (an azole-type derivative)	Inhibits SRB	Toxicity to environment has not been studied	Sheng et al. (2013)
	Aminoalcohols, calcium nitrites, sodium monofluorophosphates	Used to protect steel in concrete	Chloride susceptible	Söylev and Richardson (2008)

2009). EDDS disrupts the biofilm and exposes SRB within the biofilm, and glutaraldehyde can be used at a lower concentration than when EDDS is not present. While glutaraldehyde, formaldehyde, ozone, and chlorine dioxide are oxidizing biocides, nonoxidizing biocides such as anthraquinones, chromate, quaternary salts, and amine-containing chemicals are chemically stable and are also used as biocides for biofilms containing SRB. 2-Methylbenzimidazole has been demonstrated to inhibit corrosion of stainless steel 316 by several species of SRB (Sheng et al. 2013).

A study of biocides tested against *D. vulgaris* Hildenborough using glutaraldehyde (Glut), tetrakis (hydroxymethyl) phosphonium sulfate (THPS), and benzalkonium chloride (BAC) at subinhibitory concentrations revealed a differential impact on gene expression (Lee et al. 2010). The number of genes responding to the presence of Glut, THPS, and BAC was 256, 96, and 198 genes, respectively. Glut, which promotes protein cross linking, genes were induced to degrade inactivated proteins. BAC had an effect on genes that influenced ribosomal structure, and THPS targeted genes that had an impact on energy metabolism. Glut, THPS, and BAC had a synergy when used to inhibit hydrogen sulfide by SRB and underscore the apparent mode of action of these biocides.

8.1.6.3 Green Strategies to Control Biocorrosion

Due to the severe impact that some of the chemicals used as biocides may have on the environment, there is an interest in using "green" strategies to control biocorrosion. This topic has been reviewed (Guiamet and Gómez de Saravia 2005; Little et al. 2007; Zuo 2007; Lin and Ballim 2012), and while the results often show promise, the development of green strategies depends on the advances made in the area of biofilm control and mechanisms of biocorrosion. There are three categories that encompass these green strategies to control biocorrosion (including SRB corrosion) and these include the following: (i) using environmentally friendly biocides, (ii) changing the chemical environment, and (iii) using bacteria or bacterial products to inhibit biocorrosion.

The use of biocides with little to no toxicity to the environment includes use of plant products, chemicals readily degraded by bacteria, and residue-free biocide. Of the various chemicals used as biocides, ozone is highly attractive because it produces no residual toxicity to the environment (Guiamet and Gómez de Saravia 2005). On-site systems for generation and stabilization of ozone could be important for field applications. Natural products in water extracts from plants have been employed to test for control of established biofilms in the laboratory. The inhibition of corrosion bacteria including SRB was attributed to plant extracts from black mustard seed, garlic bulb, Egyptian lupine seeds, orange peels, and mandarin peels (Guiamet and Gómez de Saravia 2005; Omran et al. 2013). Allyl isothiocyanate is the major chemical in black mustard, which is responsible for killing the bacteria. Many different plant products have been used to control biocorrosion, and these activities have been reviewed by Amitha Rani and Basu (2012). Long-term effects of these natural product biocides on established biocorrosion at environmental sites have not yet been provided. As with traditional biocides to treat biocorrosion biofilms, "green" biocides are more effective against planktonic bacteria than attached bacteria, and it has been found that SRB attached to metallic surfaces are remarkably resistant to biocides.

8.1.6.4 Impact of O_2 on Anaerobic Biocorrosion

In an open environment where ferrous metal is submerged in water or buried in soil, the region is anaerobic where biocorrosion of metal occurs. Additionally, there is the

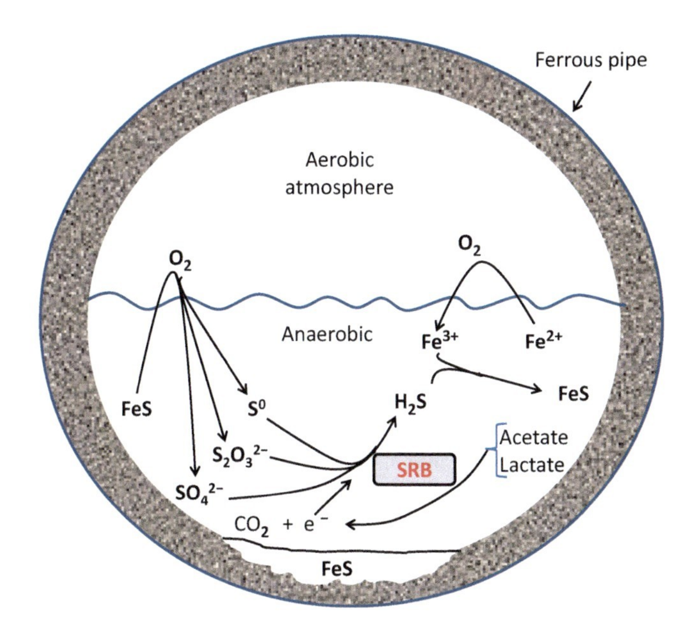

Fig. 8.5 Biocorrosion model showing the impact of atmospheric oxygen on supporting biological and chemical interactions in a corroding ferrous pipe

potential for exposure of the outer regions of corrosion to O_2 as air slowly diffuses toward the site of biocorrosion. Abiotic oxidation of hydrogen sulfide and ferrous ion results in elemental sulfur (S^0) and ferric ion, respectively. Additionally, there are microbial reactions that use O_2 as the terminal electron acceptor with the oxidation of hydrogen sulfide to elemental sulfur, thiosulfate, and sulfate and transformation of ferrous in FeS to ferric ion. The magnitude of these oxidation reactions will vary considerably with the physical and chemical environment with little impact on the corrosion rate of elemental iron (Fe^0). This holistic approach of including O_2 activity in anaerobic corrosion has been incorporated into the ferrous biocorrosion models proposed by Nielsen et al. (1993), Lee et al. (1993a, b), Hamilton (2003), and Beech and Sunner (2007). A model depicting the impact of O_2 on corrosion of ferrous pipes is given in Fig. 8.5.

Since SRB are considered to be anaerobic bacteria, there has been the suggestion that oxygen toxicity could be used to inhibit SRB. Zhang et al. (2012) examined biocorrosion under three conditions: O_2-saturated growth solutions, air-saturated growth solutions, and N_2-saturated growth solutions. They found that the rate of biocorrosion was markedly decreased under O_2-saturated environmental conditions, but the rate of corrosion under air-saturated conditions was the same as when the environment was saturated with N_2. Oxidation of hydrogen sulfide occurs rapidly with the exposure to O_2, and the impact of aerobic exposure would not only be inhibitory to SRB but would diminish the concentration of hydrogen sulfide and, thereby, reduce the rate of corrosion.

Another feature needs to be considered, and it is the capability of many SRB to be aerotolerant and not obligate anaerobes. The environment where SRB are growing apparently reflects their tolerance to oxygen. SRB isolated from oxic sediments are more tolerant to oxygen than isolates from anoxic muds (Sass et al. 1998), and SRB isolated from freshwater are more tolerant to oxygen than those from marine waters (Dolla et al. 2006). As reviewed by Marietou (2016), SRB that can grow in the presence of near atmospheric levels of oxygen include *D. cuneatus* and *D. desulfuricans* ATCC 27774, while D. *desulfuricans* NCIB 8301, *D. desulfuricans* Essex, *D. desulfuricans* CSN, and *D. oxyclinae* can tolerate 0.4–5.0% oxygen. Thus, transient treatment of corrosion biofilm with air or oxygen may not give the desired effect in controlling ferrous corrosion.

8.1.6.5 Biocompetitive Exclusion with Nitrate

While nitrate treatment to control sulfide production by SRB is currently used in controlling biocorrosion, there is a long history of using nitrate to control odors due to hydrogen sulfide from environmental waters. As mentioned by Allen (1949), the initial use of nitrate for treatment of sewage dates back to 1929, and several reports for nitrate treatment of municipal and industrial wastewaters followed (Carpenter 1932; Heukelelekian 1943; Ryan 1945; Allen 1949; Lawrance 1950). More recently, nitrate was effective in controlling sulfidogenic activity in rice paddies (Takai and Kamura 1966), lagoons (Poduska and Anderson 1981), and sewage wastes (Jennamen et al. 1986). These observations provide the background for the application of nitrate to control biocorrosion attributed to hydrogen sulfide release from SRB. Ultimately, nitrate treatment became an important approach to control sulfide production attributed to SRB in oil industry reservoirs and production waters (Jack et al. 1985; Jack and Westlake 1995; Davidova et al. 2001; Eckford and Fedorak 2004; Tabari et al. 2011).

The addition of nitrate in the seawater injection systems was found to promote the growth of indigenous NRB, which outcompeted SRB (Thorstenson et al. 2002; Hubert and Voordouw 2007). The NRB use electron donors from the environment, and in environments deficient in electron donors for bacteria, this consumption of electron donor diminishes the availability of electron donor for SRB. Additionally, the NRB establish a redox level in the environment, which is not conducive for SRB growth since the midpoint potential for the NO_3^-/NO_2^- couple is +430 mV while for the SO_4^{2-}/H_2S couple is −217 mV (Thauer et al. 2007). It has been reported that nitrate directly inhibits the metabolism and growth of SRB (Haverman et al. 2004; He et al. 2010) and that nitrite could be a competitive inhibitor for sulfite in dissimilatory sulfate reduction. To enhance the activity of nitrate reduction, it has been suggested that sites could be augmented with NRB; however, when tested in the environment, the NRB that were introduced were unable to persist (Bouchez et al. 2000). Nitrate treatment is a transient control because SRB populations quickly become reestablished once nitrate additions are terminated (Mohanakrishnan et al. 2011).

Reports of nitrate/nitrite utilization by pure cultures of SRB first appeared in the late 1950s. H_2 oxidation coupled to nitrite using cells of *D. desulfuricans* was reported (Senez and Pichinoty 1958), and a nitrate-respiring sulfate-reducing bacterium (later classified as *D. desulfuricans* strain 27774) was isolated (Coleman 1959). Nitrate reduction by SRB was also reported in the biocorrosion field where *D. desulfuricans* ATCC 27774, *D. desulfuricans* Essex 6, and "*D. multispirans*" NCIB 12078 transferred electrons from corroding steel to nitrate (Rajagopal and LeGall 1989). These reports in addition to the growing acceptance that nitrate-respiring SRB were broadly found in nature (Keith and Herbert 1983; McCready et al. 1983; Seitz and Cypionka 1986; Mitchell et al. 1986; Trinkerl et al. 1990; Dalsgaard and Bak 1994; Moura et al. 1997) supported the concept that some SRB could reduce nitrate and sulfate. Also, SRB capable of reducing nitrate to ammonium were isolated from ferrous corrosion sites including from the Danish North Sea oil field following treatment with continuous nitrate addition (Dunsmore et al. 2004).

One of the unexpected effects of nitrate addition to a sulfate-containing environment was the detection of bacteria that coupled the oxidation of sulfide to elemental sulfur with the reduction of nitrate to nitrite (Voordouw et al. 2002; Hubert et al. 2006). Nitrite production by bacteria in the nitrate-enriched environment was not detrimental to all SRB, but in fact, several strains readily utilize nitrite (Moura et al. 2007). Nitrite at 0.71–0.86 mM was shown to inhibit hydrogen sulfide production in samples from an oil field, and this inhibition of SRB was more effective than glutaraldehyde in inhibiting souring (Reinsel et al. 1996). Hitzman et al. (1995) reported that nitrate, nitrite, and molybdate had a synergetic effect in the inhibition of SRB. Using a consortium of SRB and pure cultures of SRB isolated from western Canadian oil fields, Nemati et al. (2001) reported the inhibition of SRB by nitrite and molybdate. The sensitivity of SRB to nitrite and molybdate was dependent on physiological activity with cells in early growth phase being more readily inhibited than cells growing logarithmically. Additionally, with a mixed culture containing SRB, greater concentrations of nitrite and molybdate were required to inhibit hydrogen sulfide production than with pure cultures of SRB.

The Veslefrikk and Gullfaks oil fields in the North Sea were subjected to nitrate treatment. A reduction of biocorrosion activity was observed 4 months after 0.25–0.33 mM nitrate was continuously introduced into the oil platforms. After nitrate treatment at the Veslefrikk site, the number of nitrate-reducing bacteria increased, while the number of SRB cells was diminished 20,000 fold, and there was a 50-fold decrease in SRB activity (Thorstenson et al. 2002). Over the ~8 years of nitrate treatment, the SRB activity at the Gullfaks oil field was reduced by 40% (Bødtker et al. 2008).

8.1.6.6 Bacteria to Inhibit SRB in Biofilms

Corrosion control of metals, concrete, and stone with protective biofilms produced by bacteria is reviewed by Kip and van Veen (2015). The use of bacteria to inhibit SRB is attractive because biofilm studies reveal that some microorganisms within a bacterial biofilm release peptides and amino compounds, which reduce viable bacteria in the biofilm. SRB are inhibited by antibiotics used effectively against Gram-negative bacteria; however, if the SRB biofilm is present before antibiotics are introduced, the antimicrobial chemicals are unable to penetrate the biofilm. The application of antibiotics to biofilms is environmentally unacceptable and economically prohibitive; however, establishing antibiotic-producing bacteria in biofilms has some potential. An in situ study indicates that establishing a strain of *Bacillus brevis* in biofilms overexpressed gramicidin S and inhibited the growth of SRB (Jayaraman et al. 1999a; Zuo et al. 2004). In another study, the antibacterial cationic peptides of indolicidin and bactenecin from bovine cells were expressed in strains of *Bacillus subtilis*, which had been introduced into a biofilm before *D. vulgaris* and *D. gigas* were inoculated into the biofilm (Jayaraman et al. 1999b). While this control of biocorrosion by the use of a genetically engineered biofilm is useful to demonstrate the concept, it would appear that there is little potential for it to be used in an industrial setting.

8.1.6.7 Inhibition by Disruption of Quorum Sensing

A novel approach for the control of SRB in biocorrosion involves the regulation of quorum sensing in SRB. While quorum sensing has not been explored in SRB, seven species of *Desulfovibrio* and five species of *Desulfotomaculum* were determined to have homologues to the LuxS protein in the quorum sensing system of *Vibrio harveyi* (Scarascia et al. 2016). Additionally, several homologues to several quorum sensing genes were detected in other sulfate-reducing bacteria. This reduction of SRB growth in biocorrosion biofilms by employing quorum-quenching molecules is futuristic and dependent on the understanding of quorum sensing in SRB and the regulation of biofilms by quorum sensing.

8.2 Perspective

Although the association of SRB with ferrous corrosion was made about a hundred years ago, unifying mechanisms for this biocorrosion are still being developed. The term “microbially induced corrosion” (MIC) characterizes the two models involving chemical microbially induced corrosion (CMIC) and electrical microbially induced corrosion (EMIC). These models have been studied primarily in laboratory settings, which has highly controlled environments. Utilization of H_2 produced by CMIC enhances ferrous corrosion, and SRB without hydrogenase do not produce corrosion. The movement of electrons from Fe^0 to SRB by EMIC enables sulfidogenic reactions from sulfate to occur. While contact between metallic iron and the SRB by EMIC may involve electron-conductive extensions from the cell or outer membrane cytochromes, these components have thus far been demonstrated in only a few SRB. Similarly, the use of shuttle molecules to transfer electrons from metallic iron to the

SRB is associated with only a few SRB. Future studies will enhance our understanding of the oxidative process involving biocorrosion.

Mitigation of ferrous corrosion by the use of biocides to disrupt the SRB-containing biofilm or use of chemicals to enhance competition between bacteria in the environment and SRB cultures shows some effectiveness in reducing biocorrosion, but the biocides do not target SRB specifically. Oxidizing agents such as glutaraldehyde and quaternary ammonium detergents have been used to disrupt corrosion in biofilms, and research revealed that the SRB film attached to the ferrous metal are more resistant than planktonic SRB. A search for "green" chemicals to control SRB-related corrosion has resulted in application of nitrate. Laboratory studies and field sites reveal that nitrate can effectively reduce biocorrosion, but once the continuous application of nitrate is terminated, SRB return to be the dominant bacteria. The development of biological strains to be active in competition with SRB in biofilms and the use of select chemicals including quorum sensing signals to manipulate SRB have some potential but are indeed futuristic.

References

Adams ME, Farrer TW (1953) The influence of ferrous iron on bacterial corrosion. J Appl Chem 3: 117–120

Al Abbas FM, Williamson C, Bhola SM, Spear JR, Olson DL, Mishra B (2013) Influence of sulfate-reducing bacterial biofilm on corrosion behavior of low-alloy, high-strength steel (API-5L X80). Int Biodeterior Biodegrad 78:34–42

Ali OA, Aragon E, Fahs A, Davidson S, Ollivier B, Hirschler-Rea A (2020) Iron corrosion induced by the hyperthermophilic sulfate-reducing archaeon Archaeoglobus fulgidus at 70 °C. Int Biodeterior Biodegrad 154:105056

Allen LA (1949) The effect of nitro-compounds and some other substances on production of hydrogen sulphide by sulphate-reducing bacteria in sewage. Proc Soc Appl Bacteriol 2:26–38

Alshehri ANZ (2017) Formation of electrically conductive bacterial nanowires by Desulfovibrio acetoxidans in microbial fuel cell reactor. Int J Curr Microbiol Appl Sci 6:1197–1211

Amitha Rani BE, Basu BBJ (2012) Green inhibitors for corrosion protection of metals and alloys: an overview. Int J Corros 2012: 380217, 15 p. https://doi.org/10.1155/2012/380217

Anandkumar B, George RP, Maruthamuthu S, Parvathavarthini N, Mudali KM (2015) Corrosion characteristics of sulfate-reducing bacteria (SRB) and the role of molecular biology in SRB studies: an overview. Corros Rev 2015. https://doi.org/10.1515/correv-2015-0055

Angeles-Ch C, Mora-Mendoza JL, Garcia-Esquivel R, Padilla-Viveros AA, Perez-Campos R, Flores O, Martinez L (2002) Microbiologically influenced corrosion by Citrobacter in sour gas pipelines. Mater Perform 41:50–55

Antony PJ, Singh Raman RK, Mohanram R, Kumar P, Raman R (2008) Influence of thermal aging on sulfate-reducing bacteria (SRB)-influenced corrosion behaviour of 2205 duplex stainless steel. Corros Sci 50:1858–1864

Ausset P, Lefèvre RA, Del Monte M (2000) Early mechanisms of development of sulfated black crusts on carbonate stone. In: Fassina V (ed) Proceedings of the 9th international congress on deterioration and conservation of stone. Elsevier Science, Venice, Italy: Amsterdam, The Netherlands, pp 329–337

Barr WM, Buchanan RE (1912) The production of excessive hydrogen sulfide in sewage disposal plants and consequent disintegration of the concrete. Iowa State Coll Eng Exp Station Bull 26: 1–15

Barton LL, Fauque GD (2009) Chapter 2: Biochemistry, physiology and biotechnology of sulfate-reducing bacteria. Adv Appl Microbiol 68:41–98

Beech IB, Campbell SA (2008) Accelerated low water corrosion of carbon steel in the presence of a biofilm harbouring sulphate-reducing and sulphur-oxidising bacteria recovered from a marine sediment. Electrochim Acta 54:14–21

Beech I, Cheung SC (1995) Interactions of exopolymers produced by sulphate-reducing bacteria with metal ions. Int Biodeterior Biodegrad 35:59–72

Beech IB, Coutinho CLM (2003) Biofilms on corroding materials. In: Lens P, Moran AP, Mahony T, Stoodly O, O'Flaherty V (eds) Biofilms in medicine, industry and environmental biotechnology – characteristics, analysis and control. IWA Publishing of Alliance House, London, pp 115–131

Beech IB, Gaylarde CC (1999) Recent advances in the study of biocorrosion: an overview. Rev Microbiol 30:117–190

Beech IB, Sunner JA (2004) Biocorrosion: towards understanding interactions between biofilms and metals. Curr Opin Biotechnol 15:181–186

Beech IB, Sunner JA (2007) Sulphate-reducing bacteria and their role in biocorrosion. In: Barton LL, Hamilton WA (eds) Sulphate-reducing bacteria – environmental and engineered systems. Cambridge University Press, Cambridge, UK, pp 459–482

Beech IB, Sunny Cheung CW, Chan CS, Hill MA, Franco R, Lino A-R (1994a) Study of parameters implicated in the biodeterioration of mild steel in the presence of different species of sulphate-reducing bacteria. Int Biodeterior Biodegrad 34:289–303

Beech IB, Sunny Cheung CW, Chan CS, Hill MA, Franco R, Lino A-R (1994b) Study of parameters implicated in the biodeterioration of mild steel in the presence of different species of sulphate-reducing bacteria. Int Biodeterior Biodegrad 34:289–303

Beech IB, Zinkevich V, Tapper R, Gubner R (1998) The direct involvement of extracellular components from a marine sulfate-reducing bacterium in deterioration of steel. Geomicrobiol J 15:119–132

Beech IB, Sunner JA, Hiraoka K (2005) Microbe-surface interactions in biofouling and biocorrosion processes. Int Microbiol 8:157–168

Bermont-Bouis D, Javier M, Grimont P, Dupont I, Vallaeys T (2007) Both sulfate-reducing bacteria and enterobacteriaceae take part in marine biocorrosion of carbon steel. J Appl Microbiol 102: 161–168

Bødtker G, Thorstenson T, Lillebe BP, Thorbjørnsen BE, Ulvøen RH, Sunde E, Torsvik T (2008) The effect of long-term nitrate treatment on SRB activity, corrosion rate and bacterial community composition in offshore water injection systems. J Ind Microbiol Biotechnol 35:1625–1636

Booth GH (1964) Sulphur bacteria in relation to corrosion. J Appl Bacteriol 27:174–181

Booth GH, Tiller AK (1962) Polarization studies of mild steel in cultures of sulfate-reducing bacteria. Part 3. Halophilic organisms. Trans Faraday Soc 58:2510–2516

Booth G, Shinn P, Wakerley D (1964) The influence of various strains of actively growing sulphate-reducing bacteria on the anaerobic corrosion of mild steel. In: Congres Int de la Corrosion Marine et des Salissures, Cannes, vol 336

Booth GH, Cooper PM, Wakerly DS (1966) Corrosion of mild steel by actively growing cultures of sulphate-reducing bacteria. The influence of ferrous ions. Br Corros J 1:345–349

Booth GH, Elford L, Wakerly DS (1968) Corrosion of mild steel by sulphate-reducing bacteria: an alternative mechanism. Br Corros J 3:242–245

Bouchez T, Patureau D, Dabert P, Juretschko S, Doré J, Delgenés P, Moletta R, Wagner M (2000) Ecological study of a bioaugmentation failure. Environ Microbiol 2:179–190

Bugini R, Tabasso ML, Realini M (2000) Rate of formation of black crusts on marble. A case study. J Cult Herit 1:111–116

Bunker HJ (1936) A review of the physiology and biochemistry of the Sulphur bacteria. Dep Sci Ind Res Chem Res. Special Rep. no. 3. London: His Majesty's Stationery Office, London, 48 p
Butlin KR, Adams ME, Thomas M (1949) Sulfate-reducing bacteria and internal corrosion of ferrous pipes conveying water. Nature 163:26–27
Canstein HV, Ogawa J, Shimizu S, Lloyd JR (2008) Secretions of flavins by Shewanella species and their role in extracellular electron transfer. Appl Environ Microbiol 74:615–623
Cappitelli F, Zanardini E, Ranalli G, Mello E, Daffonchio D, Sorlini C (2006) Improved methodology for bioremoval of black crusts on historical stone artworks by use of sulfate-reducing bacteria. Appl Environ Microbiol 72:3733–3737
Carpenter WT (1932) Sodium nitrate used to control nuisances. Water Works Sewerage 79:175–176
Castanier S, Le Métayer-Levrel G, Orial G, Loubière J-F, Perthuisot J-P (2000) Bacterial carbonatogenesis and applications to preservation and restoration of historic property. In: Ciferri O, Tiano P, Mastromei G (eds) Of microbes and art. Springer, Boston. https://doi.org/10.1007/978-1-4615-4239-1_14
Cervantes FJ, de Bok FA, Duong-Dac T, Stams AJM, Lettinga G, Field JA (2002) Reduction of humic substances by halorespiring, sulphate-reducing and methanogenic microorganisms. Environ Microbiol 4:51–57
Chan K-Y, Xu L-C, Fang HHP (2002) Anaerobic electrochemical corrosion of mild steel in the presence of extracellular polymeric substances produced by a culture enriched in sulfate-reducing bacteria. Environ Sci Technol 36:1720–1727
Chen S, Wang P, Zhang D (2014) Corrosion behavior of copper under biofilm of sulfate-reducing bacteria. Corros Sci 87:407–416
Chen Y, Tang Q, Senko JM, Cheng G, Newby BZ, Castaneda H, Ju L-H (2015) Long-term survival of Desulfovibrio vulgaris on carbon steel and associated pitting corrosion. Corros Sci 90:89–100
Churchill AV (1963) Microbial fuel tank corrosion: mechanisms and contributory factors. Mater Protect 2:18–23
Cobb JW (1911) Influence of impurities on the corrosion of iron. Iron Steel Inst 83:171–190
Coleman GS (1959) The isolation and some properties of a sulphate-reducing bacterium from the sheep rumen. J Gen Microbiol 21:i
Copenhagen WJ (1934) Sulfur as a factor in the corrosion of iron and steel structures in the sea. Trans Roy Soc S Afr 22:103–127
Crolet J-L (1992) From biology and corrosion to biocorrosion. Oceanol Acta 15:87–94
Dall'Agnol LT, Moura JJG (2014) Sulphate-reducing bacteria (SRB) and biocorrosion. In: Liengen T, Basséguy R, Féron D, Beech IB (eds) Understanding biocorrosion: fundamentals and applications. Elsevier, Amsterdam, pp 77–105
Dalsgaard T, Bak F (1994) Nitrate reduction in a sulphate-reducing bacterium, Desulfovibrio desulfuricans, isolated from rice paddy soil: sulfide inhibition, kinetics and regulation. Appl Environ Microbiol 60:291–297
Davidova I, Hicks MS, Fedorak PM, Suflita JM (2001) The influence of nitrate on microbial processes in oil industry production waters. J Ind Microbiol Biotechnol 27:80–86
De Muynck W, De Belie N, Verstraete W (2010) Microbial carbonate precipitation in construction materials: a review. Ecol Eng 36:118–136
Deng X, Okamoto A (2017) Energy acquisition via electron uptake by the sulfate-reducing bacterium Desulfovibrio ferrophilus IS5. J Jpn Soc Extremophil 16:67–75
Deng X, Hashimoto K, Okamoto A (2016) Iron corrosive sulfate-reducing bacteria uptake extracellular electrons via outer membrane c-type cytochromes. The Electrochemical Society Meeting Abstract MA2016-02 3245. http://ma.ecsdl.org/content/MA2016-02/44/3245.abstract
Deng X, Dohmae N, Nealson KH, Hashimoto K, Okamoto A (2018) Multi-heme cytochromes provide a pathway for survival in energy-limited environments. Sci Adv 4:eaao5682
Dhami NK, Reddy MS, Mukherjee A (2014) Application of calcifying bacteria for remediation of stones and cultural heritages. Front Microbiol 5:304. https://doi.org/10.3389/fmicb.2014.00304

Dinh HT, Kuever J, Mußmann M, Hassel AW, Stratmann M, Widdel F (2004) Iron corrosion by novel anaerobic microorganisms. Nature 427:829–832
Dolla A, Fournier M, Dermoun Z (2006) Oxygen defense in sulfate-reducing bacteria. J Biotechnol 126:87–100
Du Z, Li H, Gu T (2007) A state of the art review on microbial fuel cells: A promising technology for wastewater treatment and bioenergy. Biotechnol Adv 25:464–482
Duncan KE, Gieg LM, Parisi VA, Tanner RS, Tringe SG, Bristow J, Suflita JM (2009) Biocorrosive thermophilic microbial communities in Alaskan north slope oil facilities. Environ Sci Technol 43:7977–7984
Dunsmore BC, Whitfield TB, Lawson PA, Collins MD (2004) Corrosion by sulfate-reducing bacteria that utilize nitrate. In: Proceedings of CORROSION/2004, NACE International, Houston, TX, Paper No. 04763
Eaktasang N, Kang CS, Lim H, Kwean OS, Cho S, Kim Y, Kim HS (2016) Production of electrically-conductive nanoscale filaments by sulfate-reducing bacteria in the microbial fuel cell. Bioresour Technol 210:61–67
Eckford RE, Fedorak PM (2004) Using nitrate to control microbially-produced hydrogen sulfide in oil field waters. Stud Surf Sci Catal 151:307–340
Edyvean R, Benson I, Thomas C, Beech I, Videla H (1998) Biological influences on hydrogen effects in steel in seawater. Mater Perform 37:40–44
Elmouaden K, Jodeh S, Chaouay A, Oukhrib R, Salghi R, Bazzi L, Hilali M (2016) Sulfate-reducing bacteria impact on copper corrosion behavior in natural seawater environment. J Surf Eng Mat Adv Technol 6:36–46
Enning D, Garrelfs J (2014) Corrosion of iron by sulfate-reducing bacteria: new views of an old problem. Appl Environ Microbiol 80:1226–1236
Enning D, Venzlaff H, Garrelfs J, Dinh HT, Meyer V, Mayrhofer K, and three co-authors (2012) Marine sulfate-reducing bacteria cause serious corrosion of iron under electroconductive biogenic mineral crust. Environ Microbiol 14:1772–1787
Enos D, Taylor S (1996) Influence of sulfate-reducing bacteria on alloy 625 and austenitic stainless steel weldments. Corrosion 52:831–842
European Commission (2004) Edetic acid (EDTA) (CAS No: 60-00-4): European Union Risk Assessment Report Vol. 49. EINECS No: 200-449-4. European Chemicals Bureau, Luxembourg, online. Available at: http://ecb.jrc.ec.europa.eu/DOCUMENTS/Existing-chemicals/RISK_ASSESSMENT/REPORT/edtareport061.pdf
Ewing SP (1955) Electrochemical studies of the hydrogen sulfide corrosion mechanism. Corrosion 11:497t–501t
Fang HHP, Xu L-C, Chan K-Y (2002) Effects of toxic metals and chemicals on biofilm and biocorrosion. Water Res 36:4709–4716
Fernandes P (2006) Applied microbiology and biotechnology in the conservation of stone cultural heritage materials. Appl Microbiol Biotechnol 73:291–296
Ford T, Mitchell R (1990) Metal embrittlement by bacterial hydrogen – an overview. MTS J 24:29–35
Franklin MJ, Nivens DE, Vass AA, Mittelman MW, Jack RF, Dowling NJE, White DC (1991) Effect of chlorine, chlorine/bromine biocide treatments on the number and activity of biofilm bacteria and on carbon steel corrosion. Corrosion 47:128–134
Fu W, Li Y, Xu D, Gu T (2014) Comparing two different types of anaerobic copper biocorrosion by sulfate- and nitrate-reducing bacteria. Mater Perform 53:66–70
Gaines RH (1910) Bacterial activity as a corrosive influence in the soil. J Ind Eng Chem 2:128–130
Gardner LR, Stewart PS (2002) Action of glutaraldehyde and nitrite against sulfate-reducing bacteria biofilms. J Ind Microbiol Biotechnol 29:354–360
Gauri KL, Chowdhury AN, Kulshreshtha NP, Punuru AR (1989) The sulfation of marble and the treatment of gypsum crusts. Stud Conserv 34:201–206
Gauri K, Parks L, Jaynes J, Atlas R (1992) Removal of sulphated crust from marble using sulphate-reducing bacteria. In: Robin GM (ed) Stone cleaning and the nature, soiling and decay

mechanisms of stone, proceedings of the international conference. Donhead Publishing Ltd., Edinburgh, UK, pp 160–165
Ginter RL (1927) Interior corrosion of oil flow tanks in fields where the sulfur conditions are bad. Proc Am Petrol Inst 8:400–410
Graeber M, Boehm S, Kuever J (2014) Molecular methods for studying biocorrosion. In: Liengen T, Basséguy R, Féron D, Beech IB (eds) Understanding biocorrosion: fundamentals and applications. Elsevier, Amsterdam, pp 57–75
Grengg C, Mittermayr F, Baldermann A, Böttcher ME, Leis A, Koraimann G, Grunert P, Dietzel M (2015) Microbiologically induced concrete corrosion: A case study from a combined sewer network. Cem Concr Res 77:16–25
Gu JD (2003) Microbiological deterioration and degradation of synthetic polymeric materials: recent research advances. Int Biodeterior Biodegrad 52:69–91
Gu T (2012) New understanding of biocorrosion mechanisms and their classifications. J Microbial Biochem Technol 4:iii–vi. https://doi.org/10.4172/1948-5948.1000e107
Gu T, Xu D (2013) Why are some microbes corrosive and some not? Paper No. C2013-0002336, CORROSION/2013, Orlando, FL, March 17–21, 2013
Guan F, Zhai X, Duan J, Zhang M, Hou B (2016) Influence of sulfate-reducing bacteria on the corrosion behavior of high strength steel EQ70 under cathodic polarization. PLoS One. https://doi.org/10.1371/journal.pone.0162315
Guezennec JG (1994a) Cathodic protection and microbially induced corrosion. Int Biodeterior Biodegrad 34:275–288
Guezennec JG (1994b) Cathodic protection and microbially induced corrosion. Int Biodeterior Biodegrad 34:275–288
Guiamet PS, Gómez de Saravia SG (2005) Laboratory studies of biocorrosion control using traditional and environmentally friendly biocides: an overview. Lat Am Appl Res 35(4): 295–300
Guo J, Yuan S, Jiang W, Lv L, Liang B, Pehkonen SO (2018) Polymers for combating biocorrosion. Front Mater 5:10. https://doi.org/10.3389/fmats.2018.0001
Gutiérrez-Padilla MGD, Bielefeldt A, Ovtchinnikov S, Hernandez M, Silverstein J (2010) Biogenic sulfuric acid attack on different types of commercially produced concrete sewer pipes. Cem Concr Res 40:293–301
Hamilton WA (1985) Sulphate-reducing bacteria and anaerobic corrosion. Annu Rev Microbiol 39: 195–217
Hamilton WA (2003) Microbially influenced corrosion as a model system for the study of metal microbe interactions: a unifying electron transfer hypothesis. Biofoul. 19L65-76
Hamilton WA, Lee W, Biocorrosion. (1995) In: Barton LL (ed) Sulfate-reducing bacteria. Plenum Press, New York, pp 243–265
Hansel C, Lentini C, Tang Y, Johnston DT, Wankel SD, Jardine PM (2015) Dominance of sulfur-fueled iron oxide reduction in low-sulfate freshwater sediments. ISME J 9:2400–2412
Harris OJ (1960) Bacterial activity at the bottom of back-filled pipe line ditches. Corrosion 16:149t–154t
Haverman SA, Greene EA, Stilwell CP, Voordouw JK, Voordouw G (2004) Physiological and gene expression analysis of inhibition of Desulfovibrio vulgaris Hildenborough by nitrite. J Bacteriol 186:7944–7950
He Q, He Z, Joyner D, Joachimiak M, Price MN, Yang ZK, 12 co-authors. (2010) Impact of elevated nitrate on sulfate-reducing bacteria: a comparative study of Desulfovibrio vulgaris. ISEM J 4:1386–1397
Heggendorn FL, Gonçalves LS, Dias EP, de Oliveira Freitas Lione V, Lutterbach MTS (2015) Biocorrosion of endodontic files through the action of two species of sulfate-reducing bacteria: Desulfovibrio desulfuricans and Desulfovibrio fairfieldensis. J Contemp Dent Sci 16:665–673
Hernandez ME, Newman DK (2001) Extracellular electron transfer. Cell Mol Life Sci 58:1562–1571

Heukelelekian H (1943) Effect of the addition of sodium nitrate to sewage on hydrogen sulfide production and B.O.D. reduction. Sewage Works J 15:255–261

Hitzman DO, Sperl GT, Sandbeck KA (1995) Method for reducing the amount of and preventing the formation of hydrogen sulfide in aqueous system. United States Patent no. 5405531

Horvath J (1960) Contributions to the mechanisms of anaerobic microbiological corrosion. Acta Chem Acad Sci Hung 25:65–78

Huang G, Chan KY, Fang HHP (2004) Microbiologically induced corrosion of 70Cu-30Ni alloy in anaerobic seawater. J Electrochem Soc 151:B434

Hubert C, Voordouw G (2007) Oil field souring control by nitrate-reducing Sulfospirillum spp. that out compete sulfate-reducing bacteria for organic electron donors. Appl Environ Microbiol 73: 2644–2652

Hubert C, Arensdorf J, Voordouw G, Jenneman G (2006) Control of souring through a novel class of bacteria that oxidize sulfide as well as oil organics with nitrate. CORROSION/2006 Houston, TX: NACE International. Paper No. 06669

Hunter JB, McConomy HF, Weston RF (1948) Environmental pH as a factor in control of anaerobic bacterial corrosion. Corrosion 4:567–581

Islam S, Karr EA (2013) Examination of metal corrosion by Desulfomicrobium thermophilum, Archaeoglobus fulgidus, and Methanothermobacter thermautotrophicus. Bios 84:59–64

Iverson WP (1966) Direct evidence for the cathodic depolarization theory of bacterial corrosion. Science 151:986–988

Iverson WP (1968) Corrosion of iron and formation of iron phosphide by Desulfovibrio desulfuricans. Nature 217:1265–1267

Iverson WP (1972) Biological corrosion. In: Fontana MG, Staehle RW (eds) Advances in corrosion science and technology, vol 2. Springer, Boston, pp 1–42

Iverson WP, Olsen GJ (1984) Anaerobic corrosion of iron and steel: a novel mechanism. In: Klug MJ, Reddy CA (eds) Current perspectives in microbial ecology. American Society for Microbiology, Washington, DC, pp 623–627

Jack TR (2002) Biological corrosion failures. In: Shipley RJ, Bedker WT (eds) Failure analysis and prevention. ASM handbook, vol 11. ASM International, Materials Park, pp 881–898

Jack TR, Westlake DWS (1995) Control in industrial settings. In: Barton LL (ed) Sulfate-reducing bacteria. Plenum Press, New York, pp 265–292

Jack TR, Lee E, Mueller J (1985) Anaerobic gas production from crude oil. In: Zajic JE, Donaldson EC (eds) Microbes and oil recovery, vol 1. Bioresource Publications, El Paso, pp 167–180

Javed MA, Stoddart PR, Wade SA (2015) Corrosion of carbon steel by sulphate-reducing bacteria: initial attachment and the role of ferrous ions. Corros Sci 93:48–57

Javed MA, Neil WC, McAdam G, Wade SA (2017) Effect of sulphate-reducing bacteria on the microbiologically influenced corrosion of ten different metals using constant test conditions. Int Biodeterior Biodegrad 185:73–85

Jayanaman A, Earthman JC, Wood KY (1997) Corrosion inhibition by aerobic biofilms on SAE 1018 steel. Appl Microbiol Biotechnol 47:62–68

Jayaraman A, Hallock PJ, Carson RM et al (1999a) Inhibiting sulfate-reducing bacteria in biofilms on steel with antimicrobial peptides generated in situ. Appl Microbiol Biotechnol 52:267–275

Jayaraman A, Mansfeld F, Wood TJ (1999b) Inhibiting sulfate-reducing bacteria in biofilms by expressing the antimicrobial peptides indolicidin and bacterexin. Ind Microbiol Biotechnol 22: 167. https://doi.org/10.1038/sj.jim.2900627

Jeffrey RJ, Melchers RE (2009) Effect of vertical length on corrosion of steel in the tidal zone. Corrosion 65:695–702

Jennamen GE, McInerney MJ, Knapp RM (1986) Effect of nitrate on biogenic sulfide production. Environ Microbiol 51:1205–1211

Jhobalia CM, Hu A, Gu T, Nesic S (2005) Biochemical engineering approaches to MIC. CORROSION/2005 Houston, TX: NACE International. Paper No. 05500

Jia R, Yang D, Xu D, Gu T (2018) Carbon steel biocorrosion at 80 C by a > thermophilic sulfate reducing archaeon biofilm provides evidence for its utilization of elemental iron as electron donor through extracellular electron transfer. Corros Sci 145:47–54
Jodeh S, Khadija EM, Chaouay A, Oukhrib R, Salghi R, Bazzi L, Hilali M (2016) Sulfate-reducing bacteria impact on copper corrosion behavior in natural seawater environment. J Surf Eng Mater Adv Technol 6:36–46
Kato B (2016) Microbial extracellular electron transfer and its relevance to iron corrosion. Microbial Biotechnol 9:141–148
Keith SM, Herbert RA (1983) Dissimilatory nitrate reduction by a strain of Desulfovibrio desulfuricans. FEMS Microbiol Lett 18:55–59
King R, Miller J, Smith J (1973a) Corrosion of mild steel by iron sulphides. Br Corros J 8:137–141
King R, Miller J, Wakerly D (1973b) Corrosion of mild steel in cultures of sulphate-reducing bacteria: effect of changing the soluble iron concentration during growth. Br Corros J 8:89–93
Kip N, van Veen JA (2015) The dual role of microbes in corrosion. ISME J 9:542–551
Kusumi A, Li XS, Katayama Y (2011) Mycobacteria isolated from Angkor monument sandstones grow chemolithoautotrophically by oxidizing elemental sulfur. Front Microbiol 2:1–7
Lahav O, Lu Y, Shavit U, Loewenthal R (2004) Modeling hydrogen sulphide emission rates in gravity sewage collection systems. J Environ Eng 11:1382–1389
Lawrance WA (1950) The addition of sodium nitrate to the Androscoggin River. Sewage Ind Wastes 22:820–832
Le Métayer-Levrel G, Castanier S, Orial G, Loubière JF, Perthuisot JP (1999) Applications of bacterial carbonatogenesis to the protection and regeneration of limestones in buildings and historic patrimony. Sediment Geol 126:25–34
Lee W-C, Lewandowski Z, Okabe S, Characklis WG, Avcl R (1993a) Corrosion of mild steel underneath aerobic biofilms containing sulphate-reducing bacteria – Part I. At low dissolved oxygen concentrations. Biofouling 7:197–216
Lee W-C, Lewandowski Z, Morrison M, Characklis WG, Avci R, Nielsen PH (1993b) Corrosion of mild steel underneath aerobic biofilms containing sulphate-reducing bacteria – Part II. At high dissolved oxygen concentrations. Biofouling 7:217–239
Lee JS, Ray RI, Little BJ (2007) Comparison of Key West and Persian Gulf seawaters. CORROSION/2007 Houston, TX: NACE International, Paper No, 07518
Lee M-HP, Caffrey SM, Voordouw JK, Vordoow G (2010) Effects of gene expression in the sulfate-reducing bacterium Desulfovibrio vulgaris. Appl Microbiol Biotechnol 87:1109–1118
Lewandowski Z, Dickinson WH, Lee WC (1997) Electrochemical interactions of biofilms with metal surfaces. Water Sci Technol 36:295–302
Li SY, Kim YG, Jeon KS, Gu T, Liu H (2001) Microbiologically influenced corrosion of carbon steel exposed to anaerobic soil. Corrosion 57:815–828
Li J, Yuan W, Du Y (2010) Biocorrosion characteristics of the copper alloys BFe30- 1-1 and HSn70-1AB by SRB using atomic force microscopy and scanning electron microscopy. Int Biodeterior Biodegrad 64:363–370
Li H, Xu D, Li Y, Li Y, Liu H, Guan F, Zhai X (2015) Extracellular electron transfer is a bottleneck in the microbially influenced corrosion of C1018 carbon steel by the biofilm of sulfate-reducing bacterium Desulfovibrio vulgaris. PLoS One 10(8):e0136183. https://doi.org/10.1371/journal.pone.0126183
Li X, Duan J, Xiao H et al (2017) Analysis of bacterial community composition of corroded steel immersed in Sanya and Xiamen seawaters in China via method of Illumine MiSeq Sequencing. Front Microbiol. https://doi.org/10.3389/fmicb.2017.01737
Li Y, Xu D, Chen C, Li X, Jia R, Zhang D, three co-authors. (2018) Anaerobic microbiologically influenced corrosion mechanisms interpreted using bioenergetics and bioelectrochemistry: a review. J Mater Sci Technol 34:1713–1718
Liengen T, Féron D, Basséguy R, Beech I (2014) Understanding biocorrosion: fundamentals and applications. EFC publication no 66. Elsevier, Amsterdam

Lin J, Ballim R (2012) Biocorrosion control: current strategies and promising alternatives. Afr J Biotechnol 11:15736–15747
Little BJ, Lee JS (2015) Microbiologically influenced corrosion. In: Revie RW (ed) Oil and gas pipelines integrity and safety handbook. Wiley, Hoboken, pp 387–398
Little BJ, Wagner P, Ray RI, Jones JM (1991) Microbiologically influenced corrosion of copper alloys in saline waters containing sulfate-reducing bacteria. Paper No. 101. In: Corrosion'1991. NACE International, Houston, TX, USA
Little BJ, Wagner P, Ray RI (1992) An experimental evaluation of titanium's resistance to microbiologically influenced corrosion. Paper No. 92173. In: Corrosion'1992. NACE International, Houston, TX, USA
Little BJ, Ray RI, Pope RK (2000) Relationship between corrosion and the biological sulfur cycle: a review. Corrosion 56:433–443
Little B, Lee J, Ray R (2007) A review of 'green' strategies to prevent or mitigate microbiologically influenced corrosion. Biofouling 23:87–97
Lopes FA, Morin P, Oliveira MLF (2006) Interaction of Desulfovibrio desulfuricans with stainless steel surfaces and its impact on bacterial metabolism. J Appl Microbiol 101:1087–1095
Magot M, Ravot G, Campaignolle X, Ollivier B, Patel BK, Fardeau ML, three co-authors. (1997) Dethiosulfovibrio peptidovorans gen. nov., sp. nov., a new anaerobic slightly halophilic, thiosulfate-reducing bacterium from corroding offshore oil wells. Int J Syst Bacteriol 47:818–824
Malard E, Gueuné H, Fagot A, Lemière A, Sjogren L, Tidblad J, and four co-authors (2013) Microbiologically induced corrosion of steel structures in port environment: improving prediction and diagnosis of ALWC (MICSIPE). RFCS Publications, Luxembourg
Mand J, Park HS, Jack TR, Voordouw G (2014) The role of acetogens in microbially influenced corrosion of steel. Front Microbiol. https://doi.org/10.3389/fmicb.2014.00268
Marietou A (2016) Nitrate reduction in sulfate-reducing bacteria. FEMS Microbiol Lett 363(15): fnw155. https://doi.org/10.1093/femsle/fnw155
Marsili E, Baron DB, Shikhare ID, Coursolle D, Gralnick JA, Bond DR (2008) Shewanella secretes flavins that mediate extracellular electron transfer. Proc Natl Acad Sci U S A 105:3968–3973
Marty F, Gueune H, Malard E, Sánchez-Amaya JM, Sjögren L, Abbas B, three co-authors. (2014) Identification of key factors in accelerated low water corrosion through experimental simulation of tidal conditions: influence of stimulated indigenous microbiota. Biofouling 30:281–297
Matos JS, Aires CM (1995) Mathematical modeling of sulphides and hydrogen sulphide gas build-up in the Costa do Estoril sewerage system. Water Sci Technol 31:255–261
McCready RGL, Gould WD, Cook FD (1983) Distribution and regulation of nitrate and nitrite reduction by Desulfovibrio sp. Arch Microbiol 135:182–185
McLeod ES, MacDonald R, Brozel VS (2004) Distribution of Shewanella putrefaciens and Desulfovibrio vulgaris in sulfidogenic biofilms of industrial cooling water systems determined by fluorescent in situ hybridization. Water SA 28:123–128
Meyer FH, Riggs OL, McGlasson RL, Sudbury JD (1958) Corrosion products of mild steel in hydrogen sulfide environments. Corrosion 14:109t–115t
Miller RN, Herron WC, Krigens AG, Cameron LL, Terry BM (1964) Microorganisms cause corrosion in aircraft fuel tanks. Mater Protect 3:60–67
Miranda E, Bethencourt M, Botana FJ, Cano MJ, Sánchez-Amaya JM, Corzo A, and four co-authors (2006) Biocorrosion of carbon steel alloys by an hydrogenotrophic sulfate-reducing bacterium Desulfovibrio capillatus isolated from a Mexican oil field separator. Corros Sci 48: 2417–2431
Mitchell GJ, Jones JG, Cole JA (1986) Distribution and regulation of nitrate and nitrite reduction by Desulfovibrio and Desulfotomaculum species. Arch Microbiol 144:35–40
Mohanakrishnan J, Kofoed MVW, Barr J, Yuan Z, Schramm A, Meyeret RL (2011) Dynamic microbial response to sulfidogenic wastewater biofilm to nitrate. Appl Microbiol Biotechnol 91: 1647–1657

Moura I, Bursakov S, Costa C, Moura JJG (1997) Nitrate and nitrite utilization in sulphate-reducing bacteria. Anaerobe 3:279–290

Moura JJG, Gonzalez P, Moura I, Fauque G (2007) Dissimilatory nitrate and nitrite ammonification by sulphate-reducing eubacteria. In: Barton LL, Hamilton WA (eds) Sulphate-reducing bacteria. Cambridge University Press, Cambridge, UK, pp 241–264

Mystkowska J (2016) Biocorrosion of dental alloys due to Desulfotomaculum nigrificans bacteria. Acta Bioeng Biomech 18. https://doi.org/10.5277/ABB-00499-2015-03

Nemati M, Mazutinec TJ, Jenneman GE, Voordouw G (2001) Control of biogenic H2S production with nitrite and molybdate. J Ind Microbiol Biotechnol 26:350–355

Neville A, Hodgkiess T (2000) Corrosion of stainless steels in marine conditions containing sulphate-reducing bacteria. Br Corros J 35:60–69

Newman R, Webster B, Kelly R (1991) The electrochemistry of SRB corrosion and related inorganic phenomena. ISIJ Int 31:201–209

Nielsen PH, Lee W-C, Lewandowski Z, Morison M, Characklis WG (1993) Corrosion of mild steel in an alternating oxic and anoxic biofilm system. Biofouling 7:267–284

Nielsen AH, Yongsiri C, Hvitved-Jacobsen T, Vollertsen J (2005) Simulation of sulfide buildup in wastewater and atmosphere of sewer networks. Water Sci Technol 52:201–208

O'Connell M, McNally C, Richardson MG (2010) Biochemical attack on concrete in wastewater applications: a state of the art review. Cem Concr Compos 32:479–485

Ochynski FW, Postgate JR (1963) Some biological differences between freshwater and salt water strains of sulphate-reducing bacteria. In: Oppenheimer CH (ed) Marine microbiology. C.C. Thomas Publisher, Springfield, pp 426–441

Odom JM (1993) Industrial and environmental activities of sulfate-reducing bacteria. In: Odom JM, Singleton R Jr (eds) The sulfate-reducing bacteria: contemporary perspectives. Springer, New York, pp 189–210

Omran BA, Fatthalah NA, El-Gendy NS, Elshatoury E, Abouzeid MA (2013) Green biocides against sulphate-reducing bacteria and macrofouling organisms. J Pure Appl Microbiol 7:2219–2232

Païssé S, Ghiglione JF, Marty F, Abbas B, Gueuné H, Amaya JMS, Muyzer G, Quillet L (2013) Sulfate-reducing bacteria inhabiting natural corrosion deposits from marine steel structures. Appl Microbiol Biotechnol 97:7493–7504

Parker CD (1945) The corrosion of concrete. Aust J Exp Biol Med Sci 23:81–98

Parker CD (1951) Mechanics of corrosion of concrete sewers by hydrogen sulfide. Sewage Ind Wastes 23:1477–1485

Phan HC, Blackall LL, Wade SA (2021) Effect of multispecies microbial consortia on microbially influenced corrosion of carbon steel. Corros Mater Degrad 2:133–149

Pochon J, Coppier O, Tchan YT (1951) Role of bacteria on alterations of stone monuments. Chem Eng (Paris) 65:496–600

Poduska RA, Anderson BD (1981) Successful storage lagoon odor control. J Water Pollut Control Fed 53:299–310

Polo A, Cappitelli F, Brusetti L, Principi P, Villa F, Giacomucci L, Ranalli G, Sorlini C (2010) Feasibility of removing surface deposits on stone using biological and chemical remediation methods. Microb Ecol 60:1–14

Postgate JR (1960) The economic activities of sulphate-reducing bacteria. Progr Ind Microbiol 2: 49–69

Postgate JR (1979) The sulphate-reducing bacteria. Cambridge University Press, London

Rajagopal BS, LeGall J (1989) Utilization of cathodic hydrogen by hydrogen-oxidizing bacteria. Appl Microbiol Biotechnol 31:406–412

Ranalli G, Chiavarini M, Guidetti V, Marsala F, Matteini M, Zanardini E, Sorlini C (1997) The use of microorganisms for the removal of sulphates on artistic stoneworks. Int Biodeterior Biodegrad 40:255–261

Rao TS, Kora AJ, Anupkumar B, Narasimhan SV, Feser R (2005) Pitting corrosion of titanium by a freshwater strain of sulphate-reducing bacteria (Desulfovibrio vulgaris). Corros Sci 47:1071–1084
Reinsel MA, Sears JT, Stewart PS, PS, McInerney MJ. (1996) Control of microbial souring by nitrate, nitrite or glutaraldehyde injection in a sandstone column. J Ind Microbiol 17:128–136
Ribas-Silva M (1995) Study of biological degradation applied to concrete. In: Proc., transactions of 13th Int. Conf. on structural mechanics in reactor technology-SMiRT. Univ. Federal do Rio Grande do Sul, Porto Alegre, Brazil, pp 327–332
Ringas C, Robinson FPA (1988a) Corrosion of stainless steel by sulfate-reducing bacteria total immersion test-results. Corrosion 44:671–678
Ringas C, Robinson FPA (1988b) Corrosion of stainless steel by sulfate-reducing bacteria electrochemical techniques. Corrosion 44:386–396
Rodriguez-Navarro C, Sebastian E, Rodriguez-Gallego M (1997) An urban model for dolomite precipitation: authigenic dolomite on weathered building stones. Sediment Geol 109:1–11
Rodriguez-Navarro C, Rodriguez-Gallego M, Ben Chekroun K, Gonzalez-Muñoz MT (2003) Conservation of ornamental stone by Myxococcus xanthus-induced carbonate biomineralization. Appl Environ Microbiol 69:2182–2193
Ryan WA (1945) Experiences with sodium nitrate treatment of cannery wastes. Sewage Works J 17:1227–1230
Saleh A, Macpherson R, Miller JDA (1964) The effect of inhibitors on sulphate-reducing bacteria: a compilation. J Appl Bacteriol 27:281–293
Sand W, Bock E (1984) Concrete corrosion in the Hamburg sewer system. Environ Technol Lett 5: 517–528
Sass H, Berchold M, Branke J, Langner H, Schumann P, Kroppenstedt RM, Spring S, Rosenzweig RF (1998) Psychrotolerant sulfate-reducing bacteria from an oxic freshwater sediment description of Desulfovibrio cuneatus sp. nov. and Desulfovibrio litoralis sp. nov. Syst Appl Microbiol 21:212–219
Satoh H, Odagiri M, Ito T, Okabe S (2009) Microbial community structures and in situ sulfate-reducing and sulfur-oxidizing activities in biofilms developed on mortar specimens in a corroded sewer system. Water Res 43:4729–4739
Scarascia G, Wang T, Hong P-Y (2016) Quorum sensing and the use of quorum quenchers as natural biocides to inhibit sulfate-reducing bacteria. Antibiotics 5:39. https://doi.org/10.3390/antibiotics5040039
Seitz H-J, Cypionka H (1986) Chemolithotrophic growth of Desulfovibrio desulfuricans with hydrogen coupled to ammonification of nitrate and nitrite. Arch Microbiol 146:63–67
Senez JC, Pichinoty F (1958) Reduction of nitrite at the expense of molecular hydrogen by Desulfovibrio desulfuricans and other bacterial species. Bulletin de la Société Chimique et Biologique de Paris 40:2099–2017
Sheng X, Pehkonen SO, Ting Y-P (2013) Biocorrosion of stainless steel 316 in seawater: inhibition using an azole type derivative. Corros Eng Sci Technol 47:388–393
Sherar BWA, Power IM, Keech PG, Mitlin S, Southam G, Shoesmith DW (2011) Characterizing the effect of carbon steel exposure in sulfide containing solutions to microbially induced corrosion. Corros Sci 53:955–960
Smith M, Bardiau M, Brennan R, Burgess H, Caplin J, Santanu Ray S, Urios T (2019) Accelerated low water corrosion: the microbial sulfur cycle in microcosm. NPJ Mater Degrad 3:37. https://doi.org/10.1038/s41529-019-0099-9
Söylev TA, Richardson MG (2008) Corrosion inhibitors for steel in concrete. State-of-the-art report. Constr Build Mater 22:609–622
Spruit CJP, Wanklyn JN (1951) Iron/Sulfide ratios in corrosion by sulfate-reducing bacteria. Nature 168:951–952
Starkey RL (1957) The general physiology of the sulfate-reducing bacteria in relation to corrosion. In: Sulfate-reducing bacteria – their relation to the secondary recovery of oil. Science symposium. St. Bonaventure University, pp 25–43

Starkey RL (1958) The general physiology of the sulfate-reducing bacteria in relation to corrosion. Prod Month 22:12–30
Starkey RL (1960) Sulfate-reducing bacteria – physiology and practical significance. New Jersey Agricultural Experiment Station/The State University, Rutgers/New Brunswick
Starkey RL, Wright KM (1945) Anaerobic corrosion of iron in soil. Amer Gas Assoc Tech Report Distribution Comm. New York, 108 p
Stoodley P, Sauer K, Davies DG, Costerton JW (2002) Biofilms as complex differentiated communities. Ann Rev Microbiol 56:187–209
Stumper R (1923) La Corrosion du Fer dans la Présence du Sulfure de Fer. Compt Rend 176:1316–1317
Sure S, Ackland ML, Torriero AAJ, Adholeya A, Kochar M (2015) Microbial nanowires: an electrifying tale. Microbiology 162:2017–2028
Tabari K, Tabari M, Tabari O (2011) Application of biocompetitive exclusion in prevention and controlling biogenic H2S of petroleum reservoirs. Aust J Basic Appl Sci 5:715–718
Takai Y, Kamura T (1966) Mechanism of reduction in waterlogged paddy soil. Folia Microbiol 11: 304–313
Thauer RK, Stackebrandt E, Hamilton WA (2007) Energy metabolism and phylogenetic diversity of sulphate-reducing bacteria. In: Barton LL, Hamilton WA (eds) Sulphate-reducing bacteria. Cambridge University Press, Cambridge, UK, pp 1–38
Thorstenson T, Bodtker G, Lillebo BP, Torsvik T, Sunde E, Fores S, Beeder J (2002) Biocide replacement by nitrate in sea water injection systems. In: Paper 02033, corrosion 2002. NACE International, Denver, CO
Tiller AK, Booth GH (1962) Polarization studies of mild steel in cultures of sulfate-reducing bacteria. Part 2. Thermophilic organisms. Trans Faraday Soc 58:110–115
Trinkerl M, Breuning A, Schauder R, König H (1990) Desulfovibrio termitidis sp. nov., a carbohydrate-degrading sulphate-reducing bacterium from the hindgut of a termite. Syst Appl Microbiol 13:372–377
Updegraff DM (1955) Microbiological corrosion of iron and steel. Corrosion 11:442t–446t
Van Ommen Kloeke F, Bryant RD, Laishley EJ (1995) Localization of cytochromes in the outer membrane of Desulfovibrio vulgaris (Hildenborough) and their role in anaerobic biocorrosion. Anaerobe 1:351–358
Vasconcelos C, McKenzie JA, Bernasconi S, Grujic D, Tien AJ (1995) Microbial mediation as a possible mechanism for natural dolomite formation at low temperatures. Nature 377:220–222
Venzlaff H, Enning D, Srinivasan J, Mayrhofer KJJ, Hassel AW, Widdel F, Stratmann M (2013) Accelerated cathodic reaction in microbial corrosion of iron due to direct electron uptake by sulfate-reducing bacteria. Corros Sci 66:88–96
Vigneron A, Alsop EB, Chambers B, Lomans BP, Head IM, Tsesmetzis N (2016) Complementary microorganisms in highly corrosive biofilms from an offshore oil production facility. Appl Environ Microbiol 82:2545–2554
Vollertsen J, Nielsen AH, Jensen HS, Tove WA, Thorkild HJ (2008) Corrosion of concrete sewers-the kinetics of hydrogen sulfide oxidation. Sci Total Environ 394:162–170
von Wolzogen Kühr CAH, van der Vlugt LS (1934) The graphitization of cast iron as an electrobiochemical process in anaerobic soils. Water 18:147–165
Voordouw G, Nemati M, Jenneman GB (2002) Use of nitrate-reducing, sulfide-oxidizing bacteria to reduce souring in oil fields: interactions with SRB and effects on corrosion. CORROSION/ 2002 Houston, TX: NACE International. Paper No. 02034
Wanklyn JN, Spruit CJP (1952) Influence of sulfate-reducing bacteria on the corrosion potential of iron. Nature 169:928–929
Ward CB (1963) Corrosion resulting from microbial fuel-tank contamination. Mater Protect 2:10–17
Wei S, Jiang Z, Liu H, Zhou D, Sanchez-Silva M (2013) Microbiologically induced deterioration of concrete – a review. Braz J Microbiol 44:1001–1007

Wen J, Gu T (2007) Evaluations of a green biocide and a green biocide enhancer for the mitigation of biocorrosion using an electrochemical bioreactor. In: AIChE (American Institute of Chemical Engineers) (eds) Annual meeting conference proceedings. Salt Lake City, UT 4–9 November 2007. AIChE: New York

Wen J, Zhao K, Gu T, Raad II (2009) A green biocide enhancer for the treatment of sulfate-reducing bacteria (SRB) biofilms on carbon steel surfaces using glutaraldehyde. Int Biodeter Biodegrad 63:1102–1106

Wikiel AJ, Datsenko J, Vera M, Sand W (2014) Impact of Desulfovibrio alaskensis biofilms on corrosion behavior of carbon steel in marine environment. Bioelectrochemistry 97:52–60

Xu D, Li Y, Gu T (2014) D-methionine as a biofilm dispersal signaling molecule enhanced tetrakis hydroxymethyl phosphonium sulfate mitigation of Desulfovibrio vulgaris biofilm and biocorrosion pitting. Mater Corros 65:837–845

Xu D, Li Y, Song F, Gu T (2013) Laboratory investigation of microbiologically influenced corrosion of C1018 carbon steel by nitrate-reducing bacterium Bacillus licheniformis. Corros Sci 77:385–390

Zhang L, De Schryver P, De Gusseme B, De Muynck W, Boon N, Verstraete W (2008) Chemical and biological technologies for hydrogen sulfide emission control in sewer systems: a review. Water Res 42:1–12

Zhang Y, Pei G, Chen L, Zhang W (2016) Metabolic dynamics of Desulfovibrio vulgaris biofilm grown on a steel surface. Biofouling 32:725–736

Zhang Q, Wang P, Zhang D (2012) Stainless steel electrochemical corrosion behaviors induced by sulphate-reducing bacteria in different aerated conditions. Int J Electrochem Sci 7:11528–11539

Zhang P, Xu D, Li Y, Yang K, Gu T (2015) Electron mediators accelerate the microbiologically influenced corrosion of 304 stainless steel by the Desulfovibrio vulgaris biofilm. Bioelectrochemistry 101:14–21

Zhao K, Wen J, Gu T, Nesic S (2005) Effects of biocides and a biocide enhancer on SRB growth. In: AIChE (American Institute of Chemical Engineers) (eds) Annual meeting conference proceedings. Cincinnati, OH. 30 October–4 November, 2005. AIChE: New York

Zuo R (2007) Biofilms: strategies for metal corrosion inhibition employing microorganisms. Appl Microbiol Biotechnol 76:1245–1253

Zuo R, Ornek D, Syrell CC, Green RM, Hsu C-H, Mansfeld FB, Wood TK (2004) Inhibiting mild steel corrosion from sulfate-reducing bacteria using antimicrobial-producing biofilms in Three-Mile-Island process water. Appl Microbial Biotechnol 64:275–283

Chapter 9
Ecology of Dissimilatory Sulfate Reducers: Life in Extreme Conditions and Activities of SRB

9.1 Introduction

Reviews addressing the ecology of sulfate-reducing prokaryotes usually focus on a relatively small number of species of the *Deltaproteobacteria* phylum and of the *Clostridia* phylum, which are represented by Gram-negative and Gram-positive bacteria, respectively (Muyzer and Stams 2008). With continued examination of the environment, new isolates of sulfate-reducing prokaryotes (SRP) have been obtained, which reveal their presence in areas too extreme for most heterotrophic microorganisms. One of the environmental extremes is temperature, and while most of the SRB are anaerobic bacteria growing under mesophilic temperatures, there are a large number of SRB species that are thermophilic, a few hyperthermophilic SRP species belong to the archaea, and SRB are also found in extremely cold environments. Another environmental variable is pH, and SRP range from about pH 3 to pH 10. When examining the growth potential over a range of temperature and pH, many unique SRP have been found, and these microorganisms have adjusted to selective pressures of the environment (Fig. 9.1). With many of the SRP found in aquatic environments, these microorganisms have adjusted to acidic and alkaline environments, including those areas where the salt concentrations are extremely high or the hydrostatic pressure is intense. Thus, the SRP are polyextremophiles, and this versatility enables them to inhabit all areas of the Earth. The first part of this chapter provides information on some of the more commonly isolated SRP. The second part of the chapter examines the environmental activities attributed to these SRP and explores the community structure where these sulfate reducers are found.

L. L. Barton, G. D. Fauque, *Sulfate-Reducing Bacteria and Archaea*,
https://doi.org/10.1007/978-3-030-96703-1_9

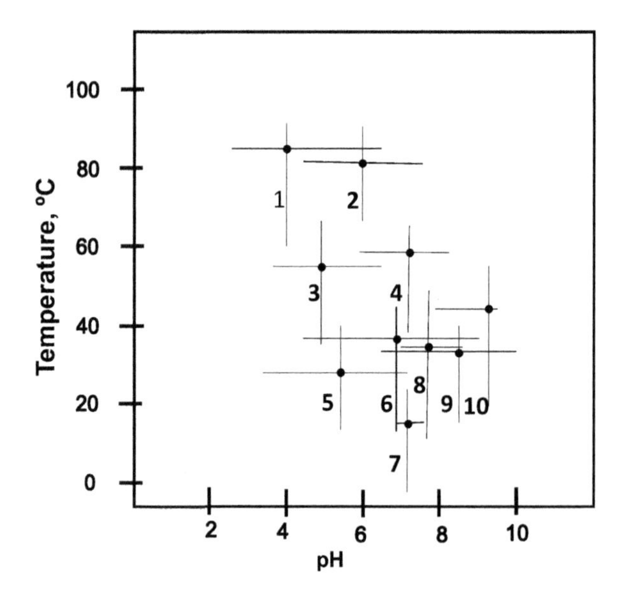

Fig. 9.1 Temperature and pH ranges associated with the growth of a selection of sulfate-reducing bacteria. The line intercepts are at the optimum growth conditions. 1 = *Caldivirga maquilingensis*, 2 = *Archaeoglobus profundus*, 3 = *Thermodesulfobium acidiphilium*, 4 = *Desulfacinum hydrothermale*, 5 = *Desulfosporosinus acididurans*, 6 = *Desulfovibrio tunisensis*, 7 = *Desulfofrigus fragile*, 8 = *Desulfovibrio desulfuricans*, 9 = *Desulfonatronum cooperativum*, and 10 = *Desulfohalophilus alkaliarsenatis*. (Data in text)

9.2 Microbes in Extreme Environments

There are several categories used to group SRP, and these are based on genetic and optimum growth characteristics. SRB grow in extreme chemical/physical environments and display halophilic, alkaliphilic, and acidophilic growth. Often the SRB are polyextremophiles that survive and grow in more than one of the extreme environments. While it is interesting to learn the mechanisms which account for the resistance of these extremophilic SRB, many strains have an extremely slow growth rate (Table 9.1), which presents a problem to cell biologists and biochemists. The following sections address some of the characteristics of the extremophilic sulfate reducers.

9.2.1 *Hyperthermophiles*

The environments of oil fields, submarine hydrothermal vents, and terrestrial hot springs harbor hyperthermophilic microorganisms, and some of these prokaryotes use sulfate as the terminal electron acceptor. The first sulfate-reducing archaeon was *A. fulgidus*, and it was isolated from a hydrothermal vent system in Italy (Stetter et al.

Table 9.1 Growth rates for representative species of SRP growing optimally in extreme environmental conditions

Groups of thermophiles[a]	Doubling time in hours under optimal growth
Hyperthermophiles (80 °C and beyond, but not <50 °C)[b]	
Archaeoglobus fulgidus	4
Thermodesulfobacterium geofontis	5
Thermodesulfovibrio yellowstonii	24
Extreme thermophiles (65–79 °C)[b]	
Thermocladium modestius	15–18
Thermodesulfobacterium commune	4
Thermodesulfobacterium hveragerdense	5.4
Thermodesulfobacterium hydrogeniphilum	3.1
Psychrophiles (7.0–18 °C)[c]	
Desulfofaba gelida	144
Desulfofrigus fragile	19
Desulfofrigus oceanense	169
Desulfotalea arctica	33
Desulfotalea psychrophila	27
Psychrotolerant bacteria (4 °C)[d]	
Desulforhopalus vacuolatus[e]	22.5
Desulfovibrio cuneatus[f]	82.75
Desulfovibrio litoralis[f]	82.75
Halophiles	
Desulfohalobium retbaense[g]	~5
Desulfocella halophile[h]	28
Desulfonauticus submarinus[i]	~12
Desulfonatronospira thiodismutans[j]	67
Desulfonatronospira delicate[j]	112
Desulfohalophilus alkaliarsenatis[k]	82.5
Acidophile	
Thermodesulfobium acidiphilum[l]	16
Neutrophilic mesophiles	
Desulfovibrio gigas[m]	7–7.5
Desulfovibrio vulgaris[n]	2.3

[a]Stetter (2006)
[b] Muyzer and Stams (2008)
[c]Knoblauch et al. (1999a)
[d]Sass et al. (1998)
[e]Isaksen and Teske (1996)
[f]Tarpgaard et al. (2006), Sass et al. (1998)
[g]Ollivier et al. (1991)
[h]Brandt et al. (1999)
[i]Audiffrin et al. (2003)
[j]Sorokin et al. (2008)
[k]Blum et al. (2012)
[l]Frolov et al. (2017)
[m]Traore et al. (1982)
[n]Traore et al. (1981)

1987). Other species of sulfate-reducing hyperthermophiles include members of the *Euryarchaeota* and *Crenarchaeota* phyla in the *Archaea* domain. *A. profundus*, *A. fulgidus*, and *A. sulfaticallidus* are classified as *Euryarchaeota*, while *Caldivirga maquilingensis* and *Thermocladium modestius* belong to the *Crenarchaeota* phylum. As reported, *T. modestius* grows with dissimilatory sulfate reduction and concomitant production of hydrogen sulfide; sulfate supports only "scanty growth" (Itoh et al. 1998). Another sulfate-reducing archaeron, "*A. lithotrophicus*," was reported by Stetter et al. (1993), and this species grows autotrophically on H_2 and sulfate as energy sources and CO_2 as the only carbon source (Vorholt et al. 1995); however, the validity of the name awaits complete description. Two additional archaea, *A. infectus* and *A. veneficus*, are dissimilatory thiosulfate reducers, but these organisms will not use sulfate as final electron acceptors (Mori et al. 2008). Several characteristics of the sulfate-reducing archaea are presented in Table 9.2. A spectrum of physiological activities is observed with the sulfate-reducing archaeal members.

While S^0 is an electron acceptor used by *C. maquilingensis* and *T. modestius*, S^0 inhibits the growth of *A. profundus*, *A. fulgidus*, and *A. sulfaticallidus.* Of the hyperthermophilic sulfate-respiring archaea, only *A. fulgidus* and *A. sulfaticallidus* produce CH_4 and have the unique enzymes and cofactors for methanogenesis. Following exposure of cells of *A. fulgidus* and *A. sulfaticallidus* to ultra violet light, the cells show a blue-green fluorescence, and this is attributed to the presence of coenzyme F_{420}, which is needed for methanogenesis. Trace levels of methane are produced by *A. fulgidus*, and since the quantity of methane produced is far below the amount produced by methanogenic *Archaea*, this organism has been called a "mini-methanogen" (Mori et al. 2008). As indicated in Table 9.2, some of the hyperthremophiles grow under acidic conditions, and others grow only at neutral pH values.

9.2.2 Thermophiles: Extreme and Moderate

The best-characterized SRP are the mesophilic and thermophilic sulfate reducers that are grouped according to rRNA sequence analysis (Castro et al. 2000) and include Gram-negative mesophiles, Gram-positive mesophile-producing spores, thermophilic bacteria, and thermophilic archaea. The Gram-negative mesophilic SRB are members of several families in the delta (δ) *Proteobacteria*: *Desulfovibrionaceae* and *Desulfobacteriaceae*. The *Desulfovibrionaceae* family contains the *Desulfovibrio* and *Desulfomicrobium* genera while the *Desulfobacteriaceae* contains members of the following genera: *Desulfobacter*, *Desulfobacterium*, *Desulfococcus*, *Desulfosarcina*, *Desulfomonile*, *Desulfonema*, *Desulfobotulus*, *Desulfoarculus*, *Desulfobacula*, *Desulfospira*, *Desulfocella*, *Desulfobacca*, and *Desulfacinum*, *Thermodesulforhabdus* which belongs to the Syntrophobacteraceae family; *Desulfobulbus*, *Desulfocapsa*, *Desulforhopalus* and *Desulfofustis* belong to the *Desulfobulbaceae* family (Castro et al. 2000). The suggestion has been raised that based on rRNA gene analysis the genus of *Pseudodesulfovibrio* should be used to

Table 9.2 Characteristics of hyperthermophilic sulfate-reducing archaea

Characteristics	*Caldivirga maquilingensis*	*Thermocladium modestius*	*Archaeoglobus sulfaticallidus*	*Archaeoglobus fulgidus*	*Archaeoglobus profundus*
	JCM 10307^T	JCM 10088^T	DSM 19444^T	DSM 4304	DSM 5631
References	Itoh et al. (1999)	Itoh et al. (1998)	Steinsbu et al. (2010)	Mori et al. (2008)	Mori et al. (2008)
	Steinsbu et al. (2010) Stetter (1988)	Steinsbu et al. (2010) Burggraf et al. (1990)			
Isolation	Acidic hot springs in the Philippines	Mud from acidic hot spring in Japan	Juan de Fuca Ridge eastern Pacific Ocean	Hydrothermal system Vulcano, Italy	Deep sea Guaymas, Mexico
Cell motility by flagella	–	–	–	+	–
Cell morphology	straight or curved rods, may bend or branch, spherical bodies at end of cell	twig-like rods, with branches or buds	irregular lobes or triangles	irregular cocci	irregular cocci
Cell size (μm)[a]	0.4–0.7 × 3–20	0.4–0.5 × 5–20	0.4 × 2.2	0.4 x 1	1.3
G + C % (DNA)	43	52	42	46/47.4	41/43.7
Growth:					
Microaerophilic	+	+	–	–	–
Autotrophic	–	–	$H_2/CO_2 + SO_4^{2-}$	$H_2/CO_2 + S_2O_3^{2-}$	–
H_2/CO_2 + carbon source (acetate)	+	+	+	+	+
Temperature (optimum, °C)	60–92 (85)	45–82 (75)	60–80 (75)	60–95 (83)	65–90 (82)
pH (optimum)	2.3–6.4 (4.0)	2.6–5.1 (4.0)	6.3–7.6 (7.0)	5.5–7.5	4.5–7.5 (6.0)
NaCl as % (optimum)	0–0.75	0–1.0	0.5–3.5 (2.0)	nr	0.9–3.6 (1.8)

(continued)

Table 9.2 (continued)

Electron acceptors:					
S^0	+	+	–	–	–
SO_4^{2-}	+	+	+	+	+
SO_3^{2-}	–	nr	+	+	+
$S_2O_3^{2-}$	+	+	+	+	+
NO_3^-	sg	sg	–	–	–
Fe^{3+}	sg	sg	–	–	–
Electron donors:					
Acetate	–	nr	–	+/–	–
Formate	–	nr	–	–	nr
Lactate	–	nr	+	+	–
Pyruvate	–	nr	+	+	–
Yeast Extract	+	+	–	+	–
Gelatin, glycogen, beef extract, or peptone	+	+	nr	nr	nr
Lipids:					
Tetraether lipids[b]	+	+	nr	nr	nr

nr not reported; *sg* slight growth

[a]width × length

[b]Cyclized glycerol–bisphytanyl–glycerol tetraethers present

group a few of the *Desulfovibrio* that have the unique property of growing under high hydrostatic pressures (Cao et al. 2016a). The Gram-positive spore-forming SRB are taxonomically arranged in the low G+C Gram-positive bacterial group according to DNA analysis (Castro et al. 2000). Cluster I is the largest and contains members of the *Desulfotomaculum* genus, Cluster II consists of *Desulfosporosinus* genus, and Cluster III contains *Dst. guttoideum.* The taxonomic association of *Desulfovirgula thermocuniculi* (Kaksonen et al. 2007) with spore-forming SRB has not yet been resolved.

The capability of SRB to grow at thermophilic temperatures (50 °C–80 °C) is found in organisms associated with several different taxonomic groups, and these SRB can be arranged in categories of thermophilic (Tables 9.3 and 9.4) and thermophilic spore-forming SRB (Table 9.5). While the sulfate-reducing archaea have the highest optimum temperature for growth, the next level of thermophilicity is found in many of the Gram-negative sulfate reducers, and the *Desulfotomaculum* species grow at the lowest thermophilic temperatures. Often these SRB with slight, moderate, or extreme heat requirements also display a preference for acidic or saline environments. Thus far, polyextremophilic SRB that are thermophiles and also alkaliphiles have not been isolated.

9.2.3 *Psychrophiles*

Psychrophilic bacteria grow optimally at <15 °C, while psychrotolerant bacteria grow at temperatures near 0 °C with maximum growth rates at 20–40 °C. Psychrophilic/psychrotolerant SRB thus far reported also grow at acid levels near neutrality (~pH 7). Examples of SRB that thrive in cold environments are listed in Tables 9.6 and 9.7. With the examination of polar sediments where the environment is permanently cold, the rates of sulfate reduction were comparable to those of mesophilic environments. Additionally, most probable number (MPN) counts indicated 4.3×10^5 cells of SRB/cm^{-3} in Artic sediments, which is comparable to the number of SRB cells in temperate marine sediments (Knoblauch et al. 1999a). The slow growth rate is evident from the observation that MPN counts of SRB with acetate as the electron donor with incubation at 4 or 10 °C required six months before positive growth was observed. With lactate in the growth medium, the generation time for SRB strains LSv54 (*Desulfotalea psychrophila*), LSv514 (*Desulfotalea arctica*), and LSv21 (*Desulfofrigus fragile*) was 4 to 6 days while the division time for SRB isolate ASv26 (*Desulfofrigus oceanense*) growing on acetate and PSv29 (*Desulfofaba gelida*) growing on propionate was 5 weeks (Knoblauch et al. 1999b). *Desulfoconvexum algidum* DSM 21856^T is a psychrophilic SRB growing optimally at 14–16 °C, and its use of a broad collection of organic compounds as electron donors (Table 9.8) reflects the capability of using alternate carbon compounds when fermentation products and H_2 are unavailable (Könneke et al. 2013). A dissimilatory SRB identified as *D. gilichinskyi* was isolated within the permafrost region from a

Table 9.3 Taxonomic groups of thermophilic sulfate-reducing bacteria of the *Thermodesulfobacteria*, *Thermodesulfobiaceae*, *Nitrospirae*, and *Proteobacteria* phyla

Electron acceptors in addition to sulfate*			Electron donors									
Organisms	G + C (%)	Temperature (optimum) °C	pH	NaCl (%)	Acetate	Lactate	S^0	SO_3^{2-}	$S_2O_3^{2-}$	H_2	Formate	Pyruvate
***Thermodesulfobacteria* Phylum**												
T[a]. *hveragerdense* JSP[T](1)	40	55–74 (70)	7.0	0	–	–	–	+	+	–	+	+
T. hydrogeniphilum (2) DSM 14290[T]	28	50–80 (75)	6.3–6.8	3.0	+	–	–	–	–	–	–	–
T. commune (3) DSM 2178[T]	34	45–85 (70)	6.0–8.0	nr	+	–	+	+	+	nr	+	+
T. thermophilum (4) DSM 2176[T]	38	44–85 (65)	7.0	0.1	–	+	+	+	+	nr	+	+
T. geofontis (5) ATCC BAA2454	30.6	70–90 (83)	6.5–7.0	0	+	–	+	–	–	+	+	+
Thermodesulfatator atlanticus (6) DSM 21156[T]	45.6	55–75	5.5–8.0	2.5	+	–	–	–	–	–	–	–
Thermodesulfatator indicus (7) DSM 15286[T]	46	55–80 (70)	6.0–6.7	1–3.5	+	nr	nr	–	–	–	–	–
***Thermodesulfobiaceae* Phylum**												
Thermodesulfobium narugense (8) DSM 14796 [T]	35.1	37–65 (50–55)	4.4–6.5	<1	+	–	+	–	–	–	–	+
Thermodesulfobium acidiphilum (9) DSM 102892 [T]	33.7	37–65 (55)	3.7–6.5	1.5	+	–	+	–	–	–	–	+
***Nitrospirae* Phylum**												
Thermodesulfovibrio aggregans (10) DSM 17283[T]	35.2	45–70 (60)	6.0–8.5	nr	+	–	–	+	+	–	–	+

Thermodesulfovibrio islandicus[b] (1) DSM 12570[T]	38	45–70 (65)	6.0–8.5	nr	nr	+	–	+	+	+	nr	nr
Thermodesulfovibrio thiophilus (10) DSM 17215[T]	34	45–60 (55)	6.0–8.5	nr	nr	+	–	+	+	+	–	+
Thermodesulfovibrio yellowstonii (11) Dsm 11347[T]	30.5	65–90 (83)	5.5–8.5	nr	+	–	+	–	–	+	–	+
Thermodesulfovibrio hydrogeniphilus (12) DSM 18151[T]	36.1	50–70 (60)	7.1	nr	+	–	+	+	+			
***Proteobacteria* Phylum**												
Thermodesulforhabdus norvegicus (13) A8444	51	44–74 (60)	6.1–7.7	1.6	+	+	+	+	–	–	+	–

References: 1= Sonne-Hansen and Ahring (1999); 2 = Jeanthon et al. (2002); 3 = Zeikus et al. (1983); 4 = Rozanova and Khudyakova (1974); Rozanova and Pivovarova (1988); 5 = Hamilton-Brehm et al. (2013); 6 = Alain et al. (2010); 7 = Moussard et al. (2004); 8 = Mori et al. (2003); 9 = Frolov et al. (2017); 10 = Sekiguchi et al. (2008); 11 = Henry et al. (1994); 12 = Haouari et al. (2008a, b); 13 = Beeder et al. (1995)

[a]T = *Thermodesulfobacterium*

[b]*Thermodesulfobivrio islandicus* reduces nitrate but other species were reported to not use nitrate, fumarate, or Fe^{3+} as electron acceptors

Table 9.4 *Desulfosoma* and *Desulfacium* species that are thermophilic sulfate-reducing bacteria

Characteristics	*Desulfosoma caldarium*[a] DSM 22027^T	*Desulfacinum infernum*[b] BaG1^T	*Desulfacinum hydrothermale*[c] DSM 13146
Site of isolation	Terrestrial hot spring Columbian Andes, Columbia	Petroleum reservoir	Hydrothermal vent north eastern Tunisia
Cell morphology	oval to rod shape	oval cells	oval to short rods
Motility	single polar flagellum	nonmotile	peritrichous flagella
Cell size (μm) $^+$	1.0–1.5 × 2.0	1.0 × 2.5–3	0.8–1 × 1.5–2.5
G + C % (DNA)	56.1	64	59.5
Growth:			
Temperature, °C	50–62	40–65	37–64
(optimum)	57	60	60
pH range (optimum)	5.7–7.7 (6.8)	7.1–7.5	5.8–8.2 (7.1)
NaCl range (optimum)	0.5–3% (2.5%)	0–5% (1%)	1.5–7.8% (3.2–3.6%)
Electron acceptors:			
SO_4^{2-}	+	+	+
SO_3^{2-}	+	+	+
$S_2O_3^{2-}$	+	+	+
Electron donors:			
H_2	+	+	+
Acetate	–	+	+
Lactate	+	+	+
Pyruvate	+	+	+

[a]Baena et al. (2011)
[b]Rees et al. (1995)
[c]Sievert and Kuever (2000)

Yamal Peninsula cryopeg (a highly mineralized ancient aquatic ecosystem) that was capable of growing at subzero temperatures (Ryzhmanova et al. 2019).

One of the physiological accommodations required for bacteria to grow at near-freezing conditions has focused on membrane lipids where molecular flexibility is required to accommodate membrane fluidity. Membrane integrity is required to maintain charge across the membrane, and molecular movement within the membrane (membrane fluidity) is needed to accomplish solute transport activities. While branched fatty acids, the hexadecenoic acid (16:0 10-me), or the heptadecenoic acid (17: lc9) are characteristic for mesophilic *Desulfovibrio, Desulfobacter*, and *Desulfobulbus* species, respectively (Taylor and Parkes 1983; Dowling et al. 1986; Vainshtein et al. 1992), unsaturated and short-chained fatty acids are characteristic of psychrophilic SRB. For example, over 70% of the fatty acids of *Desulfofaba gelida* are heptadecanoic acids (C_{15}), while hexadecenoic acids (C_{16}) are the predominate lipids of *Desulfofrigus oceanense*, *Desulfofrigus fragile*, *Desulfotalea psychrophile*, and *Desulfotalea artica* (Knoblauch et al. 1999b). In a study where

Table 9.5 Taxonomic groups of moderately thermophilic sulfate-reducing bacteria that produce spores

Electron acceptors in addition to sulfate			Oxidation of substrates in addition to H_2				
Organisms	Temperature °C	Isolation area	sugars	fatty acids[a]	alcohols	C or I[b]	References
Desulfotomaculum (Dst.)							
Dst. alcoholovirax DSM 16058[T]	33–51	Fluidized-bed reactor treatment of acidic wastewater	–	–	–	I thiosulfate, S^0	Kaksonen et al. (2008)
Dst. alkaliphilum DSM 12257[T]	30–58	Cow/pig manure mixture	–	–	–	I thiosulfate, sulfite	Pikuta et al. (2000)
Dst. arcticum DSM 117038[T]	26–46	Cold fjord sediment of Svalbard	–	–	+	I thiosulfate, sulfite	Vandieken (2006)
Dst. australicum Strain ST90[T]	40–74	Great Artesian Basin Australia, aquifer at 914 m	–	+	+	C thiosulfate, sulfite	Love et al. (1993)
Dst. carboxydivorans DSM 14880	30–68	bioreactor treating paper mill wastewater	+	nr	+	I thiosulfate, sulfite	Parshina et al. (2005)
Dst. ferrireducens GSS09T	30–55	Compost, China	+	–	nr	+ thiosulfate, sulfite, Fe^{3+}	Yang et al. (2016)
Dst. geothermicum DSM 3669[T]	37–56	Geothermal water at 2500 m, France	+	+	+	I sulfite	Daumas et al. (1988)
Dst. hydrothermale DSM 118033[T]	40–60	Hot spring, Tunisia	–	–	+	I thiosulfate, sulfite, Fe^{3+}, As^{5+}	Haouori et al. (2008a, b)
Dst. kuznetsovii DSM 6115[T]	50–85	Oil field at 3000 m in Russia	–	+	+	C thiosulfate, sulfite	Nazine et al. (1989)
Dst. luciae SLT[T]	50–70	Hot spring sediment at 2700 m in Taylorsville Triassic Basin in Virginia, US	–	–	+	C thiosulfate	Liu et al. (1997) Karnauchow et al. (1992)
Dst. nigrificans DSM 574	45–70	Soil, compost heaps, spring water	+	–	–	I thiosulfate	Liu et al. (1997) Widdel (1988)

(continued)

Table 9.5 (continued)

Electron acceptors in addition to sulfate			Oxidation of substrates in addition to H_2				
Organisms	Temperature °C	Isolation area	sugars	fatty acids[a]	alcohols	C or I[b]	References
Dst. peckii DSM 23769[T]	50–65	Abattoir waters, Tunisia	–	–	+	I thiosulfate, sulfite	Jabari et al. (2013)
Dst. putei DSM 12395[T]	50–65	Deep terrestrial subsurface Taylorsville Triassic Basin in Virginia, USA	–	–	+	I thiosulfate, sulfite	Liu et al. (1997)
Dst. salinum strain 435T	40–70	Oil field in Siberia	–	+	nr	C thiosulfate, sulfite	Nazina and Rozanova (1978)
Dst. *solfataricum* DSM 14956[T]	48–65	Hot solfataric fields in Iceland	+	+	+	C thiosulfate, sulfite	Goorissen, et al. (2003)
Dst. thermoacetoxidans DSM 5813[T]	45–60	Anaerobic reactor, benzoate enrichment	–	+	–	C thiosulfate, S^0	Min and Zinder (1990)
Dst. thermobenzoicum DSM 6193[T]	40–70	enrichment of granular sludge	–	+	+	I thiosulfate, sulfite	Tasaki et al. (1991)
Dst. thermocisternum DSM 10259	41–75	Oil field at 2600 m, North Sea	–	+	+	I thiosulfate, sulfite	Nilsen et al. (1996)
D. thermosapovorans DSM 6562[T]	35–60	Growing on rice hulls	–	+	+	I thiosulfate, sulfite	Fardeau et al. (1995)
Dst. thermosubterraneum DSM 16057[T]	50–72	Mine at 250 m, Japan	–	+	+	I thiosulfate, sulfite, S^0	Kaksonen et al. (2006)
Dst. tongense strain TGB60-1[T]	37–60	*Hydrothermal vent sediment, Tofua Arc in the Tonga Trench*	–	–		thiosulfate, sulfite	Cha et al. (2013)
Dst. varum strain RH04-3[T]	37–55	Great Artesian Basin Australia, runoff channel from borehole	+	–	–	I thiosulfate, sulfite, S^0	Ogg and Patel (2011)

Desulfovirgula thermocunicula DSM 16036T	70–80	Mine at 250 m, Japan	–	+	nr	C thiosulfate, sulfite, S^0	Kaksonen et al. (2007)
Desulfovirgula thermocuniculi DSM 16036[T]	61–80	Geothermal underground mine in Japan	–	–	–	I thiosulfate, sulfite, S^0	Kaksonen et al. (2007)
Desulfosporosinus							
orientis DSM 765[T]	30–37	Isolated from soil in Singapore	–	–	–	I thiosulfate, sulfite	Campbell and Postgate (1965) Stackebrandt et al. (1997)

[a]Long- or short-chain fatty acids; [b]C is complete and I is incomplete oxidation

Table 9.6 Characteristics of psychrophilic sulfate-reducing bacteria

Characteristics	*Desulfofaba gelida*[a] PSv29^T (DSM 12344^T)	*Desulfofrigus fragile*[a] LSv21^T (DSM 12345^T)	*Desulfofrigus oceanense*[a] ASv26^T (DSM 12341^T)	*Desulfotalea arctica*[a] LSv514^T (DSM 12342^T)	*Desulfotalea psychrophila*[a] LSv54^T (DSM 12343^T)
Isolation	Hornsund artic sediment at 2.6 °C in Svalbard	Hornsund artic sediment at 2.6 °C in Svalbard	Hornsund artic sediment at 2.6 °C in Svalbard	Artic Storfjord sediment at −1.7 °C	Artic Storfjord sediment at −1.7 °C
Cell morphology	Straight rods	slender rods	rods	slender rods	slender rod
Cell size (μm)	3.1–5.6	0.8 × 3.6	2.1–5.1	0.7 × 2.2	0.6–5.7
G + C % (DNA)	52.8	52.1	52.8	41.8	46.8
Growth:					
Temperature (optimum, °C)	−1.8 to 10 (7)	−1.8 to 27 (18)	−1.8 to 16 (10)	−1.8 to 26 (18)	−1.8 to 19 (10)
pH (range)	7.1–7.6	7.0–7.4	7.0–7.5	7.2–7.9	7.3–7.6
NaCl as % (range)	1.4–2.5	1.0–2.5	1.5–2.5	1.9–2.5	1.0
Electron acceptors:					
S^0	–	–	–	–	–
SO_4^{2-}	+	+	+	+	+
SO_3^{2-}	+	–	+	–	+
$S_2O_3^{2-}$	+	–	+	+	+
NO_3^-	nr	+	–	–	–
Fe^{3+}	–	+	+	+	+
Electron donors:					
Acetate	–	–	+	–	–
Formate	+	+	+	+	+
Lactate	+	+	+	+	+
Pyruvate	+	+	+	+	+
Ethanol	+	+	+	+	+
Propionate	+	–	+	–	–

nr not reported
[a]Knoblauch et al. (1999b)
[+] width × length

Table 9.7 Characteristics of *Desulfobacter*, *Desulfovibrio*, *Desulfoconvexum*, and *Desulforhopalus* that grow in cold environments

Characteristics	*Desulfobacter psychrotolerans*[a] akvb^T	*Desulfovibrio cuneatus* [b] STL4	*Desufovibrio litoralis*[b] STL6	*Desulfoconvexum algidum*[c] DSM 21856^T	*Desulforhopalus vacuolatus*[d] DSM 9700
Cell morphology	Straight rods	slender rods	rods	vibrioid, curves	rods with gas vacuoles
Cell size (μm)$^+$	3.1–5.6	0.8 × 3.6	2.1–5.1	1.5 × 2.0–3.5	1.5–1.8 × 3–5
G + C % (DNA)	52.8	52.1	52.8	46	48.4
Growth:					
Temperature	7	0 to 33	10	14–16	1.8 to 26
(optimum, °C)	(7)	(18)	(10)	(14–16)	(18)
pH (optimum)	7.1–7.6	7.0–7.4	7.0–7.5	7.2–7.4	6.8–7.2
NaCl as % (range)	1.4–2.5	1.0–2.5	1.5–2.5	2.0–3.0	0.5
Electron acceptors:					
S^0	–	+	+	+	–
SO_4^{2-}	+	+	+	+	+
SO_3^{2-}	+	+	+	–	+
$S_2O_3^{2-}$	+	+	+	+	+
NO_3^-	nr	+	–	nr	–
Fe^{3+}	–	+	+	nr	–
Electron donors:					
Acetate	–	–	+	–	–
Formate	+	+	+	+	–
Lactate	+	+	+	+	+
Pyruvate	+	+	+	–	+
Ethanol	+	+	+	+	+
Propionate	+	–	–	+	+

nr not reported
$^+$Width × length
[a]Tarpgaard et al. (2006)
[b]Sass et al. (1998)
[c] Könneke et al. (2013)
[d] Isaksen and Teske (1996)

Table 9.8 Electron donors for sulfate reduction by *Desulfoconvexum algidum* DSM 21856T[a]

Compounds metabolized as electron donors		Compounds not utilized as electron donors
Hydrogen	Glycerol	Acetate
Formate	Glycine	Propionate
Lactate	Alanine	Succinate
Malate	Serine	Pyruvate
Valerate	Betaine	
Caproate	Choline	
Ethanol	Proline	
Methanol	Sorbitol	
Propanol	Mannitol	
Butanol	Benzoate	

[a]Source is Könneke et al. (2013)

Desulfobacterium autotrophicum was grown at several different temperatures, the quantity of unsaturated fatty acids and short-chain fatty acids was greater when grown at 4 °C than that at 28 °C. Over 70% of the fatty acids of *Desulfobacterium autotrophicum* were unsaturated when grown at low temperatures (Rabus et al. 2002).

Another adjustment for bacteria to grow in low temperatures involves enzyme structure and production; however, proteins of psychrophilic SRB are just starting to attract attention from scientists. An evaluation of the proteome of *Desulfobacterium autotrophicum* grown at 4 °C and 28 °C revealed growth at 4 °C was not attributed to new enzymes. Two proteins were significantly greater in abundance when *Desulfobacterium autotrophicum* was grown at 28 °C as compared to 4 °C (Rabus et al. 2002).

It has been estimated that 95% of the seafloor is >4 °C (Levitus and Boyer 1994). Since sulfate-reducing bacteria are the predominant carbon mineralizers along the continental shelf and slope (Kasten and Jørgensen 2006), it would follow that a robust psychrophilic sulfate-reducing bacterial community would be there as well. When the sediments along the southwest and southeast Atlantic and cold Artic fjords were sampled, the optimal temperature for growth of sulfate reducers was 30 °C. It was concluded that these cold environments contained both psychrophilic and cold-tolerant mesophilic sulfate reducers (Sawicka et al. 2012). The mesophiles would have been swept in from temperate waters and could have adapted to persist in the cold environment. Certainly, this influx of mesophilic sulfate-reducing bacteria would have provided a continuous source of bacterial DNA for horizontal gene transfer.

9.2.4 Halophiles

Halophilic microorganisms are extremophiles requiring salt, and the most notable group of organisms in this category are members of the Archaea. There are, however, several species of bacteria that are halophilic or halotolerant. The categories of halophiles include the following: extreme halophiles that require 20–30% salt (NaCl) for growth, moderate halophiles that require 4.8–20% salt, and slight halophiles that grow in 1.7–4.7% salt (Ollivier et al. 1994). Sulfate-reducing bacteria are readily isolated from seawater (3.5% salt), salt lakes, artificial salterns, and various other brine environments. An early discussion concerning the bacterial reduction of sulfate marine environments was published by van Delden (1903), and salt tolerance by sulfate reducers was discussed in an early review by Postgate (1965). Even though there was a limited number of SRB strains isolated, salt tolerance by SRB was designated by Ovhynski and Postgate (1963) and Postgate (1984) as slightly tolerant (some strains of *D. desulfuricans*), moderately tolerant (some strains of *D. desulfuricans*), and extremely tolerant (*D. salexigens*). While fresh water strains of *D. vulgaris* are sensitive to the presence of salt, a few strains can be adapted to grow at low salt concentrations (<2%). Currently, there are several SRB isolates that are moderately tolerant to salt, and the halotolerant/halophilic SRBs are grouped as those that grow at alkaline pH levels (Table 9.9) or pH values near 7 (Table 9.10). Many halophilic microorganisms use compatible solutes such as glycerol, ecotine, and glycine-betaine as osmoprotectants to provide cell stability against osmotic stress attributed to high concentrations of environmental salt. *D. halophilus*, a moderate halophile, is unable to synthesize compatible solutes, but it readily acquires them from the environment (Ollivier et al. 1994). Cells of *D. halophilus* have been reported to contain trehalose, betaine, and glycine as osmoprotectants (Belyakova et al. 2006). It has long been considered that incomplete oxidizers are capable of tolerating higher salt levels than complete oxidizers, and this may reflect the utilization of compatible solutes in the environment that have been produced by other bacteria (Foti et al. 2007).

While over a hundred strains of sulfur-oxidizing bacteria have been isolated from saline environments, relatively few strains of SRB have been reported. Of the SRB isolates, several are polyextremophiles displaying the capability of growing optimally in both salt and alkaline environments. SRB strains that are low salt tolerant alkaliphiles include *Desulfonatronovibrio hydrogenovorans* (Zhilina et al. 1997), *Desulfonatronum lacustre* (Pikuta et al. 1998), *Desulfonatronum thiodismutans* (Pikuta et al. 2003), *Desulfonatronum cooperativum* (Zhilina et al. 2005b), and *Desulfotomaculum alkaliphilum* (Pikuta et al. 2000). SRB displaying moderate halophilic and alkaliphilic activities include *Desulfonatronospira thiodismutans* (Sorokin et al. 2008), *Desulfonatronospira sulfatiphila* (Sorokin and Chernyh 2017), *Desulfonatronospira delicata* (Sorokin et al. 2008), and *Desulfobotulus alkaliphilus* (Sorokin et al. 2010a) (Tables 9.11 and 9.12). While *Desulfonatronospira thiodismutans* (Sorokin et al. 2008) and *Desulfohalobium retbaense* (Ollivier et al. 1991) can grow slowly at NaCl concentrations above

Table 9.9 Characteristics of halophilic bacteria growing optimally under alkaline conditions

Characteristics	*Desulfonatronovibrio hydrogenovorans*[a] DSM 9292	*Desulfonatronospira thiodismutan*[b] DSM19093[T]	*Desulfonatronospira sulfatiphila*[c] DSM 100427	*Desulfonatronospira delicata*[b] DSM19491[T]	*Desulfohalophilus alkaliarsenatis*[d] DSM25765
Site of isolation	Alkaline lake sediment Lake Madagi in Kenya	Soda lakes at Kulunda Steppe, Altai, Russia	Hypersaline soda lake southwestern Siberia, Russia	Soda lakes at Kulunda Steppe, Altai, Russia	Searles Lake in Mojave Desert, California
Cell shape	motile vibrio	motile vibrio	short rods, vibrioid	vibrio form	vibrio
Motility	monopolar flagella	single polar flagella		-motile	two polar flagella
Cell size (μm)[e]	0.5 × 1.5–2	0.6–0.8 × 2–30	0.7–0.8 × 1.5–3.0	0.4–0.6 × 1.2–4.0	0.5–3
G + C % (DNA)	48.6	50.4	51.1	50.2	45.1
Growth:					
Temperature	15–43	mesophilic	mesophile	mesophilic	20–55
(optimum, °C)	(37)	(45)	(33–35)		(44)
pH range (opt.)	8.0–10.5 (9.5–9.7)	8.3–10.5 (10)	9.0–10.3 (9.7–10)	8.0–10.6 (10)	7.75–9.75 (9.25)
NaCl range (opt.)	1–12% (3%)	8.7–23.2% (11.6–14.5%)	5.8–23.2% (11.6%)	4.6–20.3% (5.8%)	5.5–33% (20%)
Autotrophy	−	+	nr	+	+
Electron acceptors:					
SO_4^{2-}	+	+	+	+	+
SO_3^{2-}	+	+	+	+	+
$S_2O_3^{2-}$	+	+	−	+	+
Disproportionation		sulfite/ thiosulfate	sulfite	sulfite/ thiosulfate	
Fermentation	pyruvate	pyruvate	+	pyruvate	−
Electron donors:					
H_2	+	+	−	+	−
Acetate	−	+ with sulfite	−	+ with thiosulfate	−
Lactate	−	+	+	+	+

Pyruvate	- (with sulfate) or sulfite	+ with thiosulfate	+	+ with thiosulfate or sulfite	nr
Major fatty	66% are $C_{16:0}$ and $C_{18:0}$,	iso$C_{15:0}$, $C_{16:0}$,	16:1, il5:0,	iso$C_{15:0}$, 10Me$C_{16:01}$,	44.8% of lipid is saturated fatty acids
acids in	11% are branched fatty acids	iso$C_{17:1}$	18:1ω7, 16:1ω7	iso$C_{17:0}$	of C-12,16 and 18.
membrane	12% monounsaturated				14% as branched (15:0 iso)
	fatty acids as $C_{16:1}$, $C_{19:1}$, $C_{18:1}$ ω9, $C_{18:1}$ ω11				24.9% as normal/ unsaturated fatty acid

nr not reported
[a]Zhilina et al. (1997)
[b]Sorokin et al. (2008)
[c]Sorokin and Chernyh (2017)
[d]Blum et al. (2012)
[e]Width × length

Table 9.10 Characteristics of halophilic bacteria growing optimally at neutral pH

Characteristics	*Desulfovibrio tunisensis*[a] DSM 19275[T]	*Desulfovibrio halophilus*[b] DSM 5663[T]	*Desulfovibrio brasiliensis*[c] DSM 15816	*Desulfovibrio gabonensis*[d] DSM 10636	*Desulfocella halophila*[e] DSM 11763T	*Deslufovibrio salexigens*[f] NCIB 2638	*Desulfohalobium retbaense*[g] DSM 5692
Site of isolation	Water from oil refinery in Tunisia	Solar Lake Sinai, Egypt	Sediment of Lagoa Vermelha, Rio de Janeiro, Brazil	Offshore oil field Galbon, West Africa	Hypersaline, thalassohaline sediment of Great Salt Lake Dakar, Senegal	From "sling mud". British Guiana	Lake Retba, a hypersaline lake near Utah
Cell morphology	vibrio	motile vibrio	vibrioid cells	motile vibrio	vibrio	vibrio	rod
Motility	motile	single polar flagellum	one subpolar flagellum	polar flagellum	polar flagellum	polar flagellum	polar flagellum
Cell size (μm)[+]	0.5 × 1.5–2	0.6 × 2.5–5	0.3–0.45 × 1–3.5	0.4 × 2–4	0.5–0.7 × 2–4	0.5 × 2.5–3.5	0.7–0.9 × 1–3
G + C % (DNA)	59.6	60.7	56.3	59.5	35	46.1	57.,1
Growth:							
Temperature	15–45	15–45	15–45	15–40	mesophilic	15–37	mesophilic
(optimum, °C)	(37)	(35–40)	(33)	(30)	(34)	(31)	(37–40)
pH range (optimum)	4.5–9 (7.0)	6.3–9 (7.6)	6.4–8.2 (6.9–7.3)	5.8–7.6 (6.5–7.3)	(7.0)	(7–7.5)	5.5–8 (6.5–7)
NaCl range (optimum)	0–7 (4)	3–18% (6–7%)	1–15% (3–10%)	2–19% (4–5%)	(4–5%)	1–8%	10%
Electron acceptors:							
SO_4^{2-}	+	+	+	+	+	+	+
SO_3^{2-}	+	+	–	+	–	nr	+
$S_2O_3^{2-}$	+	+	+	+	–	+	+
Fermentation	pyruvate, fumarate	–	pyruvate	pyruvate, malate, fumarate	–	pyruvate, fumarate	pyruvate malate

Electron donors:							
H_2	+	+	+	+	–	+	+
Acetate	–	–	–	–	–	–	–
Lactate	+	+	+	+	–	+	+
Pyruvate	+	+	+	+	+	+	+

nr not reported
[a]Ben Ali Gam et al. (2009)
[b]Caumette et al. (1991)
[c]Warthmann et al. (2005)
[d] Tardy-Jacquenod et al. (1996)
[e] Brandt et al. (1999)
[f]Postgate (1984)
[g]Ollivier et al. (1991)
[h] Width × length

Table 9.11 Sulfate-reducing bacteria that are moderate halophiles

Organisms	Characteristics	Isolation source	References
Desulfobacter postgatei	Marine strains require 2% NaCl and	brackish water and marine	Widdel and Pfennig (1981)
DSM 2034	0.3% $MgCl_2$, 10–37 °C, pH 6.22–8.5 Optimum pH 7.3	sediment	
Desulfobulbus oligotrophicus	Grows 0–18% NaCl 2.5% optimum	municipal, anaerobic sludge digester	El Houari et al. (2017)
DSM 103420[T]	20–37 °C and 35 °C optimum, pH 6.5–8.5 and 7.6 optimum	in Marrakech, Morocco	
Desulfobulbus propionicus	Marine strains grow in 0–18% NaCl	Marine mud	Widdel and Pfennig (1982)
DSM 2032	2.5% NaCl optimum 10–43 °C and 39 °C optimum, pH 6.0–8.6 and 7.2 optimum		
Desulfobulbus rhabdoformis	Grow 1.5–2% NaCl	water–oil separation system	Lien et al. (1998)
DSM 8777[T]	pH 6.8–7.2 optimum 10–40 °C and 31 °C optimum		
Desulfohalobium utahense	Grows with 8–10% NaCl,	Sediment of north arm	Jakobsen et al. (2006)
DSM 17720[T]	pH 6.5–8.3, (6.8). 15–44 °C (37 °C)	of Great Salt Lake, Utah	
Desulfonauticus submarinus	Grows optimally with 2% NaCl but	Deep-sea hydrothermal vent	Audiffrin et al. (2003)
DSM 15269[T]	tolerates 5% salt. Temperature is 30–60 °C (45 °C). Optimum pH is 7	East-Pacific Rise at depth of 2600 m	
Desulfotomaculum halophilum	Grows in 1–14% NaCl with optimum	Production fluid from oil well in	Tardy-Jacquenod et al. (1998)
DSM 11559[T]	at 6%. Optimum growth at 35 °C and pH of 7.3. Produces spores.	France	
Desulfovermiculus halophilus	Grows at 3–23% salt with optimum	Oil field water in Russia	Belyakova et al. (2006)
DSM 18834[T]	growth at 8–10% NaCl. Optimum growth at 37 °C and pH 7.2		
Desulfovibrio bastinii	Grows optimally at 4% NaCl,	production-water samples from	Magot et al. (2004)
DSM 116055[T]	pH 5.8–6.2, pH of 5.2	Emeraude Oilfield, Congo	
Desulfovibrio cavernae	Grows at 0.15–19% NaCl	subsurface sandstone at	Sass and Cypionka (2004)

(continued)

Table 9.11 (continued)

Organisms	Characteristics	Isolation source	References
Strain H1M	20–50 °C	600–1060 m deep	
Desulfovibrio gracilis	optimum growth at 5–6% NaCl,	production-water samples from	Magot et al. (2004)
DSM 16080^T	pH 6.8–7.2,	Emeraude Oilfield, Congo	
Desulfovibrio indonesiensis	Grows at 1–10% (5–6% optimum) NaCl,		Feio et al. (1998)
DSM 15121^T	pH 6.5–8.5 (6.8–7.2 optimum) 10–37 °C		
Desulfovibrio marinus	Grows at 0.5–11% (5% optimum) NaCl,	Marine sediment near Stax, Tunisia	Ben Dhia Thabet et al. (2007)
DSM 18311^T	pH 6.5–8.5 (7.0 optimum), 20–50 °C (37 °C optimum)		
Desulfovibrio oxyclinae	Grows at 2–22.5% (5–10% optimum)	hypersaline cyanobacterial mat	Krekeler et al. (1997)
DSM 11498	pH (7.0 optimum), 34–37 °C optimum	obtained from Solar Lake, Sinai	
Desulfovibrio profundus	Grows at 0.6–10% NaCl,	deep marine sediments	Bale et al. (1997)
DSM 11384	pH 4.5 to 9, (7.9 optimum)	at 80 and 500 m depths.	
	15–65 °C (25 °C optimum) Requires 100–150 atm pressure	obtained from the Japan Sea	
Desulfovibrio salexigens	Grows at 0.5–12% (2–4% optimum) NaCl,	muds	Postgate and Campbell (1966)
DSM 2638	pH (7.0 optimum), 34–37 °C optimum		Krekeler et al. (1997)

20%, *Desulfohalophilus alkaliarsenatis* can be considered an extremely halophilic SRB in that it grows optimally at 20% salt (Foti et al. 2007; Blum et al. 2012). *C.* Desulfonatronobulbus propionicus was isolated from sediments from a hypersaline soda lake in Kulunda Steppe (Altai, Russia). This SRB oxidizes propionate incompletely to acetate and grows at pH 10 with a Na^+ concentration of 1.2 M Na^+ (Sorokin and Chernyh 2016).

Desulfohalophilus alkaliarsenatis is also capable of arsenate and nitrate respiration in addition to dissimilatory sulfate reduction (Blum et al. 2012). With sulfide as the electron donor, *Desulfohalophilus alkaliarsenatis* grows by chemolithotrophy, while sulfide oxidation coupled to arsenate reduction requires acetate or other carbon structures for growth (Blum et al. 2012). With *Desulfohalophilus alkaliarsenatis*, the range of salt supporting growth with sulfate respiration is 12.5–33%, while with arsenate respiration it is 5.5–17.5% (Foti et al. 2007). This dual respiration of sulfate

Table 9.12 Sulfate-reducing bacteria that are slightly halophilic

Organisms	Characteristics	Isolation source	References
Desulfobulbus mediterraneus	Grows at 1–7% NaCl,	Deep sea sediment	Sass et al. (2002)
DSM 13871[T]	10–30 °C (25 °C optimum)	western Mediterra-nean Sea	
Desulfobacter curvatus	Grows at 0.5–2% NaCl and 0.1–0.3%	Marine sediment	Widdel (1987)
DSM 3379	$MgCl_2$, pH 6.8–7.2 opti-mum *28–31* °C optimum		
Desulfobacter latus	*Grows at 2% NaCl and 0.3%* $MgCl_2$	*Marine sediment*	Widdel (1987)
DSM3381	*pH 7–7.3 optimum, 30* °C optimum		
Desulfoplanes formicivorans	Grows at 0.5–8% (1–4) NaCl,	Sediment in Lake Harutoni,	Watanabe et al. (2015)
DSM 28890[T]	*pH 6.1–8.6 and 7.0–7.5 optimum,* *13–50* °C (42–45 °C optimum)	*Japan*	
Desulfovibrio salinus	Grows at 1–12% (3%) NaCl,	Saline lake in Tunisia	Ben Ali Gam et al. (2018)
DSM 101510[T]	15–45 °C (40 °C opti-mum) pH 6–8.5 and optimum at 6.7		
Desulfovibrio salexigens	Grows at 2.5–5.0% NaCl,	sea water, marine, and estuarine	Widdel and Pfen-nig (1984)
DSM 2638[T]	34–37 °C optimum, pH 7.5 optimum	muds, pickling brines	
Desulfomonile limimaris	Grows at 0.32–2.5% (1.25%) NaCl	Shallow marine sediments	Sun et al. (2001)
ATCC 700979[T]	37 °C optimum	Gulf Breeze, Florida	

and arsenate with lactate as the electron donor is also observed in the salt-sensitive *Desulfotomaculum auripigmentum* (current name *Desulfosporosinus auripigmenti*) (Newman et al. 1997), *Desulfomicrobium* sp. strain Ben-RB (Macy et al. 2000), and *Desulfotomaculum hydrothermale* (Haouori et al. 2008a, b).

9.2.5 Alkaliphiles

Alkaliphilic bacteria grow in environments where the pH is 8.5 to 11 and an optimum pH for growth is about pH 10. A few slightly halophilic SRB grow optimally at alkaline levels (Table 9.13), and some strains have very low tolerance to salt but grow in an alkaline environment (Table 9.14). In addition to the

Table 9.13 Alkaliphilic SRB strains that have a low salt tolerance

	Desulfotomaculum alkaliphilum[a]	*Desulfovibrio alkalitolerans*[b]	*Desulfonatronum lacustre*[c]	*Desulfonatronum thiodismutans*[d]	*Desulfonatronum cooperativum*[e]	*Desulfonatronum parangueonense*[f]
	DSM 12257^T	DSM 16529^T	DSM 10312^T	DSM 14708^T	DSM 16749^T	DSM 103602
Cell shape	curved rods	vibrio	vibrio	vibrioid	vibrioid	vibrio
Spore formation	yes	no	no	no	no	no
Size (μm)	0.6–0.7 × 3.0–3.5	0.5–0.8 × 1.4–1.9	0.7–0.9 × 2.5–3.0	0.6–0.7 × 1.2–2.7	0.4–0.5 × 1.0–2.5	0.7–0.9 × 1.2–2.3
Motility	no	single polar flagellum	single polar flagellum	single polar flagellum	single polar flagellum	single polar flagellum
pH range	8.0–9.15 (8.7)	6.9–9.9 (9.0–9.4)	8.0–10.1 (9.3–9.5)	8.0–10.0 (9.5)	6.7–10.3 (8–9)	8.3–10.4 (9.0)
NaCl (%)	0.1–5 (0.1)	0.7 (0.13)	0–10 (0)	>1.0–7.0 (3.0)	0.1–8 (0.5–1.5)	0–8.0 (5.0)
Temperature (°C)	30–60 (50–55)	16–47 (43)	22–45 (40)	15–48 (37)	15–40 (35–38)	15–40 (35)
G + C of DNA	40.9	64.7%	57.3%	63.1%	56.5%	56.1
Electron donors						
formate	+	+	+	+	+	+
lactate	+	+	–	–	+	+
pyruvate	+	+	–	–	–	+
malate	–	–	–	nr	–	nr
ethanol	+	–	+	+	–	+

(continued)

Table 9.13 (continued)

	Desulfotomaculum alkaliphilum[a]	*Desulfovibrio alkalitolerans*[b]	*Desulfonatronum lacustre*[c]	*Desulfonatronum thiodismutans*[d]	*Desulfonatronum cooperativum*[e]	*Desulfonatronum parangueonense*[f]
	DSM 12257^T	DSM 16529^T	DSM 10312^T	DSM 14708^T	DSM 16749^T	DSM 103602
Electron acceptors						
sulfate	+	+	+	+	+	+
thiosulfate	+	+	+	+	+	+
sulfite	+	+	+	+	+	+
H_2/acetate	+	+	nr	+	+	+
Fermentation or		pyruvate fermentation	disproportionation of thiosulfate and sulfite	disproportionation of thiosulfate and sulfite	disproportionation disproportionation of thiosulfate and sulfite	–

[a]Pikuta et al. (2000)
[b]Abildgaard et al. (2006)
[c]Pikuta et al. (1998)
[d]Pikuta et al. (2003)
[e]Zhilina et al. (2005b)
[f]Pérez Bernal et al. (2017)

Table 9.14 Characteristics of mesophilic and thermophilic acidophilic sulfate-reducing bacteria

Characteristics	*Desulfosporosinus acidiphilium*[c]	*Desulfosporosinus narugense*[d]	*Thermodesulfobium acididurans*[a]	*Thermodesulfobium acidiphilus*[b]
	DSM 27692^T	DSM 22704^T	DSM 102892^T	DSM 14796^T
Cell morphology	rods	rods	rods	rods
Cell size (μm)	0.7 × 3–5	0.8–1 × 4–7	0.5 × 1–5	0.5 × 2–4
G + C % (DNA)	41.8	42.3	33.7	35.1
Growth:				
Temperature (50–55) (optimum, °C)	15–40 (30)	15–40 (30)	37–65(54)	37–65
pH range (optimum)	3.8–7 (5.5)	3.6–5.6 (5.2)	3.7–6.5 (4.8–5)	4.4–6.5
NaCl as % (optimum)	0.6	1.5	<1.0	<1
Electron acceptors:				
SO_4^{2-}	−	−	−	−
$S_2O_3^{2-}$	+	+	+	+
NO_3^-	+	+	−	+
Fe^{3+}	+	−	−	−
Electron donors:				
Acetate	−	−	−	−
Formate	+	−	+	+
Lactate	+	+	−	−
Pyruvate	+	+	−	−
Ethanol	+	−	−	−
Propionate-	−	−	−	

[a]Sánchez-Andrea et al. (2015)
[b]Alazard et al. (2010)
[c]Frolov et al. (2017)
[d]Mori et al. (2003)

alkaliphilic SRB strains in Table 9.9 isolated from soda lakes, *Desulfonatronum lacustre* (Pikuta et al. 1998) and *Desulfonatronum cooperativum* (Zhilina et al. 2005a) were isolated from the alkaline Lake Khadin at Tuva in Central Asia, *Desulfonatronum thiodismutans* (Pikuta et al. 2003) from the soda Mono Lake in California, *Desulfonatronum zhilinae* (Zakharyuk et al. 2015) from soda Lake Alginskoe, Trans-Baikal Region, Russia, and *Desulfonatronum buryatense* (Ryzhmanova et al. 2013) from alkaline brackish lakes located in the Siberian region of Russia. *Dst. alkaliphilum* (Pikuta et al. 2000) was isolated from cow/pig manure on farms near Moscow, Russia, and *D. alkalitolerans* (Abildgaard et al. 2006) was from alkaline heating water in Denmark. The mechanisms accounting for the resistance of SRB cells to an extreme alkaline environment are to be determined in future studies.

Non-alkaliphilic SRB in a bacterial consortium have been found to survive extremely alkaline conditions. Using bacteria present in water from Danish district heating plants, hydrogen sulfide was not detected at pH 10.5, but if the pH of the water was reduced to pH 9.3, hydrogen sulfide was produced (Goeres et al. 1998). However, hydrogen sulfide production occurred at pH 10.2 if the nonalkaliphilic SRB were associated with a mixed biofilm containing bacteria found in the Danish heating water. It was assumed that the biofilm provided mico-niches where metabolism of the bacterial consortium accounted for localized neutralization of the alkalinity of the bulk solution.

9.2.6 Acidophiles

While SRB have been reported to be present in samples from hydrothermal vents, streams from metal mines, and acidic springs where sulfur oxidation may occur, and display optimal growth at about pH 5 (Table 9.14), none of these SRB grow optimally at pH <3. The initial report of SRB growing in an acidic environment was by Tuttle et al. (1969). They reported the presence of *Desulfovibrio* and *Desulfotomaculum* strains in the acidic mine water with the production of hydrogen sulfide from sulfate at pH 3.0 with a mixed culture, but with pure cultures of SRB, at pH 5.5, sulfate was not reduced to hydrogen sulfide. While there have been other reports of SRB in natural acidic environments, the use of inappropriate enrichment and cultivation techniques delayed isolating acidophilic SRB, and these difficulties have been well documented (Sánchez-Andrea et al. 2015). The first acidophilic SRB identified and partially characterized included *Desulfosporosinus* strains P1 and M1^T (Sen and Johnson 1999), which grew at pH 3.8 with glycerol as the electron donor (Kimura et al. 2006). An acidophilic thermophile, *Thermodesulfobium narugense*, grows at pH 4.0 to 6.5 (Mori et al. 2003), and a mesophilic acidophile, *Desulfosporosinus acidiphilus* strain SJ4(T), grows at pH 3.6 to 5.5 (Alazard et al. 2010). Several isolated bacteria have not been characterized, and they include strains CL4 (Rowe et al. 2007), CEB3 (Ňancucheo and Johnson 2012), and strains I, D, and E (Sánchez-Andrea et al. 2013). The long generation times (Table 9.1) and conditions for growth may contribute to the current availability of a few isolated strains of acidophilic SRP.

Electron donors for SRP growing under extremely acidic conditions often include diverse organic compounds. However, there are several considerations to culturing acidophilic SRP and, as reviewed by Ňancucheo et al. (2016), include (i) using glycerol or electron donors other than aliphatic acids, (ii) adding zinc ions to control the toxic accumulation of hydrogen sulfide, and (iii) using a heterotrophic acidophile to consume acetic acid produced by SRB isolates that incompletely oxidize glycerol. Under acidic conditions below the pKa of small-molecular-weight organic salts (i.e., formate, acetate, lactate, pyruvate, etc.), these organic salts become protonated to form membrane-permeable aliphatic acids, which disrupt membrane charge and prevent cell growth. However, since most of the acidophilic SRB can grow at near neutrality conditions (Table 9.6), these bacteria may have the capability of utilizing lactate and pyruvate, which are hallmark characteristics of SRB. As reported (Sánchez-Andrea et al. 2013), successful enrichments have resulted from using glycerol, methanol, or hydrogen at pH 4, lactate at pH 4.5, and succinate at pH 5.5. The addition of zinc chloride to media precipitates sulfide as ZnS, which prevents accumulation of toxic hydrogen sulfide and also functions as a pH buffer. The growth medium contains ferrous ion, which will also precipitate sulfide; however, at acidities lower than pH 5, solubility limitations of FeS limits the removal of sulfide, while ionic ZnS is an effective sulfide sink with the formation of ZnS even at low pH levels. Hydrogen sulfide is an effective reversible toxic metabolic product and at 16.1 mM hydrogen sulfide inhibits neutrophilic SRB (Reis et al. 1992). Since many SRB are incomplete oxidizers of carbon compounds, acetic acid may accumulate in the growth medium using glycerol as the carbon source. The presence of *Acidocella aromatica* or a similar heterotrophic acidophile will consume acetic acid as it is produced. Additionally, the presence of a heterotrophic acidophile is useful to deplete pyruvate, which is commonly found in agar and would be toxic to SRB at <pH 5 (Ňancucheo et al. 2016).

A common feature of acidophilic SRB is chemolithoautotrophic growth with H_2 as the electron donor, CO_2 as the carbon source, and sulfate as the electron acceptor. This is in contrast to *Thermodesulfobium narugense* and *Thermodesulfobium acidiphilun*, which use only hydrogen and formate as electron donors. *Desulfosporosinus acididurans* and *Desulfosporosinus acidiphilus* are capable of using several organic compounds as electron donors. *Caldivigra maquilingensis*, a sulfate-reducing archaeon, is unable to fix carbon dioxide but grows heterotrophically requiring sulfate, thiosulfate, or elemental sulfur as the electron acceptor using glycogen, gelatin, beef extract, peptone, tryptone, or yeast extract as the carbon source (Itoh et al. 1999).

Thermodesulfobium acidiphilum strain 3127-1^T (DSM 102892^T), isolated from geothermally heated soil in the Kamchatka peninsula of Russia, grows chemolithoautotrophically with H_2 as the electron donor, CO_2 as the carbon source, and sulfate as the electron acceptor (Frolov et al. 2017). Formate can serve as the only alternate electron donor; however, the growth is significantly less than with H_2 and only thiosulfate has been found to be used as an alternate electron acceptor.

Certainly, it is apparent that *Thermodesulfobium acidiphilum* uses a limited number of nutrients to support growth (Table 9.3). The closest relative to *Thermodesulfobium acidiphilum* DSM 102892^T is *Thermodesulfobium narugense* $Na82^T$, with an average nucleotide identity between the genomes being 87%, which is not within the 95% identity required for the same species. Several novel SRB genomes have been detected in samples from across Southeast China using metagenomics and metatranscriptomics, which are proposed to be members of a new deltaproteobacterial order designated as *Candidatus* Acidulodesulfobacterales (Tan et al. 2019). Reconstructed gene profiles suggest the presence of facultative anaerobic bacteria with genes for dissimilatory sulfate reduction (*dsrAB*, *dsrD*, *dsrL*, and *dsrEFH*) with the potential for sulfide oxidation (Tan et al. 2019). Some of these *dsr* genes are not found in other SRB. DsrL is a subunit of glutamate synthase and has been demonstrated to be required for sulfur oxidation and DsrEFH are enzymes present in sulfur oxidizers (Lübbe et al. 2006; Grimm et al. 2011; Stockdreher et al. 2012; Weissgerber et al. 2014). Thus, it was proposed that the new uncultured species of *Candidatus* Acidulodesulfobacterales oxidize sulfide to sulfate using a reversal of the dissimilatory sulfate reduction pathway (Tan et al. 2019).

9.2.7 Piezophiles

In the mid-1900s, bacteria present in deep ocean environments attracted the attention of scientists. Zobell and Johnson (1949) provided the first publication to describe marine bacteria growing optimally under high hydrostatic pressure, and they used the term "barophile" to describe their activity. Over the years, "piezophile" has replaced barophile because in Greek, "piezo" refers to pressure while "baro" refers to weight (Kato and Horikoshi 2009). Hydrostatic pressure is expressed in megapascal (MPa; 1 MPa = 10 bar or 10 atm) and the hydrostatic pressure increases 1 MPa for every 100 m increase in depth. The operating designation for an organism to be a piezophile is that optimum growth must be at 0.1 MPa or greater (Yayanos 1995) and at this time five species of *Desulfovibrio* are piezophiles (Table 9.15). The piezophilic SRB as well as many of the SRB that tolerate enhanced hydrostatic conditions are polyextremophiles in that the temperatures where the bacteria are isolated are about 3–5 °C. Therefore, the physiological response needed to enable sulfate-respiring bacteria to grow under high hydrostatic pressure must be distinguished from those physiological factors that enable growth to occur at cold temperatures.

The first report of piezophilic sulfate reducers was provided by Zobell and Morita (1957) as they reported their observations with deep-sea sediments. When inoculated samples were incubated at 3–5 °C, greater numbers of sulfate reducers were found in samples from the Philippine Trench at 1000 atm (100 MPa) as compared to incubation at 1 atm. Furthermore, mud as inoculum produced greater SRB numbers when incubated at 700 atm (70 MPa) than at 1 atm (0.1 MPa). When small mud samples from the Weber Deep were transferred into enrichment media, compressed to

Table 9.15 Sulfate-reducing bacteria that grow optimally at high pressures

Species	Growth at elevated pressures	References
Desulfovibrio brasiliense DSM 15816	Grows at pressure up to 37 MPa	Warthmann et al. (2005)
Pseudodesulfovibrio indicus J2[T]	Grows at a range of 0 to 30 MPa	Cao et al. (2016b)
DSM 101483[T]	Optimum growth is at 10 MPa	
Desulfovibrio hydrothermalis DSM 14728T	Grows better at 260 atm (26 MPa) than at 1 atm (0.1 MPa)	Alazard et al. (2003)
Desulfovibrio piezophilus[a]	Optimal growth at 10 MPa and upper limit	Khelaifia et al. (2011)
DSM 21447[T]	for growth is 25–30 MPa	
Desulfovibrio produndus[a]	Grows at pressure range of 0.1 to 40 MPa	Bale et al. (1997)
DSM 11384[T]	(optimum pressure for growth is 10 to 15 MPa and this may vary with the strain)	

[a]See text for discussion concerning placement in *Pseudodesulfovibrio* genus

700 atm (70 MPa), and incubated at 3–5 °C, the inoculated media showed no H_2S production at 60 days, but after 10 months of incubation, H_2S was detected. In comparison, mud samples incubated at 3–5 °C and 1 atm (0.1 MPa) showed no evidence of growth after 26 months of incubation. Sulfate reducers from the Weber Deep enrichment were successfully subcultured five times, and the bacterial culture consisted of nonmobile ovoid cells that were 0.3 μ wide and 0.5–0.8 μ long. Unfortunately, this culture was lost.

The response of *D. hydrothermalis* AM13 to hydrostatic pressure was examined using transcriptomics (Amrani et al. 2014). This SRB was isolated from a depth of 2600 m from a hydrothermal chimney in the East-Pacific Rise (Alazard et al. 2003). *D. hydrothermalis* readily adapts to a change in hydrostatic pressure and the genes expressed when this organism was cultivated under 0.1, 10, and 26 MPa. Using transcriptomics, 65 genes were affected by changes in hydrostatic pressures and activities of these genes belonged to the following categories: (i) aromatic amino acid and glutamate metabolism, (ii) energy metabolism, (iii) signal transduction, and (iv) unknown functions (Amrani et al. 2014). The adaptation to elevated hydrostatic pressures by *D. hydrothermalis* involves several mechanisms, with glutamate metabolism and energy production being central to this exposure. At 10 vs 0.1 MPa, there are 33 genes differentially expressed with 28 genes downregulated, while at 26 vs 10 MPa, there are 20 genes differentially expressed with 15 genes upregulated. The genes downregulated include three genes for heat shock response, three genes for anthranilate and glutamate metabolism, five genes for signal transduction mechanisms, and the gene for Fe-containing alcohol dehydrogenase. Genes that were upregulated following exposure to >10 MPa include genes for transmembrane Hmc complex, iron transport, and cobalt binding. As the cells of *D. hydrothermalis* are exposed to greater hydrostatic pressures, the amount of intracellular glutamate increases. Bacteria grown at 0.1, 10, and 26 MPa contain 82.3, 142.2, and 185.6 nmol glutamate/mg cell protein, respectively. The accumulation of glutamate in the cells reflects a decrease of glutamine required for aromatic amino acid biosynthesis and a reduction of glutamate turnover in the cells. The impact of hydrostatic pressure on energy metabolism was evaluated by examining the intracellular pools of ATP and ADP. At 26 MPa, the ATP level was 176 pmol/mg dry cells, which was significantly greater than the ATP level of 102 pmol/mg dry cells at 10 MPa. The ADP/ATP ratio in cells grown at 0.1, 10, and 26 MPa was 0.44, 0.22, and 0.10, respectively. The increased level of phosphorylation in cells exposed to high hydrostatic pressure may reflect an enhanced oxidative phosphorylation activity attributed to an increased production of transmembrane Hmc complex components.

Recently, the genomic and proteomic characterization of *D. piezophilus* C1TLV30^T was published, and several insights into the capability of this deep-sea organism have been revealed (Pradel et al. 2013). This SRB, isolated at a depth of 1700 m from wood falls in the Mediterranean Sea (Khelaifia et al. 2011), is similar to other *Desulfovibrio* in that it has seven genomic islands and few insertion sequences/transposase elements. While unsaturated fatty acids with extended chain lengths are associated with response to high hydrostatic pressures, the enzymes for membrane lipid synthesis in *D. piezophilus* are similar to non-piezophilic bacteria, and maintaining membrane fluidity under high hydrostatic pressures may be attributed to the regulation of these enzymes for lipid biosynthesis. An interesting observation was that aspartic acid, glutamic acid, lysine, asparagine, serine, and tyrosine in proteins of non-piezophilic *Desulfovibrio* were replaced by arginine, histidine, alanine, and threonine in the piezophile *D. piezophilus*. The role that these "piezophilic amino acids" may contribute to the stability and activity of proteins in *D. piezophilus* exposed to high hydrostatic pressure is unresolved at this time.

Proteome evaluation of *D. piezophilus* growing under different hydrostatic pressures revealed that 24 proteins increased and 16 proteins decreased at 0.1 MPa as compared to growth at 10 MPa (Pradel et al. 2013). The genes for these proteins influenced by hydrostatic pressure were not in a single genomic island but are randomly dispersed throughout the genome. Of the genes influenced by hydrostatic pressure, 11 proteins produced were associated with amino acid transport and metabolism. An important response to hydrostatic pressure was the regulation of genes for glutamate/glutamine transport and metabolism with both copies of the gene for ABC transporter glutamine binding proteins expressed at a high level when *D. piezophilus* was cultivated under low hydrostatic pressure. Under conditions of high hydrostatic pressure, high levels of alanine dehydrogenase, ATP phosphoribosyltransferase HisG, and ornithine carbamolyltransferase ArgF were produced, and these enzymes are considered to be important for alanine, histidine, and arginine biosynthesis. This could be the source of the "piezophilic amino acids" required for proteins synthesis. Also, the enhanced production of QmoA (quinone bound membrane flavoprotein with oxidoreductase activity), HynA-1 (the [NiFe] hydrogenase small subunit), and AprA (the adenylylsulfate reductase alpha subunit) indicates the importance of sulfate respiration by *D. piezophilus* at high hydrostatic pressures. Increased production of proteins for respiration by *D. hydrothermalis* (Amrani et al. 2014) and other piezophiles (Pradel et al. 2013) appears to be a characteristic of bacteria growing under high hydrostatic conditions.

The isolation and characterization of a piezophilic SRB from the Indian Ocean have resulted in the designation of a new genus, *Pseudodesulfovibrio* (Cao et al. 2016a). While the genus *Pseudodesulfovibrio* shares some characteristics of *Desulfobivrio* and related genera in the family *Desulfovibrionaceae*, the SRB of the *Pseudodesulfovibrio* genus are distinguished from the *Desulfovibrio* in G+C % content of DNA, electron donors for growth, and 16S rRNA gene sequence. Included in *Pseudodesulfovibrio* genus would be the newly isolated *Pseudodesulfovibrio indicus* (Cao et al. 2016a) and several SRB previously designated as *D. piezophilus* (Khelaifia et al. 2011) and *D. profundus* (Bale et al. 1997). Also, *D. portus* (Suzuki et al. 2009) and *D. aespoeensis* (Motamedi and Pedersen 1998) are placed in the new genera of *Pseudodesulfovibrio* even though their piezophilic characteristics remain to be defined. At this time, the taxonomy of *D. hydrothermalis* (Alazard et al. 2003) and *D. brasiliense* (Warthmann et al. 2005) has not been challenged.

9.3 Activities and Communities in Extreme, Unique, or Isolated Environments

9.3.1 Adaptations to the Environment

Colonization by sulfate reducers involves many stages or events including adaptations to specific environments. Important criteria defining an environment may involve pH, temperature, salt content, and available nutrients. As indicated in several of the tables in this chapter, individual species of SRB have a fairly narrow optimum range for pH, temperature, and salt in the culture media but remain viable and can grow in suspended culture at suboptimum conditions. Additionally, there is a transition into the extreme environment where SRB are exposed at the periphery of the site where induction of genetic expression provides for resistant phenotypes. As observed with one SRB species (i.e., *D. vulgaris*), regulation of several genes following exposure to physical and chemical stress enables cells to adjust (see review by Plugge et al. (2011) and references therein). The resulting SRB population reflects a community that has undergone differentiation that often produces a patchy SRB distribution (Pérez-Jiménez and Kerkhof 2005). Often, the production of bacterial endospores is viewed as a highly important mechanism for maintaining bacterial species; however, spore production by sulfate-reducing bacteria in nutrient-limited environments may be terminal because germination may require nutrients absent in that environment.

From physiological studies and genome analysis, it is apparent that SRP have considerable metabolic flexibility (Plugge et al. 2011), which enables these microorganisms to sustain themselves in the absence of sulfate as well as to persist in extreme environments (Stevens 1997). The microbial community in a deep anaerobic alkaline environment included sulfate-reducing bacteria even though the aquifer in question contained <0.05 mg/L sulfate (Fry et al. 1997). Some of the dissimilatory sulfate reducers use S^0, sulfite, thiosulfate, Fe^{2+}, nitrate, and arsenate as alternate inorganic electron acceptors, which enable sulfate reducers to grow in changing environments. Fermentation with an organic compound is a biochemical characteristic of a few SRB where pyruvate, fumarate, or malate serves as the electron acceptor. While lactate and pyruvate have been used by SRB for cultivation with sulfate, the SRB are capable of using a large array of organic compounds as electron donors. Most strains of SRB are capable of using H_2 as an electron donor, and abiotic formation may result in dissolving of ferrosilicate minerals with the formation of H_2 and magnetite (Stevens and McKinley 1995). In the absence of alternate electron donors or acceptors, a few SRB (Table 9.13) use thiosulfate or sulfite as both the electron donor and electron acceptor by a process known as disproportionation. A few SRB, including some *Desulfosporosinus* species and *Desulfohalophilus alkaliarsenatis* (Blum et al. 2012), can also grow by using nitrate, Fe^{3+}, or arsenate as terminal electron acceptors. A novel process energetics for a sulfate reducer is displayed by *Desulfobulbus propionicus*, which uses S^0 as both the electron donor and electron acceptor according to the following disproportionation reaction: $4S^0 + 4H_2O \rightarrow SO_4^{2-} + 3HS^- + 5H^+$ (Lovley and Phillips 1994). Another physiological feature of SRB is their response to O_2, with some species highly tolerant and capable of growing in the presence of low levels of O_2. Tolerance of SRB to O_2 as well as the ability to use alternate electron acceptors may vary with isolates within a species.

9.3.1.1 Impact of Temperature

The presence of SRB in various environments can be assessed by measuring rates of sulfate metabolism and using molecular methods to identify the bacteria present. *Desulforhopalus vacuolatus* was isolated from a temperate environment, has a growth optimum at 18 °C, and is considered the first psychrophilic SRB isolated (Isaksen and Teske 1996). The ocean floor represents an environment of significant size, and activities of the SRB would be important here because over 90% of the global ocean floor is <4 °C. A study of the permanently cold (~2.6 °C) Artic sediment at Hornsund and Storfjord off the coast of Svalbard resulted in the isolation of numerous psychrophilic SRB that grew optimally at <20 °C (see Table 9.7) (Knoblauch et al. 1999a,b). All strains of SRB grew at the temperatures of isolation (2.6 at Hornsund and −1.7 at Storfjord) and displayed metabolic rates of sulfate reduction that were comparable or exceeded those of mesophilic strains of SRB (Table 9.16). The psychrophilic strains of SRB listed in Table 9.6 do not grow by disproportion of S^o or thiosulfate, and all oxidize fatty acids to acetate with incomplete oxidation except *Desulfofrigus oceanense*, which uses fatty acids with complete oxidation to CO_2. The presence of lactate oxidizing SRB to acetate oxidizing

Table 9.16 Rates of sulfate reduction with SRB isolates from permanently cold environments compared to mesophilic strains of SRB[a]

Bacteria	Substrate	Sulfate reduction (fmol cell^{-1} day^{-1})
Psychrophilic strains isolated from Hornsund and incubated at 2.6 °C		
Strain LSv20	lactate	14.0
Desulfofrigus fragile	lactate	2.7
Strain LSv22	lactate	13,0
Desulfofrigus oceanense	lactate	6.9
Desulfofaba gelida	propionate	41.9
Strain ASv25	acetate	25.3
Strain ASv26	acetate	3.8
Psychrophilic strains isolated at Storfjord and incubated at −1.7 °C		
Desulfotalea arctica	lactate	3.6
Strain LSv52	lactate	7.6
Desulfotalea psychrophila	lactate	1.9
Strain LSv55	lactate	6.2
Established mesophilic strains incubated at 4 °C		
Desulfobacter postgatei	acetate	11.0
Desulfobacter hydrogenophilus	hydrogen	8.0
Desulfobulbus sp.3pr10	propionate	4.2
Desulfovibrio vulgaris	lactate	0.4
Desulfobacterium autotrophicum	lactate	1.6

[a]Knoblauch et al. (1999a, b)

SRB in the cold Hornsund and Storfjord waters is about 10 or 100 times to 1. Also, from 16 sRNA sequence analysis, there is great diversity among the isolates off the coast of Svalbard, and similar diversity is also observed with mesophilic SRB in nonextreme environments (Sahm et al. 1999).

When examining the activity and community structures of SRB in polar, temperate, and tropical marine sediments, it was determined that environmental temperature was important for the selection of SRB in microbial communities. Optimal temperatures that were determined for H_2S production in sediments from the Arctic and Antarctic, temperate zone, and tropical areas were 24–26 °C, 0–35 °C, and 38–44 °C, respectively (Robador et al. 2015). The optimal temperature for sulfate respiration in sediments from the Arctic and Antarctic was 25 °C above in situ temperatures. While each sediment site had a unique SRB community, warm or cold regions contained specific phylotypes. For example, Arctic and Antarctic sediments contained similar phylotypes even though they were geographically separated, and the Arctic and Antarctic sediments were distinct from the phylotypes of the Arabian Sea and North Sea, which are geographically closer (Robador et al. 2015). Thus, the environment selects the SRB as residents.

9.3.1.2 Extremely Acidic Environments

With respect to growth, it was generally accepted that SRB grew optimally at pH 6 to 8 (Widdel 1988). With the isolation of SRB from environments with extreme pH values, it is now apparent that there are bacteria capable of growing in highly acidic and alkaline environments. Acidophilic bacteria grow optimally in environments of pH 1–5, and there are a few extreme acidophiles that grow optimally at <pH 3. Several reports indicate the production of hydrogen sulfide by SRB in acidic environments. The pit Lake of Brunita Mine in Spain has a pH of 2.2–5.0 and high concentrations of sulfate, iron, manganese, and copper, and hydrogen sulfide production is achieved by *Desulfobacca*, *Desulfomonile*, *Desulfurispora*, and *Desulfosporosinus* (Sánchez-España et al. 2020). Hydrogen sulfide production was reported in sediments of an acidic volcanic lake where the sediment had a pH of 3 (Koschorreck et al. 2003). As evidenced by hydrogen sulfide production, SRB were actively metabolizing in permanently acidic (pH 2–3) mine tailings sites (Praharaj and Fortin 2004) and in acid mine water with a pH of 4 (Church et al. 2007). A survey of several acidic lakes and wetlands where SRB metabolism was active revealed that while the water had a pH range of 2.3 to 4.0, the sediment had a pH of ~6 and hydrogen sulfide could have been produced by SRB growing at a pH level greater than the water Koschorreck (2008).

The growth of SRB in an acidic environment has been reviewed by Koschorreck (2008) and Tran et al. (2021), and there are two fundamental concerns: the acidic environment and the impact of hydrogen sulfide. One explanation of the growth of SRB in acidic environments is that microniches of neutral pH surround the bacteria (Fortin et al. 1996); however, Koschorreck (2008) has calculated that the rate of SRB metabolism is not great enough to maintain a neutral pH microniche in a pH 3 environment.

Acetic, lactic, and other organic acids are undissociated in low pH environments, and these free organic acids become uncouplers of growth because they readily pass through the cell membrane. With 45 mmole/L acetate at pH 5, the membrane charge of *Clostridium thermoaceticum* is readily destroyed and *Sarcina ventriculi* cells in an environment of pH 3 have an internal pH of 4.25 (Baronofsky et al. 1984). Newly isolated strains of *Desulfobacter* sp., *D. desulfuricans*, and *D. sapovorans* did not grow in Postgate B medium adjusted at pH 5. When these three strains were mixed and added to the growth medium at pH 5, lactate was oxidized with the production of hydrogen sulfide. *Desulfobacter* sp. oxidized acetate to CO_2 and, thereby, prevented acetic acid toxicity to the *Desulfovibrio* cultures (Mormontoy and Hurtado 2013). While acetic acid toxicity with SRBs is readily demonstrated under laboratory growth conditions with 5 mm/L, the level of acetate in nature is insufficient to inhibit SRB. Acetate in acidic mine lake sediments is generally <10 μmol/L (Koschorreck (2008), acetate in the sediment of the acidic Lake Caviahue is <380 μmol/L (Koschorreck et al. 2003), and acetate in acidic forest soil was 6 to 51 μmol/L (Tani et al. 1993). Thus, in natural acidic environments, volatile fatty acids would not appear to be inhibitory to SRB. Since the growth of SRB was inhibited by organic acids below their pK_a, glycerol was effective as an electron donor in isolation of SRB at pH 3.8–4.2 (Kimura et al. 2006), while glycerol, methanol, and hydrogen were used at pH 4 (Sánchez-Andrea et al. 2013). With genome analysis of *Desulfosporosinus acididurans* strain $M1^T$, it was reported that this strain uses urea as a nitrogen source, and a possibility was presented that ammonia resulting from urease could provide protection against an acid environment (Petzsch et al. 2015); however, this would require an available quantity of urea in the environment.

With a low pH environment, there is a greater concentration of protons in the bulk fluid than at a neutral pH value, and the Gibbs free energy is more electronegative with an acidic environment than at a neutral pH (Tran et al. 2021). To maintain a charge on the membrane, the SRB cell would need to export protons from the cell at a high rate, and this is an increased energy demand on the cell. This expenditure of energy is observed with *D. vulgaris* under laboratory conditions and mixed SRB cultures in marine sediments of Milos Island, Greece, where the rate of H_2S production was 60% less at pH 5 than at pH 6–7 (Bayraktarov et al. 2013). With H_2 as the electron donor for sulfate reduction, the free energy expressed as ΔG at pH 3 is -198 kJ/ mol sulfite, while at pH 5 the ΔG is -175 kJ/mol sulfite (Koschorreck 2008) and the extra energy would be used for proton export. It would follow that the lowest pH SRB can tolerate is determined by the energy limitation for proton export. Stress response with SRB exposed to acidic environments is discussed in Sect. 6.3.6 of Chap. 6.

Depending on the pH, the end product of sulfate respiration by SRB may be either S^{2-}, HS^-, or H_2S, and below pH 5, H_2S is favored (Hao et al. 1996). Metal salts react with H_2S at low pH to produce highly insoluble metal-sulfide precipitates, and the result may be reduced bioavailability of essential metals for the production of metabolically active metalloproteins, removal of toxic metals from the environment, and forming a rigid layer of metal sulfides on the cell surface (Koschorreck 2008). At 2–15 mM H_2S, sulfate respiration can be reduced by about 50% (Koschorreck 2008). H_2S permeates the plasma membrane, and H_2S toxicity is proposed to be attributed to sulfide bonds between polypeptides and metabolic cofactors (Moosa and Harrison 2006). At >3 mM H_2S, the maintenance coefficient of *D. desulfuricans* was increased and growth was inhibited (Okabe et al. 1995); however, it should be noted that inhibition by H_2S is reversible and experiments indicate that after return to a neutral pH and incubation for some time, the inhibited culture will grow (Rinzema and Letting 1988).

9.3.2 Soda Lakes and Other Alkaline Environments

SRB have been detected using rRNA-based analyses in two alkaline aquifers in the Columbia River Basalt Group in Washington, where the pH was 9.9 and 8.0 (Fry et al. 1997). Twenty SRB clones from these alkaline aquifers were most similar to a *Desulfovibrio* sp. PT2 from a groundwater source and *D. longreachii* from

Table 9.17 Rates of hydrogen sulfide production in sediments and freshwater

Location		Concentration		Maximal sulfate reduction (μmol/dm^3 day^{-1})
	Depth of sample	Salinity (g/L)	sulfate (mM)	
Soda lakes from the Kulunda Steppe, Siberia Russia[a]				
Picturesque Lake	2–6 cm	405	278–584	423
Cock Lake	0–4 cm	59	6.7–8.5	326
Tanatar I Lake	0–3 cm	475	6.7–8.5	113
Narrow Lake	2–17 cm	110	235–278	162
Mongolian soda lake[b] surface of dry lake				3000
Sediments of low sulfate water				
Freshwater sediments[c]				0.5–13
Yellowstone hot spring sediment[d]				5.6–104
Marine sediment[c]				15–20
Big Soda Lake, Nevada[e]		340	58	0.74–3.3 $\times$ 10^{-3}
Mono Lake, California[f]		90		2.3 $\times$ 10^{-3}

[a]Foti et al. (2007)
[b] Gorlenko et al. (1999); Sorokin et al. (2004)
[c]Li et al. (1999)
[d]Jørgensen and Bak (1991)
[e]Smith and Oremland (1987)
[f]Oremland et al. (2004)

Australian artesian waters (Redburn and Patel 1994). In addition to SRB listed in Table 9.9, other SRB that grow at an alkaline pH include *Dst. alkaliphilum* (Pikuta et al. 2000), *Desulfonatronum zhilinae* (Zakharyuk et al. 2015), *Desulfonatronum cooperativum* (Zhilina et al. 2005b), *Desulfonatronum parangueonense* (Pérez Bernal et al. 2017), and *Desulfonatronum buryatense* (Ryzhmanova et al. 2013).

SRB in soda lakes have an important role in sulfur cycling as demonstrated by the coupling of sulfate reduction by species of *Desulfonatronovibrio* and *Desulfonatronum* to the decomposition of organic compounds produced from microbial digestion of cellulose while in hypersaline soda lakes (Sorokin et al. 2014). SRB species of the *Desulfobacteriaceae* family are important in the sulfur cycle (Zavarzin, et al. 2008). Alkaline (pH ~10) lakes in the Kulunda Steppe produce considerable H_2S even though some lakes have high salinity (Table 9.17), and these rates of sulfate reduction are exceeded only by the 3000 μmol/dm^3 day^{-1} reported for Mongolian soda lakes (Gorlenko et al. 1999; Sorokin et al. 2004) and 5000 μmol/dm^3 day^{-1} for marine mat containing *Beggiatoa* (Boetius et al. 2000). The high rate of sulfate reduction for the Mongolian soda lake can be attributed to the cyanobacterial mat that was collected on the surface of the dry lake where the sample was obtained. In comparison to the soda lakes of the Kulunda Steppe, the rate of sulfate reduction from freshwater, marine, or Yellowstone hot springs was markedly lower (Table 9.17). In fact, the rates of sulfate reduction for Big Soda Lake and Mono Lake (Table 9.17) are also low compared to lakes of the Kulunda Stepp.

To understand different rates of sulfate reduction, alkalinity, salinity, pH, and other physical/chemical factors have been examined as well as the composition of the microbial ecosystem. "*C.* Contubernalis alkalaceticum" was isolated from the sediment of Khadyn Lake (Tuva, Russia) (Zhilina et al. 2005a). This obligate syntrophic bacterium oxidizes acetate in the presence of *Desulfonatronum cooperativum* and other species of the hydrogen-consuming genera *Desulfonatronum* and *Desulfonatronovibrio*. Samples from soda lakes of the Kulunda Steppe in Siberia, Russia, contained equal numbers of SRB that grew autotrophically (H_2 + CO_2) or heterotrophically (lactate and butyrate) (Foti et al. 2007).

Not only would the presence of specific species of SRB influence rates of sulfate reduction, but the level of diversity of these SRB species may also be important. As reviewed by Dillon et al. (2007), the diversity of SRB determined from *dsrAB* phylotypes is greater in hypersaline mats of saline lakes in Sinai, Egypt and Guerrero Negro, Mexico than in cyanobacterial mats from Mushroom Spring in Yellowstone National Park (USA) where the sulfate level is low. In the soda lakes of the Kulunda Steppe, the abundance of SRB is dependent on the salinity with 10^5–10^7 viable SRB cells ml^{-1} salt is in low concentration and 10^2 to 10^3 viable SRB cells when the lake reaches soda saturation (Sorokin et al. 2010b).

9.3.3 SRB Growing on Surfaces

Early studies by Heukelekian and Heller (1940) revealed that the growth of marine bacteria was enhanced when cultivated in the presence of glass, and Zobell (1943) reported that the density of marine bacteria was greater on environmental surfaces than suspended in the water. Senez (1953) reported attached growth of *D. desulfuricans* strain Canet 41, which was proposed to be attributed to a polysaccharide produced by the SRB. The highly viscous cultures of *D. desulfuricans* strain El Agheila Z (Grossman and Postgate 1955) and of *D. vulgaris* Hildenborough (Ochynski and Postgate 1963) were found to produce exopolysaccharide material. The mucopolysaccharide (mucin) produced by *D. vulgaris* Hildenborough was identified as a polymer of mannose (Ochynski and Postgate 1963), and marine strains of *Desulfovibrio* generally produced mucin as compared to SRB from low salt environments.

9.3.3.1 Environmental Mats

In some alkaline environments, SRB display respiratory and growth activities even though these SRB are sensitive to an alkaline pH, and this is apparently attributed to protection provided by complex microbial interactions in biofilms or mats (Goeres et al. 1998). The immobilized SRB are one segment of a microbial consortium where bacterial end products produce an environment with a neutral pH and a low redox environment beneficial to SRB. Thus, unlike suspended bacterial culture, microsite (or niche) formation occurs naturally and enables SRB to grow in aerobic mats and on corroding iron surfaces (see reviews by Fauque 1995; Hamilton and Lee 1995; Baumgartner et al. 2006; Okabe 2007). The activity of SRB in environmental films has been extensively studied in wastewater treatment systems and in marine hypersaline mats. An important question has always been how the anaerobic respiring sulfate-reducing bacteria could exist in environments with SRB in environmental mats containing cyanobacteria producing O_2 by photosynthetic processes. The presence of microsites in mat strata would be one possibility. Microsites for the growth of SRB had been proposed decades ago since microsites that were 50 to 200 μm diameter had been assigned to SRB for marine sediment containing both sulfate-reducing and sulfur-oxidizing bacteria (Jørgensen 1977).

The environmental mat in wastewater systems is only a few mm thick (Ramsing et al. 1998; Okabe et al. 1999, 2003; Kühl and Jørgensen 1992; Ito et al. 2002), and atmospheric oxygen would readily diffuse into the mat. Wastewater contains 100 to 1000 μM sulfate, which is consumed by SRB, which commonly includes members of the following genera: *Desulfomicrobium*, *Desulfovibrio*, *Desulforegula*, *Desulfobacterium*, and *Desulfobulbus* (Okabe 2007). As reviewed by Okabe (2007), SRB are most abundant at the surface (0–300 μm) of wastewater mats where there is an abundance of sulfur-oxidizing bacteria. Bacteria oxidize the highly toxic H_2S to elemental sulfur (S^0) with a concentration of S^0 reaching 22 μmol cm^{-3} at the oxic/anoxic interface (Okabe et al. 1999, 2005). It has been reported by Okabe et al. (2020) that S^0 and H_2S are oxidized to sulfate by sulfur-oxidizing bacteria that use O_2 or nitrate as the electron acceptor. The turnover rate of H_2S in the sulfide-oxidizing zone was reported to take only 18 to 36 sec, which is significantly less than the abiotic oxidation of H_2S with O_2, which would take minutes to hours. Thus, there is an internal sulfur cycle in these mats. It should be emphasized that there is not a vertical separation of sulfate-reducing and sulfur-oxidizing bacteria, but sulfide oxidization occurs in the same zone as sulfate reduction.

9.3.3.2 Lithification of Mats

Sulfate reduction was reported to occur in the oxic zone of hypersaline mats at Baja, California, Sur Mexico (Canfield and Des Morais 1991). Reports of sulfate reduction in the aerobic zone of other mats supported the observation that SRB grew in microbial mats in the presence of cyanobacteria (Canfield and Des Morais 1991; Teske et al. 1998; Wierenga et al. 2000; Jonkers et al. 2003). SRB in the oxic zone of mats have been shown to precipitate carbonate, and this lithified layer of ~0.1 cm thick had a higher rate of sulfate reduction than in other areas of the mat (Visseher et al. 2000; Dupraz et al. 2004; Dupraz and Visscher 2005). The number of SRB in microbial mats was reported to be in greatest abundance in lithified layers than in nonlithified layers (Baumgartner et al. 2006). An examination of exopolymeric material from a nonlithifying microbial mat near the Eleuthera Island, Bahamas, revealed a 80% turnover of exopolymeric material due to glucosidase activity in the upper 15–20 mm of the mat (Braissant et al. 2009). Low-molecular-weight organic carbon material was released by enzymes in the microbial mat, and Ca^{2+} was tightly bound to this low-molecular-weight material. A model has been produced to describe the interaction of exopolymeric substances in the mat with Ca^{2+} (Braissant et al. 2009). Exopolymeric material produced by SRB isolated from a lithifying mat was reported to have three buffering capacities (Braissant et al. 2007), and they are as follows: (i) attributed to carboxylic acids ($pK_a = 3.0$), (ii) sulfur-containing groups including thiols and sulfonic acids ($pK_a = 7.0–7.1$), and (iii) amino acids ($pK_a = 8.4–9.2$). The calcium-binding of this exopolymeric material at pH 9.0 was 0.112 to 0.15 g calcium/g exopolymeric material (Braissant et al. 2007).

The process of $CaCO_3$ precipitation in hypersaline waters involves several steps (Baumgartner et al. 2006). Alkalinity in the mat is increased as SRBs reduce sulfate to hydrogen sulfide and use organic acids as electron donors. Production of exopolymeric substances by cyanobacteria and SRB would initially bind free Ca^{2+}, and this would prevent $CaCO_3$ precipitation. Hydrolytic enzymes released into the mat from heterotrophic bacteria and SRB degrade the exopolymeric material, and Ca^{2+} is released into the surrounding region. *D. vulgaris* produces a dispersal hexosaminidase known as DisH (DUV2239), and the genome contains seven additional genes for putative polysaccharide degrading enzymes (Zhu et al. 2018). With an elevated concentration of Ca^{2+}in localized areas, $CaCO_3$ precipitation occurs using degraded exopolymeric material as nucleation sites for mineral formation. The complexation between the partially degraded exopolymeric material and the developing precipitation of $CaCO_3$ results in amorphous mineral formation. This process described above is supported in a publication by Spring et al. (2019).

While there are reports of SRB isolated from hypersaline cyanobacterial mats, only *Desulfohalovibrio reitneri* DSM 26903^T has been fully characterized (Spring et al. 2019). This SRB was isolated from a gelatinous mat 10–15 cm thick on the Kiritimati Atoll (Kiribati, Central Pacific) and may be unique in that such mats are rarely found in other marine waters. Using 16S rRNA gene identity values, *Desulfohalovibrio reitneri* was only distantly related to *D. alkalitolerans* DSM 16529T, *Desulfovibrio* sp. X2, "*Desulfovibrio cavernae*" H1M, and *Desulfocurvus vexinense* DSM 17965T. A unique feature of *Desulfohalovibrio reitneri*, a member of the family *Desulfovibrionaceae*, is the attachment of cells to the glass walls of the culture tube, forming a biofilm, and with cell growth, aggregates of cells collect at the bottom of the culture tube. Cells become embedded in the exopolymeric substances, and this fibrillar network is seen in Fig. 9.2.

A profile of the hypersaline mat on the Kiritimati Atoll reveals the spatial distribution of major groups of SRB in a vertical profile of the lithifying mat (Fig. 9.3). The greatest number of SRB in the profile were members of *Desulfobacteraceae*, which were concentrated in layer 5 with *Desulfoarculaceae* members found in layers 7–9 (Fig. 9.3). The presence of *Desulfohalovibrio reitneri* also known as the L21-Syr-ABT species resided in layer 4 (see Fig. 9.3). Phototrophic cyanobacteria were present in layers 1–3.

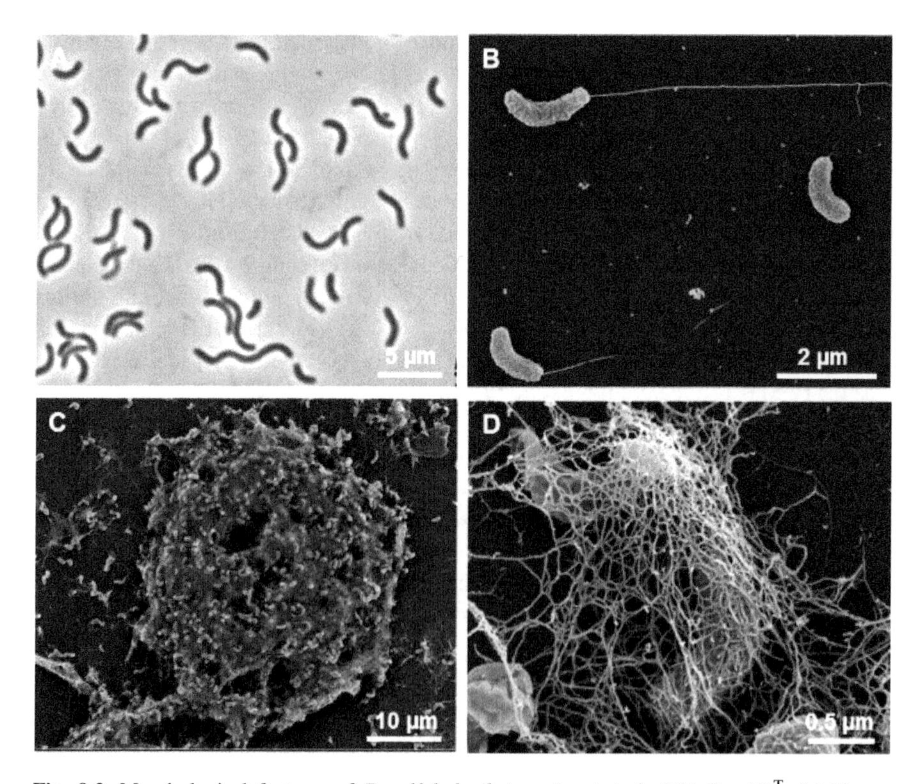

Fig. 9.2 Morphological features of *Desulfohalovibrio reitneri* strain L21-Syr-AB^T. (**a**) Phase-contrast micrograph illustrating the shape and size of cells grown in complex medium with pyruvate, sulfate, and yeast extract. (**b**) Scanning electron micrograph of single cells with monopolar monotrichous flagellation. (**c**) Scanning electron micrograph of a cellular aggregate formed in defined mineral medium with pyruvate as the sole source of carbon and energy. (**d**) Network of fibrillary EPS enveloping cells within aggregates. (Source: Spring et al. (2019).

Gypsum and other mineral aggregates were found in layer 5. *Desulfosalsimonas* (*Desulfobacteraceae*) were dominant in layer 5, and *Desulfovermiculus* (*Desulfohalobiaceae*) was in greatest abundance in layer 6 (Fig. 9.3).

9.3.3.3 Dolomite Bioformation

While the role of SRB in limestone dolomitization in early marine environments is unclear, the formation of dolomite appears influenced by bacterial sulfate reduction. The abundance of dolomite [$CaMg(CO_3)_2$] in Phanerozoic oceans compared to the relatively few new dolomite formations has been perplexing. Biogenic production of

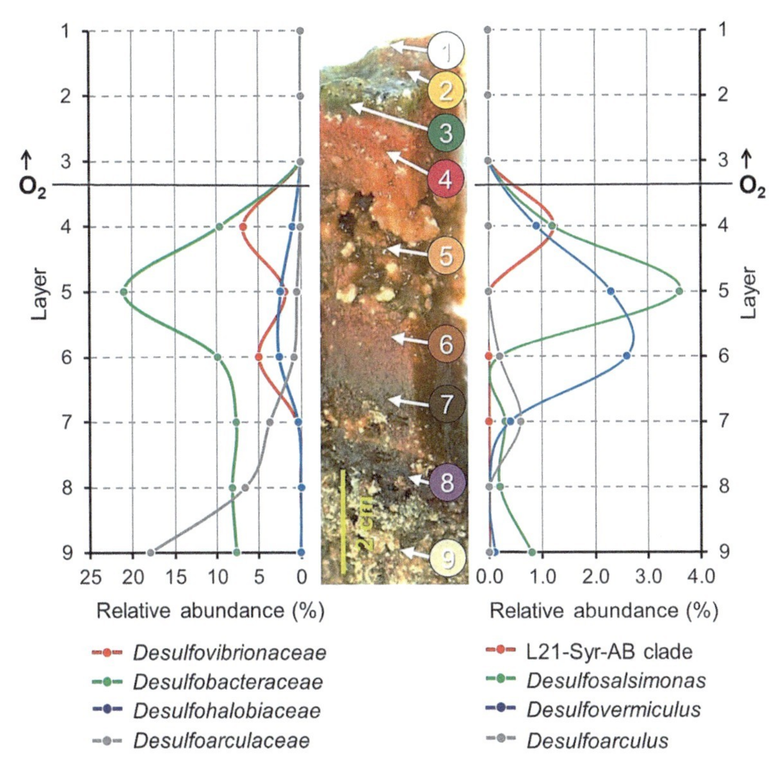

Fig. 9.3 Depth profiles of prevalent groups of sulfate-reducing bacteria in a lithifying hypersaline cyanobacterial mat. Spatial distribution patterns of distinct phylogenetic clades are based on the proportion of partial 16S rRNA gene sequences in different mat layers. A representative section of the Lake 21 mat is depicted in the middle. Numbers indicate different layers of the mat used for the generation of the corresponding 16S rRNA gene sequence libraries by high-throughput pyrosequencing of amplified DNA fragments. (Modified from Schneider et al. (2013). On the left, the proportion of sequences that could be affiliated with family-level clades of sulfate-reducing bacteria is plotted against different mat layers from top to bottom. On the right, the distribution patterns of the three most prevalent genera and a species-level clade comprising the novel isolate L21-Syr-AB^T are shown. Data points are represented by dots. Lines between the data points were extrapolated. Source: Spring et al. (2019).

dolomite has been examined at natural field sites using controlled laboratory studies, and the important conditions include (i) high sulfate-saturation index, (ii) a high Mg^{2+}/Ca^{2+} ratio, and (iii) a low sulfate concentration (Deng et al. 2010). From a series of laboratory studies using field samples, SRB have been found to contribute directly to the precipitation of dolomite (Vasconcelos and McKenzie 1997; Wright 1999; Warthmann et al. 2000; van Lith et al. 2002; van Lith et al.

2003a, b; Wright and Wacey 2005; Bontognali et al. 2010). There does not appear to be a specific SRB species for this activity because dolomite precipitation has been reported with various *Desulfovibrio* sp. (Wright and Wacey 2005; van Lith et al. 2003b), *D. brasiliensis* (Warthmann et al. 2005), *Desulfotomaculum* sp. (Wright and Wacey 2005), and *Dst. ruminis* (Deng et al. 2010). Live SRB and not dead cells are required for dolomite precipitation, and there has been the suggestion that exopolymeric material from SRB is critical for the production of calcium minerals in alkaline environments (Braissant et al. 2007). In a laboratory experiment, SRB were inactivated with antibiotics after the bacteria had grown and secreted exopolymeric substances (Bontognali et al. 2014). When Ca^{2+} and Mg^{2+} were added to the mixture containing inactivated bacteria and exopolymeric substances, Mg-calcite and Ca-dolomite were precipitated in the exopolymeric material. Using a stable carbonate platform from the Upper Jurassic in the Northern German Basin, the activity of SRB was proposed to promote dolomite formation (Baldermann et al. 2015).

Following an evaluation of biogeochemical activities of lakes of the Coorong region, South Australia, Wright (1999) reported that elevated concentrations of sulfate and carbonate anions with an increase in alkalinity was attributed to evaporation. The production of hydrogen sulfide by SRB in the lake sediment kills macrobiota, and Mg^{2+} is released from cyanobacterial cells. Dolomite formation follows this removal of sulfate and elevated concentration of Mg^{2+}, Ca^{2+}, and carbonate. Rapid dolomite precipitation results in the formation of nanosized grains with defects in a disordered lattice, and stable dolomite is not formed. Bacteria, including SRB, in the environment serve as nucleation sites for this rapid formation of unstable dolomite, and this prevention of dolomite has been attributed to SRB (Wright 1999). Thus, as reviewed by Gallagher et al. (2014), there is some criticism of the involvement of SRB in carbonate precipitation, and the reason there are no new environmental dolomite precipitations has not been explained. It appears that conditions required for dolomite formation by bacteria are complex and dependent on a series of processes.

9.3.3.4 Biofilms

Biofilms consist of bacterial cells and extracellular polymeric substances adhering onto a variety of materials including plastics, mineral crystals, clay particles, pipes, trickling filters, and other aquatic surfaces (Donlan 2002). Perhaps one of the most unexpected biocorrosion events attributed to SRB is the biocorrosion of endodontic files (steel posts used in root canals) (Heggendorn et al. 2018). While there is a natural electrostatic repulsion between the cell and the substratum, the extracellular material produced by the bacteria enables the bacterium to adsorb onto the environmental surface (Corpe 1980). As reviewed by Li et al. (2015), biocorrosion of steel is by direct contact of the bacterial cell to the metal surface or involves a biofilm where bacteria are imbedded in the exopolymeric matrix. The mechanism of biocorrosion of metals is discussed in Chap. 8. Biofilm produced by *D. vulgaris* (ATCC 7757) on

304 stainless steel is generally patchy and may vary in thickness from a few μm to several mm (Li et al. 2015). Biofilms produced by SRB on stainless steel can function as a physical barrier and contribute to biocorrosion by reducing O_2 diffusion and concentrates H_2S and FeS, which contribute to biocorrosion of steel (Li et al. 2013).

Environmental biofilms contain metal ions and minerals from the environment (Geesey et al. 1988; Beech and Sunner 2004, 2007) as well as several different bacterial types. Natural biofilms collected from the Piquette Pn-Zn deposit (Tennyson, WI) contained sphalerite (ZnS) nanoparticles, which were formed by SRB in the biofilm (Labrenz et al. 2000). The core bacteria present in the biofilm accounting for corrosion of steel placed in Sanya and Xiamen Seawaters in China were *Desulfovibrio* spp., *Desulfotomaculum* spp., and *Desulfobacter* spp. (Li et al. 2017a, b). New SRB are being evaluated for biocorrosion, and an interesting marine SRB species was reported to use direct contact to obtain electrons from metallic iron (Dinh et al. 2004).

In biofilm-producing cultures, the extracellular polymeric substances consist of either carbohydrate polymers or proteins, and both of these hydrophobic molecules are used by SRB. The formation of a biofilm of *D. vulgaris* Hildenborough on the glass surface was attributed to the presence of proteins (Clark et al. 2007). Wild-type *D. vulgaris* did not produce extensive amounts of exopolysaccharide material. Wild-type strain and a strain with the 200-kb megaplasmid deleted were examined for biofilm development, and the quantity of biofilm in the wild-type strain was significantly greater than with the strain without the megaplasmid. Proteases added to the biofilm by wild-type strain diminished the amount of biofilm. It is unresolved that protein for biofilm formation is flagellar or flagella-like structure, but it has been proposed that flagella do not adsorb onto the surface but reduce the repulsive forces between the cell and the surface (Donlan 2002). Biofilm produced by *D. alaskensis* accounts for pitting of carbon steel, and the formation of this biofilm is attributed to proteins in the exopolymeric substance (Wikie et al. 2014). As discussed in Sect. 2.2.2.2 of Chap. 2, nanowires (nanofilaments or pili) of *D. ferrophilus* are electron conductive and contribute to biocorrosion via direct cell contact with the metal surface (Deng et al. 2018).

The role of exopolymeric substances in biocorrosion is reviewed by Beech and Sunner (2004). Extracellular polysaccharides produced by *D. desulfuricans* (NCIMB 8313) was enhanced by the presence of metal coupons in the culture medium (Beech et al. 1991), and neutral sugars present in biofilm on mild steel coupons included rhamnose, mannose, glucose, galactose, xylose, allose, and rhamnose. Extracellular polysaccharide material isolated from the culture medium of *D. desulfuricans* growing in the presence of mild steel and stainless steel had molar ratios of glucose:mannose:galactose of 1:12:0.2 and 1:5.6:0.2, respectively. Additionally, uronic acids were detected in biofilm on metal coupons but not in the extracellular culture-fluid of *D. desulfuricans.*

Progress is being made to understand the molecular control for the production of biofilm by SRB. Genes encoding for the synthesis and export of exopolymeric substances are present in the megaplasmid of *D. vulgaris* strain DePue but not in the megaplasmid of *D. vulgaris* Hildenborough (Walker et al. 2009). In addition to proteins present in the exopolymeric substance of *D. vulgaris* Hildenborough (Clark et al. 2007), the biofilm matrix of *D. vulgaris* consists of mannose, fucose, and *N*-acetylgalactosamine (Poosarla et al. 2017). For biofilm production by *D. vulgaris* Hildenborough, the DVU2956 (gene for σ^{54}-dependent regulator) must be repressed, and planktonic growth occurs when DVU2956 is expressed (Zhu et al. 2019).

9.3.4 Hydrothermal Vent Sediments

Examination of prokaryotes associated with deep-sea hydrothermal chimney structures in the Central Indian Ridge in the Indian Ocean, Okinawa Trough near Japan and Izu-Bonin Arc near Japan, revealed diverse SRP populations (Nakagawa et al. 2004). DNA extracted from structures of chimneys actively venting at >250 °C was examined using dissimilatory sulfate-reducing genes *(dsrAB)* and arranged into seven phylogenic groups. The major group of SRP clones from the Central Indian Ridge chimney was related to *Archaeoglobus,* with several clones related to *Thermodesulfobacterium*. The bacteria associated with the chimneys of the Okinawa Trough formed a novel group related to the *Thermodesulfovibrio*, while the clones from the chimney of the Izu-Bonin Arc were related to *Desulfobulbus*. Since members of these SRP clones were also detected in hydrothermal sediments in Guaymas Basin, Mexico and other hydrothermal fields along the Central Indian Rige and Izu-Bonin Arc, it appears that these SRP are distributed widely in the deep-sea hydrothermal ecosystem.

Microbial activity is high in deep-sea hydrothermal ecosystems and suggests the functioning of a sulfur cycle near active chimneys. In an interesting experiment, sulfide deposits were collected from the Endeavour Segment of the Juan de Fuca Ridge in the Northwest Pacific, and the quantity of sulfate reduced/gram sediment/day (mmole g^{-1} day^{-1}) at 90 °C was calculated. Values for samples from Needles was 2.670 μmole g^{-1} day^{-1}, Chower Hill was 1.09 μmole g^{-1} day^{-1}, and Dead Dog was 0.13 μmole g^{-1} day^{-1} (Frank et al. 2013). As suggested from research at the Southwest Indian Ridge, this bioproduction of sulfide near hydrothermal chimneys would be metabolized by sulfur-oxidizing bacteria, and this would constitute a microbial sulfur cycle (Cao et al. 2014). The SRP would be present only in actively flowing hydrothermal chimneys and are absent (or in exceeding low abundance) with inactive sulfide chimneys undergoing oxidative weathering (Li et al. 2017a, b).

9.3.5 Deep Subsurface and Mines

Using *dsrB* gene analysis, SRB present in the alkaline, cold water of the nickel Kotalajti Mine in Leppävirta, Finland, was examined. SRB were present in low numbers (4×10^3 genes ml^{-1}, represented only ~0.5% of the total bacterial population present). Bacteria with *dsrB* genes included an uncultured group of sulfate reducers, *Desulfotomaculum* and several members of *Desulfobulbus* (Bomberg et al. 2015). The dominance of *Desulfobulbus* in the SRB population may reflect the capability of this bacterium to tolerate oxygen and grow with nitrate respiration. Both nitrogen and oxygen were present in waters of the Kotalajti Mine. The SRB responsible for the production of hydrogen sulfide in this cold (~7 °C), alkaline (~pH 10) mine water is unresolved because SRB characterized as psychrophiles were undetected.

A single bacterial species has been found at a depth of 2.8 kM (1.74 miles) in the Mponeng gold mine near Johannesburg, South Africa, and it has been designated *Candidatus* (*C.*) Desulforudis audaxviator (Chivian et al. 2008). After collecting 5600 L of water from a fracture in the mine wall and evaluating the DNA present, 99.9% of the DNA was from a single species, Ca. Desulforudis audaxviator that grows at 60 °C and pH 9.3. The presence of a single microbial species in this deep mine is in contrast with other South African deep mines where about 44 different archaea types and 280 different bacteria types form the rich micro-biosphere. The genomics of this Gram-positive bacterium indicates that this bacterium is capable of autotrophic growth with $H_2 + CO_2$, uses ammonia or N_2 as nitrogen sources, and can survive desiccation by forming cysts. The abiotic source of H_2 could have been attributed to the decay of radioactive uranium, potassium, or thorium. The genome of *C.* Desulforudis audaxviator has two CRISPER (clustered regularly interspaced short palindromic repeat) sections and CAS (CRISPER-associated proteins), which could function as a defense against bacterial viruses. The DNA sequence of the CRISPER system in *C.* Desulforudis audaxviator is most closely related to that of *Thermosinus carboxydivorans*, and this archaeal organism may have been the source of these genes.

Groundwater at a 450 m depth was taken from a deep granitic rock aquifer on the island of Äspö, Sweden, and was found to have ~15^5 bacterial cells ml^{-1} (Pederson 2012). Glass slides suspended in the groundwater served as a support for a biofilm, and evaluation of planktonic and biofilm bacteria indicated that about 25% of the bacterial populations were SRB. Using 16 sRNA analysis, *D. aespoeenis* and *D. ferrophilus*-like clones were observed in the biofilms. Using groundwater from the Äspö, the rate of sulfide production was determined to be 0.48×10^{-15} mol h^{-1} per cell and 1.09×10^{-15} mol h^{-1} per cell with H_2 and acetate as electron donors, respectively (Pederson 2012). In comparison, established SRB cultures have been reported to release $0.01–2 \times 10^{-15}$ mol h^{-1} per cell (Jörgensen 1978). Evaluation of the SRB present at the Äspö site resulted in identifying several isolates as *Desulfomicrobium baculatum*, some isolates related to *D. salexigens*, and some related to *D. longreachii.* Another isolate at the 600 m depth was *D. aespoeensis* DSM 10631(Motamedi and Pedersen 1998). Due to the metabolic characteristics, some consider *Pseudodesulfovibrio aespoeensis* as a more appropriate name. A lytic bacterial virus targeting *D. aespoeensis* was isolated (Eydal and Pedersen 2009). This was not the first SRB virus isolated because the virus for *D. vulgaris* (Handley et al. 1973) and *D. salexigens* (Kamimura and Araki 1989) had been reported earlier.

Since the virus attacking *D. desulfuricans* was demonstrated to transform the host cell (Rapp and Wall 1987), the presence of SRB virus in the groundwater of the Äspö site could facilitate the transfer of genes between bacteria as well as lyse the host SRB.

Through the use of culture and molecular approaches, *Desulfotomaculum* spp. have been found in many deep geological formations. Samples (16S rRNA gene sequences) from the 800-m-deep Kantaishi mine in Japan contained bacteria that were closely related to *Dst. geothermicum*, *Desulfosporosinus lactus*, *Desulfosporosinus acidiphilus*, *Desulfosporosinus youngiae*, and *C.* Desulforudis audaxviator (Ishii et al. 2000). Phylotypes related to *Dst. putei*, *Dst. geothermicum*, *Dst. arcticum*, and *Dst. kuznetsovii* were detected in the aquifer of the Paris Basin (Basso et al. 2009). Numerous other reports of spore-forming SRB found in marine seabed are in the review of Aüllo et al. (2013), and they suggest that spore-forming SRB are found primarily in the surface of the marine floor and absent at great subsurface depths.

9.3.6 Floodplains and Estuaries

The diverse ecosystem of river floodplains of three major Dutch rivers (Meuse, Rhine, and Overijsselse Vechtare) were examined using polar lipid-derived fatty acid (PLFA), *dsrB*-based denaturing gradient gel electrophoresis (DGGE), and microarray of 16S rRNA gene-based oligonucleotide analyses (Miletto et al. 2008). Based on the microbial populations in the Dutch floodplain, the following geographical patterns for SRP in the floodplain were reported and compared to results in other publications on floodplains: (i) tidal sites were readily distinguished from the non-tidal sites, (ii) the presence of *Syntrophobacter*-related microorganism throughout the floodplains supports the suggestion that these SRB are nutritionally diverse and adaptable to the changing environment, (iii) *Desulfosarcina* were present in nontidal sites where salinity is low (<0.5%) and *Desulfosarcina*-related SRB are present in oligohaline environments, and (iv) the presence of *Desulfobacteraceae*- and *Desulfomonile*-related SRPs was influenced by salinity and a low eutrophic status of the environment.

Estuaries have a highly active ecosystem where fresh and marine waters mix. Using molecular fingerprinting, the distribution of SRP communities was examined in waters flowing from the Adour estuary on the South French Atlantic coast into the Gulf of Biscay (Colin et al. 2017). The four sampling stations and the depths at which the samples were taken included inner part of the estuary (ES)—1 and 5 m; concentrated plume (CP)—1, 5, and 24 m; diluted plume (DP)—1, 13, and 55 m; and marine station (MS)—5, 22, 240, and 800 m. Using *dsrAB* probes, *D. desulfuricans*, *Desulfomicrobium baculatum*, *D. aespoeensis*, and *Citrobacter freundii* were detected at the ES site. The detection of *Citrobacter freundii* is indeed interesting because the detection of *dsrAB* genes would imply the capability of dissimilatory sulfate reduction. While this *Citrobacter freundii* was not checked for anaerobic growth on sulfate, other reports do indicate the growth of *Citrobacter freundii* using sulfate respiration (Angeles-Ch et al. 2002; Qiu et al. 2009). The CP site contained *Citrobacter freundii* and *Desulfopila aestuari* while *D. oceani* was detected at site DP. Surface waters contained SRB that were tolerant to O_2.

A biogeochemical review of SRB and methanogens in freshwater wetlands revealed that these microorganisms have an impact on carbon cycling and greenhouse emissions (Pester et al. 2012). While the concentration of sulfate in natural wetlands and rice paddy soils is at the millimolar level, which is too low to maintain the continuous production of sulfide, there are numerous reports of appreciable levels of sulfate reduction in wetlands. Following an examination of methods employed to measure sulfate reduction in wetlands, a hidden sulfur cycle was revealed in wetlands and this rate of sulfate reduction was comparable to that reported for marine surface sediments (Pester et al. 2012). Sulfide is oxidized to S^0 with the reduction of Fe III, quinone of dissolved organic matter, or O_2. Thiosulfate could be formed by microbial oxidation of sulfide with oxygen as the electron acceptor. A chemical reaction with sulfide oxidation to elemental sulfur by FeIII would also occur in environments containing iron. Since the dissimilatory sulfate reduction process is thermodynamically more favorable than fermentation or methanogenesis, the formation of methane would not be favored and the amount of methane transmitted to the atmosphere would be minimal. Thus, SRB would be important in the decomposition of organic matter in peat and wetland environments. Using quantitative *dsrAB* gene analysis, Pester et al. (2012) determined that several members of SRB were uncultivated and additional testing indicated that they had novel phylogenetic lineages. This is another environment where new strains of uncultivated SRB are present and their contribution remains to be established.

9.3.7 Low Nutrient Environment

Nutrient limitation will lead to a decline in growth rate and eventually result in cell death of cells. Cooperation between SRB cells and other anaerobes in the environment has resulted in the formation of a consortium consisting of numerous bacterial species or a syntrophic culture with a single species in addition to the SRB species. Interspecies hydrogen transfer is one of these synergisms and was first reported with a *Desulfovibrio* sp. that was able to grow on lactate and ethanol in the absence of sulfate with a methanogen acquiring H_2 produced by the SRB (Bryant et al. 1977). Also, with interspecies H_2 transfer, *Methanosaeta concilii* (an acetoclastic methanogen) produces H_2 as a byproduct, which is used as the electron donor for growth of *D. vulgaris* in a sulfate medium (Ozuolmez et al. 2015). It appears that the syntrophic activity of *Desulfovibrio* is flexible and changes with the composition of the methanogenic environment. That is, the electron carriers and cell energetics for *D. alaskensis* G20 are flexible and adjust as the shuttle compounds (hydrogen and formate) are produced for syntrophic growth with *Methanospirillum hungatei* or *Methanococcus maripaludis* (Meyer et al. 2013). Similarly, regulation of transcription was observed when *D. vulgaris* growing syntrophicaly with *Methanosarcina barkeri* is shifted to sulfidogenic metabolism (Plugge et al. 2010). Another example of syntrophic growth involves "*C.* Contubernalis alkalaceticum," which oxidizes acetate, ethanol, propanol, and a few other small-molecular-weight compounds only when grown with a H_2-utilizing sulfate reducer such as *Desulfonatronum cooperativum* (Zhilina et al. 2005a).

The ability to grow chemoautotrophically with H_2 and CO_2 enables some SRB to colonize anaerobic environments with extremely low or no organic compounds. Acetate is commonly found in anaerobic zones where organic decomposition is occurring and some of the SRB that are incapable of chemolithotrophic growth can grow on H_2 and CO_2 provided acetate is used as the carbon source. Growth of bacteria in deep marine sediments requires energy to be conserved whenever possible and perhaps SRB are capable of using uptake transport systems to acquire amino acids, sugars, and vitamins released by other bacteria and thereby saving biosynthetic energy. Lastly, for a bacterial cell to avoid bacteriophage attack, an adaptive immune system has evolved, and CRISPER/Cas systems present in sulfate-reducing bacteria are being examined (Morais-Silva et al. 2014).

9.4 Summary and Perspective

SRB are found in any environment that is anaerobic, microphilic, or transiently exposed to atmospheric oxygen. Sulfate-reducing archaea do not appear to be as highly dispersed in the environment as SRB; however, the search for these archaeal microorganisms has not been as active as for SRB. The SRP have adapted over time to basically every bioenvironment on earth through the acquisition of genes from other bacteria and archaea in the environment. This adaptability is evident as the SRB thrive in extreme physical environments and to employ diverse electron donors and electron acceptors other than sulfate. As methods develop for characterization of uncultured strains of SRP, there will be additional strains of microorganisms that are characteristic of dissimilatory sulfate reduction. Examination of syntrophic strains of SRB will provide additional insight into the diversity of these sulfate-reducing microorganisms.

References

Abildgaard L, Nielsen MB, Kjeldsen KU, Ingvorsen K (2006) *Desulfovibrio alkalitolerans* sp. nov., a novel alkalitolerant, sulphate-reducing bacterium isolated from district heating water. Int J Syst Evol Microbiol 56:1019–1024

Alain K, Postec A, Grinsard E, Lesongeur F, Prieur D, Godfroy A (2010) *Thermodesulfatator atlanticus* sp. nov., a thermophilic, chemolithoautotrophic, sulfate-reducing bacterium isolated from a Mid-Atlantic Ridge hydrothermal vent. Int J Syst Evol Microbiol 60:33–38

Alazard D, Dukan S, Urios A, Verhe F, Bouabida N, Morel F, Thomas P, Garcia J-L, Ollivier B (2003) *Desulfovibrio hydrothermalis* sp. nov., a novel sulfate-reducing bacterium isolated from hydrothermal vents. Int J Syst Evol Microbiol 53:173–178

Alazard D, Joseoh M, Battaglia-Brunet F, Cayol JL, Ollivier B (2010) *Desulfosporosinus acidiphilus* sp. nov.: a moderately acidophilic sulfate-reducing bacterium isolated from acid mining drainage sediments: new taxa: Firmicutes (Class Clostridia, Order Clostridiales, family *Peptococcaceae*). Extremophiles 14:305–312

Aüllo T, Ranchou-Peyruse A, Ollivier B, Magot M (2013) *Desulfotomaculum* spp. and related gram-positive sulfate reducing bacteria in deep subsurface environments. Front Microbiol 4: article 362. https://doi.org/10.3389/fmicb.2013.00362

Amrani A, Bergon A, Holota H, Tamburini C, Garel M, Ollivier B et al (2014) Transcriptomics reveal several gene expression patterns in the piezophile *Desulfovibrio hydrothermalis* in response to hydrostatic pressure. PLoS One 9:e106831. https://doi.org/10.1371/journal.pone.0106831

Angeles-Ch C, Mora-Mendoza JL, Garcia-Esquivel R, Padilla-Viveros AA, Perez-Campos R, Flores O, Martinez L (2002) Microbiologically influenced corrosion by *Citrobacter* in sour gas pipelines. Mater Perform 41:50–55

Audiffrin C, Cayol J-L, Joulian C, Casalot L, Thomas P, Garcia J-L, Ollivier B (2003) *Desulfonauticus submarinus* gen, nov., sp. nov., a novel sulfate-reducing bacterium isolated from a deep-sea hydrothermal vent. Int J Syst Evol Microbiol 53:1585–1590

Baena S, Perdomo N, Carvajal C, Diaz C, Patel BKC (2011) *Desulfosoma caldarium* gen. nov., a thermophilic sulfate-reducing bacterium from a terrestrial hot spring. Int J Syst Evol Microbiol 61:732–736

Bale SJ, Goodman K, Rochelle PA, Marchesi JR, Fry JC, Weightman AJ, Parkes RJ (1997) *Desulfovibrio profundus* sp. nov., a novel barophilic sulfate-reducing bacterium from deep sediment layers in the Japan Sea. Int J Syst Bacteriol 47:515–521

Baldermann A, Deditius A, Dietzel M, Fichtner F, Fischer C, Hippler D, six co-authors. (2015) The role of bacterial sulfate reduction during dolomite precipitation: implications from Upper Jurassic platform carbonates. Chem Geol 412:1–14

Baronofsky JJ, Schreurs WJA, Kashket ER (1984) Uncoupling by acetic acid limits growth of and acetogenesis by *Clostridium thermoaceticum*. Appl Env Microbiol 48:1134–1139

Basso O, Lascourreges J-F, Le Borgne F, Le Goff C, Magot M (2009) Characterization by culture and molecular analysis of the microbial diversity of a deep subsurface gas storage aquifer. Res Microbiol 160:107–116

Baumgartner LK, Reid RP, Dupraz C, Decho AW, Buckley DH, Spear JR, Przekop KM, Visscher PT (2006) Sulfate reducing bacteria in microbial mats: changing paradigms, new discoveries. Sed Geol 185:131–145

Bayraktarov E, Price RE, Ferdelman TG, Finster K (2013) The pH and pCO_2 dependence of sulfate reduction in shallow-sea hydrothermal CO_2 – venting sediments (Milos Island, Greece). Front Microbiol 2013(4):111. https://doi.org/10.3389/fmicb.2013.00111

Beech IB, Gaylarde CC, Smith JJ, Geesey GG (1991) Extracellular polysaccharides from *Desulfovibrio desulfuricans* and *Pseudomonas fluorescens* in the presence of mild and stainless steel. Appl Microbiol Biotechnol 35:65–71

Beech I, Sunner J (2004) Biocorrosion: towards understanding interactions between biofilms and metals. Curr Opin Biotechnol 15:181–186

Beech IB, Sunner JA (2007) Sulphate-reducing bacteria and their role in corrosion of ferrous metal. In: Barton LL, Hamilton WA (eds) Sulphate-reducing bacteria - environmental and engineered systems. Cambridge University Press, Cambridge, UK, pp 459–482

Belyakova EV, Rozanova EP, Borzenkov IA, Tourova TP, Pusheva MA, Lysenko AM, Kolganova TV (2006) The new facultatively chemolithoautotrophic, moderately halophilic, sulfate-reducing bacterium *Desulfovermiculus halophilus* gen. nov., sp. nov., isolated from an oil field. Microbiology 75:161–171

Ben Ali Gam Z, Oueslati R, Abdelkafi S, Casalot L, Tholozan JL, Labat M (2009) *Desulfovibrio tunisiensis* sp. nov., a novel weakly halotolerant, sulfate-reducing bacterium isolated from exhaust water of a Tunisian oil refinery. Int J Syst Evol Microbiol 59:1059–1063

Ben Ali Gam Z, Thioye A, Cayol J-L, Joseph M, Fauque G, Labat M (2018) Characterization of *Desulfovibrio salinus* sp. nov., a slightly halophilic sulfate-reducing bacterium isolated from a saline lake in Tunisia. Int J Syst Evol Microbiol 68:715–720. https://doi.org/10.1099/ijsem.0.002567. Epub 2018 Jan 12

Ben Dhia Thabet O, Fardeau M-L, Suarez-Nuñez C, Hamdi M, Thomas P, Ollivier B, Alazard D (2007) *Desulfovibrio marinus* sp. nov., a moderately halophilic sulfate-reducing bacterium isolated from marine sediments in Tunisia. Int J Syst Evol Microbiol 57:2167–2170

Blum JS, Kulp TR, Han S, Lanoil B, Saltikov CW, Stolz JF, Miller LG, Oremland RS (2012) *Desulfohalophilus alkaliarsenatis* gen. nov., sp. nov., an extremely halophilic sulfate- and arsenate-respiring bacterium from Searles Lake, California. Extremophiles 16:727–742

Boetius A, Ravenschlag K, Schubert CJ, Rickert D, Widdel F, Gieseke A, four co-authors. (2000) A marine microbial consortium apparently mediating anaerobic oxidation of methane. Nature 407:623–626

Bomberg M, Arnold M, Kinnunen P (2015) Characteristics of the sulphate-reducing community in the alkaline and constantly cold water of the closed Kotalahti Mine. Fortschr Mineral 5:452–472

Bontognali TRR, McKenzie JA, Warthmann RJ, Vasconcelos C (2014) Microbially influenced formation of Mg-calcite and Ca-dolomite in the presence of exopolymeric substances produced by sulphate-reducing bacteria. Terra Nova 26:72–77

Bontognali TRR, Vasconcelos C, Warthmann RJ, Bernasconi SM, Dupraz C, Strohmenger CJ, McKenzie JA (2010) Dolomite formation within microbial mats in the costal sabkha of Abu Dhabi (United Arab Emirates). Sedimentology 57:824–844

Braissant O, Decho AW, Dupraz C, Glunk C, Przekop MK, Visscher PT (2007) Exopolymeric substances of sulfate-reducing bacteria interactions with calcium at alkaline pH and implication for formation of carbonate minerals. Geobiology 5:401–411

Braissant O, Decho AW, Przekop KM, Gallagher KL, Glunk C, Dupraz C, Visscher PT (2009) Characteristics and turnover of exopolymeric substances in a hypersaline microbial mat. FEMS Microbial Ecol. 67:293–307

Brandt KK, Patel BK, Ingvorsen K (1999) *Desulfocella halophila* gen. nov., sp. nov., a halophilic, fatty-acid-oxidizing, sulfate-reducing bacterium isolated from sediments of the Great Salt Lake. Int J Syst Bacteriol 49:193–200

Breeder J, Torsvik T, Lein T (1995) *Thermodesulforhabdus norvegicus* gen. nov., a novel thermophilic sulfate-reducing bacterium from oil field water. Arch Microbiol 164:331–336

Bryant MP, Campbell LL, Reddy CA, Crabill MR (1977) Growth of *Desulfovibrio* in lactate or ethanol media low in sulfate in association with H_2-utilizing methanogenic bacteria. Appl Environ Microbiol 33:1162–1169

Burggraf S, Jannasch HW, Nicolaus B, Stetter KO (1990) *Archaeoglobus profundus* sp. nov., represents a new species within the sulfate-reducing archaebacteria. Syst Appl Microbiol 13:24–28

Canfield DE, Des Marais DJ (1991) Aerobic sulfate reduction in microbial mats. Science 251: 1471–1473

Caumette P, Cohen Y, Matheron R (1991) Isolation and characterization of *Desulfovibrio halophilus* sp. nov., a halophilic sulfate-reducing bacterium isolated from Solar Lake (Sinai). Syst Appl Microbiol 14:33–38

Campbell LL, Postgate JR (1965) Classification of the spore-forming sulfate-reducing bacteria. Bacteriol Rev 29:359–363

Cao J, Gayet N, Zeng X, Shao Z, Jebbar M, Alain K (2016a) *Pseudodesulfovibrio indicus* gen. nov., sp. nov., a piezophilic sulfate-reducing bacterium from the Indian Ocean and reclassification of four species of the genus *Desulfovibrio*. Int J Syst Evol Microbiol 66:3904–3911

Cao J, Maignien L, Shao Z, Alain K, Jebbar M (2016b) Genome sequence of the piezophilic, mesophilic sulfate-reducing bacterium *Desulfovibrio indicus* J2^T. Genome Announc 4(2): e00214–e00216. https://doi.org/10.1128/genomeA.00214-16
Cao H, Wang Y, Lee OO, Zeng X, Shao Z, Qiana P-Y (2014) Microbial sulfur cycle in two hydrothermal chimneys on the Southwest Indian Ridge. mBio 5(1):e00980-13. https://doi.org/10.1128/mBio.00980-13
Castro HF, Williams NH, Ogram A (2000) Phylogeny of sulfate-reducing bacteria. FEMS Microbiol Ecol 31:1–9
Cha IT, Roh SW, Kim SJ, Hong HJ, Lee HW, Lim WT, Rhee SK (2013) *Desulfotomaculum tongense* sp. nov., a moderately thermophilic sulfate-reducing bacterium isolated from a hydrothermal vent sediment collected from the Tofua Arc in the Tonga Trench. Antonie Van Leeuwenhoek 104:1185–1192
Chivian D, Broide EL, Alm EJ, Culley DE, Dehal PS, DeSantis TZ et al (2008) Environmental genomics reveals a single species ecosystem deep within the earth. Science 322:275–278
Church CD, Wilkin R, Alpers CN, Rye RO, McCleskey RB (2007) Microbial sulfate reduction and metal attenuation in pH 4 acid mine water. Geochem Trans 8:10. https://doi.org/10.1186/1467-4866-8-10
Clark ME, Edelmann RE, Duley ML, Wall JD, Fields MW (2007) Biofilm formation in *Desulfovibrio vulgaris* Hildenborough is dependent upon protein filaments. Environ Microbiol 9:2844–2854
Colin Y, Goñi-Urriza M, Gassie C, Carlier E, Monperrus M, Guyoneaud R (2017) Distribution of sulfate-reducing communities from estuarine to marine bay waters. Microbial Ecol 73:39–49
Corpe WA (1980) Microbial surface components involved in adsorption of microorganisms onto surfaces. In: Bitton G, Marshall KC (eds) Adsorption of microorganisms to surfaces. Wiley, New York, pp 105–144
Daumas S, Cord-Ruwisch R, Garcia JL (1988) *Desulfotomaculum geothermicum* sp. nov., a thermophilic, fatty acid-degrading, sulfate-reducing bacterium isolated with H_2 from geothermal ground water. Antonie Van Leeuwenhoek 54:165–178
Deng S, Dong H, Lv G, Jiang H, Yu B, Bishop ME (2010) Microbial dolomite precipitation using sulfate-reducing and halophilic bacteria: results from Qinghai Lake, Tibetan plateau. NW China Chem Geol 278:151–159
Deng X, Dohmae N, Nealson KH, Hashimoto K, Okamoto A (2018) Multi-heme cytochromes provide a pathway for survival in energy-limited environments. Sci Adv 2018(4):eaao5682
Dillon JG, Fishbain S, Miller SR, Bebout BM, Habicht KS, Webb SM, Stahl DA (2007) High rates of sulfate reduction in a low-sulfate hot spring microbial mat are driven by a low level of diversity of sulfate-reducing microorganisms. Appl Environ Microbiol 73:5218–5226
Dinh HT, Kuever J, Mussmann M, Hassel AW, Stratmann M, Widdel F (2004) Iron corrosion by novel anaerobic microorganisms. Nature 427:829–832
Donlan RM (2002) Biofilms: microbial life on surfaces. Emerg Infect Dis 8:881–890
Dowling NJE, Widdel F, White DC (1986) Phospholipid ester-linked fatty acid biomarkers of acetate-oxidizing sulphate reducers and other sulphide-forming bacteria. J Gen Microbiol 132: 1815–1825
Dupraz D, Visscher PT (2005) Microbial lithification in marine stromatolites and hypersaline mats. Trends Microbiol 13:429–438
Dupraz CD, Visscher PT, Baumgartner LK, Reid RP (2004) Microbe–mineral interactions: early carbonate precipitation in a hypersaline lake (Eleuthera Island, Bahamas). Sedimentology 51: 745–765
El Houari A, Ranchou-Peyruse M, Ranchou-Peyruse A, Dakdaki A, Guignard M, Idouhammou L, four co-authors. (2017) *Desulfobulbus oligotrophicus* sp. nov., a sulfate-reducing and propionate-oxidizing bacterium isolated from a municipal anaerobic sewage sludge digester. Int J Syst Evol Microbiol 67:275–281
Eydal HSC, Pedersen K (2009) Use of an ATP assay to determine viable microbial biomass in Fennoscandian Shield groundwater from depths of 3-1000 m. J Microbiol Meth. 70:363–373

Fardeau M-L, Ollivier B, Patel BKC, Dwived P, Ragot M, Garcia J-L (1995) Isolation and characterization of a thermophilic sulfate-reducing bacterium, *Desuvotomaculum themosapovorans* sp. nov. Int J Syst Bacteriol 95:218–221

Fauque GD (1995) Ecology of sulfate-reducing bacteria. In: Barton LL (ed) Sulfate reducing bacteria. Plenum Press, New York, pp 217–241

Feio MJ, Beech IB, Carepo M, Lopes JM, Cheung CWS, Franco R, four co-authors. (1998) Isolation and characterization of a novel sulphate-reducing bacterium of the *Desulfovibrio* genus. Anaerobe 4:117–130

Fortin D, Davis B, Beveridge TJ (1996) Role of *Thiobacillus* and sulfate-reducing bacteria in iron biocycling in oxic and acidic mine tailings. FEMS Microb Ecol 21:11–24

Foti M, Sorokin DY, Lomans B, Mussman M, Zacharova EE, Pimenov NV, Kuenen JG, Muyzer G (2007) Diversity, activity, and abundance of sulfate-reducing bacteria in saline and hypersaline soda lakes. Appl Environ Microbiol. 73:2093–2100

Frank KL, Rogers DR, Olins HC, Vidoudez C, Girguis PR (2013) Characterizing the distribution and rates of microbial sulfate reduction at Middle Valley hydrothermal vents. ISME J 7:1391–1401

Frolov EN, Kublanov IV, Toshchakov SV, Samarov NI, Novikov AA, Lebedinsky AV, Bonch-Osmolovskaya EA, Chernyh NA (2017) *Thermodesulfobium acidiphilum* sp. nov., a thermoacidophilic, sulfate-reducing, chemoautotrophic bacterium from a thermal site. Int J Syst Evol Microbiol 67:1482–1485

Fry NK, Fredrickson JK, Fishbain S, Wagner M, Stahl DA (1997) Population structure of microbial communities associated with two deep, anaerobic alkaline aquifers. Appl Environ Microbioll 63:1498–1504

Gallagher KL, Dupaz C, Visscher PT (2014) Two opposing effects of sulfate reduction on carbonate precipitation in normal marine, hypersaline, and alkaline environments. Geology Forum January 2014 Geological Society of America

Geesey GG, Jang L, Jolley JG, Hankins MR, Iwaoka T, Griffiths PR (1988) Binding of metal ions by extracellular polymers of biofilm bacteria. Water Sci Technol 20:161–165

Goeres DM, Nielsen PH, Smidt HD, Frølund B (1998) The effect of alkaline pH conditions on a sulphate-reducing consortium from a Danish district heating plant. Biofouling 12:273–286

Goorissen HP, Boschker HT, Stams AJ, Hansen TA (2003) Isolation of thermophilic *Desulfotomaculum* strains with methanol and sulfite from solfataric mud pools, and characterization of *Desulfotomaculum solfataricum* sp. nov. Int J Syst Evol Microbiol 53:1223–1229

Gorlenko VM, Namsaraev BB, Kulyrova AV, Zavarzina DG, Zhilina YN (1999) Activity of sulfate-reducing bacteria in bottom sediments of soda lakes of the southeastern Transbaikal region. Microbiology 68:580–585

Grimm F, Franz B, Dahl C (2011) Regulation of dissimilatory sulfur oxidation in the purple sulfur bacterium *Allochromatium vinosum*. Front Microbiol 2:51–62

Grossman JP, Postgate JR (1955) The metabolism of malate and certain other compounds by *Desulphovibrio desulphuricans*. J Gen Microbiol 12:429–445

Hamilton WA, Lee W (1995) Biocorrosion. In: Barton LL (ed) Sulfate- reducing bacteria. Plenum Press, New York, pp 243–264

Hamilton-Brehm SD, Gibson RA, Green SJ, Hopmans EC, Schouten S, van der Meer MT, three co-authors. (2013) *Thermodesulfobacterium geofontis* sp. nov., a hyperthermophilic, sulfate-reducing bacterium isolated from Obsidian Pool, Yellowstone National Park. Extremophiles 17: 251–263

Handley J, Adams V, Akagi JM (1973) Morphology of bacteriophage-like particles from *Desulfovibrio vulgaris*. J Bacteriol 115:1205–1207

Hao OJ, Chen JM, Huang L, Buglass RL (1996) Sulfate-reducing bacteria. Crit Rev Environ Sci Technol 26:155–187

Haouari O, Fardeau ML, Cayol JL, Casiot C, Elbaz-Poulichet F, Hamdi M, Joseph M, Ollivier B (2008a) *Desulfotomaculum hydrothermale* sp. nov., a thermophilic sulfate-reducing bacterium isolated from a terrestrial Tunisian hot spring. Int J Syst Evol Microbiol 58:2529–2535. https://doi.org/10.1099/ijs.0.65339-0

Haouari O, Fardeau M-L, Cayol J-L, Fauque G, Casiot C, Elbaz-Poulichet F, Hamdi M, Ollivier B (2008b) *Thermodesulfovibrio hydrogeniphilus* sp. nov., a new thermophilic sulphate-reducing bacterium isolated from a Tunisian hot spring. Syst Appl Microbiol 31:38–42

Heggendorn FL, Fraga AGM, Ferreira DC, Gonçalves LS, Lione VOF, Lutterbach MTS (2018) Sulfate-reducing bacteria: biofilm formation and corrosive activity in endodontic files. Int J Dent 2018:8303450. https://doi.org/10.1155/2018/8303450. PMID: 29861730; PMCID: PMC5976933

Henry HA, Devereux R, Maki JS, Gilmour CC, Woese CR, Mandelco L, three co-authors. (1994) Characterization of a new thermophilic sulfate-reducing bacterium. Arch Microbiol 161:62–69

Heukelekian H, Heller A (1940) Relation between food concentration and surface for bacterial growth. J Bacteriol 40:547–558

Isaksen MF, Teske A (1996) *Desulforhopalus vacuolatus* gen. nov., sp. nov., a new moderately psychrophilic sulfate-reducing bacterium with gas vacuoles isolated from a temperate estuary. Arch Microbiol 166:160–168

Ishii K, Takii S, Fukunaga S, Aoki K (2000) Characterization of denaturing gradient gel electrophoresis of bacterial communities in deep groundwater at the Kamaishi mine. Japan J Gen Appl Microbiol 46:85–93

Ito T, Okabe S, Satoh H, Watanabe Y (2002) Successional development of sulfate-reducing bacterial populations and their activities in a wastewater biofilm growing under microaerophilic conditions. Appl Environ Microbiol 68:1392–1402

Itoh T, Suzuki K-i, Nakasel T (1998) *Thermocladium modestius* gen. nov., sp. nov., a new genus of rod-shaped, extremely thermophilic crenarchaeote. Int J Syst Evol Microbiol 48:879–887

Itoh T, Suzuki K-i, Sanchez PC, Nakasel T (1999) *Caldivirga maquilingensis* gen. nov., sp. nov., a new genus of rod-shaped crenarchaeote isolated from a hot spring in the Philippines. J Syst Bacteriol 49:1157–1163

Jabari L, Gannoun H, Cayol J-L, Hamdi M, Ollivier B, Fauque G, Fardeau M-L (2013) *Desulfotomaculum peckii* sp. nov., a moderately thermophilic member of the genus *Desulfotomaculum*, isolated from an upflow anaerobic filter treating abattoir wastewaters. Int J Syst Evol Microbiol 63:2082–2087

Jakobsen TF, Kjeldsen KU, Ingvorsen K (2006) *Desulfohalobium utahense* sp. nov., a moderately halophilic, sulfate-reducing bacterium isolated from Great Salt Lake. Int J Syst Evol Microbiol 56:2063–2069

Jeanthon C, L'Haridon S, Cueff V, Banta A, Reysenbach A-L, Prieur D (2002) *Thermodesulfobacterium hydrogeniphilum* sp. nov., a thermophilic, chemolithoautotrophic, sulfate-reducing bacterium isolated from a deep-sea hydrothermal vent at Guaymas Basin, and emendation of the genus *Thermodesulfobacterium*. Int J Syst Evol Microbiol 52:765–772

Jonkers HM, Ludwig R, De Wit R, Pringault O, Muyzer G, Niemann H, Finke N, De Beer D (2003) Structural and functional analysis of a microbial mat ecosystem from a unique permanent hypersaline inland lake: dLa Salad de ChipranaQ (NE Spain). FEMS Microbiol Ecol 44:175–189

Jörgensen BB (1978) A comparison of methods for the quantification of bacterial sulphate reduction in costal marine sediments. Geomicrobiol J 1:46–64

Jørgensen BB (1977) Bacterial sulfate reduction within reduced microniches of oxidized marine sediments. Mar Biol 41:7–17

Jørgensen BB, Bak F (1991) Pathways and microbiology of thiosulfate transformation and sulfate reduction in a marine sediment (Kattegat, Denmark). Appl Environ Microbiol 57:847–856

Kaksonen A, Spring S, Schumann P, Kroppenstedt RM, Puhakka JA (2006) Desulfotomaculum *thermosubterraneum sp. nov., a* thermophilic *sulfate-reducer isolated from an underground mine located in a geothermally active area.* Int J Syst Evol Microbiol 56:2603–2608

Kaksonen AH, Spring S, Schumann P, Kroppenstedt RM, Puhakka JA (2007) *Desulfovirgula thermocuniculi* gen. nov., sp. nov., a thermophilic sulfate-reducer isolated from a geothermal underground mine in Japan. Int J Syst Evol Microbiol 57:98–102

Kaksonen AH, Spring S, Schumann P, Kroppenstedt RM, Puhakka JA (2008) *Desulfotomaculum alcoholivorax* sp. nov., a moderately thermophilic, spore-forming, sulfate-reducer isolated from a fluidized-bed reactor treating acidic metal- and sulfate-containing wastewater. Int J Syst Evol Microbiol 58:833–838

Kamimura K, Araki M (1989) Isolation and characterization of a bacteriophage lytic for *Desulfovibrio salexigens*, a salt-requiring sulfate-reducing bacterium. Appl Environ Microbiol 55:645–648

Kimura S, Hallberg KB, Johnson DB (2006) Sulfidogenesis in low pH(3.8 – 4.2) media by a mixed population of acidophilic bacteria. Biodegradation 17:57–65. https://doi.org/10.1007/s10532-005-3050-4

Karnauchow TM, Koval SF, Jarrell KF (1992) Isolation and characterization of three thermophilic anaerobes from a St. Lucia Hot Spring. Syst Appl Microbiol 15:296–310

Kasten S, Jørgensen BB (2006) Sulfate reductions in marine sediments. In: Zabel M, Schulz H (eds) Marine Geochemistry. Springer, Berlin, pp 271–309

Kato C, Horikoshi K (2009) Characteristics of deep-sea environments and biodiversity of piezophilic organisms. In: Gerday C, Glansdoroff N (eds) Extremophiles–Vol III Encyclopedia of Life Support Systems (EOLSS). Eolss Publishers Co, Ltd, Oxford, UK, pp 174–198

Khelaifia S, Fardeau M-L, Pradel N, Aussignargues C, Garel M, Tamburini C, four co-authors. (2011) *Desulfovibrio piezophilus* sp. nov., a piezophilic, sulfate-reducing bacterium isolated from wood falls in the Mediterranean Sea. Int J Syst Evol Microbiol 61:2706–1271. https://doi.org/10.1099/ijs.0.028670-0

Knoblauch C, Jørgensen BB, Harder J (1999a) Community size and metabolic rates of psychrophilic sulfate-reducing bacteria in Artic marine sediments. Appl Environ Microbiol 65:4230–4233

Knoblauch C, Sahm K, Jørgensen BB (1999b) Psychrophilic sulfate-reducing bacteria isolated from permanently cold Arctic marine sediments: description of *Desulfofrigus oceanense* gen. nov., sp. nov., *Desulfofrigus fragile* sp. nov., *Desulfofaba gelida* gen. nov., sp. nov., *Desulfotalea psychrophila* gen. nov., sp. nov. and *Desulfotalea arctica* sp. nov. Int J Syst Bacteriol 49:1631–1643

Könneke M, Kuever J, Galushko A, Jørgensen BB (2013) *Desulfoconvexum algidum* gen. nov., sp. nov., a psychrophilic sulfate-reducing bacterium isolated from a permanently cold marine sediment. Int J Syst Bacteriol 63:959–964

Koschorreck M (2008) Microbial sulphate reduction at a low pH. FEMS Microbiol Ecol 64:329–342

Koschorreck M, Wendt-Potthoff K, Geller W (2003) Microbial sulfate reduction at low pH in sediments of an acidic lake in Argentina. Environ Sci Technol 37:1159–1162

Krekeler D, Sigalevich P, Teske A, Cypionka H, Cohen Y (1997) A sulfate-reducing bacterium from the oxic layer of a microbial mat from Solar Lake (Sinai), *Desulfovibrio oxyclinae* sp. nov. Arch Microbiol 167:369–375

Kühl M, Jørgensen BB (1992) Microsensor measurements of sulfate reduction and sulfide oxidation in compact microbial communities of aerobic biofilms. Appl Environ Microbiol 58:1164–1174

Labrenz M, Druschel GK, Thomsen-Ebert T, Gilbert B, Welch SA, Kemner KM et al (2000) Formation of sphalerite (ZnS) deposits in natural biofilms of sulfate-reducing bacteria. Science 290:1744–1747

Levitus S, Boyer T (1994) World Ocean Atlas, volume 4: temperature. US Department of Commerce, Washington, DC

Li H, Xu D, Li Y, Feng H, Liu Z, Li X, Gu T, Yang K (2015) Extracellular electron transfer is a bottleneck in the microbiologically influenced corrosion of C1018 carbon steel by the biofilm of sulfate-reducing bacterium *Desulfovibrio vulgaris*. PLoS One 10(8):e0136183. https://doi.org/10.1371/journal.pone.0136183

Li J, Cui J, Yang Q, Cui G, Wei B, Wu Z, Wang Y, Zhou H (2017a) Oxidative weathering and microbial diversity of an inactive seafloor hydrothermal sulfide chimney. Front Microbiol 2017(8):1378. https://doi.org/10.3389/fmicb.2017.01378

Li J, Purdy KJ, Takii S, Hayashi H (1999) Seasonal changes in ribosomal RNA of sulfate-reducing bacteria and sulfate reducing activity in freshwater lake sediment. FEMS Microbiol Ecol 28:31–39

Li X, Duan J, Xiao H, Li Y, Liu H, Guan F, Zhai X (2017b) Analysis of bacterial community composition of corroded steel immersed in Sanya and Xiamen Seawaters in China via method of illumina miseq sequencing. Front Microbiol 2017(8):1737. https://doi.org/10.3389/fmicb.2017.01737

Li K, Whitfield M, Van Vliet KJ (2013) Beating the bugs: roles of microbial biofilms in corrosion. Corros Rev 321:3–6

Lien T, Madsen M, Steen IH, Gjerdevik K (1998) *Desulfobulbus rhabdoformis sp.* nov., a sulfate reducer from a water-oil separation system. Int J Syst Bacteriol 48:469–474

Liu Y, Karnauchow TM, Jarrell KF, Balkwill DL, Drake GR, Ringelberg D, Clarno R, Boone DR (1997) Description of two new thermophilic *Desulfotomaculum* spp., *Desulfotomaculum putei* sp. nov., from a deep terrestrial subsurface, and *Desulfotomaculum luciae* sp. nov., from a hot spring. Int J Syst Bacteriol 47:615–621

Love CA, Patel BKC, Nichols PD, Stackebrandt E (1993) *Desulfotomaculum australicum*, sp. nov., a thermophilic sulfate-reducing bacterium isolated from the Great Artesian Basin of Australia. Syst Appl Microbiol 16:244–251

Lovley DR, Phillips EJP (1994) Novel process for anaerobic sulfate production from elemental sulfur by sulfate-reducing bacteria. Appl Environ Microbiol 60:2394–2399

Lübbe YJ, Youn HS, Timkovich R, Dahl C (2006) Siro(haem)amide in *Allochromatium vinosum* and relevance of DsrL and DsrN, a homolog of cobyrinic acid a,c-diamide synthase, for Sulphur oxidation. FEMS Microbiol Lett 261:194–202

Macy JM, Santini JM, Pauling BV, O'Neil AH, Sli LI (2000) Two new arsenate/sulfate-reducing bacteria: mechanisms of arsenate reduction. Arch Microbiol 173:49–57

Magot M, Basso O, Tardy-Jacquenod C, Caumette P (2004) *Desulfovibrio bastinii* sp. nov. and *Desulfovibrio gracilis* sp. nov., moderately halophilic, sulfate-reducing bacteria isolated from deep subsurface oilfield water. Int J Syst Evol Microbiol 54:1693–1697

Meyer B, Kuehl JV, Deutschbauer AM, Arkin AP, Stahl DA (2013) Flexibility of syntrophic enzyme systems in *Desulfovibrio* species ensures their adaption capacity to environmental changes. J Bacteriol 195:4900–4914

Miletto M, Loy A, Antheunisse AM, Loeb R, Bodelier PLE, Laanbroek HJ (2008) Biogeography of sulfate-reducing prokaryotes in river foodplains. FEMS Microbiol Ecol 64:395–406

Min H, Zinder SH (1990) Isolation and characterization of a thermophilic sulfate-reducing bacterium *Desulfotomaculum thermoacetoxidans* sp. nov. Arch Microbiol 153:399–404

Moosa S, Harrison S (2006) Product inhibition by sulphide species on biological sulphate reduction for the treatment of acid mine drainage. Hydrometal 83:214–222

Morais-Silva FO, Rezende AM, Pimentel C, Santos CI, Clemente C, Varela-Raposo A, eight co-authors. (2014) Genome sequence of the model sulfate reducer *Desulfovibrio gigas*: a comparative analysis within the *Desulfovibrio* genus. Microbiol Open 3:513–530. https://doi.org/10.1002/mbo3.184. Epub 2014 Jul 23

Mori K, Kim H, Kakegawa T, Hanada S (2003) A novel lineage of sulfate-reducing microorganisms: *Thermodesulfobiaceae* fam. nov., *Thermodesulfobium narugense*, gen. nov., sp. nov., a new thermophilic isolate from a hot spring. Extremophiles 7:283–290

Mori K, Maruyama A, Urabe T, Suzuki K, Hanada S (2008) *Archaeoglobus infectus* sp. nov., a novel thermophilic, chemolithoheterotrophic archaeon isolated from a deep-sea rock collected at Suiyo Seamount, Izu-Bonin Arc, western Pacific Ocean. Int J Syst Evol Microbiol 58:810–816

Mormontoy J, Hurtado JE (2013) Hydrogen sulphide production at alkaline, neutral and acid pH by a bacterial consortium isolated from Peruvian mine tailing and wetland. Adv Mater Res 825: 384–387

Motamedi M, Pedersen K (1998) *Desulfovibrio aespoeensis* sp. nov., a mesophilic sulfate-reducing bacterium from deep groundwater at Aspö hard rock laboratory, Sweden. Int J Syst Bacteriol 48: 311–315

Moussard H, L'Haridon, Tindall BJ, Banta A, Schumann P, Stackebrandt E, Reysenbach A-L, Jeanthon C (2004) *Thermodesulfatator indicus* gen. nov., sp. nov., a novel thermophilic chemolithoautotrophic sulfate-reducing bacterium isolated from the Central Indian Ridge. Int J Syst Evol Microbiol 54:227–233

Muyzer G, Stams AJM (2008) The ecology and biotechnology of sulphate-reducing bacteria. Nat Rev Microbiol 6:441–454

Nakagawa T, Nakagawa S, Inagaki F, Takai K, Horikoshi K (2004) Phylogenic diversity of sulfate-reducing prokaryotes in active deep-sea hydrothermal vent chimney structures. FEMS Microbiol Lett 232:145–152

Ňancucheo I, Johnson DB (2012) Selective removal of transition metals from acidic mine waters by novel consortia of acidophilic sulfidogenic bacteria. Microb Biotechnol 5:34–44

Ňancucheo I, Rowe OF, Hedrich S, Johnson DB (2016) Solid and liquid media for isolating and cultivating acidophilic and acid-tolerant sulfate-reducing bacteria. FEMS Microbiol Lett 363(10):pii:fnw083. https://doi.org/10.1093/femsle/fnw083

Nazina TN, Ivanova AE, Kanchavli LP, Rozanova EP (1989) A new spore-forming thermophilic methylotrophic sulfate-reducing bacterium, *Desulfotomaculum kuznetsovii* sp. nov. Microbiology 57:659–663

Nazina T, Rozanova E (1978) Thermophillic sulfate-reducing bacteria from oil-bearing strata. Mikrobiologiia 47:142–148

Newman DK, Kennedy EK, Coates JD, Ahmann D, Ellis DJ, Lovley DR, Morel M (1997) Dissimilatory arsenate and sulfate reduction in *Desulfotomaculum auripigmentum* sp. nov. Arch Microbiol 168:380–388

Nilsen RK, Torsvik T, Lien T (1996) *Desulfotomaculum thermocisternum* sp. nov., a sulfate reducer isolated from a hot North Sea oil reservoir. Int J Syst Bacteriol 46:397–402

Ochynski FW, Postgate JR (1963) Some biochemical differences between fresh water and salt water strains of sulphate-reducing bacteria. In: Oppenheimer CH (ed) Symposium on marine microbiology. C. C. Thomas Publisher, Springfield. IL, pp 426–441

Okabe S, Ito T, Satoh H (2003) Sulphate-reducing bacterial reducing community structure, function. Appl Microbiol Biotechnol 63:322–334

Okabe S (2007) Ecophysiology of sulphate-reducing bacteria in environmental biofilms. In: Barton LL, Hamilton WA (eds) Sulphate-reducing bacteria environmental and engineered systems. Cambridge University Press, Cambridge, UK, pp 359–382

Okabe S, Ito T, Sugita K, Satoh H (2005) Succession of internal sulfur cycles and sulfur-oxidizing bacterial communities in microaerophilic wastewater biofilms. Appl Environ Microbiol 71: 2520–2529

Okabe S, Ito T, Sugita K, Satoh H (2020) Succession of internal sulfur cycles and sulfur-oxidizing bacterial communities in microaerophilic wastewater biofilms. Appl Environ Microbiol 71: 2520–2529

Okabe S, Itoh T, Satoh H, Watanabe Y (1999) Analysis of special distributions of sulphate reducing bacteria. Appl Environ Microbiol 65:5107–5116

Okabe S, Nielsen PH, Jones WL, Characklis WG (1995) Sulfide product inhibition of *Desulfovibrio desulfuricans* in batch and continuous cultures. Water Res 29:571–578

Ollivier B, Caumette P, Garcia J-L, Mah RA (1994) Anaerobic bacteria from hypersaline environments. Microbiol Rev 58:27–38

Ollivier B, Hatchikian CE, Guezennec PJ, Garcia J-L (1991) *Desulfohalobium retbaense* gen. nov., sp. nov. a halophilic sulfate-reducing bacterium from sediments of a hypersaline lake in Senegal. Int J Syst Bacteriol 41:74–81

Ogg CD, Patel BK (2011) *Desulfotomaculum varum* sp. nov., a moderately thermophilic sulfate-reducing bacterium isolated from a microbial mat colonizing a Great Artesian Basin bore well runoff channel. 3 Biotech 1(3):139–149. https://doi.org/10.1007/s13205-011-0017-5. Epub 2011 Aug 9. PMID: 22611525; PMCID: PMC3339622

Oremland RS, Stolz JF, Hollibaugh JT (2004) Microbial arsenic cycle in Mono Lake, California. FEMS Microbiol Ecol 48:15–27

Ovhynski FW, Postgate JR (1963) Some biochemical differences between fresh water and salt water strains of sulphate-reducing bacteria. In: Oppenheimer CH (ed) Symposium on marine microbiology. Charles C Thomas, Publisher, Springfield, pp 426–441

Ozuolmez D, Na H, Lever MA, Kjeldsen KU, Jørgensen BB, Plugge CM (2015) Methanogenic archaea and sulfate reducing bacteria co-cultured on acetate: teamwork or coexistence? Front Microbiol 6:Article 492. https://doi.org/10.3389/fmicb.2015.00492

Parshina SN, Sipma J, Nakashimada Y, Henstra AM, Smidt H, Lysenko AM, three co-authors. (2005) *Desulfotomaculum carboxydivorans* sp nov., a novel sulfate-reducing bacterium capable of growth at 100% CO. Int J Syst Evol Microbiol 55:2159–2165

Pedersen K (2012) Subterranean microbial populations metabolize hydrogen and acetate under in situ conditions in granitic groundwater at 450 m depth in the Äspö Hard Rock Laboratory, Sweden. FEMS Microbiol Ecol 81:217–229

Pérez Bernal MF, Souza Brito EM, Bartoli M, Aubé J, Fardeau ML, Rodriguez C, three co-authors. (2017) *Desulfonatronum parangueonense* sp. nov., a sulfate-reducing bacterium isolated from sediment of an alkaline crater lake. Int J Syst Evol Microbiol 67:4999–5005

Pérez-Jiménez JR, Kerkhof LJ (2005) Phylogeography of sulfate-reducing bacteria among disturbed sediments, disclosed by analysis of the dissimilatory sulfite reductase gene (*dsrAB*). Appl Environ Microbiol 71:1004–1011

Pester M, Knorr K-H, Friedrich MW, Wagner M, Loy A (2012) Sulfate-reducing microorganisms in wet-lands – fameless actors in carbon cycling and climate change. Front Microbio 3:72. https://doi.org/10.3389/fmicb.2012.00072

Petzsch P, Johnson DB, Daniel R, Schlömann M (2015) Genome sequence of the moderately acidophilic sulfate-reducing firmicute *Desulfosporosinus acididurans* (strain M1^T). Genome Announc 3(4):e00881–e00815. https://doi.org/10.1128/genomeA.00881-15

Pikuta EV, Hoover RB, Bej AK, Marsic D, Whitman WB, Cleland C, Krader P (2003) *Desulfonatronum thiodismutans* sp. nov., a novel alkaliphilic, sulfate-reducing bacterium capable of lithoautotrophic growth. Int J Syst Evol Microbiol 53:1327–1332

Pikuta E, Lysenko A, Suzina N, Osipov G, Kuznetsov B, Tourova T, Akimenko V, Laurinavichius K (2000) *Desulfotomaculum alkaliphilum* sp. nov., a new alkaliphilic, moderately thermophilic, sulfate-reducing bacterium. Int J Syst Evol Microbiol 50:25–33

Pikuta EV, Zhilina TN, Zavarzin GA, Kostrikina NA, Osipov GA, Rainey FA (1998) *Desulfonatronum lacustre* gen. nov., sp. nov.: a new alkaliphilic sulfate-reducing bacterium utilizing ethanol. Microbiology 67:105–113

Plugge CM, Scholten JCM, Culley DE, Nie I, Brockman FJ, Zhang W (2010) Global transcriptomics analysis of the *Desulfovibrio vulgaris* change from syntrophic growth with *Methanosarcina barkeri* to sulfidogenic metabolism. Microbiology 156:2746–2756

Plugge CM, Zhang W, Scholten JCM, Stams AJM (2011) Metabolic flexibility of sulfate-reducing bacteria. Front Microbiol 2:Article 81. https://doi.org/10.3389/fmicb.2011.00081

Poosarla VG, Wood TL, Zhu L, Miller DS, Yin B, Wood TK (2017) Dispersal and inhibitory roles of mannose, 2-deoxy-d-glucose and N-acetylgalactosaminidase on the biofilm of Desulfovibrio vulgaris. Environ Microbiol Rep 9:779–787

Postgate JR (1965) Recent advances in the study of the sulfate-reducing bacteria. Bacteriol Rev 29: 425–441

Postgate JR (1984) The Sulphate-reducing bacteria, 2nd edn. Cambridge University Press, Cambridge, UK

Postgate JR, Campbell LL (1966) Classification of *Desulfovibrio* species, the nonsporulating sulfate-reducing bacteria. Bacteriol Rev 30:732–738

Pradel N, Ji B, Gimenez G, Talla E, Lenoble P, Garel M, nine co-authors. (2013) The first genomic and proteomic characterization of a deep-sea sulfate reducer: insights into the piezophilic lifestyle of *Desulfovibrio piezophilus*. PLoS One 8(1):e55130. https://doi.org/10.1371/journal.pone.0055130

Praharaj T, Fortin D (2004) Indicators of microbial sulfate reduction in acidic sulfide-rich mine tailings. Geomicrobiol J 21:457–467

Qiu R, Zhao B, Liu J, Huang X, Li Q, Brewer E, Wang S, Shi N (2009) Sulfate reduction and copper precipitation by a *Citrobacter* sp. isolated from a mining area. J Hazard Mater 164:1310–1315

Rabus R, Brüchert V, Amann J, Könneke M (2002) Physiological response to temperature changes of the marine, sulfate-reducing bacterium *Desulfobacterium autotrophicum*. FEMS Microbiol Ecol 42:409–417

Ramsing NB, Kühl M, Jørgensen BB (1998) Distribution of sulfate-reducing bacteria, O_2, and H_2S in photosynthetic biofilms determined by oligonucleotide probes and microelectrodes. Appl Environ Microbiol 59:3840–3849

Rapp BJ, Wall JD (1987) Genetic transfer in *Desulfovibrio desulfuricans*. Proc Natl Acad Sci U S A 84:9128–9130

Redburn AC, Patel BKC (1994) *Desulfovibrio longreachii* sp. nov., a sulfate-reducing bacterium isolated from the Great Artesian Basin of Australia. FEMS Microbiol Lett 115:33–38

Rees GN, Grassia GS, Sheehy AJ, Dwivedi PP, Patel BKC (1995) *Desulfacinum infernurn* gen. nov., sp. nov., a thermophilic sulfate-reducing bacterium from a petroleum reservoir. Int J Syst Bacteriol 45:85–89

Reis MAM, Almeida JS, Lemos P, Carrondo MJT (1992) Effect of hydrogen sulfide growth of sulfate-reducing bacteria. Biotechnol Bioeng 40:593–600

Rinzema A, Letting G (1988) The effect of sulphide on the anaerobic degradation of propionate. Environ Technol Lett 9:83–88

Robador A, Müller AL, Sawicka JE, Berry D, Hubert C, Loy A, Jørgensen BB, Brüchert V (2015) Activity and community structures of sulfate-reducing microorganisms in polar, temperate and tropical marine sediments. ISME J 10:796–809

Rowe OF, Sánchez-España J, Hallberg KB, Johnson DB (2007) Microbial communities and geochemical dynamics in an extremely acidic, metal-rich stream at an abandoned sulfide mine (Huelva, Spain) underpinned by by two functional primary productivity systems. Environ Microbiol 9:1761–1771

Rozanova EP, Khudyakova AI (1974) A new nonspore-forming thermophilic sulphate-reducing organism, *Desulfovibrio thermophilus* nov. sp. Microbiology 43:1069–1075

Rozanova EP, Pivovarova TA (1988) Reclassification of *Desulfovibrio thermophilus* (Rozanova, Khudyakova, 1974). Microbiology 57:102–106

Ryzhmanova Y, Abashina T, Petrova D, Shcherbakova V (2019) *Desulfovibrio gilichinskyi* sp. nov., a cold-adapted sulfate-reducing bacterium from a Yamal Peninsula cryopeg. Int J Syst Evol Microbiol 69:1081–1086

Ryzhmanova Y, Nepomnyashchaya Y, Abashina T, Ariskina E, Troshina O, Vainshtein M, Shcherbakova V (2013) New sulfate-reducing bacteria isolated from Buryatian alkaline brackish lakes: description of Desulfonatronum buryatense sp. nov. Extremophiles 17:851–859

Sahm K, Knoblauch C, Aman R (1999) Phylogenic affiliation and quantitation of psychropilic sulfate-reducing isolates in marine Artic sediments. Appl Environ Microbiol 65:3876–3981

Sánchez-Andrea I, Stams AJ, Amils R, Sanz JL (2013) Enrichment and isolation of acidophilic sulfate-reducing bacteria from Tinto River sediments. Environ Microbiol Rep 5:672–678

Sánchez-Andrea I, Stams AJ, Hedrich S, Ňancucheo I, Johnson DB (2015) *Desulfosporosinus acididurans* sp. nov.:an acidophilic sulfate-reducing bacterium isolated from acidic sediments. Extremophiles 19:39–47

Sánchez-España J, Yusta I, Ilin A, van der Graaf C, Sánchez-Andrea I (2020) Microbial geochemistry of the acidic saline pit Lake of Brunita Mine (La Unión, SE Spain). Mine Water Environ 39:535–555

Sass H, Berchtold M, Branke J, König H, Cypionka H, Babenzien H-D (1998) Psychrotolerant sulfate-reducing bacteria from an oxic freshwater sediment description of *Desulfovibrio cuneatus* sp. nov. and *Desulfovibrio litoralis* sp. nov. Syst Appl Microbiol 21:212–219

Sass H, Cypionka H (2004) Isolation of sulfate-reducing bacteria from the terrestrial deep subsurface and description of *Desulfovibrio cavernae* sp. nov. Syst Appl Microbiol 27:541–548
Sass A, Rütters H, Cypionka H, Henrik SH (2002) *Desulfobulbus mediterraneus* sp. nov., a sulfate-reducing bacterium growing on mono- and disaccharides. Arch Microbiol 177:468–474
Sawicka JE, Jørgensen BB, Bruchert V (2012) Temperature characteristics of bacterial sulfate reduction in continental shelf and slope sediments. Biogeosciences 9:3425–3435
Schneider D, Arp G, Reimer A, Reitner J, Daniel R (2013) Phylogenetic analysis of a microbialite-forming microbial mat from a hypersaline lake of the Kiritimati atoll, Central Pacific. PLoS One 8:e66662. https://doi.org/10.1371/journal.pone.0066662
Sekiguchi Y, Muramatsu M, Imachi H, Narihiro T, Ohashi A, Harada H, Hanada S, Kamagata. (2008) *Thermodesulfovibrio aggregans* sp. nov. and *Thermodesulfovibrio thiophilus* sp. nov., anaerobic, thermophilic, sulfate-reducing bacteria isolated from thermophilic methanogenic sludge, and emended description of the genus*Thermodesulfovibrio*. Int J Syst Evol Microbiol 58:2541–2548. https://doi.org/10.1099/ijs.0.2008/000893-0
Sen A, Johnson B (1999) Acidophilic sulphate-reducing bacteria: candidates for bioremediation of acid mine drainage. Proc Metall 9:709–718
Senez J (1953) Sur l'activité et la croissance des bactéries anaérobies sulfato-réductrices en cultures semi-autotrophes. Ann Inst Pasteur (Paris) 84:595–604
Sievert SM, Kuever J (2000) *Desulfacinum hydrothermale* sp. nov., a thermophilic, sulfate-reducing bacterium from geothermally heated sediments near Milos Island (Greece). Int J Syst Evol Microbiol 50:1239–1246
Smith RL, Oremland RS (1987) Big Soda Lake (Nevada). Pelagic sulfate reduction. Limnol Oceanogr 32:794–803
Sonne-Hansen J, Ahring BK (1999) *Thermodesulfobacterium hveragerdense* sp.nov., and *Thermodesulfovibrio islandicus* sp.nov., two thermophilic sulfate- reducing bacteria isolated from a Icelandic Hot Spring. Syst Appl Microbiol 22:559–564
Sorokin DY, Berben T, Melton ED, Overmars L, Vavourakis CD, Muyzer G (2014) Microbial diversity and biogeochemical cycling in soda lakes. Extremophiles 18:791–809
Sorokin DY, Chernyh NA (2016) "*Candidatus* Desulfonatronobulbus propionicus": a first haloalkaliphilic member of the order *Syntrophobacterales* from soda lakes. Extremohiles 20: 895–901
Sorokin DY, Chernyh NA (2017) *Desulfonatronospira sulfatiphila* sp. nov., and *Desulfitispora elongata* sp. nov., two novel haloalkaliphilic sulfidogenic bacteria from soda lakes. Int J Syst Evol Microbiol 67:396–401
Sorokin SY, Detkova EN, Muyzer G (2010a) Propionate and butyrate dependent bacterial sulfate reduction at extremely haloalkaline conditions and description of *Desulfobotulus alkaliphilus* sp. nov. Extremophiles 14:71–77
Sorokin DY, Gorlenko VM, Namsaraev BB, Zorigto B, Namsaraev ZB, Lysenko AM et al (2004) Prokaryotic communities of the north-eastern Mongolian soda lakes. Hydrobiology 522:235–248
Sorokin SY, Rusanov II, Pimenov NV, Tourova TP, Abbas B, Muyzer G (2010b) Sulfidogenesis under extremely haloalkaline conditions in soda lakes of Kulunda Steppe (Altai, Russia). FEMS Microbial Ecol 73:278–290
Sorokin DY, Tourova TP, Henstra AM, Stams AJM, Galinski EA, Muyzer G (2008) Sulfidogenesis under extremely haloalkaline conditions by *Desulfonatronospira thiodismutans* gen. nov., sp. nov., and *Desulfonatronospira delicata* sp. nov. a novel lineage of *Deltaproteobacteria* from hypersaline soda lakes. Microbiology 154:1444–1453
Spring S, Sorokin DY, Verbarg S, Rohde M, Woyke T, Kyrpides NC (2019) Sulfate-reducing bacteria that produce exopolymers thrive in the calcifying zone of a hypersaline cyanobacterial mat. Front Microbiol. https://doi.org/10.3389/fmicb.2019.00862

Stackebrandt E, Sproer C, Rainey FA, Burghardt J, Pauker O, Hippe H (1997) Phylogenetic analysis of the genus *Desulfotomaculum*: evidence for the misclassification of *Desulfotomaculum guttoideum* and description of *Desulfotomaculum orientis* as *Desulfosporosinus orientis* gen. nov., comb. nov. Int J Syst Bacteriol 47:1134–1139

Steinsbu BO, Thorseth IH, Nakagawa S, Inagaki F, Lever MA, Engelen B, Øvreås L, Pedersen RB (2010) *Archaeoglobus sultaticallidus* sp. nov., a thermophilic and facultatively lithotrophic sulfate-reducer isolated from black rust exposed to hot ridge flank crustal fluids. Int J Syst Evol Microbiol 60:2745–2752

Stetter KO, Hube R, Blöchl E, Kurr M, Eden RD, Fielder M, Cash H, Vance I (1993) Hyperthermophilic archaea are thriving in deep North Sea and Alaskan oil reservoirs. Nature 365:743–745

Stetter KO, Lauerer G, Thomm M, Neuner A (1987) Isolation of extremely thermophilic sulfate reducers: evidence for a novel branch of archaebacteria. Science 236:822–824

Stockdreher Y, Venceslau SS, Josten M, Sahl H-G, Pereira IAC, Dahl C (2012) Cytoplasmic sulfurtransferases in the purple sulfur bacterium *Allochromatium vinosum*: evidence for sulfur transfer from DsrEFH to DsrC. PLoS One 2012(7):e40785

Stetter K (2006) History of discovery of the first hyperthermophiles. Extremophiles 10:357–362. https://doi.org/10.1007/s00792-006-0012-7

Stetter KO (1988) *Archaeoglobus fulgidus* gen. nov., sp. nov.: new taxon of extremely thermophilic *Archaebacteria*. Syst Appl Microbiol 10:172–173

Stevens T (1997) Lithoautotrophy in the subsurface. FEMS Microbiol Rev 20:327–337

Stevens TO, McKinley JP (1995) Lithoautotrophic microbial ecosystem in deep basalt aquifers. Science 270:450–454

Suzuki D, Ueki A, Amaishi A, Ueki K (2009) *Desulfovibrio portus* sp. nov., a novel sulfate-reducing bacterium in the class Deltaproteobacteria isolated from an estuarine sediment. J Gen Appl Microbiol 55:125–133

Sun B, Cole JR, Tiedje JM (2001) *Desulfomonile limimaris* sp. nov., an anaerobic dehalogenating bacterium from marine sediments. Int J Syst Evol Microbiol 51:365–371

Tan S, Liu J, Fang Y, Hedlund BP, Lian Z-H, Huang L-H, six co-authors. (2019) Insights into ecological role of a new deltaproteobacterial order *Candidatus* Acidulodesulfobacterales by metagenomics and metatranscriptomics. ISME J 13:2044–2057

Tani M, Higashi T, Nagatsuka S (1993) Dynamics of low molecular-weight aliphatic carboxylic-acids (lacas) in forest soils. 1. Amount and composition of lacas in different types of forest soils in Japan. Soil Sci Plant Nutr 39:485–495

Tardy-Jacquenod C, Magot M, Laigret F, Kaghad M, Patel BKC, Guezennec J, Witheron R, Caumette P (1996) *Desulfovibrio gabonensis* sp. nov., a new moderately halophilic sulfate-reducing bacterium isolated from an oil pipeline. Int J Syst Bacteriol 46:710–715

Tardy-Jacquenod C, Magot M, Patel BKC, Matheron R, Caumette P (1998) *Desulfotomaculum halophilum* sp. nov., a halophilic sulfate-reducing bacterium isolated from oil production facilities. Int J Syst Bacteriol 48:333–338

Tarpgaard IH, Boetius A, Finster K (2006) *Desulfobacter psychrotolerans* sp. nov., a new psychrotolerant sulfate-reducing bacterium and descriptions of its physiological response to temperature changes. Antonie Van Leeuwenhoek 89:109–124

Tasaki M, Kamagata Y, Nakamura K, Mikami E (1991) Isolation and characterization of a thermophilic benzoate-degrading, sulfate-reducing bacterium, *Desulfotomaculum thermobenzoicum* sp. nov. Arch Microbiol 155:348–352

Taylor J, Parkes RJ (1983) The cellular fatty acids of the sulphate-reducing bacteria, *Desulfobacter* sp., *Desulfobulbus* sp. and *Desulfovibrio desulfuricans*. Gen Microbiol 129:3303–3309

Teske A, Ramsing NB, Habicht KS, Fukui M, Küver J, Jørgensen BB, Cohen Y (1998) Sulfate-reducing bacteria and their activities in cyanobacterial mats of Solar Lake (Sinai, Egypt). Appl Environ Microbiol 64:2943–2951

Tran TTT, Kannoorpatti K, Padovan A, Thennadil S (2021) Sulphate-reducing bacteria's response to extreme pH environments and the effect of their activities on microbial corrosion. Appl Sci 11(5):2201. https://doi.org/10.3390/app11052201

Traore AS, Hatchikian CE, Belaich J-P, Le Gall J (1981) Microcalorimetric studies of the growth of sulfate-reducing bacteria: energetics of *Desulfovibrio vulgaris* growth. J Bacteriol 145:191–199

Traore AS, Hatchikian CE, LeGall J, Belaich J-P (1982) Microcalorimetric studies of the growth of sulfate-reducing bacteria: comparison of the growth parameters of some *Desulfovibrio* species. J Bacteriol 149:606–611

Tuttle JH, Dugan PR, Macmillan CB, Randles CI (1969) Microbial dissimilatory sulfur cycle in acid mine water. J Bacteriol 97:594–602

Vainshtein M, Hippe H, Kroppenstedt RM (1992) Cellular fatty acid composition of *Desulfovibrio* species and its use in classification of sulfate-reducing bacteria. Syst Appl Microbiol 15:554–566

van Delden A (1903) Beiträge zur Kenntnis der Sulfatreduktion durch Bakterien. Centr Bakteriol Abt II 11:81–94

van Lith Y, Vasconcelos C, Warthmann R, Martins JCF, McKenzie JA (2002) Bacterial sulfate reduction and salinity: two controls on dolomite precipitation in Lagoa Vermelha and Brejo de Espinto (Brazil). Hydrobiologia 485:35–49

van Lith Y, Warthmann R, Vasconcelos C, McKenzie JA (2003a) Microbial fossilization in carbonate sediments: a result of the bacterial surface involvement in dolomite precipitation. Sedimentology 50:327–245

van Lith Y, Warthmann R, Vasconcelos C, McKenzie JA (2003b) Sulphate-reducing bacteria induce low-temperature Ca-dolomite and high Mg-calcite formation. Andean Geol 1:77–79

Vandieken V (2006) *Desulfotomaculum arcticum* sp. nov., a novel spore-forming, moderately thermophilic, sulfate-reducing bacterium isolated from a permanently cold fjord sediment of Svalbard. Int J Syst Evol Microbiol 56:687–690

Vasconcelos C, McKenzie JA (1997) Microbial mediation of modern dolomite precipitation and diagenesis under anoxic conditions (Lagoa Vermelha, Rio de Janeiro, Brazil). J Sed Res 67: 378–390

Visscher PT, Reid RP, Bebout BM (2000) Microscale observations of sulfate reduction: correlation of microbial activity with lithified micritic laminae in modern marine stromatolites. Geology 28: 919–922

Vorholt J, Kunow J, Stetter KO, Thauer RK (1995) Enzymes and coenzymes of the carbon monoxide dehydrogenase pathway for autotrophic CO_2 fixation in *Archaeoglobus lithotrophicus* and the lack of carbon monoxide dehydrogenase in the heterotrophic *A. profundus*. Arch Microbiol 163:112–118

Walker CB, Stolyar S, Chivian D, Pinel N, Gabster JA, Dehal PS et al (2009) Contribution of mobile genetic elements to *Desulfovibrio vulgaris* genome plasticity. Environ Microbiol 11: 2244–2252

Warthmann R, van Lith Y, Vasconcelos C, McKenzie JA, Karpoff AM (2000) Bacterially induced dolomite precipitation in anoxic culture experiments. Geology 28:1991–1994

Warthmann R, Vasconcelos C, Sass H, McKenzie JA (2005) *Desulfovibrio brasiliensis* sp. nov., a moderate halophilic sulfate-reducing bacterium from Lagoa Vermelha (Brazil) mediating dolomite formation. Extremophiles 9:255–261

Watanabe M, Kojima H, Fukui M (2015) *Desulfoplanes formicivorans* gen. nov., sp. nov., a novel sulfate-reducing bacterium isolated from a blackish meromictic lake, and emended description of the family *Desulfomicrobiaceae*. Int J Syst Evol Microbiol 65:1902–1907

Weissgerber T, Sylvester M, Kröninger L, Dahl C (2014) A comparative quantitative proteomic study identifies new proteins relevant for sulfur oxidation in the purple sulfur bacterium *Allochromatium vinosum*. Appl Environ Microbiol 80:2279–2292

Widdel F (1987) New types of acetate-oxidizing sulfate-reducing *Desulfobacter* species, *D. hydrogenophilus* sp. nov., *D. lactus* sp. nov., and *D. curvatus* sp. nov. Arch Microbiol 148:286–291

Widdel F (1988) Microbiology and ecology of sulfate and sulfur-reducing bacteria. In: Zehnder AJB (ed) Environonmental microbiology of anaerobic bacteria. Wiley, New York, pp 469–585

Widdel F, Pfennig N (1981) Studies on dissimilatory sulfate-reducing bacteria that decompose fatty acids. Arch Microbiol 129:395–400

Widdel F, Pfennig N (1982) Studies on dissimilatory sulfate-reducing bacteria that decompose fatty acids II. Incomplete oxidation of propionate by *Desulfobulbus propionicus* gen. nov., sp. nov. Arch Microbiol 131:360–365

Widdel F, Pfennig N (1984) Dissimilatory sulfate-and sulfur-reducing bacteria. In: Krieg NR, Holt JG (eds) Bergey's manual of systematic bacteriology. Williams & Wilkins, Baltimore, MD, pp 663–679

Wierenga EBA, Overmann J, Cypionka H (2000) Detection of abundant sulphate-reducing bacteria in marine oxic sediment layers by a combined cultivation and molecular approach. Environ Microbiol 2:417–427

Wikie AJ, Datsenko L, Vera M, Sand W (2014) Impact of *Desulfovibrio alaskensis* biofilms on corrosion behaviour of carbon steel in marine environment. Bioelectrochemistry 97:52–60

Wright DT (1999) The role of sulphate-reducing bacteria and cyanobacteria in dolomite formation in distal ephemeral lakes of the Coorong region, South Australia. Sed Geol 126:147–157

Wright DT, Wacey D (2005) Precipitation of dolomite using sulphate-reducing bacteria from the Coorong Region, South Australia: significance and implications. Sedimentology 52:987–1008

Yang G, Guo J, Zhuang L, Yuan Y, Zhou S (2016) *Desulfotomaculum ferrireducens* sp. nov., a moderately thermophilic sulfate-reducing and dissimilatory Fe(III)-reducing bacterium isolated from compost. Int J Syst Evol Microbiol 66:3022–3028

Yayanos AA (1995) Microbiology to 10,500 meters in the deep sea. Annu Rev Microbiol 49:777–805

Zakharyuk AG, Kozyreva LP, Khijniak TV, Namsaraev BB, Shcherbakova VA (2015) *Desulfonatronum zhilinae*sp. nov., a novel haloalkaliphilic sulfate-reducing bacterium from soda lake alginskoe, Trans-Baikal region. Russia Extremophiles 19:673–680

Zavarzin GA, Zhilina TN, Dulov LE (2008) Alkaliphilic sulfidogenesis on cellulose by combined cultures. Microbiology 77:419–429

Zeikus JG, Dawson MA, Thompson TE, Ingvorsent K, Hatchikian EC (1983) Microbial ecology of volcanic sulphidogenesis: isolation and characterization of *Thermodesulfobacterium commune* gen. nov. and sp. nov. J Gen Microbiol 129:1159–1169

Zhilina T, Zavarzin G, Rainey F, Pikuta E, Osipov G, Kostrikina N (1997) *Desulfonatronovibrio hydrogenovorans* gen. nov., sp. nov., an alkaliphilic, sulfate-reducing bacterium. Int J Syst Bacteriol 47:144–149

Zhilina TN, Zavarzina DG, Kolganova TV, Yourova TP, Zavarzin GA (2005a) "*Candidatus* Contubernalis alkalaceticum" an obligately syntrophic alkaliphilic bacterium capable of anaerobic acetate oxidation in a coculture with *Desulfonatronum cooperativum*. Microbiology 74: 695–703

Zhilina TN, Zavarzina DG, Kuever J, Lysenko AM, Zavarzin GA (2005b) *Desulfonatronum cooperativum* sp. nov., a novel hydrogenotrophic, alkaliphilic, sulfate-reducing bacterium, from a syntrophic culture growing on acetate. Int J Syst Evol Microbiol 55:1001–1006

Zhu L, Gong T, Wood TL, Yamasaki R, Wood TK (2019) σ_{54} -Dependent regulator DVU2956 switches *Desulfovibrio vulgaris* from biofilm formation to planktonic growth and regulates hydrogen sulfide production. Environ Microbiol 21:3564–3576

Zhu L, Poosarla VG, Song S, Wood TL, Miller DS, Yin B, Wood T (2018) Glycoside hydrolase DisH from *Desulfovibrio vulgaris* degrades the N-acetylgalactosamine component of diverse biofilms. Environ Microbiol 20:2026–2037

Zobell CE (1943) The effect of solid surfaces on bacterial activity. J Bacteriol 46:39–56

Zobell CE, Johnson FH (1949) The influence of hydrostatic pressure on the growth and viability of terrestrial and marine bacteria. J Bacteriol 57:179–189

Zobell CE, Morita RY (1957) Barophylic bacteria in some deep-sea sediments. J Bacteriol 73: 563–568

Chapter 10
Interactions of SRB with Animals and Plants

10.1 Introduction

The general concept proposed by Bass Becking in 1934 that "Everything is everywhere *but* the environment selects" (deWit and Bouvier 2006) is supported by the various associations that SRB have with living hosts. While one can envision a variety of processes that would contribute to the global distribution of sulfate reducers, the association of SRB with eukaryotic organisms is not random but restricted and dependent on selective features. At this time, associations between eukaryotes and SRB are well documented, but there are no reported associations involving archaeal sulfate reducers. It is apparent that certain species of SRB are associated with a host, and scientists are only at the initial phase of understanding the mechanisms contributing to this symbiosis. Establishing and maintaining SRB–host interaction suggests binding between the SRB cell and host tissue. For example, it is known that SRB become established in the gut of animals and humans early after birth and SRB are retained in the gut for the life of the host. Also, a "metabolic fit" is required where nutrients for the growth of SRB and an anaerobic environment are provided. Future studies concerning bacterial community dynamics associated with SRB symbiotic states are needed to understand the microbe–host relationship.

This chapter will illustrate the array of microfauna, plants, and animals that cooperatively support the growth of SRB. As reviewed initially by Postgate (1979), the presence of SRB in the intestine of mammals and in the rumen of cattle had been known since the 1960s; however, these reports were in the category of biological curiosities than in understanding the activities of these bacteria. Since then, tremendous advancements have been made concerning the association of SRB with various life forms.

L. L. Barton, G. D. Fauque, *Sulfate-Reducing Bacteria and Archaea*,
https://doi.org/10.1007/978-3-030-96703-1_10

10.2 Symbiosis with Termites and a Gut-Residing Protist

The initial report of SRB in the gut of termites was in 1990. Using benzoate in the enrichment of gut contents from *Cubitermes speciosus*, a soil-feeding higher termite, several bacterial strains were isolated, which were similar to *Desulfovibrio* (*D.*) *desulfuricans* and *D. giganteous* (Brauman et al. 1990). A new species of SRB, *D. termitidis*, isolated from the hindgut of the termite *Heterotermes indicola* (Wasman), was found to use carbohydrates, lactate, and pyruvate as electron donors with sulfate, sulfite, thiosulfate, sulfur, and nitrate as final electron acceptors (Trinkerl et al. 1990). Another species of SRB, *D. intestinalis*, was isolated from the hindgut of the lower termite *Mastotermes darwiniensis* Froggatt (Fröhlich et al. 1999). In vivo and in vitro reduction of sulfate reduction was demonstrated using the gut contents from *Mastotermes darwiniensis* (Dröge et al. 2005). *Desulfovibrio* spp. that were closely related to *D. desulfricans* or *D. termitidis* were isolated from the termite, *Reticulitermes santonensis*, and these bacteria were able to use O_2 as the final electron acceptor with H_2, sulfide, sulfite, or thiosulfate as electron donors (Kuhnigk et al. 1996). When the termite *Reticulitermes santonensis* was fed a high sulfate diet, the gut fluid contained 10^8 SRB cells per ml.

SRB in termites appears to have a symbiotic relationship with gut-residing protists present in wood-eating termites (Dolan 2001). Bacterial symbionts on protists are either epibiotic or endobiotic. *Trichonympha agilis*, a flagellate protist in the gut of the wood-feeding termite *Reticulitermes speratus*, harbors an uncultured bacterium identified as *Candidatus* (*C.*) Desulfovibrio trichonymphae (Sato et al. 2009). About 1800 of these SRB cells are localized in the cortical layer near the anterior part of this flagellate cell. A *Desulfovibrio* symbiont was discovered buried in the plasma membrane of *Trichonympha*, the host protist, residing in the termite *Incisitermes marginipennis*, which revealed that the *Desulfovibrio* symbiont was connected to the environment outside of the protist by a small pore that has a diameter of 41.4 nm (Strassert et al. 2012; Kuwahara et al. 2017). This is a tripartite symbiotic system in *Trichonympha agilis* with *C.* D. trichonymphae, a second endosymbiotic bacterium (*C.* Endomicrobium trichonymphae) and a hydrogenosome structure. *C.* Desulfovibrio trichonymphae uses H_2 produced by the hydrogenosome and sulfate or fumarate as the electron acceptor. The pore enables sulfate and fumarate, which is in the environment outside *Trichonympha agilis* to gain access to the SRB. It is unknown if there is an interaction between the two bacterial endosymbionts. While SRB are associated with gut contents of higher and lower termites, the metabolic processes of these SRB remain to be established.

10.3 Symbiosis with Root-Feeding Larvae

The gut microbiota of the root-feeding larva of the European cockchafer (*Melolontha melolontha*) was examined by constructing 16S rRNA gene clone libraries of the midgut and hindgut contents (Egert et al. 2005). *Desulfovibrio*-related bacteria were detected in the hindgut wall and not in the lumen or the midgut. SRB in the hindgut wall account for about 10% of the bacterial gut wall community, and the presence of SRB in the hindgut wall was observed in larvae collected from different geographical regions. The role of SRB in the root-feeding larvae is unresolved, but a possible role may be to contribute to the reduction of free O_2 in the gut, as has been suggested for SRB in termites (Kuhnigk et al. 1996).

10.4 Symbiosis with Invertebrates

10.4.1 Gutless Marine Worm

Dissimilatory sulfate reduction activity is associated with the gutless marine worms that are found worldwide. One of the oligochaetes is *Olavius* (*O.*) *algarvensis*, which is only 0.2 mm in diameter and 1–2 cm in length and appears white due to the deposition of elemental sulfur by the symbiont bacterium. *O. algarvensis* is dependent on five different bacterial symbionts where two species are sulfate-reducing symbionts, two sulfur-oxidizing bacterial species, and one is a spirochete (Dubilier et al. 2001). Another oligochaete species is *O. ilvae*, which also has two sulfate-reducing symbionts and two sulfur-oxidizing symbionts but no spirochete bacteria (Ruehland et al. 2008). There is an unusual cooperative interaction of sulfate-reducing bacteria that produce sulfide, which is oxidized by the sulfur oxidizers to produce elemental sulfur. Metaproteonomics reveals that these gutless worms have unusual metabolic pathways and a novel energy-generating system (Kleiner et al. 2012).

10.4.2 Polychaete Serpulid Worm

The polychaete serpulid worms are another group of marine invertebrates, and these worms have fibrillar or lamellar wall structures. The agglutinated polychaete worms are found attached to the walls and ceiling of marine caves and have a distinctive relationship with SRB (Guido et al. 2014). It has been suggested that the metabolism of SRB produces insoluble carbonates (calcite) and that polychaetes mix the insoluble carbonates with mucus to build the irregular agglutinated worm tube wall. The sulfate reducers would obtain nutrients from organic material released from the polychaetes, and surrounding organisms would provide decaying material.

Several different species of SRB were detected in the tubes constructed by marine infaunal polychaete *Diopatra cuprea* (Matsui et al. 2004). Based on 16S rDNA sequence analysis, several clones of *Desulfobacter*, *Desulfosarcina*, and *Desulfobacterium* were present in these tubes. Based on the phospholipid fatty acid biomarkers, the quantity of *Desulfobacter* was predicted to be greater than either *Desulfovibrio* or *Desulfobulbus*. In addition to SRB, the tubes of *Diopatra cuprea* contained a great variety of other bacteria.

10.4.3 Sea Cucumber

The sea cucumber, *Apostichopus japonicas*, is one of 1200 species of Echinodermata, and it is found along the coasts of Russia, Japan, and Korea. *Apostichopus. japonicas* is a deposit feeder ingesting organic matter and microorganisms as well as inorganic materials. Analysis of the hind gut of *Apostichopus japonicas* by 16S rRNA pyrosequencing revealed that *Desulfosarcina* were the most abundant bacteria in the hindgut (Gao et al. 2014). Also, the sediment where these sea cucumbers were found contained high numbers of *Desulfobulbus* and *Desulfosarcinia.* Since SRB are important in the oxidation of carbon material in marine sediments, it is likely that *Desulfosarcina* contributes to the digestion of organic compounds in the hindgut of *Apostichopus japonicas.*

10.5 SRB Presence/Activity in Mice

The presence of SRB in mice has been reported, and the distribution of these anaerobes varies with different regions of the gastrointestinal tract. Using oligonucleotides specific for the *Desulfobacter*, *Desulfovibrio*, and *Desulfotomaculum* groups, SRB were found to be present throughout the intestinal tract in mature mice (Deplanche et al. 2000). Bacteria corresponding to each of the three SRB groups were colonized in the cecum, proximal colon, and distal colon by the time the mice were 14 days old. A relationship between SRB and *Bacteroides* spp. was observed in the region of the sulfomucin-containing goblet cells at the surface of distal gut mucosa (Deplanche et al. 2000). It was proposed by Deplanche et al. (2000) that sulfate was cleaved from sulfomucin by *Bacteroides*, and the free sulfate was used to promote SRB growth.

Using gnotobiotic mice, an artificial microbiome consisting of eight human gut bacterial species plus *D. piger* was introduced into the mice (Rey et al. 2013). Their research indicated that colonization by *D. piger* was dependent on elevated gene activity for hydrogen consumption and sulfate metabolism but not on the quantity of free sulfate in the diet. With germ-free mice co-colonized with *D. piger* and *Bacteroides theaiomicron*, the level of *D. piger* in the mouse gut was attributed to the sulfatase activity of *Bacteroides theaiomicron.* The cross-feeding of sulfate released by *Bacteroides theaiomicron* and growth of *D. piger* was confirmed by in vitro studies where the co-culture of bacteria in media containing chondroitin sulfate accounted for the increased density of *D. piger*.

As discussed by Rey et al. (2013), an inflammatory response in the distal colon is attributed to the reduced oxidative capacity of colonocytes by a high concentration of H_2S in the intestinal lumen. In the germ-free mouse study where eight gut bacteria plus *D. piger* were colonized in the intestine, there was no indication of an inflammatory response due to H_2S nor was the fat pad weigh diminished. However, the level of mRNA encoding for several subclasses of immune globulins and for the tight junction protein claudin-4 was significantly decreased in the proximal colon of mice colonized with *D. piger* as compared to animals not colonized with this SRB. The level of metalloproteinase-7 was increased in mice colonized by *D. piger*. Thus, while some differences are observed in mice colonized with and without *D. piger*, no obvious physiological problems were reported in this short experimental study (Rey et al. 2013).

Using a mouse model, Lin and colleagues observed that hydrogen sulfide produced from the metabolism of SRB had a pronounced impact on the host. When mice were tested in a maze system following administration of *D. vulgaris*, memory-linked discriminating function was impaired as compared to a control system where no SRB was dispensed to the mice (Ritz et al. 2016). Additionally, hydrogen sulfide from respiring SRB slowed intestinal transit in mice, and this activity was reversed with oral treatment using bismuth subsalicylate (Ritz et al. 2015). In the treatment of intestinal distress, bismuth compounds have been used for about a hundred years. It has been recently now demonstrated that chelated bismuth compounds inhibit the growth of SRB (Barton et al. 2016), and it is suggested that a target of bismuth inhibition is the F_1 subunit of the ATP synthase (Barton et al. 2019).

10.6 SRB Present in Rats

There are several studies concerning SRB in the gastrointestinal tract of animals, and rats have been a model system to study inflammatory activity attributed to H_2S production (Pitcher and Cummings 1996). One approach to this research has been to examine factors that influence H_2S levels in the rat colon. It was found that carrageenan (a sulfate-containing polysaccharide) fed to rats greatly enhanced the production of hydrogen sulfide, methanethiol, and dimethylsulfide in the colon, and these gases were rapidly metabolized by the cecal mucosal tissue (Suarez et al. 1998). Detoxification of H_2S in the rat colon was not due to methylation reactions, but methanethiol and dimethylsulfide were oxidized to H_2S, which was oxidized to thiosulfate (Levitt et al. 1999). Presumedly, this oxidation of hydrogen sulfide was by the enzyme rhodanese in cells of the colonic mucosa, and this accounted for the presence of thiosulfate in cecal venus blood (Levitt et al. 1999).

To control colonic sulfide production, fecal material from rats was treated with antibiotics, and the number of viable SRB and the amount of hydrogen sulfide produced were measured. Since over 95% of colonic hydrogen sulfide was detoxified in the colon and only 5% of the hydrogen sulfide was released in rat feces, in vivo tests were needed to evaluate the effect of antibiotics on SRB (Levitt et al. 1999, 2002). Administering daily doses of ciprofloxacin, metronidazole, or sulfasalazine to rats did not reduce the number of SRB in the rat colon nor was there a reduction in the amount of in vivo hydrogen sulfide produced (Ohge et al. 2003). However, when the rats treated with antibiotics were placed on a diet containing 6.5% bismuth subnitrate, the number of SRB in the colon and the amount of hydrogen sulfide produced were reduced.

The role of SRB in the development of inflammatory bowel diseases and ulcerative colitis was studied using three sets of rats (Kushkevych 2014). One group of rats received a standard diet and served as the control, a second group received a diet supplement with sulfate, which stimulated the growth of indigenous SRB in the rat intestine, and the third group received the standard diet plus a dose containing *D. piger* Vib-7 and *Desulfotomicrobium* sp. Rod-9. After 25 days, the distal colon was examined, the number of bacteria was determined, the levels of sulfide and acetate were measured, and the level of ulcer formation was assessed. In the second and third groups of rats, there was a decrease in the number of bacteria belonging to the following genera: *Bifidobacterium*, *Lactobacillus*, *Peptostreptococcus*, and *Peptococcus*. In the second and third groups of rats, the numbers of bacteria belonging to the following genera or groups were increased: *Clostridium*, *Escherichia*, *Staphylococcus*, *Proteus*, and SRB. The greatest level of colonic hydrogen sulfide and acetate was with the third group, and this group also exhibited the most severe ulceration. These observations were used to support the proposal that *D. piger* Vib-7 and *Desulfomicrobium* sp. Rod-9 contributed to colon diseases in rats (Kushkevych 2014). The amount of hydrogen sulfide produced in rodents with ulcerative colitis was higher than the amount of hydrogen sulfide produced by healthy animals (Kováč et al. 2018).

Breath test for malodor has been used to reflect inflammatory bowel diseases and was used in a rat model to indicate the production of hydrogen sulfide from microbial activity in the intestine. When SRB were introduced into rats, the excretion of rat-breath H_2 was inhibited by molybdate, and this suggested the correlating of breath H_2 to sulfate-reducing activity of SRB (Burnett et al. 2015).

10.7 SRB in Pigs

There has been interest in controlling the odors from raising pigs, including the production of hydrogen sulfide (Zahn et al. 2001). One approach to reducing the level of hydrogen sulfide in pig fecal material has been feeding pigs a diet containing low sulfur diet (Whitney et al. 1999; Poulsen et al. 2012), and another has been the use of aroma chemicals and herb extracts (Arakawa et al. 2000). When examining the gastrointestinal tract of pigs by using rRNA-targeted oligonucleotide probes, *Desulfovibrio* species were the most abundant SRB present, and the distribution of

SRB along the large intestine varied considerably (Lin et al. 1997; Kerr et al. 2010; Poulsen et al. 2012). While there was a correlation between the concentration of sulfide in the pig intestine and changes in intestinal inflammatory compounds (Kerr et al. 2010), it was not clear if SRB were responsible for the change in inflammatory activity.

An obligate intracellular bacterium has been isolated from pig intestines (Gebhart et al. 1993). Using primers specific for bacterial 16S ribosomal DNA, the DNA sequences were 91% similar to *D. desulfuricans* ATCC 27774. It would be interesting to see if this intracellular bacterium is similar to *Lawsonia intracellularis*, which is a pathogenic intracellular bacterium also related to *Desulfovibrio* (McOrist et al. 1995).

10.8 SRB Associated with Ruminates

The rumen was one of the first areas where bacterial communities were examined. Over 50 years ago, the spore-forming SRB (currently classified as *Dst. ruminis*) was isolated from the sheep rumen (Coleman 1960). Marvin P. Bryant isolated *D. desulfuricans* ATCC 27774 from the rumen of sheep, and this SRB is commonly used in research laboratories today because it grows with either nitrate or sulfate respiration. Subsequently, many SRB isolates have been demonstrated to display dissimilatory nitrate reduction. Sulfate-reducing bacteria have been demonstrated in fresh rumen samples from sheep, cattle, buffalo, deer, llama, and caecal samples from horses (Morvan et al. 1996). With the development of rRNA-targeted oligonucleotide probes for Gram-negative SRB, the rumen of cattle (steers and cows), sheep, and goats have been demonstrated to contain <1.6% of bacteria as *Desulfovibrio* species and, <0.8% as members of the *Desulfobacter* group; however, no members of the *Desulfobulbus* group were detected (Lin et al. 1997). While SRB are more commonly associated with the rumen, a significant number of unidentified *Desulfovibrio* spp. have been found in the colon of deer (Li et al. 2014). Although the Gram-negative and Gram-positive SRB are considered to be of low numbers in the rumen, they are a significant component of rumen microbiota and have an important role in the microbial community structure.

10.9 Interactions of SRB with Humans

It was not until the latter part of the twentieth century that the presence of SRB in the human gut became a focus for research. The potential SRB with the production of hydrogen sulfide to have a role in intestinal diseases has been reviewed (Macfarlane et al. 2007; Kushkevych 2017; Barton et al. 2017). SRB that inhabit the intestine and other tissues of an animal generally have the capability of metabolizing nitrate and nitric oxide (NO). The presence of intestinal nitrate is attributed to diet, while NO is produced from arginine as part of the host resistance to infections. (See Sect. 6.3.5 in Chap. 6 for nitrate, nitrite, and NO stress.) SRB that have had an association with the animal intestines are usually capable of dissimilatory nitrate reduction and produce rubredoxin:oxidoreductase, ROO, and the hybrid protein cluster, Hpc, which protect against damage due to NO (Cadby et al. 2017). ROO and Hpc have been demonstrated to protect *D. vulgaris* against macrophage infection (Figueiredo et al. 2013).

10.9.1 SRB as Flora of the Human Gastrointestinal Tract

It is well established that SRB are present in the intestinal tract of many humans, and due to the fact that H_2S, the respiratory product of SRB, has important physiological effects on human health, there is considerable interest in SRB in the human colon. The report by Moore et al. (1976) brought attention to the presence of *Desulfomonas pigra* (currently classified as *D. piger*) present in the human intestine. Beerens and Romond (1977) identified several SRB in human feces, and *Desulfovibrio* spp. were found to be present at about 10^7 bacteria per gram of feces with no difference in SRB numbers based on gender. *Desulfovibrio*, *Desulfomonas*, *Desulfobacter*, *Desulfobulbus*, and *Desulfotomaculum* are the most commonly identified SRB in the human intestine (Gibson et al. 1988). SRB become established as normal residents in the human intestine in the first few months of life where the SRB become attached to the gut wall. The abundance of SRB was found to vary with demographics, with a significantly greater number of SRB found in the feces of individuals living in the United Kingdom as compared to individuals in South Africa (Gibson et al. 1988). While cultivation of bacteria from intestinal contents or fecal material has often been used to establish the presence of SRB, culture-independent studies give a more complete evaluation of SRB present in healthy individuals as well as individuals with gastrointestinal diseases (Scanlan et al. 2009; Ding et al. 2012; Jia et al. 2012). Diet influences metabolic products in the intestine, and one of the important end products is H_2, which is used by both SRB and methanogens. Based on the evaluation of 87 healthy individuals, three different groups of intestinal microbiota were established (Gibson et al. 1993). Group 1 was characterized as having no detectable production of H_2S but appreciable production of methane. Group 2 had no detectable production of methane with high levels of H_2S produced. Group 3 had intermediate levels of both methane and H_2S produced.

10.9.2 Oral SRB

A publication over 25 years ago (van der Hoeven et al. 1995) signaled to the scientific community that SRB are present in the human mouth. Using enrichment techniques, cultures of *Desulfovibrio* and *Desulfobacter* were obtained from subgingival dental plaque of 58% of individuals tested (van der Hoeven et al.

1995). When examining the oral cavity of healthy and periodontitis patients, 10% of mucosa samples and 22% of tongue and supragingival plaque samples contained SRB with 86% of periodontitis patients having SRB in at least one of the periodontal pockets (Langendijk et al. 1999). In another study, SRB were obtained from 9 or 17 subgingival samples, and a unique strain of *Desulfovibrio* (that is, the isolate was capable of using glucose and fructose in addition to lactate as a carbon source and nitrate in addition to sulfate as a final electron acceptor) was isolated from a person with periodontal disease (Boopathy et al. 2002). In addition to SRB, periodontal pockets also contained *Treponema denticola*, *Porphytomonas gingivalis*, and *Bacteroides forsythus* (Langendijk-Genevaux et al. 2001). Two SRB species of interest with respect to periodontal disease are *Desulfomicrobium orale* and *Desulfovibrio* strain NY682, which has a 16S rDNA sequence similar to *D. fairfieldensis* (Langendijk et al. 2001).

Studies concerning the bacterial etiology of periodontitis prompted the use of various molecular techniques for the global characterization of oral bacterial ecology. From 16S clonal analysis of periodontal biofilms (Kumar et al. 2005), real-time quantitative PCR (Vianna et al. 2008), and RNA-oligonucleotide quantification technique (Teles et al. 2011), *Desulfobulbus* was more abundant in individuals with periodontitis than in healthy subjects. Additionally, phylogenic analysis of subgingival biofilm from individuals with implants had greater levels of *Desulfobulbus* as compared to biofilm from healthy individuals (da Silva et al. 2014).

When genomic amplicons of single cells of uncultured *Desulfovibrio* and *Desulfobulbus* were sequenced and compared to available deltaproteobacterial genomes, several virulence components were encoded in these uncultured SRB cells (Campbell et al. 2013). Based on a comparison of the oral *Desulfobulbus* strain to other SRB, it is most closely related to *Desulfobulbus propionicus*. Also, based on gene content, the oral *Desulfobulbus* is capable of both sulfate respiration and nitrate respiration and has genes encoding for glycolysis. When comparing oral host-associated *Desulfovibrio* and *Desulfobulbus* to environmental isolates, Campbell et al. (2013) found significant differences in gene content (see discussion in Sect. 10.9.5).

Since sulfate-reducing bacteria are Gram-negative cells, the endotoxin potential of the lipopolysaccharide (LPS) segment of these bacteria has been examined. The immunological response of gingival fibroblasts to LPS from *D. desulfuricans* demonstrated an enhanced secretion of IL-6 and IL-8. In a related experiment, co-culture of *D. desulfuricans* and *D. fairfieldiensis* with KB cells indicated that the SRB were internalized by the tissue cells by microtubule rearrangement and an enhanced production of IL-6 and IL-8 (Bison-Boutellienz et al. 2010). This stimulation of cytokine release by LPS supports the suggestion that SRB could be involved with the initiation and/or progression of periodontitis (Dzierżewicz et al. 2010; Bison-Boutellienz et al. 2010).

With SRB associated with periodontal disease, the presence of oral hydrogen sulfide may be suggestive of anaerobic plaque formation with developing gingivitis and periodontitis. Detection of hydrogen sulfide using micro-sulfide sensors may be an early indicator of gingival disease (Pavolotskaya et al. 2006). Various sources from the environment may contribute to the presence of oral SRB, and dental units not meeting the American Dental Association's limit are of concern because SRB may be present on high-speed drills and in compressed air and water syringes used in dental practice (Dogruöz et al. 2012).

10.9.3 Are Desulfovibrio *Human Pathogens?*

In the last couple of decades, there has been a lot of interest concerning the possible ill effects of SRB on humans. Perhaps the initial observation about SRB in a clinical setting was mentioned by John Postgate in his book (1979), where he mentions that a desulfoviridin-positive bacterium was isolated from the blood of a patient "under treatment for a variety of symptoms" and "the patient spontaneously recovered." Subsequently, there have been numerous instances where SRB have been isolated from humans, and these bacteria may be considered opportunists growing in immune-compromised individuals (Table 10.1). While there are several examples of high SRB levels in individuals with ulcerative colitis, definitive evidence for hydrogen sulfide toxicity has not been presented (Pitcher et al. 2000). Practices proposed to control the growth of intestinal SRB have included the use of bacteriophages (Kushkevych et al. 2020) and use of the prebiotic glycomacropeptide (Sawin et al. 2015). While most of the SRB isolated from humans are also routinely found in the environment, *D. fairfieldensis* has been isolated from hospitalized

Table 10.1 *Desulfovibrio* spp. reported in human infections under hospital treatment

Organism	Infection	Source	Reference
Desulfovibrio sp.	Appendicitis *Appendicitis* *Brain abscess*	Peritoneal fluid *Blood* *Pus*	Beerens and Romond (1977) Loubinoux et al. (2000) Lozniewski et al. (1999) Loubinoux et al. (1999, 2000)
D. desulfuricans	Sepsis	Blood	Porschen and Chan (1977) Goldstein et al. (2003) Tanamachi et al. (2011)
	Bacteremia	Blood	Liderot et al. (2010) Verstreken et al. (2012)
	Cerebral infarction	Blood	Hagiwara et al. (2014)
	Liver abscess	Blood	Koyano et al. (2015)
D. fairfieldensis	Liver abscess Polyps Meningoencephalitis Choledocholithiasis Bacteremia Acute sigmoiditis	Pus Blood Urine Blood Blood Blood	Tee et al. (1996) McDougall et al. (1997) McDougall et al. (1997) Pimentel and Chan (2007) Urata et al. (2008) Gaillard et al. (2011)
D. legallii	Joint infection	Synovial fluid	Vasoo et al. (2014)
D. intestinalis	Vaginatis	Vagina	Ichiishi et al. (2010)

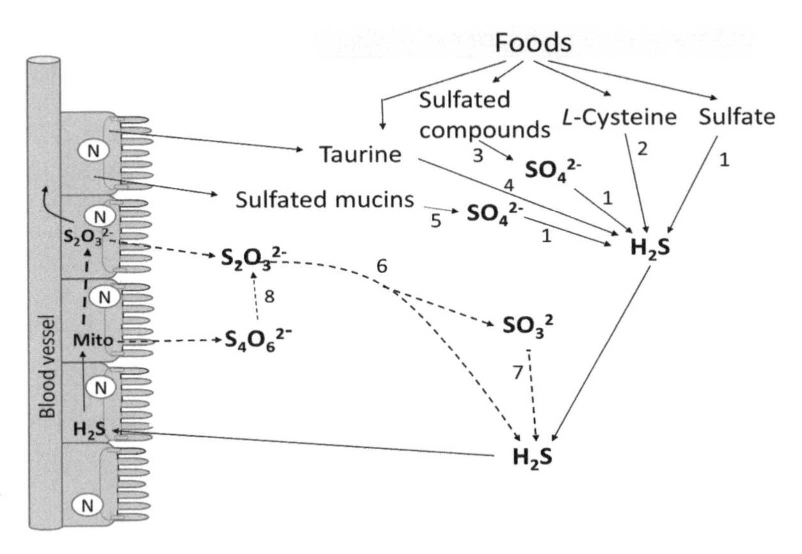

Fig. 10.1 A simplified model depicting H_2S production from major substrates in the healthy colon. A more detailed model is available (Barton et al. 2017). Solid lines indicate the production of H_2S, while broken lines indicate thiosulfate shunt. "Mito" refers to mitochondria and "N" refers to the nucleus in colonic cells. Enzymes are as follows: 1. dissimilatory sulfate reduction by SRB, 2. cysteine desulfhydrase, 3. alkyl sulfatase and aryl sulfatase, 4. taurine dehydrogenase and taurine:pyruvate aminotransferase, 5. glycosulfatase, 6. thiosulfate reductase, 7. dissimilatory sulfite reductase, and 8. tetrathionate reductase

patients, and it has not been reported from environmental studies where metagenomic analysis was used. It has been suggested that *D. piger* is associated with inflammatory bowel disease because the presence of this SRB was 55% greater in patients with inflammatory bowel disease than in healthy individuals (Loubinoux et al. 2002). Recently, it has been reported that SRB increase the expression of the proinflammatory pathway and the mechanism of this induction remains to be established (Singh et al. 2020). The toxicity of H_2S to humans is well recognized, and within the past decade, H_2S has gained attention as it was proposed to contribute to inflammatory bowel diseases, including ulcerative colitis (Pitcher and Cummings 1996) and colorectal cancer (Attene-Ramos et al. 2006). As reviewed by Loubinoux et al. (2002), Singh and Lin (2015), and Barton et al. (2017), hydrogen sulfide produced by SRB is of great concern in producing inflammatory bowel diseases. The intestine is a unique environment in that it is anaerobic in the center and aerobic at the periphery. When hydrogen sulfide enters the aerobic cells along the exterior of the intestine, mammalian cells oxidize sulfide, with thiosulfate being one of the major sulfur compounds produced. As indicated in Fig. 10.1, thiosulfate generated in the intestine is subsequently reduced to sulfide, and this forms a thiosulfate shunt in the intestinal sulfur cycle (Barton et al. 2017). A thiosulfate shunt in the sulfur cycle was proposed by Jørgensen (1990) for sulfur cycling in marine sediments.

Other publications suggest that SRB have a role in regressive autism (Finegold et al. 2012). Recently, there have been several publications evaluating the potential role of SRB and autism-like behaviors (Srikantha et al. 2019; Bezawada et al. 2020; Johnson et al. 2020). Children with autism spectrum disorder (ASD) have an elevated level of intestinal SRB as compared to healthy individuals (Finegold et al. 2012; De Angelis et al. 2013; Liu et al. 2019), and the abundance of SRB is correlated with the severity of ASD (Tomova et al. 2015). Although specific activities of SRB that may contribute to ASD have not been reported, *D. desulfuricans* AT5 cultivated from a patient with ADS was capable of immobilizing iron as greigite and pyrite in the proximal part of the intestinal tract (Karnachuk et al. 2021). Additionally, SRB have also been implicated as an endogenous neuromodulator (Abe and Kimura 1996) and contributing to Parkinson's disease (Murros et al. 2021; Lin et al. 2018; Murros 2020, 2021).

For many years, physicians were not routinely looking for SRB in samples, and the presence of SRB in clinical samples is believed to be greatly underestimated. SRB in samples from patients grow slowly and can be readily overgrown by other bacteria present in the sample. To isolate SRB from a clinical sample, commercially available anaerobic blood culture bottles are inoculated and incubated at 37 °C for several days (Loubinoux et al. 2003; Pimentel and Chan 2007; Koyano et al. 2015). Attempts to initially cultivate SRB from a clinical sample using media described by Postgate (1979) for environmental samples is not satisfactory. A new growth medium was developed by Zinkevich and Beech (2000), which indicated that all biopsy samples from ulcerative colitis patients and samples from non-colitis controls contained SRB while the growth of colitis specimens and non-colitis samples was 92% and 52%, respectively, and were positive for SRB when using Postgate medium B.

Samples are subcultured from the blood bottles onto blood agar plates incubated under anaerobic conditions. After 5 to 7 days, very small, translucent, nonhemolytic colonies are observed on the blood agar plates, and often these SRB display capability for nitrate reduction as well as sulfate reduction. Identification of isolated colonies as SRB is accomplished using multiplex PCR with molecular biology techniques or MALDI-TOF (matrix-assisted laser desorption/ionization time-of-flight) mass spectrometry. Broad implication of SRB with respect to regressive autism and human health has been raised (Finegold et al. 2012), and this may be an important research area in the future. The presence of virulence factors in SRB would suggest the possibility for specific SRB to be an opportunistic pathogen.

10.9.4 Antibiotic Susceptibility and Resistance

Several different antibiotics have been used successfully to inhibit the growth of clinical isolates of *Desulfovibrio* spp. (Table 10.2). Several *Desulfovibrio* spp. are resistant to penicillin, and this has been shown to be due to the production of β-lactamase (Morin et al. 2002; Lozniewski et al. 2001). Inactivation of numerous antibiotics has been demonstrated at elevated concentrations of H_2S (Shatalin et al. 2011). SRB shown to be resistant to ciprofloxacin are inhibited if bismuth subnitrate

Table 10.2 Antibiotic susceptibility and resistance of clinical isolates of *Desulfovibrio*

Clinical isolate	Susceptible	Resistance	Reference
Desulfovibrio sp.	Metronidazole, ciprofloxacin, *imipenem, amoxicillin-clavulanate ticarcillin-clavulnate, azithromycin, clindamycin*	Penicillin, ampicillin, *cephalothin, cofactor, ticarcillin, trimethoprim, sulfamethoxazole-trimethoprin, gentamicin, cefotaxime,* vancomycin	McDougall et al. (1997)
Desulfovobirio sp.	Cefotaxime piperacilin		Lozniewski et al. (1999)
Desulfovibrio spp.	Imipenem, metronidazole, clindamycin, chloramphenicol	Penicillin, piperacillin, piperacillin- tazobactam, cefoxitin, cefotetan	Lozniewski et al. (2001)
D. legallii	Metronidazole, ertapenem, meropenem	Piperacillin-tazobactam	Vasoo et al. (2014)
D. fairfieldensis	Ticarcillin-clavulanate		Pimentel and Chan (2007)
D. fairfieldensis	Ampicillin, rifampin		La Scola and Raoult (1999)
D. fairfieldensis	Ciprofloxacin		McDougall et al. (1997)
D. fairfieldensis	Cefotaxime, metronidazole, ampicillin, ciprofloxacin		Tee et al. (1996)
D. desulfuricans	Piperacillin-tazobactam amoxicillin-cavulanate		Koyano et al. (2015)
D. desulfuricans	Doxycycline		Goldstein et al. (2003)

is added to the antibiotic regimen (Ohge et al. 2003). While bismuth (Bi^{3+}) salts are only slightly soluble in water, inhibition of *D. desulfuricans* ATCC 27774 has been demonstrated when using deferiprone as a chelator for bismuth (Barton et al. 2016).

10.9.5 Do SRB Have Virulence Factors?

With the possibility that SRB could be opportunistic pathogens, there was interest in determining if SRB had virulence factors to enhance their growth in human tissues. Bacteria belonging to the *Desulfovibrio* genus have a Gram-negative cell wall, and scientists have explored the possibility that they could produce endotoxins. The optimal procedure for the isolation of lipopolysaccharide (LPS) from the outer membrane of *D. desulfuricans* was established (Pawtowska et al. 2004), and it was determined that the endotoxin consisted of three segments: lipid A, core polysaccharide, and O-specific polysaccharide (Wolny et al. 2011; Lodowska et al. 2012). The core polysaccharide for *D. desulfuricans* is glucosamine. The endotoxic polysaccharide core contains rhamnose, fucose, mannose, glucose, galactose, heptose, and 2-keto-3deoxyoctonic acid (KDO). The fatty acids present in lipid A included dodecanoic acid, tetradecanoic acid, 3-hydroxytetradecanoic acid, and hexadecanoic acid. An evaluation of the fatty acids in lipid A of several different isolates of *D. desulfuricans* revealed that a variation of fatty acid content and the endotoxic activity was observed with the different isolates (Zhang-Sun et al. 2015). In terms of the effect of LPS from *D. desulfuricans* on cells, endotoxin was found to induce secretion of IL-6, IL-8, and E-selectin by endothelial cells (Weglarz et al. 2003). LPS is important in modulating the activities of human gingival fibroblasts (Dzierżewicz et al. 2010), secretion of TNF-alpha by human mononuclear cells (Weglarz et al. 2006), and influence IL-6 secretion and IL-6 receptor genes in colon cancer (Cholewa et al. 2010). SRB that inhabit the intestine and other animal tissues generally have the capability of metabolizing nitrate and nitric oxide (NO). The presence of nitrate in animal intestines is attributed to diet, while NO is produced from arginine as part of the host resistance to infections. (See Sect. 6.3.5 in Chap. 6 for nitrate, nitrite, and NO stress.) SRB that have had an association with the animal intestine are usually capable of dissimilatory nitrate reduction and produce rubredoxin:oxidoreductase, ROO, and the hybrid protein cluster, Hpc, which protect against damage due to NO (Cadby et al. 2017). ROO and Hpc have been demonstrated to protect *D. vulgaris* against macrophage infection (Figueiredo et al. 2013).

In a study of uncultured oral SRB, genomic analysis of single cells of *Desulfobulbus* sp. and *Desulfovibrio* sp. revealed that these SRB contained numerous genes typically associated with environmental SRB, and they also have genes that enable them to survive within a host (Campbell et al. 2013). Seven single cells of *Desulfobulbus* sp. and five single cells of *Desulfovibrio* sp. were analyzed, and the estimated genome sizes of these bacteria were 2.48 Mbp and 2.63 Mbp, respectively. Genome analysis of cells identified as Dsb1-5 (*Desulfobulbus* sp.) and Dsv1 (*Desulfovibrio* sp.) revealed the presence of putative virulence factors that could be assigned to seven different categories. These categories are listed as follows and the number of genes annotated in each category is given in parenthesis: iron acquisition including siderophores (6), secretion (4), stress response (10), evasion (20), defense mechanism including multidrug export system (5), protease/peptidase (13), and adhesion (5) (Campbell et al. 2013). Clearly, additional evaluation of uncultured SRB is critical to evaluate their capability in functioning as oral pathogens and also in general human pathogens.

10.10 SRB Interactions with Plants

It is well documented that SRB are found in the rhizosphere, and a few studies have examined the interaction of SRB with plants. A black band has often been observed in golf courses and especially in the root zone of the putting greens of "Penncross" creeping bentgrass (*Agrostis palustris*). Berndt and Vargas (2006) demonstrated that this black zone is attributed to sulfate-reducing bacteria producing sulfide, which reacts with metal ions in the soil to produce the black precipitate. Since dissimilatory sulfate reduction is an anaerobic process, the low redox potential required for the growth of SRB could be achieved by the high respiratory activity of soil microorganisms, restricted diffusion due to water-logged soil, formation of soil salt crusts, or biofilm formation.

In rice fields of Senegal, sulfide accumulation is responsible for injurious effects on rice plants, with hydrogen sulfide being produced by 10^7 to 10^9 SRB in each gram of dry soil (Ouattara and Jacq 1992). Seven isolates were identified with six strains similar to *D. vulgaris* and one strain of *D. desulfuricans*. Since rice plants secrete high quantities of lactate and pyruvate from their roots (Ouattara and Jacq 1992), this would be an appropriate carbon source for the *Desulfovibrio*. Acetate or palmitate-oxidizing SRB were found in Senegal rice fields, but they were significantly less abundant than the lactate-oxidizing SRB. By using different electron donors in isolation media, cultures characteristic of *Desulfovibrio*, *Desulforhabdus*, *Desulfotomaculum*, *Desulfobulbus*, *Desulfobotulus*, and *Desulfosarcina* were obtained from rice field soil and roots of rice plants (Wind et al. 1999). The significance of SRB activities in rice patty soil has been toxicity to rice plants resulting in diminished grain yield, early plant death, and inhibition of seed germination. In many countries including Senegal and Burkina Faso (Otoidobiga et al. 2015), rice fields have been abandoned due to sulfide toxicity which has greatly disrupted regional agricultural economies.

Salt marshes constitute important ecosystems, and SRB are known to be present in the rhizosphere of *Spartina alterniflora*, an invasive plant in estuaries of China (Bahr et al. 2005). Using a probe for the *dsrAB* gene, Bahr and colleagues found considerable diversity in the SRB population. While 80% of the clone sequences were similar to genes from *Desulfosarcina* and *Desulfobacterium*, other sequences were similar to *Desulfotomaculum* sp. and *Syntrophobacteraceae* of the δ-proteobacterial group. Of considerable interest is the observation that a bacterial cluster contained the *dsrAB* gene but did not affiliate with any of the cultured SRB. It appears that the organic compounds secreted by *Spartina alterniflora* selects for the type of SRB present (Bahr et al. 2005). In another study, the invasion of *Spartina alterniflora* in (reed) *Phragmites australis*-vegetated sediments in the Yangtze River estuary produced an increase in the abundance of SRB, which were primarily members of the families *Desulfobacteraceae* and *Desulfobulbaceae* (Zeleke et al. 2013). With the invasion by *Spartina alterniflora*, the community composition of the SRB was not significantly changed; however, a unique cluster that consisted of about 11% of *dsrB* gene sequences did decline with the invasion (Zeleke et al. 2013). The rhizosphere of *Phragmites australis* in a noninvaded environment (Lake Velence, Hungary) was reported to contain *Desulfovibrio alcoholivorans*, various species of *Desulfovibrio*, *Desulfotomaculum*, and *Desulfobulbus* and uncultivated strains of *Desulfovibrio* and *Desulfobublus* as suggested from partial *dsrAB* sequences (Vladár et al. 2008). With the invasion of *S. alterniflora* in *Kandelia obovata*-dominated estuary of Jiulong River, rhizoplane bacteria (including SRB) were not as sensitive as endotrophic bacteria (Hong et al. 2015).

A diversity of SRB was reported to be associated with floating macrophytes in a lake in the Amazon basin of Bolivia (Achá et al. 2005). Members of the SRB subgroups detected on the macrophytes and roots included *Desulfotomaculum*, *Desulfobulbus*, *Desulfobacter*, *Desulfococcus–Desulfonema–Desulfosarcina*, and *Desulfovibrio–Desulfomicrobium. Desulfobacterium* was not detected on macrophytes in this plant study, which included a survey of C_4 plants (*Polygonum densiflorum* and *Hymenachne donacifolia)* and C_3 plants *(Ludwigia helminthorrhiza* and *Eichhornia crassipes*). In sediments containing rooted *Elodea nuttallii*, an invasive submerged macrophyte, the methylation of mercury was attributed to rhizosphere bacteria belonging to the *Desulfuromonadales* and *Desulfobacteraceae* taxonomic units, which were stimulated by root secretions (Regier et al. 2012).

The bacterial communities present in sea grass have been the subject of several studies, with SRB being involved with root decomposition or mutualistic activity. There is a difference in the bacterial community between the bulk sediment and the rhizosphere. *D. zosterae* was isolated from the root surface of the macrophyte *Zostera marina*, and it is unusual in that it is not only able to use fumarate in addition to short-chain organic acids as the electron donor but that it can grow by fermentation of fructose, pyruvate, or fumarate (Nielsen et al. 1999). While nitrogen fixation by root and rhizome-associated bacteria was several times greater than with bulk sediment, sulfate reduction on plant roots and rhizomes was only slightly elevated as compared to overall sulfate reduction (Nielsen et al. 2001). When using molecular techniques to study bacteria in sediments colonized by *Zostera noltii*, enrichments from a lactate-combined nitrogen media produced members of the *Desulfovibrionaceae* while lactate media without combined nitrogen selected *Desulfobacteriaceae* (Cifuentes et al. 2003). Not only did Cifuentes and colleagues suggest the presence of nondescribed species of SRB, but they also reported that an archaeon, nonextreme Crenoarchaeota, was present with the SRB in some of the enrichments.

With the sea grass *Zostera marina* (Jensen et al. 2007), oxygen loss is at the root tip, and this aeration inhibits adjacent SRB with SRB colonization occurring on older parts of the root and in the bulk sediment. With the sea grass *Halodule wrightii* (Küsel et al. 1999), SRB are located in the rhizoplane as well as in cortex cells deep within the root. While aerobic bacteria are commonly associated with plants as endophytes, the role of anaerobic bacteria such as SRB as endophytes merits additional study.

10.11 SRB on Surfaces of Living Marine Organisms

SRB become associated with the surface of different forms of marine life, and it is often difficult to classify this activity as ectosymbiotic or epizoic. Clams such as *Artica islandica* survive transient anoxia in marine sediments that have SRB on their surface, and it is considered that these SRB metabolize acetate, propionate, and succinate released from the bivalve (Bussmann and Reichardt 1991). Free-living anaerobic marine ciliates (*Metopus contortus* and *Caenomorpha levanderi*) were found to have SRB as ectosymbionts (Fenchel and Ramsing 1992).

Black band disease (BBD) is a serious problem attributed to a microbial consortium consisting of photosynthetic and nonphotosynthetic bacteria. *Desulfovibrio* spp. are found at the base of the black band, which is found between the living coral tissue and the coral skeleton (Bourne et al. 2011). Sulfide produced by the SRB enhances the growth of the black mat containing hundreds of bacterial species, which collectively destroy the healthy coral tissue.

10.12 SRB with Other Animals

As broad screening molecular techniques are used to identify environmental populations, SRB are being reported in various environmental samples. The broad distribution of SRB in various animal species indicates the ability of SRB to compete and grow in environments containing highly diverse microbial populations. The following list provides preliminary information about the distribution of SRB in animals, and these examples were not discussed earlier in this chapter:

- Based on 16S RNA sequences, the mucus shed by the medicinal leech (*Hirudo verbena*) contains *Desulfovibrio* spp., which account for about 3.6% of all bacteria with the shed mucus (Ott et al. 2015).
- A bacterial isolate from a dog with a fever of 105°F and a tense abdomen was shown by 16S rDNA analysis to be 99% identical to *D. desulfuricans* ATCC 27774 (Shukla and Reed 2000).
- A SRB was isolated from mesenteric lymph nodes of ferrets and hamsters with proliferative bowel diseases, and the sequence of the amplified 16S rRNA gene from this bacterium showed 87.5% similarity to *D. desulfuricans* (Fox et al. 1994).
- Cockchafers are European beetles of the genus *Melolontha*, which live both above and below ground. The adult lives above ground eating leaves, and the larvae eat roots below ground. It has been reported that 10–15% of the bacterial numbers in the hind gut of larvae and adult *Melolontha melolonthia* consist of *Desulfovibrio* spp. In the larvae of *Melolontha hippocastani*, a significant number of *Desulfovibrio* are found on the epithelium in the mid gut and on food in the hindgut. Since acetate is commonly found in the hind gut of *Melolontha melolonthia*, it has been proposed that a role of SRB in the gut is to modulate the acetate level (Arias-Cordero et al. 2012).
- In a survey of dogs with periodontal diseases, 70% of the periodontal pockets contained SRB, and higher levels of SRB were found on the dorsal surface of the tongue and supragingival plaque in dogs with periodontal disease as compared to healthy dogs (Costinar et al. 2010).
- Black howler monkeys (*Alouatta pigra*) growing in the wild have a diet of plant leaves, which are high in fiber and low in energy content and have a diversity of

intestinal *Desulfovibrionales* than captive monkeys fed commercial food (Nakamura et al. 2011).

- There are 26 different families of fecal microbiota in chickens, and the dominant bacteria belonged to *Campylobacteraceae* and *Desulfovibrioaceae* (Videnska et al. 2014).
- In Pekin and Muscovy ducks, the abundance of *Desulfovibrioaceae* in the ceca increased when the ducks were overfed (Vasaï et al. 2014).
- Evaluation of microbiota in the gastrointestinal tract of deer revealed that the greatest abundance of *Desulfovibrio* spp. (0.49%) was in the colon (Li et al. 2014).

10.13 Summary and Perspective

The adaptability of SRB is apparent with the broad distribution of these bacteria throughout the biosphere as they associate with both animals and plants. Requirements for successful association of SRB with animals appear to be a requirement for sulfate and a suitable anaerobic environment. However, these requirements are subject to some variation because SRB have highly flexible respiratory processes that enable them to grow in the absence of sulfate, and a complex oxygen stress response enables SRB to persist (at least transiently) to exposure to the aerobic atmosphere. Therefore, it should not be a great surprise that SRB are found in the termite gut and gut of root-feeding larvae. The association of SRB with the gutless marine worm, polychaete serpulid worms, and sea cucumbers indicates an adaptability to associate with invertebrate organisms. SRB are also found in the oral cavity and gut of humans and various animals. Historically, the demonstration of SRB in a sample from an animal has relied on the growth of the SRB, but the availability of molecular techniques to detect SRB may reveal many more associations of SRB with animal hosts. While hydrogen sulfide is an unpleasant end product of SRB metabolism, it is generally unknown what benefit, if any, SRB may provide to animals as a host or in the immediate environment. Additional research is needed in this area of SRB–host interaction.

References

Abe K, Kimura H (1996) The possible role of hydrogen sulfide as an endogenous neuromodulator. J Neurosci 1996(16):1066–1071

Achá D, Iñiguez V, Roulet M, Davée Guimarães JR, Luna R, Alanoca L, Sanchez S (2005) Sulfate-reducing bacteria in floating macrophyte rhizospheres from an Amazonian floodplain lake in Bolovia and their association with Hg methylation. Appl Environ Microbiol 71:7531–7535

Arakawa T, Ishikawa Y, Ushida K (2000) Volatile sulfur production by pig cecal bacteria in batch culture and screening inhibitors of sulfate-reducing bacteria. J Nutr Sci Vitaminol (Tokyo) 46: 193–198

Arias-Cordero E, Ping L, Reichwald K, Delb H, Platzer M, Boland W (2012) Comparative evaluation of the gut microbiota associated with the below- and above- ground life stages (larvae and beetles) of the forest cockchafer, *Melolontha hippocastani*. PLoS One. http://dx.doi.org/10.1371)journal.pone.0051557

Attene-Ramos MS, Wagner ED, Plewa MJ, Gaskins HR (2006) Evidence that hydrogen sulfide is a genotoxic agent. Mol Cancer Res 4:9–14

Bahr M, Crump BC, Klepac-Ceraj V, Teske A, Sogin ML, Hobbie JE (2005) Molecular characterization of sulfate-reducing bacteria in a New England salt marsh. Environ Microbiol 7:1175–1185

Barton LL, Granat AS, Lee S, Xu H, Ritz NL, Hider R, Lin HC (2019) Bismuth (III) interactions with *Desulfovibrio desulfuricans*: inhibition of cell energetics and nanocrystal formation of Bi_2S_3 and Bi^0. Biometals 32:803–811

Barton LL, Lyle DA, Ritz NL, Granat AS, Khurshid AN, Kherbik N, Hider R, Lin HC (2016) Bismuth(III) deferiprone effectively inhibits growth of *Desulfovibrio desulfuricans* ATCC 27774. Biometals 29:311–319

Barton LL, Ritz NL, Fauque GD, Lin HC (2017) Sulfur cycling and the intestinal microbiome. Dig Dis Sci 62:2241–2257

Beerens H, Romond C (1977) Sulfate-reducing anaerobic bacteria in human feces. Am J Clin Nutr 30:1770–1776

Berndt WL, Vargas JM Jr (2006) Dissimilatory reduction of sulfate in black layer. Hort Sci 41:815–817

Bezawada N, Phang TH, Hold GL, Hansen R (2020) Autism spectrum disorder and the gut microbiota in children: a systematic review. Ann Nutr Metab 76:16–29

Bison-Boutellienz C, Massin F, Dumas D, Miller N, Lozniewski A (2010) *Desulfovibrio* spp. survive within KB cells and modulate inflammatory responses. Mol Oral Microbiol 25:226–235

Boopathy R, Robichaux R, LaFont D, Howell M (2002) Activity of sulfate-reducing bacteria in human periodontal pocket. Can J Microbiol 48:1099–1103

Bourne DG, Muirhead A, Sato Y (2011) Changes in sulfate-reducing bacterial populations during the onset of black band disease. ISME J 5:559–564

Brauman A, Koenig JF, Dutreix J, Garcia JL (1990) Characterization of two sulfate-reducing bacteria from the gut of the soil-feeding termite, *Cubitermes speciosus*. Antonie Van Leeuwenhoek 58:271–275

Burnett BJ, Ritz NL, Barton LL, Wilson M, Singh SB, Lin HL (2015) Excretion of breath hydrogen is reduced by inhibition of sulfate reduction by molybdate in rats. Gastroenterology 148(4): S-924

Bussmann I, Reichardt W (1991) Sulfate-reducing bacteria in temporarily toxic sediments with bivalves. Mar Ecol Prog Ser 78:987–102

Cadby I, Faulkner M, Chèneby J, Long J, Van Helden J, Dolla A, Cole J (2017) Coordinated response of the *Desulfovibrio desulfuricans* 27774 transcriptome to nitrate, nitrite and nitric oxide. Sci Rep Nature Publishing Group 7(1). https://doi.org/10.1038/s41598-017-16403-4

Campbell AG, Campbell JH, Schwientek P, Woyke T, Allman S, Sczyrba A et al (2013) Multiple single-cell genomes provide insight into functions of uncultured deltaproteobacteria in the human oral cavity. PLoS One. https://doi.org/10.1371/journal.pone.0059361

Cholewa K, Weglarz L, Parfiniewicz B, Lodowska J, Jaworska-Kik M (2010) The influence of *Desulfovibrio desulfuricans* endotoxin on IL-6 and IL-6 receptor genes expression in colon cancer Caco-2 cells. Farn Przegl Nauk 3:27–32

Cifuentes A, Antón J, de Wit R, Rodriguwz-Valera F (2003) Diversity of bacteria and archaea in sulphate reducing enrichment cultures inoculated from serial dilution of *Zostera noltii* rhizosphere samples. Environ Microbiol 5:754–764

Coleman GS (1960) A sulphate-reducing bacterium from the sheep rumen. J Gen Microbiol 22: 423–436

Costinar L, Herman V, Pascu C (2010) The presence of sulfate-reducing bacteria in dog's oral cavity. Lucrări Stiintifice Medicină Veterinară Timişoara XLIII:128–131 da Silva ESC, Feres M, Figueiredo LC, Shibli JA, Ramiro FS, Faven M (2014) Microbiological diversity of peri-implantitis biofilm by Sanger sequencing. Clinical Oral Implants Res 25:1192–1199

De Angelis M, Piccolo M, Vannini L, Siragusa S, De Giacomo A, Serrazzanetti DI, Cristofori F, Guerzoni ME, Gobbetti M, Francavilla R (2013) Fecal microbiota and metabolome of children with autism and pervasive developmental disorder not otherwise specified. PLoS One 2013(8): e76993

Deplanche B, Hristova KR, Oakley HA, McCracken VJ, Aminov R, Mackie RI, Gaskins HR (2000) Molecular ecological analysis of the succession and diversity of sulfate-reducing bacteria in the mouse gastrointestinal tract. Appl Environ Microbiol 66:2166–2174

de Wit R, Bouvier T (2006) 'Everything is everywhere, but, the environment selects'; what did Baas Becking and Beijerinck really say? Environ Microbiol 8:755–756

Ding J, Zhang Q, Liu X, Tian F, Zhang H, Chen W (2012) Quantity of desulfovibrios and analysis of intestional microbiota diversity in healthy and intestinal disease people in Wuxi, Jiangsu province. Comparative Study, Wei Sheng Wu Xue Bao 52:1033–1039

Dogruöz N, Ilhan-Sungur E, Göksay D, Türetgen I (2012) Evaluation of microbial contamination and distribution of sulphate-reducing bacterial dental units. Environ Monitor Assess 184:133–139

Dolan MF (2001) Speciation of termite gut protists: the role of bacterial symbionts. Int Microbiol 4: 203–208

Dröge S, Limper U, Emtiazi F, Schönig I, Pavlus N, Drzyzga O, Fischer U, König H (2005) In vivo and in vitro sulfate reduction in the gut of the termite *Mastotermes darwiniensis* and the rose-chafer *Pachnoda marginata*. J Gen Appl Microbiol 51:57–64

Dubilier N, Mülders C, Ferdelman T, de Beer D, Pernthaler A, Klein M et al (2001) Endosymbiotic sulfate-reducing and sulfide-oxidizing bacteria in an oligochaete worm. Nature 411:298–302

Dzierżewicz Z, Szczerba J, Lodowska J, Wilny D, Gruchlik A, Orchel A, Weglarz L (2010) The role of *Desulfovibrio desulfuricans* lipopolysaccharides in modulation of periodontal inflammation through stimulation of human gingival fibroblasts. Arch Oral Biol 55:515–522

Egert M, Stingl U, Bruun LD, Pommerenke B, Brune A, Frieddrich MW (2005) Structure and topology of microbial communities in the major gut compartments of *Melolontha melolontha* larvae (Coleoptera: Scarabaeidae). Appl Environ Microbiol 71:4556–4566

Fenchel T, Ramsing NB (1992) Identification of sulphate-reducing ectosymbiotic bacteria from anaerobic ciliates using 16S rRNA binding oligonucleotide probes. Arch Microbiol 158:394–397

Figueiredo MCO, Lobo SAL, Sousa SH, Pereira FP, Wall JD, Nobre LS, Saraiva LM (2013) Hybrid cluster proteins and flavodiiron proteins afford protection to *Desulfovibrio vulgaris* upon macrophage infection. J Bacteriol 195:2684–2690

Finegold SM, Downes J, Summanen PH (2012) Microbiology of regressive autism. Anaerobe 18: 260–262

Fox JG, Dewhirst FE, Fraser GJ, Paster BJ, Shames B, Murphy JC (1994) Intracellular *Campylobacter*-like organisms from ferrets and hamsters with proliferative bowel disease is a *Desulfovibrio* sp. J Clin Microbiol 32:1229–1237

Fröhlich J, Sass H, Babenzien HD, Huhnigk T, Varma A, Saxena S et al (1999) Isolation of *Desulfovibrio intestinalis* sp. nov. from the hindgut of the lower termite *Mastotermes darwiniensis*. Can J Microbiol 45:145–152

Gaillard T, Pons S, Darles C, Beausset O, Monchal T, Brisou P (2011) Bactériémie à *Desulfovibrio fairfieldensis* au cours d'une sigmoïdite aiguë. [*Desulfovibrio fairfieldensis* bacteremia associated with acute sigmoiditis]. Med Mal Infect 41(5):267–8 French. https://doi.org/10.1016/j.medmal.2010.11.016. Epub 2011 Jan 3. PMID: 21208758

Gao F, Li F, Tan J, Yan J, Sun H (2014) Bacterial community composition in the gut content and ambient sediment of sea cucumber *Apostichopus japonicas* revealed by 16S rRNA gene pyrosequencing. PLoS One. https://doi.org/10.1371/journal.pone.0100092

Gebhart CJ, Barns SM, McOrist S, Lin GF, Lawson GH (1993) Ileal symbiont intracellularis, an obligate intracellular bacterium of porcine intestines showing a relationship to *Desulfovibrio* speces. Int J Syst Bacteriol 43:533–538

Gibson GR, Macfarlane GT, Cummings JH (1988) Occurrence of sulphate-reducing bacteria in human faeces and the relationship of dissimilatory sulphate reduction to methanogenesis in the large gut. J Appl Bacteriol 65:103–111

Gibson GR, Macfarlane GT, Cummings JH (1993) Metabolic interactions involving sulphate-reducing and methanogenic bacteria in the human large intestine. FEMS Microbiol Ecol 12: 117–125

Goldstein EJC, Citron DM, Peraino VA, Cross SA (2003) *Desulfovibrio desulfuricans* bacteremia and review of human *Desulfovibrio* infections. J Clin Microbiol 41:2752–2754

Guido A, Mastandrea A, Rosso A, Sanfilippo R, Tosti F, Riding R, Russo F (2014) Commensal symbiosis between agglutinated polychaetes and sulfate-reducing bacteria. Geobiology 12:265–275

Hagiwara S, Yoshida A, Omata Y, Tsukada Y, Takahashi H, Kamewada H et al (2014) *Desulfovibrio desulfuricans* bacteremia in a patient hospitalized with acute cerebral infarction: case study and review. J Infect Chemother 20:274–277

Hong Y, Liao D, Hu A, Wang H, Chen J, Khan S, Su J, Li H (2015) Diversity of endotrophic and rhizoplane bacterial communities associated with exotic *Spartina alternifora* and native mangrove using illumine amplicon sequencing. Can J Microbiol 61:723–733

Ichiishi S, Tanaka K, Nakao K, Izumi K, Mikamo H, Watanabe K (2010) First isolation of desulfovibrio from the human vaginal flora. Anaerobe 16:229–233

Jensen SI, Kühl M, Prieme A (2007) Different bacterial communities associated with the roots and bulk sediment of the seagrass *Zostera marina*. FEMS Microbiol Ecol 62:108–117

Jia W, Whitehead RN, Griffiths L, Dawson C, Bai H, Waring RH et al (2012) Diversity and distribution of sulfate-reducing bacteria in human faeces from healthy subjects and patients with inflammatory bowel disease. FEMS Immunol Med Microbiol 65:55–68

Johnson D, Letchumanan V, Thurairajasingam S, Lee LH (2020) A revolutionizing approach to autism spectrum disorder using the microbiome. Nutrients 12(7):1983. https://doi.org/10.3390/nu12071983. PMID: 32635373; PMCID: PMC7400420

Jørgensen BB (1990) A thiosulfate shunt in the sulfur cycle of marine sediments. Science 249:152–154

Karnachuk OV, Ikkert OP, Avakyan MR, Knyazev YV, Volochaev MN, Zyusman VS, four co-authors. (2021) *Desulfovibrio desulfuricans* AY5 Isolated from a patient with autism spectrum disorder binds iron in low-soluble greigite and pyrite. Microorganisms 9:2558. https://doi.org/10.3390/microorganisms9122558

Kerr BJ, Weber TE, Ziemer CJ, Spence C, Cotta MA, Whitehead TR (2010) Effect of dietary inorganic sulfur level on growth performance, fecal composition and measures of inflammation and sulfate-reducing bacteria in the intestine of growing pigs. J Anim Sci 89:426–437

Kleiner M, Wentrup C, Lott C, Teeling H, Wetzel S, Young J et al (2012) Metaproteomics of a gutless marine worm and its symbiotic microbial community reveal unusual pathways for carbon and energy use. Proc Natl Acad Sci U S A 109:7148–7149

Kováč J, Vítězová M, Kushkevych I (2018) Metabolic activity of sulfate-reducing bacteria from rodents with colitis. Open Med 13:344–349

Koyano S, Tatsuno K, Okazaki M, Ohkusu K, Sasaki T, Saito R, Okugawa S, Moriya K (2015) A case study of liver abscess with *Desulfovibrio desulfuricans* bacteremia. Case Rep Infect Dis 2015:Article ID 354168.,4 pages. https://doi.org/10.1155/2015/354168

Kuhnigk T, Branke J, Krekeler D, Cypionka H, König H (1996) A feasible role of sulfate-reducing bacteria in the termite gut. Syst Appl Microbiol 19:139–149

Kumar PS, Griffen AL, Moeschberger ML, Leys EJ (2005) Identification of candidate periodontal pathogens and beneficial species by quantitative 16S clonal analysis. J Clin Microbiol 43:3944–3955

Küsel L, Pinkart HC, Drake HL, Devereux R (1999) Acetogenic and sulfate-reducing bacteria inhabiting the rhizoplane and deep cortex cells of the sea grass *Halodule wrightii*. Appl Environ Microbiol 65:5117–5123

Kushkevych IV (2014) Etiological role of sulfate-reducing bacteria in the development of inflammatory bowel disease and ulcerative colitis. Am J Infect Dis Microbiol 2:63–73

Kushkevych IV (2017) Intestinal sulfate-reducing bacteria. Masarky University, Brno, p 320

Kushkevych I, Dordević D, Vítězová M, Rittmann SKR (2020) Environmental impact of sulfate-reducing bacteria, their role in intestinal bowel diseases, and possible control by bacteriophages. Appl Sci 11:735. https://doi.org/10.3390/app11020735

Kuwahara H, Yuki M, Izawa K, Ohkuma M, Hongoh Y (2017) Genome of '*Ca. Desulfovibrio trichonymphae*', an H_2-oxidizing bacterium in a tripartite symbiotic system within a protist cell in the termite gut. ISME J 11:766–776

La Scola B, Raoult D (1999) Third human isolate of a *Desulfovibrio* sp. identical to the provisionally named *Desulfovibrio fairfieldensis*. J Clin Microbiol 37:3076–3077

Langendijk PS, Hagemann J, van der Hoeven JS (1999) Sulfate-reducing bacteria in periodontal pockets and in healthy oral sites. J Clin Periodontol 26:596–299

Langendijk-Genevaux PS, Grimm WD, van der Hoeven JS (2001) Sulfate-reducing bacteria in relation with other potential periodontal pathogens. J Clin Periodontol 28:1151–1157

Langendijk PS, Kulik EM, Sandmeier H, Meyer J, van der Hoeven JS (2001) Isolation of *Desulfomicrobium orale* sp. nov. and *Desulfovibrio* strain NY682, oral sulfate-reducing bacteria involved in human periodontal disease. Int J Syst Evol Microbiol 51:1035–1044

Levitt MD, Furne J, Springfield J, Suarez F, DeMaster E (1999) Detoxification of hydrogen sulfide and methaethiol in the cecal mucosa. J Clin Invest 104:1107–1114

Levitt MD, Springfield J, Furne J, Koenig T, Suarez FL (2002) Physiological sulfide in the rat colon: use of bismuth to assess colonic sulfide production. J Appl Physiol 92:1655–1660

Li Z, Zhang Z, Xu C, Zhao J, Liu H, Fan Z, Yang F, Wright A-DG, Li G (2014) Bacteria and methanogens differ along the gastrointestinal tract of Chinese roe deer (*Capreolus pygargus*). PLoS One. https://doi.org/10.1371/journal.pone.0114513

Liderot K, Larsson M, Boräng S, Özenci V (2010) Polymicrobial bloodstream infection with *Eggerthella lenta* and *Desulfovibrio desulfuricans*. J Clin Microbiol 48:3810–3812

Lin C, Raskin L, Stahl DA (1997) Microbial community structure in gastrointestinal tracts of domestic animals: comparative analysis using rRNA-targeted oligonucleotide probes. FEMS Microbiol Ecol 22:281–294

Lin A, Zheng W, He Y, Tang W, Wei X, He R et al (2018) Gut microbiota in patients with Parkinson's disease in Southern China. Parkinsonism Relat Disord. https://doi.org/10.1016/j.parkreldis.2018.05.007

Liu F, Li J, Wu F, Zheng H, Peng Q, Zhou H (2019) Altered composition and function of intestinal microbiota in autism spectrum disorders: a systematic review. Transl Psychiatry 9:43. https://doi.org/10.1038/s41398-019-0389-6

Lodowska J, Wolny D, Jaworska-Kik M, Kurkiewicz S, Dzierżewicz Z, Weglarz L (2012) The chemical composition of endotoxin isolated from intestinal strain of *Desulfovibrio desulfuricans*. Sci World J 2012:Article ID 647352

Loubinoux J, Bronowicki JP, Pereira IA, Mougenel JL, Faou AE (2002) Sulfate-reducing bacteria in human feces and their association with inflammatory bowel diseases. FEMS Microbiol Ecol 40:107–112

Loubinoux J, Maurer R, Schumacher H, Carlier JP, Mory F (1999) First isolation of *Desulfovibrio* species as part of a polymicrobial infection from a brain abscess. Eur J Clin Microbiol Infect Dis 18:602–603

Loubinoux J, Mory F, Pereira IAC, LeFaou AE (2000) Bacteremia caused by a strain of *Desulfovibrio* related to the provisionally named *Desulfovibrio fairfieldensis*. J Clin Microbiol 38:931–934

Loubinoux J, Jaulhac B, Piemont Y, Monteil H, Le Faou AE (2003) Isolation of sulfate-reducing bacteria from human thoracoabdominal pus. J Clin Microbiol 41:1304–1306

Lozniewski A, Labia R, Haristoy X, Mory F (2001) Antimicrobial susceptibilities of clinical *Desulfovibrio* isolates. Antimicrob Agents Chemother 45:2933–2935

Lozniewski A, Maurer P, Schumacher H, Carlier JP, Mory F (1999) First isolation of *Desulfovibrio* species as part of a polymicrobial infection from a brain abscess. Eur J Clin Microbiol Infect Dis 18:602–603

Macfarlane GT, Cummings JH, Macfarlane S (2007) Sulphate-reducing b acteria and the human large intestine. In: Barton LL, Hamilton WA (eds) Sulphate-reducing bacteria: environmental and engineered systems. Cambridge University Press, Cambridge, UK, pp 503–521

Matsui GY, Ringelberg DB, Lovell CR (2004) Sulfate-reducing bacteria in tubes constructed by the marine infaunal Polychaete *Diopatra cuprea*. Appl Environ Microbiol 70:7053–7065

McDougall R, Robson J, Paterson D, Tee W (1997) Bacteremia caused by a recently described novel *Desulfovibrio species*. J Clin Microbiol 35:1805–1808

McOrist S, Gebhart CJ, Boid R, Barns SM (1995) Characterization of *Lawsonia intracellularis* gen. nov., sp. nov., the obligately intracellular bacterium of porcine proliferative enteropathy. Int J Sys Bacteriol 45:820–825

Moore WEC, Johnson JL, Holderman LV (1976) Emendation of *Bacteriodaceae* and *Butyrivibrio* and descriptions of *Desulfomonas* gen. nov. and ten new species in the genera *Desulfomonas, Butyrivibrio, Eubacterium, Clostridium* and *Ruminococcus*. Int J Bacteriol 26:238–252

Morin A-S, Poirel L, Mory F, Labia R, Nordmann P (2002) Biochemical-genetic analysis and distribution of DES-1, an amber class A extended-spectrum β-lactamase from *Desulfovibrio desulfuricans*. Antimicrob Agents Chemother 46:3215–3222

Morvan B, Bonnemoy F, Fonty G, Gouet P (1996) Quantitative determination of H_2 – utilizing acetogenic and sulfate-reducing bacteria and methanogenic archaea from the digestive tract of different mammals. Curr Microbiol 32:129–133

Murros KE (2020) Sulfate-reducing gut bacteria and Parkinson's disease. Eur J Neurol. https://doi.org/10.1111/ene.14626

Murros KE (2021) Sulfate reducing gut bacteria and Parkinson's disease. Eur J Neurol 28(3):e21. https://doi.org/10.1111/ene.14626

Murros KE, Huynh VA, Takala TM, Saris PEJ (2021) *Desulfovibrio* bacteria are associated with Parkinson's disease. Front Cell Infect Microbiol 11:652617. https://doi.org/10.3389/fcimb.2021.652617

Nakamura N, Amato KR, Garber PA, Gaskins R (2011) Analysis of the hydrogenotrophic microbiota of wild and captive black howler monkeys (*Alouatta pigra*) in Palenque national park, Mexico. Am J Primatol 73:909–919

Nielson JT, Liesack W, Finster K (1999) *Desulfovibrio zosterae* sp. nov., a new sulfate reducer isolated from surface-sterilized roots of the seagrass *Zostera marina*. Int J Syst Bacteriol 49:859–865

Nielson LB, Finster K, Welsh DT, Donelly A, Herbert RA, de Wit R, Lomstein BA (2001) Sulphate reduction and nitrogen fixation rates associated with roots, rhizomes and sediments from *Zostera noltii* and *Spartina maritime* meadows. Environ Microbiol 3:63–71

Ohge H, Furne JK, Springfield J, Sueda T, Madoff RD, Levittet MD (2003) The effect of antibiotics and bismuth on fecal hydrogen sulfide and sulfate-reducing bacteria in the rat. FEMS Lett 228:137–142

Otoidobiga CH, Keita A, Yacouba H, Traore AS, Dianou D (2015) Dynamics and activity of sulfate-reducing bacterial populations in patty soil under subsurface drainage: case study of Kamcoinse in Burkina Faso. Agric Sci 6:1393–1403

Ott BM, Rickards A, Gehrke L, Rio RV (2015) Characterization of shed medicinal leech mucus reveals a diverse microbiota. Front Microbiol. https://doi.org/10.3389/fmicb.2014.00757

Ouattara AS, Jacq VA (1992) Characterization of sulfate-reducing bacteria isolated from Senegal rice fields. FEMS Microbiol Ecol 101:217–228

Pavolotskaya A, McCombs G, Darby M, Marinak K, Dayanand NN (2006) Sulcular sulfide monitoring: an indicator of early dental plaque-induced gingival disease. J Dental Hygiene 80:1–12

Pawtowska K, Lodowska J, Jaworska-Kik M, Dzierzewicz Z, Weglarz L, Wilczok T (2004) Choice of isolation procedure for endotoxins from the outer membrane of the bacteria *Desulfovibrio desulfuricans*. Med Dosw Mikrobiol 56:293–300

Pimentel JD, Chan RC (2007) *Desulfovibrio fairfieldensis* bacteremia associated with cholidocholithiasis and endoscopic retrograde cholangiopancreatography. J Clin Microbiol 45:2747–2750

Pitcher MCL, Beatty ER, Cummings JH (2000) The contribution of sulphate-reducing bacteria and 5-aminosalicylic acid to faecal sulphide in patients with ulcerative colitis. Gut 2000(46):64–72

Pitcher MCL, Cummings JH (1996) Hydrogen sulfide: a bacterial toxin in ulcerative colin? Gut 39: 1–4

Porschen RK, Chan P (1977) Anaerobic vibrio-like organisms cultured from blood: *Desulfovibrio desulfuricans* and *Succinivibrio* species. J Clin Microbiol 5:444–447

Postgate JR (1979) The Sulphate-reducing bacteria. Cambridge Press, Cambrodge. UK

Poulsen HV, Jensen BB, Finster K, Spence C, Whitehead TR, Cotta MA, Canibeet N (2012) Microbial production of volatile sulphur compounds in the large intestine of pigs fed two different diets. J Appl Microbiol 113:143–154

Regier N, Frey B, Converse B, Roden E, Grosse-Honebrink A, Bravo AG, Cosio C (2012) Effect of *Elodea nuttallii* roots on bacterial communities and MMHg proportion in a Hg polluted sediment. PLoS One 7(9):e45565. https://doi.org/10.1371/journal.pone0045565

Rey FE, Gonzalez MD, Cheng J, Wu M, Ahern PP, Gordon JI (2013) Metabolic niche of a prominent sulfate-reducing human gut bacterium. Proc Natl Acad Sci U S A 110:13582–13587

Ritz NL, Burnett BJ, Barton LL, Wilson M, Singh SB, Lin HC (2015) Live but not killed sulfate-reducing bacteria slow intestinal transit in a bismuth-reversible fashion in mice. Gastroenterology 148 Supplement1:89-S-90

Ritz NL, Burnett BJ, Setty P, Reinhart KM, Wilson M, Alcock J et al (2016) Sulfate-reducing bacteria impairs working memory in mice. Physiol Behav 157:281–287

Ruehland C, Blazejak A, Lott C, Loy A, Erséus C, Dubilier N (2008) Multiple bacterial symbionts in two species of co-occurring gutless oligochaete worms from Mediterranean Sea grass sediments. Environ Microbiol 10:3404–3416

Sato T, Hongoh Y, Noda S, Hattori S, Ui S, Ohkuma M (2009) *Candidatus* Desulfovibrio trichonymphae, a novel intracellular symbiont of the flagellate *Trichonympha agilis* in termite gut. Environ Microbiol 11:1007–1015

Sawin EA, De Wolfe TJ, Aktas B, Stroup BM, Murali SG, Steele JL, Ney DM (2015) Glycomacropeptide is a prebiotic that reduces *Desulfovibrio* bacteria, increases cecal short-chain fatty acids, and is anti-inflammatory in mice. Am J Physiol Gastrointest Liver Physiol 309: G590–G601

Scanlan PD, Shanahan F, Marchesi JR (2009) Culture-independent analysis of desulfovibrios in the human distal colon of healthy, colorectal cancer and polypectomized individuals. FEMS Microbiol Ecol 69:213–221

Shatalin K, Shatalina E, Mironov A, Nudler E (2011) H2S: a universal defense against antibiotics in bacteria. Science 334:986–990

Shukla SK, Reed KD (2000) *Desulfovibrio desulfuricans* bacteremia in a dog. J Clin Microbiol 38: 1701–1702

Singh SB, Lin HC (2015) Hydrogen sulfide in physiology and diseases of the digestive tract. Microorganisms 3:866–889

Singh S, Coffman C, Carroll-Portillo A, Lin HC (2020) Sulfate-reducing bacteria increase expression of proinflammatory pathway via notch signaling in macrophages. Gastrroenterology 158:S-1049

Srikantha P, Mohajeri MH (2019) The possible role of the microbiota-gut-brain-axis in autism spectrum disorder. Int J Mol Sci 20(9):2115. https://doi.org/10.3390/ijms20092115. PMID: 31035684; PMCID: PMC6539237

Strassert JFH, Köhler T, Wienemann THG, Ikeda-Ohtsubo W, Faivre N, Franckenberg S, three co-authors. (2012) '*Candidatus* Ancillula trichonymphae', a novel lineage of endosymbiotic

Actinobacteria in termite gut flagellates of the genus *Trichonympha*. Environ Microbiol 14: 3259–3270

Suarez F, Furne J, Springfield J, Levitt M (1998) Production and elimination of sulfur-containing gasses in the rat colon. Am J Phys 274:G727–G733

Tanamachi C, Hashimoto K, Itoyama T, Horita R, Yano T, Tou K et al (2011) A case of *Desulfovibrio desulfuricans* cultured from blood in Japan. Rinsho Byori 59:466–469

Tee W, Dyall-Smith M, Woods W, Eisen D (1996) Probable new species of *Desulfovibrio* isolated from a pyrogenic liver abscess. J Clin Microbiol 34:1760–1764

Teles FRF, Teles RP, Siegelin Y, Paster B, Haffajee AD, Socransky SS (2011) RNA-oligonucleotide quantification technique (ROQT) for the enumeration of uncultivated bacterial species in subgingival biofilms. Mol Oral Microbiol 26:127–139

Tomova A, Husarova V, Lakatosova S, Bakos J, Vlkova B, Babinska K, Ostatnikova D (2015) Gastrointestinal microbiota in children with autism in Slovakia. Physiol Behav 138:179–187

Trinkerl M, Breunig A, Schauder R, König H (1990) *Desulfovibrio termitidis* sp. nov., a carbohydrate-degrading sulfate-reducing bacterium from the hindgut of a termite. Syst Appl Microbiol 13:372–377

Urata T, Kikuchi M, Hino T, Yusuke Yoda Y, Kiyoko Tamai K, Kodaira Y, Hitomi S (2008) Bacteremia caused by *Desulfovibrio fairfieldensis*. J Infect Chemother 14:368–370

van der Hoeven JS, van den Kieboom CWA, Schaeken MJM (1995) Sulfate-reducing bacteria in the periodontal pocket. Oral Microbiol Immunol 10:288–290

Vasaï F, Brugirard Ricaud K, Bernadet MD, Cauquil L, Bouchez O, Combes S, Davail S (2014) Overfeeding and genetics affect the composition of intestinal microbiota in *Anas platyrhynchos* (Pekin) and *Cairina moschata* (Muscovy) ducks. FEMS Microbiol Ecol 87:204–216

Vasoo S, Mason EL, Gustafson DR, Cunningham SA, Cole NC, Vetter EA, five co-authors. (2014) *Desulfovibrio legallii* prosthetic shoulder joint infection and review of antimicrobial susceptibility and clinical characteristics of desulfovibrio infections. J Clin Microbiol 52:3105–3110

Verstreden I, Laleman W, Waulters G, Verhaegen J (2012) *Desulfovibrio desulfuricans* bacteremia in a immunocompromised host with a liver graft and ulcerative colitis. J Clin Microbiol 50:199–201

Vianna ME, Holtgraewe S, Seyfarth I, Conrads G, Horz HP (2008) Quantitative analysis of three hydrogenotrophic microbial groups, methanogenic archaea, sulfate-reducing bacteria, and acetogenic bacteria, within plaque biofilms associated with human periodontal disease. J Bacteriol 190:3779–3785

Videnska P, Rahman M, Faldynova M, Babak V, Matulova ME, Prukner-Radovcic E et al (2014) Characterization of egg laying hen and broiler fecal microbiota in poultry farms in Croatia, Czech Republic, Hungary and Slovenia. PLoS One 9. https://doi.org/10.1371/journal.pone.0110076

Vladár P, Rusynyák A, Máfialigen K, Borsodi AK (2008) Diversity of sulfate-reducing bacteria inhabiting the rhizosphere of *Phragmites australis* in Lake Velencer (Hungary) revealed by a combined cultivation-based and molecular approach. Microbial Ecol 56:64–75. https://doi.org/10.1007/s00248-007-9324-0

Weglarz L, Dzierzewicz Z, Shop B, Orchel A, Parfiniewicz B, Wisniowska B, Swiatkowska L, Wilczok T (2003) *Desulfovibrio desulfuricans* lipopolysaccharides induce endothelial cell IL-6 and IL-8 secretion and E-selectin and VCAM-1 expression. Cell Mol Lett 8:991–1003

Weglarz L, Parfiniewicz B, Mertas A, Kondera-Anasz Z, Jaworska-Kik M, Dzierżiewicz Z, Swiatkowska L (2006) Effect of endotoxins isolated from *Desulfovibrio desulfuricans* soil and intestinal strain on the secretion of TNF-alpha by human mononuclear cells. Polish J Environ Stud 15:615–622

Wind T, Stubner S, Conrad R (1999) Sulfate-reducing bacteria in rice field soil and on rice roots. Syst Appl Microbiol 22:269–279

Whitney MH, Nicolai R, Shurson GC (1999) Effects of feeding low sulfur starter diets on growth performance of early weaned pigs and odor, hydrogen sulfide, and ammonia emissions in nursery rooms. J Anim Sci 77(Suppl):70 (Abstr.) 10064029

Wolny D, Lodowska J, Jaworska-kik M, Kurkiewicz S, Węglarz L, Dzierżewicz Z (2011) Chemical composition of *Desulfovibrio desulfuricans* lipid A. Arch Microbiol 193:15–21

Zahn JA, Hatfield JL, Laird DA, Hart TT, Do YS, DiSpirito AA (2001) Functional classification of swine manure management systems biased on effluent and gas emission characteristics. J Environ Qual 30:635–647

Zeleke J, Sheng Q, Wang JG, Huang MY, Xia F, Wu JH, Quan Z-X (2013) Effects of *Spartina alterniflora* invasion on the communities of methanogens and sulfate-reducing bacteria in the estuarine marsh sediments. Front Microbiol 4:423. https://doi.org/10.3389/fmicb.2013.00243. eCollection 2013

Zhang-Sun W, Augusto LA, Zhao L, Caroff M (2015) *Desulfovibrio desulfuricans* isolates from the gut of a single individual: structural and biological lipid A characterization. FEBS Lett 589:165–171

Zinkevich V, Beech IB (2000) Screening of sulfate-reducing bacteria in colonoscopy samples from healthy and colitic human gut mucosa. FEMS Microbiol Ecol 34:147–155

Index

A
Accelerated low water corrosion (ALWC), 438
Acetate kinase, 303
Acetobacterium woodii, 432
Acid mine drainage (AMD), 388–390
Acid mine waters, 355
Acid rock drainage (ARD), 388–390
Acidification, 100
Acidithiobacillus sp., 428
Acidophiles, 490–492
Acidophilic thermophile, 490
Adaptations to the environment
  extremely acidic environments, 498–499
  impact of temperature, 496–497
  soda lakes and other alkaline environments, 499–501
Adenosine 5'-phosphosulfate (APS), 100–102
Aerobic oxidative phosphorylation, 282
Aerotaxis, 17
AhpC-like enzyme, 62
Alcohol dehydrogenase, 257
Alkaliphilic SRB strains, 487
Allyl isothiocyanate, 446
Alternate energy-yielding systems, 255, 275
Amino acid, 311
Aminolevulinic acid (ALA), 219, 220
Anaerobic bacteria, 20, 331
Anaerobic bacterium, 37
Anaerobic corrosion, 12
Anaerobic decomposition, 33
Anaerobic heterotrophic bacteria, 358
Anaerobic methane oxidation (AMO), 360–362
Anaerobic methanotrophic archaea (ANME), 32, 361, 366
Anaerobic mineralization, 359
Anaerobic oxidation of methane (AOM), 30, 366
Anaerobic oxidative phosphorylation, 282
Anionic depolarization, 431
Antibodies, 71
Antisera-based reactions, 60
APS reductase, 326
Archaea, 79
Archaea domain, 68, 83, 85, 86
Archaeoglobus, 430
Archaeoglobus fulgidus, 126
Archaeoglobus fulgidus VC-16 DSM 4304, 83
Assimilatory sulfite reductase (aSiR), 131–133
Atmospheric pollutants, 428
Atomic force microscopy (AFM), 97
ATP sulfurylase (ATPS), 123, 125, 326
ATPase (F-type and V-5ype), 269–270
ATP-binding cassette (ABC) system, 142
Autotrophy, 33

B
Bacillus brevis, 450
Bacillus megaterium, 132
Bacillus subtilis, 450
Bacteria, 449, 450
Bacterial cell, 437
Bacterial ferredoxins, 191
Benzalkonium chloride (BAC), 445
Biochemical characteristics
  cell architecture, 60–63
  cell morphology and anatomy, 57–59
  cytoplasmic structures, 64–69
  gram-positive, 57
  microorganisms, 74

L. L. Barton, G. D. Fauque, *Sulfate-Reducing Bacteria and Archaea*,
https://doi.org/10.1007/978-3-030-96703-1

Biochemical characteristics (*cont.*)
spore-forming bacteria, 69, 70
sulfate-reducing prokaryotes, 57, 58
Biocorrosion
artistic stoneworks, 428
biofilms, 449, 450
chemical control, 444, 446
concrete, 427, 428
destructive effects, 427
ferrous metals, 431–436, 445
ferrous wire, 431
green strategies, 446
nanowires, 441, 442
nitrate, 448, 449
O_2, 447, 448
oil facility, 440
outer membrane cytochromes, 441, 442
physical methods, 443, 444
seawater, 441
shuttle molecules, 443
steel alloys, 436
stone, 427, 428
sulfate reduction process, 439
Biofilm cells, 334
Biofilms, 322, 323, 334, 336, 337, 437
Biologically induced minerals
biogenic metals, 377
carbonate minerals and dolomite, 381
Co sulfides, 380–381
Cu sulfide deposits, 378
iron sulfide mineral precipitation, 377–378
Mo sulfide minerals, 380
Ni sulfide formations, 379–380
Zn sulfide deposits, 379
Biomarkers, 70, 71
Bioremediation, 355, 388–392
Black band disease (BBD), 545
Bohr effect, 211

C
Cable bacteria, 59, 95–100
Cadmium resistance, 62
Calcium carbonate ($CaCO_3$), 428
Calcium sulfate dihydrate, 428
Caldivirga maquilingensis, 85
"*Candidatus* Magnetoglobus multicellularis", 373
Carbon cycling, 356
anaerobic methane oxidation (AMO), 360–362
decomposition, organic matter, 357–360
Carbon dioxide
laboratory experiments, 18
reduction, 18
strains, 18
Carbonyl cyanide m-chlorophenylhydrozone (CCCP), 138
Catalase, 218, 219
Cathodic depolarization, 431
Cell architecture
nanowires, 62, 63
surfaceome and outer membrane protein complexes, 60–62
Cell shaving, 60
Cell wall, 59
Chemical microbially influenced corrosion (CMIC), 433, 436
Chemotaxonomy
biomarkers, 70, 71
cytochromes, 72
DNA G + C content, 74
FISH technologies, 71, 72
GeoChip, 71, 72
lipids, 72–74
PhyloChip, 71, 72
quinones, 72–74
Chlorate decomposition, 391
Chloropseudomonas ethylica, 27
Citrobacter amalonaticus, 436
Clostridia, 441
Co-culture biofilm, 31
CO cycling, 277
Co^{III}EDTA (ethylenediaminetetraacetic acid), 380
Conserved hypothetical proteins, 62
Coo hydrogenase, 183, 185
CO oxidation, 9, 10
Crenarchaeota organisms, 3
Crisper, 337
CRISPR-Cas systems, 337
in *Desulfovibrio* spp., 337, 338
Cryptic plasmid (pDMC1), 87
Cryptic sulfur cycle, 100
Culture media, 1
Cytochrome *b*, 173, 211, 221
Cytochrome c_{553}, 203
Cytochrome-rich SRP, 72
Cytochromes, 62, 72
characteristic, 202
c-type cytochromes, 202
electron transport complexes, 212
heme *b* distribution in sulfate reducers, 211–212
hexadecaheme cytochrome *c*3 (HmcA), 208–209
homodimeric diheme split-soret cytochrome *c*, 205–206

molecular docking, Bohr effect and proton thrustor, 209–210
monoheme cytochrome *c*553, 203–205
nonaheme cytochrome *c*3 (NhcA), 208–209
octaheme cytochrome *c*3 (Mr 26,000), 207–208
physiological activities, 203
tetraheme cytochrome *c*3 (TpI-*c*3/TpII-*c*3), 206–207
Cytoplasmic hydrogenases, 259
Cytoplasmic proteins
with high redox potentials
desulfoferrodoxin (Dfx), 200
desulforedoxin (Dx), 201
neelaredoxin (Nlr), 201, 202
rubredoxins, 197–199
rubrerythrin (Rr), 199, 200
with low redox potentials
ferredoxin, 191–194
flavodoxins, 194–197
Cytoplasmic structures
gas vacuoles, 69
iron inclusions, 63, 65–67, 85
magnetosomes, 63, 65–67, 85
PHA, 64, 65
phosphorus inclusions, 67, 69
polyglucose, 64

D

Decolorization, 396
Dehalorespiration, 20
Deltaproteobacteria, 81, 83, 85–87, 93, 373, 441
Desulfobacter postgatei, 140
Desulfobacterium autotrophicum, 147
Desulfobulbus propionicus, 126, 141
Desulfocapsa sulfoexigens, 24
Desulfocapsa thiozymogenes, 150
Desulfococcus multivorans, 126, 141
Desulfoferrodoxin (Dfx), 200
Desulfomicrobium baculatum, 136
Desulforedoxin (Dx), 201
Desulfosarcina variabilis, 126
Desulfotomaculum, 450
Desulfovibrio alaskensis
electron flow, 254
energy metabolism, 249
Desulfovibrio desulfuricans, 35, 375
Desulfovibrio sp., 57, 59, 122, 428, 450
distribution of proviruses and CRISPR arrays, 339
energy balance, 247
Desulfovibrio vulgaris, 123
acetate oxidation, 310–312
butyric and propionic acid oxidation, 312–313
cell surface
cell size, 297
outer membranes, 296
porins, 296
vesicles, 296, 297
CO as a metabolite, 316
coculture, 335
CO cycle, 317
formate oxidation, 317–318
fumarate respiration
fumarate disproportionation, 304
fumarate reductase, 304
fumarate-sulfate system, 304
genome analysis, 295
glycolytic capability, 307
growth with CO, 315–316
identification of genes, 295
inhibition by CO, 317
intermediary carbon metabolism, 309
intermediary metabolism
fueling substrates, 307
TCA cycle, 305, 307
lactate dehydrogenase proteins, 298
lactate oxidation
cellular energy, 297
lactate fermentation, 300
lactate-sulfate metabolism, 299
operon luo, 298
oxidation of alcohols
ethanol, 313
methanol, 314–315
pyruvate oxidation
acetate kinase, 303
phosphotransacetylase, 303
pyruvate as an electron donor, 300–302
pyruvate fermentation, 302–303
pathway, 299
stress response, 327
sugars and amino acids, 307–310
Desulfoviridin test, 70
Desulfuromonas acetoxidans, 133, 143
Detoxification, 372, 376, 387, 389
Dimethylsulfoniopropionate (DMSP), 18, 150
Dimethylsulfoxide (DMSO), 21
Direct electron transfer (DET), 434, 435
Disproportionation, 16, 21–24
Dissimilatory (bi)sulfite reductase (DSR), 100–102
Dissimilatory sulfate reduction, 270–272, 531
anaerobic bacteria and archaea, 134

Dissimilatory sulfate reduction (*cont.*)
- *aprAB* and *qmoABC* genes, 132
- APS reductase, 130
- biological sulfur cycle, 146
- desulfoviridin, 135
- enzyme, 134
- mechanism of bisulfite reduction, 138–140
- oxidation of H_2, 134
- polyacrylamide gel electrophoresis, 136
- porphyroprotein, 134
- Qmo and Apr proteins, 125
- QmoABC-AprAB interaction, 129
- reaction cycle, 131
- sirohemes, 135
- sulfide production, 124
- tetrahydroporphyrin prosthetic group, 134
- thiosulfate reductase, 137, 138
- trithionate metabolism, 138
- trithionate pathway, 136

Dissimilatory sulfite reductase Dsr (dSiR) genes, 135
Dissimilatory-sulfate reducers (DSR), 57
Dithionate, 13
DNA-directed RNAS polymerase, 83
DNA G + C content, 74
DNA maintenance, 69
Dolomite, 381
Dsr complex, 264
Dyes, 174

E

Ecology, 97
Electrical microbially induced corrosion (EMIC), 433, 450
Electroautotrophs, 396
Electrochemical pathways, 437
Electron acceptors, 87
- anaerobes, 17
- capability, 17
- in cyanobacterial marine, 17
- elemental sulfur, 13, 14
- fumarate, 16
- H_2 oxidation, 16
- microorganisms, 14
- organic compounds, 16
- oxidized compounds, 15
- respiration, 15
- SRB, 13
- sulfate, 13
- thiosulfate, 13

Electron-bifurcating hydrogenases, 179
Electron-conductive nanofilaments, 62
Electron-dense bodies, 69
Electron donors, 126
- hydrogenase activity, 10
- phosphite, 12
- sulfide oxidation, 11
- sulfate-reducing bacteria, 10
- sulfur pathway enzymology, 11

Electron microscopy, 97
Electron paramagnetic resonance (EPR), 125
Electron transport complexes, 261
Electron transport proteins, 248
Electron transport system, 85
Electronema, 99
Elemental sulfur reduction, 272
Endospore-producing SRB, 75, 79, 80
Endospores, 69, 70
Energy currency, 245
Entner-Doudoroff pathway, 64
Escherichia coli, 133
Ethylenediamine disuccinate (EDDS), 444
Ethylenediaminetetraacetic acid (EDTA), 444
Extended X-ray absorption fine structure (EXAFS), 124
Extracellular electron transfer (EET), 434
Extracellular electron transport, 12
Extracellular polymeric substance (EPS), 431, 437
Extracellular uptake, 62
Extremophiles, 479

F

Fatty acids, 72
Fe-P organelle, 66
Fermentation, 21, 25
Ferredoxin, 191–194
Ferredoxin-dependent oxidoreductases, 83
Ferric iron uptake, 62
Ferrous metals, 431–436
Ferrous sulfide, 436
Filamentous, 95–100
FISH technologies, 71, 72
Flavin adenine dinucleotide (FAD), 435, 443
Flavin mononucleotide (FMN), 194
Flavodoxins, 194–197
Fluorescence, 70
Focused ion beam scanning electron microscopy (FIB-SEM), 99
Formate, 173
Formate cycling, 278
Formate dehydrogenase (Fdh/FDh), 173, 255, 256
- characterization, 189

cytoplasmic, 190–191
FdhAB, 189
membrane-associated Fdh, 188
periplasmic and membrane, 188–190
sulfate-reducing microorganisms, 189
Formate dehydrogenase/hydrogenase (FhcABCD), 190, 191
Formate hydrogen lyase (FHL) complex, 191
Fumarate, 217
as an electron acceptor, 273
disproportionation, 281, 304
reductase, 304
Fumarate-sulfate system, 304

G
Gammaproteobacteria, 441
Gas vacuoles, 69
Genetic manipulations, 102, 103
Genome analysis, 63, 99
D. vulgaris, 295
Genomic analysis, 87
Genomic island (GEI[H]), 325, 326
GeoChip, 71, 72
Glutaraldehyde (Glut), 445
Glycosidic linkages, 4
Gram-negative, 57
Gram-negative bacteria, 62
Gram-negative SRB, 59, 72, 258
Gram-negative stain, 69
Gram-negative sulfate reducers, 245, 266
Gram-positive bacterium, 60
Gram-positive cell wall, 61
Gram-positive sulfate reducers, 266
Gram-positive-type cell wall, 60
Growth coefficients, 245
Gulf of Mexico, 440

H
H_2 oxidation, 19, 246, 260, 270, 272, 273, 275, 277, 282
H_2S pollution, 362–364
H_2S production, 539
Halophiles, 465, 479, 484
Heme biosynthesis, 274
Heme-containing enzymes
catalase, 218
molybdopterin oxidoreductase (Mop), 218
nitrate reductase, 214, 215
nitrite reductase, 213, 214
oxygen reductases, 215
cytochrome *cc*(*b*/*o*)o3 oxidase, 216–217
quinol *bd*-type oxidase, 216
quinol:fumarate reductase (QFR), 217–218
sirohemes, 215
Heme proteins, 62
Heterodisulfide reductase (Hdr), 268
Heteromeric protein-protein interactions, 62
Hexadecaheme cytochrome *c*3 (HmcA), 209–210
Hexavalent chromium, 382
Holocene dolomite, 381
Human infections, 538
Hydrocarbon degradation, 28
Hydrocarbons, 437
Hydrogenase, 10, 173, 183
bifurcating hydrogenases, 180
catalytic activity, 176
in cells of Desulfovibrio, 175
characteristics, 176
discovery, 174
distribution, in *Desulfovibrio*, 178
D. norvegicum ferredoxin I, 204
in *Dst. gibsoniae*, 177
electron-bifurcating hydrogenases, 179–181
features, 174
general physiochemical and catalytic properties, 177
growth medium, 187
H_2 oxidation with sulfate reducers, 174
H_2-sensing hydrogenase, 181
H_2-sensory, 181
HsfB-type, 179
hydAB genes, 177
hydrogenase maturation and architecture, 181–182
maturation, 181
“metal-free”, 174
molecular characteristics, 174
Mtt pathway, 175
and [NiFe], 186
vs. [NiFeSe] hydrogenase, 187
species of *Desulfovibrio* carry genes, 177
in SRB, 176
standard cell-fractionation techniques, 175
Tat pathway, 175
types in *Desulfotomaculum* spp., 179
Hydrogen atoms, 436
Hydrogen cycling, 275–277
Hydrogen peroxide, 327, 328
Hydrogen sulfide, 2, 427, 435
Hydroxylamine, 151
Hydroxyvaleric acid, 64
HynAB, 86
Hyperthermophiles, 466

I
Industrial applications
dye decolorization, 396
energy technology, 395–396
production of metallic nanoparticles, 393–395
Inorganic nanoparticles, 393
Insights
archaea domain, 68, 83, 85, 86
Deltaproteobacteria, 85–87, 93
Desulfovibrio spp. genes, 83
plasmids, 93, 94
uncultured bacteria, 86, 94, 95
Ion-translocating NADH dehydrogenase complexes, 267–268
Iron inclusions, 63, 65–67, 85
Iron mineralization, 371–373
Iron sulfide, 428, 436, 438
Iron sulfide–induced corrosion, 431

K
Knockout mutants, 321

L
Lactate oxidation, 5, 252, 254, 255, 257
Lactate-starved cells, 62
Lactate-sulfate medium, 251
Lateral gene transfer (LGR), 100–102
Lipids, 72–74

M
Magnetic nanoparticles, 68
Magnetosome membrane proteins, 372
Magnetosomes, 63, 65–67, 85, 371–373, 378
Magnetotactic bacteria, 372, 373
mamABEKQMOPT genes, 87
Marine microbiology, 3
Mediated electron transfer (MET), 434, 435
Membrane-associated ATPase, 269
Membrane-associated Fdh, 188
Membrane heme proteins, 62
Membrane lipids, 73
Membrane targeting and translocation (Mtt) pathway, 175
Menaquinones (MK), 72
Mercury methylation, 373–377
Mercury resistance, 375, 376
Mesophilic and thermophilic acidophilic sulfate-reducing bacteria, 489
Metabolic pathways, 305, 310, 339
Metabolism, 16
Metabolism of hydrocarbons
oxidation of environmentally relevant organic compounds, 366–367
reductive dehalogenation, 371
Metagenomics, 99
Metalloids, 383
Metalloproteins, 201, 210
Metal-reducing bacteria (MRB), 382
Methanobacterium, 433
Methanococcus maripaludis, 433
Methanosarcina barkeri, 133
Methyl viologen (MV), 126
Methylation of mercury, 374, 375
Methylmercury, 373–376
Microbes
in acidophiles, 490–492
in alkaliphiles, 486–490
in halophiles, 479–486
in hyperthermophiles, 464–466
in piezophiles, 492–495
in psychrophiles, 469–478
in thermophiles, 466–469
Microbial corrosion, 427
Microbially induced concrete corrosion (MICC), 428
Microbially induced corrosion (MIC), 450
Microbially influenced corrosion (MIC), 430
Microbiologically induced corrosion (MIC), 427
Microbiologically induced deterioration (MID), 427
Microbiologically influenced corrosion (MIC), 437
Microorganisms, 1, 57, 121, 427, 431
with anaerobic sulfate respiration, 57
Model organisms, 375
Molecular biology techniques, 204
Molecular docking, 210, 211
Molybdate transport system (Mod ABC), 142
Molybdenite (MoS_2), 380
Molybdenum, 380
Molybdenum sulfide, 189
Molybdopterin guanine dinucleotide (MGD), 152
Molybdopterin oxidoreductase (Mop), 218
Multi-culture metabolism, 28

N
NADH:quinone oxidoreductase, 184
NADPH:rubredoxin oxidoreductase (NRO), 202
Nanofilaments, 62, 63
Nanowires, 62, 63, 97, 441, 442

NapA, 215
Neelaredoxin (Nlr), 201
Neurospora, 122
Nfn (NADH-dependent reduced ferredoxin: NADP+ oxidoreductase) protein complex, 268
[NiFe] Hydrogenases
  Desulfovibrio, 183
  [FeNiSe] hydrogenases, 186–188
  dimeric soluble, 182
  maturation, 182, 185
  model of the active site, 182
  NADH:quinone oxidoreductase, 184
  X-ray crystal analysis and related studies, 182
Nigerythrin (Ngr), 202
Nitrate reductase, 214, 215
Nitrate-reducing bacteria (NRB), 437
Nitrite
  as an electron acceptor, 273–274
  reduction, 214
Nitrite reductase, 213, 214
Nitrogen cycles, 25
Nitrogen fixation, 155, 156
Nitrogen respiration
  ammonium, 152
  enzymes, 156
  half-cell reactions, 151
  nitrite reduction, 153–155
  nitrogen fixation, 155, 156
  periplasmic nitrate reductase, 151
  role of SRP, 156, 157
  sulfate-reducing microorganism, 151
  terminal electron acceptor, 151, 152
NMR spectroscopy, 67
Nonaheme cytochrome *c*3 (NhcA), 208–209
Non-redox active protein (NapD), 153
Non-sporing bacteria, 81, 83

O
Octaheme cytochrome *c*3 (Mr 26,000), 207–208
Organic carbon compounds, 355
Outer membrane proteins, 60–62
Oxidative phosphorylation, 245, 257, 259, 269–271, 274
Oxygen stress, 327

P
Phosphoadenosine-5'-phosphosulfate (PAPS), 122
Phosphorus inclusions, 67, 69
Phosphotransacetylase, 303
PhyloChip, 71, 72
Physiology, 97
Piezophiles, 492–495
Plasma membrane, 61
Plasmids, 93, 94
Polar lipid-derived fatty acids (PLFA), 73
Pollutants and bioremediation processes
  acid mine and ARD bioremediation, 389–390
  biogenic hydrogen sulfide production, 388–389
  bioremediation, petroleum hydrocarbons, 392
  perchlorate reduction and use, 391, 392
  uranium remediation, 390–391
Poly-3-hydroxybutyric acid (3HB), 64
Polychaete serpulid worms, 531
Polyextremophiles, 463, 464, 479, 492
Polyglucose, 64
Polyhydroxyalkanoate (PHA), 64, 65
Polysaccharides, 437
Polysulfide reductase (PSR), 99
Proteins, 437
Proteobacteria, 372, 373
Proteomics, 296, 320, 334, 339
Protodolomite, 381
Proton motive force (Pmf), 258, 259, 267, 275, 278
Proton thruster, 211
Proton translocation experiments, 260
Proton-translocating pyrophosphatase, 260–261
Proton tunneling, 210
Pseudomonas zelinskii, 132
Psychrophiles, 469–478
Pyruvate, 279
Pyruvate phosphoroclastic reaction, 279–280

Q
QmoABC complex, 253, 254, 263
Quinol:fumarate reductase (QFR), 217, 218
Quinone-interacting membrane-bound oxidoreductase complex (Qmo), 262
Quinone oxidoreductase complex, 262–264
Quinone reductase complex (Qrc), 267
Quinones, 72–74
Quorum sensing
  biological strains, 451
  ferrous corrosion, 451
  nitrate, 451
  oxidizing agents, 451

R
Radionuclides, 383
Raman microscopy, 97
Redox complexes, 266–267
Reduction of metal(loid)s
 arsenic, 387–388
 chromium, 382–386
 molybdenum, 386–387
 selenium, 387
Reductive dehalogenation, 371
Riboflavin, 435
rRNA (16S) gene, 81, 82, 95
Rubredoxin gene (*rub*), 198
Rubredoxin-2, 198
Rubredoxins, 197–199
Rubrerythrin (Rr), 199, 200

S
S-adenosylmethionine production, 324
Screening molecular techniques, 545
Sea cucumber, 532
Selenium, 387
Sequence of -Cys-XX-Cys-, 201
Single-cell genomes (SAGs), 95
Sirohemes, 215
Species-specific and genus-specific markers, 70
Spectral near-edge structure (XANES) analysis, 66
Spherical cells, 59
Split-Soret cytochrome (SSC), 205, 206
Spore-forming SRB, 80
Sporulation, 79
SRP-PhyloChip, 71
Steel alloys, 436
STEM-EDS analysis, 68
Storage polymer, 64
Stress responses, 320, 322, 326
 in *D. vulgaris*
  cold shock, 329–330
  heat shock, 329
  nitrate stress, 332
  nitrite stress, 332, 333
  NO stress, 333
  oxidative stress, 327–328
  salt adaptation, 330–331
  starvation response and CO_2 stress, 328–329
 expression of genes, 326
Substrate-level phosphorylation fermentation, 278, 279
Succinate-fumarate reactions, 280–282
Sulfate reducers, 495
Sulfate reduction, 502
Sulfate-reducing bacteria (SRB), 1, 8, 22, 121, 295, 355, 486, 493
 accelerated low water corrosion (ALWC), 438
 bacterial cell, 437
 biofilms, 437
 cultures, 36
 cytoplasmic activity, 258–259
 to detoxify/remediate polluted environments, 393
 diversity, 356
 electrogenic activity, 437
 electron transport, 173
 energy metabolism, 248
 eukaryotic organisms, 529
 genus, 2
 geographical regions, 2
 with global carbon cycle, 357
 gram-negative SRB, 258
 growth, 4
 with humans, 535, 536
  antibiotic Susceptibility snd resistance, 540
  as flora of human gastrointestinal tract, 536
  Desulfovibrio human pathogens, 538–540
  oral SRB, 536–538
  virulence factors, 541–542
 H_2S pollution, 362–364
 hydrothermal system, 3
 interactions with plants, 542–544
 magnetosomes and iron mineralization, 371–373
 in marine microbiology, 2
 metal corrosion, 429, 430
 metal ions, 19
 metals, 429
 in mice, 532–533
 nomenclature and classification, 3
 in nutrient-limited environments, 21
 O_2 respiration, 174
 and oil technology, 364–365
 organic substrates, 5, 369
 with other animals, 545–546
 oxidized minerals and compounds, 355
 periplasmic formate dehydrogenases, 258
 in pigs, 534–535
 in rats, 533–534
 reduction of metals with proteins, 385
 reports, 36
 with ruminates, 535

species, 34
spore-forming, 36
strains, 15, 24
on surfaces
biofilms, 506–508
deep subsurface and mines, 508–510
dolomite bioformation, 504–506
environmental mats, 501–502
floodplains and estuaries, 510–511
hydrothermal vent sediments, 508
lithification of mats, 502–503
living marine organisms, 544
low nutrient environment, 511–512
sulfate, 2
sulfate reduction pathway, 3
sulfate-respiring archaea, 355
sulfur and nitrogen cycling (*see* Sulfur cycling and nitrogen cycling)
Sulfate-reducing microorganisms (SRMs), 1, 251, 279, 306, 308, 314
Sulfate-reducing prokaryotes (SRP), 1, 6, 121, 174–176, 186, 215, 220, 463, 464, 491, 495, 508, 510, 512
alcohol dehydrogenase, 257
energetics, 245
energy balance, 247
formate cycling, 278
formate dehydrogenases, 255–257
lactate oxidation, 250–254
metabolic pathway, 246
pyruvate oxidation, 254–255
taxonomy (*see* Taxonomy)
Sulfate transport, 140–143
Sulfide/sulfur oxidation
bacterium and archaea, 30
culture, 27
fermentation, 29
microorganisms, 27
nutrients, 28
polyploidy, 27
reduction, 26
syntrophic growth, 28
syntrophic interactions, 30
syntrophy, 26
Sulfite reduction, 215
Sulfite reduction complex, 264
Sulfonates, 19
Sulfur and nitrogen cycling
carbon cycling, 356 (*see also* Carbon cycling)
Sulfur reduction
anaerobic environments, 144
APS reductase, 126, 129, 131
assimilatory sulfite reductase (aSiR), 131–133
ATP sulfurylase (ATPS), 123, 125
bacterial metabolism, 149
biotic and abiotic action, 147
components, 144
cryptic sulfur cycle, 144
electron acceptor, 121, 144
electron donors, 150
elemental, 143, 144
geochemical reactions, 149
inorganic pyrophosphatase, 125, 126
marine sediments, 147, 148
metabolism, 144
microbial physiology, 121
microorganisms, 121
nitrate, 121
phosphate, 122
polysulfides, 148
salt marshes, 150
sulfate transport, 140–143
sulfur-oxy anions, 121
sulfur redox cycle, 149
thiosulfate, 148
Supplemental energy, 245
Surface-exposed proteins, 60
Surfaceome, 60–62
Symbiosis, 529
with invertebrates
gutless marine worm, 531
polychaete serpulid worms, 531
sea cucumber, 532
with root-feeding larvae, 531
with termites and gut-residing protist, 530
Synthesis of heme, 219–220
Syntrophobacter, 312
Syntrophobacter fumaroxidans, 87
Syntrophy, 26, 33, 300, 312

T

Taxonomic markers, 70
Taxonomy
APS, 100–102
cable bacteria, 95–100
chemotaxonomy, 70–74
classification
archaea, 79
Deltaproteobacteria, 81–83
endospore-producing SRB, 75, 79, 80
non-sporing bacteria, 81, 83
sulfate-reducing prokaryotes and characteristics, 74, 75

Taxonomy (*cont.*)
DSR, 100–102
endospores, 69, 70
filamentous, 95–100
genetic manipulations, 102, 103
insights, 83–95
LGT, 100–102
phenotypic characteristics (*see* Biochemical characteristics)
TCA cycle, 305, 307
Tetrachloroethene (TCE), 371
Tetraheme cytochrome *c*3 (TpI-*c*3/TpII-*c*3), 206–207
Tetrakis (hydroxymethyl) phosphonium sulfate (THPS), 445
Thermodesulfobacterium commune, 136
Thermodesulfovibrio sp., 430
Thermodynamics, 20
Thermophiles, 466–469
Thermophilic SRB endospores, 80
Thiosulfate, 21
Thiosulfate reductase, 137, 138
Tindallia texcoconensis, 441
TolC-like protein, 62
Tooth-shaped magnetite crystals, 66
Transcriptional analysis, 30
Transcriptome analysis, 62
Transmembrane complex (TMC), 253, 262, 275
Transmission electron micrograph, 4
Transmission electron microscopy, 67
Trithionate metabolism, 138
Truncated rubrerythrin gene, 200
TupABC transport system, 325
Twin arginine transport (TAT), 153

U
Ulcerative colitis, 534, 538–540
Uncultured bacteria, 86, 94, 95

V
Vertical zones, 357
Vibrio harveyi, 450
Virulence factors, 540–542

W
Wood-Ljungdahl pathway, 36, 99, 374